Pollutants, Human Health and the Environment

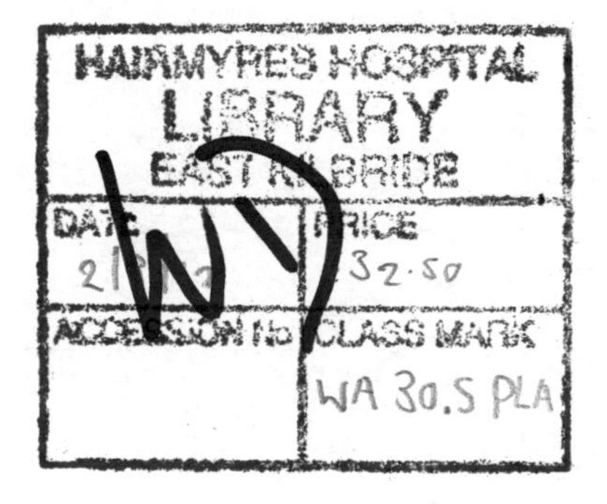

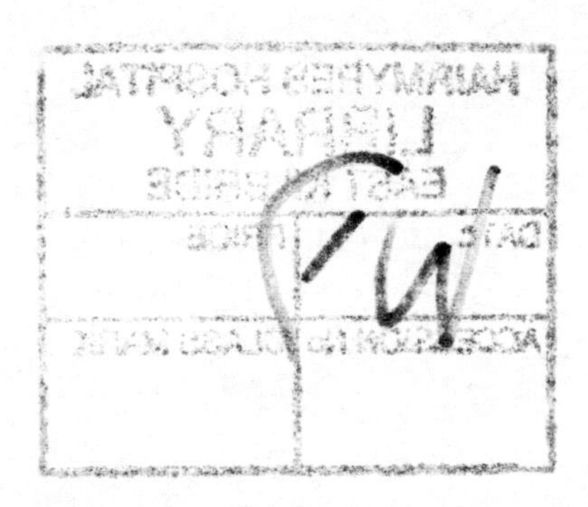

Pollutants, Human Health and the Environment

A Risk Based Approach

Editors

Jane A. Plant
Imperial College London

Nikolaos Voulvoulis
Imperial College London

K. Vala Ragnarsdottir
University of Iceland

A John Wiley & Sons, Ltd., Publication

Wiley-Blackwell is an imprint of John Wiley & Sons, formed by the merger of Wiley's global Scientific, Technical and Medical business with Blackwell Publishing.

Registered office: John Wiley & Sons, Ltd, The Atrium, Southern Gate, Chichester, West Sussex, PO19 8SQ, UK

Editorial offices: 9600 Garsington Road, Oxford, OX4 2DQ, UK
The Atrium, Southern Gate, Chichester, West Sussex, PO19 8SQ, UK
111 River Street, Hoboken, NJ 07030-5774, USA

For details of our global editorial offices, for customer services and for information about how to apply for permission to reuse the copyright material in this book please see our website at www.wiley.com/wiley-blackwell.

Library of Congress Cataloging-in-Publication Data

Pollutants, human health, and the environment : a risk based approach / [edited by] Jane A. Plant, Nikolaos Voulvoulis, and K. Vala Ragnarsdottir.
p. cm.
Includes bibliographical references and index.
ISBN 978-0-470-74261-7 (cloth) – ISBN 978-0-470-74260-0 (pbk.)
1. Pollution–Environmental aspects. 2. Pollution–Health aspects. 3. Nature–Effect of human beings on. I. Plant, Jane A. II. Voulvoulis, Nikolaos III. Ragnarsdottir, Kristin Vala, 1954-.
TD196.C45P65 2011
363.73–dc23

2011025642

A catalogue record for this book is available from the British Library.

This book is published in the following electronic formats: ePDF 9781119950110; Wiley Online Library 9781119950127; ePub 9781119951063; Mobi 978111951070

Set in 9/11pt, Times Roman by Thomson Digital, Noida, India
Printed and bound in Singapore by Markono Print Media Pte Ltd

First Impression 2012

The book is dedicated to Charlotte Rich, a PhD student and friend of many of the authors of the book, who died tragically in 2007 and Michael Gillott who died of cancer. Michael had the utmost respect for scientific research and the quest for truth. He persuaded us of the need to inform a wide audience, including health professionals, of the hazards and risks of pollutants.

Contents

Forewords

A Foreword by Lord Selborne

Over the past century a large number of man-made chemical substances have been widely dispersed into our environment, both by accident and design, raising concerns about their adverse effects on human health and the environment. There is no doubt that there has been a worrying increase in health problems that are related, at least in part, to these substances and their release from manufacturing processes, spills, inadequate handling, storage and use, and careless after-use disposal.

Over the past 30 years or so there has been a plethora of legislation at international and national levels aimed at controlling and regulating the production, use and disposal of chemicals. The introduction of the REACH (Registration, Evaluation, Authorisation and restriction of Chemicals) by the EU in 2007 has provided the innovative concept that the burden of proof is now on manufacturers to provide evidence of the safety of their products before supplying them: a practical example of the so-called precautionary principle. It also provides rules for the phasing out and substitution of the most dangerous substances already in circulation, though this, unfortunately, is likely to be a protracted process. The main objective of REACH is to improve risk management of industrial chemicals by banning their manufacture or importation into Europe.

One concern about the REACH legislation is that it will drive research and development and manufacturing involving chemicals to parts of the world where legislation is weak or non-existent. This would be highly unsatisfactory, since chemical pollution is a global issue. For example, the use of arsenical groundwater to grow crops in south-east Asia has resulted in health warnings on rice, especially warning against feeding it to babies in the USA and Europe; persistent organic pollutants such as pesticides and plasticisers, despite being used mainly at low latitudes, are accumulating in marine fish and mammals in polar and sub-polar regions; and the manufacture of pharmaceuticals in countries where environmental legislation fails to require the clean-up of waste water has the potential to increase the antibiotic resistance of pathogens.

This book provides a balanced view of the risks and benefits of several groups of substances: essential, toxic trace and radioactive elements; synthetic organic agricultural and industrial chemicals and pharmaceuticals; and particulates and nanomaterials. Most of these substances are important to modern industrialised societies but can have adverse impacts on the environment and human health. It also deals with risk reduction and the future role of chemicals in achieving sustainable development. The issue of sustainability in a world of finite resources is likely to become ever more important in considering the use of chemicals in the twenty-first century.

The book uses a risk-based approach for industrial and other chemicals. It includes a discussion of the potential use of chemicals in sustainable development and suggests that in the future there will be more emphasis on green chemistry and biomimicry, which involve learning from nature, with industry developing clean cycles of production that are built on natural processes whereby waste from one process is feed for another – the cradle-to-cradle concept.

The book includes well-researched material, with references to the latest published work. It is written in accessible English and provides an excellent introduction to anyone wishing to know more about the increasingly important subject of chemicals in the environment. The information it contains will be particularly useful to everyone affected by recent legislation including the REACH legislation.

The Earl of Selborne GBE FRS
Chair of the Council of Science and Technology

A Foreword by Professor Karol Sikora

Global warming has captured the attention of the world's media, public and politicians, but although the dramatically increasing carbon footprint we are leaving is important it is not the only effect of the twenty-first century on the environment. Many diverse pollutants have the potential to cause lasting damage to our environment. Processes that may make a quick buck today could cause untold difficulties for our successors who will inherit the challenge. We have already seen an inexorable rise in cancer incidence in the world. Although an increasingly aging population is the main driver, there is no doubt that other more subtle influences are at work. The incidence of cancer acts as a litmus test for the deleterious effects of the environment and lifestyle changes on our bodies.

It was announced this year that the lifetime risk of breast cancer in the UK has gone up from 1 in 9 women to 1 in 8 women, a rate comparable to that in the USA. Rates of this and many other types of cancer have risen dramatically since reliable cancer registries were first developed in the 1950s and they are projected to continue to do so in the future – according to the World Cancer Research Fund Report, cancer rates worldwide could increase by a further 50 per cent to 15 million new cases a year by the year 2020. Rates of many other chronic diseases, from Alzheimer's disease to Parkinsonism, as well as mood disorders such as anxiety and depression are also increasing across the globe following industrialisation and development, at a time when many of the costs of health care are becoming unaffordable. It is clear that we must learn how to reduce the risks of such diseases and prevent the human and economic toll that they are taking on society worldwide.

Effective prevention requires a detailed understanding of the pathogenesis of disease and the dissection of the positive and negative drivers that influence the process. Cancer is a disease of cumulative somatic mutations leading to disruption in cellular growth control. It is not surprising that many pollutants can influence this process. By understanding the detailed factors involved in the aetiology of cancer it may be possible to devise public-health strategies to minimise the overall burden of disease. Furthermore, our increasing knowledge of the molecular mechanisms involved in the interplay of the environment and our genetic background may make it possible to personalise prevention strategies at some future point. This individualisation is far more likely to achieve wider compliance amongst the population rather than bland generic messages.

While it is widely acknowledged that the causes of many chronic conditions are multifactorial, attention is being directed increasingly to the role of chemicals in our diet and the wider environment – especially following the understanding of the important role of epigenetics in disease progression. For example, the role of endocrine-disrupting substances in hormone-dependent cancer, asbestiform particulates in mesothelioma and particulates generally in chronic lung disease has been established. It has even been suggested that chemicals found in certain plastics, as well as in cigarette smoke, may increase the risks of obesity or diabetes. It is clear that greater efforts should be made to reduce exposure and hence the risk to human health of potentially hazardous substances beyond those associated simply with smoking and alcohol consumption.

This book, *Pollutants, Human Health and the Environment,* equips health professionals with an up-to-date knowledge of hazardous substances to help them limit the risks to human health and prevent many chronic diseases. It explains clearly the difference between hazard and risk, and goes on to discuss groups of hazardous substances, chapter by chapter. It includes discussions of many controversial issues, including: toxic trace elements such as arsenic, cadmium and mercury; radiation and radioactive elements such as naturally occurring radon gas and other natural and artificial fission products; industrial chemicals such as benzene and trichloroethylene; pesticides and pharmaceuticals, which enter water supplies and the wider environment; and particulates, including asbestos. It also includes chapters on engineered nanomaterials, essential and beneficial trace elements such as selenium, copper and zinc, and natural oestrogens.

The book contains some striking information, for example: the numbers of people at risk of skin or bladder cancer from increased exposure to arsenic; the number of conditions for which there is evidence that selenium deficiency is a cause or a factor; the fact that the greatest exposure of US citizens to ionising radiation is from medical diagnostics and treatment and that this is 500 times the dose from the nuclear industry; the increased amount of oestrogen in our food because of changes in farming practices; our increased exposure to neurotoxic substances used as pesticides or preservatives; and the fact that mercury levels have increased by a factor of four over the last 100 years.

This book is highly recommended to all health professionals who wish to play an effective role in reducing the risk to human health of chemicals in the environment. For the sake of our children's children we all need to understand the footprint we are leaving for them. The knowledge, understanding and information in this book are the key to developing effective action plans across the globe.

Karol Sikora

Professor of Cancer Medicine, Hammersmith Hospital and
Dean, University of Buckingham Medical School

Tribute

Professor Stanley Bowie FRS, 1917 to 2008

Stanley Hay Umphray Bowie FRS, FRSE FEng, FIMM

Stanley Bowie was a scientist of international standing who, as Chief Geochemist, established and led the highly successful Geochemical Division of the British Geological Survey (BGS, formerly the Institute of Geological Sciences, IGS), which became a model for similar divisions in geological surveys throughout the world. He and his staff made major contributions in isotope geology, fluid-inclusion studies, trace-element geochemistry (including high-resolution geochemical mapping), ore mineralogy, economic geology and analytical chemistry. The first inductively coupled plasma mass spectrometer was developed by Alan Gray of the University of Surrey and Alan Date in the IGS with funding from the European Commission, negotiated by Stanley. Later he was involved in further instrument development, including the portable XRF analyser and the first towed seabed gamma spectrometer.

A Shetlander by birth, Stanley Bowie graduated in 1941 with a first-class honours degree in geology from the University of Aberdeen where he had also studied chemistry and physics. He was awarded the Mitchell prize for the best Honours Geology student and the Senior Kilgour Research Scholarship.

In January 1942, during the Second World War, he joined the Meteorological Branch of the Royal Air Force and was commissioned flying officer a year later. He was stationed with Bomber Command in East Anglia, which was later the base for the first American B17 squadron stationed in Britain.

In 1946 he joined the Geological Survey of Great Britain (GSGB) with the Special Investigations Unit (renamed the Atomic Energy Division, AED, in 1951). This was the Unit which had been responsible for advising the British Government on the availability of uranium supplies for the Manhattan Project during the war and subsequently provided geological information for the UK's atomic-weapons and nuclear-energy programmes. It was Britain's knowledge and ownership of uranium reserves that ensured that the country remained in the American-led nuclear club after 1945. Stanley worked on autoradiography studies of uranium and thorium minerals in thin and polished sections and, in collaboration with the Atomic Energy Research Establishment (AERE) at Harwell, began a programme of instrument development for uranium exploration that helped to develop Geiger-Müller counters for use in uranium exploration, borehole logging and aero-radiometric surveys. He also developed an index of radioactive minerals, which remained classified until 1976.

In 1955 Stanley was promoted to Chief Geologist of the AED and represented the UK at international conferences on atomic energy, helping to develop advanced radiometric instrumentation. He also developed, with Ken Taylor, a new system of opaque-mineral identification based on the measurement of indentation hardness and reflectance – a major advance over previous complex systems – which gave Britain an important lead in economic geology. He used the system to investigate and document uranium deposits throughout the free world, and it remained in use by most ore mineralogists until the advent of the electron microprobe.

In 1968 he was appointed Chief Geochemist, in charge of the analytical, mineralogy and isotope-geology units as well as the field geochemistry programmes. From 1968 to 1973 he led a uranium reconnaissance programme on behalf of the UKAEA using many of the instrumental methods developed earlier in his career as well as newer geochemical methods based on the delayed-neutron method of analysis. In 1970 he was appointed by NASA as a principal investigator for returned lunar samples.

His work with Peter Simpson on the ore mineralogy of these samples and with Clive Rice on the distribution of uranium using fission-track analysis made an important contribution to understanding the lunar surface.

In 1972 he obtained funding for a programme of systematic geochemical mapping of Great Britain. This programme developed, for the first time, quantitative reproducible methods for the preparation of geochemical maps of similar standing to gravity, magnetic and other geophysical maps prepared by geological surveys. Led by Dr Jane Plant, this programme became the model for geochemical databases worldwide, and many of the sampling, analytical, quality-control and quality-assurance techniques and the methods of data processing form the basis of recommendations of the IUGS/IAGC Task Group on 'Global Geochemical Baselines' initiated by Dr Arthur Darnley, a former colleague.

In 1975 he established and led a Royal Society Working Party on Environmental Geochemistry and Health, which included other notable scientists such as Professor John Webb of Imperial College, Dr Colin Mills of the Rowett Institute of Nutrition and Health and Dr Gerry Shaper of the Royal Free and University College Medical School. The Working Party was in contact with national coordinating committees in the USA and USSR, the Academies of Science of the five Scandinavian countries and individuals elsewhere. The proceedings of a Royal Society discussion meeting in 1977 entitled 'Environmental Geochemistry and Health' continue to be regarded as a key scene-setting volume covering geochemistry and the health of man, animals and plants. He collaborated in 1990 with Dr Cameron Bowie (not a relative) of the Somerset Health Authority in a book on radon and health and in a paper on the same topic, published in the Lancet in 1991.

In 1959 he was awarded the Silver Medal of the Royal Society of Arts. In 1963 he was elected a Fellow of the Royal Society of Edinburgh. He was elected a Fellow of the Royal Society in 1976 and in the same year he became president of the Institution of Mining and Metallurgy.

In 1984 a new platinum-group mineral was named bowieite by the United States Geological Survey in recognition of Stanley's contribution to ore mineralogy. He was visiting Professor at Strathclyde University until 1985 and visiting Professor at Imperial College from 1985 until 1989, and he served on the Commission of Ore Mineralogy of the International Mineralogical Association until 1987.

This book is a tribute to Professor Stanley Bowie FRS, honouring him as one of the pioneers of geochemistry as applied in the real world, recognising especially his role in establishing high-quality geochemical mapping, researching radioactivity and radio-elements in the Earth's crust, and applying these studies to the exploration and development of mineral resources and to the improvement of human health.

The Editors

Professor Jane Plant CBE, DSc, FRSM, FRSE, FRSA, FRGS, FIMMM, FGS, CEng, CGeol holds the Anglo American chair of Geochemistry at Imperial College London. She was formerly Chief Geochemist and later Chief Scientist of the British Geological Survey. She has been awarded seven honorary doctorates and many prizes and distinctions for her contribution to science, including the prestigious Lord Lloyd of Kilgerran Award of the Foundation of Science and Technology, the Coke Medal of the Geological Society and the Tetleman Fellowship of Yale University. She formerly chaired the Government's Advisory Committee on Hazardous Substances and was a member of the Royal Commission on Environmental Pollution 2000–2006. She is presently on the Council of the UK All Party Parliamentary and Scientific Committee and the College of Medicine and is patron of several cancer charities, including the famous Penny Brohn Centre in Bristol.

She supervises and undertakes research in environmental geochemistry with particular reference to human health. She is an international expert on environmental pollution, specialising in understanding and modelling the sources and behaviour of essential, beneficial and toxic trace elements, radioelements and radioactivity in the environment. She established the world-renowned geochemical baseline programme of the UK (G-BASE), and subsequently co-led the global geochemical baseline International Union of Geological Sciences/International Association of GeoChemistry (IUGS/IAGC) Programme with Dr David Smith of the United States Geological Survey.

Professor Plant is the author of the internationally best-selling book *Your Life in Your Hands*, on overcoming breast cancer, and several other books on health including ones on osteoporosis and prostate cancer. Her latest popular health book, entitled *Beating Stress, Anxiety and Depression*, was published in 2009 and has been described as ground-breaking. These popular health books aim to empower sufferers by making available the latest scientific information on diet and lifestyle, as well as conventional medical treatments. She has played a leading role in developing this volume in order to help to communicate to others the significant health problems caused by chemicals in the environment.

Dr Nikolaos Voulvoulis is a Reader in Environmental Technology, leader of the Environmental Quality Research theme at the Centre for Environmental Policy and Director of the world-renowned MSc in Environmental Technology at Imperial College London. He supervises and undertakes research in the area of environmental analysis and assessment for environmental quality management. This focuses on the development of methods for assessing emerging environmental contaminants and their sources, pathways and fate in the environment, with emphasis on waste and waste-water-treatment processes. He is an international expert in environmental pollution by hazardous substances such as biocides, pesticides, endocrine-disrupting chemicals and pharmaceuticals, and on the associated policy and management issues. His research activities also involve the development and application of environmental-analysis tools, multi-criteria assessment, risk management and sustainability assessment. This research aims to develop methodologies that establish the influence of different parameters of environmental quality, process performance, and indicators of effects. His research has been having an impact on environmental decision-making and policy on environmental quality, climate change and human health nationally and internationally. Through surveys, environmental monitoring, modelling, laboratory experiments and lab-scale trials, he delivers high-quality research that has been published in some of the top journals in the field.

Dr Voulvoulis engages in a number of high-profile external teaching and research activities. Through such activities, he has developed strong links with industry, regulators, research organisations and NGOs. He is a member of the Steering Group of the Global Contaminated Land Network of the Chartered Institution of Water and Environmental Management and Director of the Opal Soil Centre responsible for the National Soil and Earthworm survey. This survey was recently included as an example of a science-based education programme and data-collection method in the European Atlas of Soil Biodiversity launched by the European Commission's Joint Research Centre in September 2010 as part of the International Year of Biodiversity. In addition, he has recently been in charge of the evaluation of over 1000 environmental projects that were co-financed by the Instrument for Structural Policies

Pre-Accession or Cohesion Fund by the European Commission, assessing the effectiveness of these projects and their contribution to the acquis communautaire in the field of the environment – specifically in the fields of water quality and management and waste collection and treatment.

Professor Kristín Vala Ragnarsdóttir is the Dean of Engineering and Natural Sciences at the University of Iceland. She was a Professor of Environmental Geochemistry and Environmental Sustainability at the University of Bristol, UK until 2008. Educated in Geochemistry and Petrology at the University of Iceland, Reykjavík (BSc) and Geochemistry at Northwestern University, Evanston, Illinois (MS, PhD) she changed her focus a decade ago from Earth Sciences to cross-disciplinary Sustainability Science. Her research pertains to sustainability in its widest context, including nature protection, economics, society and the wellbeing of citizens. She is currently developing a framework for the establishment of sustainable communities. Vala is also working on soil-sustainability indicators for land management and undertaking a comparative study of the relative fertility of conventionally versus organically managed land to ensure future food security. Her activities also include the establishment of a framework for a sustainable financial system and natural-resource use, and she is investigating the factors involved in complex multi-factorial disease development. Previously she studied the behaviour of pollutants in the natural environment and the link between environment and health.

Professor Ragnarsdóttir was a member of the Scientific Advisory Board for Framework 7 Environment Programme from 2006 to 2008. She has been a member of grant research panels for the EC (Brussels), NERC (UK), NSF (USA) and ESA (Netherlands). Vala is a past Director of the Geochemical Society and was a member of the Board of the European Association for Geochemistry and the Geological Society of Great Britain. She was the chair of the Schumacher Society and is a current board member of the Balaton Group. Professor Ragnarsdottir is a past Associate Editor of *Geochimica Cosmochimica Acta*, *Chemical Geology* and *Geochemical Transactions*. She is a current Guest Editor of *Solutions*.

Contributors

E. Louise Ander
British Geological Survey, Kingsley Dunham Centre, Keyworth, Nottingham, NG12 5GG

Aldo R. Boccaccini
University of Erlangen-Nuremberg, Department of Materials Science and Engineering, 91058 Erlangen, Germany

Pamela Castle
Former Chair of the Environmental Law Foundation

Mark R. Cave
British Geological Survey, Kingsley Dunham Centre, Keyworth, Nottingham, NG12 5GG

Ho-Sik Chon
Centre for Environmental Policy, Imperial College London, Prince Consort Road, London SW7 2AZ

Alexandra Collins
Centre for Environmental Policy, Imperial College London, Prince Consort Road, London SW7 2AZ

Jason Dassyne
Centre for Environmental Policy, Imperial College London, Prince Consort Road, London SW7 2AZ

Edward Derbyshire
Centre for Quaternary Research, Royal Holloway, University of London, Egham, Surrey, TW20 0EX

Danelle Dhaniram
Centre for Environmental Policy, Imperial College London, Prince Consort Road, London SW7 2AZ

Mustafa B. A. Djamgoz
Department of Life Sciences, Imperial College London, Prince Consort Road, London SW7 2AZ

Sally Donovan
Centre for Environmental Policy, Imperial College London, Prince Consort Road, London SW7 2AZ

Richard M. Evans
Centre for Toxicology, School of Pharmacy, University of London, 29–39 Brunswick Square, London

Claire J. Horwell
Institute of Hazard, Risk and Resilience, Department of Earth Sciences, Durham University, Science Laboratories, South Road, Durham, DH1 3LE

Timothy P. Jones
School of Earth and Ocean Sciences, Main Building, Cardiff University, Cardiff, Wales, CF10 3YE

Qin-Tao Liu
Current address: Dow Corning (China) Holding Co., Ltd.

Olwenn V. Martin
Institute for the Environment, Brunel University, Kingston Lane, Uxbridge, Middlesex, UB8 3PH

Rebecca McKinlay
Centre for Toxicology, University of London School of Pharmacy, 29–39 Brunswick Square, London

Superb K. Misra
Natural History Museum, Mineralogy, London SW7 5BD

Christopher J. Oates
Applied Geochemistry Solutions, 49 School Lane, Gerrards Cross, Buckinghamshire, SL9 9AZ

Dieudonné-Guy Ohandja
Centre for Environmental Policy, Imperial College London, Prince Consort Road, London SW7 2AZ

Richard Owen
University of Exeter Business School, Streatham Court, Rennes Drive, Exeter, EX4 4PU; European Centre for Environment and Human Health, Peninsula College of Medicine and Dentistry, Royal Cornwall Hospital, Truro, TR1 3HD

Jilang Pan
Centre for Environmental Policy, Imperial College London, Prince Consort Road, London SW7 2AZ

Xiyu Phoon
Centre for Environmental Policy, Imperial College London, Prince Consort Road, London SW7 2AZ

Jane A. Plant
Centre for Environmental Policy and Department of Earth Science and Engineering, Imperial College London, Prince Consort Road, London SW7 2AZ

K. Vala Ragnarsdottir
Faculty of Earth Sciences, School of Engineering and Natural Sciences, Askja, University of Iceland, Reykjavik 101, Iceland

Khareen Singh
Centre for Environmental Policy, Imperial College London, Prince Consort Road, London SW7 2AZ

Barry Smith
Intelliscience Ltd, 38A Station Rd, Nottingham, NG4 3DB

Teresa D. Tetley
Section of Pharmacology and Toxicology, National Heart and Lung Institute, Imperial College London, London SW3 6LY

Andrew Thorley
National Heart and Lung Institute, Imperial College London, London SW3 6LY

James Treadgold
Centre for Environmental Policy, Imperial College London, Prince Consort Road, London SW7 2AZ

Eugenia Valsami-Jones
Natural History Museum, Mineralogy, London SW7 5BD

Nikolaos Voulvoulis
Centre for Environmental Policy, Imperial College London, Prince Consort Road, London SW7 2AZ

Acknowledgements

The editors wish to thank Henry Haslam for his invaluable help in writing this book and Xiyu Phoon for acting as research assistant. They also wish to thank those who reviewed the manuscripts anonymously and made constructive, helpful suggestions which have improved the overall quality of the book. They thank colleagues at Imperial College London, especially Professor John Cosgrove, Professor Jan Gronow and Claire Hunt for their help and advice.

They acknowledge financial assistance from the charity Cancer P Prevent administered by Professor Jane Plant, Diana Patterson-Fox and Sandie Bernor on behalf of Imperial College Trust. A particularly generous financial donation was received from The Stanley Foundation Ltd.

Introduction

Jane A. Plant[1*], Nikolaos Voulvoulis[2] and K. Vala Ragnarsdottir[3]
[1]Centre for Environmental Policy and Department of Earth Science and Engineering, Imperial College London, Prince Consort Road, London SW7 2AZ
[2]Centre for Environmental Policy, Imperial College London, Prince Consort Road, London SW7 2AZ
[3]Faculty of Earth Sciences, School of Engineering and Natural Sciences, Askja, University of Iceland, Reykjavik 101, Iceland
*Corresponding author, email jane.plant@imperial.ac.uk

This book is concerned with current and emerging issues associated with the impact of pollutants on human health and the environment. Public concern and the media are presently focused on just one pollutant – carbon dioxide – and this has distracted attention from the wider issue of chemicals in the environment generally. Carbon dioxide has become the centre of international political and media attention because of its high levels in the atmosphere, mainly as a result of the burning of fossil fuels, and its potential to cause global climate change and acidification of the oceans – changes that could not be corrected on human timescales. However, this is just one of the many significant changes in the chemistry of the Earth System caused by the burgeoning human population and the increasing demand for material goods. The changes made to the Earth by humans are now so great that many scientists propose that a new geological period known as the Anthropocene be added to the geological column (Crutzen and Stoermer, 2000). Some people date the beginning of this period at 8000 years ago, with the start of forest clearing and settled agriculture; others date it at 1784, the date of Stevenson's steam engine, 'The Rocket'. In this book we examine the impact that humanity has had on the chemistry of the Earth System during the Anthropocene, beginning from 8000 years ago (van Andel, 1995; Ruddiman, 2005).

There are many books dealing with aspects of environmental geochemistry: the carbon cycle, for example, or heavy metals such as lead or mercury, or the chemistry of particular environmental compartments such as soil or water. This book is different. It aims to give information on the many different groups of substances that are having an impact on the surface environment with the potential to cause harm to human health.

The contribution of chemicals to improvements in life expectancy, human health and material living standards for most people in Western-style democracies is widely acknowledged. Over the past 50 years, however, there has been mounting unease about the widespread use of synthetic chemicals and their risk to human health, including via environmental pathways (RCEP, 2003). There is also concern about the damaging effects of chemicals on the ecosystems upon which, ultimately, all life on Earth depends. Society's fears about the risks of chemicals and other hazardous substances in the environment have been propagated by the media and, in the late twentieth century, by an information-technology revolution that has globalised communication (Gardner, 2008). Hazardous substances in the environment, from carbon dioxide, radioactivity and oil pollution to arsenic and polychlorinated biphenyls (PCBs), now occupy a central place in social, economic and political debates and developments worldwide (Beck, 1986) and will probably continue to do so in the future.

Concerns about chemicals in the environment date back at least to the early 1920s, when scientists and doctors such as Sir Robert McCarrison and Sir Albert Howard in the UK and Rudolf Steiner in Austria expressed concern about the change from traditional biologically based to chemical-based agriculture. One of the pioneers of organic farming in the UK was Lady Eve Balfour, and in 1946 she helped to found the Soil Association. The first half of the twentieth century had witnessed a growing number of early warnings of the hazards and risks of chemicals (EEA, 2001), but it was the publication in 1962 of Rachel Carson's landmark book *Silent Spring* that is widely credited with launching the modern environmental movement (Glausiusz, 2007). The book catalysed public concern about synthetic chemicals in the environment and their unintended effects by documenting the damaging effects of pesticides, particularly on birds; it stated, for example, that DDT caused eggshell thinning, reproductive problems and death. Carson also accused the chemical industry of spreading disinformation, and public officials of accepting industry claims uncritically.

The concerns about the effects of pesticides raised by Carson have since been confirmed and extended to include a wide range of other health effects, from carcinogenicity, mutagenicity and reprotoxicity to endocrine disruption, genotoxicity, neurotoxicity

Pollutants, Human Health and the Environment: A Risk Based Approach, First Edition. Edited by Jane A. Plant, Nikolaos Voulvoulis and K. Vala Ragnarsdottir.

and immunotoxicity. These effects have been observed in both wildlife and humans (Ecobichon and Joy, 1993; Colborn *et al.*, 1997; Cadbury, 1998; Ecobichon, 2001; Stenersen, 2004). Carson also highlighted the broader ecological impacts of pesticides, which are increasingly recognised as a major factor affecting biodiversity and habitat loss. In the short term they can have toxic effects on directly exposed organisms, as well as causing changes in habitat and the food chain over the long term (Isenring, 2010).

In the US, Carson's work led to the establishment of the Environmental Protection Agency in December 1970 – a model which has been followed by many individual countries as well as the European Union. Such agencies have developed formal processes of risk assessment and risk-management practices, as discussed throughout this book, as well as improved regulation.

We now know that there are many other chemicals of concern besides pesticides. In 1965, Clair Patterson, the scientist who in 1956 used mass spectrometry to establish that the Earth was 4.5 billion years old, published *Contaminated and Natural Lead Environments of Man,* which drew attention to the increased levels of lead from industrial sources in the environment and food chain (see Chapter 4). His work led to a ban on the use of tetraethyl lead (TEL) in gasoline (petroleum) in the US in 1986 and later on the use of lead in food cans. By the late 1990s, lead levels in the blood of Americans were reported to have dropped by up to 80 per cent (CDC, 1997; HHS, 1997). Lead in petrol and food cans is now banned in most of the developed world, but manufacturers of TEL initially shifted their sales to developing countries and leaded petrol was not banned in some African countries until 2006, 20 years later than in the US.

Concern about the chronic effects of toxic trace elements other than lead on the environment and human health has also mounted. For example, according to some authorities mercury levels in the biosphere have increased by up to four times since the 1880s (Monteiro and Furness, 1997). A high proportion of mercury emissions is from the burning of fossil fuels, especially in China and India. This contaminates the oceans, bioaccumulating up the food chain as fat-soluble methylmercury to top predators such as oily fish and, ultimately, to those who eat them (Walsh *et al.*, 2008). As a result, the Food and Drugs Administration in the US and the Food Standards Agency in the UK now recommend limiting the amount of oily fish consumed, especially by pregnant women.

One of the most dramatic poisoning events in recent years, however, has been due to another trace element, arsenic (As), as a result of the exploitation of groundwater resources in Bangladesh. It has been estimated that between 33 and 77 million people in Bangladesh, out of the total population of 125 million, are at risk of arsenicosis from drinking As-contaminated water (Smith *et al.*, 2000). Mercury and arsenic are discussed in Chapter 4.

Meanwhile, there has been evidence for decades (Mills, 1969) that many people have diets that are lacking essential trace elements such as selenium (see Chapter 3), iodine, and lithium, deficiencies that may contribute to chronic ill health. This problem is caused, at least in part, by the development of intensive chemical agriculture. In the early nineteenth century, scientists such as Sir Humphrey Davy and J. von Liebig in Germany argued that since humus simply supplied N, P and K to plants it could be replaced by chemicals such as superphosphate. Justus van Liebig (van Liebig *et al.*, 1841; van Liebig, 1843) defined the principles for sustainable agriculture, which is today known as 'Liebig's law' which is based on adding superphosphate to soils. This reductionist approach failed to recognise the other functions of humus, such as acting as a carbon sink, buffering water levels, improving soil texture and capturing conservative elements such as iodine from rainwater and sea spray and recycling essential and beneficial trace elements.

The range of environmental chemical contaminants now includes a vast range of synthetic and natural substances, from plasticisers to pharmaceuticals and industrial chemicals such as benzene and PCBs (see Chapters 6 and 7). The potentially additive or synergistic effects of such chemicals, as mixtures, or 'chemical cocktails', is now a major concern, as are the chronic effects of exposure, even at low or very low levels (less than parts per trillion). Moreover, some of the chemicals (such as PCBs) have been shown to be highly persistent and to have remained in the environment long after their hazardous properties have been identified and restrictions on their use put in place. Over the last few decades it has also been shown that some persistent chemicals travel great distances (EEA, 1998). Some of these chemicals, such as halogenated (chlorinated, brominated) organic chemicals, which include some pesticides and fire retardants, bioaccumulate up the food chain, so that foods such as fatty meat and dairy produce are contaminated. There is also increasing evidence of widespread chemical contamination of human tissues with synthetic chemicals, in some cases to levels higher than those known to cause adverse health effects in other species, even in remote populations of the globe (such as the Inuit of Northern Canada and Greenland and inhabitants of remote islands such as the Faroes) (Walsh *et al.*, 2008). Chemical pollution is a global issue, crossing national boundaries.

Problems caused by chemicals can arise quite unexpectedly, despite the development of extensive toxicity testing and risk-assessment procedures based on sophisticated science and technology. Many such problems – such as the ozone-depleting action of the chlorofluorocarbons (CFCs) and the endocrine-disrupting effects of the birth-control pill released to the aquatic environment – had not been anticipated (RCEP, 2003; Jobling and Owen, 2010; EEA, 2001). The often unpredictable and uncertain nature of innovation and technology and the wider impacts of new developments (Collingridge, 1980; RCEP, 2008) are also demonstrated by the identification in 1992 of the transgenerational endocrine-disrupting properties of common chemicals, including some plasticisers, detergents, pesticide formulations and pharmaceuticals. These studies, led by the distinguished biologist Theo Colborn (TEDX, 2010), showed that such substances were being transferred from top predator females to their offspring, resulting in chemically induced alterations in sexual and functional development. The problems resulting from exposure to low doses and/or low ambient levels of endocrine-disrupting chemicals have been described in *Our Stolen Future* (Colborn *et al.*, 1997) and *The Feminization of*

Nature (Cadbury, 1996). The risks and benefits of natural oestrogens are considered in Chapter 9.

In addition to the chronic effects of chemicals, there have also been several high-profile chemical accidents such as those at Flixborough in the UK in 1974, Seveso in Italy in 1976, Bhopal in India in 1984, Toulouse in France in 2001, and at the Hertfordshire Oil Storage Terminal at Buncefield in the UK in 2005.

Concerns about pollution are not restricted to chemicals. Fears of radioactivity (see Chapter 5) date back to the bombing of Hiroshima and Nagasaki in 1945 at the end of the Second World War. Radiation has been and continues to be an emotive issue (Fisk, 1996), and psychometric studies of public perception of risk have shown that the dangers associated with radioactivity are the most dreaded and least understood hazards (Slovic, 1987).

There have been several serious nuclear accidents in the former Soviet Union, which have resulted in major explosions and/or the release of radioactive materials, with many deaths and hundreds of thousands of people evacuated. The most recent accident, at the uncontained civil nuclear reactor at Chernobyl in the Ukraine in 1986, caused severe local contamination as well as regional contamination, which continues to affect livestock farming in parts of Europe. There is concern about further explosions at the site. Outside the countries of the former Warsaw Pact and Soviet Union, however, there had been only two partial meltdowns in civil nuclear power stations until the Fukushima incident in March 2011 (see Chapter 5). The first, in the Fermi Reactor in Michigan in 1966 required the reactor to be repaired; the other, the Three Mile Island accident in Pennsylvania, led to the permanent shutdown of the reactor. The high level of safety of Western reactors, however, meant there were no deaths or serious injuries.

The Three Mile Island accident ended nuclear power development in the US for more than 30 years. The irrationality of our approach to risk (Gardner, 2008) is perhaps best illustrated by our concerns about radioactivity. For example, the average US citizen now receives approximately 50 per cent of their dose of ionising radiation from medical diagnostics compared to 0.1 per cent from the nuclear industry, and yet public concern continues to focus on nuclear energy and nuclear waste repositories.

There have been many environmentally damaging accidental oil spills, some of which were associated with considerable loss of human life. These include the Torrey Canyon disaster in 1967, the Amoco Cadiz spill in 1978 and, most recently, the BP deep-drilling rig accident in the Gulf of Mexico in 2010. In July 1988, the Piper Alpha North Sea production platform exploded, and the resulting fire destroyed the platform and killed 167 men. It has recently been revealed that more than 5 tonnes of highly toxic carcinogenic PCBs were lost in the disaster, and other oil spills have resulted in the release of toxic chemicals. These accidents highlight the dangers to the Earth's fragile ecosystems of servicing our heavily energy- and resource-dependent societies.

As well as chemicals and radiation, environmental exposure to many other substances, including particulates (see Chapter 10), is now known to cause health problems. For example, asbestiform minerals used in the construction industries are associated with lung disease including mesothelioma, a serious type of cancer. More than 3000 people a year now die in the UK as a result of inhaling asbestos dust, and one US government estimate suggests it may eventually kill 5.4 million Americans (Tang *et al.*, 2010). Silica dust has long been known to cause silicosis and other lung diseases in miners and quarry workers. It has also been suggested that airborne particulate matter may cause between 100 000 and 300 000 deaths in Europe each year (Tainio *et al.*, 2007). These findings have given rise to concerns about the potential effects of manufactured or engineered nanoparticles such as carbon nanotubes (see Chapter 11), one of the most important groups of emerging contaminants (Royal Society and Royal Academy of Engineering, 2004).

Looking back over the past two and a half centuries since the industrial revolution, and especially since the second world war, there has been a change in the developed world from the industrial societies of the 'dark satanic mills' and smogs where risks of pollution were more localised and immediately perceptible, to a globalised, risk-conscious and even risk-averse society which is no longer simply concerned with 'making nature useful but with problems resulting from techno-economic development itself' (Beck, 1986). Such risks are now known to be more complex and to have global impacts, although they are often hidden and undetectable through the senses (Gardner, 2008). Many of the risks may remain poorly understood, or unknown.

Over the past 50 years, the rapid development of disciplines such as toxicology, ecotoxicology and environmental chemistry have played an important role in improving our understanding of the hazards, sources, behaviour, fate and health effects of pollutants in the environment. Environmental data in these and related disciplines, including geochemical databases, on regional, national and international scales, are increasingly available in the scientific literature and beyond. These developments have been accompanied by great improvements in analytical methods, such as inductively coupled plasma (ICP) mass spectrometry for inorganic elements and isotopes, and gas chromatography for separating chemicals in complex environmental matrices with low detection limits. New disciplines have also emerged, including genomics, proteomics, metabolomics and now epigenetics, which help to explain the biological effects of hazardous substances. These fundamental scientific developments have been matched, notably through organisations such as the Organisation for Economic Co-operation and Development, by the development of standardised tests for the hazardous effects of chemicals and effluents entering the environment. Many of these issues and developments are discussed in the chapters which follow.

The book is based on a risk-analysis framework, using the source–pathway–target model of organisations such as the US Environmental Protection Agency and the European Environment Agency. This model is explained in Chapters 1 and 2. Each of the remaining chapters follows a similar format, to ensure comprehensiveness of cover and easy cross reference. The introduction to each chapter describes the relevance and discovery of the health and environmental impacts of the substances to be discussed. This is followed by sections on the known hazardous properties; the principal natural and artificial sources; the main environmental pathways and exposure routes; the effects on receptors, especially humans, with particular reference to health; and methods of risk reduction. The final chapter of the book proposes ways in which human impact on the chemistry of the Earth System can be reduced. The book as a whole provides an integrated overview of the impact of chemicals in the environment, but each of the chapters can stand alone as a complete work in its own right. Each chapter has been peer reviewed.

The book is intended for students of the environment, especially environmental geochemistry, engineering and medical geography, and also for health professionals, for whom it provides essential information for epidemiological studies and public health. One of our main aims is to engage the medical profession in the increasingly important issue of the chemical disturbance of the environment by humanity, threatening not only our own health but also that of future generations and the Earth itself.

References

Beck, U. (1986) *Risk Society: towards a new modernity*. Sage Publications, London, 260 pp.

Cadbury D. (1998) *The Feminization of Nature. Our Future at Risk*. Penguin Books, London.

CDC (1997) CDC (1997) Blood Lead Levels Keep Dropping; New Guidelines Proposed for Those Most Vulnerable. Centre for Disease Control. 20th February 1997. Available online [http://www.cdc.gov/nchs/pressroom/97news/bldlead.htm] Accessed 1 November 2010.

Clifton, J.C. 2nd (2007) Mercury exposure and public health. *Pediatric Clinician North America* **54**, 237–69.

Colborn T., Dumanoski D., and Peterson Meyers J. (1997) *Our Stolen Future. Are we Threatening our Fertility, Intelligence and Survival?* Abacus Books, New York.

Collingridge, D. (1980) *The Social Control of Technology*. Francis Pinter (Publishers) Ltd, London, 200 pp.

Crutzen, P. J., Stoermer, E. F. (2000) The Anthropocene. *Global Change Newsletter* **41**, 17–18.

Ecobichon, D. J. (2001) Toxic effects of pesticides. In: Klaassen, C. (ed) *Cassarett and Doull's Toxicology. The Basic Science of Poisons*. New York, McGraw-Hill, pp. 763–810.

Ecobichon, D. J., Joy, R. M. (1993) *Pesticides and Neurological Diseases*, 2nd edition. Boca Raton, CRC Press.

EEA (European Environment Agency) (2001). *Late Lessons from Early Warnings: the precautionary principle 1896-2000* pp. 135–143.

European Parliament (2008) Export-ban of mercury and mercury compounds from the EU by 2011. Press release. http://www.europarl.europa.eu/news/expert/infopress_page/064-29478-140-05-21-911-20080520IPR29477-19-05-2008-2008-false/default_en.htm. Accessed 4 January 2011

Fisk, D. J. (1996) Opening address on behalf of the Secretary of State for the Environment. *Radiation Protection Dosimetry* **68**, 1–2.

Gardner, D. (2008) *Risk: The Science and Politics of Fear*. Virgin Books, UK.

Glausiusz, J. (2007) Better Planet: Can a Maligned Pesticide Save Lives? *Discover Magazine* 20th November 2007.

HHS (1997) *Blood Lead Levels Keep Dropping; New Guidelines Proposed for Those Most Vulnerable,* 1997-02-20, http://www.hhs.gov/news/press/1997pres/970220.html. Retrieved 31 July 2007

Isenring, R. (2010) Pesticides and the loss of biodiversity. Pesticide Action Network Report.

Jobling S., Owen R. (2010) Ethinyl oestradiol: bitter pill for the precautionary principle, In: Gee. D (ed). Late Lessons from Early Warnings II, European Environment Agency.

Lindh, U., Hudecek, R., Danersund, A., Eriksson, S., Lindvall A. (2002) Removal of dental amalgam and other metal alloys supported by antioxidant therapy alleviates symptoms and improves quality of life in patients with amalgam-associated ill health. *Neuroendocrinology Letters* **23**, 459–482.

Lorscheider, F., Vimy, M. (1993) Evaluation of the safety of mercury released from dental fillings. *Journal of the Federation of American Societies for Experimental Biology* **7**, 1432–1433.

Monteiro, L.R., Furness, R.W. (1997) Accelerated increase in mercury contamination in north Atlantic mesopelagic food chains as indicated by time series of seabird feathers. *Environmental Toxicology and Chemistry* **16**, 2489–2493.

RCEP (2003) Royal Commission on Environmental Pollution, Report no 24: Chemicals in Products: Safeguarding the Environment and Human Health. Available at www.rcep.org.uk

RCEP (2008) Royal Commission on Environmental Pollution, Report no 27: Novel Materials in the Environment: The case of Nanotechnology. Available at www.rcep.org.uk

Royal Society and Royal Academy of Engineering (2004) Nanosciences and Nanotechnologies: Opportunities and Uncertainties. Available at www.nanotec.org.uk/finalReport.htm

Ruddiman W.F. (2005) *Plows, Plagues, and Petroleum: How Humans Took Control of Climate*. Princeton, New Jersey, Princeton University Press.

Slovic, P. (1987) Perception of risk. *Science* **236**, 280–285.

Stenersen, J. (2004) *Chemical Pesticides: mode of action and toxicology*. Florida, CRC Press.

Tainio, M., Tuomisto, J. T., Hänninen, O., Ruuskanen, J., Jantunen, M. J., and Pekkanen, J. (2007) Parameter and model uncertainty in a life-table model for fine particles (PM2.5): a statistical modeling study. *Environmental Health* **6**, 24.

Tang, N., Stein, J., Hsia, R.Y., Maselli, J.H., Gonzales, R. (2010) Trends and Characteristics of US Emergency Department Visits, 1997-2007. *JAMA* **304** (6), 664–670.

TEDX (2010) The Endocrine Disruption Exchange. Available online [http://www.endocrinedisruption.org] Accessed 1 November 2010

van Andel, T. H. (1994) *New Views on an Old Planet: A History of Global Change*, 2nd edition. Cambridge, Cambridge University Press.

von Liebig, J., Playfair, L., Webster, J. W. (1841) *Organic Chemistry and its Applications to Agriculture and Physiology*. Cambridge, J. Owen Publishing House.

von Liebig, J. (1843) *Familiar Letters of Chemistry and its Relation to Commerce, Physiology and Agriculture*. London, J. Gardener.

Walsh, P. J., Smith, S. L., Fleming, L. E., Solo-Gabriele, H. M. and Gerwick, W. H. (2008) *Oceans and Human Health. Risk and Remedies from the Seas*. Elsevier.

Windham, B. (2010) The results of removal of amalgam fillings: 60,000 documented cases. DAMS International. Available from: www.flcv.com/dams. www.thenaturalrecoveryplan.com, 23pp.

1

The scientific appraisal of hazardous substances in the environment

Olwenn V. Martin[1]* and Jane A. Plant[2]

[1]*Institute for the Environment, Brunel University, Kingston Lane, Uxbridge, Middlesex, UB8 3PH*
[2]*Centre for Environmental Policy and Department of Earth Science and Engineering, Imperial College London, Prince Consort Road, London SW7 2AZ*
**Corresponding author, email olwenn.martin@brunel.ac.uk*

1.1 Introduction

This book describes a wide range of non-living toxins that are present in the environment and are potentially harmful to human health or the environment. It covers both organic and inorganic substances, natural as well as manufactured substances. It deals mostly with substances that are hazardous because of their chemical properties, but also includes some where the hazard derives from their physical properties, for example, particulates and nanoparticles. This chapter summarises some important concepts of basic toxicology and environmental epidemiology relevant to an understanding of the possible effects of pollutants in the environment.

A common misconception is that chemicals made by nature are intrinsically good and, conversely, those manufactured by man are bad (Ottoboni, 1991). However, there are many examples of toxic compounds produced by algae or other micro-organisms, venomous animals and plants. There are even examples of environmental harm resulting from the presence of relatively benign natural compounds, either in unexpected places or in unexpected quantities. It is therefore of prime importance to define what is meant by 'chemical' when referring to chemical hazards in this chapter and the rest of this book. The correct term for a chemical compound to which an organism may be exposed, whether of natural or synthetic origins, is xenobiotic, i.e. a substance foreign to an organism (the term has also been used for transplants). A xenobiotic can be defined as a chemical which is found in an organism but which is not normally produced or expected to be present in it. It can also cover substances that are present in much higher concentrations than are usual.

1.2 Fundamental concepts of toxicology

Toxicology is the science of poisons. A poison is commonly defined as 'any substance that can cause an adverse effect as a result of a physicochemical interaction with living tissue' (Duffus, 2006). The use of poisons is as old as the human race, as a method of hunting or warfare as well as murder, suicide or execution. The evolution of this scientific discipline cannot be separated from the evolution of pharmacology, or the science of cures. Theophrastus Phillippus Aureolus Bombastus von Hohenheim, more commonly known as Paracelsus (1493–1541), a physician contemporary of Copernicus, Martin Luther and da Vinci, is widely considered as the father of toxicology. He challenged the ancient concepts of medicine based on the balance of the four humours (blood, phlegm, yellow and black bile) associated with the four elements and believed that illness occurred when an organ failed and poisons accumulated. This use of chemistry and chemical analogies was particularly offensive to the contemporary medical establishment. He is famously credited with the quotation that still underlies present-day toxicology.

Pollutants, Human Health and the Environment: A Risk Based Approach, First Edition. Edited by Jane A. Plant, Nikolaos Voulvoulis and K. Vala Ragnarsdottir.

'All substances are poisons; there is none which is not a poison. The right dose differentiates a poison from a remedy.'

Paracelsus

In other words, all substances are potential poisons, since all can cause injury or death following excessive exposure. Conversely, this statement implies that all chemicals can be used safely if handled with appropriate precautions and exposure is kept below a defined limit below which risk is considered tolerable (Duffus, 2006). The concepts of tolerable risk and adverse effect illustrate the value judgements embedded in an otherwise scientific discipline relying on observable, measurable empirical evidence. What is considered abnormal or undesirable is dictated by society rather than science. Any change from the normal state is not necessarily an adverse effect even if statistically significant. An effect may be considered harmful if it causes damage, irreversible change or increased susceptibility to other stresses, including infectious disease. The stage of development or state of health of the organism may also have an influence on the degree of harm.

1.2.1 Routes of exposure

Toxicity will vary depending on the route of exposure. There are three routes by which exposure to environmental contaminants may occur:

- Ingestion.
- Inhalation.
- Skin adsorption.

In addition, direct injection may be used in testing for toxicity. Toxic and pharmaceutical agents generally produce the most rapid response and greatest effect when given intravenously, directly into the bloodstream. A descending order of effectiveness for environmental exposure routes would be inhalation, ingestion and skin adsorption.

Oral toxicity is most relevant for substances that might be ingested with food or drinks. It could be argued that this is generally under an individual's control, but people often don't know what chemicals there are in their food or water and are not well informed about the current state of knowledge about their harmful effects.

Inhalation of gases, vapours, dusts and other airborne particles is generally involuntarily (with the notable exception of smoking). The destination of inhaled solid particles depends upon their size and shape. In general, the smaller the particle, the further into the respiratory tract it can go. A large proportion of airborne particles breathed through the mouth or cleared by the cilia of the lungs can enter the gut.

Dermal exposure generally requires direct and prolonged contact with the skin. The skin acts as a very effective barrier against many external toxicants, but because of its large surface area (1.5–2 m^2) some of the many and diverse substances it comes in contact with may elicit topical or systemic effects (Williams and Roberts, 2000). If dermal exposure is often most relevant in occupational settings, it may nonetheless be pertinent in relation to bathing waters (ingestion is also an important route of exposure in this context). The use of cosmetics raises the same questions regarding the adequate communication of current knowledge about potential effects as those related to food.

1.2.2 Duration of exposure

The toxic response will also depend on the duration and frequency of exposure. The effect of a single dose of a chemical may be severe whilst the same total dose given at several intervals may have little or no effect: the effect of drinking four beers in one evening, for example, is very different from that of drinking four beers in four days. Exposure duration is generally divided into four broad categories: acute, sub-acute, sub-chronic and chronic. Acute exposure to a chemical usually refers to a single exposure event or repeated exposures over a duration of less than 24 hours. Sub-acute exposure to a chemical refers to repeated exposures for 1 month or less, sub-chronic exposure to continuous or repeated exposures for 1 to 3 months or approximately 10 per cent of the lifetime of an experimental species, and chronic exposure to continuous or repeated exposures for more than 3 months, usually 6 months to 2 years in rodents (Eaton and Klaassen, 2001). Chronic exposure studies are designed to assess the cumulative toxicity of chemicals with potential lifetime exposure in humans. The same terms are used in real-life situations, though it is generally very difficult to ascertain with any certainty the frequency and duration of exposure.

For acute effects, the time component of the dose is not important, as it is the high dose that is responsible for the effects. However, the fact that acute exposure to agents that are rapidly absorbed is likely to induce immediate toxic effects does not rule out the possibility of delayed effects, and these are not necessarily similar to those associated with chronic exposure (e.g. latency between the onset of certain cancers and exposure to a carcinogenic substance). The effect of exposure to a toxic agent may depend on the timing of exposure. In other words, long-term effects as a result of exposure to a toxic agent during a critically sensitive stage of development may differ markedly from those seen if an adult organism is exposed to the same substance. Acute effects are almost always the result of accidents, or, less commonly, criminal poisoning or self-poisoning (suicide). Chronic exposure to a toxic agent is generally associated with long-term low-level chronic effects, but this does not preclude the possibility of some immediate (acute) effects after each administration. These concepts are closely related to the mechanisms of metabolic degradation and excretion of ingested substances, as illustrated in Figure 1.1.

1.2.3 Mechanisms of toxicity

The interaction of a foreign compound with a biological system is two-fold: there is the effect of the organism on the compound

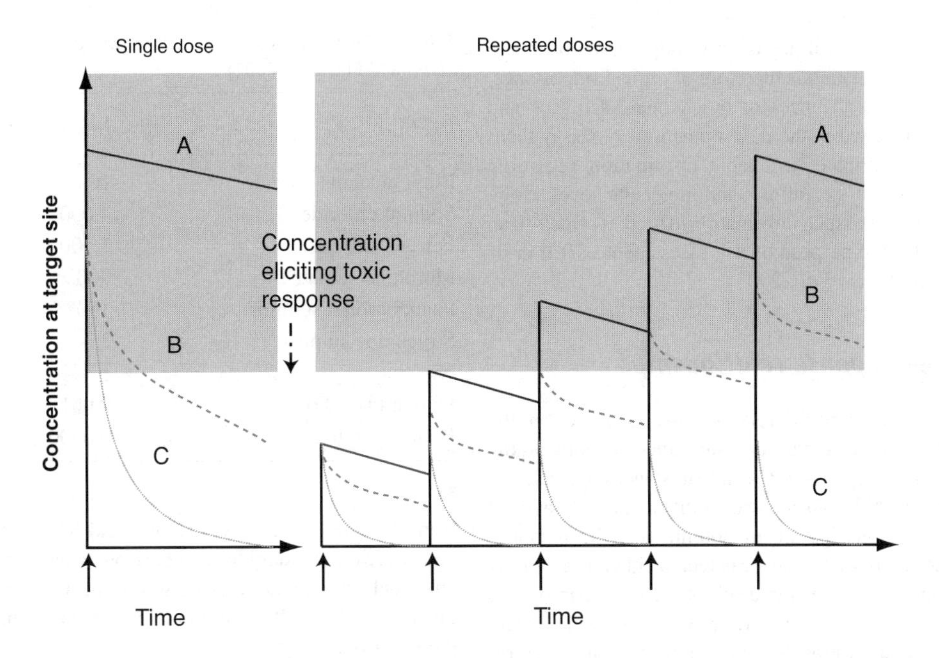

Figure 1.1 Relationship between dose and concentration at the target site under different conditions of dose frequency and elimination rate (reproduced from Eaton and Klaassen, 2001). *Line A.* Chemical with very slow elimination. *Line B.* Chemical with a rate of elimination slower than frequency of dosing. *Line C.* Rate of elimination faster than the dosing frequency. Shaded area represents the concentration range at the target site exhibiting a toxic response

(toxicokinetics) and the effect of the compound on the organism (toxicodynamics).

Toxicokinetics relate to the delivery of the compound to its site of action, including absorption (transfer from the site of administration into the general circulation), distribution (via the general circulation into and out of the tissues), and elimination (from general circulation by metabolism or excretion). The target tissue refers to the tissue where a toxicant exerts its effect, and is not necessarily where the concentration of the toxic substance is highest. Many halogenated compounds such as polychlorinated biphenyls (PCBs) or flame retardants such as polybrominated diphenyl ethers (PBDEs) are known to bioaccumulate in body fat. Whether such sequestration processes are actually protective to the individual organisms (by lowering the concentration of the toxicant at the site of action) is not clear (O'Flaherty, 2000). In an ecological context however, such bioaccumulation may serve as an indirect route of exposure for organisms at higher trophic levels, thereby potentially contributing to biomagnification through the food chain.

Absorption of any compound that has not been intravenously injected will entail transfer across membrane barriers before it reaches the systemic circulation, and the efficiency of absorption processes is highly dependent on the route of exposure.

It is also important to note that distribution and elimination, although often considered separately, take place simultaneously. Elimination itself comprises two kinds of processes, excretion and biotransformation, which also take place simultaneously. Elimination and distribution are not independent of each other, as effective elimination of a compound will prevent its distribution in peripheral tissues, whilst, conversely, wide distribution of a compound will impede its excretion (O'Flaherty, 2000). Kinetic models attempt to predict the concentration of a toxicant at the target site from the administered dose. The ultimate toxicant, i.e. the chemical species that induces structural or functional alterations resulting in toxicity, may be the compound administered (parent compound), but it can also be a metabolite of the parent compound generated by biotransformation processes, i.e. toxication rather than detoxication (Timbrell, 2000; Gregus and Klaassen, 2001). The liver and kidneys are the most important excretory organs for non-volatile substances, whilst the lungs excrete volatile compounds and gases. Other routes of excretion include the skin, hair, sweat, nails and milk. Milk may be a major route of excretion for lipophilic chemicals due to its high fat content (O'Flaherty, 2000).

Toxicodynamics is the study of toxic response at the site of action, including the reactions with and binding to cell constituents, and the biochemical and physiological consequences of these actions. Such consequences may therefore be manifested and observed at the molecular or cellular levels, at the target organ or on the whole organism. Therefore, although toxic responses have a biochemical basis, the study of toxic response is generally subdivided, either depending on the organ on which toxicity is observed, including hepatotoxicity (liver), nephrotoxicity (kidney), neurotoxicity (nervous system), pulmonotoxicity (lung) or depending on the type of toxic response, including teratogenicity (abnormalities of physiological development), immunotoxicity (immune system impairment), mutagenicity (damage of genetic material), carcinogenicity (cancer causation or promotion). The choice of the toxicity endpoint to observe in experimental toxicity testing is therefore of critical importance. In recent years, rapid advances of biochemical sciences and technology have resulted in the development of bioassay techniques that can contribute invaluable information regarding toxicity mechanisms at the cellular and molecular level. However, the extrapolation of such information to predict effects in an intact organism for the purpose of risk assessment is still in its infancy (Gundert-Remy *et al.*, 2005).

1.2.4 Dose–response relationships

The theory of dose–response relationships is based on the assumptions that (1) the activity of a substance depends on the dose an organism is exposed to (i.e. all substances are inactive below a certain threshold and active over that threshold), and (2) dose–response relationships are monotonic (i.e. the response rises with the dose). Toxicity may be detected either as an all-or-nothing phenomenon such as the death of the organism or as a graded response such as the hypertrophy of a specific organ. Dose–response relationships for all-or-nothing (quantal) responses are typically S-shaped and this reflects the fact that sensitivity of individuals in a population generally exhibits a normal or Gaussian distribution (bell-shaped curve). When plotted as a cumulative frequency distribution, a sigmoid dose–response curve is observed (Figure 1.2).

Studying dose response and developing dose–response models are central to determining 'safe' and 'hazardous' levels.

The simplest measure of toxicity is lethality, and determination of the median lethal dose, the LD_{50} is usually the first toxicological test performed with new substances. The LD_{50} is the dose at which a substance is expected to cause the death of half of the experimental animals and it is derived statistically from dose–response curves (Eaton and Klaassen, 2001). LD_{50} values are the standard for comparison of acute toxicity between chemical compounds and between species. Some values are given in Table 1.1. It is important to note that the higher the LD_{50}, the less toxic the substance.

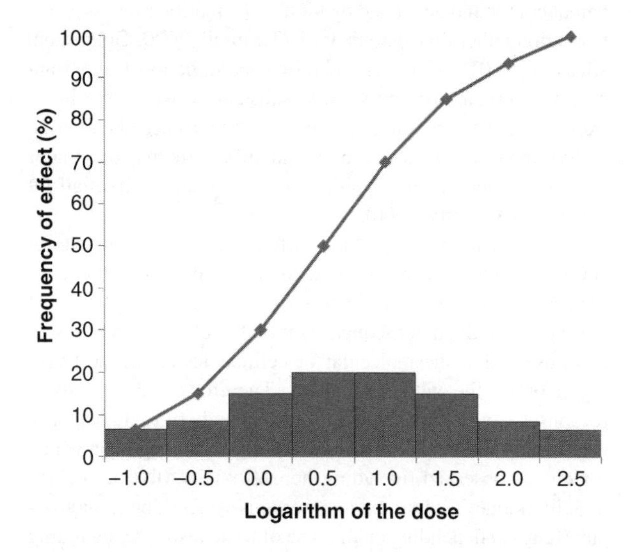

Figure 1.2 Quantal dose–response relationship. Bar chart shows the proportion of individuals affected at each dose and the line shows the cumulative frequency

Table 1.1 Acute LD_{50} of some well-known substances (adapted from Eaton and Klaassen, 2001)

Agent	LD_{50}, mg/kg body weight
Ethyl alcohol	10,000
Sodium chloride	4,000
Ferrous sulphate	1,500
Morphine sulphate	900
Phenobarbital sodium	150
Strychnine sulphate	2
Nicotine	1
Dioxin (TCDD)	0.001
Botulimum toxin	0.00001

Similarly, the EC_{50}, the median effective dose, is the quantity of the chemical that is estimated to have an effect in 50 per cent of the organisms. However, median doses alone are not very informative, as they do not convey any information on the shape of the dose–response curve. This is best illustrated by Figure 1.3. While toxicant A seems (always) more toxic than toxicant B on the basis of its lower LD_{50}, toxicant B will start affecting organisms at lower doses (lower threshold) while the steeper slope for the dose–response curve for toxicant A means that once individuals become overexposed (exceed the threshold dose) the increase in response occurs over much smaller increments in dose.

1.2.4.1 Low dose responses

The classic paradigm for extrapolating dose–response relationships at low doses is based on the concept of threshold for non-carcinogens, whereas for carcinogens it is assumed that there is no threshold and a linear relationship is hypothesised (Figures 1.4 and 1.5).

The NOAEL (No Observed Adverse Effect Level) is the exposure level at which there is no statistically or biologically significant increase in the frequency or severity of adverse effects between exposed population and its appropriate control. The NOEL for the most sensitive test species and the most sensitive indicator of toxicity is usually employed for regulatory purposes. The LOAEL (Lowest Observed Adverse Effect Level) is the lowest exposure level at which there is a statistically or biologically significant increase in the frequency or severity of adverse effects between exposed population and its appropriate control. The main criticism of NOAEL and LOAEL is that they are dependent on study design, i.e. the dose groups selected and the number of individuals in each group. Statistical methods of deriving the concentration that produces a specific effect EC_x,

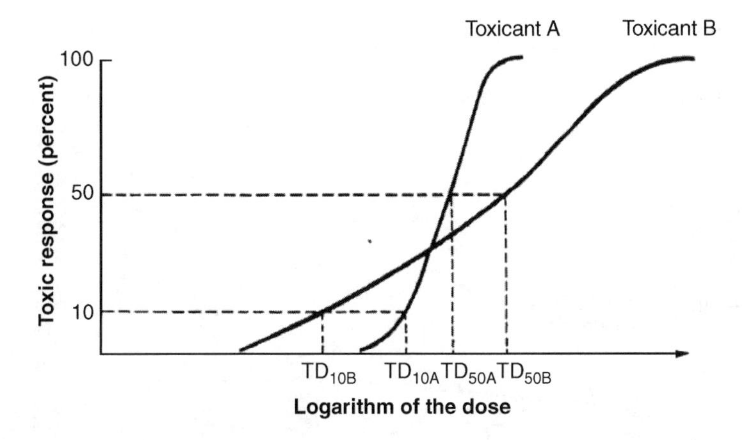

Figure 1.3 Importance of the dose–response relationship (Reproduced from *Principles of Toxicology: environmental and industrial applications*, R. C. James, © 2000 by John Wiley & Sons, with permission from John Wiley & Sons Inc.)

or a benchmark dose (BMD), the statistical lower confidence limit on the dose that produces a defined response (the benchmark response or BMR), are increasingly preferred.

To understand the risk that environmental contaminants pose to human health requires the extrapolation of limited data from animal experimental studies to the low doses critically encountered in the environment. Such extrapolation of dose–response relationships at low doses is the source of much controversy. Recent advances in the statistical analysis of very large populations exposed to ambient concentrations of environmental pollutants have not observed thresholds for cancer or non-cancer outcomes (White *et al.*,2009). The actions of chemical agents are triggered by complex molecular and cellular events that may lead to cancer and non-cancer outcomes in an organism. These processes may be linear or non-linear at an individual level. A thorough understanding of critical steps in a toxic process may help refine current assumptions about thresholds (Boobis *et al.*, 2009). The dose–response curve, however, describes the response or variation in sensitivity of a population. Biological and statistical attributes such as population variability, additivity to pre-existing conditions or diseases induced at background exposure will tend to smooth and linearise the dose–response relationship, obscuring individual thresholds.

1.2.4.2 Hormesis

Dose–response relationships for substances that are essential for normal physiological function and survival are actually U-shaped. At very low doses, adverse effects are observed due to a deficiency. As the dose of such an essential nutrient is increased, the adverse effect is no longer detected and the organism can function normally in a state of homeostasis. Abnormally high doses, however, can give rise to a toxic

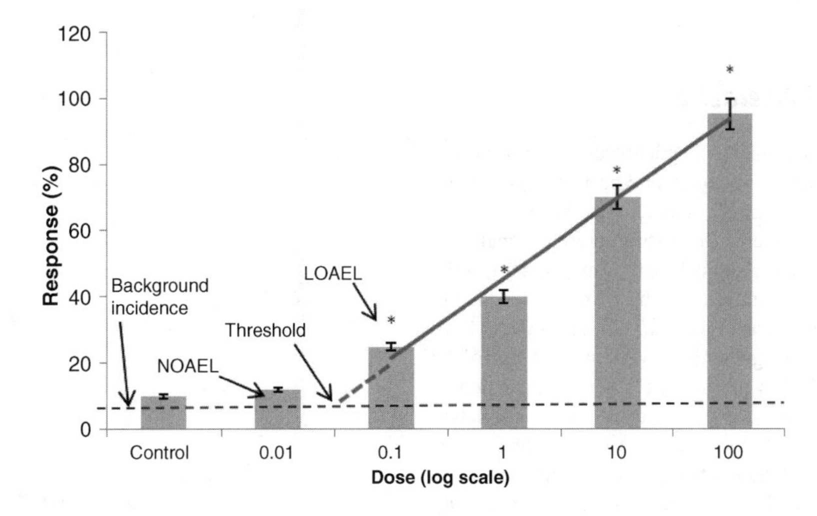

Figure 1.4 Extrapolation of the dose–response relationship at low doses for non-carcinogens (threshold concept). Vertical lines represent the standard error and * denotes statistical significance

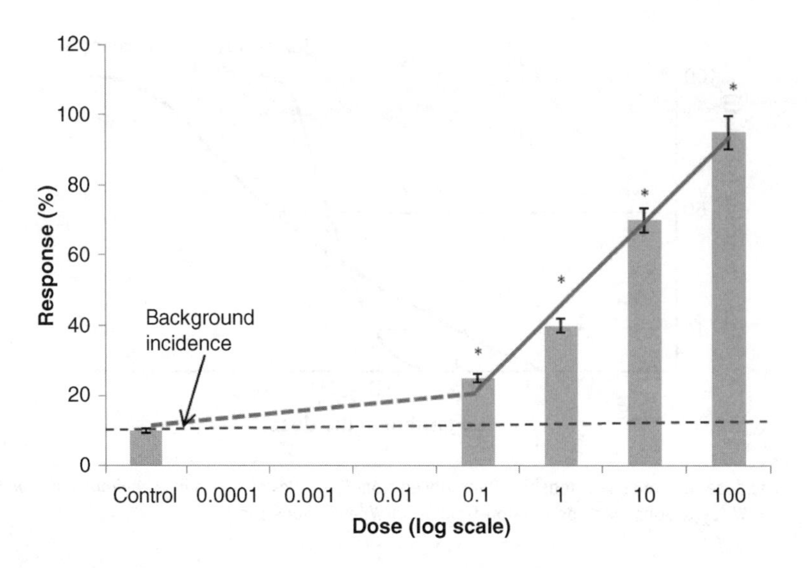

Figure 1.5 Extrapolation of the dose–response relationship at low doses for carcinogens (assumes no threshold). Vertical lines represent the standard error and * denotes statistical significance

response. This response may be qualitatively different and the toxic endpoints measured at very low and very high doses are not necessarily the same.

There is evidence that non-essential substances may also impart an effect at very low doses (Figure 1.6). Some authors have argued that hormesis, defined as beneficial stimulatory effects of toxins at low doses, ought to be the default assumption in the risk assessment of toxic substances (Calabrese and Baldwin, 2003). Whether such low dose effects should be considered stimulatory or beneficial is controversial. Further, potential implications of the concept of hormesis for the risk management of the combinations of the wide variety of environmental contaminants present at low doses that individuals with variable sensitivity may be exposed to are at best unclear.

1.2.5 *Chemical interactions*

In regulatory hazard assessment, chemical hazards are typically considered on a compound-by-compound basis, the possibility of chemical interactions being accounted for by the use of safety or uncertainty factors. Mixture effects still represent a challenge for the risk management of chemicals in the environment, as the presence of one chemical may alter the response to another chemical. The simplest interaction is *additivity*: the effect of two or more chemicals acting together is equivalent to the sum of the effects of each chemical in the mixture when acting independently. *Synergism* is more complex and describes a situation when the presence of both chemicals causes an effect that is greater than the sum of their effects when acting alone. In *potentiation,* a substance that does not produce specific toxicity on its own increases the toxicity of another substance when both are present. *Antagonism* is the principle upon which antidotes

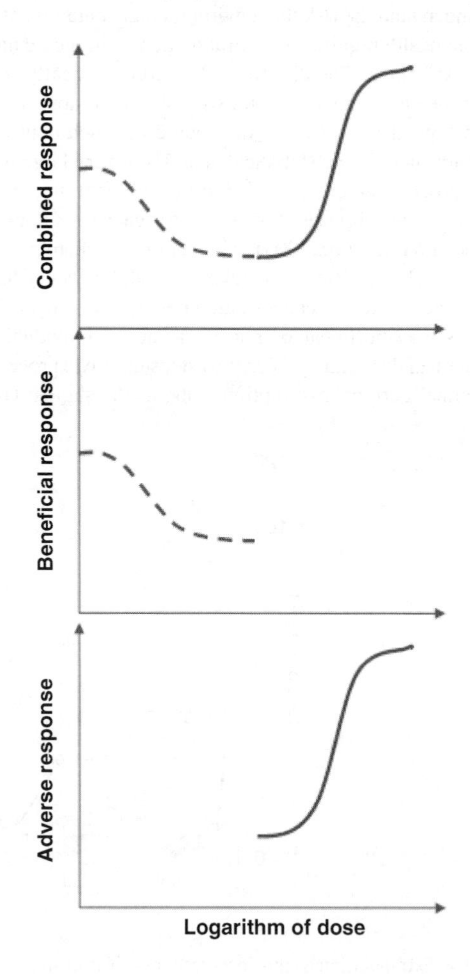

Figure 1.6 Hypothetical hormetic dose–response

Table 1.2 Mathematical representations of chemical interactions (reproduced from James *et al.*, 2000)

Effect	Hypothetical mathematical illustration	Example
Additive	$2 + 3 = 5$	Organophosphate pesticides
Synergistic	$2 + 3 = 20$	Cigarette smoking + asbestos
Potentiation	$2 + 0 = 10$	Alcohol + carbon tetrachloride
Antagonism	$6 + 6 = 8$ or	Toluene + benzene
	$5 + (-5) = 0$ or	Caffeine + alcohol
	$10 + 0 = 2$	Dimercaprol + mercury

are based, whereby a chemical can reduce the harm caused by a toxicant (James *et al.*, 2000; Duffus, 2006). Mathematical illustrations and examples of known chemical interactions are given in Table 1.2.

There are four main ways in which chemicals may interact (James *et al.*, 2000);

1. Functional: both chemicals have an effect on the same physiological function.
2. Chemical: a chemical reaction between the two compounds affects the toxicity of one or both compounds.
3. Dispositional: the absorption, metabolism, distribution or excretion of one substance is increased or decreased by the presence of the other.
4. Receptor-mediated: when two chemicals have differing affinity and activity for the same receptor, competition for the receptor will modify the overall effect.

1.2.6 Relevance of animal models

A further complication in the extrapolation of the results of toxicological experimental studies to humans, or indeed other untested species, is related to the anatomical, physiological and biochemical differences between species. This requires some previous knowledge of the mechanism of toxicity of a chemical and the comparative physiology of different test species. When adverse effects are detected in screening tests, these should be interpreted with the relevance of the chosen animal model in mind. For the derivation of safe levels, safety or uncertainty factors are again usually applied to account for the uncertainty surrounding inter-species differences (James *et al.*, 2000; Sullivan, 2006).

1.2.7 A few words about doses

When discussing dose–response, it is also important to understand which dose is being referred to and to differentiate between concentrations measured in environmental media and the concentration that will elicit an adverse effect at the target organ or tissue. The exposure dose in a toxicological testing setting is generally known or can be readily derived or measured from concentrations in media and average consumption (of food or water for example) (Figure 1.7). Whilst toxicokinetics help to develop an understanding of the relationship between the internal dose and a known exposure dose, relating concentrations in environmental media to the actual exposure dose, often via multiple pathways, is in the realm of exposure assessment.

1.2.8 Other hazard characterisation criteria

It is important to understand the difference between hazard and risk. Hazard is defined as the potential to produce harm; it is therefore an inherent qualitative attribute of a given chemical substance. Risk, on the other hand, is a quantitative measure of the magnitude of the hazard and the probability of it being realised. Hazard assessment is therefore the first step of risk assessment, followed by exposure assessment and finally risk characterisation.

'Carcinogenic, mutagenic or reprotoxic' (CMR) is a designation applied by manufacturers and legislative bodies in the EU to chemicals identified as hazardous substances capable of: initiating cancer; increasing the frequency of changes in an organism's genetic material above their natural background level; and/or harm the ability of organisms to successfully reproduce. Chemical producers wishing to produce or import chemicals in quantities greater than 1 t per year in the EU are now obliged to carry out standardised toxicological tests to identify

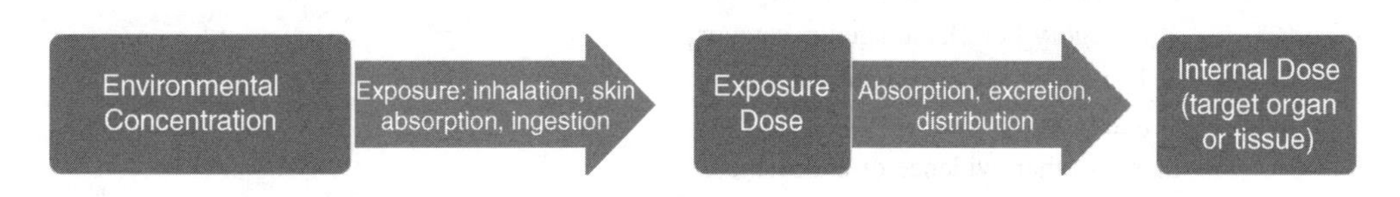

Figure 1.7 Relationships between environmental concentration, exposure dose and internal dose

CMR properties. Formerly, this information was forwarded to EU designated member-state authorities to be classified by the European Chemicals Bureau (Langezaal, 2002). Since 2007, however, regulatory functions have been passed over to the European Chemicals Agency, which has replaced the European Chemicals Bureau and plays a central role in coordinating the implementation of the REACH Directive. REACH is the legislation now in place for the **R**egistration, **E**valuation, **A**uthorisation and Restriction of **Ch**emicals. Substances which are found to have CMR properties are classified alongside persistent, bioaccumulative and toxic substances as Substances of Very High Concern. Their sale and use are strictly regulated (ECHA, 2007).

Toxicity is not the sole criterion evaluated for hazard characterisation purposes. Some chemicals have been found in the tissues of animals in the arctic, for example, where they have never been used or produced. This realisation that some persistent pollutants are able to travel considerable distances and bioaccumulate through the food web has led researchers to take account of such inherent properties of organic compounds, as well as their toxicity, for the purpose of hazard characterisation.

Persistence is the result of resistance to environmental degradation mechanisms such as hydrolysis, photodegradation and biodegradation. Hydrolysis only occurs in the presence of water, photodegradation in the presence of UV light and biodegradation is primarily carried out by micro-organisms. Degradation is related to water solubility, itself inversely related to lipid solubility, so persistence tends to be correlated with lipid solubility (Francis, 1994). The persistence of inorganic substances has proved more difficult to define as they cannot be degraded to carbon and water.

Chemicals may accumulate in environmental compartments and constitute environmental sinks that could be remobilised and lead to toxic effects on organisms. Further, some substances can accumulate in one species without adverse effects but be toxic to its predator(s). *Bioconcentration* refers to accumulation of a chemical from its surrounding environment rather than specifically through food uptake. *Biomagnification* refers to uptake from food without consideration for uptake through the body surface. *Bioaccumulation* integrates both paths, surrounding medium and food. *Ecological magnification* refers to an increase in concentration through the food web from lower to higher trophic levels. Accumulation of organic compounds generally involves transfer from a hydrophilic to a hydrophobic phase and correlates well with the n-octanol/water partition coefficient (Herrchen, 2006).

Persistence and bioaccumulation of a substance is evaluated by standardised OECD tests. Criteria for the identification of persistent, bioaccumulative and toxic substances (PBT), and very persistent and very bioaccumulative substances (vPvB) as defined in Annex XIII of the European Directive on the Registration, Evaluation, Authorisation and Restriction of Chemicals (REACH) (European Union, 2006) are given in Table 1.3. To be classified as a PBT or vPvB, a given substance must fulfil each criterion.

Table 1.3 REACH criteria for identifying PBT and vPvB chemicals

Criterion	PBT criteria	vPvB criteria
Persistence	Either: Half-life > 60 days in marine water Half-life > 60 days in fresh or estuarine water Half-life > 180 days in marine sediment Half-life > 120 days in fresh or estuarine sediment Half-life > 120 days in soil	Either: Half-life > 60 days in marine, fresh or estuarine water Half-life > 180 days in marine, fresh or estuarine sediment Half-life > 180 days in soil
Bioaccumulation	Bioconcentration factor (BCF) > 2000	Bioconcentration factor (BCF) > 2000
Toxicity	Either: Chronic no-observed effect concentration (NOEC) < 0.01 mg/l substance is classified as carcinogenic (category 1 or 2), mutagenic (category 1 or 2), or toxic for reproduction (category 1, 2 or 3) there is other evidence of endocrine-disrupting effects	

> ***Key points***
>
> - Toxicological studies generate experimental data to further our understanding of the mode of action and/or toxicity of a xenobiotic.
> - Dose–response relationships are central to the prediction of toxicity in regulatory risk assessment.
> - There are many uncertainties in the extrapolation of such results to the prediction of actual risks from environmental exposure to low levels of numerous pollutants via multiple routes.

1.3 Some notions of environmental epidemiology

Epidemiology is an observational approach to the study of associations between environment and disease. It can be defined as 'the study of how often diseases occur and why, based on the measurement of disease outcome in a study sample in relation to a population at risk.' (Coggon *et al.*, 2003). Environmental epidemiology refers to the study of distribution patterns of disease and health related to exposures that are exogenous and involuntary. Such exposures generally occur in the air, water, diet or soil and include physical, chemical and biological agents. The extent to which environmental epidemiology is considered to include social, political, cultural and engineering or architectural factors affecting human contact with such agents varies according to authors. In some contexts, the environment can refer to all non-genetic factors, although dietary habits are generally excluded, despite the facts that some deficiency diseases are environmentally determined and nutritional status may also modify the impact of an environmental exposure (Steenland and Savitz, 1997; Hertz-Picciotto, 1998).

Most of environmental epidemiology is concerned with endemics (acute or chronic disease occurring at relatively low frequency in the general population due partly to a common and often unsuspected exposure) rather than epidemics (acute outbreaks of disease affecting a limited population shortly after the introduction of an unusual known or unknown agent). Measuring such low-level exposure to the general public may be difficult – if not impossible – particularly when seeking historical estimates of exposure to predict future disease. Estimating very small changes in the incidence of health effects of low-level common multiple exposure on common diseases with multifactorial aetiologies is particularly difficult, because often greater variability may be expected for other reasons and environmental epidemiology has to rely on natural experiments which, unlike controlled experiments, are subject to other, often unknown, risk factors and variables. However, environmental epidemiology may still be important from a public-health perspective, as small effects in a large population can have large attributable risks if the disease is common (Steenland and Savitz, 1997; Coggon *et al.*, 2003).

1.3.1 Definitions

1.3.1.1 What is a case?

The definition of a case generally requires a dichotomy, i.e. for a given condition people can be divided into two discrete classes – the affected and the non-affected. It increasingly appears that diseases exist in a continuum of severity within a population rather than an all-or-nothing phenomenon. For practical reasons, a cut-off point to divide the diagnostic continuum into 'cases' and 'non-cases' is therefore required. This can be done on a statistical, clinical, prognostic or operational basis. On a statistical basis, the 'norm' is often defined as within two standard deviations of the age-specific mean, thereby arbitrarily fixing the frequency of abnormal values at around 5 per cent in every population. Moreover, it should be noted that what is usual is not necessarily good. A clinical case may be defined by the level of a variable above which symptoms and complications have been found to become more frequent. On a prognostic basis, some clinical findings may carry an adverse prognosis, yet be symptomless. When none of the other approaches is satisfactory, an operational threshold will need to be defined, e.g. based on a threshold for treatment (Coggon *et al.*, 2003).

1.3.1.2 Incidence, prevalence and mortality

The *incidence* of a disease is the rate at which new cases occur in a population during a specified period or frequency of incidents.

$$\text{Incidence} = \frac{\text{Number of new cases}}{\text{Population at risk} \times \text{time during which cases were ascertained}}$$

The *prevalence* of a disease is the proportion of the population that are cases at a given point in time. This measure is appropriate only in relatively stable conditions and is unsuitable for acute disorders. Even in a chronic disease, the manifestations are often intermittent and a point prevalence will tend to underestimate the frequency of the condition. A better measure when possible is the period prevalence, defined as the proportion of a population that are cases at any time within a stated period.

$$\text{Prevalence} = \text{incidence} \times \text{average duration}$$

In studies of aetiology, incidence is the most appropriate measure of disease frequency, as different prevalences result from differences in survival and recovery as well as incidence.

Mortality is the incidence of death from a disease (Coggon *et al.*, 2003).

1.3.1.3 Interrelation of incidence, prevalence and mortality

Each incident case enters a prevalence pool and remains there until either recovery or death:

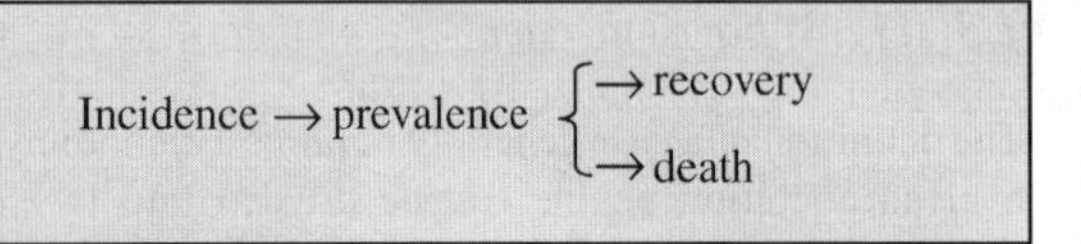

A chronic condition will be characterised by both low recovery and low death rates, and even a low incidence will produce a high prevalence (Coggon *et al.*, 2003).

1.3.1.4 Crude and specific rates

A crude incidence, prevalence or mortality is one that relates to results for a population taken as a whole, without subdivisions or refinement. To compare populations or samples, it may be helpful to break down results for the whole population to give rates specific for age and sex (Coggon *et al.*, 2003).

1.3.1.5 Measures of association

Several measures are commonly used to summarise association between exposure and disease.

Attributable risk is most relevant when making decisions for individuals and corresponds to the difference between the disease rate in exposed persons and that in unexposed persons. The *population attributable risk* is the difference between the rate of disease in a population and the rate that would apply if all of the population were unexposed. It can be used to estimate the potential impact of control measures in a population.

Population attributable risk
= attributable risk × prevalence of exposure to risk factor

The *attributable proportion* is the proportion of disease that would be eliminated in a population if its disease rate were reduced to that of unexposed persons. It is used to compare the potential impact of different public-health strategies.

The *relative risk* is the ratio of the disease rate in exposed persons to that in people who are unexposed.

Attributable risk
= rate of disease in unexposed persons × (relative risk − 1)

Relative risk is less relevant to risk management but is nevertheless the measure of association most commonly used because it can be estimated by a wider range of study designs. Additionally, where two risk factors for a disease act in concert, their relative risks have often been observed empirically to come close to multiplying.

The *odds ratio* is defined as the odds of disease in exposed persons divided by the odds of disease in unexposed persons (Coggon *et al.*, 2003).

1.3.1.6 Confounding

Environmental epidemiological studies are observational, not experimental, and compare people who differ in various ways, known and unknown. If such differences happen to determine risk of disease independently of the exposure under investigation, they are said to confound its association with the disease and the extent to which observed association are causal. It may equally give rise to spurious associations or obscure the effects of a true cause (Coggon *et al.*, 2003). A confounding factor can be defined as a variable which is both a risk factor for the disease of interest, even in the absence of exposure (either causal or in association with other causal factors), and is associated with the exposure but not a direct consequence of the exposure (Rushton, 2000).

In environmental epidemiology, nutritional status suggests potential confounders and effect modifiers of associations between environment and disease. Exposure to environmental agents is also frequently determined by social factors: where one lives, works, socialises or buys food, and some argue that socio-economic context is integral to most environmental epidemiology problems (Hertz-Picciotto, 1998).

Standardisation is usually used to adjust for age and sex, although it can be applied to account for other confounders. Other methods include mathematical modelling techniques such as logistic regression and are readily available. They should be used with caution, however, as the mathematical assumptions in the model may not always reflect the realities of biology (Coggon *et al.*, 2003).

1.3.1.7 Standardisation

Direct standardisation is suitable only for large studies, and entails the comparison of weighted averages of age and sex-specific disease rates, the weights being equal to the proportion of people in each age and sex group in a reference population.

In most surveys the indirect method yields more stable risk estimates. Indirect standardisation requires a suitable reference population for which the class-specific rates are known for comparison with the rates obtained for the study sample (Coggon *et al.*, 2003).

1.3.2 Measurement error and bias

1.3.2.1 Bias

Bias is a systematic tendency to underestimate or overestimate a parameter of interest because of a deficiency in the design or execution of a study. In epidemiology, bias results in a difference between the estimated association between exposure and disease and the true association. Three general types of bias can be identified: selection bias, information bias, and confounding bias. *Information bias* arises from errors in measuring exposure or disease, and the information is wrong to the extent that the relationship between the two can no longer be correctly estimated. *Selection bias* occurs when the subjects studied are not representative of the target population about which conclusions are to be drawn. It generally arises because of the way subjects are recruited or the way cases are defined (Bertollini *et al.*, 1996; Coggon *et al.*, 2003).

1.3.2.2 Measurement error

Errors in exposure assessment or disease diagnosis can be important sources of bias in epidemiological studies, and it is therefore important to assess the quality of measurements. Errors may be differential (different for cases and controls) or non-differential. Non-differential errors are more likely to occur than differential errors and have until recently been assumed to tend to diminish risk estimates and dilute exposure-response gradients (Steenland and Savitz, 1997). Non-differential misclassification is related to both the precision and the magnitude of the differences in exposure or diagnosis within the population. If these differences are substantial, even a fairly imprecise measurement would not lead to much misclassification. A systematic investigation of the relative precision of the measurement of the exposure variable should ideally precede any study in environmental epidemiology (Bertollini *et al.*, 1996; Coggon *et al.*, 2003).

1.3.2.3 Validity

The validity of a measurement refers to the agreement between this measure and the truth. It is potentially a more serious problem than a systematic error, because in the latter case the power of a study to detect a relationship between exposure and disease is not compromised. When a technique or test is used to dichotomise subjects, its validity may be analysed by comparison with results from a standard reference test. Such analysis will yield four important statistics: sensitivity, specificity, systematic error and predictive value. It should be noted that both systematic error and predictive value depend on the relative frequency of true positives and true negatives in the study sample (prevalence of the disease or exposure being measured) (Bertollini *et al.*, 1996; Coggon *et al.*, 2003).

1.3.2.4 Repeatability

When there is no satisfactory standard against which to assess the validity of a measurement technique, then examining the repeatability of measurements within and between observers can offer useful information. Whilst consistent findings do not necessarily imply that a technique is valid, poor repeatability does indicate either poor validity or that the measured parameter varies over time. When measured repeatedly in the same subject, physiological or other variables tend to show a roughly normal distribution around the subject's mean. Misinterpretation can be avoided by repeat examinations to establish an adequate baseline, or by including a control group. Conversely, conditions and timing of an investigation may systematically bias subjects' response and studies should be designed to control for this.

The repeatability of measurements of continuous variables can be summarised by the standard deviation of replicate measurements or by their coefficient of variation. Within-observer variation is considered to be largely random, whilst between-observer variation adds a systematic component due to individual differences in techniques and criteria to the random element. This problem can be circumvented by using a single observer or, alternatively, allocating subjects to observers randomly. Subsequent analysis of results by observers should highlight any problem and may permit statistical correction for bias (Coggon *et al.*, 2003).

1.3.3 Exposure assessment

The quality of exposure measurement underpins the validity of an environmental epidemiology study. Assessing exposure on an ever/never basis is often inadequate because the certainty of exposure may be low and a large range of exposure levels with potentially non-homogeneous risks are grouped together. Ordinal categories provide the opportunity to assess dose–response relations, whilst quantified measures, where possible, also allow researchers to assess comparability across studies and can provide the basis for regulatory decision making. Instruments for exposure assessment include (Hertz-Picciotto, 1998):

- interviews, questionnaires, and structured diaries;
- measurement in the macro-environment, either conducted directly or obtained from historical records;
- concentration in the personal micro-environment;
- biomarkers of physiological effect in human tissues or metabolic products.

All questionnaires and interview techniques rely on human knowledge and memory, and hence are subject to error and recall bias. Cases tend to report exposure more accurately than controls and this biases risk estimates upwards and could lead to false positive results. There are techniques that can be applied to detect this bias, such as including individuals with a disease unrelated to the exposure of interest, probing subjects about the understanding of the relationship between the disease and exposure under study, or attempting to corroborate information given by a sample of the cases and controls through records, interviews, or environmental or biological monitoring. Interviews either face-to-face or on the phone may also elicit under-reporting of many phenomena subject to the 'desirability' of the activity being reported. Self-administered questionnaires or diaries can avoid interviewer influences but typically have lower response rates and do not permit the collection of complex information (Bertollini *et al.*, 1996; Hertz-Picciotto, 1998).

A distinction has been made between exposure measured in the external environment, at the point of contact between the subject and its environment, and measurements made in human tissue or sera. Measurements in external media yield an ecological measure and are useful when group differences outweigh inter-individual differences. Macro-environment measures are also more relevant to the exposure context rather than to individual pollutants. Sometimes, the duration of contact (or potential contact) can be used as a surrogate quantitative measure, the implicit assumption being that duration correlates with cumulative exposure. When external measurements are available, they can be combined with duration and timing of residence and activity-pattern information to assign quantitative exposure estimates for individuals. Moreover, many pollutants are so dispersed in the environment that they can reach the body through a variety of environmental pathways (Bertollini *et al.*, 1996; Hertz-Picciotto, 1998).

The realisation that human exposure to pollutants in micro-environments may differ greatly from those in the general environment was a major advance in environmental epidemiology. It has led to the development of instrumentation suitable for micro-environmental and personal monitoring and sophisticated exposure models. Nonetheless, these estimates of individual absorbed doses still do not account for inter-individual differences due to breathing rate, age, sex, medical conditions, and so on (Bertollini *et al.*, 1996; Hertz-Picciotto, 1998).

The pertinent dose at the target tissue depends on toxicokinetics, metabolic rates and pathways that could either produce the active compound or detoxify it, as well as storage and retention times, and elimination. Measuring and modelling of integrated exposure to such substances are difficult at best, and, when available, the measurement of biomarkers of internal doses will be the preferred approach. Whilst biomarkers can account for individual differences in pharmacokinetics, they do not tell us which environmental sources and pathways are dominating exposure and in some situations they could be poor indicators of past exposure. Moreover, many pollutants are so dispersed in the environment that they can reach the body through a variety of environmental pathways (Bertollini *et al.*, 1996; Hertz-Picciotto, 1998).

To study diseases with long latency periods, such as cancer, or those resulting from long-term chronic insults, exposures or residences at times in the past are more appropriate. Unfortunately, reconstruction of past exposures is often fraught with problems of recall, incomplete measurements of external media, or inaccurate records that can no longer be validated, and retrospective environmental exposure assessment techniques are still in their infancy (Bertollini *et al.*, 1996; Hertz-Picciotto, 1998).

1.3.4 Types of studies

1.3.4.1 Ecological studies

In ecological studies, the unit of observation is the group, a population or a community, rather than the individual. The relation between disease rates and exposures in each of a series of populations is examined. Often the information about disease and exposure is abstracted from published statistics such as those published by the World Health Organisation (WHO) on a country-by-country basis. The populations compared may be defined in various ways (Steenland and Savitz, 1997; Coggon *et al.*, 2003):

- Geographically. Care is needed in the interpretation of results, due to potential confounding effects and differences in ascertainment of disease or exposure.
- Time trends or time series. Like geographical studies, analysis of secular trends may be biased by differences in the ascertainment of disease. However, validating secular changes is more difficult as it depends on observations made and often scantily recorded many years ago.
- Migrants studies. These offer a way of discriminating genetic from environmental causes of geographical variation in disease, and may also indicate the age at which an environmental cause exerts its effect. However, the migrants may themselves be unrepresentative of the population they leave, and their health may have been affected by the process of migration.
- By occupation or social class. Statistics on disease incidence and mortality may be readily available for socio-economic or occupational groups. However, occupational data may not include data on those who left this employment, whether on health grounds or not, and different socio-economic groups may have different access to healthcare.

1.3.4.2 Longitudinal or cohort studies

In a longitudinal study subjects are identified and then followed over time with continuous or repeated monitoring of risk factors and known or suspected causes of disease and subsequent morbidity or mortality. In the simplest design, a sample or cohort of subjects exposed to a risk factor is identified along with a sample of unexposed controls. By comparing the incidence rates in the two groups, attributable and relative risks can be estimated. Case-response bias is entirely avoided in cohort studies where exposure is evaluated before diagnosis. Allowance can be made for suspected confounding factors, either by matching the controls to the exposed subjects so that they have similar patterns of exposure to the confounder, or by measuring exposure to the confounder in each group and adjusting for any difference in statistical analysis. One of the main limitations of this method is that when it is applied to the study of chronic diseases a large number of people must be followed up for long periods before sufficient cases accrue to give statistically meaningful results.

When feasible, the follow-up could be carried out retrospectively, as long as the selection of exposed people is not influenced by factors related to their subsequent morbidity. It can also be legitimate to use the recorded disease rates in the national or regional population for control purposes, when exposure to the hazard in the general population is negligible (Bertollini *et al.*, 1996; Coggon *et al.*, 2003).

1.3.4.3 Case-control studies

In a case-control study, patients who have developed a disease are identified and their past exposure to suspected aetiological factors is compared with that of controls or referents that do not have the disease. This allows the estimation of odds ratio but not of attributable risks. Allowance is made for confounding factors by measuring them and making appropriate adjustments in the analysis. This adjustment may be rendered more efficient by matching cases and controls for exposure to confounders, either on an individual basis or in groups. Unlike a cohort study, however, matching does not on its own eliminate confounding, and statistical adjustment is still required (Coggon *et al.*, 2003).

1.3.4.3.1 Selection of cases and controls In general, selecting incident rather than prevalent cases is preferred. The exposure to risk factors and confounders should be representative of the population of interest within the constraints of any matching criteria. It often proves impossible to satisfy both those aims. The exposure of controls selected from the general population is likely to be representative of those at risk of becoming cases, but assessment of their exposure may not be comparable with that of cases due to recall bias, and studies will tend to overestimate risk. Recall bias can be addressed by including a control group composed of patients with other diseases, but their exposure may be unrepresentative, and studies will tend to underestimate risk if the risk factor under investigation is involved in other pathologies. It is therefore safer to adopt a range of control diagnoses rather than a single disease group. Interpretation can also be helped by having two sets of controls with different possible sources of bias. Selecting equal numbers of cases and controls generally makes a study most efficient, but the number of cases available can be limited by the rarity of the disease of interest. In this circumstance, statistical confidence can be increased by taking more than one control per case. There is, however, a law of diminishing returns, and it is usually not worth going beyond a ratio of four or five controls to one case (Coggon *et al.*, 2003).

1.3.4.3.2 Exposure assessment Many case-control studies ascertain exposure from personal recall, using either a self-administered questionnaire or an interview. Exposure can sometimes be established from existing records such General Practice notes. Occasionally, long-term biological markers of exposure can be exploited, but they are only useful if not altered by the subsequent disease process (Coggon *et al.*, 2003).

1.3.4.4 Cross-sectional studies

A cross-sectional study measures the prevalence of health outcomes or determinants of health, or both, in a population at a point in time or over a short period. The risk obtained is disease prevalence rather than incidence. Such information can be used to explore aetiology, but associations must be interpreted with caution. Bias may arise because of selection into or out of the study population, giving rise to effects similar to the healthy-worker effect encountered in occupational epidemiology. A cross-sectional design may also make it difficult to establish what is cause and what is effect. Because of these difficulties, cross-sectional studies of aetiology are best suited to non-fatal degenerative diseases with no clear point of onset and to the pre-symptomatic phases of more serious disorders (Rushton, 2000; Coggon *et al.*, 2003).

1.3.5 Critical appraisal of epidemiological reports

1.3.5.1 Design

A well-designed study should state precisely formulated, written objectives and the null hypothesis to be tested. This should in turn demonstrate the appropriateness of the study design for the hypothesis to be evaluated. Ideally, a literature search of relevant background publications should be carried out in order to explore the biological plausibility of the hypothesis (Elwood, 1998; Rushton, 2000; Coggon *et al.*, 2003).

In order to be able to appraise the selection of subjects, each study should first describe the target population that the study participants are meant to represent. The selection of study participants affects not only how widely the results can be applied but also, more importantly, their validity. The internal validity of a study relates to how well a difference between the two groups being compared can be attributed to the effects of exposure rather than to chance or confounding bias. In contrast, the external validity of a study refers to how well the results can be applied to the general population. Whilst both are desirable, design considerations that help increase the internal validity of a study may decrease its external validity. However, the external validity of a study is only useful if the internal validity is acceptable. The selection criteria should therefore be appraised by considering the effects of potential selection bias on the hypothesis being tested and the external and internal validity of the study population. The selection process itself should be effectively random (Elwood, 1998; Rushton, 2000; Coggon *et al.*, 2003).

The sample size should allow the primary purpose of the study, formulated in precise statistical terms, to be achieved, and its adequacy should be assessed. If it is of particular interest that certain subgroups are relatively over-represented, a stratified random sample can be chosen by dividing the study population into strata and then drawing a separate random sample from each. Two-stage sampling may be adequate when the study population is large and widely scattered but there is some loss of statistical efficiency, especially if only a few units are selected at the first stage (Rushton, 2000; Coggon *et al.*, 2003).

To be able to appraise a study, a clear description of how the main variables were measured should be given. The choice of method needs to allow a representative sample of adequate size to be examined in a standardised and sufficiently valid way. Ideally, observers should be allocated to subjects in a random manner to minimise bias due to observer differences. Importantly, methods and observers should allow rigorous standardisation (Rushton, 2000; Coggon *et al.*, 2003).

1.3.5.2 Bias

Almost all epidemiological studies are subject to bias, and it is important to allow for the probable impact of biases in drawing conclusions. In a well-reported study, this question would already have been addressed by the authors, who may even have collected data to help quantify bias (Coggon *et al.*, 2003).

Selection bias, information bias and confounding have all been discussed in some detail in previous sections, but it is worth mentioning the importance of accurately reporting response rates, as selection bias can also result if participants differ from non-participants. The likely bias resulting from incomplete response can be assessed in different ways: subjects who respond with and without a reminder could be compared, or a small random sample can be drawn from the non-responders and particularly vigorous efforts made to collect some of the information that was originally sought and findings then compared with those of the earlier responders; or differences based on available information about the study population such as age, sex and residence could give an indication of the possibility of bias, and making extreme assumptions about the non-responders can help to put boundaries on the uncertainty arising from non-response (Elwood, 1998).

1.3.5.3 Statistical analysis

Even after biases have been taken into account, study samples may be unrepresentative just by chance. An indication of the potential for such chance effects is provided by statistical analysis and hypothesis testing. There are two kinds of errors that one seeks to minimise. A type-I error is the mistake of concluding that a phenomenon or association exists when in truth it does not; by convention, the rate of such errors is usually set at 5 per cent. A result is therefore called statistically significant, when there is a less than 5 per cent probability to have observed an association in the experiment when such an association does not actually exist. A type-II error, failing to detect an association that actually does exist, is, also by convention, often set at 20 per cent, although this is in fact often determined by practical limitations of sample size (Armitage and Berry, 1994). It is important to note that failure to reject the null hypothesis (i.e. no association) does not equate with its acceptance but only provides reasonable confidence that if any association exists it would be smaller than an effect size determined by the power of the study. The issues surrounding power and effect size should normally be addressed at the design stage of a study, although this is rarely reported (Rushton, 2000).

1.3.5.4 Confounding versus causality

If an association is found and not explained by bias or chance, the possibility of unrecognised residual confounding still remains. Assessment of whether an observed association is causal depends in part on the biological plausibility of the relation. Certain characteristics of the association, such as an exposure–response gradient, may encourage causal interpretation, though in theory it may still arise from confounding. Also important is the magnitude of the association as measured by the relative risk or odds ratio. The evaluation of possible pathogenic mechanisms and the importance attached to exposure–response relations and evidence of latency are also a matter of judgement (Coggon *et al.*, 2003).

1.3.6 Future directions

Some progress has been made in the area of exposure assessment, but more work is needed in integrating biological indicators into exposure assessment, and much remains to be done with respect to timing of exposures as they relate to induction and latency issues.

An obstacle to analysis of multiple exposures is the near impossibility of separating induction periods, dose–response, and interactive effects from one another. These multiple exposures include not only the traditional chemical and physical agents, but should also be extended to social factors as potential effect modifiers.

An emerging issue for environmental epidemiologists is that of variation in susceptibility. This concept is not new: it constitutes the element of the 'host' in an old paradigm of epidemiology that divided causes of disease into environment, host and agent. It has, however, taken on a new dimension with the current technology that permits identification of genes implicated in many diseases. The study of gene–environment interactions as a mean of identifying susceptible subgroups can lead to studies with a higher degree of specificity and precision in estimating effects of exposures.

Key points

- The type of studies required to obtain reliable estimates of long-term risks following chronic exposures are expensive and time-consuming.
- Again, assessing environmental exposure to low levels of pollutants via multiple routes is an issue.
- Inter-individual variability and interactions between environment and disease can also obscure results.

1.4 Scientific evidence and the precautionary principle

1.4.1 Association between environment and disease

Scientific evidence on associations between exogenous agents and health effects is derived from epidemiological and toxicological studies. As discussed previously, both types of methods have advantages and disadvantages, and much uncertainty and controversy stem from the relative weights attributed to different types of evidence. Environmental epidemiology requires the estimation of often very small changes in the incidence of common diseases with multifactorial aetiologies following low-level multiple exposures. For ethical reasons, it is necessarily observational, and natural experiments are subject to confounding and to other, often unknown, risk factors (Steenland and Savitz, 1997; Coggon *et al.*, 2003). Some progress has been made in the development of specific biomarkers, but this is still hindered by issues surrounding the timing of exposures as they relate to induction and latency. Toxicology, on the other hand, allows the direct study of the relationship between the quantity of chemical to which an organism is exposed and the nature and degree of consequent harmful effect. Controlled conditions, however, limit the interpretation of toxicity data, as they generally differ considerably from those prevailing in the natural environment.

1.4.1.1 Bradford-Hill criteria

Since 1965, evaluations of the association between environment and disease have often been based on the nine 'Bradford-Hill criteria' (Hill, 1965).

Bradford-Hill criteria

- Strength
- Consistency
- Specificity
- Temporality
- Biological gradient
- Plausibility
- Coherence
- Experiment
- Analogy

Results from cohort, cross-sectional or case-control studies of not only environmental but also accidental, occupational, nutritional or pharmacological exposure, as well as toxicological studies, can inform all the Bradford-Hill tenets of association between environment and disease. Such studies often include some measure of the *strength of the association* under investigation and its statistical significance. Geographical studies and migrant studies provide some insights into the *consistency of observations*. Consistency of observations between studies of different chemicals exhibiting similar properties, or between studies of different species, should also be considered. Whilst *specificity* provides evidence of specific environment–disease association, the lack of it, or association with multiple endpoints, does not constitute proof against a potential association. Time-trend analyses are directly related to the *temporality* aspect of a putative association, whether trends in environmental release of the chemical agents of interest precedes similar trends in the incidence of disease. This is also particularly relevant in the context of the application of the precautionary principle, as the observation of intergenerational effects in laboratory animals (Newbold *et al.*, 1998, 2000) may raise concerns of 'threats of irreversible damage'. Occasionally, studies are designed to investigate the existence of a *biological gradient* or dose–response. *Plausibility* is related to the state of mechanistic knowledge underlying a putative association, while *coherence* can be related to what is known of the aetiology of the disease. *Experimental evidence* can be derived both from toxicological studies and from natural epidemiological experiments following occupational or accidental exposure. Finally, *analogy,* where an association has been shown for analogous exposure and outcomes, should also be considered.

1.4.2 Precautionary principle

A common rationale for the precautionary principle is that increasing industrialisation and the accompanying pace of technological development and widespread use of an ever-increasing number of chemicals exceed the time needed to test those chemicals adequately and collect sufficient data to form a clear consensus among scientists as to their potential to do harm (Burger, 2003).

The precautionary principle became European Law in 1992 when the Maastricht Treaty modified Article 130r of the treaty establishing the European Economic Community, and in just over a decade it has also been included in several international environmental agreements (Marchant, 2003). The precautionary principle is nonetheless still controversial and lacks a definitive formulation. This is best illustrated by the differences between two well-known definitions of the principle, the Rio Declaration produced in 1992 by the United Nations Conference on Environment and Development and the Wingspread Statement formulated by proponents of the precautionary principle in 1998 (Marchant, 2003). One interpretation of the principle is therefore that uncertainty is not justification to delay the prevention of a potentially harmful action, whilst the other implies that no action should be taken unless it is certain that it will do no harm

(Rogers, 2003). Definitions also differ in the level of harm necessary to trigger action, from 'threats of serious or irreversible damage' to 'possible risks' (Marchant, 2003). Whilst there are situations where risks clearly exceed benefits and vice versa, there is a large grey area in which science alone cannot decide policy (Kriebel *et al.*, 2001), and the proponents of a strong precautionary principle advocate public participation as a means to make environmental decision-making more transparent. This will require the characterisation and efficient communication of scientific uncertainty to policy-makers and the wider public, and scientific uncertainty is a well-known 'dread factor', increasing the public's perception of risk (Slovic, 1987).

> 'When there are threats of serious and irreversible damage, lack of full scientific certainty shall not be used as a reason for postponing cost-effective measures to prevent environmental degradation'
>
> *Rio Declaration*
>
> 'When an activity raises threats of harms to human health or the environment, precautionary measures should be taken even if some cause and effect relationships are not fully established scientifically'
>
> *Wingspread Statement*

1.4.2.1 Sufficiency of evidence

The European Environment Agency *Late Lessons* report (2001) provided a working definition of the precautionary principle, and this was improved following further discussions and legal developments.

> 'The Precautionary Principle provides justification for public policy actions in situations of scientific complexity, uncertainty and ignorance, where there may be a need to act in order to avoid, or reduce, potentially serious or irreversible threats to health or the environment, using an appropriate level of scientific evidence, and taking into account the likely pros and cons of action and inaction.'
>
> *(Gee, 2006)*

It specifies complexity, uncertainty and ignorance as contexts where the principle may be applicable and makes explicit mention that precautionary actions need to be justified by a sufficiency of scientific evidence. The report also offers a clarification of the terms Risk, Uncertainty and Ignorance and corresponding states of knowledge with some examples of proportionate actions (Table 1.4).

1.5 Uncertainty and controversy: the endocrine disruption example

More than 10 years after the publication of Theo Colborn's *Our Stolen Future* (Colborn *et al.*, 1996), endocrine disruption probably remains one of the most controversial current environmental issues. News stories about the potential effects of 'gender-bending chemicals' on unborn male fetuses are still being printed in some sections of the general media, while by virtue of the precautionary principle, the term 'endocrine disrupters' can be found in emerging European environmental legislation, such as the Water Framework Directive or the REACH proposal (European Community, 2000; Commission

Table 1.4 Examples of precautionary actions and the scientific evidence justifying them (reproduced from European Environment Agency, 2001)

Situation	State and dates of knowledge	Examples of action
Risk	'Known impacts'; 'known probabilities'; e.g. asbestos causing respiratory disease, lung and mesothelioma cancer, 1965–present	Prevention: action taken to reduce known hazards; e.g. eliminate exposure to asbestos dust
Uncertainty	'Known' impacts; 'unknown' probabilities; e.g. antibiotics in animal feed and associated human resistance to those antibiotics, 1969–present	Precautionary prevention: action taken to reduce potential risks; e.g. reduce/eliminate human exposure to antibiotics in animal feed
Ignorance	'Unknown' impacts and therefore 'unknown'probabilities; e.g. the 'surprises' of chlorofluorocarbons (CFCs) and ozone layer damage prior to 1974; asbestos mesothelioma cancer prior to 1959	Precaution: action taken to anticipate, identify and reduce the impact of 'surprises'; e.g. use of properties of chemicals such as persistence or bioaccumulation as 'predictors' of potential harm; use of the broadest possible sources of information, including long-term monitoring; promotion of robust, diverse and adaptable technologies and social arrangements to meet needs, with fewer technological 'monopolies' such as asbestos and CFCs

of the European Communities, 2003). It is therefore of interest to consider here what makes endocrine disruption such a challenging topic for environmental toxicologists.

1.5.1 Emergence of the 'endocrine disruption' hypothesis

The realisation that human and animal hormonal function could be modulated by synthetic variants of endogenous hormones is generally attributed to a British scientist, Sir Edward Charles Dodds (1889–1973), a professor of biochemistry at the Middlesex Hospital Medical School at the University of London, who had won international acclaim for his synthesis of the oestrogen diethylstilbestrol (DES) in 1938, subsequently prescribed for a variety of gynaecologic conditions, including some associated with pregnancy (Krimsky, 2000). By then, it was also known that the sexual development of both male and female rodents could be disrupted by prenatal exposure to sex hormones (Greene *et al.*, 1938). It was not until 1971, however, that an association was made between DES exposure *in utero* and a cluster of vaginal clear-cell adenocarcinoma in women under 20, an extremely rare type of cancer for this age group (Herbst *et al.*, 1971). It took another 10 years to link DES prescription to pregnant women to other genital-tract abnormalities in their progeny.

Meanwhile, Rachel Carson famously associated the oestrogenic pesticide *o,p*-dichlorodiphenyltrichloroethane (DDT) with eggshell thinning in her book *Silent Spring* (Carson, 1962). Until then, overexploitation and habitat destruction were considered the most significant causes of declining wildlife populations. Pesticides were subsequently found in tissues of wildlife from remote parts of the world, and Carson's observation that these concentrations increased with trophic levels, a process called biomagnification, was verified. Nevertheless, it took a further 30 years for the endocrine disruption hypothesis to emerge as a result of the convergence of several separate lines of enquiry.

In 1987, Theo Colborn began an extensive literature search on toxic chemicals in the Great Lakes. Wildlife toxicology had previously concentrated on acute toxicity and cancer, but Colborn found that reproductive and developmental abnormalities were more common than cancer and effects were often observed in the offspring of exposed wildlife (Colborn *et al.*, 1996). Another path to the generalised endocrine hypothesis originated from studies of male infertility and testicular cancer. The advent of artificial insemination was accompanied by the development of techniques to assess sperm quality and such information began to be recorded. In the early 1970s, Skakkebaek, a Danish paediatric endocrinologist, noticed a group of cells resembling fetal cells in the testes of men diagnosed with testicular cancer and began to suspect that testicular cancer had its origin in fetal development. A study of 'normal' subjects in the mid-1980s found that 50 per cent of these males had abnormal sperm; it was suggested that environmental factors may be at work, and oestrogenic compounds were suspected (Carlsen *et al.*, 1992).

Although the concept of endocrine disruption first developed when it was observed that some environmental chemicals were able to mimic the action of the sex hormones (oestrogens and androgens), it has now evolved to encompass a range of mechanisms involving the many hormones secreted directly into the blood circulatory system by the glands of the endocrine system and their specific receptors and associated enzymes (Harvey *et al.*, 1999).

1.5.2 Definitions

Many national and international agencies have proposed their definitions for endocrine disrupters. One of the most commonly used definitions is referred to as the 'Weybridge' definition and was drafted at a major European Workshop in December 1996.

> 'The Weybridge definition: An endocrine disrupter is an exogenous substance that causes adverse health effects in an intact organism, or its progeny, subsequent to changes in endocrine function.'
>
> *(MRC Institute for Environment and Health, 1997)*
>
> 'An endocrine disruptor is an exogenous agent that interferes with the synthesis, secretion, transport, binding, action, or elimination of natural hormones in the body that are responsible for the maintenance of homeostasis, reproduction, development and/or behaviour.'
>
> *(EPA, 1997)*
>
> 'An endocrine disruptor is an exogenous substance or mixture that alters function(s) of the endocrine system and consequently causes adverse health effects in an intact organism or its progeny or (sub)populations.'
>
> *(IPCS, 2002)*

A major issue with the Weybridge definition is the use of the term 'adverse'. For a chemical to be considered an endocrine disrupter, its biological effect must amount to an adverse effect on the individual or population and not just a change that falls within the normal range of physiological variation (Barker, 1999).

The US Environmental Protection Agency Risk Assessment Forum's definition focuses more on any biological change regardless of amplitude.

The International Programme on Chemical Safety (IPCS) modified the Weybridge definition to clarify the fact that endocrine disruption is a mechanism that explains a biological effect.

1.5.3 Modes of action

1.5.3.1 The endocrine system

There are two main systems by which cells of metazoan organisms communicate with each other.

The nervous system serves for rapid communication using chains of interconnected neurones transmitting transient impulses and also producing chemicals called neurotransmitters, which are rapidly destroyed at the synapses. Such responses are generally associated with sensory stimuli.

The endocrine system uses circulating body fluids such as blood to carry chemical messengers secreted by ductless glands to specific receptors non-uniformly distributed on target organs or tissues that are physicochemically programmed to react and respond to them (Highnam and Hill, 1977; Bentley, 1998; Hale *et al.*, 2002). These messengers, referred to as hormones, have a longer biological life and are therefore suited to control long-term processes within the body, such as growth, development, reproduction and homeostasis. Recently, the number of endogenous chemicals found to have hormonal activity has increased dramatically. Many are local hormones (paracrine or autocrine), delivered to their target organ by non-endocrine routes (Harvey *et al.*, 1999).

Nerves and hormones are often mutually interdependent. Central nervous activity in most animals is likely to be strongly affected by hormones, and hormone production and release are dependent on nervous activity (Highnam and Hill, 1977; Bentley, 1998). Similarly, the endocrine system is known to influence and be influenced by the immune system.

1.5.3.2 Levels of effect

To understand the significance of endocrine disruption, it is necessary to determine whether there is a causal relationship between an environmental factor and an observed effect. Endocrine disruption is not a toxicological endpoint *per se* but a functional change that may or may not lead to adverse effects. Endocrine disruption can be observed at different levels, and each level of observation gives a different insight into the mode of action of an endocrine disrupter. At the cellular level, the information is gained regarding the potential mechanism of action of a contaminant, whilst at the population level, a greater understanding of the ecological significance of such mechanism is gained. A classification of the different levels at which endocrine disruption can be observed is proposed in Figure 1.8.

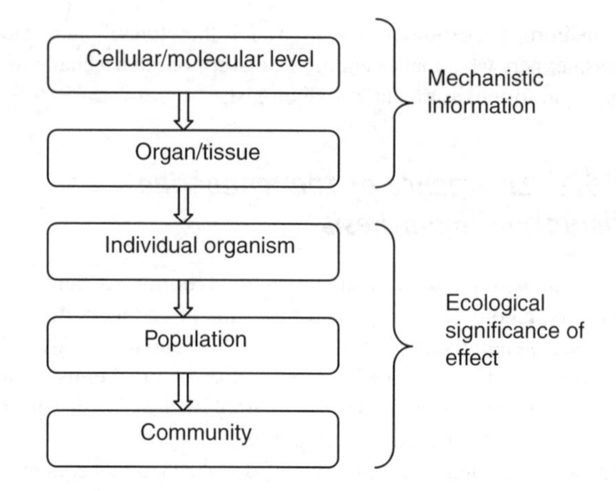

Figure 1.8 Hierachy of levels of observation for endocrine disruption effects

It is then clear that any effect observed at any one level cannot constitute evidence of endocrine disruption in itself.

Harvey *et al.* have suggested a classification scheme to cover the main types of endocrine and hormonally modulated toxicity (Harvey *et al.*, 1999):

- Primary endocrine toxicity involves the direct effect of a chemical on an endocrine gland, manifested by hyperfunction or hypofunction. Because of the interactions between endocrine glands and their hormones and non-endocrine target tissues, direct endocrine toxicity often results in secondary responses.
- Secondary endocrine toxicity occurs when effects are detected in an endocrine gland as a result of toxicity elsewhere in the endocrine axis. An example would be castration cells that develop in the pituitary as a result of testicular toxicity.
- Indirect toxicity involves either toxicity within a non-endocrine organ, such as the liver, resulting in an effect on the endocrine system or the modulation of endocrine physiology as a result of the stress response to a toxicant.

There is a general consensus that indirect endocrine toxicity should not be described as endocrine disruption and that the term itself may have been sometimes misused to include toxicological effects better described in terms of classical toxicology (Eggen *et al.*, 2003).

1.5.4 Mechanisms

Endocrine disruption was first recognised when it was found that certain environmental contaminants were able to mimic the actions of endogenous hormones. Some chemicals were subsequently shown to be able to block such actions, and other mechanisms involved in the control of circulating hormone levels were identified.

Contaminants have been shown (Cheek *et al.*, 1998; Folmar *et al.*, 2001; Guillette and Gunderson, 2001) to:

- act as hormone receptor agonists or antagonists;
- alter hormone production at its endocrine source;
- alter the release of stimulatory or inhibitory hormones from the pituitary or hypothalamus;
- alter hepatic enzymatic biotransformation of hormones;
- alter the concentration or functioning of serum-binding proteins, altering free hormone concentrations in the serum; and
- alter other catabolic pathways of clearance of hormones.

Receptor-mediated mechanisms have received the most attention, but other mechanisms have been shown to be equally important.

1.5.4.1 Hormone–receptor agonism and antagonism

The current focus for concerns about endocrine-mediated toxicity has mostly been on chemicals interacting with the steroid hormone receptor superfamily, receptors for oestrogens, androgens, thyroid hormones, etc. These receptors are predominantly involved in changing gene transcription (Barton and Andersen, 1997). According to the accepted paradigm for receptor-mediated mechanisms, a compound binds to a receptor forming a ligand-receptor complex with high binding affinity for specific DNA sequences or responsive elements. Once bound to this responsive element, the ligand-receptor complex induces gene transcription followed by translation into specific proteins which are the ultimate effectors of observed responses (Zacharewski, 1997).

Whilst hormone agonists are not only able to bind to the receptor under consideration, but also induce gene transcription thereby amplifying endogenous hormonal response, antagonists bind to the receptor but are unable to effect increased gene transcription, but rather competitively inhibit it by their occupancy of receptor binding sites.

1.5.5 Dose–response relationships

The theory of dose–response relationships of xenobiotics generally assumes that they are monotonic, the response rising with the dose. However, endogenous hormones are already present at physiological concentrations and are therefore already beyond the threshold (Andersen *et al.*, 1999b). Additionally, most endocrine processes are regulated by feedback controls such as receptor autoregulation and control of enzymes involved in synthesis of high-affinity ligands. This is expected to give rise to highly non-linear dose–response characteristics and abrupt changes from one biological condition to another over a very small change in concentration. While many of these non-linear switching mechanisms are expected to produce non-linear dose–response curves for the action of endogenous hormones, the dose response for effects of exogenous compounds still depends on the combination of effects of the native ligand and the endocrine-disrupting chemical (Andersen *et al.*, 1999a). Evidence of low-dose effects has proved very controversial, mainly due to lack of reproducibility, and it has been suggested that this may be related to the natural variability between individuals (Ashby *et al.*, 2004).

There are also important time-dependent variations in normal endogenous hormone levels such as circadian rhythms, puberty, oestrous or menstrual cycles, and reproductive senescence and aging. This introduces the additional problems of critical life stages: exposure at sensitive developmental stages can result in irreversible changes and latency (the time between exposure and the observed effects) as was exemplified by DES exposure *in utero* (Barlow *et al.*, 2002).

1.6 Concluding remarks

This chapter intends to illustrate the complex issues surrounding attempts to predict the effects of environmental contaminants and equip the reader with some basic concepts that may aid a critical understanding of evidence for such effects. It should encourage rather than deter the reader from reading and referring to the authoritative works cited.

References

Andersen, H. R., A. M. Andersson, S. F. Arnold, H. Autrup, M. Barfoed, N. A. Beresford, P. Bjerregaard, L. B. Christiansen, B. Gissel and R. Hummel *et al.* (1999a). Comparison of short-term estrogenicity tests for identification of hormone-disrupting chemicals. *Environmental Health Perspectives* **107** (Supplement 1): 89–108.

Andersen, M. E., R. B. Conolly, E. M. Faustman, R. J. Kavlock, C. J. Portier, D. M. Sheehan, P. J. Wier and L. Ziese (1999b). Quantitative mechanistically based dose–response modeling with endocrine-active compounds. *Environmental Health Perspectives* **107**: 631–638.

Armitage, P. and G. Berry (1994). *Statistical Methods in Medical Research*. Cambridge, Cambridge University Press.

Ashby, J., H. Tinwell, J. Odum and P. Lefevre (2004). Natural variability and the influence of concurrent control values on the detection and interpretation of low-dose or weak endocrine toxicities. *Environmental Health Perspectives* **112**(8): 847–853.

Barker, J. (1999). Liquid chromatography-mass spectrometry (LC/MS) *Mass Spectrometry*. D. J. Ando. Chichester, John Wiley & Sons: 320–334.

Barlow, S. M., J. B. Greig, J. W. Bridges, A. Carere, A. J. M. Carpy, G. L. Galli, J. Kleiner, I. Knudsen, H. B. W. M. Koeter, L. S. Levy, C. Madsen, S. Mayer, J. F. Narbonne, F. Pfannkuch, M. G. Prodanchuk, M. R. Smith and P. Steinberg (2002). Hazard identification by methods of animal-based toxicology. *Food and Chemical Toxicology* **40**(2–3): 145–191.

Barton, H. A. and M. E. Andersen (1997). Dose–response assessment strategies for endocrine-active compounds. *Regulatory Toxicology and Pharmacology* **25**(3): 292–305.

Bentley, P. J. (1998). *Comparative Vertebrate Endocrinology*. Cambridge, Cambridge University Press.

Bertollini, R., M. D. Lebowitz, R. Saracci and D. A. Savitz, Eds. (1996). *Environmental Epidemiology – Exposure and Disease*. Copenhagen, Lewis Publishers on behalf of the World Health Organisation.

Boobis, A., G. Daston, R. Preston and S. Olin (2009). Application of key events analysis to chemical carcinogens and noncarcinogens. *Critical Reviews in Food Science and Nutrition* **49**(8): 690–707.

Burger, J. (2003). Making decisions in the 21(st) century: Scientific data, weight of evidence, and the precautionary principle. *Pure and Applied Chemistry* **75**(11–12): 2505–2513.

Calabrese, E. and L. Baldwin (2003). Hormesis: The dose–response revolution. *Annual Review of Pharmacology and Toxicology* **43**: 175–197.

Carlsen, E., A. Giwervman, N. Keiding and N. E. Skakkebaek (1992). Evidence for decreasing quality of semen during the past 50 years. *British Medical Journal* **305**: 609–613.

Carson, R. (1962). *Silent Spring*. New York, Penguin Books.

Cheek, A. O., P. M. Vonier, E. Oberdorster, B. C. Burow and J. A. McLachlan (1998). Environmental signaling: A biological context for endocrine disruption. *Environmental Health Perspectives* **106**: 5–10.

Coggon, D., G. Rose and D. J. P. Barker (2003). *Epidemiology for the Unititiated.* BMJ Publishing Group.

Colborn, T., D. Dumanoski and M. J. Peterson (1996). *Our Stolen Future*. London, Abacus.

Commission of the European Communities (2003). Proposal for a Regulation of the European Parliament and of the Council concerning the Registration, Evaluation, Authorisation and Restriction of Chemicals (Reach), establishing a European Chemicals Agency and amending Directive 1999/45/EC and Regulation (EC) {on Persistent Organic Pollutants} {SEC(2003 1171}/* COM/2003/0644 final—COD 2003/0256 */**I**.

Duffus, J. H. (2006). Introduction to Toxicology. *Fundamental Toxicology*. J. H. Duffus and H. G. J. Worth. Cambridge, Royal Society of Chemistry: 1–17.

Eaton, D. L. and C. D. Klaassen (2001). Principles of Toxicology. *Casarett & Doull's Toxicology: the basic science of poisons*. C. D. Klaassen, McGraw-Hill: 11–34.

Eggen, R. I. L., B. E. Bengtsson, C. T. Bowmer, A. A. M. Gerritsen, M. Gibert, K. Hylland, A. C. Johnson, P. Leonards, T. Nakari, L. Norrgren, J. P. Sumpter, M. J. F. Suter, A. Svenson and A. D. Pickering (2003). Search for the evidence of endocrine disruption in the aquatic environment: Lessons to be learned from joint biological and chemical monitoring in the European Project COMPREHEND. *Pure and Applied Chemistry* **75**(11–12): 2445–2450.

Elwood, M. (1998). *Critical Appraisal of Epidemiological Studies and Clinical Trials*. Oxford, Oxford University Press.

ECHA (2007) Guidance for the Preparation of an Annex XV Dossier on the Identification of Substances of Very High Concern. Guidance for the Implementation of REACH. Helsinki, European Chemicals Agency: 58.

EPA (1997). Special report on environmental endocrine disruption: an effects assessment and analysis. Washington D.C., United States Environmental Protection Agency (EPA).

European Community (2000). Directive 200/60/EC of the European Parliament and of the Council of 23 October 2000 establishing a framework for Community action in the field of water policy: 72.

European Environment Agency (2001). Late Lessons from Early Warnings: The Precautionary Principle 1896–2000. Copenhagen, European Environment Agency.

European Union (2006). Regulation (EC) No 1907/2006 of the European Parliament and of the Council of 18 December 2006 concerning the Registration, Evaluation, Authorisation and Restriction of Chemicals (REACH), establishing a European Chemicals Agency, amending Directive 1999/45/EC and repealing Council Regulation (EEC) No 793/93 and Commission Regulation (EC) No 1488/94 as well as Council Directive 76/769/EEC and Commission Directives 91/155/EEC, 93/67/EEC, 93/105/EC and 2000/21/EC. *Regulation (EC) No 1907/2006.* Brussels: 278.

Folmar, L. C., N. D. Denslow, K. Kroll, E. F. Orlando, J. Enblom, J. Marcino, C. Metcalfe and L. J. J. Guillette (2001). Altered serum sex steroids and vitellogenin induction in Walleye (Stizostedion vitreum) collected near a metroplitan sewage treatment plant. *Archives of Environmental Contamination and Toxicology* **40**: 392–398.

Francis, B. M. (1994). Water pollution, persistence, and bioaccumulation. *Toxic Substances in the Environment*. New York, John Wiley & Sons: 93–120.

Gee, D. (2006). Late lessons from early warnings: toward realism and precaution with endocrine-disrupting substances. *Environmental Health Perspectives* **114** (Suppl. 1): 152–160.

Greene, R. R., M. W. Burrill, and A. C. Ivy (1938) Experimental intersexuality: the production of feminized male rats by antenatal treatment with estrogens. *Science* **188**(2275): 130–131.

Gregus, Z. and C. D. Klaassen (2001). Mechanisms of toxicity. *Casarett & Doull's Toxicology: the Basic Science of Poisons*. C. D. Klaassen, McGraw-Hill: 35–82.

Guillette, L. J. and M. P. Gunderson (2001). Alterations in development of reproductive and endocrine systems of wildlife populations exposed to endocrine-disrupting contaminants. *Reproduction* **122**(6): 857–864.

Gundert-Remy, U., P. Kremers, A. Renwick, A. Kopp-Schneider, S. G. Dahl, A. Boobis, A. Oberemm and O. Pelkonen (2005). Molecular approaches to the identification of biomarkers of exposure and effect – report of an expert meeting organized by COST Action B15. November 28, 2003. *Toxicology Letters* **156**(2): 227–40.

Hale, R. C., M. J. La Guardia, E. Harvey and T. M. Mainor (2002). Potential role of fire retardant-treated polyurethane foam as a source of brominated diphenyl ethers to the US environment. *Chemosphere* **46**(5): 729–35.

Harvey, P. W., K. C. Rush and A. Cockburn, Eds. (1999). *Endocrine and Hormonal Toxicology*. Chichester, John Wiley & Sons.

Herbst, A. L., H. Ulfelder and D. C. Peskanzer (1971). Adenocarcinoma of the vagina: association of maternal stilbestrol therapy with tumor appearances in young women. *New England Journal of Medicine* **284**: 878–881.

Herrchen, M. (2006). Pathways and behaviour of chemicals in the environment. *Fundamental Toxicology*. J. H. Duffus and H. G. J. Worth. Cambridge, The Royal Society of Chemistry: 238–256.

Hertz-Picciotto, I. (1998). Environmental epidemiology. *Modern Epidemiology*. K. J. Rothman and S. Greenland. Philadelphia, Lippincott-Raven Publishers: 555–583.

Highnam, K. C. and L. Hill (1977). *The Comparative Endocrinology of the Invertebrates*. London, William Clowes & Sons.

Hill, A. B. (1965). The environment and disease: association or causation? *Proceedings of the Royal Society of Medicine* **58**: 295–300.

International Programme on Chemical Safety (2002). *Global Assessment of the State-of-the-Science of Endocrine Disruptors*. World Health Organisation: 180.

James, R. C., S. M. Roberts and P. L. Williams (2000). General principles of toxicology. *Principles of Toxicology: Environmental and Industrial Applications*. R. C. James, S. M. Roberts and P. L. Williams. New York, John Wiley & Sons: 3–34.

Kriebel, D., J. Tickner, P. Epstein, J. Lemons, R. Levins *et al.* (2001). The precautionary principle in environmental science. *Environmental Health Perspectives* **109**(9): 871–876.

Krimsky, S. (2000). *Hormonal Chaos. The Scientific and Social Origins of the Environmental Endocrine Hypothesis*. Baltimore, The Johns Hopkins University Press.

Langezaal, I. (2002) *The Classification and Labelling of Carcinogenic, Mutagenic, Reprotoxic and Sensitising Substances*. Ispra, European Chemicals Bureau, Joint Research Centre: 193.

Marchant, G. (2003). From general policy to legal rule: aspirations and limitations of the precautionary principle. *Environmental Health Perspectives* **111**(14): 1799–803.

MRC Institute for Environment and Health (Editor) (1997). *European Workshop on the Impact of Endocrine Disrupters on Human Health and Wildlife*. Brussels, European Commission: 125.

Newbold, R. R., R. B. Hanson, W. N. Jefferson, B. C. Bullock, J. Haseman and J. A. McLachlan (1998). Increased tumors but uncompromised fertility in the female descendants of mice exposed developmentally to diethylstilbestrol. *Carcinogenesis* **19**(9): 1655–63.

Newbold, R. R., R. B. Hanson, W. N. Jefferson, B. C. Bullock, J. Haseman and J. A. McLachlan (2000). Proliferative lesions and reproductive tract tumors in male descendants of mice exposed developmentally to diethylstilbestrol. *Carcinogenesis* **21**(7): 1355–63.

O'Flaherty, E. J. (2000). Absorption, Distribution, and Elimination of Toxic Agents. *Principles of Toxicology: Environmental and Industrial Applications*. P. L. Williams, R. C. James and S. M. Roberts. New York, John Wiley & Sons: 35–56.

Ottoboni, M. A. (1991). *What are chemicals? The dose makes the poison—A plain language guide to toxicology*. New York, Van Nostrand Reinhold: 5–18.

Rogers, M. D. (2003). Risk analysis under uncertainty, the Precautionary Principle, and the new EU chemicals strategy. *Regulatory Toxicology and Pharmacology* **37**(3): 370–381.

Rushton, L. (2000). Reporting of occupational and environmental research: use and misuse of statistical and epidemiological methods. *Occupational and Environmental Medicine* **57**(1): 1–9.

Slovic, P. (1987). Perception of risk. *Science* **236**(4799): 280–5.

Steenland, K. and D. A. Savitz, Eds. (1997). *Topics in Environmental Epidemiology*. New York, Oxford University Press.

Sullivan, F. M. (2006). Reproductive toxicity. *Fundamental Toxicology*. J. H. Duffus and H. G. J. Worth. Cambridge, Royal Society of Chemistry: 142–153.

Timbrell, J. (2000). Factors affecting toxic responses: metabolism. *Principles of Biochemical Toxicology*. London, Taylor & Francis: 65–112.

White, R., I. Cote, L. Zeise, M. Fox, F. Dominici and T. Burke (2009). State-of-the-Science Workshop Report: issues and approaches in low-dose–response extrapolation for environmental health risk assessment. *Environmental Health Perspectives* **117**(2): 283–287.

Williams, F. S. and S. M. Roberts (2000). Dermal and ocular toxicology: toxic effects of the skin and eyes. *Principles of Toxicology: Environmental and industrial Applications*. P. L. Williams, R. C. James and S. M. Roberts. New York, John Wiley & Sons: 157–168.

Zacharewski, T. (1997). In vitro bioassays for assessing estrogenic substances. *Environmental Science and Technology* **31**(3): 613–623.

2

Regulatory systems and guidelines for the management of risk

Dieudonné-Guy Ohandja[1]*, Sally Donovan[1], Pamela Castle[2], Nikolaos Voulvoulis[1] and Jane A. Plant[3]

[1]*Centre for Environmental Policy, Imperial College London, Prince Consort Road, London SW7 2AZ*
[2]*Former Chair of the Environmental Law Foundation*
[3]*Centre for Environmental Policy and Department of Earth Science and Engineering, Imperial College London, Prince Consort Road, London SW7 2AZ*
**Corresponding author, email d.ohandja@imperial.ac.uk*

2.1 Introduction

Over the past century, there has been a significant increase in the production and use of chemicals in areas as diverse as manufacturing, transport, agriculture and health, as well as in the home (Hudson, 1983). Since the 1960s, however, especially after the publication of Rachel Carson's book *Silent Spring* in 1962, the ever-increasing use of synthetic chemicals, coinciding with the development of increasingly sensitive analytical equipment, has raised serious concerns about the adverse effects of many of these substances on human health and the environment. Although harmful chemicals are usually properly used and disposed of, especially in advanced industrial nations, incidents involving exposure to humans and/or the environment during one or more stages of a product's lifecycle are well documented. Examples include chemical manufacture, spills, inadequate handling, storage and use, as well as incorrect disposal. An example of such incidents is the recent explosion of the Deepwater Horizon oil rig operated by BP, Transocean and Halliburton in April 2010. The explosion which killed 11 platform workers and injured 17 others resulted in thousands of tonnes of crude oil being poured into the Gulf of Mexico. Crude oil, of course, is a natural substance, but most of the concern relates to synthetic chemicals. Some chemicals such as pesticides are deliberately distributed widely through the environment, and many others are commonly used in everyday household products (RECP, 2003). Hence, a wide range of synthetic chemicals with diverse properties are presently found in air, water, soil and sediments. These include inorganic and radioactive substances, persistent organic pollutants (POPs), pharmaceuticals and personal-care products, chemicals used by agribusiness to enhance growth or health of livestock (USEPA, 2008a), surfactants, plasticisers, pesticides, flame retardants, particulates and engineered nanomaterials, as well as associated degradation products (Yan, Subramanian *et al.*, 2010). More details of each chemical class can be found in the following chapters of this book.

The inorganic chemicals of most concern continue to be toxic trace elements such as arsenic, cadmium and mercury, which have no known biochemical value and are harmful to humans and most animals at low concentration levels as explained in Chapter 4. Elements such as fluorine, selenium and zinc (Chapter 3) are needed for normal biological function but can also cause problems for human and animal health.

POPs are another group of chemicals of concern that are widespread in the environment. They include chemicals such as polychlorinated biphenyls (PCBs), organochlorines, organophosphates, dioxins and furans (Ragnarsdottir, 2000; Plant, Korre *et al.*, 2004). Common characteristics of POPs are persistence (they can remain in the environment for decades after use), the ability to bioaccumulate and a semi-volatile nature, meaning that

Pollutants, Human Health and the Environment: A Risk Based Approach, First Edition. Edited by Jane A. Plant, Nikolaos Voulvoulis and K. Vala Ragnarsdottir.

they slowly evaporate in the air and travel long distances through repeated short hops before being redeposited on land or water. Their widespread presence in the environment is the result of multiple emission-transport-deposition cycles, known as the 'grasshopper' effect (Semeena and Lammel, 2005). Many POPs have or have had a range of industrial applications historically and some, such as organophosphates, continue to be used widely in agriculture (Ragnarsdottir, 2000). There are a large number of chemicals in this group, and the risks associated with some of them, such as the alkylphenols, phthalates, bisphenols, dioxins, polybrominated flame retardants, polyethoxylates, polyaromatic hydrocarbons (PAHs), PCBs, furans (which are known carcinogens), neurotoxins and reprotoxins, have not been fully established (Plant, Korre *et al.*, 2004). The long-term effects of some POPs on human health have been demonstrated by accidents such as the explosion at a pesticide factory in Bhopal, India where the leakage of methyl-isocyanate gas and other toxins resulted in the death of 2500–6000 people. Survivors of the accident continue to experience a high incidence of health problems, including febrile illnesses and respiratory, neurological, psychiatric and ophthalmic symptoms (Mishra, Samarth *et al.*, 2009).

Chemical risk management aims to minimise the risks associated with the handling and distribution of potentially hazardous chemicals and to prevent or reduce any potentially adverse effects on human health or the environment. In many cases, risk-management results from a comprehensive risk assessment (RA) exercise. In most countries where it is applied, RA is initially based on the source-pathway-target framework as set out by the United States Environmental Protection Agency (USEPA) and the European Union (EU). Within this framework, for any risk to cause harm there should be a source of the substance, an exposure pathway and a target (receptor) in place. This receptor can be a living organism, an ecosystem or the environment. Risk assessment involves various steps including hazard identification, risk data evaluation, exposure assessment, effect assessment, risk characterisation and risk evaluation (Diderich and Ahlers, 1995; Vermeire and Van Der Zandt, 1995; Ahlers, 1999).

Depending on the results of the risk characterisation, a risk-reduction strategy can be developed and subsequent remedial action (risk reduction) can be based on any or all three elements of the contamination chain. Risk assessment is thus based on scientific evidence and has been widely used to manage risk around the world.

2.1.1 The precautionary principle

In cases where there is not enough evidence to quantify risk, the precautionary principle (PP) is generally adopted to find alternatives to any potentially destructive practices or to avoid environmental catastrophes. The PP derives from the German concept 'Vorsorgeprinzip', meaning foresight-planning, which has been the basis of German environmental policy since the 1970s (Boehmer-Christiansen, 1994). The PP gained prominence in the 1980s when it began to be incorporated into many national and international environmental policies (Lokke and Christensen, 2008). The PP states that when an activity has the potential to cause harm to human health or the environment, precautionary measures should be taken, even in the absence of conclusive scientific evidence that such harm would occur (Tickner and Raffensperger, 2001). The PP applies where there is considerable scientific uncertainty about causality, magnitude, probability and nature of harm – though some form of scientific analysis is required for the PP to be applied (UNESCO, 2005). The Vienna Convention on Ozone Depleting Substances was the first international environmental treaty to evoke precautionary measures to be taken at national and international levels (Cameron, 2001). In 1992, the PP was universally recognised as a legal tool during the United Nations Conference on Environment and Development (Mann, 1992; Hohmann, 1994; Post, 2006) and it was integrated into the EU Maastricht Treaty of 1993, which states that all Community policies on the environment shall be based on the PP. Since then, it has been incorporated into many EU and international environmental agreements and treaties, such as the 2001 Stockholm Convention on POPs and the EU Communication on the Precautionary Principle (Europa, 2001).

The application of the PP varies from country to country, and can be a subject of contention (Post, 2006). In the EU, it is a key element in the regulatory framework, where it has been incorporated into many regulations and treaties in the area of food, chemicals and biotechnology (Hohmann, 1994; Sand, 2000; Eckley and Selin, 2004). It is also increasingly accepted as an important principle in international environmental law; for example, in the provisional measures of the United Nations Convention on the Law of the Sea, which were invoked in the case of Ireland's opposition to the UK operating a mixed oxide (MOX) nuclear plant on the Irish Sea in 2004. This case recommended that the two parties cooperate for the prevention of pollution of the marine environment as a result of the operation of the MOX plant. Application of the PP remains especially contentious in the United States of America (USA) and Canada, where it is often regarded as an obstacle to progress and risk management is based entirely on risk assessment. Many opponents of the PP in these countries argue that the unrealistic zero risk of the PP can result in the banning of useful chemicals (Tickner and Raffensperger, 2001).

In order to manage risks associated with chemicals effectively, many countries have developed guidelines, standards and regulations and are signatories to international treaties. Such risk-management tools, which are either preventive or remedial measures, are based on risk assessment, the PP, or a combination of both. This chapter focuses on current approaches to managing the potentially adverse impacts of hazardous chemicals on the environment and/or human health around the world.

2.2 Current regulation on chemicals

2.2.1 The Organisation for Economic Co-operation and Development (OECD) and the International Council of Chemical Associations (ICCA) High Production Volume (HPV) Chemicals Programme

In 1990, the Organisation for Economic Co-operation and Development (OECD) initiated a system for evaluating the hazards of High Production Volume (HPV) chemicals, which refers to chemicals produced or imported into OECD member countries or into the EU in quantities in excess of 1000 tonnes per year (ICCA, 2000; OECD, 2010). To develop the system, a wide range of industries were encouraged to provide characterisation, effects and exposure information from government and public sources in order to complete a Screening Information Data Set (SIDS) with an agreed number of initial toxicological and ecotoxicological end points. Following completion of the SIDS, an initial assessment of the data is made, the potentially hazardous properties of chemicals are identified and areas for future work recommended. In 1998, the ICCA began to provide internationally harmonised data sets and hazard assessments for approximately 1000 HPV chemicals. By 2007, there were 4638 HPV chemicals on the ICCA list (OECD, 2010). The OECD uses the ICCA list of HPV chemicals to establish a list of priority chemicals for action.

2.2.2 United Nations Environment Programme (UNEP)

The United Nations coordinates a wide range of international activities to control hazardous chemicals.

The *Rotterdam Convention* was adopted on 10 September 1998 in Rotterdam, the Netherlands. It promotes shared responsibility and careful use of hazardous chemicals among parties involved in their trade. Between 12 September 1998 and 10 September 1999, the convention was signed by 72 states and one regional economic integration organisation and came into force on 24 February 2004 (Garrod, 2006). The Rotterdam Convention recognises the Prior Informed Consent (PIC) procedure, which has been adopted by many countries. The PIC procedure prevents the export of a group of chemicals that have been banned or restricted, or are known to be harmful to humans and the environment (EEA, 2005). Currently, 40 chemicals subject to the PIC procedure are listed in Annex III of the convention. These chemicals include 25 pesticides, 4 severely hazardous pesticide formulations and 11 industrial chemicals. It is expected that many other chemicals will be added to this list. The Rotterdam Convention encourages importing countries to withhold importation of potentially hazardous chemicals that they are unable to manage safely. It also enforces compliance on exporters to meet the requirements of PIC.

The *Stockholm Convention* on POPs is a worldwide treaty aimed at protecting the environment and human health. It was adopted in May 2001 in Stockholm, Sweden. It was ratified by the EU in November 2004 and the UK in January 2005. The objectives of the convention are to eliminate or restrict the production and use of some POPs, pesticides and industrial chemicals; it also proposes to control and reduce emission of by-products resulting from combustion processes while ensuring the safe disposal of such compounds. The convention applied initially to aldrin, chlordane, dieldrin, endrin, heptachlor, mirex, toxaphene, DDT, hexachlorobenzene (HCB), polychlorinated biphenyls (PCBs), polychlorinated dibenzodioxins (PCDDs) and polychlorinated dibenzofurans (PCDFs) (Lallas, 2001). It also laid the framework for adding new chemicals to the list, and in May 2009, nine new POPs were added, including chlordecone; hexabromobiphenyl; hexabromodiphenyl ether and heptabromodiphenyl ether; alpha hexachlorocyclohexane; beta hexachlorocyclohexane; lindane; pentachlorobenzene (PeCB); perfluorooctane sulphonic acid (PFOS), its salts and perfluorooctane sulphonyl fluoride (PFOS-F) (UNEP, 2010a).

The Strategic Approach to International Chemicals Management (SAICM) was adopted by the UNEP Governing Council in February 2002 (Garrod, 2006). Under this approach, the UNEP works with member governments and other stakeholders to review current actions to advance the sound international management of chemicals, to further advance the approach of the Intergovernmental Forum on Chemical Safety (IFCS) and to propose specific projects to be given priority.

2.2.3 Regulations in Europe

2.2.3.1 The Oslo and Paris Convention for the Protection of the Marine Environment of the North East Atlantic (OSPAR)

The OSPAR Convention came into force in March 1998 with the aim of safeguarding the marine ecosystem of the North East Atlantic and protecting human health, including from pollution by hazardous chemicals (OSPAR, 1992). A list of substances of concern was drawn up, and individual chemicals prioritised for risk assessment and action as necessary. The list was prepared according to a Dynamic Selection and Prioritisation Mechanism for Hazardous Substances (DYNAMEC) developed by the convention. Industries are invited to bring forward data for each prioritised chemical and each chemical is adopted by a sponsor country, which, together with OSPAR, produces a background document covering the properties of the chemical and likely pathways to the marine environment. The list is revised regularly. OSPAR adopts decisions and recommendations, which are legally binding on the contracting countries. Examples of such decisions include the OSPAR Decision 2007/1 to prohibit the storage of carbon dioxide streams in the water column or on the seabed, the OSPAR Decision 2005/1 amending OSPAR Decision 2000/2 on a harmonised mandatory control system for the use and reduction of the discharge of offshore chemicals, and the OSPAR Decision 98/4 on emission and discharge limit values for the manufacture of vinyl chloride monomer (VCM) including the manufacture of 1, 2-dichloroethane (EDC).

2.2.3.2 The Restriction of the Use of Certain Hazardous Substances in Electrical and Electronic Equipment regulations (The RoHS regulations)

The RoHS regulations aim to restrict the use of some hazardous chemicals in electrical and electronic equipment (EEE). These regulations implement EU Directive 2002/95/EC which bans the placing of new EEE containing more than agreed levels of lead, cadmium, mercury, hexavalent chromium, polybrominated biphenyl (PBB) and polybrominated diphenyl ether (PBDE) flame retardants onto the EU market. The regulations do not apply to large-scale stationary industrial tools, EEE and spare parts used for repair, which were placed on the market before July 2006. The regulation further lists other exemptions in its Annex and amendments. These exemptions include specific uses and chemical states of metals such as lead, mercury, cadmium, chromium and their compounds (EC, 2003).

2.2.3.3 New European Regulation (EC) No 1272/2008 on Classification, Labelling and Packaging of Substances and Mixtures (CLP regulation)

The CLP regulation (EC, 2008a) was adopted in December 2008 and came into force in the EU on 20 January 2009. It replaces the Dangerous Substance Directive 67/548/EEC and the Dangerous Preparations Directive 1999/45/EC. The main objective of the CLP regulation is to identify the properties of substances and mixtures that lead to their classification. These properties include physical hazards to both human health and the environment. Although this regulation clearly states that manufacturers, importers and downstream users of those substances or mixtures are all responsible for hazard identification, it also highlights that none of these parties should be obliged to produce new toxicological and/or ecotoxicological data needed for such classification. To achieve effective hazard identification and management, manufacturers are required to label and package substances and mixtures according to the CLP regulation before placing them on the market. Distributors and end-users should follow the classification of the substances and the mixtures derived from this regulation.

2.2.3.4 Registration, Evaluation and Authorisation of Chemicals (REACH)

The REACH regulations are possibly the most ambitious piece of environmental legislation ever attempted. They were initiated on 1 June 2007 and will be fully implemented in the EU by 31 May 2018 (Ruden and Hansson, 2010). The main objective of REACH is to improve risk management of industrial chemicals in Europe, with the following goals (Williams, Panko *et al.*, 2009):

- Compile a suite of physicochemical, toxicological, and ecotoxicological data for each substance.
- Establish safe-usage parameters by conducting chemical safety assessments (CSAs).
- Allow for regulatory evaluation of substances to determine potential hazards based on the compiled data.
- Prevent the use of substances of very high concern (SVHCs) without approval of the European Chemicals Agency (ECHA).
- Restrict the use of chemicals for which no safe-usage parameters can be established.

2.2.3.4.1 Registration Chemicals produced or imported in quantities greater than 1 tonne per annum (tpa) by any single producer or importer must be registered in a central database and are subject to much more demanding legislation (Ruden and Hansson, 2010). There are four tonnage bands, grouping chemicals manufactured or imported in quantities greater than 1 tonne per annum but less than 10 tpa, 10–100 tpa, 100–1000 tpa, and more than 1000 tpa. However, Annexes IV and V of the REACH regulations indicate that some chemicals are exempt from registration regardless of the amount produced or imported, including medicinal products and foodstuffs, polymers, radioactive substances, waste materials and naturally occurring substances.

A key element of the REACH process is the stepwise increase in toxicity data required as the market or production volume of a chemical increases. Information on physicochemical, toxicological and ecotoxicological properties and environmental impact is required for each substance. The requirements are different for each tonnage band, and they also depend on the tonnage manufactured or imported – in general the higher the tonnage, the more information required (ECHA, 2008). As a result, REACH is having a major impact on chemical-risk-assessment practices in Europe, including abolishing the different approaches to 'existing' and 'new' chemicals and shifting responsibility for risk assessment from government authorities to the producer or importer. Criteria used for assessment of the required information for substance registration include relevance, reliability, and adequacy (ECHA, 2008; Williams, Panko *et al.*, 2009). As defined by ECHA (2008), relevance is the 'extent to which data and tests are appropriate for a particular hazard identification or risk characterisation'; reliability refers to 'the inherent quality of a test report or a publication relating to preferably standardised methodology and the way the experimental procedure and results are described to give evidence of the clarity and plausibility of the findings'; and adequacy is 'the usefulness of the data for hazard and risk assessment purposes'.

Moreover, the producer/importer is obliged to make toxicity and exposure data and safe-handling advice publicly available. For substances within the lowest (1–10 tpa) tonnage band, the manufacturer or importer is required to produce a chemical safety report (CSR) with the main objective of assessing and characterising the risk associated with the use of each substance, and potential measures to manage those risks. The chemical safety report provides a detailed chemical safety assessment (CSA). Each CSA includes human assessment; environmental hazard assessment; human-health assessment for physicochemical properties; persistent, bioaccumulative assessment; exposure assessment; and risk characterisation. Information collected in a chemical safety report is the basis of a safety data sheet (SDS), and all exposure scenarios from the CSR and SDS are compiled to produce extended safety data sheets (eSDS). These eSDSs will therefore summarise all information offered by suppliers. The European Chemical Agency (ECHA) assigns a registration number after checking that the dossier is complete.

2.2.3.4.2 Evaluation The evaluation stage includes three independent processes: compliance to check whether the information submitted by registrants complies with the legal requirements; examination of testing proposals to avoid unnecessary animal testing; and substance evaluation to clarify whether the use of a substance may cause harm to human health or the environment (ECHA, 2009). It is anticipated that all dossiers submitted in the high-tonnage bands will be evaluated for testing proposals and that testing proposals for the highest tonnage will be prioritised for carcinogenic, mutagenic and reprotoxic (CMR); persistent, bioaccumulative, and toxic (PBT); and very persistent and very bioaccumulating (vBvP) substances (Williams, Panko *et al.*, 2009).

Carcinogenic, mutagenic or reprotoxic (CMR) is a designation applied by manufacturers and legislative bodies in the EU to substances identified as hazardous substances capable of initiating cancer, increasing the frequency of changes in an organism's genetic material above the natural background level, and/or harming the ability of organisms to reproduce successfully. Since the likelihood of cancer and reproductive outcomes are dependent on genetic stability, many substances given this designation show more than one of these effects. Chemical producers wishing to market products in the EU are obliged to carry out toxicological tests to identify whether their products have CMR properties. Formerly this information was forwarded to EU-designated member-state authorities to be classified by the European Chemicals Bureau (Langezaal *et al.*, 2002). Since 2007, regulatory functions have been passed to the European Chemicals Agency, which has replaced the European Chemicals Bureau and plays a central role in coordinating and implementing REACH legislation.

2.2.3.4.3 Authorisation The authorisation process consists of four steps. Industry has obligations in the third step. However, all interested parties have the opportunity to provide input in steps 1 and 2 (ECHA, 2010).

- Step 1: Identification of substances of very high concern by Member State Competent Authorities or the ECHA, by preparing a dossier compliant with Annex XV. This will result in compiling a list of identified substances, which are candidates for prioritisation (the 'candidate list').
- Step 2: Prioritisation process by relevant authorities to determine which substances on the candidate list should be subject to authorisation. Interested parties are invited to submit comments during this process.
- Step 3: Application for authorisation (by industries) for each use not exempted from the authorisation requirement. Such applications need to be made within the set deadlines accompanied by relevant information such as a chemical safety report and an analysis of possible alternative substances or technologies.
- Step 4: Granting of authorisations by the European Commission upon proof of adequate means to control the substance.

The REACH legislation will move the management of hazardous products to earlier in their life cycle and place the burden on those who profit from the manufacture or use of chemicals, rather than the end-user or taxpayer. The REACH regulation will therefore require the industry to manage risks from chemicals and publish safety information in a central database before the chemicals are used. It is hoped that the implementation of REACH will result in a reduction of human health problems associated with exposure to chemicals. Although REACH was developed to control chemicals management in the EU, it will also have significant implications on chemical management globally, since countries trading chemicals with the EU will need to comply with REACH regulation. Development of REACH might also have triggered modifications of chemical management elsewhere. For example in the USA, the federal government adopted the Kid-safe Chemicals Act (amending the Toxic Substances Control Act, TSCA), which ensures that all the chemicals used in baby bottles and children's toys are proven safe before being put on the market (US Senate, 2008). This regulation shows a clear shift from the common USA approach of 'innocent until proven guilty' towards a precautionary approach as prescribed by REACH. Other countries such as Japan, Switzerland, Turkey and Canada are also modifying their existing chemical regulations to include features similar to those proposed by REACH (ICIS, 2010).

2.2.3.5 Hazardous waste legislation

Council Directive (91/689/EC) on hazardous waste was released in 1991 (EC, 1991a). This piece of legislation provided a precise definition of 'hazardous' waste and methods of managing such waste with the least possible harm to the environment. The definition of hazardous is provided through a list of hazardous properties. Key components of this legislation are that it forbids the mixing of different types of hazardous waste or of hazardous and non-hazardous waste, and that any treatment that the waste goes through must be recorded up to final disposal.

The Waste Framework Directive (2008/98/EC) (EPC, 2008a) now incorporates hazardous waste, overriding Directive 91/689/EC. This simplifies the process for waste management. The key elements of hazardous-waste management are still valid, but a new category has been added to the list of substances, namely 'Sensitising'.

In the past, hazardous wastes have been transported to other countries, with less stringent regulations and dumped untreated. The Basel Convention, which came into force in 1992 and has

173 parties, prohibits this action. In the EU it was initially regulated through 93/98/EEC (EC, 1993), and this was updated in 1997 with 97/640/EC (EC, 1997). The key parts of the legislation are: exporting or importing hazardous wastes or other wastes to or from countries outside the EU is prohibited; export between states is prohibited unless the receiving country has given specific written consent; the involved member states must be informed and given an opportunity to perform a risk assessment on the waste; the transportation may not occur unless no danger is predicted from the movement and disposal; any transported wastes must be packaged, labelled and transported in conjunction with the treaty, and written records must be kept from the initial movement of waste until its disposal; and additional requirements may be imposed by the involved member states, in accordance with the convention.

Further pieces of legislation have also been passed which apply to specific waste types with hazardous components.

The Directive on End of Life Vehicles in 2000 (EPC, 2000) aimed to minimise waste production and organise collection and delivery to appropriate treatment facilities. The treatment must involve dismantling the vehicle, and any hazardous components must then be treated in compliance with the Waste Framework Directive (EPC, 2008a). Manufacturers are required to provide information on dismantling any vehicles they place on the market within six months. The directive sets quantitative recycling and recovery targets. Member states are further required to provide three-yearly reports to the council on the progress of its implementation.

In 2002 two pieces of legislation relating to electrical and electronic equipment (EEE) were published. The aforementioned restriction on hazardous components (EC, 2003) and another relating to waste EEE (EPC, 2003) require improvements in product design with a view to end of life, so that it is easy to dismantle the equipment and recover and recycle the component parts. Provision must be made for separate collection of waste electrical and electronic equipment (WEEE) and delivery to appropriate treatment facilities. Treatment must employ the best available technology and quantitative material recovery rates, which are also specified. The treatment may take place overseas provided it is compliant with the waste-shipment regulations. The financial responsibility for treatment and recovery may fall partially to the producers, and further products must be labelled to raise awareness of the requirement to separate WEEE from the rest of the waste streams, particularly for households. Finally, a report on progress of implementation must be provided to the EU every three years.

The directive on Waste Batteries and Accumulators (EC, 1991b), which aimed to inhibit marketing of certain types of batteries with a high content of hazardous components, significantly increased recycling rates and shifted some responsibility onto producers to improve labelling and removability of their products. This legislation was updated in 2006 (EPC, 2006). The original legislation applied only to batteries containing mercury, lead or cadmium and excluded button cells. The updated version applies to all types. It sets clear quantitative targets for collection and recycling and shifts part of the responsibility for this to producers by requiring clear labelling and safe removability from products. Disposal by incineration is prohibited and landfilling is only permitted as a last resort.

2.2.4 UK approaches to chemicals in the environment

2.2.4.1 UK Chemicals Strategy

As well as complying with EU law, the UK has independently developed several additional regulations. In 1999, the UK government set out its proposals for a Chemical Strategy on the Sustainable Production and Use of Chemicals (DEFRA, 2002a), aimed at avoiding harm to the environment and human health from environmental exposure to chemicals. The strategy aimed to make information about the environmental risks of chemicals publicly available and to continue to reduce the risks to the environment from chemicals, while maintaining the competitiveness of industry. The strategy is also aimed at phasing out the manufacture and use of chemicals that pose an unacceptable risk to the environment and human health. A key feature was the establishment of the UK Chemicals Stakeholder Forum (CSF) in 2000 to promote better understanding between stakeholders on issues of chemicals and the environment, and to provide advice to the government about chemicals in the environment to guide the development of policy (DEFRA, 2002b). The Advisory Committee on Hazardous Substances (ACHS), made up of eminent scientists, advises the government and the CSF on risk management of chemicals (DEFRA, 2009a, 2009b).

2.2.4.2 Royal Commission on Environmental Pollution (RCEP)

The RCEP, which closed in March 2011, was established in 1970 as an independent body to advise the Queen, the government and members of the public on environmental issues. It sought to challenge past approaches to the regulation of chemicals and to make recommendations for a future regulatory system (RCEP, 2010). Some of the RCEP recommendations to the REACH proposals included a stepwise system for handling chemicals, commencing with the compilation of a list of all marketed chemicals, sorting, selecting and evaluating chemicals of concern and introducing risk-management action where necessary (RCEP, 2010).

2.2.4.3 The Chemicals Hazard Information and Packaging for Supply (CHIPS) regulations

The CHIPS regulations implemented in 2002 deal with the marketing of dangerous substances (chemical elements or compounds such as mercury or sulphuric acid, and preparations, mixtures, solutions or substances such as paints or inks) within the UK. These regulations are amended regularly and require the suppliers of such chemicals to classify their hazards, provide information in the form of labels and material safety data sheets (MSDS), and package the chemicals safely (OPSI, 1991, 2002).

2.2.4.4 Hazardous-waste regulation in the UK

In England and Wales, the 1991 hazardous waste directive was initially implemented by the 'Special Waste' regulations of 1996 (DETR, 1996). These were replaced by the Hazardous Waste Regulations of 2005 (DEFRA, 2005a) and the list of waste regulations (DEFRA, 2005b). The list contains 20 chapters of waste types, based on their origins, which indicates whether or not a waste should be considered hazardous. Subchapters within each of these can then be used to identify the classification of the waste. Wastes that are listed with an asterisk are hazardous and appear in either blue or red. Blue entries are 'absolute hazardous' meaning they are definitely to be considered hazardous. Red entries are mirror wastes and need further consideration to determine their classification. In general, waste producers should know precisely what their waste contains and can therefore determine the classification by consulting a material safety data sheet. The list of hazardous properties in the appendix of the Waste Framework Directive must be consulted at this point. If the waste is shown to contain harmful concentrations of substances with one or more of these properties then it must be considered hazardous. If the producer does not know the content of its waste it must be tested. There is some difficulty in this assessment, particularly with the recent addition of 'Sensitising' as a hazardous waste property, as there is not yet a clear-cut method for testing this property.

In 2009 the Hazardous Waste Regulations (DEFRA, 2009b) were updated, in accordance with the 2008 waste framework directive incorporating hazardous wastes and amendments therein. One of the key changes is towards moving the management of hazardous substances further up the production line, to reduce the amount of hazardous waste produced. The new regulations also incorporate changes to the classification and list of hazardous substances.

Mixed household waste that contains hazardous components is still treated as non-hazardous waste, unless it is collected separately. Many local authorities now offer this service, but there is a lack of public awareness on what products contain hazardous substances. A recent survey (Slack, Zerva *et al.*, 2005) found that while 93 and 92 per cent of householders surveyed could identify batteries and motoring products, respectively, as hazardous, only 18 per cent identified fluorescent lights as hazardous and 60 per cent considered the hazard to be broken glass rather than the mercury content. Therefore better education on hazardous products is going to be an important part of achieving reductions in uncontrolled hazardous waste, but focusing on phasing out the use of hazardous chemicals as far as possible will be the ultimate objective.

2.2.5 Chemicals management in the USA

In the USA, many laws and regulations are in place to manage chemicals at the federal level. These laws and regulations include the Toxic Substances Control Act (TSCA) of 1976 and the Pollution Prevention Act (PPA) of 1990 and a host of voluntary initiatives (USEPA, 2010a). These laws and regulations cover various aspects of chemicals management such as pollution prevention, risk assessment, hazard and exposure assessment and characterisation, and risk management. This part will mainly focus on the TSCA and the PPA.

Through the TSCA, the USEPA controls the production, importation, use, and disposal of specific chemicals including PCBs, asbestos, radon and lead-based paint (USEPA, 2010b). The TSCA requires any party involved in chemicals manufacture, processing and distribution to inform the USEPA of any substantial risk to health or the environment that chemicals or mixtures could pose. In 2008, the USEPA initiated the Chemical Assessment and Management Programme (ChAMP) with the aim of making all the basic screening-level toxicity information on them publicly available (USEPA, 2009a). The main objectives of the ChAMP were to 'develop screening-level characterisations for an estimated 6,750 chemicals produced or imported in quantities of 25,000 pounds or more a year and to prioritise the chemicals for the collection of additional data or the consideration of control measures' (USEPA, 2010c). In December 2009, the ChAMP was superseded by 'action plans on phthalates, long-chain perfluorinated chemicals (PFCs), polybrominated diphenyl ethers (PBDEs) in products, and short-chain chlorinated paraffins. These action plans outline the risks that each chemical may present and identify the specific steps the Agency is taking to address those concerns' (USEPA, 2010c).

The PPA is a national policy, which stipulates that pollution should be prevented or reduced at source. The PPA requires business operators filing a toxic chemical release form to include a toxic reduction and recycling report, and mandates the USEPA to develop and implement strategies to reduce pollution at source and establish databases of relevant information. The PPA also encourages the individual states to promote source reduction by businesses (USEPA, 2006).

While many policies, laws and regulations are made at the federal level, their enactment varies from one state to another, and several other chemical policies are developed at state levels. Development of chemical laws at the state and local levels has been prompted by the inability of Congress to amend TSCA since 1976, and the fact that federal regulations provided little information on chemical hazards and did not sufficiently address the risks of chemicals to humans and the environment. For example, the states of Maine, Washington and California have adopted several 'Green Chemistry' laws very similar to the REACH regulation, which apply only to these states. These laws will restrict the presence of hazardous chemicals in consumer products and establish mandatory priority chemical notification requirements. Many other laws and regulations are proposed in other states, including Oregon, New York, Illinois and Minnesota (BV, 2010).

In 2007, the state of Maine adopted the Toxic Chemicals in Children's Products Law, which aims at reducing exposure to 'chemicals of high concern' or prohibiting their use where safer alternatives exist. In 2008, the state of Washington adopted the Children's Safe Products Act, which requires the Department of Ecology and Department of Health to develop a list of chemicals of high concern for children and to identify children's products or product categories that may contain such chemicals. Under this Act, manufacturers and distributors of children's products are required to provide annual notice to the Department of Ecology if their product contains a high-priority chemical. Senate Bill No. 509 enacted by the state of California in 2010 requires the California Environmental Protection Agency Department of Toxic Substances Control (DTSC) to create an internet database to collect and disseminate chemical hazard information (CSS, 2010). The Assembly Bill No. 1879 requires the DTSC to adopt regulations identifying and prioritising chemicals of concern in consumer products and reduce public exposure to those chemicals (CL, 2010).

2.2.5.1 *Hazardous-waste regulation in the USA*

The first piece of federal legislation relating to solid-waste management in the USA was the 1965 Solid Waste Disposal Act (Wang, 2010). The first significant amendment to this act was the 1976 Resource Conservation and Recovery Act (RCRA) (USEPA, 1976) which substantially altered the regulations in the initial act and extended it to cover hazardous waste. In 1984 the Hazardous and Solid Waste amendments were enacted and are relatively close to the regulations in force today. A significant component of this amendment was incorporating waste minimisation into the regulations, including reducing or eliminating particularly hazardous wastes. In 1992 the Federal Facility Compliance Act (USEPA, 1992a) strengthened the authority to enforce the act at federal facilities and in 1996 the Land Disposal Program Flexibility Act (USEPA, 1996) provided regulatory flexibility for land disposal of certain wastes (Shammas, 2010).

The Hazardous and Solid Waste Act is divided into 10 sections covering most wastes. Those with relevance to hazardous waste will be discussed here. *Section D: Solid waste* applies to non-hazardous waste and to households and industries that produce hazardous waste in quantities less than 100 kg per month, known as 'conditionally exempt small-quantity generators'. It follows a 'reduce, recycle, energy recovery, landfill disposal' hierarchy, similar to the EU waste framework directive. Segregated collection of hazardous household waste is not required, but local government can choose to provide services. These are usually either as bring sites or kerbside collection, but their success is dependent on public awareness of what wastes are hazardous (USEPA, 2008b).

Section C regulations are divided into small-quantity generators (producing between 100 and 1000 kg per month) and large-quantity generators (producing more than 1000 kg per month). This section contains criteria to determine whether waste must be classified as hazardous (USEPA, 2009b). The regulations then follow a 'cradle to grave' approach, divided into three sections: generators; transporters; and treatment, storage or disposal facilities (TSDFs). The regulations applicable to each of these parties are enforced through permitting. States may opt to have their own regulations, as long as they are more stringent than the federal regulations.

Separate legislation exists for certain widely used hazardous wastes including used oil (USEPA, 2008c). The act was designed to encourage the recycling of used oil, and thus reduce the amount requiring disposal. The regulation means that handlers of waste oil do not need to apply for hazardous-waste handling permits unless they are going to dispose of it, and it regulates energy recovery. In 1995 Universal Waste program (USEPA, 1995) was promulgated, which similarly eased the burden of recycling certain common hazardous wastes including batteries, pesticides, thermostats and lamps.

The USA is a signatory on the Basel convention and has similar policies regarding the import and export of hazardous waste (USEPA, 2008d). Further to this they have special agreements with Canada, Mexico, Costa Rica, Malaysia and the Philippines. The agreements are similar to those between European countries. First, consent must be obtained from the receiving country. In the case of USA to Canada (USEPA, 2007), for example, the exporter must send USEPA a detailed report of what they intend to export, and USEPA will pass this on to Environment Canada. If both parties agree, then export must take place with detailed records of the track of waste to the facility, and, finally, reports must be presented to the USEPA, especially where the export is taking place over a period of time, in which case reports must be provided biannually.

2.3 Guideline values

A guideline value represents the concentration of a constituent that does not result in any significant risk to health over a lifetime of consumption. Guideline values for chemicals in the environment are aimed at standardising the evaluation of contaminated water, soils and air.

Guideline values are available for many chemical pollutants. Several provisional guideline values have been established at concentrations that are reasonably achievable through standard treatment approaches. There are two principal sources of information on health effects resulting from exposure to chemicals, which can be used in deriving guideline values. The first (and preferred) source is epidemiological studies on human populations (WHO, 2006). The value of such studies for many substances is limited, however, by the lack of quantitative information on the concentration to which people have been exposed or on simultaneous exposure to other agents (Pan, 2009). The second (and most frequently used) source of information is toxicological studies using laboratory animals (WHO, 2006). In order to derive a guideline value to protect human health, it is necessary to select the most suitable study or studies. Expert judgement is always exercised in the selection of the most appropriate study from the range of information available. Different countries use different guideline values for contaminants. Here we review guideline values for water, soil and air, mainly those recommended by the WHO, the USA, Australia and the EU, with specific legislation from various member states especially the UK.

2.3.1 Water-quality guideline values (WGVs)

Water-quality guidelines have been developed mainly to protect human health, and they are used in the development of risk-management strategies to control the amounts of hazardous substances. The aim of drinking-water guidelines is to ensure the safety of drinking water, and this can be achieved by a combination of adequate management and the effective control of treatment processes, distribution, storage and subsequent handling (IPCS, 2010a). Various risk-assessment exercises have been used to develop drinking-water guidelines, including those from the Environmental Health Criteria monographs (IPCS, 2010a), the Concise International Chemical Assessment Documents, the International Agency for Research on Cancer (WHO, 2010), the Joint Meeting on Pesticide Residues and the Joint Expert Committee on Food Additives (IPCS, 2010b, 2010c). WHO encourages guidelines to be adjusted according to national, regional and local circumstances, and many countries have developed their own national guideline values. The guideline values can be relaxed in certain conditions such as emergencies and disasters (Pan, 2009). The WHO drinking-water guidelines are the most widely used globally, with most variations reflecting differences in country-specific background levels. Table 2.1 summarises the WGVs for heavy metals and Table 2.2 those for some other chemicals, set by the UK, the USA, Australia and the WHO. It is clear that Australia has the most stringent values.

Whilst most countries have developed water guidelines for drinking and surface waters, there remains a lack of strict guidelines for water used for irrigation, which in some countries is used for many additional purposes including small-scale industries, livestock and inland fisheries (Jensen, Matsuno *et al.*, 2001). Irrigation water has various origins, including surface water, groundwater and treated waste water. The WHO has issued health guidelines on the use and reuse of waste water in agriculture and aquaculture. Where guideline values for irrigation exist, they are developed for one particular use and do not take account of other potential uses of the water or local area-specific conditions such as soil, climate or land use. Irrigation water guidelines assume that threshold values can be used to protect crops. However, there is also a need to include parameters such as accumulation of water-borne pollutants in soils and crops. For example, in Bangladesh where arsenic-contaminated groundwater is used for agricultural irrigation, high levels of arsenic have been found in agricultural land and rice (Rahman, 2009).

The quality of fresh water is affected by many factors including run-off from mining, agriculture, sediments and landfill, and influx from sewage and treated waste water effluents, storm water and atmospheric depositions (Chon, Ohandja *et al.*, 2009). These are all potential sources of organic and inorganic pollutants. In some cases it is difficult to regulate sources of contaminants because of their distributed nature. Nevertheless, the quality of treated effluents from waste water treatment plants (WTPs) is strictly regulated throughout the world in an attempt to reduce contamination of the receiving waters. Incidents of accidental discharges of untreated waste water have been widely reported and their adverse impact on the environment and human health acknowledged (Nyenje, Foppen *et al.*, 2010; UNEP, 2010b). Discharge of effluents containing high concentrations of nutrients such phosphorus and nitrogen can cause eutrophication of the receiving surface-water bodies, leading to reduction of oxygen levels, toxic algal blooms, and, potentially, death of the aquatic fauna and flora. Table 2.3 shows typical concentrations of selected metals from conventionally treated effluents across the world.

Conventional biological sewage treatment – which is the most commonly used method of treating municipal waste water around the world – was designed mainly to remove suspended solids (SS), and reduce biochemical oxygen demand (BOD), chemical oxygen demand (COD) and nutrients (Gagnon and Saulnier, 2003; Oliveira, Bocio *et al.*, 2007). If a conventional biological waste-water treatment plant removes metals as well, that is an added benefit (Karvelas, Katsoyiannis *et al.*, 2003); metals that are not removed are either adsorbed onto sludge or discharged with treated waste-water effluents into receiving water bodies. Effluent quality depends mostly on the discharge consents set by regulators at national and regional levels.

In the UK, for example, effluent quality is regulated by two key legal frameworks: the Water Resources Act 1991 (a piece of UK legislation which applies to all sewage treatment works, STWs), and the Urban Waste Water Treatment Directive 91/271/EEC (a slice of EU legislation that applies to larger STWs in agglomerations of more than 2000, with more advanced treatment required for agglomerations of more than 10 000; EC, 1991c). Other pieces of legislation setting site-specific standards include the EC Habitats Directive (EC, 1992), the EC Shellfish Waters Directive (EC, 2006a) and the EC Bathing Water Directive (EC, 2006b).

The implementation of the EU Water Framework Directive (WFD) aims to restore the quality of surface waters and ground waters based on river-basin management plans. The WFD requires EU member states not only to prevent the deterioration of their water quality but also to achieve 'good' chemical, ecological and quantitative status of their waters by 2015 (EC, 2000). The WFD identifies 33 dangerous substances also known as 'priority substances'. The environmental quality standards for these substances have been derived from various assessments of their risks to the environment. The substances are divided in two main groups – priority substances, which need to be controlled to achieve a progressive reduction of discharges, and priority hazardous substances, which should be removed or phased out of discharges, emissions and losses by 2020 (EC, 2000, 2001, 2008b).

Table 2.1 Guideline values for selected metals in water (adapted from Pan, 2009)

Metal		Water guideline values		
		Drinking water (mg/l)	Fresh water (μg/l)	Irrigation water (mg/l)
As	UK (Anon, 2000)	0.01	–	–
	USA (USEPA, 2010d)	0.01	150–340	–
	WHO (WHO, 2006)	0.01	–	–
	Australia (NEPC, 1999a)	0.007	50	0.1
Cd	UK	0.005	0.08–0.25	–
	USA	0.005	0.25–2	–
	WHO	0.003	–	–
	Australia	0.002	0.2–2	0.01
Cu	UK	2	–	–
	USA	1.3	9–13	–
	WHO	2	–	–
	Australia	2	2–5	0.2
Hg	UK	0.001	0.05	–
	USA	0.002	0.77–1.4	–
	WHO	0.006	–	–
	Australia	0.001	0.1	0.002
Mo	UK	–	–	–
	USA	–	–	–
	WHO	0.07	–	–
	Australia	0.05	–	0.01
Ni	UK	0.02	20	–
	USA	–	52–470	–
	WHO	0.07	–	–
	Australia	0.02	15–150	0.02
Pb	UK	0.025	7.2	–
	USA	0.015	2.5–6.5	–
	WHO	0.01	–	–
	Australia	0.01	1–5	0.2
Sb	UK	0.005	–	–
	USA	0.006	–	–
	WHO	0.02	–	–
	Australia	0.003	30	–
Se	UK	0.01	–	–
	USA	0.05	5	–
	WHO	0.01	–	–
	Australia	0.01	5	0.02
Tl	UK	–	–	–
	USA	0.002	–	–
	WHO	–	–	–
	Australia	–	4	–
Zn	UK	–	–	–
	USA	5	120	–
	WHO	15	–	–
	Australia	3	5–50	2

Table 2.2 Guideline values for other chemicals in water

Substance		Water guideline values	
		Drinking water (μg/l)	Fresh water (μg/l)
Atrazine	England and Wales	–	0.6
	USA	3	–
	WHO	2	–
	Australia	1	–
Benzene	England and Wales	1	10
	USA	5	–
	WHO	10	–
	Australia	1	10
Carbon tetrachloride	England and Wales	–	12
	USA	0	–
	WHO	5	–
	Australia	3	3
Aldrin	England and Wales	0.03	$\Sigma = 0.001$ for aldrin + dieldrin + endrin + isodrin
	USA	–	–
	WHO	$\Sigma = 0.03$ for aldrin + dieldrin	–
	Australia	0.01	1
Dieldrin	England and Wales	0.030	See aldrin
	USA	–	–
	WHO	See aldrin	–
	Australia		1
Endrin	England and Wales	–	See aldrin
	USA	2	–
	WHO	0.6	–
	Australia	–	1
Dichloromethane	England and Wales	–	20
	USA	5	–
	WHO	20	–
	Australia	4	–
Hexachlorobutadiene	England and Wales	–	0.1
	USA	–	–
	WHO	0.6	–
	Australia	0.7	–
Pentachlorophenol	England and Wales	–	0.4
	USA	1	–
	WHO	9	–
	Australia	–	–
Tetrachloroethene	England and Wales	10	10
	USA	5	–
	WHO	40	–
	Australia	50	10
Tricholorethylene	England and Wales	10	10
	USA	5	–
	WHO	20	–
	Australia	Insufficient data to set guideline	30

Table 2.3 Concentrations of selected metals in conventionally treated waste-water effluent in the world

Mean concentration (μg/l)				Reference
Cd	Pb	Hg	Ni	
12.00–20.00	65.0–180.0	0.50–1.60	–	Brown, Hensley *et al.*, 1973[a]
1.00	15.00	<1.00	270.00	Oliver and Cosgrove, 1974
8.50	48.50	0.14	116.00	Chen, Young, *et al.*, 1974
0.20	2.60	–	7.80	Brombach, Weiss *et al.*, 2005
1.50	27.00	–	430.00	Karvelas *et al.*, 2003
0.06	22.57	0.05	–	Oliveira, Bocio *et al.*, 2007
0.01–0.25	2.50–16.40	0.60–2.00	1.93–11.70	Carletti, Fatone *et al.*, 2008[a]

[a]A range of mean values from different sites.

In the USA, the Clean Water Act lays out the regulatory framework for the discharge of pollutants, and the National Pollutant Discharge Elimination System permit program controls water pollution by setting standards for point sources discharging pollutants into waters such as wetlands, lakes, rivers, estuaries, bays and oceans (USEPA, 2010e).

2.3.2 Soil Guideline Values (SGVs)

2.3.2.1 UK

The CLEA (Contaminated Land Exposure Assessment) model published by the Department for Environment, Food and Rural Affairs (DEFRA) and the Environment Agency (EA) in March 2002 sets out a framework for the appropriate assessment of risks to human health from contaminated land, as required by Part IIA of the Environmental Protection Act 1990. As part of this framework, generic SGVs have been derived for 10 contaminants, to be used as 'intervention values'. These values should not be considered as remedial targets but values above which further detailed assessment should be considered. Three sets of CLEA SGVs have been produced for residential land (with and without plant uptake), allotments and commercial/industrial land.

2.3.2.2 USA

Soil Screening Levels (SSLs) are concentrations of contaminants in soil that are designed to protect people from exposure in a residential setting and are derived from Soil Screening Guidance. They aim to standardise and accelerate the evaluation and clean-up of contaminated soils at sites on the National Priorities List with future land use (USEPA, 2010f). SSLs are risk-based, derived from equations combining estimates of exposure with USEPA toxicity data. SSLs alone do not trigger the need for response actions or define unacceptable levels of contaminants in soil. In the USA SSLs, 'screening' refers to the process of identifying and defining areas, contaminants, and conditions, at a particular site which do not require further Federal attention. Generally, at sites where contaminant concentrations fall below SSLs, no further action or study is warranted under the Comprehensive Environmental Response, Compensation and Liability Act. Generic SSLs can be used in place of site-specific screening levels; however, in general, they are expected to be more conservative than site-specific levels. To calculate SSLs, the exposure equations and pathway models are run in reverse to back-calculate an 'acceptable level' of a contaminant in soil.

2.3.2.3 Australia

The assessment levels used in Australia were compiled from several sources, which were in turn compiled using different methods and studies including the National Environment Protection Council (NEPC) study (NEPC, 1999b). They are therefore generic and provide guidance only.

2.3.2.4 Comparison between different countries

Table 2.4 lists SGVs that apply in different countries for several important elements.

The variations in the guideline values for given contaminants and land use (i.e. residential, recreational, industrial and ecological) reflect the influence of politics in setting SGVs in different countries. The SGVs are often based on political considerations rather than scientific evidence, and the potential (carcinogenic or non-carcinogenic) human-health risks often remain high even where there is full compliance with the guideline values.

2.3.3 Application of biosolids to agricultural land

Biosolids is a generic term used to refer to the mix of residual organic and inorganic materials from sewage and municipal waste-water treatment processes after most of the pathogens have been killed. Biosolids contain nutrients and are excellent

Table 2.4 Soil guideline values (adapted from Pan, 2009)

Metal		Soil guideline values (mg/kg)			
		Residential	Recreational	Industrial	Ecological
As	UK (Environment Agency, 2010)	20	–	–	–
	USA (USEPA, 2010g)	0.39 (ca) 22 (nc)	–	990000	–
	The Netherlands (Swartjes, Dirven-Van Breemen *et al.*, 2007)	55	55	–	–
	Australia (Department of Environment, 2003)	100	200	–	–
Cd	UK	1 (pH=6) 2 (pH=7) 8 (pH =8)	–	1400	–
	USA	70	–	810	–
	The Netherlands	12	12	12	12
	Australia	20	40	100	3
Cu	UK	–	–	–	–
	USA	3100	–	41000	–
	The Netherlands	190	190	190	190
	Australia	1000	2000	5000	60
Hg	UK	8	–	480	–
	USA	6.7–23	–	28–310 (ca)	–
	WHO	10	10	10	10
	Australia	15	30	75	1
Mo	UK	–	–	–	–
	USA	390	–	5100	–
	The Netherlands	200	200	200	200
	Australia	390	–	10220	40
Ni	UK	50	–	5000	–
	USA	1600	–	20000	–
	The Netherlands	210	210	210	210
	Australia	600	600	3000	60
Pb	UK	450	–	750	–
	USA	400	–	800	–
	WHO	530	530	530	530
	Australia	300	600	1500	300
Sb	UK	–	–	–	–
	USA	31	–	410	–
	The Netherlands	15	15	15	15
	Australia	30	–	820	20
Se	UK	35	–	8000	–
	USA	390	–	5100	–
	The Netherlands	100	100	100	100
	Australia	–	–	–	–
Tl	UK	–	–	–	–
	USA	5.1	–	66	–
	The Netherlands	15	15	15	15

(*continued*)

Table 2.4 *(Continued)*

Metal		Soil guideline values (mg/kg)			
		Residential	Recreational	Industrial	Ecological
	Australia	–	–	–	–
Zn	UK	–	–	–	–
	USA	23000	–	310000	–
	The Netherlands	720	720	720	720
	Australia	7000	14000	3500	200

ca – carcinogenic.
nc – noncarcinogenic.

fertilisers and soil conditioners, so they are widely used on agricultural land in many countries. However, they often contain residual pathogens, organic contaminants (OCs) and potentially toxic elements (PTEs). Application of sludge to farmland should therefore comply with criteria set at national and regional levels, and some countries have established clear guideline values. After application to land, contaminants in biosolids undergo biological and chemical reactions, and physical transfer processes within the soil, thus presenting a theoretical risk for the environment and human health (van Leeuwen and Hermens, 1995). Various pathways of exposure to contaminants in biosolids have been reported, including pathways to the environment, to livestock, to humans via livestock, to humans via direct exposure, and to humans through dispersion to other environmental media (Schowanek, Carr *et al.*, 2004). Table 2.5 shows the permissible levels of OCs in biosolids in various EU member states.

The concentrations of pollutants in biosolids vary according to geographical areas and the processes used to treat waste waters, but concerns have been raised on potential risks from their use on agricultural land. There is concern that high concentrations of OCs and PTEs in biosolids could be taken up by plants and reach humans through the food chain. In Switzerland, for example, concerns about potential exposure to contaminants found in biosolids have led to a ban on the disposal of biosolids to agricultural land despite scientific evidence that this is not a significant exposure route. Research in the UK, the USA and Canada has also shown that the risks associated with biosolids application to farmland are of little or no significance for the dietary intake of contaminants via soil, aquatic ecosystems or animals (Sauerbeck and Leschber, 1992; USEPA, 1992b; Smith, 2000; Bright and Healey, 2003; USEPA, 2003).

While the presence of OCs is widely acknowledged, there is no common approach to controlling or regulating their concentrations in biosolids (Schowanek, Carr *et al.*, 2004). Despite scientific evidence that the levels of OCs in biosolids pose no threat to the environment or humans, many European countries have used the precautionary principle to set limit values for OCs in biosolids. For example, Germany has set limits on polychorophenyls, France on polyaromatic hydrocarbons and Denmark on chemicals such as linear alkylbenzene sulphonate (Smith, 2009). Table 2.5 shows the list of OCs regulated by some EU member states. Other countries, such as the USA and the UK, argue that since the concentrations of OCs currently found in sludge are not a risk to the environment or human health there is no need to set limits (USEPA, 1992b; Blackmore, Davis *et al.*, 2006).

Concentrations of PTEs in biosolids are also of concern. One study of biosolids used for agricultural land in the EU found mean concentrations of zinc, copper, nickel, lead, cadmium, chromium, arsenic and mercury at 802, 565, 59, 221, 3.4, 163, 6 and 2.3 mg/kg respectively of dry solid (Gendebien, Davis *et al.*, 2001). Many researchers agree that the levels of metals in biosolids are low and do not require regulation. In the USA, however, the USEPA has developed limit values for metals in biosolids applied to land, as outlined in Table 2.6.

In the UK, the Code of Practice for Sludge Use on Agricultural Land (1996) regulates the concentrations of toxic trace elements in soils following the application of biosolids to agricultural land, assuming that there is no atmospheric deposition and no loss of compounds over time (ADAS, 2001). This code of practice does not set limits for heavy metals in sludge but recommends the maximum permissible concentrations in soils after the application of biosolids to agricultural land as well as the maximum permissible average annual rate of application. These concentrations are shown in Table 2.7. Table 2.8 shows the maximum permissible concentrations of PTEs in soil after application of sewage sludge and maximum annual rates of addition in the UK.

2.3.4 Application to land of treated biodegradable components of municipal solid waste

The biodegradable components of municipal solid waste (MSW) are increasingly being diverted from landfill via either thermal or biological treatment processes. The residues from biological treatments, such as composting and anaerobic digestion, contain nutrients that can improve soil quality. Currently in the UK, treated biological waste that is applied to land must meet the Publicly Available Specification (PAS) 100 (British Standards Institute, 2005). Part of the requirement is that the incoming

Table 2.5 Standards for maximum concentrations of organic contaminants in sewage sludge (Reproduced, with permission, from Organic contaminants in sewage sludge (biosolids) and their significance for agricultural recycling, Smith SR (2009). *Philosophical Transactions Of The Royal Society A-Mathematical Physical And Engineering Sciences* **367**(1904): 4005–4041)

Countries	(mg/kg dry solids (DS) except PCDD/F: ng toxic equivalents (TEQ) per kg DS)						
	Absorbable organic halogen (AOX)	Di(2-ethylhexyl) phthalate (DEHP)	Linear alkylbenzene sulphonates (LASs)	Nonylphenols and nonylphenol ethoxylates (NP/NPEs)	Polycyclic aromatic hydrocarbons (PAHs)	Polychlorinated biphenyls (PCBs)	Polychlorinated dibenzo-p-dioxins and dibenzo-p-furans (PCDD/Fs)
EC (2000)[a]	500	100	2600	50	6[b]	0.8[c]	100
EC (2003b)[a]	–	–	5000	450	6[b]	0.8[c]	100
Denmark	–	50	1300	10	3[b]	–	–
Sweden	–	–	–	50	3[d]	0.4[c]	–
Lower Austria	500	–	–	–	–	0.2[e]	100
Germany	500	–	–	–	–	0.2[e]	100
France	–	–	–	–	9.5[f]	0.8c[g]	–
USA	–	–	–	–	–	–	300[h]

[a]Proposed but withdrawn and basis subject to review.
[b]Sum of nine congeners: acenaphthene, fluorene, phenanthrene, fluoranthene, pyrene, benzo(b + j + k)fluoranthene, benzo(a)pyrene, benzo(ghi)perylene, indeno(1,2,3-c,d)pyrene.
[c]Sum of seven congeners: PCB 28, 52, 101, 118, 138, 153, 180.
[d]Sum of six congeners.
[e]Each of the six congeners: PCB 28, 52, 101, 138, 153, 180.
[f]Sum of three congeners: fluoranthene, benzofluoranthen(b), benzo(a)pyrene.
[g]For pasture the limit is 0.5 mg/kg DS.
[h]Following detailed risk assessment, USEPA's final decision was not to regulate PCDD/Fs (USEPA, 2003).

Table 2.6 PTE limits and loading rates for biosolids applied to land, from the USEPA (USEPA 2004)

Pollutant	Ceiling concentration limits (mg/kg)[a]	Pollutant concentration (mg/kg)[a]	Cumulative pollutant loading rate limits for biosolids (kg/ha)	Annual pollutant loading rate limits for biosolids (kg/ha per annum)
As	75	41	41	2.0
Pb	840	300	300	15
Hg	57	17	17	0.85
Mo	75	–	–	–
Ni	420	420	420	21
Se	100	100	100	5.0
Zn	7500	2800	2800	140
Applies to:	All biosolids that are land applied	Bulk biosolids and bagged biosolids[c]	Bulk biosolids	Bagged biosolids[c]
From Part 503	Table 1 Section 503.13	Table 3 Section 503.13	Table 2 Section 503.13	Table 4 Section 503.13

[a]Dry-weight basis.
[b]As a result of the February 25, 1994, Amendment to the rule, the limits for Mo were deleted from the Part 503 rule pending EPA consideration.
[c]Bagged biosolids are sold or given in a bag or other container.

waste stream is restricted to suitable biodegradable components, although strictly controlled trials are investigating the potential use of outputs from a mixed waste stream, separated using mechanical techniques (Environment Agency, 2008). Moreover, biological components known to have been treated with potentially toxic chemicals are banned from entering the process (Environment Agency, 2008). Hence, the concentrations of these elements should be lower than in sewage sludge. MSW is collected from a large number of sources and relies on widespread public participation so there is a risk that hazardous wastes such as pesticides used in gardening could enter through the process. The hazardous nature of the products might not be known to the end user, and their presence in the waste may not be apparent to the acceptor at the waste-management facility.

2.3.5 Air quality

Air-quality legislation has a long history, dating back to ancient times. There are records of odour-nuisance legislation in ancient Greece, and in ancient Rome damage from smoke due to wood burning resulted in law suits (Jacobson, 2002). However, the massive increase in coal burning since the industrial revolution has resulted in many health problems attributable to poor air quality. While these have been largely overcome in developed countries, with improvements in smokestack-emission abatement, the WHO estimates there are still 2 million deaths per year caused by poor air quality (WHO, 2005). In less developed countries this is often due to indoor air quality associated with cooking and heating practices, while in urban areas of both

Table 2.7 Maximum permissible concentrations of metals in biosolids for agricultural application (Reproduced, with permission, from Occurrence and fate of heavy metals in the wastewater treatment process. Karvelas, M., A. Katsoyiannis, *et al.* (2003). *Chemosphere* **53**(10): 1201–1210)

Country	Maximum permissible metal concentrations (mg/kg dry weight)					
	Cd	Cr	Cu	Pb	Ni	Zn
Greece	20–40	–	1000–1750	750–1200	300–400	2500–4000
Austria	3	250	500	250	100	1200
Belgium	5	200	500	1000	100	1500
Denmark	0.4	–	–	120	30	–
Germany	10	900	800	900	200	2500
France	20	1000	1000	800	200	3000
Luxembourg	1.5	100	100	150	50	400
Holland	1	50	75	100	30	200
Spain	10	400	50	300	120	1100
Italy	10	600	600	500	200	2500
Slovenia	5	500	600	500	80	2000

Table 2.8 Maximum permissible concentration of potentially toxic elements (PTE) in soil after application of sewage sludge and maximum annual rates of addition in the UK (DEFRA, 2011)

PTE	Maximum permissible concentration of PTE in soil (mg/kg dry solids)				Maximum permissible average annual rate of PTE addition over a 10-year period (kg/ha)[b]
	pH 5.0–5.5	pH 5.5–6.0	pH 6.0–7.0	pH >7.0[c]	
Zn[a]	200	200	200	300	15
Cu[a]	80	100	135	200	7.5
Ni	50	60	75	110	3
	For pH 5.0 and above				
Cd[a]	3				0.15
Pb[a]	300				15
Hg	1				0.1
*Cr	400				15
*Mo	4				0.2
*Se	3				0.15
*As	50				0.7
*F	500				20

*These parameters are not subject to the provisions of Directive 86/278/EEC. (In 1993 the European Commission withdrew its 1988 proposal to set limits for addition of chromium from sewage sludge to agricultural land).
[a]The permitted concentrations of zinc, copper, cadmium and lead are provisional and will be reviewed when current research into their effects on soil fertility and livestock is completed. The pH qualification of limits will also be reviewed with the aim of setting one limit value for copper and one for nickel across pH range 5.0<7.0 and therefore ensuring consistency with the approach adopted for zinc in response to the recommendations from the Independent Scientific Committee.
[b]The annual rate of application of PTE to any site shall be determined by averaging over the 10-year period ending with the year of calculation.
[c]The increased permissible PTE concentrations in soils of pH greater than 7.0 apply only to soils containing more than 5% calcium carbonate.

developing and developed countries it is a result of motor-vehicle emissions.

2.3.5.1 WHO

The WHO first published air quality guidelines in 1987. The most recent update to this was in 2005 (WHO, 2005). The pollutants of concern are: particulate matter, which is split into two categories, less than 10 μm aerodynamic diameter (PM_{10}) and less than 2.5 μm aerodynamic diameter ($PM_{2.5}$); and ozone, nitrogen dioxide (NO_2) and sulphur dioxide (SO_2), as shown in Table 2.9. For PM, the value is based on studies that showed, with 95 per cent confidence, an increase in total, cardiopulmonary and lung-cancer mortality in response to long-term exposure.

The ozone level is based on a combination of epidemiological evidence (which showed that exposure to greater than this level of ozone leads to an estimated 1–2 per cent increase in daily mortality), and extrapolation from chamber and field studies. The results of the latter studies were adjusted to account for the likelihood that day-to-day exposure tends to be repetitive and chamber studies exclude highly sensitive or clinically compromised individuals. While the guideline limit is considered to provide adequate protection of public health, some health effects may occur below this level.

The NO_2 limit is based on toxicological studies, which showed significant health impacts at both short and long-term exposures. The impacts were more difficult to identify in epidemiological studies as the gas is formed during combustion and is always accompanied in the atmosphere by other potentially harmful chemicals. NO_2 is one of the main precursors of tropospheric ozone, so reducing its concentrations in the atmosphere may also reduce ozone levels.

Toxicological studies have shown respiratory system impacts due to SO_2 after exposures of only 10 minutes. The health impacts of long-term exposures based on epidemiological studies are harder to identify, as other pollutants present in the atmosphere may have been the cause.

Implementation of legislation to improve air quality is down to individual countries and should take into account the economic situation and any background concentrations. Even with concentrations of these pollutants below recommended threshold values, health effects may occur as there are still many gaps in the research.

2.3.5.2 EU

The EU Air Quality Directive specifies limit values for atmospheric concentrations of benzene, 1, 3-butadiene, carbon monoxide, lead, ozone, NOx, PM_{10} and SO_2, as shown in Table 2.9.

Table 2.9 Ambient air quality guidelines

Country	Pollutant	Limit value (μg/m^3)	Time period
WHO	Benzene	–	–
EU		5	annual mean
UK		16.25	running annual mean
USA		–	–
WHO	1,3 Butadiene	–	–
EU		–	–
UK		2.25	running annual mean
USA		–	–
WHO	Carbon monoxide	–	–
EU		10	8-hour mean
UK		10	running 8-hour mean
USA		10	8-hour mean
		40	1-hour mean
WHO	Lead	–	
EU		0.5	annual mean
UK		0.25	annual mean
USA		0.15	rolling 3-month mean
		1.5	quarterly average
WHO	NO_2	40	annual mean
		200	1-hour mean
EU		200	1-hour mean (not to be exceeded more than 18 times a year)
		40	annual mean
UK		200	1-hour mean (not to be exceeded more than 18 times a year)
		40	annual mean
USA		53 (ppb)	annual mean
		100 (ppb)	1-hour mean
WHO	PM_{10}	20	annual mean
		50	24-hour mean
EU		50	24-hour mean (not to be exceeded more than 35 times a year)
		40	annual mean
UK		50	24-hour mean (not to be exceeded more than 35 times a year)
		18	annual mean
USA		15	annual mean
		35	24-hour mean
WHO	SO_2	20	24-hour mean
		500	10-minute mean
EU		350	1-hour mean
		125	24-hour mean (not to be exceeded more than 3 times a year)
UK		350	1-hour mean (not to be exceeded more than 24 times a year)
		125	24-hour mean (not to be exceeded more than 3 times a year)
		266	15-minute mean (not to be exceeded more than 35 times a year)
USA		0.03 (ppm)	annual mean
		0.14 (ppm)	24-hour mean
		75 (ppb)	1-hour mean

Table 2.9 *(Continued)*

Country	Pollutant	Limit value ($\mu g/m^3$)	Time period
WHO	$PM_{2.5}$	10	annual mean
		25	24-hour mean
EU			
UK		25	annual mean[a]
USA		15	annual mean
		35	24-hour mean
WHO	Ozone	100	8-hour mean
EU		–	
UK		–	
USA		0.075 (ppm)	8-hour mean
		0.12 (ppm)	1-hour mean

UK values do not apply to Scotland, which has its own, more stringent limit values.
[a]At the moment this is under consultation and is not a regulation.

Ozone is the most difficult to tackle due to its long range, so that any control requires cooperation across borders (EPC, 2008b). Emission limits on $PM_{2.5}$, which are believed to cause more severe health problems than the larger PM_{10}, are included in the most recent revision of the directive. The limit values are defined over a particular time period and depend on the health effects as defined by the WHO. Limit values apply only in areas of relevant exposure, that is to say where receptors are likely to be present for the time period of the limit value. For example, there are two limits for NO_2, an annual mean of 40 $\mu g/m^3$ and a 1 hour mean of 200 $\mu g/m^3$ not to be exceeded more than 18 times a year. The annual mean is applicable in residential areas, where the same people are likely to be exposed constantly over the year, while the one hour mean would be relevant in public places, such as parks, where people are likely to be regularly present for an hour or so at any one time or several times a year if jogging, walking their dog etc. on a regular basis.

2.3.5.3 UK

In the UK the health impacts of poor air quality were highlighted in 1952 when the severe four-day 'pea soup' smog occurred. Londoners were used to heavy fogs and not concerned at first, but 4000 deaths and 100,000 illnesses resulted. In the months following the event, elevated mortality levels continued to occur. At the time these deaths were attributed to influenza, but retrospective analysis (Bell, Davis *et al.*, 2004) indicates that an influenza epidemic three times larger than the most severe epidemic on record would have to have occurred – which is extremely unlikely. Elevated level of pollutants is a much more likely explanation. The incident led to the 1956 Clean Air Act (OPSI, 1956), which reduced the use of coal for domestic heating and introduced 'smokeless zones'. Although air quality improved dramatically, there is some belief this would have occurred anyway due to a reduction in the use of coal for domestic heating and movement of large industrial plants away from densely populated areas.

The update to the act in 1968 (OPSI, 1968) introduced the use of tall chimney stacks aimed at diluting SO_2 concentrations to safe levels by the time the gas reached ground level. However, in the 1970s it came to light that the long distance travelled by these pollutants was leading to acidification of surface waters in countries such as Norway and Sweden. It was not until 1983 – when R. W. Batterbee, Professor of Environmental Change at University College, London, proved the link between these emissions and acidification in lakes in Scotland (Flower and Battarbee, 1983) – that the UK Government acknowledged the problem and began to take measures to abate emissions such as fitting coal-burning facilities with air filters to prevent the release of SO_2 and other harmful pollutants to the atmosphere. Industrial pollution is now heavily countered throughout Europe under the Large Combustion Plant Directive (EPC, 2001). The most problematic air pollutants in the UK today are NO_2 and PM_{10}, which are both mainly sourced from road vehicles. These emissions are much harder to regulate than industrial sites.

The UK air-quality strategy incorporates all of the EU limit values in their air-quality objectives (DEFRA, 2007), where they are the same or more stringent than those independently assessed for the UK (see Table 2.9). The enforcement of the objectives is the responsibility of local authorities, who are required to produce regular review and assessment reports (DEFRA, 2009c). Following an initial review and assessment to identify emission sources in the area, the likelihood of a breach of air-quality regulations is assessed. If any are identified, authorities are required to produce a detailed assessment within 12 months, incorporating monitoring data and modelling to determine the likelihood of failure to meet the objectives and estimate the magnitude and geographical extent of any exceedence. Where an exceedence is considered likely, an air-quality management area (AQMA) must be declared and a further assessment completed within 12 months. This must reconfirm exceedence and define the required improvement in air quality, and then identify source contributions and how their emissions can be reduced. An Air Quality Action Plan must be developed, outlining the authority's strategy for reducing air-quality impacts in their local area. This must be regularly updated in order to prove its effectiveness, and where it is not shown to be effective must be revised to determine a better method. Where no likely exceedence is predicted, a regular review and assessment report must still be produced in order to identify any new potential sources of pollutants, such as industrial sites or developments that might lead to a significant increase in traffic.

2.3.5.4 USA

Records of air-quality legislation in the USA at a municipal level date back as far as 1869 (Jacobson, 2002), but much of the early legislation was not enforced and was largely ineffective. The first attempt at enforcing regulation, in St Louis in 1893, was overruled by the state Supreme Court, but municipal-level air-pollution legislation continued to be enacted and approximately 200 laws had been passed by 1940. The first instance of state legislation was in Massachusetts in 1910, when regulations were introduced to control air pollution in the city of Boston. Federal-level involvement also occurred in the early 1900s, with the establishment of the Office of Air Pollution; however, this body was largely inactive and only survived a short time.

The first federal legislation was the Air Pollution Control Act of 1955 (Federal Register, 1955). This provided five-years worth of funding for research into the effects of air pollution and abatement methods. Funding was extended in 1959 and 1960, when one of the main reasons was to study the health effects of vehicle-exhaust emissions.

The first air-quality federal regulation was the Clean Air Act of 1963 (Federal Register, 1963), which imposed limits on smokestack emissions. Many amendments have been made since the original legislation was passed, some of the most significant of which are as follows.

In 1965 the Motor Vehicle Air Pollution Control Act (Federal Register, 1965) was passed, which set standards for emissions of carbon monoxide and hydrocarbons from motor vehicles. The emission standards were largely unmet.

In 1967 Air Quality Act (Federal Register, 1967) was passed, which divided the country into Air Quality Control Regions. Now relevant authorities were required to determine the extent of air pollution in their area, collect air-quality data and develop an emissions inventory. It also led to the development of air-quality criteria reports with details of potential health impacts and suggested acceptable concentrations. On the basis of these, states were then required to set and enforce air-quality standards and produce a State Implementation Plan (SIP) explaining how their standards would be achieved, for submission to the federal government.

In 1970 the USEPA was established to enforce air-quality regulations and another amendment (Federal Register, 1970) was passed. SIPs were still relevant, and states had to form action plans detailing how they would combat air pollution. The amendment included emission standards for aircrafts and regulation on the emission of hazardous substances. Perhaps the most significant development was the National Ambient Air Quality Standards (NAAQS), which defined certain substances known as criteria pollutants, based on health impacts. The six original pollutants were CO, NO_2, SO_2, total suspended particulates (TSPs), hydrocarbons and photochemical oxidants (see Table 2.9). Significant updates to this occurred in 1976 (when lead was added), in 1979 (when ozone replaced photochemical oxidants), and in 1983 when HCs were removed and TSPs were replaced with PM_{10} (Jacobson, 2002).

In 1977 another amendment was passed (Federal Register, 1977) which required any state with non-attainment areas, i.e. areas where the ambient air quality exceeded the NAAQS, to produce a revised SIP. A permitting programme was established to prevent deterioration of air quality in attainment areas, i.e. areas compliant with NAAQS. Any new developments with polluting potential were required to install the best available emissions control technologies and use computer modelling to show that their emissions wouldn't lead to an exceedence. This piece of legislation was also the first to impose regulations on CFCs.

In 1990 a further amendment was passed (Federal Register, 1990), largely because there was still widespread exceedence of NAAQS. Cities were given deadlines for reducing ambient concentrations of pollutants, based on the severity of their exceedence. Further to this, any new sources of air pollution in non-attainment areas were required to achieve the lowest possible emissions, and existing sources were required to improve their emissions as far as reasonably possible. Authorities with non-attainment areas were required to produce emissions inventories for key pollutants and prove their action plans would achieve NAAQS in the required time, using computer modelling. The 1990 amendment also contained more stringent regulations on car and truck emissions, acid deposition precursors, hazardous pollutants and CFCs. The new CFC regulations aimed to phase out the use of these substances altogether. Regulation enforcement was moved from state to federal level.

In 1997 an additional amendment (Jacobson, 2002) altered standards for ozone and PM_{10}, and instituted a new standard for $PM_{2.5}$.

2.3.5.5 Australia

Australia does not have ambient air quality standards to the same extent as European countries and the USA as they have not suffered the same air-quality issues (Jacobson, 2002). The cities and urban areas are less densely populated, energy generation plants are generally located away from populated areas and have always been so, and it does not receive pollution from other countries.

Cars built since 1986 have been required to contain catalytic converters to reduce the toxicity of emissions. The government is offering financial incentives to industry to reduce their emissions – though the aim of this is more to reduce the emission of gases that contribute to the greenhouse effect than to improve local air quality.

2.4 Conclusions and recommendations

In the past, the introduction of new chemicals into the environment has sometimes had devastating consequences for public health and the environment, and legislation to protect the public from the health risks has been largely reactive. However, incidences such as the pea-soup smogs of the 1950s in both the UK and the USA, which resulted in tens of thousands of deaths, have prompted the development of a more precautionary approach, where the potential risks of chemicals must be established before they are released. More recently, this has been taken further to make producers, rather than end-users, responsible for the fate of chemicals they produce, in order to discourage the use of hazardous substances where safer alternatives are available. This new approach has caused controversy in some countries, especially the USA and Canada where it is seen as an obstacle to progress, while the European Union has taken it into policy. The recent REACH legislation is the most ambitious example of this. The WHO plays an important role in establishing 'safe' concentrations of chemicals in the environment through a combination of epidemiological and toxicological studies. Governments can introduce the WHO's guidelines into policy, but enforcing these can be difficult. It is particularly difficult to improve air quality, as improvements in emissions from vehicles are counteracted by the ever-increasing number of vehicles on the road. Thus, although pushing producers to reduce the quantities of hazardous substances in their products and take responsibility for the management of the waste is going some way towards protecting public health, greater public awareness of risks is also important.

References

Anon (2000). The water supply (Water quality) regulations. Statutory Instrument No. 3184. UK, Her Majesty's Stationery Office. **No. 3184**.

ADAS (2001). The safe sludge matrix. Guidelines for the application of sewage sludge to agricultural land. Retrieved 8 February 2010, from http://www.adas.co.uk/Home/Publications/DocumentStore/tabid/211/Default.aspx.

Ahlers, J. (1999). The EU existing chemicals regulation – A suitable tool for environmental risk assessment and risk management? *Environmental Science and Pollution Research* **6**(3): 127–129.

Bell, M. L., D. L. Davis, *et al.* (2004). A retrospective assessment of mortality from the London smog episode of 1952: The role of influenza and pollution. *Environmental Health Perspectives* **112**(1): 6–8.

Blackmore, K., L. Davis, *et al.* (2006). Accommodating the implications of the revised EU Sludge Directive. *L. UKWIR.*

Boehmer-Christiansen, S. (1994). The precautionary principle in Germany-enabling government. *Interpreting the Precautionary Principle*. T. O'Riordan and J. Cameron. London, Earthscan.

Bright, D. A. and N. Healey (2003). Contaminant risks from biosolids land application: contemporary organic contaminant levels in digested sewage sludge from five treatment plants in Greater Vancouver, British Columbia. *Environmental Pollution* **126**(1): 39–49.

British Standards Institute (2005). PAS 100:2005 Specification for composted materials. Bristol, British Standards Institute.

Brombach, H., G. Weiss, *et al.* (2005). A new database on urban runoff pollution: comparison of separate and combined sewer systems. *Water Science and Technology* **51**(2): 119–128.

Brown, H. G., C. P. Hensley, *et al.* (1973). Efficiency of heavy metals removal in municipal sewage treatment plants. *Environmental Letters* **5**(2): 103–114.

BV (2010). US Regulatory Update: Summary of Recent State Bills Related to Priority Chemical Lists. Retrieved 25 August 2010, from http://www.bureauveritas.com/wps/wcm/connect/fbd9e30041e44057b493b638d772f879/Bulletin_10B-120.pdf?MOD=AJPERES&CACHEID=fbd9e30041e44057b493b638d772f879.

Cameron, J. (2001). The precautionary principle in international law. *Reinterpreting the Precautionary Principle*. T. O'Riordan and J. Cameroon. London, Cameron May: 113–142.

Carletti, G., F. Fatone, *et al.* (2008). Occurrence and fate of heavy metals in large wastewater treatment plants treating municipal and industrial wastewaters. *Water Science and Technology* **57**(9): 1329–1336.

Chen, K. Y., C. S. Young, *et al.* (1974). Trace metals in wastewater effluents. *Journal Water Pollution Control Federation* **46**(12): 2663–2675.

Chon, H.-S., D.-G. Ohandja, *et al.* (2009). Implementation of E.U. Water Framework Directive: source assessment of metallic substances at catchment levels. *Journal of Environmental Monitoring* **12**(1): 36–47.

CL (2010). Assembly Bill No. 1879. Retrieved 25 August 2010, from http://www.chemicalspolicy.org/legislationdocs/California/CA_AB1879.pdf.

CSS (2010). Senate Bill No. 509. Retrieved 28 August 2010, from http://info.sen.ca.gov/pub/09-10/bill/sen/sb_0501-0550/sb_509_bill_20091011_chaptered.pdf.

DEFRA (2002a). *The Government's chemicals strategy: sustainable production and use of chemicals – A strategic approach*. Department for Environment, Food and Rural Affairs.

DEFRA (2002b). UK Chemicals Stakeholder Forum. Retrieved 8 February 2010, from http://www.defra.gov.uk/environment/quality/chemicals/csf/index.htm.

DEFRA (2005a). The Hazardous Waste (England and Wales) Regulations. *Statutory Instrument 2005 No. 894*. Norwich, Her Majesty's Stationery Office.

DEFRA (2005b). The List of Wastes (England) Regulations. *Statutory Instrument 2005 No. 1673*. Norwich, Her Majesty's Stationery Office.

DEFRA (2007). Air Quality Strategy for England, Scotland, Wales and Northern Ireland. Retrieved 23 March 2010, from http://www.official-documents.gov.uk/document/cm71/7169/7169_i.asp.

DEFRA (2009a). Advisory Committee on Hazardous Substances. Retrieved 9 March 2010, from http://www.defra.gov.uk/environment/quality/chemicals/achs/.

DEFRA (2009b). The Hazardous Waste (England and Wales) (Amendment) Regulations 2009. *Statutory Instrument 2009 no. 507*. Norwich, Her Majesty's Stationery Office.

DEFRA (2009c). Local air quality management: Technical Guidance (LAQM:TG09). *Part IV of the Environment Act 1995*. London, Department for Environment, Food and Rural Affairs.

DEFRA (2011). Code of Practice For Agriculture Use Of Sewage Sludge. Retrieved 17 June 2011 from: http://archive.defra.gov.uk/

environment/quality/water/waterquality/sewage/documents/sludge-cop.pdf

Department of Environment (2003). Assessment levels for soil, sediment and water: Draft for public comment. *Contaminated Sites Management Series.* Perth, Department of Environment (DoE).

DETR (1996). The Special Waste Regulations. *Statutory Instrument no. 972.* Norwich, Her Majesty's Stationery Office.

Diderich, R. and J. Ahlers (1995). Risk assessment of existing chemicals—Technical guidance documents adopted by the EU. *Environmental Science and Pollution Research International* **2**(2): 116–116.

EC (1991a). Council Directive 91/689/EC of 12 December 1991 on hazardous waste. *Official Journal of the European Communities* **L 377**: 20–27.

EC (1991b). Council Directive 91/157/EEC of 18 March 1991 on batteries and accumulators containing certain dangerous substances. *Official Journal of the European Communities* **L 78**: 38–41.

EC (1991c). Council Directive 91/271/EEC of 21 May 1991 concerning urban waste-water treatment. *Official Journal of the European Communities* **L 135**: 40–52.

EC (1992). Council Directive 92/43/EEC of 21 May 1992 on the conservation of natural habitats and of wild fauna and flora. *Official Journal of the European Communities* **L 206** 7–50.

EC (1993). Council Decision 93/98/EEC of 1 February 1993 on the conclusion, on behalf of the Community, of the Convention on the control of transboundary movements of hazardous wastes and their disposal (Basel Convention). *Official Journal of the European Communities* **L 39**: 1–2.

EC (1997). Council Decision of 22 September 1997 (97/640/EC) on the approval, on behalf of the Community, of the amendment to the Convention on the control of transboundary movements of hazardous wastes and their disposal (Basel Convention), as laid down in Decision III/1 of the Conference of the Parties. *Official Journal of the European Communities* **L 272**: 45–46.

EC (2000). Directive 2000/60/EC of the European Parliament and of the Council of 23 October 2000 establishing a framework for Community action in the field of water policy. *Official Journal of the European Communities* **L 327**: 321–372.

EC (2001). Decision No 2455/2001/EC of the European Parliament and the Council of 20 November 2001 on establishing the list of priority substances in the field of water policy and amending Directive 2000/60/EC. *Official Journal of the European Communities* **L 331**: 1–5.

EC (2003a). Directive 2002/95/EC of the European Parliament and of the Council of 27 January 2003 on the restriction of the use of certain hazardous substances in electrical and electronic equipment. *Official Journal of the European Union* **L 37**: 19–23.

EC (2003b) Proposal for a Directive of the European Parliament and of the Council on spreading of sludge on land. 30 April2003. Brussels, Begium European Commisssion.

EC (2006a). Directive 2006/113/EC of the European Parliament and of the Council of 12 December 2006 on the quality required of shellfish waters. *Official Journal of the European Union* **L 376**: 14–20.

EC (2006b). Directive 2006/7/EC of the European Parliament and of the Council. *Official Journal of the European Union* **L 64**: 37–51.

EC (2008a). Regulation (EC) No 1272/2008 of the European parliament and of the Council of 16 December 2008 on classification, labelling and packaging of substances and mixtures, amending and repealing Directives 67/548/EEC and 1999/45/EC, and amending Regulation (EC) No 1907/2006. *Official Journal of the European Union* **L 356**: 1–1355.

EC (2008b). Directive 2008/105/EC of the European Parliament and of the Council of 16 December 2008 on environmental quality standards in the field of water policy, amending and subsequently repealing Council Directives 82/176/EEC, 83/513/EEC, 84/156/EEC, 84/491/EEC, 86/280/EEC and amending Directive 2000/60/EC of the European Parliament and of the Council. . *Official Journal of the European Union*, **L 348**: 84–97.

ECHA (2008). Guidance on information requirements and chemical safety assessment Part B: Hazard Assessment. Retrieved 23 August 2010, from http://guidance.echa.europa.eu/docs/guidance_document/information_requirements_part_b_en.pdf?vers=20_10_08.

ECHA (2009). Evaluation under REACH Progress Report 2009. Retrieved 24 August 2010, from http://echa.europa.eu/doc/progress_report_2009.pdf.

ECHA (2010). REACH and CLP guidance: authorisation. Retrieved 24 August 2010, from http://guidance.echa.europa.eu/authorisation_en.htm.

Eckley, N. and H. Selin (2004). All talk, little action: precaution and European chemicals regulation. *Journal of European Public Policy* **11**(1): 78–105.

EEA (2005). Rotterdam Convention on the Prior Informed Consent Procedure for certain hazardous Chemicals and Pesticides in international trade. Retrieved 8 February 2010, from http://www.eea.europa.eu/themes/chemicals/links/international-organisations-conventions-and-agreements/rotterdam-convention-on-the-prior-informed-consent-procedure-for-certain-hazardous-chemicals-and-pesticides-in-international-trade.

Environment Agency. (2008). Position Statement: Sustainable management of biowastes; Compost-like output from mechanical biological treatment of mixed source municipal waste. Retrieved 14 May 2010, from http://www.environment-agency.gov.uk/static/documents/mbt_2010727.pdf.

Environment Agency. (2010). Contaminated Land Exposure Assessment (CLEA). Retrieved 9 March 2010, from http://www.environment-agency.gov.uk/research/planning/33714.aspx.s

EPC (2000). Directive 2000/53/EC of the European Parliament and of the Council of 18 September 2000 on end-of life vehicles—Commission Statements. *Official Journal of the European Communities* **L 269**: 34–43.

EPC (2001). Directive 2001/80/EC on the limitation of emissions of certain pollutants into the air from large combustion plants. *Official Journal of the European Communities* **L 309**: 1–21.

EPC (2003). Directive 2002/96/EC of the European Parliament and of the Council of 27 January 2003 on waste electrical and electronic equipment (WEEE) – Joint declaration of the European Parliament, the Council and the Commission relating to Article 9. *Official Journal of the European Communities* **L 37**: 24–39.

EPC (2006). Directive 2006/66/EC of the European Parliament and of the Council of 6 September 2006 on batteries and accumulators and repealing Directive 91/157/EEC. *Official Journal of the European Communities* **L 266**: 1–14.

EPC (2008a). Directive 2008/98/EC of the European Parliament and of the Council of 19 November 2008 on waste and repealing certain Directives. *Official Journal of the European Communities* **L 312**: 3–30.

EPC (2008b). Directive 2008/50/EC of the European Parliament and of the Council of 21 May 2008 on ambient air quality and cleaner air for Europe. *Official Journal of the European Communities* **L 152**: 1–44.

Europa (2001). Communication from the Commission on the precautionary principle COM (2000)1. Retrieved 10 March 2010, from http://ec.europa.eu/environment/docum/20001_en.htm.

Federal Register (1955). Air pollution control act of 1955. *Public Law 84–159.* Washington D. C., US Government.

Federal Register (1963). Clean Air Act 163. *Public Law 88–206.* Washington D. C., US Government.

Federal Register (1965). Motor vehicle air pollution control act. *Public Law 89-272.* Washington D. C., US Government.

Federal Register (1967). Air Quality Act. *Public Law 90–148.* Washington D. C., US Government.

Federal Register (1970). Clean Air Act Extension. *Public Law 91–604.* Washington D. C., US Government.

Federal Register (1977). Clean air act amendments. *Public Law 95–96.* Washington D. C., US Government.

Federal Register (1990). Clean air act amendments *Public Law 101–549.* Washington D. C., US Gsovernment.

Flower, R. J. and R. W. Battarbee (1983). Diatom evidence for recent acidification of two Scottish lochs. *Nature* **305**(5930): 130–133.

Gagnon, C. and I. Saulnier (2003). Distribution and fate of metals in the dispersion plume of a major municipal effluent. *Environmental Pollution* **124**(1): 47–55.

Garrod, J. (2006). *The Current Regulation of Environmental Chemicals.* Cambridge, RSC Publishing.

Gendebien, A. H., R. D. Davis, *et al.* (2001). *Survey of wastes spread on land in the European Union.* Technology transfer. Proceedings of the 9th International Conference on the FAO ESCORENA Network on recycling of agricultural, municipal and industrial residues in agriculture, Gargano, Italy, 6–9 September 2000.

Hohmann, H. (1994). *Precautionary Legal Duties and Principles of Modern International Environmental Law: The Precautionary Principle: International Environmental Law between Exploitation and Protection* (International Environmental Law and Policy Series). London, Graham & Trotman; The Hague, Martinus Nijhoff, Kluwer Academic Publishers Group.

Hudson, R. (1983). Capital accumulation and chemicals production in Western Europe in the postwar period. *Environment and Planning A* **15**(1): 105–122.

ICCA (2000). High production volume. Retrieved 10 March 2010, from http://www.icca-chem.org/Home/ICCA-initiatives/High-production-volume-chemicals-initiative-HPV/.

ICIS (2010). Reach-like regulations enacted globally. From http://www.icis.com/Articles/2010/05/31/9362538/reach-like-regulations-enacted-globally.html.

IPCS (2010a). Environmental Health Criteria Monographs (EHCs) Retrieved 8 February 2010, from http://www.inchem.org/pages/ehc.html.

IPCS (2010b). JMPR: Joint FAO/WHO meeting on pesticide residues. Retrieved 8 February 2010, from http://www.who.int/ipcs/food/jmpr/en/.

IPCS (2010c). Joint FAO/WHO expert committee on food additives (JECFA). Retrieved 8 February 2010, from http://www.who.int/ipcs/food/jecfa/en/.

Jacobson, M. Z. (2002). *Atmospheric Pollution: History, Science and Regulation.* Cambridge, Cambridge University Press.

Jensen, P. K., Y. Matsuno, *et al.* (2001). Limitations of irrigation water quality guidelines from a multiple use perspective. *Irrigation and Drainage Systems* **15**(2): 117–128.

Karvelas, M., A. Katsoyiannis, *et al.* (2003). Occurrence and fate of heavy metals in the wastewater treatment process. *Chemosphere* **53**(10): 1201–1210.

Lallas, P. (2001). The Stockholm Convention on persistent organic pollutants. *The American Journal of International Law* **95**(3): 692–708.

Langezaal, I., S. Hoffmann, T. Hartung and S. Coecke (2002) Evaluation and prevalidation of an immunotoxicity toxicity test based on human whole-blood cytokine release. *ATLA Alternatives to Laboratory Animals* **30**: 581–595.

Lokke, S. and P. Christensen (2008). The introduction of the precautionary principle in Danish environmental policy: the case of plant growth retardants. *Journal of Agricultural & Environmental Ethics* **21**(3): 229–247.

Mann, H. (1992). The Rio Declaration. *American Society of International Law* **86**: 405–411.

Mishra, P., R. Samarth, *et al.* (2009). Bhopal gas tragedy: review of clinical and experimental findings after 25 years. *International Journal of Occupational Medicine and Environmental Health* **22**(3): 193–202.

NEPC (1999a). National environment protection (Assessment of site contamination) measure: Schedule B(1) Guideline on the investigation levels for soil and groundwater. Canberra, National Environment Protection Council.

NEPC (1999b). Schedule B(1) – Guideline on investigation levels for soil and groundwater. Retrieved 25 August 2010, from http://www.ephc.gov.au/sites/default/files/ASC_NEPMsch__01_Investigation_Levels_199912.pdf.

Nyenje, P. M., J. W. Foppen, *et al.* (2010). Eutrophication and nutrient release in urban areas of sub-Saharan Africa – a review. *Science of the Total Environment* **408**(3): 447–455.

OECD (2010). Description of OECD Work on Investigation of High Production Volume Chemicals. Retrieved 10 March 2010, from http://www.oecd.org/document/21/0,3343,en_2649_34379_1939669_1_1_1_1,00.html.

Oliveira, A.d.S., A. Bocio, *et al.* (2007). Heavy metals in untreated/treated urban effluent and sludge from a biological waste water treatment plant. *Environmental Science and Pollution Research* **14**(7): 483–489.

Oliver, B. G. and E. G. Cosgrove (1974). The efficiency of heavy metal removal by a conventional activated sludge treatment plant. *Water Research* **8**(11): 869–874.

OPSI (1956). Clean Air Act 1956. Retrieved 31 August 2010, from http://www.opsi.gov.uk/acts/acts1956/pdf/ukpga_19560052_en.pdf.

OPSI (1968). Clean Air Act 1968. Retrieved 31 August 2010, from http://www.opsi.gov.uk/RevisedStatutes/Acts/ukpga/1968/cukpga_19680062_en_1.

OPSI (1991) Water Resources Act 1991. Retrieved 23 June 2011 from http://www.legislation.gov.uk/ukpga/1991/57/data.pdf.

OPSI (2002) Statutory Instrument 2002 No. 1689. The Chemicals (Hazard Information and Packaging for Supply) Regulations 2002. Retrieved 23 June 2011 from http://www.legislation.gov.uk/uksi/2002/1689/pdfs/uksi_20021689_en.pdf.

OSPAR (1992). *Convention for the Protection of the Marine Environment of the North-East Atlantic.* London. http://www.ospar.org/v_publications.

Pan, J. (2009). Environmental Risk Assessment of Inorganic Chemicals in the Mining Environment London, PhD Thesis. Imperial College London.

Plant, J., A. Korre, *et al.* (2004). Chemicals in the environment. *Geochimica et Cosmochimica Acta* **68**(11): A527–A527.

Post, D. (2006). The precautionary principle and risk assessment in international food safety: How the world trade organization influences standards. *Risk Analysis* **26**(5): 1259–1273.

Ragnarsdottir, K. V. (2000). Environmental fate and toxicology of organophosphate pesticides. *Journal of the Geological Society* **157**: 859–876.

Rahman, M. (2009). Arsenic levels in rice grain and assessment of daily dietary intake of arsenic from rice in arsenic-contaminated regions of Bangladesh – implications to groundwater irrigation. *Environmental Geochemistry and Health* **31**: 179–187.

RCEP (2003). 24th report: Chemicals in products: safeguarding the environment and human health. Retrieved 22 February 2010, from http://www.rcep.org.uk/reports/24-chemicals/documents/24-chemicals.pdf.

RCEP (2010). The commission and its work. Retrieved 2 February 2010, from http://www.rcep.org.uk/.

Ruden, C. and S. O. Hansson (2010). Registration, Evaluation, and Authorization of Chemicals (REACH) is but the first step – how far will it take us? Six further steps to improve the European chemicals legislation. *Environmental Health Perspectives* **118**(1): 6–10.

Sand, P. (2000). The precautionary principle: a European perspective. *Human and Ecological Risk Assessment* **6**(3): 445–458.

Sauerbeck, D. R. and R. Leschber (1992). *German proposals for acceptable contents of inorganic and organic pollutants in sewage sludges and sludge amended soils*. Luxembourg: Commission of of the European Communities.

Schowanek, D., R. Carr, *et al.* (2004). A risk-based methodology for deriving quality standards for organic contaminants in sewage sludge for use in agriculture—Conceptual framework. *Regulatory Toxicology and Pharmacology* **40**(3): 227–251.

Semeena, V. S. and G. Lammel (2005). The significance of the grasshopper effect on the atmospheric distribution of persistent organic substances. *Geophysical Research Letters* **32**(7): 5.

Shammas, N. K. (2010). Legislation and regulations for hazardous wastes. *Handbook of Advanced Industrial and Hazardous Wastes Treatment*. L. K. Wang, Y. Hung and N. K. Shammas. Boca Raton, CRC Press.

Slack, R., P. Zerva, *et al.* (2005). Assessing quantities and disposal routes for household hazardous products in the United Kingdom. *Environmental Science & Technology* **39**(6): 1912–1919.

Smith, S. (2000). Are controls on organic contaminants necessary to protect the environment when sewage sludge is used in agriculture? *Progress in Environmental Science* **2**: 129–146.

Smith, S. R. (2009). Organic contaminants in sewage sludge (biosolids) and their significance for agricultural recycling. *Philosophical Transactions of the Royal Society A –Mathematical Physical and Engineering Sciences* **367**(1904): 4005–4041.

Swartjes, F. A., E. M. Dirven-Van Breemen, *et al.* (2007). Human health risks due to consumption of vegetables from contaminated sites: Towards a protocol for site-specific assessment. Bilthoven, RIVM/Laboratory for Ecological Risk Assessment.

Tickner, J. and C. Raffensperger (2001). The politics of precaution in the United States and the European Union. *Global Environmental Change – Human and Policy Dimensions* **11**(2): 175–180.

UNEP (2010a). The 9 new POPs under the Stockholm Convention Retrieved 3 February 2010, from http://chm.pops.int/Programmes/NewPOPs/The9newPOPs/tabid/672/language/en-US/Default.aspx.

UNEP (2010b). How Bad is Eutrophication at Present? Retrieved 3 February 2010, from http://www.unep.or.jp/ietc/publications/short_series/lakereservoirs-3/2.asp.

UNESCO (2005). The Precautionary Principle. Retrieved 22 August 2010, from http://unesdoc.unesco.org/images/0013/001395/139578e.pdf.

USEPA (1976). Resource Conservation and Recovery Act. Retrieved 24 August 2010, from http://www.epa.gov/lawsregs/laws/rcra.html.

USEPA (1992a). Federal Facility Compliance Act. Retrieved 24 August 2010, from http://www.epa.gov/fedfac/documents/ffc92.htm.

USEPA (1992b). Technical Support Document for Land Application Of Sewage Sludge, Volume II. Eastern Research Group, Lexington.

USEPA (1995). Universal Waste Rule (Hazardous Waste Management System; Modification of the Hazardous Waste Recycling Regulatory Program; Final Rule). Retrieved 24 August 2010, from http://www.epa.gov/wastes/hazard/downloads/uwr_fr.pdf.

USEPA (1996). Land Disposal Flexibility Act. Retrieved 24 August 2010, from http://www.epa.gov/osw/inforesources/pubs/flex-act.pdf.

USEPA (2003). Final Action not to Regulate Dioxins in Land-Applied Sewage Sludge. Retrieved 4 February 2010, from http://www.epa.gov/waterscience/biosolids/dioxinfs.html.

USEPA (2004) A Plain English Guide to the EPA Part 503 Biosolids Rule. Retrieved 17 June 2011 from http:/water.epa.gov/scitech/wastetech/biosolids/503pe_index.cfm.

USEPA (2006). Mandates in Federal Statutes. Retrieved 9 March 2010, from http://www.epa.gov/p2/pubs/p2policy/provisions.htm.

USEPA (2007). Waste Shipments between the United States and Canada. Retrieved 24 August 2010, from http://www.epa.gov/osw/hazard/international/us-canada.pdf.

USEPA (2008a). Pharmaceuticals and Personal Care Products (PPCPs). Retrieved 22 August 2010, 2010, from http://www.epa.gov/ppcp/.

USEPA (2008b). Household Hazardous Waste. Retrieved 24 August 2010, from http://www.epa.gov/epawaste/conserve/materials/hhw.htm.

USEPA (2008c). Used oil management program. Retrieved 24 August 2010, from http://www.epa.gov/epawaste/conserve/materials/usedoil/index.htm.

USEPA (2008d). Basel Convention. Retrieved 24 August 2010, from http://www.epa.gov/osw/hazard/international/basel.htm.

USEPA (2009a). Chemical Assessment and Management Program (ChAMP). Retrieved 9 March 2010, from http://www.epa.gov/champ/.

USEPA (2009b). Hazardous waste regulations. Retrieved 31 August 2010, from http://www.epa.gov/osw/laws-regs/regs-haz.htm.

USEPA (2010a). Laws that We Administer. Retrieved 25 August 2010, from http://www.epa.gov/lawsregs/laws/.

USEPA (2010b). Summary of the Toxic Substances Control Act. Retrieved 9 March 2010, from http://www.epa.gov/lawsregs/laws/tsca.html.

USEPA (2010c). Existing Chemicals. Retrieved 9 March 2010, from http://www.epa.gov/oppt/existingchemicals/index.html.

USEPA (2010d). Regional Screening Levels (formerly PRGs). Retrieved 14 May 2010, from http://www.epa.gov/region9/superfund/prg/index.html.

USEPA (2010e). Summary of the Clean Water Act. Retrieved 8 February 2010, from http://www.epa.gov/lawsregs/laws/cwa.html.

USEPA (2010f). Soil Screening Guidance. Retrieved 27 August 2010, from http://www.epa.gov/superfund/health/conmedia/soil/index.htm.

USEPA (2010g). Soil Screening Guidance. Retrieved 24 May 2010, from http://www.epa.gov/superfund/health/conmedia/soil/#fact.

US Senate (2008). S.3040 – Kid-Safe Chemicals Act of 2008. Retrieved 25 August 2010, from http://www.opencongress.org/bill/110-s3040/show.

van Leeuwen, C. J. and J. L. M. Hermens (1995). *Risk Assessment of Chemicals: An Introduction*. London, Kluwer Academic Publishers.

Vermeire, T. and P. Van Der Zandt (1995). *Procedures of Hazards and Risk Assessment*. London, Kluwer Academic Publishers.

Wang, L. K. (2010). *Handbook of Advanced Industrial and Hazardous Wastes Treatment*. Boca Raton, Florida, CRC Press.

WHO (2005). WHO air quality guidelines for particulate matter, ozone, nitrogen dioxide and sulfur dioxide: Global update 2005: summary of risk assessment. Geneva, World Health Organization.

WHO (2006). *Guidelines for drinking-water quality—Recommendations*. World Health Organisation.

WHO (2010) IARC Monographs on the Evaluation of Carcinogenic Risks to Humans. Retrieved 23 June 2011 from http://www.iarc.fr/en/publications/list/monographs/

Williams, E., J. Panko, *et al.* (2009). The European Union's REACH regulation: a review of its history and requirements. *Critical Reviews in Toxicology* **39**(7): 553–575.

Yan, S., S. Subramanian, *et al.* (2010). Emerging contaminants of environmental concern: source, transport, fate, and treatment. *Practice Periodical of Hazardous, Toxic, and Radioactive Waste Management* **14**(1): 2–20.

3

Essential and beneficial trace elements

Xiyu Phoon[1], E. Louise Ander[2] and Jane A. Plant[3]*

[1]*Centre for Environmental Policy, Imperial College London, Prince Consort Road, London SW7 2AZ*
[2]*British Geological Survey, Kingsley Dunham Centre, Keyworth, Nottingham, NG12 5GG*
[3]*Centre for Environmental Policy and Department of Earth Science and Engineering, Imperial College London, Prince Consort Road, London SW7 2AZ*
**Corresponding author, email jane.plant@imperial.ac.uk*

3.1 Introduction

Essential trace elements are required by humans in amounts ranging from 50 µg to 18 mg per day. They act as enzyme cofactors and/or structural components of larger molecules, which have specific functions and are essential for life (Mertz, 1981).

Approximately 96 per cent of the human body is composed of the light elements hydrogen (H), carbon (C), nitrogen (N) and oxygen (O). The remaining 4 per cent includes calcium (Ca), potassium (K) and phosphorus (P) (Rose, 1999). Heavier elements which are essential for health in trace amounts include the first-row transition elements vanadium (V), chromium (Cr), manganese (Mn), iron (Fe), cobalt (Co), nickel (Ni), copper (Cu) and zinc (Zn), together with molybdenum (Mo), selenium (Se) and iodine (I) (Underwood, 1981) (Figure 3.1). Other light elements such as lithium (Li) and boron (B) may also be essential (Nielsen, 1991). The essentiality of fluorine (F) has been questioned (Fordyce *et al.*, 2007), although it is generally accepted to be beneficial at levels less than 1.5 mg/l in water (Dean *et al.*, 1942). With the exception of Si, all the essential and beneficial trace elements occur in the Earth's crust at average concentrations of less than 1000 mg/kg (Nielsen, 1984). The bulk chemistry of the human body has been suggested to reflect that of primitive life forms that evolved at spreading centres underlain by basic rocks (basalts) through which sea water was convecting (McDowell, 1992).

Deficiencies in one or more essential trace elements in humans and other animals have for some time been reliably linked to ill health and a range of subclinical disorders such as general malaise (Selinus, 2005; Mills, 1969). All essential and beneficial trace elements are toxic if ingested or inhaled at sufficiently high levels for a long enough period of time (Figure 3.2). Selenium, F and Mo are examples of trace elements which show a relatively narrow concentration range (of the order of a few µg/g) between essential or beneficial levels and toxic levels. High concentrations of these elements can have detrimental consequences to human and animal health and the environment.

Many trace-element deficiencies, such as those for Cr, Cu and Zn, were first established in domesticated ruminant animals such as cows and sheep. In the 1940s coast disease and wasting disease in livestock in Australia were shown to be preventable and treatable by adding Co to the animals' diet (Underwood and Suttle, 1999) (Figure 3.3). Cobalt forms part of the vitamin B12 molecule and the essentiality of Co to humans was demonstrated in 1948 when injection with vitamin B12 was shown to prevent pernicious anaemia (Vanderpas *et al.*, 1990). The relationship between I deficiencies in soil water and crops and I-deficiency diseases such as cretinism and goitre is also well established (Fuge, 2005; Vanderpas *et al.*, 1990). In recent years the effects of Cu, Cr and Zn deficiencies in humans have been reported mainly from developing countries, especially at low latitudes where surface soils have been deeply leached (Plant *et al.*, 1996, 2000) and agriculture is mainly subsistence farming. These micronutrients are usually present in adequate amounts in varied Western diets, which normally include foods from many sources, so that a deficiency in one food is likely to be balanced by an adequate amount in another, although the essential micronutrients I and Se are thought to be widely deficient in diets of some European countries (Andersson *et al.*, 2007; Johnson *et al.*, 2010).

Pollutants, Human Health and the Environment: A Risk Based Approach, First Edition. Edited by Jane A. Plant, Nikolaos Voulvoulis and K. Vala Ragnarsdottir.

Exposure Limit

Key: element name / atomic number / symbol / atomic weight (mean relative mass)

hydrogen 1 H 1.0079																		helium 2 He 4.0026
lithium 3 Li 6.941	beryllium 4 Be 9.0122												boron 5 B 10.811	carbon 6 C 12.011	nitrogen 7 N 14.007	oxegen 8 O 15.999	fluorine 9 F 18.998	neon 10 Ne 20.180
sodium 11 Na 22.990	magnesium 12 Mg 24.305												aluminium 13 Al 26.982	silicon 14 Si 28.086	phosphorus 15 P 30.974	sulfur 16 S 32.065	chlorine 17 Cl 35.453	argon 18 Ar 39.948
potassium 19 K 39.098	calcium 20 Ca 40.078		scandium 21 Sc 44.956	titanium 22 Ti 47.867	vanadium 23 V 50.942	chromium 24 Cr 51.996	manganese 25 Mn 54.938	iron 26 Fe 55.845	cobalt 27 Co 58.933	nickel 28 Ni 58.693	copper 29 Cu 63.546	zinc 30 Zn 65.39	gallium 31 Ga 69.723	germanium 32 Ge 72.61	arsenic 33 As 74.922	selenium 34 Se 78.96	bromine 35 Br 79.904	krypton 36 Kr 83.80
rubidium 37 Rb 85.468	strontium 38 Sr 87.62		yttrium 39 Y 88.906	zirconium 40 Zr 91.224	niobium 41 Nb 92.906	molybdenum 42 Mo 95.94	technetium 43 Tc [98]	ruthenium 44 Ru 101.07	rhodium 45 Rh 102.91	palladium 46 Pd 106.42	silver 47 Ag 107.87	cadmium 48 Cd 112.41	indium 49 In 114.82	tin 50 Sn 118.71	antimony 51 Sb 121.76	tellurium 52 Te 127.60	iodine 53 I 126.90	xenon 54 Xe 131..29
caesium 55 Cs 132.91	barium 56 Ba 137.33	57-70 *	lutetium 71 Lu 174.97	hafnium 72 Hf 178.49	tantalum 73 Ta 180.95	tungsten 74 W 183.84	rhenium 75 Re 186.21	osmium 76 Os 190.23	iridium 77 Ir 192.22	platinum 78 Pt 195.08	gold 79 Au 196.97	mercury 80 Hg 200.59	thallium 81 Tl 204.38	lead 82 Pb 207.2	bismuth 83 Bi 208.98	polonium 84 Po [209]	astatine 85 At [210]	radon 86 Rn [222]
francium 87 Fr [223]	radium 88 Ra [226]	89-102 **	lawrencium 103 Lr [262]	rutherfordium 104 Rf [261]	dubnium 105 Db [262]	seaborgium 106 Sg [266]	bohrium 107 Bh [264]	hassium 108 Hs [269]	meitnerium 109 Mt [268]	ununnilium 110 Uun [271]	unununium 111 Uuu [272]	ununbium 112 Uub [277]		ununquadium 114 Uuq [289]				

*lanthanoids	lanthanum 57 La 138.91	cerium 58 Ce 140.12	praseodymium 59 Pr 140.91	neodymium 60 Nd 144.24	promethium 61 Pm [145]	samarium 62 Sm 150.36	europium 63 Eu 151.96	gadolinium 64 Gd 157.25	terbium 65 Tb 158.93	dysprosium 66 Dy 162.50	holmium 67 Ho 164.93	erbium 68 Er 167.26	thulium 69 Tm 168.93	ytterbium 70 Yb 173.04
**actinoids	actinium 89 Ac [227]	thorium 90 Th 232.04	protactinium 91 Pa 231.04	uranium 92 U 238.03	neptunium 93 Np [237]	plutonium 94 Pu [244]	americium 95 Am [243]	curium 96 Cm [247]	berkelium 97 Bk [247]	californium 98 Cf [251]	einsteinium 99 Es [252]	fermium 100 Fm [257]	mendelevium 101 Md [258]	nobelium 102 No [259]

Figure 3.1 The periodic table of elements. Major elements in the body are the darkest shade of grey, with white lettering. All other shaded elements are beneficial trace elements, of which the darker (mid) grey are the elements discussed in this chapter

The extent to which essential trace elements can cause harm to human health or the environment depends on their concentration, speciation, pathways and exposure. In this chapter we are concerned mainly with the first-row transition elements Cu, Cr and Zn, the metalloid Se and the halogen F, all of which occur naturally in the environment or as a result of pollution. Some of the key properties of these elements are listed in Table 3.1.

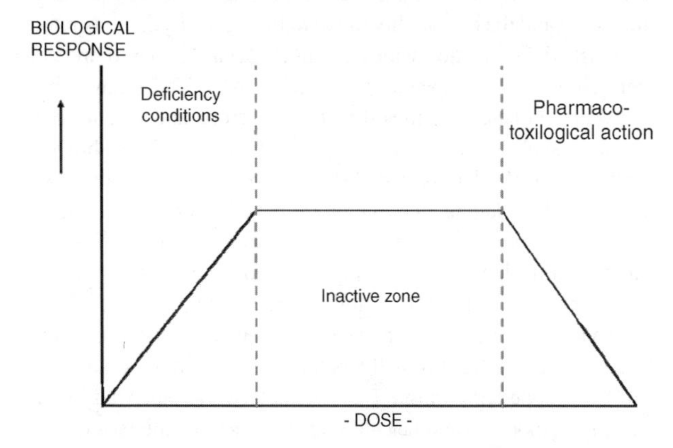

Figure 3.2 Dose–response of essential trace elements (Reproduced from Venchikov, A.E. (1974) Zones of display of biological and pharmacotoxicological action of trace elements. In: Trace Element Metabolism in Animals 2. (Hoekstra, W.G., Suttie, I.W., Ganther, H.E. & Mertz, W., eds.), pp. 295–310, Univ. Park Press, Baltimore)

In the 1960s, trivalent Cr was found to be present in enzymes involved in glucose, fat and protein metabolism. Trivalent Cr is the most stable oxidation state of Cr in living organisms and is unable to cross cell membranes easily. It is biologically active in the form of chromodulin, assisting insulin binding to cell surface receptors. Unlike trivalent Cr, hexavalent Cr, when bound to O, can cross membranes easily and react with proteins and nucleic acids in cells, causing carcinogenic effects (Pechova and Pavlata, 2007). The essentiality of Cu was established in 1928 in relation to haematology and later in the prevention of cardiovascular disease (Klevay, 2000). Copper is present in several metalloenzymes involved in physiological processes such as haemoglobin formation, drug metabolism, carbohydrate metabolism, cross-linking of collagen, elastin and keratin in hair, as well as for antioxidant defence (ATSDR, 2004). Clinical manifestations of Cu deficiency in humans include anaemia as a result of impaired haemoglobin synthesis, bone malformations and altered immunity (WHO, 1998). Copper toxicity usually reflects the presence of free Cu in the blood, where it generates free radicals and damages proteins, lipids and DNA (Brewer, 2010).

The essentiality of Zn in humans was first suggested in 1963, when growth retardation in the Middle East was found to be due to Zn deficiency (Prasad, 1991). Zinc is present in more than 200 enzymes and is second only to Fe in its importance as a trace nutrient for humans (Sanstead, 1984). A lack of Zn can cause impaired neuropsychological function, growth retardation, poor wound healing, immune problems, anorexia, anaemia and

Table 3.1 Key properties of Cr, Cu, Zn, Se and F

	Cr	Cu	Zn	Se	F
Atomic number	24	29	30	34	9
Atomic weight	51.996	63.546	65.39	78.96	18.996
Melting point (°C)	1857	1083.4	419.58	221	−219.6
Boiling point (°C)	2672	2567	907	685	−188.1
Vapour pressure (Pa)	990 at 18576 °C	0.0505 at 1084.6 °C	19.2 at 419.736 °C	0.695 at 2216 °C	–
Most important oxidation states in environment	+3, +6	+1, +2	+2	+4, +6	−1

alteration of taste perception (WHO, 2001; Barceloux, 1999a). Zinc deficiency is one of the major 'global burden of disease' risk factors according to the WHO Disability-Adjusted Life Years calculations and most seriously affects low-income countries (WHO, 2009). Zinc toxicity can occur in both acute and chronic forms. Acute adverse effects of high Zn intake include nausea, vomiting, loss of appetite, abdominal cramps, diarrhoea and headaches.

Selenium deficiency emerged as a serious cause of concern for human health when, in 1935, Keshan disease (KD) in north-eastern China was attributed to low Se in the diet; symptoms of KD include blood-circulation disorders, endocardium abnormality and myocardium necrosis. In 1971, it was reported that glutathione peroxidase, which plays an important role in the immune system, is a selenoenzyme (Rotruck *et al.*, 1972; Johnson *et al.*, 2010). There is also evidence that a selenoenzyme is involved in the synthesis of thyroid hormones (Beckett *et al.*, 1989; Combs and Combs, 1986). More recently, Se deficiency has been suggested as a cause or contributory factor in a wide range of diseases including cancer, AIDS, heart disease, muscular dystrophy, multiple sclerosis, osteoarthropathy and immune-system and reproductive disorders in humans and white-muscle disease in animals (Levander 1986; WHO 1996; Johnson *et al.*, 2010) and a recent comprehensive review of the role and state of knowledge of Se in human health and disease has been undertaken (Fairweather-Tait *et al.*, 2010). In the 1970s and 1980s the diet of people in the affected regions of China was supplemented with Se, and Na selenite was used to treat growing crops. This has resulted in a decline in the incidence of Se deficiency disease generally (Liu *et al.*, 2002). On a global scale, overt Se toxicity in human subjects is far less widespread than deficiency (Fordyce, 2005).

Figure 3.3 Cobalt deficiency in sheep. Animals fed the same experimental diet, with that of the smaller animal lacking cobalt (Reproduced by permission of CF Mills)

The first reported observation of disease thought to reflect Se toxicity was by Marco Polo in 1295 when he described 'hoof rot' in cattle and horses in the Nan Shan and Tien Shan mountains of Turkestan, where grains and forage have since been shown to contain high levels of Se (5–50 mg/kg). This also causes dystrophic changes and loss of tail switch in the animals in that region (Pais and Jones, 1977; Barceloux, 1999b)[1]. Hoof disease in livestock was reported in Colombia in 1560 and near the south coast of the US in the mid-nineteenth century.

Low concentrations of fluoride (0.5–1.5 mg/l) were thought to promote dental health, whereas excess intake may lead to dental fluorosis and, most seriously, skeletal fluorosis (Figure 3.4) (Edmunds and Smedley, 2005). The first concerns about fluoride were related to the effects of high levels on dental health. During the late 1800s and early 1900s the cause of brown staining of otherwise healthy teeth in people living in certain areas was investigated (Richmond, 1985) and in 1931 fluoride in water was identified as the cause of 'mottled enamel' (Churchill, 1932). In the 1940s, fluoride levels in water supplies in the USA were reduced from 1.5–10 mg/l to around 1.0 mg/l (Richmond, 1985).

[1] Shao and Yang (2008) suggest that the horses' symptoms may not have been due to selenosis.

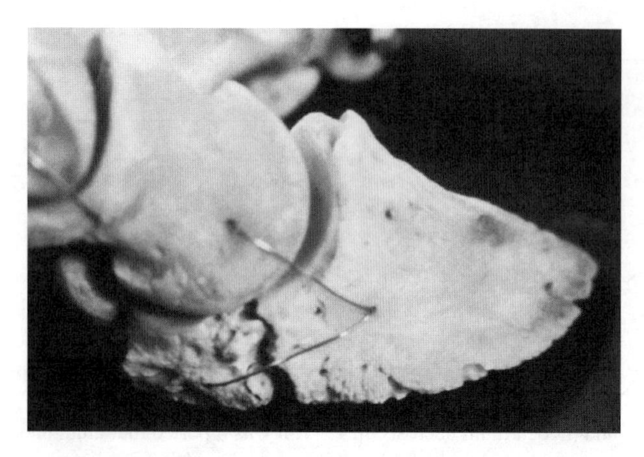

Figure 3.4 Fluorosis in livestock. Spontaneous fracture of cows foot bone (Reproduced by permission of CF Mills)

Other serious problems of fluoride toxicity associated with serious bone disease, including *genu valgum* (knock knees), have been reported from India and China. In other countries, however, there has been a major push to fluoridate water, toothpaste and other dental products as a means of preventing tooth decay (ATSDR, 2003a; Joiner *et al.*, 2009; Gibson-Moore, 2009; Meler and Meler, 2006; Edmunds and Smedley, 2005).

Recommended daily allowances for these elements have been set by various government bodies, as exposure above certain thresholds can result in symptoms of toxicity.

3.2 Hazardous properties

The essential trace elements are crucial for human health, and deficiency causes subclinical health effects such as malaise and in some cases overt health problems, but they are also potentially toxic in excess. Their toxicity depends on their concentration, chemical species, route of exposure and bioavailability. The LD_{50} values for some important environmental species are given in Table 3.2. The carcinogenic, mutagenic and reproductive properties of some compounds of Cr, Cu, Se and Zn are listed in Table 3.3.

Exposure can be either acute or chronic. Brief exposure to large quantities of an element (e.g. Cu) can be as harmful as chronic exposure to lower levels of the element over the longer term.

Chromium provides a good example of the importance of chemical speciation. Chromium, as the pure metal, has no reported health effects. The trivalent form, Cr^{3+}, is relatively insoluble and poorly absorbed in the gastrointestinal tract (HPA, 2007); and it helps to prevent adult-onset diabetes and cardiovascular disease (Anderson and Kozlovsky, 1985). Hexavalent Cr, Cr^{6+}, on the other hand, is relatively soluble; it can irritate the skin and mucous membranes and is a known carcinogen (Table 3.3) (ATSDR, 2008). It has been suggested that the hazardous properties of hexavalent Cr relate to its highly reactive state and its ability to form biological ligands, primarily nucleic acids, proteins and free sulphydryl groups in organisms (Fasulo *et al.*, 1983).

Copper is toxic to many biological species, including sheep and goats, marine invertebrates, many algae, bacteria and fungal spores. Humans and pigs are generally less sensitive. Much of the information on Cu toxicity is from studies of Bordeaux mixture (CaOH and $CuSO_4$), used to spray apples, potatoes, vines and hops. Such Cu salts owe their toxicity to the free-radical cascades generated when Cu^{2+} is reduced to Cu^{+}, destroying membrane lipids (Briggs, 1992). In humans, signs of toxicity can occur after ingestion of as little as 1 g of Cu and consuming foods or beverages with more than 25 mg/l has been linked to acute gastroenteritis (Pizarro *et al.*, 1999).

The process of Zn poisoning is currently unknown, but it results in haemolysis, regenerative anaemia, or renal failure. Hypocupremia may be present, due to the Zn–Cu antagonism of absorption in the gastrointestinal tract (ATSDR, 2005).

Selenium (as Se^{2-}) has a similar ionic radius (0.184 nm) to that of sulphur (S^{2-}) (0.198 nm) and a similar electronegativity (Se 2.4; S 2.5) (Pauling, 1970). Selenium can destroy H peroxide in the body, thus increasing tissue peroxide levels by the action of glutathione peroxidase enzyme, which contains Se (Battin and Brumaghim, 2009). The cytotoxicity of Se occurs as a result of pro-oxidant catalytic activity by selenide anions producing super-oxide anions, H peroxide and other reactive metabolites. In plants and animals, methylation can counteract the toxicity of Se species (Barceloux, 1999b; Battin and Brumaghim, 2009).

Most forms of Se have generally low toxicity. Selenium sulphide is currently classed as a probable human carcinogen, since oral ingestion in rats has been shown to induce hepatocellular carcinomas. However, when a shampoo containing 2.5 per cent Se sulphide was applied topically to rats, no carcinogenic effects were observed (NIEHS, 2005). In another study, intraperitoneal injection of dimethyl selenide in rats was found to be far less harmful than oral ingestion of Na selenite (Hsieh and Ganther, 1977).

Fluoride is the most electronegative and reactive of all the chemical elements. It has an ionic radius (35 Å) and valency (−1) closely similar to those of the hydroxyl ion. Human and animal bones are composed of hydroxyapatite, which is an end member with fluorapatite of the apatite solid-solution series. Fluoride exchanges readily with the hydroxyl ion in the apatite lattice, increasing the brittleness and solubility of bone structure at high concentrations (Fordyce *et al.*, 2007; Dissanayake and Chandrajith, 1999; Skinner, 2000). There is evidence that the adverse health effects of fluoride are enhanced by a lack of Ca, vitamins and protein in the diet (Jacks *et al.*, 1993; Li *et al.*, 1996; Zheng *et al.*, 1999).

Up to a certain level, fluoride helps to stabilise teeth and bones, but at higher levels (>1.5 mg/l in blood) it increases crystallinity and causes degeneration of bones and teeth (Bhussry *et al.*, 1970). Long-term exposure to high levels of F (Fordyce *et al.*, 2007) initially causes mottling of teeth and, in more serious cases, tooth crumbling and bone deformities, including *genu valgum*. Neurological problems that occur as a result of exposure to very high levels of fluorides are often as a consequence of fluoride binding to Ca, causing hypocalcaemia (Eichler *et al.*, 1982).

Table 3.2 Acute toxicity (LD_{50}) values of various compounds (compiled from HPA, 2007; WHO, 1998; Domingo *et al.*, 1988; WHO, 1986)

Compound	Route of exposure	Model organism	LD_{50} (mg/kg bodyweight)
Na chromate, Na dichromate, K dichromate, ammonium dichromate	Oral	Rat (male)	21–28
		Rat (female)	13–19
Cr trioxide		Rat (male)	29
		Rat (female)	25
Ca chromate		Rat (male)	249
		Rat (female)	108
Sr chromate		Rat (male)	811
		Rat (female)	21–28
Cu (II) acetate		Rat	595
Cu (II) carbonate		Rat	159
Cu (II) carbonate hydroxide		Rat (male)	1350
		Rat (female)	1495
		Rabbit	317
Cu (II) chloride		Rat	140
		Mouse	190
		Guinea pig	32
Cu (II) hydroxide		Rat	1000
Cu (I) oxide		Rat	470
Cu (II) oxychloride		Rat	1440
Cu (II) sulphate		Rat	120
		Rabbit	50
Zn acetate	Oral	Rats	794
		Mice	287
	Intraperitoneal	Rats	162
		Mice	108
Zn nitrate	Oral	Rats	1330
		Mice	926
	Intraperitoneal	Rats	133
		Mice	110
Zn chloride	Oral	Rats	1100
		Mice	1260
	Intraperitoneal	Rats	58
		Mice	91
Zn sulphate	Oral	Rats	1710
		Mice	926
	Intraperitoneal	Rats	200
		Mice	316
Na selenite	Oral	Rats	10.5–13.2
Diselenodipropionic acid	Intraperitoneal		25–30
Trimethylselenonium chloride			49.4
Dimethyl selenide			1600

No comparable data are avaliable for fluoride.

Table 3.3 Carcinogenic, mutagenic and reproductive properties of selected Cr, Cu, Zn and Se compounds

Chemical element	Species	Authorising or recommending body	Carcinogenic	Mutagenic	Reproductive	Other comments
Cr	CrO_3	ACGIH (2004)	A1; BEI issued	–	–	–
		MAK (DFG, 2004)	Carcinogen category: 2	–	–	Sensitisation of skin
	Na_2CrO_4; $Na_2Cr_2O_7$	ACGIH (2005)	A1; BEI issued	–	–	–
		MAK (DFG 2004)	Carcinogen category: 2	–	–	Sensitisation of skin (Inhalable fraction)
Cu	$CuAsHO_3$	ACGIH (2004)	A1	–	–	–
		MAK (DFG 2004)	Carcinogen category: 1	Germ-cell mutagen group: 3A	–	–
	$As_2Cu_3H_8O_{12}$/ $Cu_3(AsO_4)_2 \cdot 4H_2O$	ACGIH (2004)	A1; BEI issued;	–	–	–
		MAK (DFG, 2004)	Carcinogen category: 1	Germ cell mutagen group: 3A	–	–
	Cu	ACGIH (1992–1993)	A4 (not classifiable as a human carcinogen)	–	–	–
		MAK (DFG, 2005)	–	–	Pregnancy-risk group: D	
Cu	Cu_2O; $CuSO_4.5H_2O$	MAK (DFG, 2004)	–	–	Pregnancy-risk group: D	–
Zn	$ZnCrO_4$	ACGIH (1999)	A1 (confirmed human carcinogen)	–	–	–
		MAK (DFG, 1999)	–	–	–	Sensitisation of skin
Se	Na_2SeO_3	MAK (DFG, 2005)	–	–	Pregnancy-risk group: C	(Inhalable fraction)
	Se; H_2SeO_3; $SeOCl_2$; SeO_2; SeF_6; H_2Se; SeO_3	MAK (DFG 2004)	Carcinogen category: 3B	–	Pregnancy risk group: C	(Inhalable fraction)

ACGIH = American Conference of Industrial Hygienists, MAK = Maximale Arbeitsplatzkonzentration, DFG = Deutsche Forschungsgemeinschaft. Currently, there is no conclusive evidence that fluoride compounds are carcinogenic to humans (COT, 2003), but inhalation of vinyl containing F causes cancer in rats and mice at concentrations 25 mg/m^3 (Cantoreggi and Keller, 1997).

3.3 Sources

3.3.1 Natural sources

Chromium, Cu, Zn, Se and F occur naturally in rocks, soils, water and air; typical concentrations in different environmental compartments are listed in Table 3.4.

The natural concentrations of essential trace elements vary greatly from area to area, influenced by the local geology, the presence of any mineralisation, the soil type, pH and soil redox potential, the presence of oxides, the soil cation-exchange capacity, the type and distribution of organic matter and the rate of litter decomposition. The average concentration of the trace elements considered here in different rocks types is shown in Table 3.5 and the main minerals in which they occur are listed in Table 3.6.

Chromium is a lithophile metallic element which occurs mainly as the Cr spinel, chromite ($FeCr_2O_4$) (Widatallah *et al.*, 2005) and in magnetite, ilmenite and pyroxene formed in the early stages of crystal fractionation (FOREGS, 2005). During weathering it can accumulate in detrital minerals or in secondary oxides and clays, since the valency and ionic radius of Cr^{3+} are similar to those of Fe^{3+} and Al^{3+}.

The average crustal abundance of Cr is 122 mg/kg (Kabata-Pendias and Pendia, 2001). The highest concentrations (>1600 mg/kg) are in ultrabasic igneous rocks, with lower values in basalts and, especially, in granites, which contain 170 and 4–22 mg/kg Cr respectively (Kabata-Pendias and Pendia, 2001). Sedimentary rocks such as shale, sandstone and limestone contain on average 90, 35 and 11 mg/kg Cr respectively (Mielke, 1979).

Chromite, the main Cr ore mineral, is generally mined from ultramafic rocks. It can also be found in basaltic magmas with magnetite and ilmenite, which may also be enriched in Cr (Wedepohl, 1978). Most of the world's chromite reserves occur in the Bushveld Igneous Complex in South Africa and the Great Dyke in Zimbabwe (Schoenberg *et al.*, 2003; Nex, 2004).

Copper is a chalcophile metallic element and occurs in chalocopyrite ($CuFeS_2$), covellite (CuS) and bornite (Cu_5FeS_4) in sulphide ore deposits. Malachite ($Cu_2CO_3(OH)_2$), azurite ($2CuCO_3.Cu(OH)_2$) and, less commonly, chrysocolla ($CuSiO_3.2H_2O$) are common weathering product of these deposits. It can also occur as native copper.

The average crustal abundance of Cu is 60 mg/kg. Typical concentrations in igneous and sedimentary rocks are 4–200 mg/kg and 2–90 mg/kg, respectively (WHO, 1998; Reimann and de Caritat, 1998). Basalts and gabbros generally contain 40–60 mg/kg Cu and ultrabasic rocks around 5–40 mg/kg. Intermediate and granitic rocks have around 12–20 mg/kg Cu (BGS, 2007). It is widely dispersed in trace concentrations in rock-forming silicate minerals such as biotite, olivine, pyroxene and amphibole. Typical levels are 115 mg/kg in olivine, 120 mg/kg in pyroxene, 78 mg/kg in amphibole, 86 mg/kg in mica and 62 mg/kg in plagioclase (FOREGS, 2005). Copper in these minerals is generally not held in the silicate lattice and is more likely to be present as small sulphide inclusions.

Copper concentrations in unmineralised sediments are influenced by the presence of mafic detritus, secondary Fe and Mn oxides, clay minerals and organic matter. Black shales are enriched in Cu, reaching ore grade at some localities. Levels in sandstones and limestones are typically 5–15 mg/kg (FOREGS, 2005).

Copper-sulphide deposits are the most economically important source of Cu, accounting for around 90 per cent of Cu mined globally, about half of which is from chalcopyrite (BGS, 2007).

Table 3.4 Natural abundances of Cr, Cu, F, Se and Zn (WHO, 2001, 1998, 1988, 1986, 1984)

	Cr	Cu	Zn	Se	F
Soil (mg/kg)	2–60	2–250	10–300	0.059–0.318	200–300 (mineral soils)
Ambient air ($\mu g/m^3$)	<0.1	0.005–0.05	0.003–0.027	<0.00004	1
Fresh water (mg/l)	<0.001	0.001–0.02	<0.0001–0.05	<0.01–0.003	0.01–3 (<1–25 in groundwater)
Sea water (mg/l)	<0.001	0.00015	0.000002–0.0001	0.00006–0.00012	1.3

Table 3.5 Concentrations of Cr, Cu, Zn, F and Se in various rock types (He *et al.*, 2005)

Element	Basaltic igneous	Granitic igneous	Shales and clays	Limestone	Sandstone
	Concentration (mg/kg)				
Cr	40–600	2–90	30–590	10	35
Cu	30–160	4–30	18–120	4	2
Zn	48–240	5–140	18–180	20	2–41
Se	0.05–0.11	0.05–0.06	–	0.08	0.05
F	360	–	800	220	180

Table 3.6 Principal minerals of Cr, Cu, Zn, Se and F (FOREGS, 2005)

Element	Mineral	Chemical formula
Cr	Chromite	$FeOCr_2O_3$
	Tongbaite	Cr_3C_2
	Zincochromite	$ZnCr_2O_4$
	Chromatite	$CaCrO_2$
Cu	Chalcocite	Cu_2S
	Cuprite	Cu_2O
	Covellite	CuS
	Bornite	Cu_5FeS_4
	Malachite	$Cu_2CO_3(OH)_2$
	Azurite	$2CuCO_3.Cu(OH)_2$
	Antlerite	$Cu_3SO_4(OH)_4$
	Enargite	Cu_3AsS_4
	Chrysocolla	$CuSiO_3.2H_2O$
	Chalcopyrite	$CuFeS_2$
Zn	Sphalerite	ZnS
	Hemimorphite	$Zn_4Si_2O_7(OH)_2H_2O$
	Smithsonite	$ZnCO_3$
	Hydrozincite	$Zn_5(OH)_6(CO_3)_2$
	Zincite	ZnO
	Willemite	Zn_2SiO_4
	Franklinite	$(Zn,Fe,Mn)(Fe,Mn)_2O_4$
Se	Crookesite	$(Cu,Tl,Ag)_2Se$
	Berzelianite	Cu_2Se
	Tiemannite	HgSe
F	Fluorite	CaF_2
	Apatite	$Ca_5[PO_4]_3(Cl,F,OH)$
	Topaz	$Al_2F_2[SiO_4]$
	Cryolite	Na_3AlF_6
	Carobbite	KF
	Muscovite	$KAl_2(OH,F)_2[AlSi_3O_{10}]$

Chile has the highest identified reserves of Cu, with 150 million tonnes – 30 per cent of the world's Cu. This is followed by the USA (35 million tonnes) and Indonesia (35 million tonnes) (USGS, 2008).

Zinc is another chalcophile metallic element. It is ubiquitous in the environment, although it is rarely found in the metallic form and usually exists in the +2 oxidation state in minerals such as sphalerite (ZnS), smithsonite (ZnCO3) and zincite (ZnO). Sphalerite is the commonest Zn mineral in hydrothermal ore deposits; it is often mined in association with Pb, Cu, silver (Ag) and other metals (ILZSG, 2007). Zn^{2+} is readily partitioned into oxide and silicate minerals by substituting for Fe^{2+} and Mg^{2+}, which have similar ionic radii to Zn^{2+}. As a result, it is often present as a trace constituent of pyroxene, amphibole, mica, garnet and magnetite (FOREGS, 2005; USEPA, 2005).

The average crustal abundance of Zn is 70 mg/kg. Ultrabasic rocks contain around 50 mg/kg, while basalts, granites and syenites contain around 105, 39–60 and 130 mg/kg, respectively. The concentration and distribution of Zn in sedimentary rocks reflects the presence of clay minerals, ferromagnesian silicates and detrital oxides such as magnetite. Non-mineralised carbonate rocks (around 50 mg/kg) and quartzofeldspathic rocks (30–50 mg/kg) usually contain less Zn than greywacke (70–100 mg/kg) and shale (50–90 mg/kg) (FOREGS, 2005).

In the absence of Fe, Zn is often associated with carbonate and silicate phases. Up to 1 per cent Zn has been recorded in oolitic ironstone and recent ferromanganese nodules (FOREGS, 2005).

The main types of Zn ore deposits are volcanic-hosted massive sulphides, sediment-hosted massive sulphides, Mississippi Valley Type carbonate-hosted deposits, intrusion-related Zn ore deposits and Broken Hill type ore deposits. Lead, Cu, Au, Ag and other metals are often found in association with Zn ores. It is rare for the ore to be rich enough to be directly used by smelters and it needs to be concentrated. Zinc ores usually contain 3–10 per cent Zn, while Zn concentrates contain around 55 per cent (USGS, 2006). Zinc ores are widely dispersed geologically and geographically. Important historical deposits from the UK include those of the Mendip Hills, Central Wales, Shropshire, the Lake District and Cornwall (BGS, 1998).

Selenium is a chalcophile element and substitutes for S in ore minerals such as chalcopyrite and pyrite. Following its discovery in 1817 by JJ Berzelius, Se attracted little attention because its concentration in the environment was, until recently, below analytical detection limits. Recent improvements in analytical methods, however, mean that even low levels of Se can be determined routinely in geological and environmental samples (Johnson *et al.*, 2010). The average crustal abundance of Se is generally very low (0.05–0.09 mg/kg) (Taylor and McLennan, 1985). Intrusive igneous and volcanic rocks generally have low concentrations, although the ash and gases from volcanic activity can contain high concentrations (Fergusson, 1990). Shales are generally rich in Se, whereas limestones and sandstones rarely contain as much as 0.1 mg/kg (Neal, 1995). Very high concentrations (>300 mg/kg Se) occur in some phosphatic rocks because of the substitution of the SeO_4^{2-} anion for PO_4^{3-} (Flemming, 1980; Jacobs, 1989; Nriagu, 1989; Neal, 1995). Selenium is associated with organic matter and is high in coal (1–20 mg/kg) and black shales (up to >600 mg/kg) (Jacobs, 1989). There are no Se mines, but Se is produced as a by-product of refining other metals such as Pb or Cu and from sulphuric acid manufacture.

Fluorine is the 13th most abundant element in the Earth's crust, with an average crustal abundance of 300 mg/kg (Tebbut, 1983). Unlike the other halogens, fluoride is mostly derived from rocks and minerals rather than air, sea water or anthropogenic activities.

Fluoride concentrations determined in Tibetan sandstones, plutonic igneous rocks, shale and limestone are between 389 and 609 mg/kg, while in volcanic rocks the range is 80–1000 mg/kg

(Zhang *et al.*, 2002). Fluoride is concentrated in minerals such as fluorite, cryolite and apatite. The highest concentrations occur in acid igneous rocks such as granite, rhyolite, pegmatites and mineralised veins, as well as organic-rich sediments (Davison *et al.*, 2004; Edmunds and Smedley, 2005).

3.3.1.1 Trace element availability in equatorial, tropical and sub-tropical regions

Levels of essential trace elements tend to be low in equatorial, tropical or sub-tropical regions, especially in areas of low relief (Plant *et al.*, 1996, 2000) where the regolith reflects the cumulative effects of subaerial weathering over many millions of years (Butt and Zeegers, 1992). Some soils have been deeply leached from Tertiary times or even since the late Proterozoic, 500 million years ago (Daniels, 1975; Butt 1989). Such soils tend to be poorer, thinner and more fragile than soils in temperate regions (Murdoch, 1980) and are particularly characteristic of areas of crystalline basement in Africa and Latin America away from the Andes. Laterites which have been forming for at least 100 million years (i.e. since Jurassic times) in parts of Africa, South America and Asia sometimes have profiles extending to depths of 150 m (Figure 3.5) (Trecases, 1992) and there is very little fresh supply of chemicals from bedrock. Indeed, the chemical compositions of such soils, consisting mostly of insoluble hydroxides of aluminium (Al) and ferric Fe, together with high levels of Ti and Mn, have little relationship with bedrock composition (Nahon and Tardy, 1992). They generally lack organic matter, soluble cations such as K^+, Ca^{2+} and Mg^{2+}, anions such as Cl^-, PO_4^{3-} and NO_3^- and available trace elements such as Se. Many deserts may have had a similar history, and most now comprise a regolith of detrital quartz and other resistant minerals carried by the wind. In contrast, tropical soils developed on volcanic rocks and alluvium (e.g. in south-east Asia) are generally richer in essential trace elements. Some generalised relationships between different morphoclimatic zones are shown in Figure 3.6 (from Plant *et al.*, 1996).

3.3.2 Anthropogenic sources

Activities such as base-metal mining, agriculture and fossil-fuel combustion can increase the levels of Cr, Cu, Zn, Se and F to values higher than the recommended guideline values in both terrestrial and aquatic environments. Adding essential or beneficial trace elements to fertilisers can lead to an improved intake in populations with low dietary values, such as for Se (reviewed in Johnson *et al.*, 2010), but care must be taken not to reach unsafe levels in the environment (He *et al.*, 2005) or in foodstuffs.

Chromium is used in a wide range of industries, such as textiles, tanning and the production of heat-resistant steels, tool steels and alloy cast irons. Metallic Cr is also used as protective and decorative coatings on metal products (Chernousov *et al.*, 2003). Chromium levels of 26–150 mg/kg have been found in soils in the Linares region in southern Spain, due to intensive mining and urban industrial activity (Martínez *et al.*, 2008).

Chromium, Cu and As were, in the past, widely used in the wood preservative chromated Cu arsenate (CCA), but the USEPA and the timber industry agreed to phase out the use of CCA by the end of 2003 due to concerns over its safety (USEPA, 2003). Leaching tests on wood samples treated with the preservative showed that total Cr and Cu levels in soils rose to 191–825 mg/kg and 283–568 mg/kg from background levels of 16.7–35.6 mg/kg and 23.5–32.8 mg/kg, respectively (Balasoiu *et al.*, 2001).

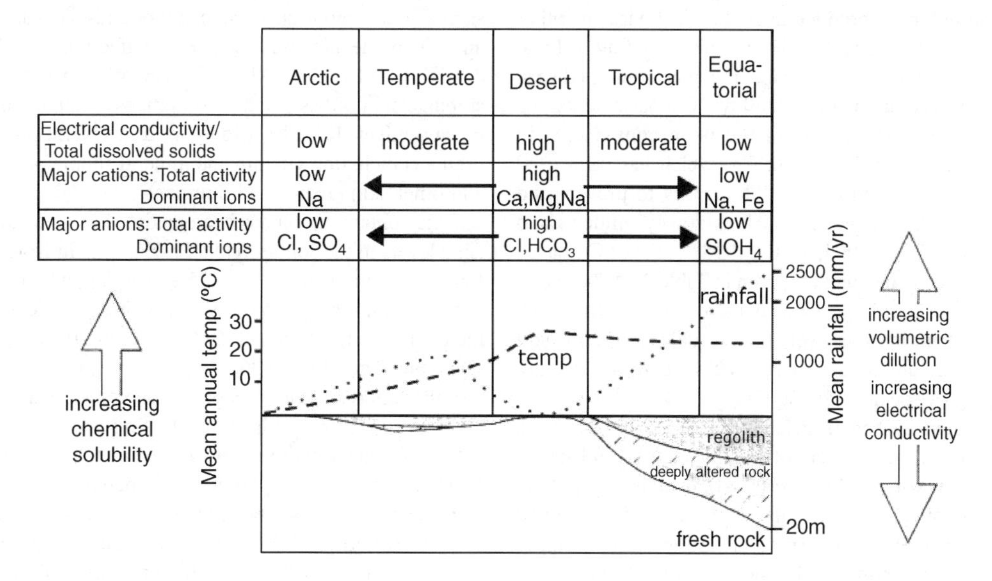

Figure 3.5 Soils, types of duricrust and thickness of the weathering zone and relationship to annual temperature and mean annual rainfall (modified after Pedro, 1985)

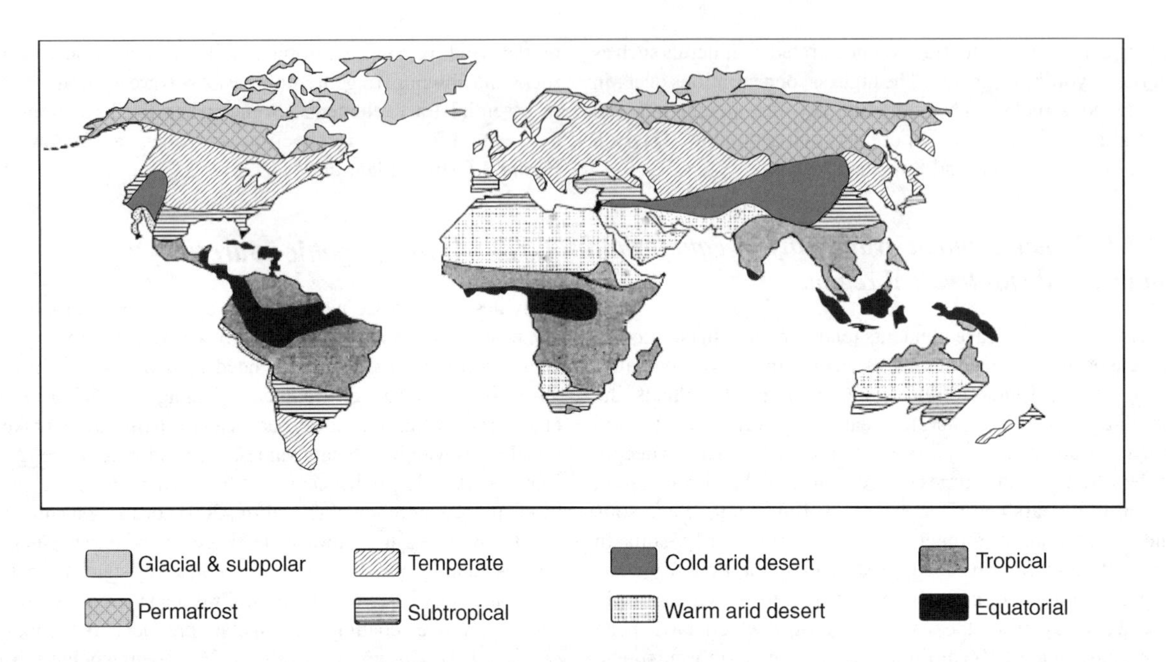

Figure 3.6 Morphoclimatic zones of the world (Reproduced from Pedro, G. (1985) Grandes tendances des sols mondiaux. Cultivar 184:78–81)

Levels of 0.004–0.007 mg/l Cr, very much higher than the reference value of <0.0001 mg/l, have been reported in groundwater samples around a scrap-iron and metal recycling facility in Denmark (Jensen *et al.*, 2000). In Hezhang County, China, soil Cr levels ranged up to 240 mg/kg (control site 71 mg/kg) as a result of atmospheric deposition around a number of Zn smelting sites (Bi *et al.*, 2006).

Industrial waste water and mining discharge have contaminated marine sediments in south-west Spain, with levels of up to 92 mg/kg Cr (Morillo *et al.*, 2004).

Copper is used in the production of electrical wire and other electrical applications, as well as in plumbing, roofing, vehicle radiators, machined parts and coinage. It can be combined with other metals to make high-strength alloys. Copper can also be used as a fungicide and is one of the main components in Bordeaux mixture (0.05–2 per cent Cu) which has been used in vineyards since the end of the 19th century to prevent and eliminate fungal plant pathogens such as downy mildews and rusts (Loland and Singh, 2004; WHO, 1998). A list of Cu-based fungicides and their Cu content is given in Table 3.7. The use of such chemicals in the Bento Gonçalves regions of Brazil has been found to result in excessively high levels of Cu in the soil (3200 mg/kg) in an area where background levels are 3.6–51.7 mg/kg (Mirlean *et al.*, 2007).

Moderately acid leachate from landfills (pH 5.0–6.0) has been reported to contain high levels of organic matter and heavy metals, including Cu (27 000–43 000 g/l), which can pollute groundwater supplies. Heavy metals in leachate and/or polluted groundwater form colloids or are present in solution. The affinity of heavy metals for humic substances is influenced by the dissociation constants of carboxylic acid and phenolic groups. A reduction in the pH, which depends mainly on the amount of putrescible waste in the landfill and O availability, can enhance heavy-metal migration (Qu *et al.*, 2008).

Copper sulphate has been used as an algicide, molluscicide and aquatic-plant herbicide in ponds and it has been suggested that Cu may accumulate in the sediment (de Oliveira-Filho *et al.*, 2004; Han *et al.*, 2001). The most toxic forms of Cu to aquatic organisms are the free Cu^{2+} ion and some hydroxyl and carbonate complexes (Mastin and Rodgers, 2000).

Zinc is used for galvanising steel and Fe to protect them from rust and corrosion. It is also used with other metals to form alloys such as brass and bronze. Zinc compounds such as Zn sulphide and Zn oxide are used in the manufacture of white paints, ceramics and rubber, whilst Zn acetate, Zn chloride and Zn sulphate are used as wood preservatives and in the manufacture of fabric dyes. It is also widely used in several pharmaceutical products including vitamin and mineral supplements, sun block and other skin creams and ointments.

Zinc contamination has been recorded around several mines. Zinc levels in the soil around a Pb-Zn mine in Korea have been found to be in the range 329–25 800 mg/kg, where control sites had concentrations of 56–119 mg/kg. Tobacco leaves grown in the contaminated area had Zn levels of 1620 mg/kg dry weight (Jung and Thornton, 1996).

Other important anthropogenic sources of Zn are Zn batteries and leaching of galvanised metals. Soil and groundwater contamination can occur in scrap-iron and metal recycling facilities. In Denmark, soil samples near car-battery salvaging centres were found to contain 6600 mg/kg Zn, against a reference value of only 16 mg/kg, while the areas with scrap cutting resulted in soil Zn levels of 4700 mg/kg (Jensen *et al.*, 2000).

Selenium Anthropogenic activity is a major source of Se, with an estimated release of 76 000–88 000 tonnes per year – compared

Table 3.7 Copper-based fungicides used for treating citrus diseases in the USA (Reproduced from Timmer, L.W., Childers, C.C. & NIGG, H.N. (2004) Pesticides registered for use on Florida citrus. Gainesville, FL: 2004 Florida Citrus Pest Management Guide, SP-43, University of Florida)

Trade name/common fungicide name	Cu, per cent	Diseases treated			
		Melanose	Greasy spot	Alternaria	Scab
Kocide 101/Cu hydroxide	50	+	+	+	+
Kocide 2000/Cu hydroxide	35	+	+	+	+
Champ dry prill/Cu hydroxide	38	+	+	+	+
Cuprofix disperses/basic Cu sulphate	20	+	+	+	−
Copper-count-N/Cu ammonium complex	8	+	−	+	−

with the natural release of Se estimated to be 4500 tonnes per year. Common anthropogenic sources of Se are shown in Table 3.8. The most important sources include fossil-fuel combustion, metal extraction, phosphate fertilisers and sewage sludge. The mining and processing of gold, base metal and coal deposits can also be important sources. For example, contamination of the Chayanta River by mine water leached from the nearby near Potosi mine in Bolivia resulted in water levels of 0.005–0.020 mg/l Se, which exceed the guideline value for freshwater aquatic organisms (0.001 mg/l) (Rojas and Vandecasteele, 2007).

The manufacture of detergents and shampoos can cause Se levels to increase locally. There is now an increased use of Se in the pharmaceuticals, glazing, photocopying, ceramics, paints and electronic industries.

Fluoride levels in the environment can be increased from many anthropogenic sources, including the application of phosphate-based fertilisers, which can increase soil fluoride levels from 217 to 454 mg/kg (Loganathan *et al.*, 2001). Other anthropogenic sources of fluoride include glass manufacture, brickworks, steelworks, pharmaceuticals and Al smelting (Debackere and Delbeke, 1978; Kabata-Pendias and Pendia, 2001). Aluminium smelting is an important source of fluorides, especially in countries such as Iceland, Canada and Norway, which have cheap sources of geothermal or hydroelectric energy (Bell and Treshow, 2002). More than two thirds of the emissions comprise gaseous fluorides (Hocking, 2005).

Table 3.8 Common anthropogenic sources of Se in the environment (Reproduced from Fordyce, F.M. (2005) Selenium deficiency and toxicity in the environment. In: Essentials of Medical Geology, pp 373–415. Eds Selinus, O., Alloway, B., Centeno, J.A., Finkelman, R.B., Fuge, R., Lindh, U. and Smedley, P. Elselvier, London)

Sources	Comments
Metal-processing industries	Important source
Burning of fossil fuels	Important source
Disposal of sewage sludge to land	Typical Se contents 1–17 mg/kg
Agricultural use of pesticides	$[K(NH_4)S]_5Se$
Agricultural use of lime	Typical Se content 0.08 mg/kg
Agricultural use of manure	Typical Se content 2.4 mg/kg
Agricultural use of phosphate fertilisers	Typical Se content 0.08 mg/kg

In the USA, a five-year study of the Prebake Primary Aluminium Production Plant near Charleston, South Carolina showed that fluoride emissions of 0.4–0.5 kg per tonne of Al produced were associated with an increase in skeletal fluorosis in the native white-tailed-deer population (Suttie *et al.*, 1987). Fluoride contamination covering 4500 km^2 of surface snow from two Al smelters has been reported in Canada (Quellet, 1987). Emissions from Al plants there have been reduced by the installation of scrubber systems and now range from 12.08 to 21.14 million tonnes per annum (Hocking, 2005).

The largest single anthropogenic source of fluoride into the atmosphere is coal combustion (0.2 Mtpa) (Ando *et al.*, 2001). In addition to releases from power plants, an estimated 400 million people in China depend on coal for domestic energy. In China each kg of coal burnt introduces about 100 mg of fluoride into the atmosphere, depending on plant efficiency (Finkleman *et al.*, 2002; Luo *et al.*, 2002).

Fluorine may also be released into the environment as H fluorocarbons (HFCs), which can be found in plastic foams, refrigerants and as an aerosol propellant. HFCs are greenhouse gases, but they are less damaging than the chlorofluorocarbons (CFCs), which they replaced and which were responsible for depleting the ozone layer (Howard and Hanchett, 1975). The use of HFCs has been phased out internationally under the terms of the 1987 Montreal Protocol.

3.4 Environmental pathways

In temperate environments, levels of Cr, Cu, Zn, Se and F in soils generally reflect parent material, especially bedrock, whereas water levels also reflect factors such as the residence time and composition of water. Atmospheric deposition, which reflects mainly anthropogenic activity, can cause dispersed as well as

Table 3.9 Mean concentrations of Cr, Cu, Zn and Se in various foods (modified from Ysart *et al.*, 2000) (Reproduced by permission of Taylor & Francis Group). F data from the US Department of Agriculture

Food group	Mean concentration (mg/kg fresh weight)				
	Cr	Cu	Zn	Se	F
Bread	0.1	1.6	9.0	0.04	39
Miscellaneous cereals	0.1	1.8	8.6	0.03	17–72
Carcass meat	0.2	1.4	51	0.12	4–166
Offal	0.1	40	43	0.42	5
Meat products	0.2	1.5	25	0.12	16–65
Poultry	0.2	0.73	16	0.15	15–21
Fish	0.2	1.1	9.1	0.39	18–210
Nuts	0.7	8.5	31	0.29	16
Eggs	0.2	0.59	11	0.19	1–5
Oils and fats	0.4	0.05	0.6	0.01	1–25
Sugars and preserves	0.2	1.5	3.8	0.01	1–19
Green vegetables	0.2	0.84	3.4	0.01	1–38
Potatoes	0.1	1.3	4.5	0.008	49
Other vegetables	0.1	0.91	2.6	0.02	1–7
Canned vegetables	0.1	1.5	3.9	0.02	1–26
Fresh fruit	< 0.1	0.94	0.8	0.004	1–12
Fruit products	< 0.1	0.73	0.7	0.005	28–104
Beverages	< 0.1	0.1	0.3	0.003	9–322
Milk	0.3	0.05	3.5	0.02	3
Dairy produce	0.9	0.45	14	0.05	3–35

more local contamination of soil and water. Food is an important pathway for essential elements, but levels vary markedly between different food types and according to where the food is produced (Table 3.9).

In tropical and arid terrains, chemical elements tend to undergo more pronounced separation depending on their mobility (Trecases, 1992), so there is greater potential for deficiency or toxicity conditions to occur. The most deeply leached regions may have only Al, Fe and Mn oxides and elements such as Zr in resistate minerals. On the other hand in arid conditions the most mobile elements such as the alkalis, alkaline earths, halogens and elements such as Se which are soluble at high pH, tend to accumulate in areas of inland drainage or at the base of soil profiles. Elements such as Cr, Cu and Zn tend to be strongly bound on ferric or manganic oxides, so, despite high total levels, little may be bioavailable. Fluorine is usually deeply leached, becoming concentrated in groundwater. The mobility of the essential trace element in different pH and Eh conditions is shown in Figure 3.7.

In Europe, the systematic coverage of the FOREGS study (FOREGS, 2005) has shown that concentrations of trace elements are highly variable and in great measure depend upon the underlying geological controls on soil and sediment formation, or water-rock interaction. Copper, Cr, Zn and Se are generally highest over soils around the Mediterranean,

RELATIVE MOBILITIES	ENVIRONMENTAL CONDITIONS			
	oxidising	acid	neutral-alkaline	reducing
VERY HIGH	I	I	I Mo *U* *Se*	I
HIGH	Mo *U* Se F *Ra* Zn	Mo *U* Se F *Ra* Zn Cu Co Ni *Hg*	F *Ra*	F *Ra*
MEDIUM	Cu Co Ni *Hg* *As Cd*	*As Cd*	*As Cd*	
LOW	*Pb Be Bi Sb Ti*	*Pb Be Bi Sb Ti* Fe Mn	*Pb Be Bi Sb Ti* Fe Mn	Fe Mn
VERY LOW TO IMMOBILE	Fe Mn *Al* Cr	*Al* Cr	*Al* Cr Zn Cu Co Ni *Hg*	*Al* Cr Mo *U* Se Zn Co Cu Ni *Hg* *As Cd* *Pb Be Bi Sb Tl*

Figure 3.7 Mobility of selected elements in different pH and Eh conditions (Reproduced, with permission, from Plant, J.A., Baldock, J.W. and Smith, B. (1996). The role of geochemistry in environmental and epidemiological studies in developing countries: a review. From Environmental Geochemistry and Health 113, 7–22)

especially Greece and parts of Italy, and lowest over the Precambrian shields of Northern Ireland, Scotland and Fennoscandia and their cover sequences in north Germany, Denmark and Poland (FOREGS, 2005).

3.4.1 Chromium

Soil. Chromium levels are particularly high in soils developed from ultrabasic rocks such as serpentinites, where they can be as high as 3500 mg/kg, while soils derived from granites and sandstones generally contain 10–40 mg/kg (WHO, 1988).

Over Europe, high topsoil levels of Cr (>60 mg/kg) occur over northern Scandinavia, central Norway, south-west Britain, north-east Northern Ireland, areas of western Germany, Italy, much of Croatia and strikingly across much of Greece, also. High subsoil levels of Cr are found in Sicily, southern Spain and eastern Slovakia (FOREGS, 2005). In contrast, low levels of Cr (<34 mg/kg) in subsoil occur in southern Scandinavia, Estonia, parts of northern Germany and Poland, central Portugal and parts of central and eastern Spain (FOREGS, 2005).

Sediments. Chromium levels in stream sediments are highest in Greece and northern Italy, probably reflecting bedrock composition. Chromium levels across Europe are <3–3324 mg/kg. The lowest values occur over Quarternary sediments in Denmark, north-east Germany, central Germany, Estonia, Latvia, southern Norway, part of south Finland. The Variscan of central and eastern Spain, north and west Portugal, south-west France, central Austria and northern Sardinia and southern Corsica. The highest values occur over the alpine ophiolite belt across Greece and Albania, high levels also occur over the Appenines of north Italy, northern Fennoscandia, the Southern Uplands of Scotland over the Caledonian suture zone and over the Tertiary basalts of Northern Ireland.

Water. In Europe, stream water contains <0.01–43.3 μg/l (median 0.38 μg/l). The lowest levels are in western, central and northern Norway, central Sweden, central England and Wales on Precambrian Shield and Caledonian terrains, north-western Spain, Corsica, Sardinia, in most of central mainland Europe from central France, Germany and Czech Republic and in most of northern and south-central Italy, Switzerland, Austria and Slovenia, western Slovakia and northern Hungary (FOREGS, 2005). Higher Cr levels are found in south-western Sweden, Finland, Lithuania, Latvia, the north-east tip of Poland, north-central and southern Spain, north-western (Liguria), central (Roman-Tuscan province) and southern (Apulia) Italy, eastern Czech Republic, Slovakia, eastern Croatia, Albania and Greece (FOREGS, 2005). The distribution in some areas of Europe is different to that in soils and sediments. The dominant influence on the latter is bedrock geology, whereas stream-water concentrations are also influenced by hydrogeochemical conditions. In Greece, there are high concentrations in all sampled media.

Air. Levels of Cr in air are determined largely by local geology. Typical background levels of Cr range between 2.5 and 10 pg/m^3 (WHO, 1988). Ambient air concentrations of Cr in the United States have been found to average 0.015 μg/m^3 with a maximum recorded value of 0.35 μg/m^3 (WHO, 1988). In the vicinity of coal-fired power stations, Cr levels can be as high as 100 mg/m^3 (USEPA, 1978).

A study of heavy metal deposition in Poland using the moss *Pleurozium schreberi* reported the highest levels of Cr (>6 mg/kg) in the south, where there are more sources of industrial emission (Grodzińska *et al.*, 1999).

3.4.2 Copper

Soil. Copper levels in subsoils across Europe range from 0.86 to 125 mg/kg and 0.81 to 256 mg/kg for topsoils. Areas with the lowest recorded subsoil Cu are in northern central Europe, parts of Scandinavia and the Baltic states, north-east Italy, central Hungary and central Portugal, whereas areas with the highest subsoil Cr have been reported in north-western and southern Spain, the western Pyrenees, across France from Brittany to the Massif Central, the southern tip of Italy, Greece, Croatia, south-central Austria, north-eastern Slovakia, the Harz Mountains and Thuringia, south-west England, northern Wales, eastern Ireland, western Norway and central Sweden (FOREGS, 2005).

Much of the Mediterranean is enriched in Cu, as weathering in dry climates causes more precipitation of (hydr)oxides, sulphate and carbonate and adsorption by Fe-Mn oxides. Bedrock composition is also significant: there are (ultra)basic rocks in Greece and mineralisation and mining in the Iberian Pyrite Belt (FOREGS, 2005).

Sediment. Total Cu values in sediments across Europe range between 2 and 495 mg/kg, with a median of 17 mg/kg. Low values (<11 mg/kg) occur over the granitic, granodioritic and gneissic areas of Scandinavia, northern Germany, Poland and the Baltic countries, the sandstone, metamorphic and granitic rocks of northern Scotland, the calcareous, clastic and crystalline rocks of central and eastern Spain and the molasse basin of Austria (FOREGS, 2005). High total Cu values in flood-plain sediment (>28 mg/kg) have been reported over central and northern Norway (attributed to mineralisation and mining), in the Stockholm region in Sweden, the mineralised areas of south-west Finland and the shales and mafic volcanics in the Scottish Midland Valley (FOREGS, 2005). High Cu levels in Wales, the Midlands and south-west England reflect lithology, base-metal and Fe mineralisation and possibly industrial pollution; the same applies to an extensive arcuate belt extending from Belgium through central Germany to the Harz Mountains, Thuringian Forest, Erzgebirge, the Bohemian Massif, the Czech Republic, the Slovakian Ore Mountains, almost the whole of Hungary (e.g. Recsk Cu-Zn mineralisation), Slovenia, Croatia, the southern Alps of Austria and northern Italy (FOREGS, 2005).

Water. Copper can be introduced to both surface and groundwater by processes such as atmospheric deposition, surface run-off, contaminant leaching and effluent discharge. In natural waters, Cu is transported either adsorbed onto suspended particles or complexed with ligands. Only a small proportion is present as free Cu^{2+} ions (WHO, 1998).

Copper values in stream water over Europe range from 0.08 to 14.6 μg/l, with a median value of 0.88 μg/l (FOREGS, 2005). Lowest Cu values in stream water (<0.38 μg/l) are found in northern Scotland, most of Scandinavia, eastern Poland, Spain, western Austria, southern Germany, Albania and most of Greece. These low concentrations in Greece indicate that the primary lithological source (as indicated by Cu concentrations in soil and sediment) is poorly soluble. Copper concentrations are higher in stream water over parts of western, southern and eastern Finland, Denmark, the Netherlands, south-east Ireland, eastern England, parts of Brittany, areas of southern Spain and southern Italy. Very high values are found in Switzerland and areas in France and Germany. Enhanced Cu values in Brittany might have an anthropogenic source and reflect the use of liquid manure from pigs and chicken farms (FOREGS, 2005).

Air. The distribution and transport of Cu from major natural and anthropogenic emission sources is mainly via the atmosphere, with most Cu deposited near the source. Typical concentrations of Cu in air in remote regions are around 5–50 ng/m^3 (WHO, 1998) while levels in more industrial areas can be as high as 920 ng/m^3 (Beavington *et al.*, 2004).

Atmospheric transportation can redistribute elements from air to land. Atmospheric levels of Cu and Zn in the Deonar region of Bombay, India are in the range 38–783 ng/m^3 and 110–1800 ng/m^3 respectively. Bulk deposition of Cu was calculated to be 21.1 kg/km^2/yr in 1988 and 12.9 kg/km^2/yr in 1989. For Zn, this was 92.3 kg/km^2/yr and 102.1 kg/km^2/yr in two years. Possible sources of the metals were from nearby chemical, petrochemical and paint industries (Tripathi *et al.*, 1993).

3.4.3 Zinc

Soil. Zinc normally accumulates in the surface horizon, as it is easily adsorbed by mineral and organic materials present in the soil, and regional soil profile studies (FOREGS, 2005; BGS, 2010) have shown that the proportion of extractable Zn declines with increasing depth, while total Zn is uniformly distributed. The concentration of extractable Zn often has a positive correlation with total Zn, organic matter, clay content and cation exchange capacity (CEC). Conversely, there is a negative correlation between extractable Zn and $CaCO_3$ and pH.

Zinc from atmospheric deposition can migrate down soil profiles. When unpolluted spodosol profiles from Houthalen, Belgium were compared with profiles polluted by a nearby Zn smelter in Balen, the average Zn concentrations in the upper horizons were found to be 9 mg/kg in the unpolluted soils and 71 mg/kg in the polluted soils (Degryse and Smolders, 2006). In polluted soils in Kempen on the Dutch-Belgian border, Zn levels decreased from 70.7 to 10.8 mg/kg with increasing depth. Most of the Zn was in the upper horizon where it is mainly in a non-soluble form that prevents it leaching into the subsoil. Deposition of Zn contaminants between 1889 and 1935 was not enough to mobilise all of the non-soluble Zn (Degryse and Smolders, 2006).

Sediment. Stream sediments with high Zn concentrations (≥110 mg/kg) in Europe are found, particularly, in some areas of Great Britain; the highest concentrations, up to 80 mg/kg, reflect areas of metalliferous mineralisation and historical mining. Some lake sediments can contain over 20 000 mg/kg Zn (FOREGS, 2005; WHO, 2001) in fine precipitates (Barceloux, 1999a).

Zinc may be present in suspended and bedload sediment in rivers. More Zn is associated with suspended sediment than dissolved in water. In an uncontaminated area of the Mississippi River, USA it was estimated that around 90 per cent of Zn was transported as suspended particles, whereas in a contaminated area it was around 40 per cent, the remainder being in the bed load (WHO, 2001).

Water. Zinc in stream water across Europe ranges from <0.001 to 0.181 mg/l. The lowest values are in northern Finland, Sweden, western Norway, Scotland, western and southern England, Wales, southern Germany and areas of France, Spain and Corsica. There is a belt of low Zn values over south-east Poland over Quaternary glacial deposits. Low levels of Zn in stream water are also found in a region that extends from south-east France across Italy into central Austria. Southern Hungary, north-east Croatia and Greece also have low Zn content in stream water (FOREGS, 2005). Mining in south Morocco has resulted in contamination of nearby rivers. Water samples there contain up to 7.34 mg/l of Zn, which is above the USEPA limit of 5 mg/l (Khalil *et al.*, 2008).

Highest levels (>8.3 μg/l) of Zn are found over the Precambrian shield in south Sweden and part of south-east Norway as well as over shield-derived glacial drift in Denmark, Estonia, Latvia, Lithuania and north Poland. Levels are also high over the Variscides of east-central Germany, Brittany, north Portugal, southern Spain and Sardinia where it also reflects the presence of mineral deposits. Zinc levels in stream waters are also high over Sicily and Quarternary deposits in the Netherlands (FOREGS, 2005).

Air. Average background levels of Zn in ambient air in remote areas are in the range <3–27 ng/m^3 (Barceloux, 1999a). Volcanic activity can emit Zn into the atmosphere, from which it is subsequently deposited onto the Earth's surface. Wind can also move soil particles containing Zn over land surfaces and into the atmosphere, from which it can be redeposited onto land via wet and dry deposition (Salomons, 1986). Anthropogenic sources of Zn particles, such as from smelting Fe scrap, are around 0.3–0.4 μm in diameter while Zn from natural sources are >0.1 μm. Levels of Zn in air can be over 1000 ng/m^3 from anthropogenic sources (WHO, 2001). Background levels of Zn are <3–27 ng/m^3 in most uncontaminated areas (Barceloux, 1999a).

3.4.4 Selenium

The main natural flux of Se is via the marine system. This is an important sink, but the main source to humans is thought to be from bedrock, which affects soil concentrations and availability for plant uptake (Fordyce, 2005).

Soil. Soil is a fundamental control on Se concentrations in the food chain and is a major control of the Se status of crops and livestock (Fordyce, 2005). The world mean value for soils is 0.4 mg/kg, with a range of 0.01–2.00 mg/kg (Fordyce, 2005). Except where there is contamination, there is usually a strong correlation between the concentration of Se in soils and that in bedrock or regolith. The biological and physicochemical properties of soils also control the mobility and uptake of Se by plants and animals and hence its concentration in food. These factors include pH and redox conditions, speciation of Se, soil texture and mineralogy, organic matter content and the presence of competing ions (Fordyce, 2005).

In the northern Midwest USA, Se toxicity affects livestock in areas where soils contain 1–10 mg/kg Se because ≥60 per cent is readily bioavailable in the semi-arid alkaline environmental conditions. On the other hand, soils in Hawaii containing as much as 20 mg/kg Se have no toxicity problems because the Se is complexed with Fe and Al in lateritic soils (Oldfield, 1999). In areas with Keshan disease in China, Se levels are around 0.112 mg/kg compared to an average of 0.234 mg/kg in non-endemic areas of the world (WHO, 1986) where it may occur in its inorganic form as selenide, selenate and selenite. Selenium can also replace sulphur to produce a range of organic Se compounds such as dimethylselenide and trimethylselenium (Meister, 1965). Also in China, Se deficiency related to KD was shown to reflect the presence of black organic-rich soils from which Se was poorly mobile rather than low total Se content (Johnson *et al.*, 2010). Essential elements from soil can leach into surface and groundwater sources. High levels of Se often occur in surface waters in farm areas where irrigation water drains from soils containing high Se. In water it exists as selenic and selenious acids (Barceloux, 1999b). In alkaline soils, Se is often present as selenite, which is bioavailable to plants and could prevent Se deficiency in people who eat them.

The oxidation state of Se is critical in determining its availability in the food chain. For example, in neutral-to-alkaline soils Se^{6+} (selenate) is the dominant state. This form of Se is generally more soluble and mobile in soils and is readily available for plant uptake. Selenite (Se^{4+}) has lower solubility and greater affinity for adsorption on soil particle surfaces than Se^{6+} and hence selenites are less bioavailable to agricultural crops (Mikkelsen *et al.*, 1989).

Sediment. Selenium levels in sediments of the Lewis and Clark Lake near the Missouri River are in the range 0.012–9.62 mg/kg, far higher than the toxic effect threshold of 2 mg/kg. The Se is thought to be derived from the erosion of shale bluffs containing high levels of the element (Lemly, 2002; Pracheil *et al.*, 2010; Johnson *et al.*, 2010).

Water. In water samples taken from the Republican River Basin of Colorado in the US, nine sites contained levels above 0.005 mg/l Se, which is considered a high hazard for Se accumulation in the planktonic food chain (May *et al.*, 2001). Typical levels of Se in groundwater and surface water range from 0.00006 to 0.400 mg/l, with some areas having as much as 6 mg/l (WHO, 1996; Fordyce, 2005).

Air. In air, Se is mostly bound to particles, with most urban air having concentrations of 0.1–10 ng/m^3 (WHO, 1996). Higher levels may occur in the vicinity of coal-fired thermal power stations (90 ng/m^3; background level, 10 ng/m^3) with Se in an amorphous state (Giere *et al.*, 2006; Jayasekher, 2009). In the UK, soil samples collected between 1861 and 1990 by the Rothamstead Agricultural Experimental Station show that the highest concentrations of Se were between 1940 and 1970, coinciding with a period of intensive coal use. The decline in Se in herbage more recently is thought to reflect a switch to fuel sources such as nuclear, oil and gas (Haygarth, 1994).

3.4.5 Fluoride

Soil. Average soil fluoride levels are in the range 100–1000 mg/kg. Most mineral soils contain 20–500 mg/kg of fluoride, which is usually associated with micas and clay minerals. Soils that have developed from fluoride-containing minerals can have as much as 38 000 mg/kg fluoride (WHO, 1984; Kabata-Pendias and Pendias, 1984). Soils with fluoride concentrations of 24–220 mg/kg have been reported in Argentina and uncontaminated soils with 113–5250 mg/kg in the UK (Lavado and Reinaudi 1979; Fuge and Andrews, 1988).

Sediments. There are few data on F levels in stream sediments. It is generally determined in water samples.

Water. Typically, levels of fluoride in fresh water are less than 1 mg/l (FOREGS, 2005). The concentration of fluoride in most waters is controlled by the low solubility of the main fluoride-bearing mineral, fluorite (CaF_2); hence, waters that are rich in Na, K and Cl and poor in Ca tend to contain high fluoride concentrations. In general, groundwaters contain more fluoride than surface waters because they have more time in contact with fluoride-bearing minerals (Edmunds and Smedley, 2005; Hem, 1992; WHO, 2000; Fordyce, 2005).

Fluoride levels in surface waters over Europe vary widely, with levels below 0.050 µg/l over the Caledonides of Scotland, Ireland, Wales and Norway as well as northern Italy and parts of Spain with higher background values (>0.36 µg/l) over south-east England, Poland and parts of Italy, Spain and the Baltic States (FOREGS, 2005).

Levels of fluoride in lakes are likely to be governed by the calcium-carbonate-phosphate-fluoride system, which tends to maintain uniform fluoride levels with depth (Kramer, 1964).

Endemic dental and/or skeletal fluorosis occurs in the East African Rift Valley, associated with volcanic rocks and thermal waters (Frencken *et al.*, 1980); in India and Sri Lanka in areas of fluoride-rich alkali groundwaters; and in China from high-fluoride groundwater and in coal smoke (Zheng *et al.*, 1999).

Groundwater from Battuvani Palli, India contains up to 3.8 mg/l fluoride, causing skeletal and dental fluorosis as a result of fluoride dissolved from fluorspar, cryolite and fluorapatite minerals (Butterworth *et al.*, 2005).

In central Europe, in countries such as Ukraine, Moldova, Hungary and Slovakia, groundwater that exceeds the WHO guidelines of 1 mg/l are widespread, with detrimental effects on human health (Fordyce *et al.*, 2007). Triassic sandstones in Shropshire, Lancashire and Cumbria show an average concentration of <0.1 mg/l and a range of <0.1–0.26 mg/l (Edmunds *et al.*, 1989).

Air. Natural background levels of fluoride in air are around 0.5 ng/m^3, the main sources being from weathering, volcanic emissions and marine aerosols (WHO, 2002). Volcanoes represent the major source of natural inorganic fluorides in the atmosphere, with an annual input estimated to be 60 000 to 6 million tonnes per annum in the form of H fluoride (HF) and fluorosilicates, depending on the amount and type of volcanic activity in the year (Symonds *et al.*, 1988). The average HF emission rates from Mt Etna have been estimated at about 75 Gg/a (Aiuppa *et al.*, 2004), making it the largest known point source of atmospheric fluoride (Francis *et al.*, 1998).

Coal combustion is the largest anthropogenic input of fluoride to the atmosphere, with an estimated annual total of 0.2 Mt/pa. The combustion of F-rich coals for heating and cooking makes a direct input of particulate and gaseous fluoride into the atmosphere and environment (Ando *et al.*, 2001). Luo *et al.* (2002) suggest that burnt coal contributes an estimated 100 mg of F per kg of coal to atmospheric F, though the amount of F released will depend on the efficiency of the plant and the treatment of the exhaust gases.

3.4.6 Food

Food is an important route of exposure for essential and beneficial trace elements. The mean concentrations in some foodstuffs are listed in Table 3.9.

Chromium can be found in processed meats, whole grains and particularly in pulses and spices. It is present in its trivalent form (EVM, 2003). The estimated daily Cr intake has increased from 0.25 mg to 0.34 mg/day between 1991 and 1994 due to increased Cr levels in oil, fats, milk, dairy and nuts, according to the 1997 Total Diet Study; the reason for this is unknown (Ysart *et al.*, 2000).

The Cu content of food is generally in the range of 0.2–44 mg/kg fresh weight (Barceloux, 1999c). Liver, seafood, nuts and seeds are particularly high in Cu (ATSDR, 2004). The UK Reference Nutrient Intake of adults is 1.2 mg per day. According to the 1997 UK Total Diet Study, the average daily intake of Cu from food is just below 1.8 mg or 0.03 mg/kg of body weight for a 60 kg adult (Ysart *et al.*, 2000). In the USA, safe consumption of Cu is considered to be 1.5–3.0 mg daily in adults (EVM, 2003).

Seafood, muscle meat, dairy products, nuts, legumes and whole grains are particularly high in Zn. Meat, poultry and fish contain around 25 mg/kg of Zn (Barceloux, 1999a). The solubility of Zn increases in weak acid solutions, so ingestion of acid foods such as citrus fruits stored in containers that contain Zn can increase overall Zn uptake in the diet (Barceloux, 1999a).

In the case of Se, the UK Total Diet Study shows a fall in intake from 60 µg/day in the 1980s to approximately half that value in 2000 (SACN, 2007) which is well below the reference nutrient intake value of 60 µg/day for adult females and 75 µg/day for adult males (Department of Health, 1991). One factor in the decrease of Se in the UK diet is the change from importing wheat from the prairies of North America to Europe.

Fresh fruits and vegetables may contain higher levels of essential elements if they have been irrigated with contaminated water and/or have been growing on contaminated soils, for example lettuce grown on contaminated soils has been reported to contain 301 mg/kg of Zn, while lettuce from uncontaminated soils has 77 mg/kg Zn (WHO, 2001).

Tea plants (*Camellia sinensis*) tend to accumulate fluoride from the soil, especially in the leaves where it is present in the form of an aluminium-fluoride complex. Some teas can contain as much as 1175 mg/kg fluoride. As a result, long-term drinking of tea either grown on fluoride-rich soils or irrigated with water rich in fluorides can cause chronic intoxication and fluorosis (Yi and Cao, 2008). Drinking such teas can account for 20 per cent of the daily intake of fluoride (Messaïtfa, 2008). Fluoride contents in several teas are shown in Table 3.10. However, drinking water is the main cause of excess dietary intake. More than 200 million people worldwide are thought to be drinking water with excess fluoride (Edmunds and Smedley, 2005).

A particularly important example of essential elements in consumer products is in toothpaste, where sodium fluoride is added to prevent dental cavities. Drinking water that is naturally low in fluoride is purposefully fluoridated with fluorosilicic acid and sodium fluorosilicate for the same reasons (ATSDR, 2003a).

3.5 Effects on human receptors

3.5.1 Human health exposure pathways

The main routes of exposure to essential elements in humans are through food and water. Inhalation can be important in occupational settings such as in metal refining and manufacture. A chemical is considered a risk when a source, pathway and receptor can be identified (Figure 3.8).

3.5.1.1 Oral exposure

Some types of foods are particularly enriched in certain elements. Shellfish, muscle meat, nuts and legumes, for example, are particularly high in Zn (Barceloux, 1999a). The average concentrations of Cr, Cu, Zn, Se and F in various foods are shown in Table 3.9.

The levels of essential elements in foods, especially fruit and vegetables, can vary greatly in relation to the conditions in which they are grown. For example, lettuce grown on contaminated soils has been found to contain 301 mg/kg Zn while the control plants contained only 77 mg/kg (WHO, 2001).

The proportion of essential elements absorbed from food and water is influenced by bioaccessibility. In a study to compare absorption between organic and inorganic forms of Se, 84 per cent of inorganic Se from selenite was absorbed and 97 per cent

Table 3.10 Fluoride content in teas and tea beverages (modified from Yi and Cao, 2008) (Reproduced by permission of Elselvier)

Tea types	F content (mg/kg)
Tea leaves	
Green tea	2.1–550.0 (China); 217.0–336.0 (Hong Kong); 71.1–180.0 (Japan)
Oolong tea	170–224 (Hong Kong); 8–176 (China)
Black tea	23.6–385.0 (China); 322.0–423.0 (Hong Kong); 30.0–340.0 (Poland); 87.6–289.2 (Turkey); 35.0–182.0 (Iran)
Flower tea	31.5–636.4 (China)
Brick tea	52.5–1175.0 (China); 680.0–878.0 (Hong Kong)
Instant tea	
Green tea	260.3–597.5 (China)
Oolong tea	248.2 (China)
Black tea	0.49–3.35[a] (Germany); 1.0 6.5[a] (USA); 151.0–631.3 (China)
Flower tea	196.3 (China)
Tea beverage	
Green tea	0.53–0.90[a] (Japan); 0.21–1.47[a] (China); 22.6–25.3[a] (Taiwan)
Black tea	0.47–2.19[a] (Japan); 0.25–1.79[a] (Germany); 0.176–2.195[a] (China); 19.9–33.4[a] (Taiwan)
Oolong tea	0.75–1.50[a] (Japan); 0.0565–4.1068[a] (China); 21.8–28.5[a] (Taiwan)
Flower tea	0.19–2.28 (China)
Fruit tea	0.03–0.09[a] (Germany); 0.671–4.452[a] (China)

[a]Values in mg/l.

of organic Se from selenomethionine (Swanson *et al.*, 1991). Bioaccessibility of different chemical species of essential elements is discussed in Section 3.5.2.

Most epidemiological studies on fluoride exposure are community based, in areas where drinking water contains naturally high levels of fluorides (ATSDR, 2003a). Fluoride is rapidly absorbed in the gastrointestinal tract by diffusion. Within 30–60 minutes following ingestion of sodium fluoride, elevated levels of fluoride are seen in the plasma (ATSDR, 2003a). Inhalation of fluoride as HF and F gas can occur in occupational and laboratory settings.

In areas where water fluoride levels are particularly low, it is fluoridated at 1 mg/l with fluorosilicic acid and sodium fluorosilicate to prevent dental cavities. The daily intake of fluoride from food and water is approximately 1 mg (ATSDR, 2003a)

Exposure assessment for fluoride has not been examined by the Expert group on Vitamins and Minerals (EVM), because fluoride fortification is a public health measure (EVM, 2003).

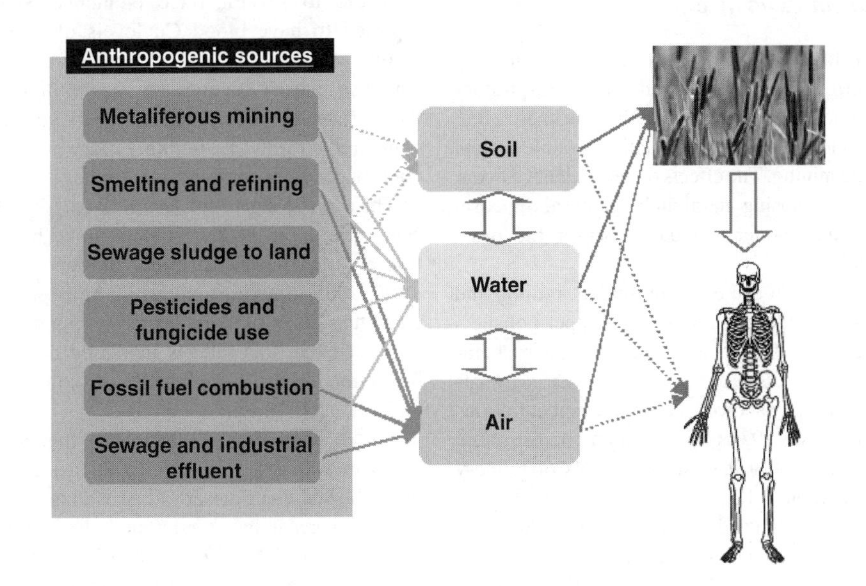

Figure 3.8 Source–pathway–receptor model of anthropogenic sources of essential elements

Table 3.11 Total oral exposure to Cr, Cu, Zn and Se (EVM, 2003)

Exposure (mg/day)	Cr	Cu	Zn	Se
Food:				
Mean	0.10	1.4	9.8	0.039
97.5 percentile	0.17	3.0	17.0	0.1
Water	0.002	6.0	10.0	–
Supplements	0.6	2.0	50.0	0.3
Total intake	0.77	11	77.0	0.4

No equivalent data avaliable for fluoride.

Levels of essential elements in drinking water depend on pH, hardness and, for Cr, Cu and Zn, the water-distribution system, especially in areas where water is soft and acid. Copper levels in drinking water usually increase with lengthy passage through Cu piping, or if the water is stagnant (EVM, 2003; WHO, 2004). Studies of Cu in drinking water in Europe, Canada and the USA have shown levels in the range $\leq$0.005 to >30 mg/l, with Cu piping the main source (WHO, 2004). Copper piping can contribute up to 1 mg of daily Cu intake (Barceloux, 1999c). Zinc can also be leached from fittings. The median level of Zn in surface water in Finland has been found to be <0.02 mg/l, while in tap water levels of up to 1.1 mg/l have been reported (WHO, 2003).

Most essential elements are available as dietary supplements. It is estimated that multivitamins and minerals taken at the recommended dosage result in daily intakes of up to 2 mg Cu per day (EVM, 2003). The average oral exposure of Cr, Cu, Zn and Se from food and drink is shown in Table 3.11.

3.5.1.2 Inhalation exposure

Inhalation exposure to Cr, Cu and Zn compounds is usually occupational. The effects of Cr compounds on the respiratory tract have been studied in workers from the chrome-plating industry, chromate and dichromate production, stainless-steel welding and chromite mining. The effects of hexavalent Cr occur at the site of contact, causing nasal itching, nasal mucosal atrophy, bronchitis, pneumonia and ulceration of the nasal septum (ATSDR, 2008).

Inhalation exposure to Cu can occur in mining, milling and smelting, as well as in welders and brass and copper polishers and carpenters using Cu-treated wood. Farmers using agrochemicals containing Cu can also be exposed to Cu aerosols (WHO, 1998). A recent study showed that serum concentrations of Cu are higher in welders (97 µg/dl) than in a control group (82 µg/dl) (Wang *et al.*, 2008). Urinary Cu levels of refinery workers are shown in Table 3.12.

Table 3.12 Levels of Cu in the urine of pyrometallurgical workers at the Monchegorsk copper refinery (modified from Nieboer *et al.*, 2007) (Reproduced by permission of The Royal Society of Chemistry)

Job category	Cu concentration (µg/l)
Anode-casting furnace: crane operator	15.0
Anode-casting furnace: pouring worker	12.8
Anode-casting furnace: smelter	14.3
Converter-furnace: crane operator	29.3
Converter worker	21.0
Crane operator assistant at the floor; anode casting and reverberatory furnace, smelter	18.8
Cu concentrate loader (feeder)	47.3
Filter worker	35.3
Reverberatory furnace, smelter	12.0
Others (gas scrubbing; laboratory chemistry worker; tractor driver; anode casting and reverberatory furnaces; quality control)	12.0
All female workers	30.8
All male workers	18.8
All	21.0

Around 0.7 µg of Zn is inhaled per day (WHO, 2001), depending on the environment. Ambient air concentrations of Zn in remote areas are in the range <3 to 27 ng/m^3 (Barceloux, 1999a) while urban areas usually contain 100–500 ng/m^3 of Zn (WHO, 2003). A range of 0.190–0.287 mg/m^3 of Zn in air was reported from recycling scrap iron and steel, while galvanising of tubes released 0.076 mg/m^3 Zn into the working environment (WHO, 2001).

Tobacco smoking is an important source of inhalation exposure to several trace elements. Smokers have been reported to have blood Cu levels of around 1.31 mg/l Cu, compared with 1.10 mg/l Cu in non-smokers There is also evidence that Cr accumulates in the lungs of smokers, with tissue samples containing around 4.3 mg/kg Cr dry weight compared to only 1.3 mg/kg Cr dry weight in non-smokers (Bernhard *et al.*, 2005).

The average cigarette contains 24 mg/kg Zn, of which 70 per cent is present in the mainstream smoke, but there is no correlation between serum Zn levels and smoking status (Bernhard *et al.*, 2005). Interestingly, while Se is present in tobacco and cigarette smoke, smoking decreases blood Se levels. It has been suggested that this reflects increased demand for antioxidant protection involving glutathione-peroxidase, which contains Se (ATSDR, 2003b).

Little information is available on inhalation exposure to Se compounds. Exposure to Se dioxide in a rectifier manufacturing plant showed that Se is absorbed via inhalation (Glover, 1970).

In humans it has been found that plasma fluoride levels increase following exposure to HF gas for 60 minutes, with a peak in concentrations 60–90 minutes after initial exposure (Lund *et al.*, 1997). Workers involved in the production of phosphate rock and triple superphosphate were exposed to

2–4 mg/m^3 of airborne fluoride, comprising around 60 per cent dust and 40 per cent HF gas. Some 2–3 hours after initial exposure, urinary fluoride levels increased from 0.5 to 4.0 mg/l, the highest levels (7–8 mg/l) being 10 hours later, after exposure had ceased (Rye, 1961).

3.5.1.3 Dermal exposure

Dermal absorption of hexavalent and trivalent Cr compounds is limited, but exposure to hexavalent Cr can cause burns (ATSDR, 2008). Potassium dichromate (VI) and Cr chloride (III) can be absorbed through the skin, but there is no evidence that Cr sulphate can be (Barceloux, 1999d). Dermal exposure to 4–25 mg/l Cr can cause allergic dermatitis, although repeated exposure can lead to decreased sensitivity (ATSDR, 2008).

Dermal exposure to some metals can cause contact dermatitis, with itching and inflammation of the skin. A patch test using 0.5–5.0 per cent Cu sulphate in water or petroleum jelly for 24–48 hours on 1190 eczema patients found that 13 per cent reacted to 2 per cent Cu in petroleum jelly (WHO, 1998). Metallic Cu and Cu (II) compounds are not generally considered to be a skin irritant and are used in dental amalgams and Cu bracelets for therapeutic use (Hostynek and Maibach, 2004).

Zinc has been used since ancient Egyptian times to enhance wound healing, although the usefulness of this approach is only partially confirmed by recent clinical data. The effects of Zn compounds on the skin are strongly influenced by its speciation. Zinc oxide is applied topically to promote wound healing (WHO, 2001), whereas Zn chloride and Zn sulphate are caustic (ATSDR, 2005). Dermal absorption of Zn can occur; burns patients treated with gauze impregnated with Zn oxide were found to have serum Zn levels of 28.3 µmol/l 3–18 days of treatment (WHO, 2001).

There is little information on dermal absorption of Se in humans. Selenium sulphide, which is present in some antidandruff shampoos, does not appear to be absorbed (ATSDR, 2003b).

Hydrogen fluoride is a strong irritatant to skin, causing itching, burning and rashes; in one study, 44 per cent of those most exposed to the gas were found to have severe skin problems, compared with 5 per cent of non-exposed residents (Dayal *et al.*, 1992).

3.5.1.4 Injection

Few studies of the effects of Cr, Cu, Zn, Se and F injection in humans are available, except for studies of intravenous administration of Cr trichloride and a few Se compounds.

In one study it was found that over 50 per cent of administered ^{51}Cr trichloride was distributed to various organs within hours following injection. The spleen and liver were found to contain the highest levels of trivalent Cr (Lim *et al.*, 1983).

In another study, the injection of Na selenite, sodium selenate, or selenomethionine was found to increase the accumulation of Se in the liver and kidneys. Injection with radiolabelled Se showed that it accumulated to high levels in the pancreas within hours of administration, before disappearing. Following injection with Na selenite or Na selenate,, it was found that within the first 2 hours 50 per cent of the Se was bound to proteins; at 4–6 hours, this increased to 85 per cent and after 24 hours, 95 per cent (ATSDR, 2003b).

3.5.2 Bioaccessibility

The proportion of trace elements absorbed by the body is influenced by speciation, interactions with other elements, the dietary status of the individual and the type of food. In studies of urban soils, Cr, Cu and Zn of anthropogenic origin were found to be more soluble and therefore more bioavailable (Ljung *et al.*, 2007).

The bioaccessibility of metals in soil can decrease exponentially over time. In one study, the amount of bioavailable Cr in soil from the A horizon was less than 0.2 mg/kg within 30 days of incubation, while in the B horizon this was 0.3 mg/kg after 100 days. The decrease in bioaccessibility was found to be more rapid in the A than the B horizon and it was concluded by the authors that the finer texture of the lower horizon decreased chemical reaction rates through diffusion limitations (Fendorf *et al.*, 2004).

Geophagy, which is the ingestion of soil, has been used as a mineral supplementation. An *in vitro* soil-ingestion simulation test revealed that soil ingestion can reduce the absorption of nutrients which are already bioavailable, particularly in Cu and Zn however (Hooda *et al.*, 2004).

Selenates are more bioavailable than selenides or elemental Se. Selenium in yeast, wheat or other plant sources is generally more bioavailable than that from animal sources; seafoods generally have the lowest bioavailable Se (Barceloux, 1999b). Radish extracts grown in soil enriched in inorganic Se (IV) or Se (VI) in a simulated-digestion experiment contained around 70 per cent Se that was soluble in gastric extracts, increasing to 90 and 100 per cent in plants that were exposed to selenate and selenite respectively (Pedrero *et al.*, 2006).

There is a marked variation in Se intake and status (as measured in blood, plasma or serum, toenails or hairs) from one part of the world to another and even between different parts of the same country (Rayman, 2000, 2005). Table 3.15 shows serum or plasma Se levels from a compilation of European studies and puts these data in the context of the optimal levels required for glutathione peroxidise and selenoprotein P activity. These data for Europe, especially compared with US serum Se levels of 125–137 µg/l (as measured in recent US National Health and Nutrition Examination Surveys (Niskar *et al.*, 1991; Bleys *et al.*, 2008), demonstrate that European populations are generally of low Se status (Fairweather-Tait *et al.*, 2010).

Fluoride is more bioavailable as sodium fluoride than in calcium fluoride, the latter being less soluble. Fluoride has been found to accumulate in the skeletal tissue of animals that consume foliage containing fluoride; so it is unlikely to bioaccumulate up the food chain (ATSDR, 2003a).

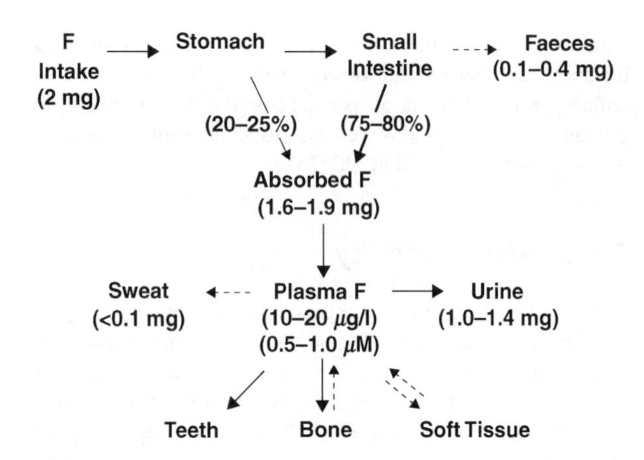

Figure 3.9 Fate of fluoride following ingestion (Cerklewski, 1997) (Reproduced by permission of Elselvier)

Numerous clinical and experimental studies show a variety of influences of fluoride on human health, depending on the content in drinking water (Gnatyuk, 1988; Grigoryeva *et al.*, 1993; Rozier, 1999). Indeed, approximately 90 per cent of fluoride ingested in water is absorbed in the gastrointestinal tract, compared with only 30–60 per cent of fluoride in food (WHO, 1996; Fordyce *et al.*, 2007).

It has been reported that low pH such as that in the stomach favours fluoride absorption (Cerklewski, 1997). Moreover 50–80 per cent of the F is absorbed when consumed with a meal, compared with 100 per cent in a fasted state. The fate of fluoride in the body is shown in Figure 3.9.

Bioavailability of an essential element can be dose dependent. For example, ingestion of 10 μg/day of dietary Cr results in 2 per cent absorption, while at 40 μg/day only 0.5 per cent is absorbed (Anderson and Kozlovsky, 1985).

Amiard *et al.* (2008) showed that cooking some types of shellfish influenced the bioaccessibility of Cu and Zn. The bioaccessibility of Zn in the digestive glands of frozen mussels decreased following cooking while in frozen oysters, the converse was true. Analysis of edible portions of other shellfish showed that the Cu and Zn in Pacific oysters (*Crassostrea gigas*) was bioavailable (97.4 and 82 per cent respectively), while Asian green mussels (*Perna viridis*) had the lowest amount of bioavailable Zn (34 per cent) and Noble scallops had the lowest amount of bioavailable Cu (38 per cent). Such differences in bioavailability have also been reported in plants: more Zn in pulses was bioavailable (27–56 per cent) than in the cereals tested (5.5–21.4 per cent) (Table 3.13). There was a negative correlation between phytate content and Zn dialyseability (Hemalatha *et al.*, 2007).

Copper bioavailability ranges from 65 to 70 per cent, depending on its chemical form and its interaction with other metals and other dietary components (Barceloux, 1999c). The presence of sugars, animal proteins and amino acids can all inhibit Cu absorption (Vitali *et al.*, 2008). Elevated levels of Mo have been found to reduce Cu uptake and induce hypocupriosis (Ward, 1978).

Table 3.13 Bioavailability of Zn in various grains (Hemalatha *et al.*, 2007) (Reproduced by permission of Elselvier)

Food grain	Zn content (mg/100g)	Zn bioaccessibility (per cent)
Cereals		
Rice	1.08	21.4
Wheat	1.62	8.93
Finger millet	1.73	8.31
Sorghum	2.24	5.51
Maize	1.48	7.82
Pulses		
Chick pea (whole)	2.03	44.9
Chick pea (decorticated)	2.68	56.5
Green gram (whole)	2.40	27.0
Green gram (decorticated)	2.19	40.8
Red gram	2.35	45.7
Black gram	2.30	33.4
Cow pea	2.57	53.0
French bean	2.18	52.5

3.5.3 Human health effects

Deficiencies of essential elements are associated with diseases mainly because of their role in enzymes that have important functions in man and other animals (Table 3.14).

The toxicity of different chemical species for the five elements considered in this chapter are indicated in Table 3.2 which lists their LD_{50} values in rodents.

Excessive amounts of these elements can cause adverse health effects to the neurological, reproductive, developmental, respiratory, cardiovascular, hepatic, renal, gastrointestinal, haematological, musculoskeletal and dermal systems of the human body. These are discussed below.

3.5.3.1 Chromium

Chromodulin, the biologically active form of Cr is an oligopeptide complex with four chromic ions (ATSDR, 2008). It is extremely rare for people with normal diets to suffer from Cr deficiency, but signs of deficiency have been observed in patients fed intravenously for long periods of time on infusates low in Cr, resulting in hyperglycaemia, weight loss, ataxia and peripheral neuropathy (Offenbacher and Pi-Sunyer, 1988). The role of Cr in humans has not been studied as extensively as that of Cu or Zn. It has been suggested that it is involved in glucose tolerance (Mertz, 1981), but no active substance has been identified (Lindh, 2005).

Metal-fume fever (MFF) can occur as a result of exposure to Cr, Cu, Se and Zn fumes from welding, electroplating and metal cutting. It has been reported that a 25-year-old male welder not wearing a respirator suffered from MFF while working in an area where galvanised steel was being cut in a poorly ventilated

Table 3.14 Enzymes and their functions

Enzyme	Function
Copper	
2-furoyl-CoA dehydrogenase	Catalyses the reaction 2-furoyl-CoA + H_2O + acceptor $\rightleftharpoons$ S-(5-hydroxy-2-furoyl)-CoA + reduced acceptor
Cu-containing amine-oxidase	Urea cycle and metabolism of amino groups, histidine metabolism, tyrosine metabolism, tryptophan metabolism
Galactose oxidase	Galactose metabolism
Hexose oxidase	Pentose phosphate pathway
Indole 2,3-dioxygenase	Tryptophan metabolism
L-ascorbate oxidase	Ascorbate metabolism
Zinc	
Alcohol dehydrogenase	Alcohol metabolism
Carbon-monoxide dehydrogenase	Methane metabolism
Galactitol-1-phosphate 5-dehydrogenase	Galactose metabolism
Methylmalonyl-CoA carboxytransferase	Propanoate metabolism
Phosphonoacetate hydrolase	Aminophosphonate metabolism
Stizolbate synthase	Tyrosine metabolism
Selenium	
Glutathione peroxidases	Catalyses reduction of peroxidases
Phospholipid hydroperoxide glutathione peroxidase	Reduces fatty acid hydroperoxides esterified to phospholipids
Thioredoxin reductase	Reduces protein disulphides
Thyroid hormone deiodinases	Formation and regulation of thyroid hormone triiodothyronine
Selenophosphate synthase	Incorporates selenocysteine into selenoproteins

No human enzyme that uses Cr or F as co-factors has been identified.

work space. He experienced headaches, breathing difficulty, coughing, a stiff neck and a metallic taste in the mouth (Hassaballa *et al.*, 2005).

Ingestion of 29 mg/kg hexavalent Cr as K dichromate by a 17-year-old male led to death by cardiac arrest. An autopsy revealed haemorrhages in the anterior papillary muscle of the left ventricle (Clochesy, 1984). There are no known cardiovascular effects associated with trivalent Cr.

Ingestion of 4.1 mg/kg hexavalent Cr as chromic acid has been reported to cause gastrointestinal haemorrhage and extensive necrosis of all digestive mucous membranes (Loubieres *et al.*, 1999). Anaemia has also been reported in a chrome-plating worker who accidentally ingested 300 g/l Cr trioxide (Fristedt *et al.*, 1965).

Hexavalent Cr can accumulate in erythrocytes, where it is reduced to trivalent Cr and binds to haemoglobin and other ligands. This stable Cr-haemoglobin complex remains in the cell over its lifespan, releasing around 1 per cent of its Cr daily (ATSDR, 2008).

Dermal exposure to some Cr substances can result in blisters and skin ulcers, which are also called chrome holes or chrome sores (ASTDR, 2008).

3.5.3.2 Copper

Copper deficiency is rare in humans, but a diet that does not provide adequate amounts of Cu can result in anaemia and neurodegeneration, with symptoms of ataxia and neuropathy. Low levels of Cu can also be caused by excessive Zn intake (Kumar, 2006).

There have been several reports from the Indian subcontinent of Cu toxicity causing childhood cirrhosis due to a combination of genetic factors and a high Cu diet (Barceloux, 1999c).

Acute exposure to 3 g of Cu sulphate ingested by an 18-month-old child caused haemolytic anaemia and renal damage. In another case, an adult who had consumed around 175 g of the same chemical had massive haemolysis and renal dysfunction (Barceloux, 1999c).

Workers sieving Cu for electrolytic processes were exposed to 434 mg/m^3 of Cu in their first year and 111 mg/m^3 in their third year. Radiographs of their lungs showed fibrosis and some had signs of nodulation (ATSDR, 2004).

There is some evidence of an increased risk of heart disease with either low or increased serum Cu levels. A sample of 4574 individuals showed that 151 who died from coronary heart disease had on average 5 per cent higher serum Cu levels than those who died from other causes. The extent to which Cu affects atherogenesis remains to be established (Ford, 2000).

Chronic exposure to Cu seriously affects those with Wilson's disease. The disease reflects an aberration of gene ATP7D on chromosome 13, which leads to defective Cu transport with Cu accumulating in the liver because sufferers are unable to excrete excess Cu (Kim *et al.*, 2008; ATSDR, 2004; Barceloux, 1999c). The disease incidence varies between 1 in 30 000 and 1 in 100 000 individuals.

Gastrointestinal problems can result from acute ingestion of copper. Concentrations (1, 3 and 5 mg/l) of Cu as Cu sulphate were used to study the acute gastrointestinal effects of Cu in drinking water. Twenty-one of the individuals exposed suffered gastrointestinal problems, including diarrhoea, abdominal pain, vomiting and nausea at levels $\geq$3 mg/l Cu (Pizarro *et al.*, 1999).

Workers exposed to airborne Cu in the range 0.64–1.05 mg/m^3 were reported to have lower haemoglobin and erythrocyte levels, but hair analysis showed that they were also exposed to Fe, Pb and Cd (ATSDR, 2004).

3.5.3.3 Zinc

Zinc deficiency is one of the major global risk factors when burdens of disease are considered and is particularly prevalent in the lowest income countries, where it has severe health outcomes (WHO, 2004, 2009).

Many problems have been reported in occupational situations. In the late 1800s, foundry workers were reported to be suffering from fever, headaches, nausea and vomiting as a result of exposure to Zn-oxide fumes (Barceloux, 1999a). Ingestion of household products containing Zn oxide can cause poisoning (ATSDR, 2005).

Hypocupraemia can reflect Zn-induced or Mo-induced Cu deficiency. A 28-month-old infant with atopic dermatitis treated with 45 mg/day elemental Zn as gluconate for 11 months developed thin, dry hypopigmented skin and brittle, sparse hair. His serum Cu level was <1.7 μmol/l, compared with the normal range (10.7–20.1 μmol/l). Once the Zn supplement was stopped, hair and skin quality improved and serum Cu levels increased to 15.2 μmol/l (Sugiura *et al.*, 2005).

Occupational asthma has been reported in welders using solder fluxes containing Zn chloride or hexavalent Cr. It is, difficult, however to determine the extent to which the effects are due to one particular metal when other potentially toxic chemicals are also present (WHO, 2001; ATSDR, 2008).

Suicide as a result of ingestion of Zn phosphide (Zn_3P_2), a widely used rodenticide has been reported on rare occasions. Ingestion of 7 g of 10 per cent Zn phosphide powder has been fund to result in tachycardia, sweaty skin and abdominal pain (Frangides and Pneumatikos, 2001).

Serum lipid profiles can reflect chronic exposure to Zn sulphate or Zn gluconate. Consumption of 2.3–4.3 mg/kg of Zn per day for 5–6 weeks or 0.71 mg/kg of Zn per day for 12 weeks can also decrease beneficial high-density lipoprotein (HDL) cholesterol levels. In people over 60 years old, Zn supplements decreased HDL and increased harmful low-density lipoprotein (LDL) cholesterol, particularly in those who exercised regularly (ATSDR, 2005).

Acute inhalation of gases from solder fluxes containing Zn chloride has led to asthma and other respiratory problems in welders (WHO, 2001). Ingestion of Zn chloride can also cause severe upper pharyngeal corrosion, emphysema and stomach damage (Chew *et al.*, 1986).

3.5.3.4 Selenium

In the body Se is used as part of a group of molecules known as selenoproteins, which contain the amino acid selenocysteine. These selenoproteins have a variety of functions, including: glutathione peroxidase, antioxidant defence; thioredoxin reductase, cell redox control; selenoprotein SEPS1, anti-inflammatory; selenoprotein P, Se transport in the plasma (Papp *et al.*, 2007). A summary of human health conditions related to Se deficiency is given in Table 3.15.

Selenium deficiency is associated with Kashin-Beck disease (KBD) and Keshan disease (KD). KBD is endemic in China, Russia and Korea; symptoms include cartilage atropy, degeneration and necrosis, as well as shortened stature from focal necrosis in the growth plate of tubular bones (Peng *et al.*, 1992). Recent studies have demonstrated a high prevalence of the Coxsackie B virus in patients with KBD and this is now thought to be a cofactor in the disease (Li *et al.*, 2000).

KBD is a permanent and disabling osteoarticular bone disease and shows a similar distribution to that of KD in China. Although the aetiology of KBD is now thought to be multifactorial, Se deficiency is considered to be an important underlying factor that predisposes the target cells (chondrocytes) to oxidative stress from free-radical carriers such as mycotoxins in stored grain. In Tibet, epidemiological studies carried out in 1995–1996 showed that KBD was also associated with I deficiency and with fungal contamination of grains (Chasseur *et al.*, 1997) and possibly organic matter in drinking water. A severe Se deficiency was documented, but it was suggested that this alone could not explain the occurrence of KBD in the villages under study (Moreno-Reyes *et al.*, 1998).

The serious cardiovascular symptoms of KD, which affects mainly young women and children in parts of China are also thought to reflect Se deficiency. Clinical symptoms include necrosis of the myocardium associated with cardiac enlargement, congestive heart failure, cardiogenic shock and death (Chen *et al.*, 1980).

The relationship between Se deficiency and viruses such as Coxsackie B has led to suggestions that Se deficiency may be an important factor in HIV developing into aids (Baum *et al.*, 1997; Cermelli *et al.*, 2002). Selenite inhibits replication of the Coxsackie B virus by interacting with thiol groups in the virus (Cermelli *et al.*, 2002). Foster (2003, 2004) has claimed that HIV can be prevented from turning into AIDS by supplementa-

Table 3.15 Summary of human health conditions related to Se deficiency (adapted from Johnson *et al.*, 2010) (Reproduced by permission of Cambridge University Press)

Health condition	Description	References
Overt Se deficiency diseases	Keshan disease (KD), an endemic cardiomyopathy (heart disease); occurs in Se-deficient areas and responds to Se supplementation. Coxsackie B virus also implicated	Fordyce, 2005; Li *et al.*, 2000
	Kashin-Beck disease, an endemic chronic disabling osteoarthropathy; multifactorial, with Se deficiency suggested to be a factor	Fordyce, 2005
Antiviral effects and HIV	The antioxidant selenoenzyme GPx1 inhibits the mutation of virus to more virulent forms	Beck *et al.*, 1998
	Low Se status increases the risk of primary liver cancer in hepatitis B/C positive patients. Risk reduced by Se supplementation	Yu *et al.*, 1999; Bjelakovic *et al.*, 2004
	HIV infection progresses more rapidly to AIDS, with higher mortality in patients with <85 μg/l Se in plasma	Foster, 2002
	In Tanzania, mortality in HIV-positive pregnant women decreased with increased plasma Se (>85 μg/l)	Kupka *et al.*, 2004
Cancer	Evidence that Se helps prevent cancer of the prostate, oesophagus and gastric cardia, bladder, liver, lung, thyroid and colorectal adenoma; from geographical, animal, prospective and intervention studies; Inverse relationship between crop and human Se status and cancer incidence in North America	Rayman, 2008; Rayman, 2005; Shamberger and Frost, 1969
Cognitive function	Lower toenail Se associated with lower cognitive score in rural elderly Chinese	Gao *et al.*, 2007
Coronary heart disease (CHD)	A meta-analysis shows a 50 per cent increase in Se concentrations is associated with 24 per cent reduction in CHD	Flores-Mateo *et al.*, 2006
Fertility and reproduction	Adequate intake improves male fertility, reduces the risk of miscarriage and possibly the risk of pre-eclampsia	Rayman, 2008; Rayman *et al.*, 2003
Immune function	Se supplementation of UK adults enhances cellular response; supplementation of elderly restores the age- related decline in immune response	Broome *et al.*, 2004; Peretz *et al.*, 1991
Mortality	Increasing Se associated with decreased mortality up to a serum concentration of 130 μg/l	Bleys *et al.*, 2008
	Low plasma Se concentration significantly associated with higher mortality	Akbaraly *et al.*, 2005
	Se supplementation tends to reduce mortality	Bjelakovic *et al.*, 2007
Thyroid effects	Enzymes that produce the active form of thyroid hormone are selenoenzymes, while the antioxidant selenoenzyme GPx is required for the protection of the thyroid from the H_2O_2 produced there	Arthur and Beckett, 1999
	Inverse association between Se status and thyroid volume, thyroid tissue damage and goitre	Derumeaux *et al.*, 2003

tion with Se and three essential amino acids which are needed to form glutathione peroxidase which is important for immune function. However, recent reviews have shown that further work is needed to clarify whether a relationship exists (Fairweather-Tait *et al.*, 2010). High rates of AIDS in Africa, where there are many crystalline rocks likely to be low in Se, in an environment which has been deeply leached, would be consistent with a link between low selenium levels and high rates of AIDS.

Long-term exposure to high levels of Se (1270 μg, 10–20 times higher than normal exposure) can cause seleniosis associated with numbness, paralysis and occasional hemiplegia (ATSDR, 2003b). Listlessness and lack of mental alertness were reported in a family who drank well water containing 9 mg/l Se for about three months, but their symptoms disappeared when consumption of the water ceased (ATSDR, 2003b). In humans, high incidence of gastrointestinal problems, poor dental health, diseased nails and skin discolouration are recognised as symptoms of Se poisoning (Smith and Westfall, 1937).

Selenium supplementation during pregnancy has been shown to reduce pregnancy-induced hypertension (Li and Shi-Mei, 1994). Contact with Se fumes and Se dioxide can cause skin rashes (Barceloux, 1999b).

Subjects fed on a high Se diet (0.0006 mg per kg of body weight each day for 21 days, followed by 0.004 mg per kg each day for 99 days) were found to have a 5 per cent decrease in white blood cells, while those fed on a low Se diet (0.0006 mg/kg each day for 21 days, followed by 0.0002 mg/kg each day for 99 days) had a 10 per cent increase due to the changes in number of granulocytes (Hawkes *et al.*, 2001).

Occupational exposure to H selenide can result in a metallic taste in the mouth, bronchopneumonia, pulmonary oedema and reduced expiratory flow rate (Barceloux, 1991b). Other adverse effects of chronic exposure to Se are brittle nails and hair (Barceloux, 1999b).

3.5.3.5 Fluoride

Fluoride deficiency can cause dental cavities and weakened tooth enamel as well as brittle bones due to bone demineralisation (Pizzo *et al.*, 2007).

During tooth formation in children up to 12 years old, fluoride is thought to accelerate mineralisation, forming fluorapatite which is less soluble than hydroxyapatite. Fluoride also acts as an antibacterial agent in the mouth, helping to minimise acid attack on teeth (Pterovich *et al.*, 1995; Pashayev *et al.*, 1990).

The detrimental effects of fluoride on dental health were discovered in the late 1800s and early 1900s, when the cause of brown staining of otherwise healthy teeth in people living in certain areas was investigated (Richmond, 1985). In 1931, F in water was identified as the cause of 'mottled enamel' (Churchill, 1932) and in the 1940s, fluoride levels in water supplies in the USA were reduced from 1.5–10 mg/l to around 1.0 mg/l (Richmond, 1985). Ingestion of excessive fluoride (above 10 mg/l) can cause more serious problems such as skeletal fluorosis, which can become crippling (Butterworth *et al.*, 2005). Excess fluoride in drinking water is thought to affect over 200 million people worldwide (Edmunds and Smedley, 2005). Groundwater can contain dangerously high amounts of fluoride. In a rural village in southern India, inhabitants have suffered from chronic skeletal fluorosis, discoloured teeth and aching joints because of drinking such water (Butterworth *et al.*, 2005). Fluorosis in parts of south-western China has been associated with inhaling fluoride from coal smoke (Zheng *et al.*, 1999). Children living in areas of China with naturally high levels of fluorides in drinking water have been reported to have low intelligence (Lu *et al.*, 2000).

Hydrogen fluoride gas can irritate the eyes, nose and skin and in large amounts cause death; F gas itself is even more dangerous, even at amounts below 0.4 mg/m^3, leading to pain in the eyes and nose (ATSDR, 2003a). Inhalation of Na fluoride can irritate and damage the respiratory tract (HPA, 2008). Currently, there is no conclusive evidence that fluoride compounds are carcinogenic to humans (COT, 2003), but inhalation of vinyl containing F causes cancer in rats and mice at concentrations $\geq$25 mg/m^3 (Cantoreggi and Keller, 1997).

In vivo studies have shown fluoride to be mutagenic, but this is often in response to fluoride levels that are much higher than those found in human blood. Exposure to 1.95–2.2 mg/l of fluoride in drinking water in those suffering from fluorosis has been shown to increase the frequency of sister chromatid exchange than in normal individuals exposed to 0.6–1 mg/l fluoride (Sheth *et al.*, 1994).

Exposure to excess fluoride can cause it to bind to serum calcium resulting in hypocalcaemia and subsequent tetany, decrease in myocardial contractility and possible cardiovascular collapse. It can also cause ventricular fibrillation and, in serious cases of fluoride poisoning, death (Bayless and Tinanoff, 1985). Abnormally high levels of white blood cells are also seen in those exposed to excessive levels of fluoride (8 mg/l) in drinking water (ATSDR, 2003a).

Workers involved in crushing cryolite (a mineral containing Na, Al and fluoride) experienced nausea followed by loss of appetite, vomiting and chronic indigestion (ATSDR, 2003a). Inhalation of HF irritates the airways and can be fatal (Lund *et al.*, 2005).

While trace amounts of fluoride are beneficial in preventing dental cavities and building bones, musculoskeletal effects are most pronounced in those exposed to high levels of fluorides. Dental fluorosis can occur whereby the porosity of subsurface enamel is increased. As teeth become more fluorosed, they become increasingly discoloured and are likely to fracture from normal mechanical stress (Fordyce *et al.*, 2007). Skeletal fluorosis may also occur as the accumulation of fluorides in the bones causes them to become brittle. In the most severe cases, kyphosis or lordosis can occur.

3.5.4 Organs of accumulation

Chromium, Cu, Zn, Se and F accumulate preferentially in different organs in the body. Under normal conditions, most Cu is present in the bile, liver, brain, heart and kidneys. Due to the mass of bone, most Cu is located there (Barceloux, 1999c).

In certain conditions, Cu accumulates mainly in the liver, particularly in sufferers of Wilson's disease. Copper can also accumulate in the brain and other organs (ATSDR, 2004).

Hexavalent Cr is mostly converted to trivalent Cr in the body, which is distributed to all organs (Barceloux, 1999d). The average dry weights of Cr in lung, aorta, pancreas, heart, kidney, liver and spleen are 15.6, 9.1, 6.5, 3.8, 2.1, 1.8 and 1.7 mg/kg respectively (WHO, 1988).

Typical liver Cu concentrations of adults are in the range 18–45 mg/g dry weight (Barceloux, 1999c). Normal Cu concentrations in kidneys are around 2.23 mg/kg dry weight (Cerulli *et al.*, 2006). Blood Cu levels in individuals with Wilson's disease are around 0.62 μg/ml while in autopsy cases of normal individuals this is 0.85 μg/ml (Lech and Sadlik, 2007).

Children with Indian childhood cirrhosis (ICC) have liver Cu concentrations of 790–6654 mg/kg dry weight; those without ICC have much lower concentrations of 8–118 mg/kg dry weight (Bhave *et al.*, 1982).

Copper concentrations in the brain, small intestine, large intestine, spleen, lung and heart in normal individuals are 3.40, 1.26, 1.24, 1.50, 2.23 and 3.30 mg/kg wet weight respectively (Lech and Sadlik, 2007).

Zinc is present in all organs in the body but it is concentrated in muscle tissue. Around 5 per cent of the body's burden of Zn is found in the liver (WHO, 2001). Individuals living near or working at mines, waste sites and metal-smelting plants may have higher levels of Zn. Tissue samples of former Cu-smelter workers were found to contain 58.9 and 31.5 mg/kg Zn wet weight in their liver and kidneys respectively, higher than the 47.2 and 23.3 mg/kg Zn in the same organs in controls. The controls, however, were found to have higher hair Zn concentrations (233 mg/kg) than the smelter workers (212 mg/kg) (Gerhardsson *et al.*, 2002).

The total body concentration of Se is between 3 and 20 mg. Around 40–50 per cent of the body's Se is in skeletal muscle as selenomethionine (Swanson *et al.*, 1991). High levels can be found in the liver (0.39–1.73 μg/g) and kidneys (0.63–0.89 μg/g) (ATSDR, 2003b).

Fluoride accumulates mainly in the bone (around 99 per cent of the body's burden) where an excess can result in brittle bones and reduced tensile strength (ATSDR, 2003a). The bones of the human body are constantly resorbed and remineralised during a lifetime and high fluoride intakes increase the accretion, resorption and Ca-turnover rates of bone tissue, affecting the homeostasis of bone-mineral metabolism (Krishnamachari, 1986). Calcification of soft tissues such as ligaments can also occur. Although approximately 80 per cent of fluoride entering the body is excreted, mainly in the urine, the remainder is adsorbed into body tissues from where it is released very slowly (WHO, 1996; Fordyce *et al.*, 2007). Thus, fluoride is a cumulative toxin and although skeletal fluorosis commonly affects older people following long exposure, crippling forms of the disease are also seen in children in endemic areas (WHO, 1996). Typical bone concentrations of fluoride are between 500 and 1500 mg/kg, levels generally increasing with age. Those with fluorosis can have over 10 000 mg/kg fluoride in their bones (Franke *et al.*, 1975). Unlike the other essential elements considered here, fluoride does not accumulate in soft tissues (ATSDR, 2003a).

3.6 Risk reduction

While trace amounts of essential elements are beneficial and prevent symptoms of deficiency, exposure to excessive levels of some of these elements can cause severe adverse health effects. Daily recommended amounts have been established (Table 3.16) and these should not be exceeded.

Table 3.16 Recommended dietary intake for some essential elements (modified from EVM, 2003; National Academy of Sciences, 2004)

Country	Demographic	Cr (III)	Cu	Zn	Se	F
		Recommended intake (mg/day)				
UK (Reference nutrient intakes)	Adults	0.025	1.2	5.5–9.5 (Males); 4.0–7.0 (Females);	0.075 (Males); 0.060 (Females); 0.075 (Lactating females)	–
	Children	0.0001–0.001	0.2–0.7	–	–	–
USA (Dietary reference intakes)	Adults	0.02–0.03	0.7–0.9	8.0–11.0	0.04–0.06	2.0–4.0
	Infants (0–0.5 years)	0.0002	0.2	2.0	0.015	0.01

Table 3.17 Drinking water guideline values (NHMRC, 2004; WHO, 2008; DWI, 2010; Health Canada, 2010)

Element	Guideline limits for drinking water (mg/l)					
	WHO	UK	Australia		Canada	
			MAC	Aesthetics[a]	MAC	Aesthetics[a]
Cr	0.05	0.05	0.05	–	0.05	–
Cu	2.0	2.0	2.0	1.0	–	1.0
Zn	3.0	0.5	–	3.0	–	5.0
Se	0.01	0.01	0.01	–	0.01	–
F	1.5	1.5	1.5	–	1.5	–

MAC = maximum accepted concentration.
[a]Aesthetic limits given for drinking water to appear potable.

Risk to humans and the environment can also be reduced by defining guideline limits for the various environmental compartments (soil, air, water). Drinking-water guidelines are particularly important and those for Cr, Cu, Zn, Se and F are listed in Table 3.17.

Soil guideline values are based on land use (Table 3.18). These are important, especially for small children. They are designed to prevent toxic levels entering the food chain, as some plants accumulate trace elements such as Zn. In areas

Table 3.18 Soil guideline values (DEFRA and EA, 2009, 2002; VROM, 2009; NEPC, 1999)

Country	Land use	Soil guideline values (mg/kg)				
		Cr	Cu	F	Se	Zn
UK	Residential with plant uptake	130	–	–	35	–
	Residential without plant uptake	200	–	–	260	–
	Allotments	130	–	–	35	–
	Commercial	5000	–	–	8000	–
Dutch intervention values		100	36	500	0.7	140
Dutch target values		380	190	–	100	720
Australia	Standard, residential with garden (less than 10 per cent vegetables consumed)	2 per cent Cr (III) 100 Cr(VI)	1000	–	–	7000
	Residential with minimal opportunities for soil access	48 per cent Cr (III) 400 Cr (VI)	4000	–	–	28000
	Recreational areas	24 per cent Cr (III) 200 Cr (VI)	2000	–	–	14000
	Commercial	60 per cent Cr (III) 500 Cr (VI)	5000	–	–	35000

Table 3.19 Occupational air standards (OSHA, 2010; ACGIH, 2001; NIOSH, 2010)

	Occupational air standards (mg/m^3)			
	OSHA PEL		ACGIH TLV	NIOSH REL
	General Industry	Construction		
Hexavalent Cr	0.005	0.005	0.05 (soluble) 0.01 (insoluble)	0.001
Trivalent Cr	0.5	0.5	0.5	0.5
Cu dusts and mists	1.0	1.0	1.0	1.0
Cu fumes	0.1	0.1	0.2	0.1
Zn chloride fumes	1.0	1.0	1.0	1.0
Zn oxide fumes and respirable dusts	5.0	5.0	5.0	5.0
Zn oxide total dust	15.0	15.0	10.0	5.0
Se and compounds	0.2	0.2	0.2	0.2
Se hexafluoride	0.4	0.16	0.16	0.4
H selenide	0.2	–	0.2	0.2
F	0.2	0.2	1.0	0.2
Fluoride	2.5	2.5	2.5	–
HF	2.0	2.0	2.6	2.5
Na fluoride	2.5	–	2.5	2.5

OSHA PEL = Occupational Safety and Health Administration Permissible Exposure Limits; ACGIH TLV = American Conference of Governmental Industrial Hygienists Threshold Limit Value; NIOSH REL = National Institute for Occupational Safety and Health Recommended Exposure Limit.

where the levels of essential elements are above the guideline values, the contaminated soil may be removed and disposed of in a licensed landfill and replaced with uncontaminated soil, although this tends to be expensive. Remediation techniques such as soil washing and phytoremediation using hyperaccumulating plants may be employed and trace elements may be immobilised by the application of organic materials like composted sewage sludge, bark chips, humus or peat (Kiikkila *et al.*, 2002). The use of clays, carbonates and phosphates can also limit migration of many trace elements (Kumpiene *et al.*, 2008).

Guideline limits for the trace elements in air tend to be for occupational exposure (Table 3.19). A guideline limit of 0.001 mg/m^3 has been set for both total Cr and F in ambient air by the WHO but no limits have been established for Cu, Zn or Se (WHO, 2000).

Occupational exposure to those substances listed in Table 3.19 can be reduced by improving ventilation in the workspace as well as by workers using respirators to prevent inhalation of toxic amounts of chemicals.

References

ACGIH (2001) Chemical substances 7th edition. American Conference of Governmental Industrial Hygienists. Avaliable online from http://www.acgih.org/store/BrowseProducts.cfm?type=catandid=16. Accessed 14 January 2011.

Aiuppa, A., Federico, C., Franco, A., Giudice, G., Gurrieri, S., Inguaggiato, S., Liuzzo, M., McGonigle, A.J.S. and Valenza, M. (2005) Emission of bromine and iodine from Mount Etna volcano. *Geochemistry, Geophysics, Geosystems* 8/6.

Akbaraly, N.T., Arnaud, J., Hininger-Favier, I., *et al.* (2005) Selenium and mortality in the elderly: results from the eva study. *Clinical Chemistry* **51**: 2117–2123.

Amiard, J.-C., Amiard-Triquet, C., Charbonnier, L., Mesnil, A., Rainbow, P.S. and Wang, W-X. (2008) Bioaccessibility of essential and non-essential metals in commercial shellfish from Western Europe and Asia. *Food and Chemical Toxicology* **46**: 2010–2022.

Anderson, R.A. and Kozlovsky, A.S. (1985) Chromium intake, absorption and excretion of subjects consuming self-selected diets. *American Journal of Clinical Nutrition* **41**: 1177–1183.

Andersson, M., de Benoist, B., Darnton-Hill, I. and Delange, F.Eds (2007) *Iodine in Europe: A Continuing Public Health Problem.* (Zurich: World Health Organisation; UNICEF.).

Ando, M., Tadano, M., Yamamoto, S., Tamura, K., Asanuma, S., Watanabe, T., Kondo, T., Sakurai, S., Ji, R., Liang, C., Chen, X., Hong, Z. and Cao, S. (2001). Fluoride pollution from coal burning in China. *Science of the Total Environment* **271**: 107–16.

Arthur, J.R. and Beckett, G.J. (1999) Thyroid function. *British Medical Bulleti* **55**: 658–668.

ATSDR (2003a) *Toxicological Profile for Fluoride.* Agency for Toxic Substances and Disease Registry, Atlanta, Georgia.

ATSDR (2003b) *Toxicological Profile for Selenium.* Agency for Toxic Substances and Disease Registry, Atlanta, Georgia.

ATSDR (2004) *Toxicological Profile for Copper.* Agency for Toxic Substances and Disease Registry, Atlanta, Georgia.

ATSDR (2005) *Toxicological Profile for Zinc.* Agency for Toxic Substances and Disease Registry, Atlanta, Georgia.

ATSDR (2008) *Toxicological Profile for Chromium*. Agency for Toxic Substances and Disease Registry, Atlanta, Georgia.

Balasoiu, C.F., Zagury, G.J. and Deschênes, L. (2001) Partitioning and speciation of chromium, copper and arsenic in CCA-contaminated soils: influence of soil composition. *The Science of the Total Environment* **280**: 239–255.

Barceloux, D.G. (1999a) Selenium. *Clinical Toxicology* **37**(2): 145–172.

Barceloux, D.G. (1999b) Copper. *Clinical Toxicology* **37**(2): 217–230.

Barceloux, D.G. (1999c) Zinc. *Clinical Toxicology* **37**(2): 279–292.

Barceloux, D.G. (1999d) Chromium. *Clinical Toxicology* **37**(2): 173–194.

Battin, E.E. and Brumaghim, J.L. (2009) Antioxidant activity of sulfur and selenium: A review of reactive oxygen species scavanging, glutathione peroxidase and metal-binding antioxidant mechanisms. *Cell Biochemistry and Biophysics* **55**: 1–23.

Baum, M.K., Shor-Posner, G., Lai, S., Zhang, G., Lai, H., Fletcher, M. A., Sauberlich, H. and Page, J.B. (1997) High risk of HIV-related mortality is associated with selenium deficiency. *Journal of Acquired Immune Deficiency Syndromes and Human Retrovirology* **15**: 370–374.

Bayless, J.M. and Tinanoff, N. (1985) Diagnosis and treatment of acute fluoride toxicity. *Journal of the American Dental Association* **110**: 209–211.

Beavington, F., Cawse, P.A. and Wakenshaw, A. (2004) Comparative studies of atmospheric trace elements: improvements in air quality near a copper smelter. *Science of the Total Environment* **332**, 39–49.

Beck, M.A., Esworthy, R.S., Ho Y.-S., *et al.* (1998) Glutathione peroxidase protects mice from viral-induced myocarditis. *FASEB J* **12**: 1143–1149.

Beckett, G.J., Macdougall, D.A., Nicol, F. and Arthur, R. (1989) Inhibition of type I and type II iodothyronine deiodinase activity in rat liver, kidney and brain produced by selenium deficiency. *Biochemical Journal* **259**: 887–892.

Bell, J.N.B. and Treshow, M. (2002). *Air Pollution and Plant Life*. 2nd edition. John Wiley and Sons, Chichester, pp.480.

Bernhard, D., Rossmann, A. and Wick, G. (2005) Metals in cigarette smoke. *IUBMB Life* **57**(12): 805–809.

BGS (2007) *Copper mineral profile*. British Geological Survey.

BGS (1998) *Minerals in Britain. Past production, future potential. Lead and zinc*. British Geological Survey.

BGS (2010) Geochemical baseline survey of the environment (G-base). Avaliable online from http://www.bgs.ac.uk/gbase/. Accessed 8 December 2010.

Bhave, S.A., Pandi, A.N., Pradhan, A.M., Sidhaye, D.G., Kantarjian, A., Williams, A., Talbot, I.C. and Tanner, M.S. (1982) Liver disease in India. *Archives of Disease in Childhood* **57**: 922–928.

Bhussry, B.R., Demole, V., Hodge, H.C., Jolly, S.S., Singh, A. and Taves, D.R. (1970) *Toxic Effects of Larger Doses of Fluoride and Human Health*. WHO, Geneva, pp. 225–272.

Bi, X., Feng, X., Yang, Y., Oiu, G., Li, G., Li, F., Liu, T., Fu, Z. and Jin, Z. (2006) Environmental contamination of heavy metals from zinc smelting areas in Hezhang County, western Guizhou, China. *Environmental International* **32**: 883–890.

Bjelakovic, G., Nikolova, D., Simonetti, R.G. *et al.* (2004) Antioxidant supplements for prevention of gastrointestinal cancers: a systematic review and meta-analysis. *Lancet* **364**: 1219–1228.

Bjelakovic, G., Nikolova, D, Gluud, L.L. *et al.* (2007) Mortality in randomised trials of antioxidant supplements for primary and secondary prevention: systematic review and metaanalysis. *JAMA* **297**: 842–857.

Bleys, J., Navas-Acien, A. and Guallar, E. (2008) Serum selenium levels and all-cause, cancer and cardiovascular mortality among US adults. *Archives of Internal Medicine* **168**: 404–410.

Brewer, G.J. (2010) Copper toxicity in the general population. *Clinical Neurophysiology* **121**: 459–460.

Briggs, S.A. (1992) *Basic Guide to Pesticides, Their Characteristics and Hazards*. Taylor and Francis. New York.

Broome, C.S., Mcardle, F., Kyle, J.A. *et al.* (2004). An increase in selenium intake improves immune function and poliovirus handling in adults with marginal selenium status. *American Journal of Clinical Nutrition* **80**: 154–162.

Butt, C.R.M. and Zeegers, H. (editors) (1992) *Regolith Exploration Geochemistry in Tropical and Subtropical Terrains. Handbook of Exploration Geochemistry, 4*. Elsevier, Amsterdam, 607 pp.

Butt, C.R.M. (1989) Genesis of supergene gold deposits in the lateritic regolith of the Yilgarn Block, Western Australia. In: Keays, R. R, Ramsay, W.R. H, Groves,and D.I. (Eds) The geology of gold deposits: the perspective in 1988. *Economic Geology, New Haven. Economic Geology Monograph* **6**: 460–470.

Butterworth, J., Reddy, Y.V.M., Renuka, B. and Reddy, G.V. (2005) Avoiding fluoride in drinking water, andhra Pradesh, India. *Waterlines* **24**(1): 24–26.

Cantoreggi, S. and Keller, D.A. (1997) Pharmacokinetics and metabolism of vinyl fluoride *in vivo* and *in vitro*. *Toxicology and Applied Pharmacology* **143**: 130–139.

Cerklewski, F.L. (1997) Fluoride bioavailability – nutritional and clinical aspects. *Nutritional Research* **17**: 907–929.

Cermelli, C., Vinceti, M., Scaltriti, E., Bazzani, E., Beretti, F., Vivoli, G., Portolani, M. (2002) Selenite inhibition of Coxsackie virus B5 replication: implications on the etiology of Keshan disease. *Journal of Trace Elements in Medicine and Biology* **16**: 41–46.

Cerulli, N., Campanella, L., Grossi, R., Politi, L., Scandurra, R., Soda, G., Soda, G., Gallo, F., Damiani, S., Alimonti, A., Alimonti, A., Petrucci, F. and Caroli, S. (2006) Determination of Cd, Cu, Pb and Zn in neoplastic kidneys and in renal tissue of fetuses, newborns and corpses. *Journal of Trace Elements in Medicine and Biology* **23**: 157–166.

Chasseur, C., Suetens, C., Nolard, N., Begaux, F. and Haubruge, E. (1997) Fungal contamination in barley and Kashin-Beck disease in Tibet. *Lancet* **350**: 1074.

Chen, X., Yang, G., Chen, J. *et al.* (1980) Studies on the relations of selenium and Keshan disease. *Biological Trace Elements Research* **2**: 91–107.

Chernousov, P.I., Golubev, O.V. and Yusfin, Y.S. (2003) Analysis of the movement of chromium in natural and anthropogenic media. *Metallurgist* **47**: 226–231.

Chew, L.S., Lim, H.S., Wong, C.Y., Htoo, M.M. and Ong, B.H. (1986) Gastric stricture following zinc chloride ingestion. *Singapore Medical Journal* **27**: 163–166.

Churchill, H.V. (1932) Occurrence of fluorides in some waters of the United States. *Industrial and Engineering Chemistry* **23**(9): 996–998.

Clochesy, J.M. (1984) Chromium ingestion: A case report. *Journal of Emergency Nursing* **10**: 281–282.

Combs, G.F. and Combs, S.B. (1986) *The Role of Selenium in Nutrition*. Academic Press Orlando, USA.

COT (2003) Committee on Toxicity of Chemicals in Food Consumer Products and the Environment. COT Statement on Fluorine in the 1997 Total Diet Study.

Daniels, J.L. (1975) Palaeogeographic development of Western Australian-Precambrian. In: *Geology of Western Australia.*

Geological Survey of Western Australia, Memoirs **Volume 2**: 437–445.

Davison, A., Howard, G., Stevens, M., Callan, P., Fewtrell, L., Deere, D. and Bartram, J. (2004) *Water Safety Plans: Managing Drinking-Water Quality from Catchment to Consumer*. IWA Publishing, London.

Dayal, H.H., Brodwick, M., Morris, R., Baranowski, T., Trieff, N., Harrison, J.A., Lisse, J.R. and Ansari, G.A. (1992) A community-based epidemiological study of health sequelae of exposure to hydrofluoric acid. *Annals of Epidemiology* **2**: 213–230.

Dean, H., Arnold, F. and Elvove, E. (1982) Domestic water and dental caries. Part 5 – Additional studies of the relation of fluoride in domestic waters and dental caroes. *Public Health Report* **57**: 1155–1179.

Degryse, F. and Smolders, E. (2006) Mobility of Cd and Zn in polluted and unpolluted Spodosols. *European Journal of Soil Science* **57**: 122–133.

Debackere, M. and Delbeke F.T (1978) Fluoride pollution caused by a brickworks in Flemish country side of Belgium. *International Journal of Environmental Studies* **11**: 245–252.

DEFRA and EA (2009) Soil guideline values for selenium in soil. Science Report SC050021/selenium SGV. Department for Environment, Food and Rural Affairs, Environment Agency.

DEFRA and EA (2002) Contaminants in soils: Collation of toxicological data and intake values for humans. Consolidated Main Report, CLR Report No 9. DEFRA, London.

de Oliveira-Filho, E.C., Lopes, R.M. and Paumgartten, F.J.R. (2004) Comparative study on the susceptibility of freshwater species to copper-based pesticides. *Chemosphere* **56**: 369–374.

Department of Health (1991) Dietary reference values for food energy and nutrients for the UK. Report on Health and Social Subjects no. 41. London: H.M. Stationery Office.

Derumeaux, H., Valeix, P., Castetbon, K. *et al.* (2003) Association of selenium with thyroid volume and echostructure in 35- to 60-year-old French adults. *European Journal of Endocrinology* **148**: 309–315.

Dissanayake, C.B. and Chandrajith, R. (1999) Medical geochemistry of tropical environments. *Earth Science Reviews* **47**: 219–258.

Domingo, J.L., Llobet, J.M., Patermain, J.L. and Corbella, J. (1988) Acute zinc intoxication: comparison of the antidotal efficacy of several chelating agents. *Veterinary and Human Toxicology* **30**(3): 224–228.

DWI (2010) What are the drinking water standards? Drinking Water Inspectorate. Avaliable online [http://www.dwi.gov.uk/consumers/advice-leaflets/standards.pdf] Accessed 14 January 2011.

Eichler, H.G., Lenz, K., Fuhrmann, M. and Hruby, K. (1982) Accidental ingestion of NaF tablets by children – reported by a poison center and one case. *International Journal of Clinical Pharmacology, Therapy and Toxicology* **20**: 334–338.

Edmunds, M., and Smedley, P. 2005. *Fluoride in natural waters*. In Selinus, O. (Ed.), *Essentials of Medical Geology. Impacts of the natural environment on public health*, pp. 301–329. Elsevier, Amsterdam.

Edmunds, W.M., Cook, J.M., Kinniburgh, D.G., Miles, D.L. and Trafford, J.M. (1989) *Trace-element occurrence in British groundwaters*. British Geological Survey Research Report SD/89/3.

EVM (2003) Expert Group on Vitamins and Minerals. Safe Upper Levels for Vitamins and Minerals. http://cot.food.gov.uk/pdfs/vitmin2003.pdf.

Fairweather-Tait, S., Bao, Y., Broadley, M., Collings, R., Ford, D., Hesketh, J. and Hurst, R. (2010). Selenium in Human Health and Disease. Antioxidants and Redox Signalling, Vol. doi:10.1089/ars.2010.3275.

Fasulo, M.P., Bassi, M. and Donini, A. (1983) Cytotoxic effects of hexavalent chromium in *Euglena gracilis*. II. *Physiological and ultra structural studies. Protoplasma* **114**: 35–43.

Fendorf, S., La Force, M.J. and Li, G. (2004) Heavy metals in the environment. temporal changes in soil partitioning and bioaccessibility of arsenic, chromium and lead. *Journal of Environmental Quality* **33**: 2049–2055.

Fergusson, J.E. (1990) *The Heavy Elements: Chemistry, Environmental Impact and Health Effects*. Pergamon Press, London.

Finkleman, R. B, Orem, W., Castranova, V., Tatu, C.A., Belkin, H. E, Zheng, B., Lerch, H.E. and Maharaj, S.V. (2002) Health impacts of coal and coal use: possible solutions. *International Journal of Coal Geology* **50**: 425–443.

Flemming, G.A. (1980) Essential micronutients II: iodine and selenium. In Davis, B.E. (editor) *Applied Soil Trace Elements*, pp. 199–234. John Wiley & Sons, New York.

Flores-Mateo, G., Navas-Acien, A., Pastor-Barriuso, R. *et al.* (2006) Selenium and coronary heart disease: a meta-analysis. *American Journal of Clinical Nutrition* **84**: 762–773.

Ford, E.S. (2000) Serum copper concentration and coronary heart disease among US adults. *American Journal of Epidemiology* **151**: 1182–1188.

Fordyce, F.M. (2005) Selenium deficiency and toxicity in the environment. In Selinus, O., Alloway, B., Centeno, J.A., Finkelman, R.B., Fuge, R., Lindh, U. and Smedley, P. (Editors) *Essentials of Medical Geology*, pp. 373–415. Elselvier, London.

Fordyce, F.M., Vrana, K., Zhovinsky, E., Povoroznuk, V., Toth, G., Hope, B.C., Iljinsky, U. and Baker, J. (2007) A health risk assessment for fluoride in Central Europe. *Environmental Gechemical Health* **29**: 83–102.

FOREGS (2005) Geochemical Atlas of Europe. EuroGeoSurveys. Available via http://www.gsf.fi/publ/foregsatlas/.

Foster, H.D. (2002) *What really causes AIDS*. Trafford Publishing, Victoria.

Foster, H.D. (2003) Why HIV-1 has diffused so much more rapidly in Sub-Saharan Africa than in North America. *Medical Hypotheses* **60**: 611–614.

Foster, H.D. (2004) How HIV-1 causes AIDS: implications for prevention and treatment. *Medical Hypotheses* **62**: 549–553.

Francis, P., Burton, M., and Oppenheimer, C. (1998) Remote measurements of volcanic gas compositions by solar occultation spectroscopy. *Nature* **396**: 567–570.

Frangides, C.Y. and Pneumatikos, I.A. (2001) Persistent severe hypoglycemia in acute zinc phosphide poisoning. *Intensive Care Medicine* **28**: 223.

Franke, I., Rath, F., Runge, H., Fengler, F., Auermann, E. and Lenart, G. (1975) Industrial Fluorosis. *Fluoride* **8**: 61–85.

Frencken, J.E., Truin, G.J., Van't Hof, M.A., Konig, K.G., Mabelya, L., Mulder, J. and Ruiken, H.M. (1980) Prevalence of dental caries in 7–13 year old children in Morogoro District, Tanzania, in 1984, 1986 and 1988. *Community Dental Oral Epidemiology* **18**: 2–8.

Fristedt, B., Lindqvist, B., Schuetz, A. and Ovrum, P. (1965) Survival in a case of acute oral chromic acid poisoning with acute renal failure treated by haemodialysis. *Acta Medica Scandinavica* **177**: 153–159.

Fuge, R. 2005. *Soils and iodine deficiency*. In Selinus O,(editor) *Essentials of Medical Geology. Impacts of the natural environment on public health* pp. 417–433. Elsevier, Amsterdam.

Fuge, R. and Andrews, M.J. (1988) Fluorine in the UK environment. *Environmental Geochemistry and Health* **10**: 96–104.

Gao, S., Jin, Y., Hall, K.S. *et al.* (2007) Selenium level and cognitive function in rural elderly Chinese. *American Journal of Epidemiology* **165**: 955–965.

Gerhardsson, L., Englyst, V., Lundström, N.G., Sandberg, S. and Nordberg, G. (2002) Cadmium, copper and zinc in tissues of deceased copper smelter workers. *Journal of Trace Elements in Medicine and Biology* **16**: 261–266.

Gibson-Moore, H. (2009) Water fluoridation for some – should it be for all? *Nutrition Bulletin* **24**: 291–295.

Giere, R., Blackford, M. and Smith, K. (2006) TEM study of $PM_{2.5}$ emitted from coal and tire combustion in a thermal power station. *Enivronmental Science and Technology* **40**: 6235–6240.

Glover, J.R. (1970) Selenium and its industrial toxicology. *Industrial Medicine and Surgery* **39**: 50–54.

Gnatyuk, P. (1988) Fluorosis and caries of temporal teeth. *Stomatology* **67**: 67–68.

Grigoryeva, L., Golovko, N., Nikolishiyn, A. and Pavlyenko, L. (1993) Fluoride influence on prevalence and intensity of stomatological disease in adolescents of Poltava Oblast. In Conference Proceedings – Fluoride Problems of Ecology, Biology, Medicine and Hygiene. Poltava, pp.25–26.

Grodzińska, K., Szarek-Łukaszewska, G. and Godzik, B. (1999) Survey of heavy metal deposition in Poland using mosses as indicators. *Science of the Total Environment* **229**: 41–51.

Han, F.X., Hargreaves, J.A., Kingery, W.L., Huggett, D.B. and Schlenk, D.K. (2001) Accumulation, distribution and toxicity of copper in sediments of catfish ponds receiving periodic copper sulfate applications. *Journal of Environmental Quality* **30**: 912–919.

Hassaballa, H.A., Lateef, O.B., Bell, J., Kim, E. and Casey, L. (2005) Metal fume fever presenting as aseptic meningitis with pericarditis, pleuritis and pneumonitis. *Occupational Medicine* **55**: 638–641.

Hawkes, W.C., Kelley, D.S. and Taylor, P.C. (2001) The effects of dietary selenium on the immune system in healthy men. *Biological Trace Elements Research* **81**: 189–213.

Haygarth, P.M. (1994) *Global Importance and Cycling of Selenium*. In: Frankenberger, W.T. and Benson, S. (editors) *Selenium in the Environment* pp. 1–28. Marcel-Dekker, New York.

He, Z.L., Yang, X.E. and Stoffella, P.J. (2005) Trace elements in agroecosystems and impacts on the environment. *Journal of Trace Elements in Medicine and Biology* **19**: 125–140.

Health Canada (2010) Guidelines for Canadian Drinking Water Quality. Summary Table. Federal-Provincial-Territorial Committee on Drinking Water.

Hem, J. (1992) *Study and Interpretation of the Chemical Characteristics of Natural Water*. US Geological Survey, Reston.

Hemalatha, S., Platel, K. and Srinivasan, K. (2007) Zinc and iron contents and their bioaccessibility in cereals and pulses consumed in India. *Food Chemistry* **102**: 1328–1336.

Hocking, M. B. (2005). *Handbook of Chemical Technology and Pollution*, 3rd edition. Academic Press, Burlington, MA USA.

Hooda, P.S., Henry, C.J.K., Seyoum, T.A., Armstrong, L.D.M. and Fowler, M.B. (2004) The potential impact of soil ingestion on human mineral nutrition. *Science of the Total Environment* **333**: 75–87.

Hostynek, J.J. and Maibach, H.I. (2004) Skin irritation potential of copper compounds. *Toxicology Mechanisms and Methods* **14**: 205–213.

Howard, P.H. and Hanchett, A. (1975) Chlorofluorocarbon Sources of Environmental Contamination. *Science* **189**, 217–219.

HPA (2007) *Health Protection Agency*. Chromium. Toxicological Overview.

HPA (2008) Health Protection Agency. Sodium fluoride. Toxicological Overview.

Hsieh H.S. and Ganther, H.E. (1977) Biosynthesis of dimethyl selenide from sodium selenite in rat liver and kidney cell-free systems. *Biochimica Biophysica Acta* **497**(1): 205–17.

Jacks, G., Rajagopalan, K., Alveteg, T. and Jonsson, M. (1993) Genesis of high-F groundwaters, southern India. *Applied Geochemistry S2*: 241–244.

Jacobs, L.W. (editor) (1989) *Selenium in agriculture and the environment*. Soil Science Society of America Special Publication no. 23. American Society of Agronomy and Soil Science Society of America, Madison, WI.

Jayasekher, T. (2009) Aerosols near by a coal fired thermal power plant: Chemical composition and toxic evaluation. *Chemosphere* **75**: 1525–1530.

Jensen, D.L., Holm, P.E. and Christensen, T.H. (2000) Soil and groundwater contamination with heavy metals at two scrap iron and metal recycling facilities. *Waste Management Research* **18**: 52–63.

Johnson, C.C., Fordyce, F.M. and Rayman, M.P. (2010). Factors controlling the distribution of selenium in the environment and their impact on health and nutrition. *Proceedings of the Nutrition Society* **69**: 119–132.

Joiner, A., Schafer, F., Hornby, K., Long, M., Evans, M., Beasley, T. and Abraham, P. (2009) Enhanced enamel benefits from a novel fluoride toothpaste. *International Dental Journal* **50**: 244–253.

Jung, M.C. and Thornton, I. (1996) Heavy metal contamination of soils and plants in the vicinity of a lead-zinc mine, *Korea. Applied Geochemistry* **11**: 53–59.

Kabata-Pendias, A. and Pendia, H. (editors) (2001). *Trace Element in Soils and Plants*, 3rd edition. CRC Press, Boca Raton, FL, USA.

Khalil, H.E., Hamiani, O.E., Bitton, G., Ouazzani, N. and Boularbah, A. (2008) Heavy metal contamination from mining sites in South Morocco: Monitoring metal content and toxicity of soil runoff and groundwater. *Environmental Monitoring and Assessment* **136**: 147–160.

Kiikkila, O., Pennanen, T., Perkiömäki, J., Derome, J. and Fritze, H. (2002) Organic material as a copper immobilising agent: a microcosm study on remediation. *Basic and Applied Ecology* **3**: 245–253.

Klevay, L.M. (2000) Cardiovascular disease from copper deficiency – a history. *Journal of Nutrition* **130**: 489s–492s.

Kim, B-E., Nevitt, T. and Thiele, D.J. (2008) Mechanisms for copper acquisition, distribution and regulation. *Nature Chemical Biology* **4**: 176–185.

Kramer, J. R. (1964) Theoretical model for the chemical composition of fresh water with application to the Great Lakes. In: Proceeding of the 7th Conference of Great Lakes Research, April 6–7, Toronto, Ontario, pp. 47–160.

Krishnamachari, K.A.V. R. (1986). Skeletal fluorosis in humans: A review of recent progress in the understanding of the disease. *Progression in Food and Nutrition Science* **10**: 279–314.

Kumar, N. (2006) Copper deficiency myelopathy (human swayback). *Mayo Clinic Proceedings* **81**: 1371–1384.

Kumpiene, J., Lagerkvist, A. and Maurice, C. (2008) Stabilization of As, Cr, Cu, Pb and Zn in soil using amendments – A review. *Waste Management* **28**: 215–225.

Kupka, R., Msamanga, G.I. and Spiegelman, D. (2004) Se status is associated with accelerated HIV disease progression among HIV-infected pregnant women in Tanzania. *Journal of Nutrition* **134**: 2556–2560.

Lavado, R.S. and Reinaudi, N. (1979) Fluoride in salt affected soils of La Pampa (Republica Argentina). *Fluoride* **12**: 28–32.

Lech, T. and Sadlik, J.K. (2007) Contribution to the data on copper concentration in blood and urine in patients with Wilson's disease and in normal subjects. *Biological Trace Element Research* **118**: 16–20.

Lemly, A.D. (2002) *Selenium Assessment in Aquatic Ecosystems: A Guide for Hazard Evaluation and Water Quality Criteria*. Springer, New York.

Levander, O. (1986) Selenium. In Mertz, W.;(editor) *Trace Elements in Human and Animal Nutrition*, pp 209–280. Academic Press, Orlando, FL.

Li, Y., Peng, T., Yang, Y. *et al.* (2000) High prevalence of enteroviral genomic sequences in myocardium from cases of endemic cardiomyopathy (Keshan Disease) in China. *Heart* **83**: 696–701.

Li, H. and Shi-Mei, Z. (1994) Selenium supplementation in the prevention of pregnancy induced hypertension. *Chinese Medical Journal* **107**: 870–871.

Li, Y., Liang, C.K., Katz, B.P., Niu, S., Cao, S. and Stookey, G.K. (1996) Effect of fluoride exposure and nutrition on skeletal fluorosis. *Journal of Dental Research* **75**: 2699.

Lindh, U. (2005). Biological functions of the elements. In Selinus, O, (editor) *Essentials of Medical Geology. Impacts of the natural environment on public health* pp. 115–160. Elsevier, Amsterdam.

Lim, T.H., Sargent, T. and Kusbov, N. (1983) Kinetics of trace element chromium (III) in the human body. *American Journal of Physiology* **224**(4): 445–454.

Liu, Y., Chiba, M., Inaba, Y. and Kondo, M. (2002) Keshan disease – a review from the aspect of history and etiology. *Nippon Eiseigaku Zasshi* **56**(4): 641–648.

Ljung, K., Oomen, A., Duits, M., Selinus, O. and Berglund, M. (2007) Bioaccessibility of metals in urban playground soils. *Journal of Environmental Science and Health* **42**: 1241–1250.

Logananathan, P., Hedley, M.J., Wallace, G.C. and Roberts, A.H.C. (2001) Fluoride accumulation in pasture forages and soils following long-term applications of phosphorus fertilisers. *Environmental Pollution* **115**: 275–282.

Loland, J.Ø. and Singh, B.R. (2004) Copper contamination of soil and vegetation in coffee orchards after long-term use of Cu fungicides. *Nutrient Cycling in Agroecosystms* **69**: 203–211.

Loubieres, Y. De Lassence, A., Bernier, M., Viellard-Baron, A., Schmitt, J.M., Page, B. and Jardin, F. (1999) Acute, fatal, oral chromic acid poisoning. *Journal of Toxicology-Clinical Toxicology* **37**: 333–336.

Lu, Y., Sun, Z.R., Wu, L.N., Wang, X., Lu, W. and Liu, S.S. (2000) Effect of high-fluoride water on intelligence in children. *Fluoride* **33**: 74–78.

Luo, K.L., Xu L.R., Li, R.B. and Xiang, L.H. (2002) Fluorine Emission from Combustion of Steam Coal of North China Plate and Northwest China. *Chinese Science Bulletin* **47**: 1346–1350.

Lund, K., Ekstrand, J., Boe, J., Søstrand, P. and Kongerud, J. (1997) Exposure to hydrogen fluoride: an experimental study in humans of concentrations of fluoride in plasma, symptoms and lung function. *Occupational and Environmental Medicine* **54**: 32–37.

Lund, K., Dunster, C., Ramis, I., Sandstrom, T., Kelly, F.J., Sostrand, P., Schwarze, P., Skovlund, E., Boe, J., Kongerud, J. and Refsnes, M. (2005) Inflammatory markers in bronchoalveolar lavage fluid from human volunteers 2 hours after hydrogen fluoride exposure. *Human and Experimental Toxicology* **24**: 101–108.

Martínez, J., Llamas, J.F., De Miguel, E., Rey, J. and Hidalgo, M.C. (2008) Soil contamination from urban and industrial activity: example of the mining district of Linares (southern Spain). *Environmental Geology* **54**: 669–677.

Mastin, B.J. and Rodgers, J.H. (2000) Toxicity and bioavailability of copper herbicides (Clearigate, Cutrine-plus and copper sulfate) to freshwater animals. *Archives of Environmental and Contaminant Toxicology* **39**: 445–451.

May, T.W., Walther, M.J., Petty, J.D., Fairchild, J.F., Lucero, J., Delvaux, M., Manring, J., Armbruster, M. and Hartman, D. (2001) An evaluation of selenium concentrations in water, sediment, invertebrates and fish from the Republican River Basin: 1997–1999. *Environmental Monitoring and Assessment* **72**: 179–206.

McDowell, L.R. (1992) *Minerals in Animal and Human Nutrition*. Academic Press, London.

Meister. A. (1965) *A Biochemistry of the Amino Acids*, **Vol. 1**, pp. 338–369. Academic Press, New York.

Meler, J. and Meler, G. (2006) Fluoridation of drinking water – advantages and disadvantages. *Journal of Elementology* **11**: 379–387.

Mertz, W. (1981) The essential trace elements. *Science* **213**: 1332–1338.

Messaïtfa, A. (2008) Fluoride contents in groundwaters and the main consumed foods (dates and tea) in Southern Algeria region. *Environmental Geology* **55**: 377–383.

Mielke, J.E. (1979) Composition of the Earth's crust and distribution of the elements. In Siegel, F. R. (editor), *Review of Research on Modern Problems in Geochemistry*. International Association for Geochemistry and Cosmochemistry. Earth Science Series No. 16.

Mikkelsen, R.L., Page, A.L. and Bingham, F.T. (1989) Factors affecting selenium accumulation by agricultural crops. In Jacob, L.W. (editor), *Selenium in Agriculture and the Environment*. Soil Science Society of America Special Publication no. 23, pp. 65–94. American Society of Agronomy and Soil Science Society of America, Madison, WI.

Mills, C.F. (1969) Trace element metabolism in animals and man. *British Medical Journal* **3**: 352–353.

Mirlean, N., Roisenberg, A. and Chies, J.O. (2007) Metal contamination of vineyard soils in wet subtropics (southern Brazil). *Environmental Pollution* **149**: 10–17.

Moreno-Reyes, R., Suetens. C., Mathieu, F., Begaux, F., Zhu, D., Rivera, M.T., Boelaert, M., Neve, J., Perlmutter, N., and Vanderpas, J. (1998) Kashin–Beck osteoarthropathy in rural Tibet in relation to selenium and iodine status. *New England Journal of Medicine* **339**: 1112–1120.

Morillo, J., Usero, J. and Gracia, I. (2004) Heavy metal distribution in marine sediments from the southwest coast of Spain. *Chemosphere* **55**: 431–442.

Murdoch, W.M. (1980) *The Poverty of Nations. The Political Economy of Hunger and Population*. John Hopkins University Press, Baltimore.

National Academy of Sciences (2004) *Dietary Reference Intakes (DRIs): Recommended Intakes for Individuals, Elements*. National Academies Press.

Nahon, D. and Tardy, Y. (1992) The ferruginous laterites. In: Butt, C.R. M, Zeegers,and H. (editors). *Regolith Exploration Geochemistry in*

Tropical and Subtropical Terrains. Handbook of Geochmistry, pp. 25–40. Elsiever, Amsterdam.

Neal, R.H. (1995) Selenium. In: Alloway, B.J. (editor), *Heavy Metals in Soils*, pp. 260–283. Blackie Academic and Professional, London.

NEPC (1999) Schedule B (1) Guideline on the investigation levels for soil and groundwater. National Environment Protection Council, Australia.

NEX, P. (2004) Formation of bifurcating chromite layers of the UG1 in the Bushveld igneous complex, an analogy with sand volcanoes. *Journal of the Geological Society* **161**: 903–909.

NHMRC (2004) Australian drinking water guidelines 6. National Health and Medical Research Council, Canberra.

Nieboer, E., Thomassen, Y., Romanova, N., Nikonov, A., Odland, J.Ø. and Chaschin, V. (2007) Multi-component assessment of worker exposures in a copper refinery. *Journal of Environmental Monitoring* **9**: 695–700.

NIEHS (2005) Report on Carcinogens, 11th edition. National Institute of Environmental Health Sciences, US Department of Health and Human Services.

Nielsen, F.H. (1984) Ultratrace elements in nutrition. *Annual Review of Nutrition* **4**: 21–41.

Nielsen, F.H. (1991) Nutritional requirements for boron, silicon, vanadium, nickel and arsenic: current knowledge and speculation. *FASEB Journal* **5**: 2661–2667.

NIOSH (2010) *Pocket Guide to Chemical Hazards*. National Institute for Occupational Safety and Health. Avaliable online: [http://www.cdc.gov/niosh/npg/pgintrod.html] Accessed 14 January 2011.

Niskar, A.S., Paschal, D.C., Kieszak, S.M. *et al.* (2003) Serum selenium levels in the US population: Third National Health and Nutrition Examination Survey, 1988–1994. *Biological Trace Elements Research* **91**: 1–10.

Nriagu, J.O. (1989) *Occurrence and Distribution of Selenium*. CRC Press, Boca Raton, FL.

Offenbacher, E.G. and Pi-Sunyer, F.X. (1988) Chromium in human nutrition. *Annual Review of Nutrition* **8**: 543–563.

Oldfield, J.E. (1999) *Selenium World Atlas*. Selenium–Tellurium Development Association, Grimbergen, Belgium.

OSHA (2010) Permissible exposure limits. Occupational Safety and Health Administration. Avaliable online: [http://www.osha.gov/SLTC/pel/index.html#standards] Accessed 14 January 2010.

Pais, I. and Jones, J.B. (1997) *The Handbook of Trace Elements*. St Lucie Press, Boca Raton, Florida.

Papp, L.V., Lu, J., Holmgren, A. *et al.* (2007) From selenium to selenoproteins: synthesis, identity and their role in human health. *Antioxidant Redox Signal* **7**: 775–806.

Pashayey, C., Akhmyedov, R. and Halifa-Zade, C. (1990) Fluoride and other biogeochemical factors influence on microstrength of enamel and dentin. *Stomatology* **69**: 10–12.

Pauling, L. (1970) *General Geochemistry*, 3rd edition. W.H. Freeman & Co., San Francisco.

Pechova, A. and Pavlata, L. (2007) Chromium as an essential nutrient: a review. *Veterinarni Medicina* **52**: 1–18.

Pedrero, Z., Madrid, Y. and Cmara, C. (2006) Selenium species bioaccessibility in enriched radish (*Raphanus sativus*): A potential dietary source of selenium. *Journal of Agricultural and Food Chemistry* **54**: 2412–2417.

Pedro, G. (1985) Grandes tendances des sols mondiaux. *Cultivar* **184**: 78–81.

Peng, A., Yang, C., Rui, H., and Li, H. (1992) Study on the pathogenic factors of Kashin-Beck disease. *Journal of Toxicology and Environmental Health* **35**: 79–90.

Peretz, A., Neve, J., Desmedt, J. *et al.* (1991) Lymphocyte response is enhanced by supplementation of elderly subjects with selenium-enriched yeast. *American Journal of Clinical Nutrition* **53**: 1323–1328.

Petrovich, Y., Podorozhnaya, R., Dmitriyeva, L., Knavo, O. and Vasyukova, O. (1995) Glutamate and organic phosphates metabolic ferments under fluorosis. *Stomatology* **74**: 26–28.

Pizarro, F., Olivares, M., Uauy, R., Contreras, P., Rebelo, A. and Gidi, V. (1999) Acute gastrointestinal effects of graded levels of copper in drinking water. *Environmental Health Perspectives* **107**: 117–121.

Pizzo, G., Piscopo, M.R., Pizzo, I. and Giuliana, G. (2007) Community fluoridation and caries prevention: a critical review. *Clinical Oral Investigations* **11**: 1436–3771.

Plant, J.A., Baldock, J.W. and Smith, B. (1996) The role of geochemistry in environmental and epidemiological studies in developing countries: a review. *Environmenal Geochemistry and Health* **113**: 7–22.

Plant, J.A., Smith, D., Smith, B. and Williams, L. (2000) Environmental geochemistry at the global scale. *Journal of the Geological Society, London* **157**: 837–849.

Pracheil, B.M., Snow, D.D. and Pegg, M.A. (2010) Distribution of selenium, mercury and methylmercury in surficial Missouri River sediments. *Bulletin of Environmental Contamination and Toxicology* **84**: 331–335.

Prasad, A.S. (1991) Discovery of human zinc deficiency and studies in an experimental human model. *American Journal of Clinical Nutrition* **53**: 403–412.

Quellet, M. (1987). Reduction of Airborne Fluoride Emission from Canadian Aluminium Smelters. *Science of the Total Environment* **66**: 65–72.

Qu, X., He, P.-J., Shao, L.-M. and Lee, D.-J. (2008) Heavy metals mobility in full-scale bioreactor landfill: Initial stage. *Chemosphere* **70**: 769–777.

Rayman, M.P., Bode, P. and Redman, C.W. (2003) Low selenium status is associated with the occurrence of the pregnancy disease preeclampsia in women from the United Kingdom. *American Journal of Obstetrics and Gynecology* **189**: 1343–1349.

Rayman, M.P. (2000) The importance of selenium to human health. *Lancet* **356**: 233–241.

Rayman, M.P. (2005) Se in cancer prevention: a review of the evidence and mechanism of action. *Proceedings of the Nutrition Society* **64**: 527–542.

Rayman, M.P. (2008) Food-chain selenium and health: emphasis on intake. *British Journal of Nutrition* **100**: 254–268.

Reimann, C. and de Caritat, P. 1998. *Chemical Elements in the Environment*. Springer, Berlin, ISBN 3-540-63670-6.

Richmond, V.L. (1985) Thirty years of fluoridation: a review. *American Journal of Clinical Nutrition* **41**: 129–138.

Rojas, J.C. and Vandecasteele, C. (2007) Influence of mining activities in the North of Potosi, Bolivia on the water quality of Chayanta River and its consequences. *Environmental Monitoring and Assessment* **132**: 321–330.

Rye, W.A. (1961) Fluorides and phosphates – clinical observations of employees in phosphate operation. International Congress on Occupational Health, July 25–29, 1960, pp. 361–364.

Rose, S. (1999) *The Chemistry of Life*, 4th edition. Penguin Press Science, London.

Rotruck, J.T., Pope, A.L., Ganther, H.E. *et al.* (1972) *Prevention of oxidative damage to rat erythrocytes by dietary selenium. Journal of the American* Chemical Society **79**: 3292–3293.

Rozier, R.G. (1999) The prevalence and severity of enamel fluorosis in North American children. *Journal of Public Health and Dentistry* **59**: 239–46.

SACN (2007) Paper for approval: Selenium and Health paper. Scientific Advisory Committee on Nutrition.

Salomons, W. (1986) Impact of atmospheric inputs on the hydrospheric trace metal cycle. In *Toxic Metals and the Atmosphere*, pp. 406–466. John Wiley & Sons, Chichester.

Sanstead, H.H. (1984) Zinc: essentiality for brain development and function. *Nutrition Today* **19**: 26–30.

Schoenberg, R., Nagler, F., Gnos, E., Kramers, J.D. and Kamber, B.S. (2003) The source of the Great Dyke, Zimbabwe and its tectonic significance; evidence from Re-Os isotopes. *Journal of Geology* **111**: 565–578.

Selinus, O. (editor) 2005. *Essentials of Medical Geology. Impacts of the natural environment on public health.* Elsevier, Amsterdam.

Shamberger, R.J. and Frost, D.V. (1969). Possible protective effect of Se against human cancer. *Canadian Medical Association Journal* **100**: 682.

Sheth, F.J., Multani, A.S. and Chinoy, N.J. (1994) Sister chromatid exchanges: A study in fluorotic individuals of North Gujarat. *Fluoride* **27**: 215–219.

Skinner, C. (2000) In praise of phosphates, or why vertebrates choose apatite to mineralise their skeletal elements. *International Geology Review* **42**: 232–240.

Smith, M.I. and Westfall, B.B. (1937) Further field studies on the selenium problem in relation to public health. *Public Health Report* **52**: 1375–1384.

Suttie, J.S., Dickie, R., Clay, A.B., Nielsen, P., Mahan, W.E., Baumann. D.P. and Hamilton, R. J. (1987). Effects of fluoride emissions from a modern primary aluminium smelter on local population of white tailed deer. Available from. http://www.jwildfredid.org/cgi/reprint/23/1/135.pdf Accessed 7 June, 2010.

Sugiura, T., Goto, K., Ito, K., Ueta, A., Fujimoto, S. and Togari, H. (2005) Chronic zinc toxicity in an infant who received zinc therapy for atopic dermatitis. *Acta Paediatrica* **94**: 1333–1335.

Swanson, C.A., Patterson, B.H., Levander, O.A., Veillon, C., Taylor, P., Helzsouer, K., McAdam, P.A. and Zech, L.A. (1991) Human (^{74}Se) selenomethionine metabolism: a kinetic model. *American Journal of Clinical Nutrition* **54**: 917–926.

Symonds, R.B., Rose, W.I. and Reed, M.H. (1988) Contribution of Cl- and F-bearing gases to the atmosphere by volcanoes. *Nature* **334**: 415–418.

Taylor, S.R. and McLennan, S.M. (1985) *The Continental Crust: Its Composition and Evolution.* Blackwell, Oxford.

Tebutt, T.H.Y. (1983) Relationship between natural water quality and health. UNESCO, Paris.

Trecases, J.J. (1992) Chemical weathering. In: Butt, C.R.M. and Zeegers, H. (editors), *Regolith Exploration Geochemistry in Tropical and Subtropical Terrains. Handbook of Geochemistry*, pp. 25–40. Elsevier, Amsterdam.

Timmer, L.W., Childers, C.C. and Nigg, H.N. (2004) Pesticides registered for use on Florida citrus. Gainesville, FL: 2004 Florida Citrus Pest Management Guide, SP-43. University of Florida.

Tripathi, R.M., Ashawa, S.C. and Khandekar, R.N. (1993) Atmospheric deposition of Pb, Cd, Cu and Zn in Bombay, India. *Atmospheric Environment* **27B**: 269–273.

Underwood, E.J. (1981) The incidence of trace element deficiency diseases. Philosophical Transactions of the Royal Society of London, Series B, *Biological Sciences* **294**: 3–8.

Underwood, E.J. and Suttle, N.F. (1999) *The Mineral Nutrition of Livestock*, 3rd edition. CABI Publishing, New York.

USEPA (2005) Toxicological review of zinc compounds (EPA/635/R-05/002). United States Environment Protection Agency, Washington, D.C.

USEPA (2003) Response to Requests to cancel certain chromated copper arsenate (CCA) wood preservative products and amendments to terminate certain uses of other CCA products, Vol. 68, No. 68. United States Environment Protection Agency, Washington, D.C.

USEPA (1978) Reviews of the environmental effects of pollutants. III. Chromium., 285 pp (ORNL/EIS-80; EPA 600/1–78–023). US Environmental Protection Agency, Washington, D.C.

USGS (2008) Copper statistics and information. United States Geological Survey. Available online: http://minerals.usgs.gov/minerals/pubs/commodity/copper/ Accessed 3 April 2008.

USGS (2006) Minerals Handbook: Zinc. Available online: http://minerals.usgs.gov/minerals/pubs/commodity/zinc/myb1–2006-zinc.pdf Accessed 4 August 2009.

Vanderpas, J.B., Contempre, B., Duale, N.L., Goossens, W., Bebe, N., Thorpe, R., Ntambue, K., Dumont, J., Thilly, C.H. and Diplock, A.T. (1990) Iodine and selenium deficiency associated with cretinism in northern Zaire. *American Journal of Clinical Nutrition* **52**: 1087–1093.

Venchikov, A.E. (1974) Zones of display of biological and pharmacotoxicological action of trace elements. In Hoekstra, W.G., Suttie, I.W., Ganther, H.E. and Mertz, W. (editors), *Trace Element Metabolism in Animals 2*, pp. 295–310. University Park Press, Baltimore.

Vitali, D., Vendrina Dragojević, I. and Šebečić, B. (2008) Bioaccessibility of Ca, Mg, Mn and Cu from whole grain tea-biscuits: Impact of proteins, phytic acid and polyphenols. *Food Chemistry* **110**: 62–68.

VROM (2009) Soil Remediation Circular 2009. Ministry of Housing, Spatial Planning and the Environment (Ministerie van Volkshuisvesting, Ruimtelijke Ordening en Milieubeheer), The Hague, Netherlands.

Wang, D., Du, X. and Zheng, W. (2008) Alteration of saliva and serum concentrations of manganese, copper, zinc, cadmium and lead among career welders. *Toxicology Letters* **176**: 40–47.

Ward, G.M. (1978) Molybdenum toxicity and hypocupriosis in ruminants: A review. *Journal of Animal Science* **46**: 1078–1085.

Wedepohl, K.H., (1978) *Handbook of Geochemistry*. Springer-Verlag, Berlin-Heidelberg.

WHO (2004) Copper in drinking-water: Background document for development of WHO Guidelines for drinking-water quality (WHO/SDE/WSH/03.04/88). World Health Organization. Geneva.

WHO (2003) Zinc in drinking-water: Background document for development of WHO Guidelines for drinking-water quality (WHO/SDE/WSH/03.04/17). World Health Organization. Geneva.

WHO (1996) Selenium in drinking-water: Background document for development of WHO Guidelines for drinking-water quality (WHO/SDE/WSH/03.04/17). World Health Organization. Geneva.

WHO (2000) Air Quality Guidelines for Europe, 2nd edition. World Health Organization Regional Office for Europe, Copenhagen.

WHO (2001) Zinc – Environmental aspects, 221. World Health Organisation, Geneva.

WHO (1998) Copper – Environmental aspects, 200. World Health Organisation, Geneva.
WHO (1988) Chromium – Environmental aspects, 61, World Health Organisation, Geneva.
WHO (1986) Selenium – Environmental aspects, 58. World Health Organisation, Geneva.
WHO (1984) Fluorine and fluorides – Environmental aspects, 36. World Health Organisation, Geneva.
WHO (2002) Fluorides – Environmental health criteria, 227. World Health Organisation, Geneva.
WHO (2008) Guidelines for Drinking-water Quality, 3rd Edition. World Health Organisation, Geneva.
WHO (2009) Global health risks: mortality and burden of disease attributable to selected health risks. World Health Organisation, Zurich. ISBN 978-92-4-156387-1.
Widatallah, H.M., Johnson, C., Berry, F.J., Jartych, E., Gismelseed, A. M., Pekala, M. and Grabski, J. (2005) On the synthesis and cation distribution of alumiumium-substituted spinel-related lithium ferrite. *Materials Letters* **59**: 1105–1109.
Yi, J. and Cao, J. (2008) Tea and fluorosis. *Journal of Fluorine Chemistry* **129**: 76–81.
Ysart, G., Miller, P., Croasdale, M., Crews, H., Robb, P., Baxter, M., De L'argy, C. and Harrison, N. (2000) 1997 UK total diet study – dietary exposures to aluminium, arsenic, cadmium, chromium, copper, lead, mercury, nickel, selenium, tin and zinc. *Food Additives and Contaminants* **17**: 775–786.
Yu, M.W., Horng, I.S., Hsu, K.H. *et al.* (1999) Plasma selenium levels and the risk of hepatocellular carcinoma among men with chronic hepatitis virus infection. *American Journal of Epidemiology* **150**: 367–374.
Zhang, X.P., Deng, W. and Yang, X.M. (2002) The background concentrations of 13 soil trace elements and their relationships to parent materials and vegetation in Xizang (Tibet). *China. Journal of Asian Earth Sciences* **21**: 167–174.

4

Toxic trace elements

Jilang Pan[1], Ho-Sik Chon[1], Mark R. Cave[2], Christopher J. Oates[3] and Jane A. Plant[4*]

[1]*Centre for Environmental Policy, Imperial College London, Prince Consort Road, London SW7 2AZ*
[2]*British Geological Survey, Kingsley Dunham Centre, Keyworth, Nottingham, NG12 5GG*
[3]*Applied Geochemistry Solutions, 49 School Lane, Gerrards Cross, Buckinghamshire, SL9 9AZ*
[4]*Centre for Environmental Policy and Department of Earth Science and Engineering, Imperial College London, Prince Consort Road, London SW7 2AZ*
**Corresponding author, email jane.plant@imperial.ac.uk*

4.1 Introduction

Some of the earliest writings about toxic trace elements (TTEs) are included in the Chinese *Materia Medica* written about 2737 BC by the mythical emperor of China, Shen Nung, who is also known as the Father of Chinese agriculture and of Chinese medicine (Robinson *et al.*, 1994). The treatise documents the properties of animal substances, plants such as aconite, opium and cannabis, and minerals such as alum (hydrated aluminium potassium sulphate), iodine, iron, sulphur and mercury, which were variously used as poisons, antidotes and drugs.

The first understanding of accidental poisoning by a mineral substance was probably in relation to lead (Pb), which was mined in Asia Minor (present-day Turkey) from about 6500 BC. Lead's properties of easy workability, low melting point and corrosion resistance made it attractive for the manufacture of jewellery, drinking vessels and, later, plumbing. By Roman times, Pb salts were also used for make-up (white lead $[(PbCO_3)_2{\cdot}Pb(OH)_2]$) and medicinal ointments, and to sweeten wine (Pb acetate $[Pb(C_2H_3O_2)_2]$).

Lead poisoning can cause irreversible neurological damage as well as renal, cardiovascular and reproductive problems (ATSDR, 2007). Some authors attribute many of the ills of the last years of the Roman Empire to Pb poisoning, though this has been disputed and remains controversial (Brewster and Perazzella, 2004; Lewis, 1985).

More recently, in the middle of the twentieth century, there was considerable concern about Pb affecting children, especially their neurological development and intelligence (Patterson, 1965). The main problems were caused by Pb paints, the use of lead solder in food cans and the use of tetraethyl Pb (TEL) in petrol as an anti-knock agent. Patterson's paper, *Contaminated and Natural Lead Environments of Man,* documented the many problems caused by the high levels of exposure to Pb in the American population. Patterson also compared the Pb, barium (Ba) and calcium (Ca) levels in 1600-year-old Peruvian skeletons and showed a 700 to 1200 fold increase in Pb levels in modern human bones, but no comparable increases in the other trace elements. His efforts led eventually to the US Environmental Protection Agency (US EPA) banning Pb from all petrol in the United States by 1986. Control measures were also introduced on Pb in food containers, paints, glazes and water distribution systems, and similar controls were introduced by many other countries. By the late 1990s Pb levels in the blood of Americans were reported to have dropped by up to 80 per cent. Lead continues to be used for roofing and batteries. It is also used for radiation shielding in the nuclear industry, where there is less likelihood of human exposure.

Several TTEs other than Pb have been a cause of concern in relation to human health over the past few decades. For example, a crippling bone disease first reported in Japan in 1912 has been attributed, since the 1950s, to cadmium (Cd) pollution (Uetani, 2007). Cadmium had been transported from a historical zinc-lead (Zn-Pb) mining and processing plant by the Jinzu River and deposited on paddy fields downstream. People who had eaten polluted rice and drunk the river water over a period of thirty years were found to have accumulated a large amount of Cd in their bodies, and this caused the serious osteoporosis-like bone disease known as 'itai-itai' (ouch-ouch) disease.

The neurological damage caused by mercury (Hg) has been recognised since at least Victorian times in England, when Lewis

Pollutants, Human Health and the Environment: A Risk Based Approach, First Edition. Edited by Jane A. Plant, Nikolaos Voulvoulis and K. Vala Ragnarsdottir.

Carroll described the odd behaviour of the Mad Hatter in his classic book *Alice's Adventures in Wonderland*; at the time, hatters used mercurous nitrate to preserve the felt used for making hats. More recently, in the 1950s, a well-studied case of Hg poisoning was reported from Japan. Mercuric chloride from a chemical and plastic plant had been discharged over a period of years into Minamata Bay (Mailman, 1980). Mercury was converted to methyl Hg chloride by bacteria in the sediments; it was taken up by fish and shellfish, which were subsequently consumed by the local population (Levi, 1997). Symptoms of Minamata disease included ataxia, numbness in the hands and feet, general muscle weakness, disturbed vision, damage to hearing and speech, and, in extreme cases, insanity, paralysis, coma and death.

Arsenic (As) was known and used as a poison in Persia and other ancient civilisations; it has been called the 'King of Poisons' (Bentley and Chasteen, 2002). In recent times, chronic As poisoning in the general population has been reported from many areas of the world, including Argentina, Bangladesh, Chile, China, Hungary, India, Mexico, Nepal, Pakistan, Sweden, Taiwan, Thailand and the USA (Hall, 2002; Tseng *et al.*, 2002). Endemic As poisoning is associated mostly with naturally high concentrations of As in drinking water, although in China it is as a result of burning coal rich in As (Ng *et al.*, 2003). Long-term exposure to As can result in chronic As poisoning (arsenicosis), skin lesions (melanosis and keratosis), and skin and bladder cancer. It has been estimated that between 33 and 77 million people in Bangladesh, out of the total population of 125 million, are at risk of arsenicosis from drinking As-contaminated water (Smith *et al.*, 2000).

Unlike organic pollutants, TTEs such as Pb and Hg are persistent and do not break down in the environment, although their species can be transformed to other species which may be less toxic. High doses of many trace elements are toxic – even those essential for human health. For the purpose of this chapter, TTEs are defined as elements with the following properties:

- They have one or more soluble species that has a level of the acute toxicity, including LC_{50}, lower than 1 mg/l or the level of chronic toxicity, including NOEL, lower than 0.1 mg/l; copper (Cu) sulphate has a lower LC_{50} than 1 mg/l but Cu is also an essential trace element and has been considered in Chapter 3.
- They have no known physiological functions; some TTEs actually damage essential physiological functions by replacing an essential element in the body – as Pb substitutes for Zn and Ca.
- They occur naturally, although they are often used in man-made products and may also be released adventitiously – as a result of coal burning, for example.

Most TTEs are heavier than carbon by atomic weight, though beryllium (Be), a light element with an atomic weight of 9, is highly toxic, especially if it is inhaled. Many of the radioactive elements as well as being radiotoxic also have high chemical toxicities. They are considered further in Chapter 5. Organometallic complexes such as TEL and tributyl tin, which are also highly toxic, are discussed in Chapter 7.

In this chapter, we discuss As, Cd, Hg, Pb and thallium (Tl) (highlighted in Figure 4.1) as examples of TTEs which occur

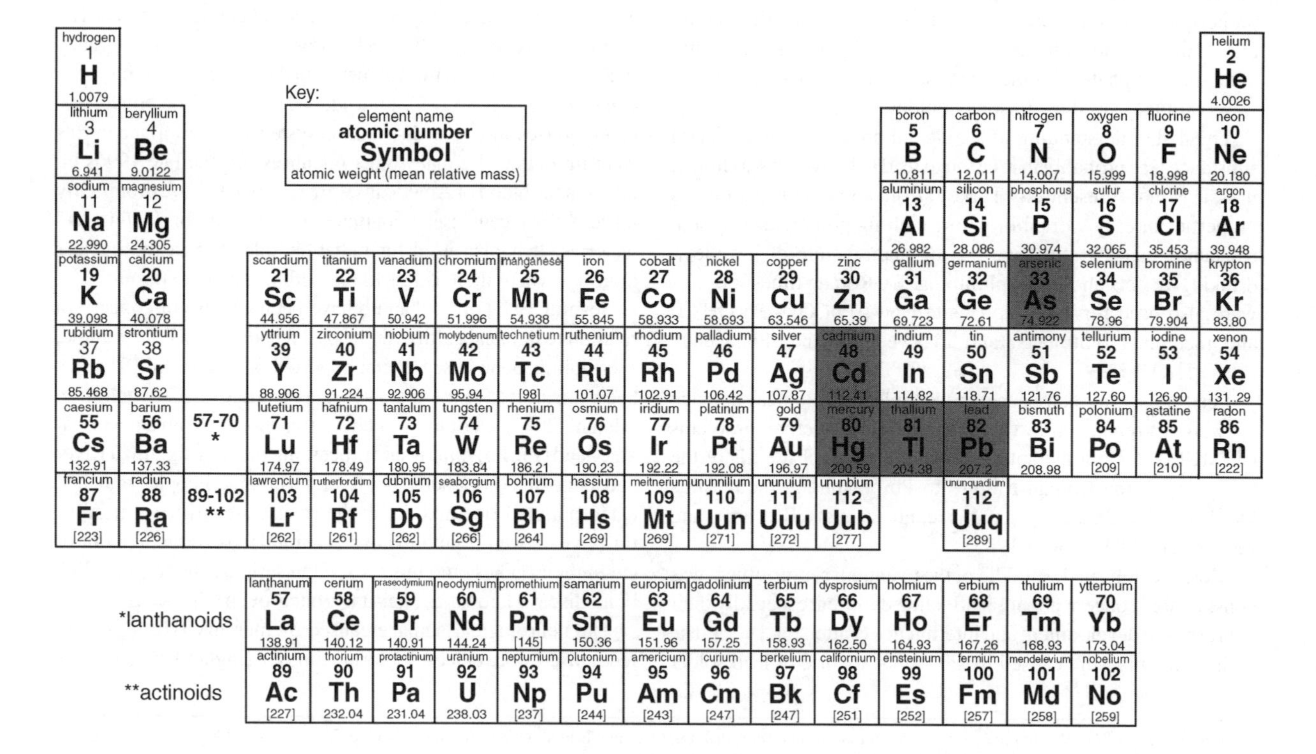

Figure 4.1 Periodic table of the elements

naturally or as a result of pollution and which, at high concentrations, have a harmful impact on human health and the environment. Except for As, they are all metals and are generally referred to as heavy metals. They are all chalcophile and hence enriched in sulphide ore deposits, although Tl and to a lesser extent Pb also have lithophile properties. Moreover, their affinity for sulphur also affects their behaviour in biological systems. In this chapter we discuss the chemical properties, natural and anthropogenic sources, behaviour in the environment and key human-health impacts of these five elements.

4.2 Hazardous properties

The hazardous properties of many TTEs reflect their ability to substitute for other metals with similar properties (such as valency and ionic radii) in enzymes and other vital biochemical substances. Many TTEs are difficult to remove from the body (some of them, like Pb, because of their unreactivity) and are cumulative poisons.

Arsenate (As^{5+}) compounds are less toxic than arsenite (As^{3+}) compounds. Arsenate has a similar structure to that of inorganic phosphate, with which it can compete, replacing the phosphate required for the production of adenosine triphosphate (ATP). This can lead to the uncoupling of oxidative phosphorylation by an unstable arsenate ester, which then hydrolyses (arsenolysis) (ATSDR, 1990; Hughes, 2002). Arsenite, on the other hand, interacts with thiol groups, inhibiting essential sulphydryl groups of enzymes and proteins (ATSDR, 1990; Rodriguez *et al.*, 2003). The main chemical species of As in nature are presented in Figure 4.2.

Cadmium can replace Zn, Ca and magnesium (Mg) in many biological systems, in particular systems that contain soft ligands such as compounds containing sulphur (S). Because of the ability to bind with sulphydryl groups, Cd can disturb the biological functions of animals, such as the formation of disulphide bridges and consequent conformational changes in the proteins (Walker *et al.*, 2006). Cadmium can bind up to ten times more strongly than Zn in certain biological systems and is notoriously difficult to remove. In the aquatic environment, Cd toxicity increases with decreasing water hardness, and soils with low pH values tend to have high Cd toxicity (Herber, 2004).

Mercury is highly toxic to most forms of life and is viewed as a priority pollutant (Fitzgerald and Lamborg, 2004; Parsons and

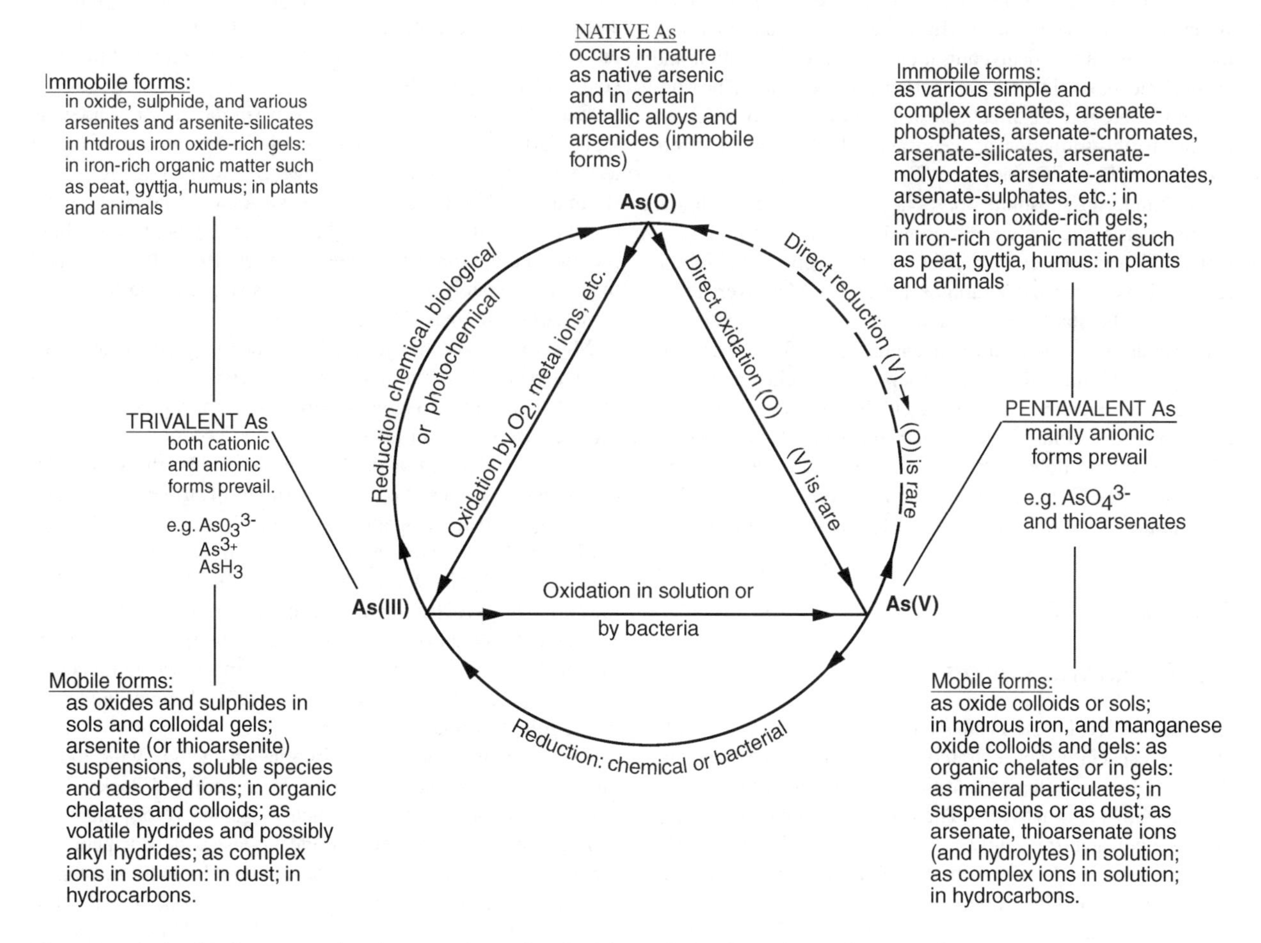

Figure 4.2 Cycle of As interconversions in nature (Reproduced, with permission, from The geochemistry of arsenic and its use as an indicator element in geochemical prospecting, Boyle, R.W. & Jonasson, I.R., Journal of Geochemical Exploration/2/3, © 1973/Elsevier)

Percival, 2005). In contrast to As, it is the organic species of Hg, for example methylmercury, that are the most toxic. Metallic Hg (Hg^0) is relatively inert and not readily taken up by organisms. However, it exists as a vapour phase at room temperature due to its high vapour pressure, and this can cause intoxication through inhalation (Drasch *et al.*, 2004; Simon *et al.*, 2006). Metallic Hg can dissolve easily in fatty compounds because of its high lipophilicity, while divalent Hg (Hg^{2+}) has high solubility in water (Drasch *et al.*, 2004). Mercury cations have a high affinity for S; this causes the interaction of Hg with biological ligands (Simon *et al.*, 2006). Mercury can bind with sulphydryl groups of proteins and disulphide groups in amino acids, resulting in inactivation of S and blocking of enzymes, cofactors and hormones (Markovich and James, 1999; Mathieson, 1995).

Lead is highly toxic, and it is the most studied trace element in relation to its effects on human health. The toxicity of Pb reflects its ability to substitute for other biologically important metals such as Ca, Zn and iron (Fe) which act as cofactors in many reactions involving enzymes. Lead interferes with the enzyme's ability to catalyse a reaction normally. The oxides and sulphides of Pb and divalent Pb (Pb^{2+}) salts are insoluble in water, whereas Pb acetate, Pb chlorate and Pb chloride are relatively soluble (Ewers and Schlipköter, 1991). Lead is not reactive in biological systems and is removed slowly from the body (mainly in urine). Inorganic Pb causes demyelinisation, axonal degeneration and presynaptic block in both the peripheral and central nervous systems. Moreover, Pb affects the steps in the chain of reactions leading to the formation of haem and inhibits enzymes from synthesising haem (Gerhardsson, 2004).

Thallium can substitute for potassium (K) due to similar chemical properties between monovalent K (K^+) and monovalent Tl (Tl^+) and therefore interferes with K-dependent biological reactions (Peter and Viraraghavan, 2005). Moreover, Tl^+ is reported to be stable with soft ligands such as S-containing compounds (Galván-Arzate and Santamaría, 1998). The high affinity of Tl for sulphydryl groups of proteins and other biomolecules enables Tl to bind at sulphydryl sites of enzymes (Aoyama *et al.*, 1988; Chandler and Scott, 1986). Compounds of Tl^+, including sulphate (Tl_2SO_4), acetate (CH_3COOTl) and carbonate (Tl_2CO_3), are soluble and therefore more toxic than Tl sulphide (Tl_2S) and Tl iodide (TlI) (Moeschlin, 1980).

4.3 Sources

4.3.1 Natural sources

Arsenic, Cd, Hg, Pb and Tl all occur naturally in rocks and soils, and are especially concentrated in certain types of mineral deposits; they are released into the environment both naturally and through human activities. Their concentration in rocks, soils and groundwater frequently reflects the tectonic setting. For example, As levels are naturally high in groundwaters in Bangladesh, Hungary (Smedley and Kinniburgh, 2002) and Holland (Van der Veer, 2006), associated with subsiding Holocene deltaic sediments near recently emergent Himalayan or Alpine mountain ranges. It has been suggested that Hg is higher over recent orogenic belts such as the Variscides, the Alps and the 'Ring of Fire' around the Pacific (Peterson, 1972).

4.3.1.1 Concentrations in rocks

The concentrations of TTEs in various rock types are shown in Table 4.1. Except for Pb, their average abundance in the continental crust is in the range 0.02 to 3.4 mg/kg. The highest concentrations are found primarily in organic-rich and sulphide-rich shales, sedimentary ironstones, phosphatic rocks and some coals; average As level, for example, can be as high as 20–200 mg/kg in such rocks (Plant *et al.*, 2004).

Arsenic levels can be particularly high in certain greywacke–shale sequences in orogenic belts (Plant *et al.*, 1998). Arsenical greywackes are frequently host to gold (Au) deposits in the south-west England orefield and Jiangxi province in China, or to tin (tungsten)-uranium (Sn(W)-U) ore deposits associated with later granite intrusions. All Au and many Cu deposits contain high concentrations of As, which frequently occurs in sulphide and sulpharsenide minerals, notably arsenopyrite (FeAsS), realgar (AsS) and orpiment (As_2S_3). Some coals, especially some Gondwana coals, are especially high in As (and Hg and Tl). Typical As concentrations in coal range from <1 to 17 mg/kg, but concentrations of up to 35 000 mg/kg are reported in some coals in China (Sun, 2004).

Cadmium has properties closely similar to Zn, and it is particularly concentrated in Zn-sulphide ore deposits with an average concentration in the range 200 to 14 000 mg/kg Cd (Reimann and de Caritat, 1998). Like As, Cd tends to be most enriched in shales, oceanic and lacustrine sediments, and phosphorites (Fergusson, 1990; Fleischer *et al.*, 1974). It is estimated that approximately 8100 t of Cd per year enters the biosphere from natural sources (Nriagu, 1980b).

Mercury, in addition to its affinity for S (e.g. cinnabar, HgS), has a high affinity for organic carbon, and it is often concentrated in coal. Transport of Hg with petroleum from source rocks into sedimentary traps can lead to significant accumulations in oil (White, 1967). Much of the Hg introduced into the superficial environment comes from the burning of fossil fuels. The Almadén deposit in Spain is the world's largest Hg deposit, until recently producing a third of the total world Hg (Ferrara *et al.*, 1998; Martín-Doimeadios *et al.*, 2000); the Hg orebodies are hosted by pyroclastic rocks and organic-rich shale, and native Hg is the dominant ore mineral (USGS, 1995). The concentration of Hg in soils sampled around this deposit was estimated to be 0.13 mg/kg, which is considerably higher than normal levels of Hg soils (Molina *et al.*, 2006).

Lead. Natural sources of airborne Pb include wind-blown dust, forest fires and volcanoes. In mineralised rocks, it is concentrated in ore minerals such as galena (PbS), anglesite ($PbSO_4$) and cerussite ($PbCO_3$), and it is also widely dispersed in other minerals, including silicates such as K-feldspar, plagioclase and mica, and the accessory minerals, zircon and magnetite (WHO, 1995). Carbonaceous shales from the USA and Europe contain 10–70 mg/kg of Pb, while phosphatic rocks can contain much higher concentrations, sometimes more than 100 mg/kg of

Pb (WHO, 1977). Sedimentary exhalative deposits (SEDEX deposits) are major producers of Zn and Pb, and include the Sullivan mine in British Columbia, Red Dog in Alaska and Mount Isa and Broken Hill in Australia. Historically, Mississippi-Valley-Type (MVT) Pb-Zn deposits (epigenetic SEDEX deposits), such as those of the famous Lead Belt and Viburnum trend of south-east Missouri and the English Pennines, have been among the most important Pb deposits. They are hosted mainly by dolostone or by limestone and sandstone, and have similar mineral assemblages, isotopic compositions and textures. Most MVT ore deposits are associated with regional hydrological processes rather than igneous activity (USGS, 1995). The Red Dog Pb-Zn deposits in Alaska, USA have been estimated to contain average Pb concentrations of 46 000 mg/kg (Jennings and King, 2002).

Lead isotopes other than ^{204}Pb, 'common Pb', are produced by the radioactive decay of U and thorium (Th), and these decay systems can be used to date rocks and model their genesis. The Pb isotopes ^{206}Pb and ^{207}Pb are produced by the decay of U^{238} and U^{235} respectively. Radiometric U-Pb dating reflects the decay of ^{238}U to ^{206}Pb ($t_{1/2} = 4.47$ billion years) and ^{235}U to ^{207}Pb ($t_{1/2} = 704$ million years) (Romer, 2003). Such dating is often performed on crystals of zircon ($ZrSiO_4$) in which U and Th and their decay products (including radon gas) are retained in the crystalline structure so that there is no disequilibrium (Mattinson, 2005). Deposits of Pb ore normally have a distinctive isotopic signature, which can be used to identify different sources of Pb in the environment.

Thallium in thallous (Tl^+) form has a similar ionic radius to K and rubidium (Rb) (Sahl *et al.*, 1978) and therefore shows similar geochemical behaviour to these elements. It is an incompatible element, becoming most concentrated in highly fractionated granite. It is widely distributed in K minerals such as K-feldspar (0.4–610 mg/kg). Like other TTEs discussed, Tl also shows chalcophile behaviour and is frequently concentrated in sulphur-containing ores. Average levels of Tl in the natural environment are low, with the highest values up to 3 mg/kg in acid igneous rocks and shales. Much higher concentrations (25 mg/kg Tl) can occur in organic-rich shales such as the Pierre Shale in the USA. Most coals contain 0.5–3 mg/kg of Tl, but levels in the range 100–1000 mg/kg have been reported in the Jurassic coals of Tadzhikistan (Smith and Carson, 1977).

4.3.1.2 Concentrations in soils

The chemical composition of soil varies in relation mainly to parent materials, climatic factors, topography, soil organisms and the age of the soil (Jenny, 1941; Kabata-Pendias and Sadurski, 2004). At high latitudes, soils comprise mainly residual phases from the parent rocks, such as quartz, feldspars and accessory minerals such as zircon (Curtis, 2003), so concentrations of TTEs in soils reflect that of the parent rock. At low latitudes, soil composition may reflect the age of the soil and the extent of chemical weathering (see Chapter 3). The fate and distribution of TTEs in soils depend on complex reactions between the chemical species and different components of solid, aqueous and gaseous phases of soils. For example, the presence of Fe and manganese (Mn) hydroxides, on which TTEs are readily sorbed, can increase their concentrations in soils (Kabata-Pendias and Sadurski, 2004).

Arsenic. The average abundance of As in soils ranges from 4.4 to 9.3 mg/kg (Table 4.1). Sandy soils have lower concentrations of As, while higher As concentrations are found in alluvial soils and soils with high levels of organic matter (Kabata-Pendias and Pendias, 1992). Acid sulphate soils contain high concentrations of As, up to 30–50 mg/kg (Dudas, 1987), and in Europe particularly high As levels occur over Variscan rocks

Table 4.1 Average abundance of As, Cd, Hg, Pb and Tl in rocks and soils (mg/kg)

Rock and soil	As	Cd	Hg	Pb	Tl	Reference
Continental crust	3.4	0.10	0.02	15	0.49	*a*
Oceanic crust	1.5*	0.13	0.02*	0.89	0.013	*a*
Ultramafic	0.5	0.05	0.01	0.1	0.01	*b*
Basalt	1.5	0.10	0.02	3.5	0.08	*a*
Andesite	0.8	0.09	0.015	35.1	1	*b*
Granite	1.5	0.09	0.03	32	1.1	*a*
Shale	10	0.13	0.45	22	0.68	*a*
Greywacke	8	0.09	0.11	14	0.20	*a*
Limestone	2.5	0.16	0.03	5	0.05	*a*
Gneiss, mica schist	4.3	0.10	0.02	16	0.65	*a*
Granulite	4*	0.10	0.02*	9.8	0.28	*a*
Soil**	4.4–9.3	0.37–0.78	0.05–0.26	22–44	–	*c*

[a]Wedepohl (1991).
[b]Govett (1983).
[c]Kabata-Pendias and Pendias (1992).
*Estimated concentration.
**The range of mean concentrations in different types of soils (podzols, cambisols, rendzinas, kastanozems and chernozems, and histosols).

of the south-western Armorican shear zone, extending from Brittany to the Massif Central, which is associated with Sb (antimony)-As-Au mineralisation (Plant *et al.*, 1998).

Cadmium in soils is generally associated with Zn, for which it substitutes isomorphically in rocks and minerals (Herber, 2004). In arid and semi-arid climates, Cd levels can be particularly high in soils, including in farming regions (McBride, 1994). Concentrations of Cd in uncontaminated soils are in the range 0.37–0.78 mg/kg (Table 4.1), with the highest concentration in clay-rich soils (Kabata-Pendias and Pendias, 1992; Page and Bingham, 1973).

Mercury concentrations in soils are typically in the range 0.05–0.26 mg/kg (Table 4.1), and surface soil values tend not to exceed 0.4 mg/kg (Kabata-Pendias, 2001). Soils developed on intrusive magmatic rocks near subaerial and submarine volcanism and hydrothermal activity can contain higher concentrations. For example, Hg levels of 1.2–14.6 mg/kg have been reported from Steamboat Springs, Nevada, USA (Gustin *et al.*, 1999) and represent a significant source of Hg to the atmosphere. Wang *et al.* (2003) also demonstrated a positive correlation between the concentrations of Hg in the atmosphere and the Hg contents in soils.

Lead concentrations in uncontaminated soils often correlate with the amount of clay minerals, Fe-Mn oxides and organic matter present (WHO, 1977; Wixon and Davies, 1993). In some soils, Pb may be highly concentrated in Ca carbonate or phosphate (Kabata-Pendias, 2001). A baseline Pb value for surface soils on the global scale is the range 22–44 mg/kg (Table 4.1); levels with values above the range are likely to reflect anthropogenic influences (Kabata-Pendias, 2001).

Thallium concentrations in soils typically range from 0.1 to about 1.0 mg/kg, with an average concentration of 0.25 mg/kg (Bowen, 1979; Brumsack, 1977; Chattopadhyay and Jervis, 1974; Schoer, 1984; Smith and Carson, 1977).

4.3.2 Anthropogenic sources

The five TTEs considered can be released into the environment as a result of activities such as mining, smelting and fossil-fuel combustion. The last of these is especially significant where low- grade coals are used, as those of China or India. The main anthropogenic sources of As, Cd, Hg, Pb and Tl, including inadvertent sources, are summarised in Table 4.2 and discussed below.

Arsenic and its compounds are used as pesticides, especially herbicides and insecticides. The high As levels in soils over the cotton-growing areas of the USA reflect the use of such pesticides in the past (US EPA, 2006). It is also used in various alloys. Other anthropogenic sources of As include coal combustion, sulphide-ore roasting and smelting, and the use of arsenical growth promoters in pig and poultry rearing (Reimann and de Caritat, 1998). Arsenic contamination of the environment as a result of historical mining and smelting is quite common (Breward *et al.*, 1994; Thornton, 1996), but local (Plant *et al.*, 2004). Oil spills and leakages increase concentrations of As in both fresh water and marine water. Moreover, oil prevents underlying sediments from adsorbing the As and removing it from the aqueous phase (Wainipee *et al.*, 2010). Some recent uses of As include its use in light-emitting diodes (LEDs) (Ryu *et al.*, 2006).

Cadmium. Anthropogenic sources of Cd in soils include rock-phosphate fertiliser used in arable farming, ash from fossil-fuel combustion, waste from cement manufacture and metallurgical works, municipal refuse and sewage sludge. About 40 per cent of anthropogenic Cd is deposited from the atmosphere. Concentrations of Cd can be as high as 500 mg/kg in phosphorites which are used for the manufacture of fertilisers; and soils which have received heavy applications of sewage sludge have been found to have total Cd soil contents of up to 64.2 mg/kg (with values of Pb up to 938 mg/kg).

It has been estimated that, historically, mining and smelting of (mainly) Pb-Zn ores have been the largest anthropogenic source of Cd to the aquatic environment (Fergusson, 1990; Fleischer *et al.*, 1974; Merrington and Alloway, 1994), including mine drainage, ore-processing waste water, overflows from tailings ponds, and run-offs from the general mine area.

Anthropogenic emissions now contribute most Cd to the atmosphere, resulting in a complex pattern of Cd deposition into soils and waters. Approximately 60 per cent of anthropogenic Cd emissions to the atmosphere are from Zn and Cu production, and about 22 per cent from indirect sources including fossil-fuel burning. The magnitude of Cd atmospheric deposition varies from 0.02–0.06 kg/km^2/yr in France (Thévenot *et al.*, 2007) to 15.30 kg/km^2/yr in Turkey (Muezzinoglu and Cizmecioglu, 2006). Globally, Cd inputs to land from atmospheric fallout, mining and fertiliser applications are in the range 2200–8400 t/yr, 4300–7300 t/yr, and 30–250 t/yr respectively (Nriagu and Pacyna, 1988).

Mercury. The global cycle of Hg has many similarities to that of carbon (C). The rising concentration of Hg in the environment is mainly a result of the combustion of fossil fuels and solid waste, as well as mining and smelting. It has been estimated that 300–8800 t of Hg per year enter the aquatic ecosystems (Nriagu and Pacyna, 1988). It has also been estimated that approximately 20 kg of Hg were released daily into the atmosphere from Hg mining and smelting in Idrija, Slovenia (Biester *et al.*, 1999). Concentrations of Hg in soils in this area ranged from 0.005 mg/kg to over 100 mg/kg (Biester *et al.*, 1999), and those in sediments of the rivers Idrijca and Soça affected by mining and smelting operations were reported to be up to 1000 mg/kg (Gosar *et al.*, 1997). Mercury concentrations in soils of the Almadén mining area in Spain are in the range 0.13–2695 mg/kg (Molina *et al.*, 2006).

Mercury has been widely used in the manufacture of thermometers, barometers, diffusion pumps, mercury-vapour lamps, low-energy light bulbs, advertising signs, mercury switches and other electronic apparatus. It continues to be used as thimerosal as a preservative in some vaccines, including influenza vaccines (US FDA, 2010) (see also page 106).

In spite of the well-known neurotoxic effects of Hg (Clifton, 2007), amalgam, a mixture of silver (Ag) and Hg, is still permitted for use in dentistry throughout the world. Due to

Table 4.2 Summary of anthropogenic sources of As, Cd, Hg, Pb and Tl

Anthropogenic sources	As	Cd	Hg	Pb	Tl
Uses	Animal-feed additives, fertilisers, pesticides, fungicides, herbicides, Timber treatment (chromated copper arsenate, CCA*), LEDs, various alloys	Ni-Cd batteries, pigments, plating, stabilisers, electronics, communications, power generation, aerospace industries, nanotechnology	Pesticides, dental preparation, antifouling paint, batteries, catalysts, new light bulbs	Water pipes, cisterns and pewter containers antiknock agents in petrol (TEL and TML)*, paint*, cable sheathing and paint pigments, Pb-acid batteries in cars, solder, ammunition, radiation shielding, metal alloys	Pesticides (rodenticides), electrical and electronic industries, special glass
Inadvertent sources	Mining, smelting, fossil fuel combustion				
	Sulphide-ore roasting	Phosphate fertilisers, waste from cement manufacture, metallurgical works, municipal waste and sewage sludge applied to land	Solid-waste combustion	Vehicle emissions in countries with no legislation on Pb in petrol	Brickworks, cement plants

*The use of these products has been phased out in most countries

the reported toxicity and effect on health of Hg amalgam (Lindh *et al.*, 2002; Lorscheider and Vimy, 1993), Norway banned amalgam in 2008, and it was banned in Sweden in 2009. A positive effect on the health following removal of Hg amalgam fillings has been reported (Windham, 2010). As of 15 March 2011 an export ban on Hg and its alloys came into force in Europe (European Parliament, 2008).

Other uses of Hg include pesticides, biocides for timber treatment, antifouling paint, batteries and catalysts (Sörme and Lagerkvist, 2002; Q. Wang *et al.*, 2004).

Lead contamination in soils can reflect mining, especially historical mining and smelting, the use of agrochemicals, the application of sewage sludge, and coal fly ash, which can contain up to 1500 mg/kg (Al-Khashman and Shawabkeh, 2006; Ansorena *et al.*, 1995; Block and Dams, 1975; Schwab *et al.*, 2007; Singh and Agrawal, 2007; Wong *et al.*, 2002). Around MVT ore deposits in Upper Silesia in southern Poland, the concentrations of Pb in soils ranged from 4.0 to 8200 mg/kg with a mean value of 102.3 mg/kg, reflecting historical mining and smelting operations and subsequent atmospheric deposition (Dudka *et al.*, 1995). The concentrations of Pb in topsoils sampled alongside roads and at port facilities around the Red Dog Pb-Zn deposits in northern Alaska were in the range of 4600– 5240 mg/kg (Kelley and Hudson, 2007). Irrigation of crops using water with high Pb levels can also increase soil concentrations. High Pb levels near roads reflect tyre and brake-lining abrasion (Sörme and Lagerkvist, 2002), the past use of TEL (which has a distinctive isotopic signature) (Preciado *et al.*, 2007) and, in older towns, the former use of Pb-rich paints. Mining, effluents from sewage works and surface runoffs can all contribute to Pb contamination of water bodies. A high Pb content has also been found in some inexpensive imported toys (Weidenhamer, 2009).

Lead emissions from vehicles used to be one of the most important sources of atmospheric Pb pollution in urban areas, but levels have declined dramatically over the last 40 years since TEL was phased out (Pacyna *et al.*, 2007). The contribution of leaded petrol to atmospheric Pb levels has been shown to have gone down (Farmer *et al.*, 2000).

Thallium contamination of soils is caused mainly by emissions and solid wastes from coal combustion and ferrous and non-ferrous smelting (Peter and Viraraghavan, 2005) as well as by dust emitted by power-generating plants, brickworks and cement plants. Historically, large amounts of contaminated waste materials from mining Hg ore and coal containing 25 to 106 mg/kg Tl resulted in chronic Tl poisoning in China, mainly via water (WHO, 1996a; Zhou and Liu, 1985). In northern Germany, historical mining activity resulted in the enrichment of Tl in soils by factors of between 3 and 35 (WHO, 1996a). Emissions from a cement plant in Lengerich, Germany increased Tl concentrations in dry soils by up to 6.9 mg/kg (Mathys, 1981).

Less than 15 t of Tl per year are produced worldwide; however, about 2000–5000 t per year are discharged by industrial activities. In industrial areas, the concentrations of Tl in surface waters was reported to be in the range of 1–100 µg/l (Kazantzis, 2000). Emissions of Tl to air are mainly from (sulphide) mineral smelters, power-generating plants and cement plants (ATSDR, 1992). Thallium compounds are volatile at high temperatures and pass through most emission-control facilities. Thallium is concentrated in the smallest particles in fly ash from coal-fired power plants and passes through conventional filters. Such particles remain suspended in the atmosphere and are respirable (Kazantzis, 2000; Manzo and Sabbioni, 1988).

4.4 Environmental pathways

Some of the properties that affect the abundance and behaviour of the TTEs are presented in Table 4.3.

4.4.1 Arsenic

The main pathways of As in the environment are summarised in Figure 4.3.

Soil. The behaviour of As in soils is similar to that of phosphorus (P). One of the main controls on As solubility in oxidising neutral to alkaline conditions is the presence of ferric

Table 4.3 Properties of As, Cd, Hg, Pb and Tl

Property		Arsenic	Cadmium	Mercury	Lead	Thallium
Symbol		As	Cd	Hg	Pb	Tl
Atomic number		33	48	80	82	81
Atomic mass		74.9	112.4	200.59	207.2	204.38
Classification		Metalloid	Metal	Metal	Metal	Metal
Density, g/cm^3		5.73	8.65	13.456	11.34	11.85
Partition	Soil-water	0.3–4.3	0.1–5.0	2.2–5.8	0.7–5.0	–1.2–1.5
coefficient[a]	Sediment-water	1.6–4.3	0.5–7.3	3.8–6.0	2.0–7.0	–0.5–3.5
(log K_d)	SPM[b] -Water	2.0–6.0	2.8–6.3	4.2–6.9	3.4–6.5	3.0–4.5
CAS registry number		7440-38-2	7440-43-9	7439-97-6	7439-92-1	7440-28-0

[a]US EPA (2005).
[b]Suspended particulate matter.

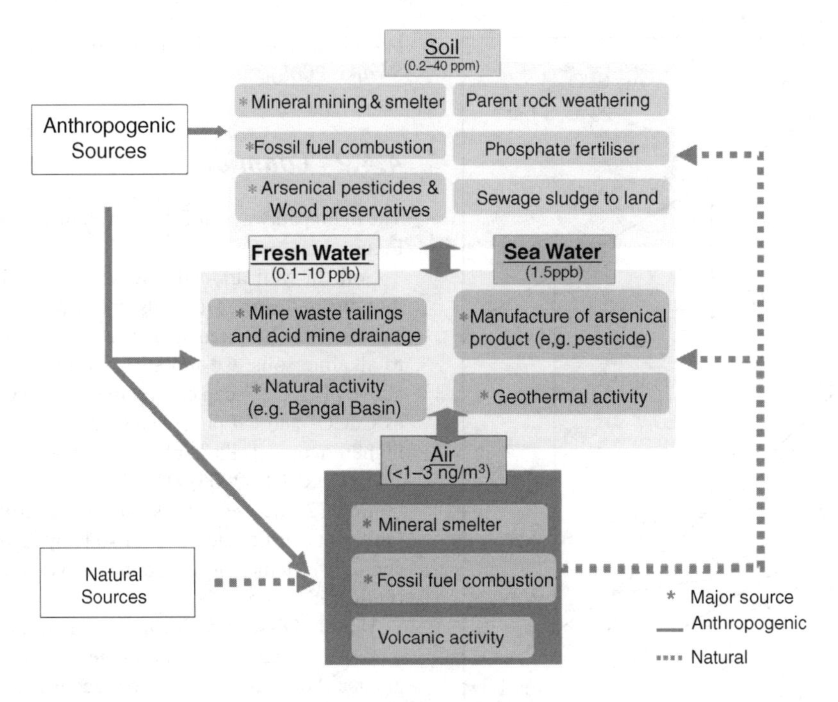

Figure 4.3 Pathways of As

iron, on which it becomes sorbed (Plant *et al.*, 2004). It is also sorbed by hydrous oxides of Fe, Mn and Al, noncrystalline aluminosilicates and to a lesser extent layer-silicate clays. The mobility of As is fairly low in acid conditions in soils containing clay and oxides. Arsenic may be more mobile in neutral to alkaline soils, especially those that are sodic, but its mobility is generally limited in anaerobic conditions. Biological transformation of As species in soil is presented in Figure 4.4.

Water. The dominant As species in water depends on the pH and redox potential (Eh), with As^{5+} being predominant under oxidising conditions. The main As species in natural waters under different pH and Eh conditions are presented in Figure 4.5. The relatively small amount of As released into stream waters during weathering is mobile only if the pH and Eh are sufficiently low to favour its persistence in trivalent form. Otherwise, dissolved As is rapidly oxidised to insoluble As^{5+} and it becomes sorbed as the arsenate ion (AsO_4^{3-}) by hydrous Fe and Mn oxides, clays and organic matter (Cheng *et al.*, 2009). Surface waters (i.e. lakes and rivers) and drinking waters indicate generally lower As concentrations, with surface waters rarely exceeding a few μg/l except in the most acid conditions such as those associated with acid mine drainage (Hem, 1992). Arsenic in groundwaters is frequently of natural origin, and it may cause health problems (see Section 4.1). In Bangladesh, As is released from the deltaic sediments, which are intercalated with peat, into the groundwaters due to anoxic conditions in which iron is soluble.

Microbial agents can affect the oxidation state of As in waters and mediate the methylation of inorganic As to organic As compounds. Microorganisms can oxidise arsenite to arsenate, reduce arsenate to arsenite, or reduce arsenate to arsine (AsH_3). Bacterial action also oxidises minerals such as orpiment (As_2S_3),

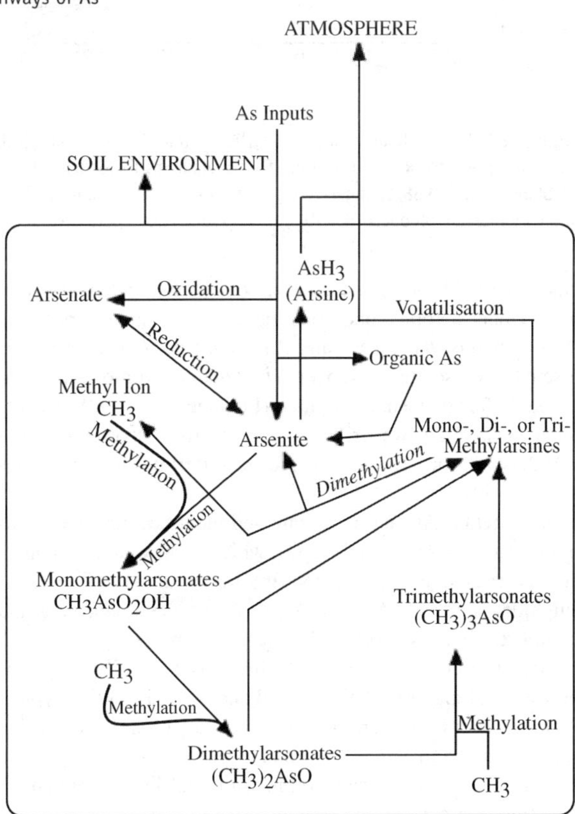

Figure 4.4 Biological transformation of As in soils (Reproduced, with permission, from Arsenic chemistry in soils: An overview of thermodynamic predictions and field observations, Sadiq, M., Water, Air, and Soil Pollution/93/1, © 1997/Springer)

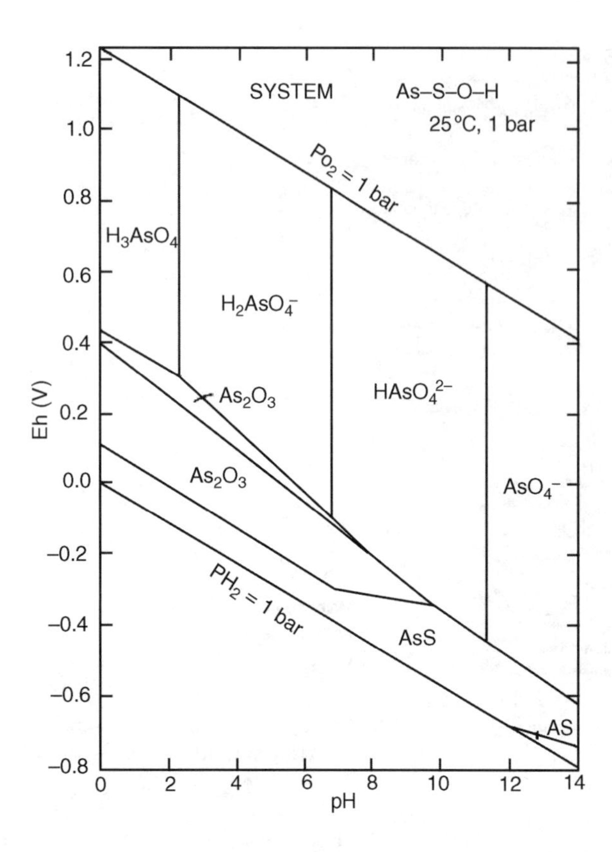

Figure 4.5 Eh-pH diagram for As in the system As-S-O-H at 25 °C and 1 bar pressure (Reproduced from Eh-pH Diagrams for Geochemistry, Brookins, D.G., © 1988/Springer-Verlag). Arsenite species including H_3AsO_3 and its ionisation products ($H_2AsO_3^-$, $HAsO_3^{2-}$ and AsO_3^{3-}) are excluded

arsenopyrite (FeAsS) and enargite (Cu_3AsS_4), releasing arsenate. Under aerobic conditions, the common aquatic bacteria, *Pseudomonas fluorescens* and *Anabaena oscillaroides* reduce arsenate to arsenite. Moreover, *Thiobacillus* promotes precipitation of ferric arsenate sulphate (Leblanc *et al.*, 1996), while *Shewanella alga* BrY releases arsenate from scorodite ($FeAsO_4 \cdot 2H_2O$) through reduction of ferric to ferrous ion (Cummings *et al.*, 1999).

Sediments. Arsenic in stream sediments occurs mainly as oxides such as As_2O_3 and As_2O_5, which form soluble arsenites and arsenates in acid waters (Irgolic *et al.*, 1995), and as sulphides such as FeAsS and As_2S_3, heavy-metal arsenates and complexes with Fe oxides (Baur and Onishi, 1978).

Air. The concentration of As in air is generally very low, between 0.4 and 30 ng/m^3 (Mandal and Suzuki, 2002). Typical As levels over Europe are currently between 0.2 and 1.5 ng/m^3 in rural areas, 0.5 to 3 ng/m^3 in urban areas, and up to 50 ng/m^3 in industrial areas (European Commission, 2001). Arsenic enters the atmosphere as a result of wind erosion, volcanic emissions, low-temperature volatilisation from soils, marine aerosols and pollution. In air, As occurs predominantly absorbed on particulate matter, usually as a mixture of arsenite and arsenate, with only negligible amounts of organic As except in areas of arsenic pesticide application or biotic activity (Mandal and Suzuki, 2002).

4.4.2 Cadmium

The main pathways of Cd in the environment are summarised in Figure 4.6.

Soil. In well-drained acid soils, Cd^{2+} has medium to high mobility, reflecting weak adsorption by organic matter, clays and oxides. It can be taken up by plants in such conditions. In neutral to alkaline soils, Cd mobility and bioavailability are low, and above pH 7 Cd^{2+} can co-precipitate with $CaCO_3$, or precipitate as $CdCO_3$ and Cd phosphates. In Europe, the levels of Cd are higher in topsoils than subsoils, with a ratio of topsoil/subsoil of 1.67 (FOREGS, 2005). This is thought to reflect (1) historical mining, (2) the use of rock-phosphate fertiliser, (3) the upward movement of the elements in soils by repeated reprecipitation and/or (4) the increased amount of organic substances in topsoils (FOREGS, 2005).

Water. Dissolved Cd in sea waters is mostly in the range 0.004 to 0.07 µg/l, but the range in rivers and lakes is much greater (Forstner, 1980). In surface waters, Cd is most soluble under oxidising conditions at pH less than 8 (Figure 4.7). Cadmium can also occur in solid phases such as carbonate ($CdCO_3$) and hydroxide ($Cd(OH)_2$) at pH values between 8 and 13, sulphide (CdS) under reducing conditions and Cd bound on organic matter and co-precipitated with hydrous oxides of Mn and possibly Fe. Since Cd^{2+} is the only stable oxidation state, the solubility of Cd is scarcely affected by changes in Eh (Fergusson, 1990). The low solubility of $CdCO_3$ restricts Cd^{2+} mobility in carbonate-rich sediments and soils (Bowen, 1982). Cadmium sulphate is also relatively insoluble, and concentrations in surface water are generally well below saturation (Hem, 1992). Cadmium readily forms complexes in solution with halides, cyanides and ammonium species, and it has a strong affinity for organic matter. Rivers contaminated with Cd can pollute the surrounding land through irrigation, the dumping of dredged sediments or flooding (Johnson and Eaton, 1980), as the Jinzu River region of Japan (see page 87). Irrigation using groundwaters high in Cd levels can also contaminate soils.

Sediments. Humic substances in sediments bind Cd^{2+} to a greater extent than the major inorganic ligands, especially at high pH (Reuter and Perdue, 1977), and clay minerals as well as hydrous ferric and manganic oxide coatings adsorb Cd (Hem, 1992). This Cd can be mobilised into readily bioavailable forms by changes in the physicochemical properties of sediment-water systems such as Eh, pH and salinity. A survey of the Lake District and the western North Pennines in England reported up to 14.3 mg/kg Cd with a mean value of 0.96 mg/kg Cd in stream sediments over mineralised Ordovician lavas, with some of the highest values occurring around former mine workings (BGS, 1992). Cadmium concentrations in stream sediments over Europe have been found to be in the range <0.02–43.1 mg/kg, with a mean value of 0.53 mg/kg (FOREGS, 2005).

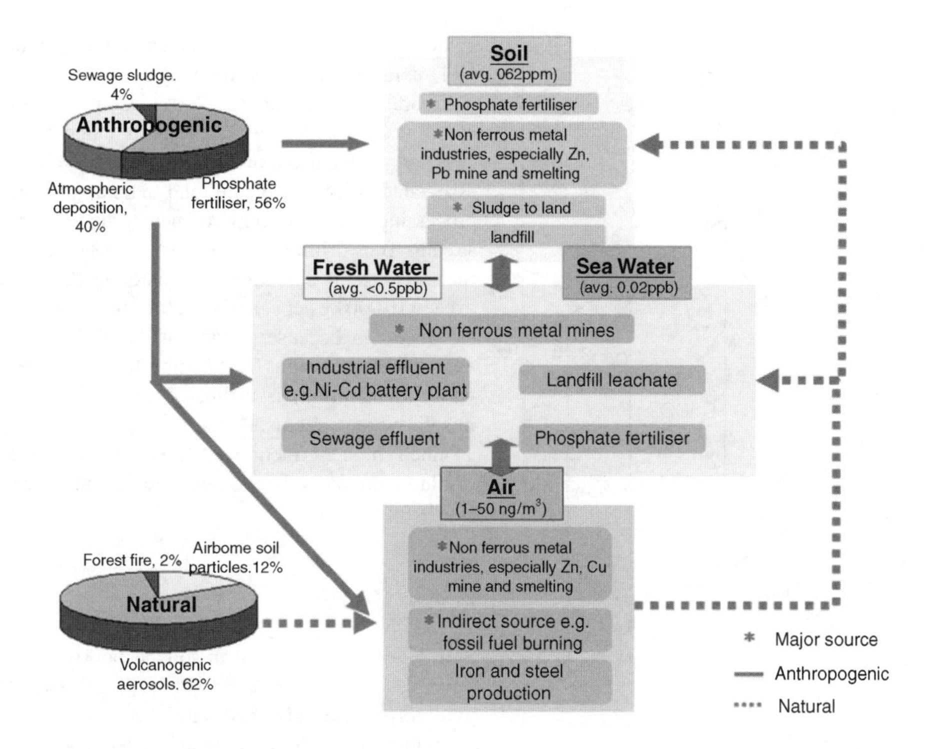

Figure 4.6 Pathways of Cd (Reproduced, with permission, from Cadmium levels in Europe: Implications for human health, Pan, J., Plant, J.A., Voulvoulis, N., Oates, C.J. & Ihlenfeld, C., Environmental Geochemistry and Health/32/1, © 2010/Springer)

Air. The concentration of Cd in air ranges from 0.1 to 150 ng/m^3 in monitored sites with different land use (ATSDR, 2008). Worldwide, annual emissions from natural sources are estimated to be 843 t Cd (Nriagu, 1980b): 12 per cent is in airborne soil particles, 62 per cent in volcanogenic aerosols, about 25 per cent originating from vegetation, and 2 per cent from forest fires. The release of Cd as a result of degassing crustal rocks to the atmosphere has not been quantified (Chernyak and Nussinov, 1976; Goldberg, 1976; Nriagu, 1980a). Anthropogenic sources contribute 7300 t/yr of Cd in the atmosphere. Nriagu (1980b) estimated that 76 per cent of such emissions were from the non-ferrous metal industries, while about 22 per cent were from indirect sources including fossil-fuel burning.

4.4.3 Mercury

The main pathways of Hg in the environment are summarised in Figure 4.8.

Soil. Mercury can exist in three states in soils under normal conditions of temperature and pressure: Hg^0, the mercurous state (Hg^+) and Hg^{2+} (Schuster, 1991). The behaviour of Hg in soils depends on its concentration, the presence of other ions such as Cl^-, OH^- and S^{2-}, and the presence of organic matter which can complex it. The presence of excess Cl^- ions decreases the sorption of Hg^+ onto mineral particles, Mn oxides, and organic matter because stable and highly soluble Hg-Cl complexes are formed. The high affinity of Hg^{2+} for Cl^- and OH^- also results in the formation of $HgCl_2$ and $Hg(OH)_2$, which are highly soluble, increasing the mobility of Hg. The amount of organic matter is, however, a critical factor in determining the status and distribution of Hg in soils (Schuster, 1991). The sorption of Hg by clay is limited, depending on pH values; the highest sorption occurs between pH 4 and 5. In acid soils, however, HgS and even metallic Hg may precipitate (Kabata-Pendias, 2001).

The mobility of organometallic Hg compounds generally depends on biological and chemical processes (Kabata-Pendias, 2001). The formation of organomercury compounds, especially the methylation of elemental Hg, plays the most important role in the Hg cycle in the environment. The accumulation of Hg in soils depends mainly on the levels of organic C and S, so higher concentrations occur in surface soils than subsoils, and organic soils have a higher Hg content than mineral soils (FOREGS, 2005).

Water. Mercury occurs in several species in natural waters, including metallic Hg, ionic Hg (Hg^+ and Hg^{2+}) and methylated Hg (CH_3Hg^+ and $(CH_3)_2Hg$) (Ravichandran, 2004). The cycle of Hg species in waters is described in Figure 4.9. Most important inorganic Hg species behave differently in different pH and Eh conditions. In oxidising acid conditions, Hg^+ and Hg^{2+} are predominant, whereas metallic Hg occupies a broad field on Eh-pH diagrams (Figure 4.10). Natural organic materials form strong complexes with Hg via functional sulphide

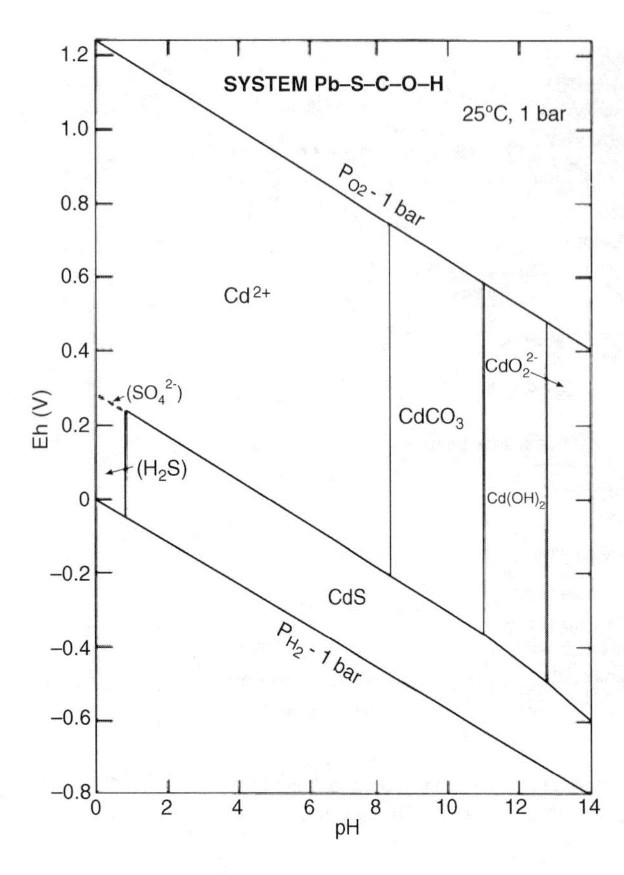

Figure 4.7 Eh-pH diagram for Cd in the system Cd-C-S-O-H at 25 °C and 1 bar pressure (Reproduced from Eh-pH Diagrams for Geochemistry, Brookins, D.G., © 1988/Springer-Verlag)

groups, which influence the concentration, speciation and subsequent bioaccessibility of Hg in the aquatic environment (Ravichandran, 2004). Levels of Hg in uncontaminated surface waters are <0.07 µg/l (US EPA, 1997). Mercury minerals such as cinnabar (HgS) and meta-cinnabar ((Hg, Zn, Fe)S, Se) are insoluble under normal conditions, and metallic Hg does not react with stream waters directly, enabling metallic Hg to persist in stream sediments for many years. When Hg is mobilised, the agent is often microbial methylation.

Sediments. Average Hg levels in organic-rich sediments are relatively high (Moore and Ramamoorthy, 1984). Mercury is particularly concentrated in sediments developed in Zn-Pb, hydrothermal and mineralised ore regions over Europe. Anomalous Hg contents in sediments also reflect anthropogenic influences such as coal-fired power plants and Hg and Ag mining. In Europe, the level of Hg in stream sediments is in the range <0.001–13.6 mg/kg, with a mean of 0.08 mg/kg (FOREGS, 2005).

Air. Mercury enters the atmosphere as a result of volcanic emissions and pollution, and is returned to the earth's surface by wet and dry deposition. Mercury levels in air are in the range 2–10 ng/m^3 (WHO, 2005). The most important anthropogenic emissions of Hg are from smelter operations and fossil-fuel combustion – particularly coal-fired power stations (Fitzgerald and Lamborg, 2004). Concentrations of Hg in cloud water have been reported to range between 7.5 and 71.8 ng/l, with a mean of 24.8 ng/l (Zahir *et al.*, 2005). Mercury associated with petroleum and natural-gas production and processing enters the environment primarily via solid waste streams (i.e. drilling and refinery waste) and combustion of fuels. The total amount of Hg entering the atmosphere may exceed 10 000 kg per year (Wilhelm, 2001), but current estimates are uncertain because of lack of data. The amount of Hg in atmospheric emissions from combustion (fossil-fuel burning and waste incineration) is estimated to be roughly equal to that in solid-waste streams, though the latter are thought to contain a much higher fraction of mercuric sulphides and other insoluble compounds and hence to be much less bioavailable (Wilhelm, 2001).

4.4.4 Lead

Soil. Lead can be present in soils as poorly soluble secondary minerals such as anglesite ($PbSO_4$), cerussite ($PbCO_3$) and pyromorphite, or adsorbed onto clay particles, hydrous Mn or Fe oxides and colloidal organic matter. The mobility of Pb in soils increases at low pH, including as a result of the concentration of organic acids (WHO, 1995). Lead concentration is usually greater in the smaller size fractions of soils (<0.075–0.25 mm), in which it is associated with carbonates and Fe-Mn oxides (Yarlagadda *et al.*, 1995). In Europe, Pb concentrations are in the range <3–938 mg/kg in subsoils and 5.3–970 mg/kg in topsoils (FOREGS, 2005).

Water. In acid conditions, Pb in the aquatic environment is usually present as the divalent ion (Pb^{2+}) and forms complexes with organic ligands, carbonates and hydroxides (Moore and Ramamoorthy, 1984). The stability of Pb species in the system Pb-S-C-O-H under different Eh and pH conditions is indicated in Figure 4.11. Lead sulphides, sulphates, carbonates, phosphates and oxides are sparingly soluble in water, while Pb nitrates, acetate and chlorides are more soluble. The concentration of Pb in stream waters over Europe ranges from <0.005 to 6.37 µg/l, with a mean of 0.224 µg/l (FOREGS, 2005). WHO (1995) reported 0.02 µg/l lower levels of Pb in surface waters. In open sea water Pb concentration is in the range <0.001–0.004 µg/l. Higher concentrations are detected in the near-surface water, possibly reflecting input from atmospheric deposition of Pb and wind-borne dust (WHO, 1995).

Sediments. Lead in sediments has a high affinity for Fe oxides (Pacifico *et al.*, 2007). On the south-west coast of Spain, 60 per cent of Pb in sediments has been found to be in the residual fraction (Morillo *et al.*, 2004); Song *et al.* (1999) reported about 35 per cent of Pb in the sand fraction, with the reminder in the silt and clay fractions. Over Europe, Pb concentrations in stream sediments range from <1 to 5760 mg/kg, with a mean of 38.6 mg/kg, and are

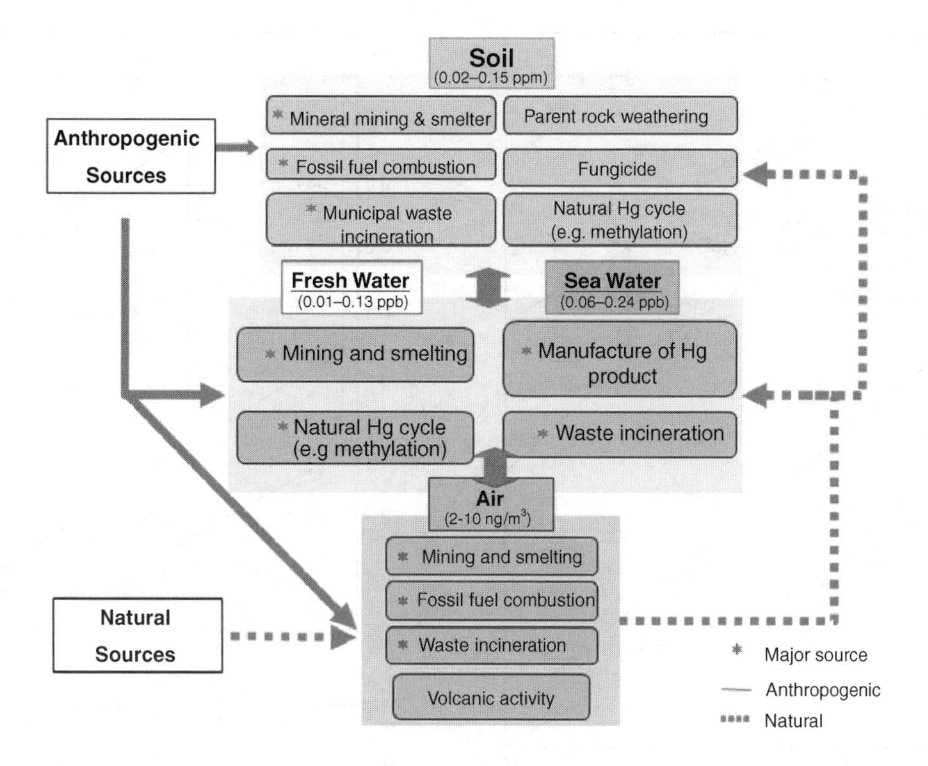

Figure 4.8 Pathways of Hg

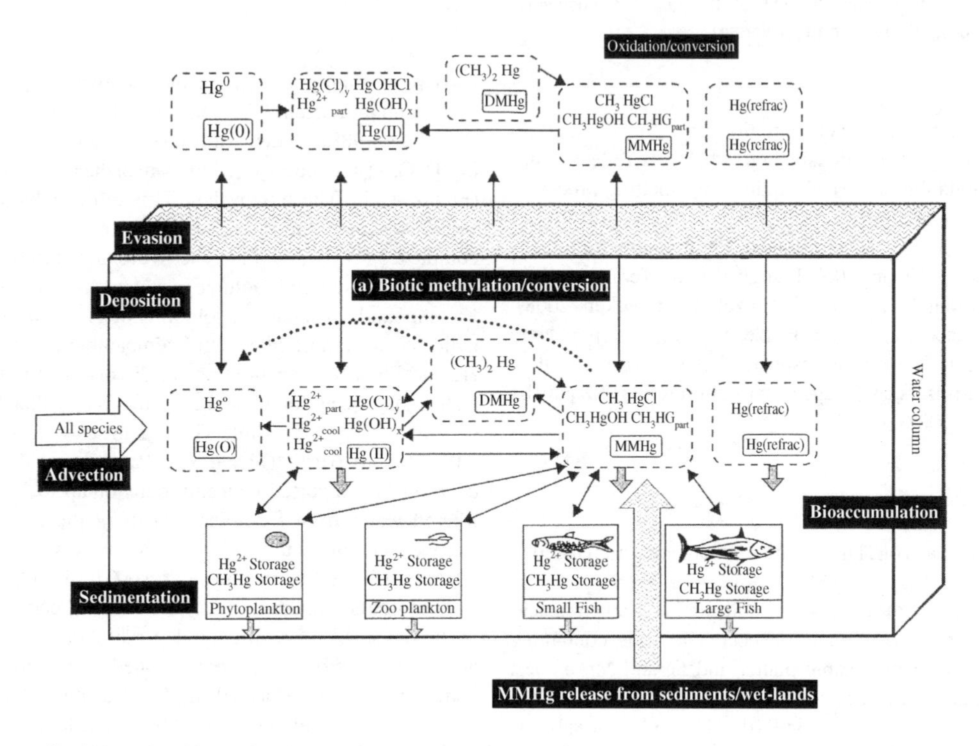

Figure 4.9 Generalised view of Hg biogeochemistry in the aquatic environment (Reproduced from Mercury pollution: Integration and synthesis, Watras, C.J. & Huckabee, J.W., © 1994/CRC Press)

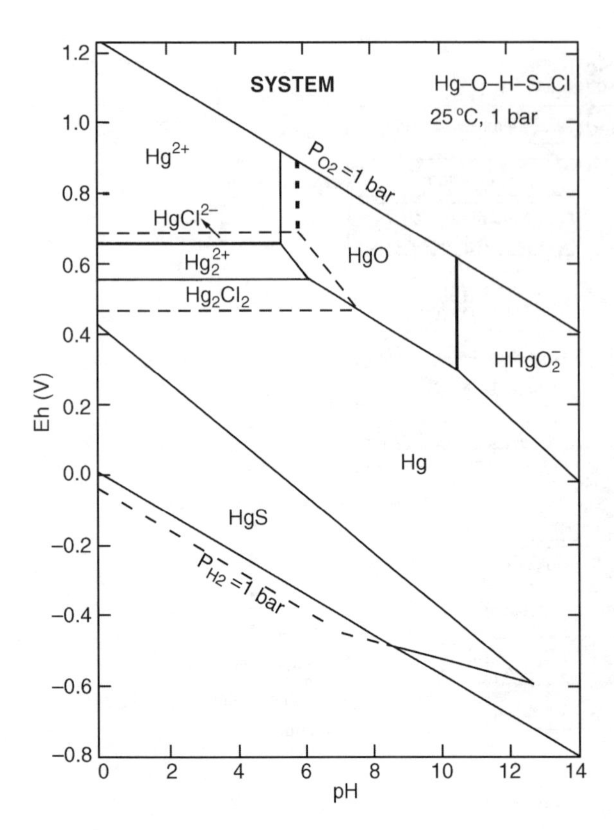

Figure 4.10 Eh-pH diagram for Hg in the system Hg-O-H-S-Cl at 25 °C and 1 bar pressure (Reproduced from Eh-pH Diagrams for Geochemistry, Brookins, D.G., © 1988/Springer-Verlag)

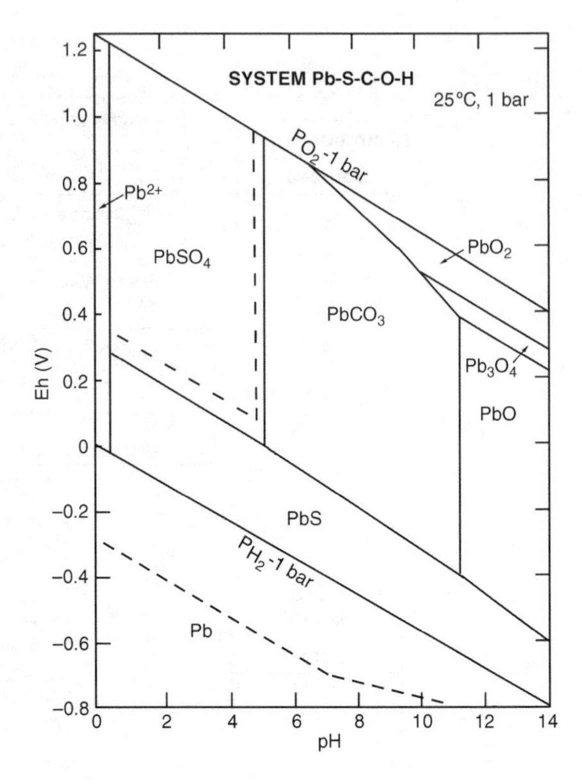

Figure 4.11 Eh-pH diagram for Pb in the system Pb-S-C-O-H at 25 °C and 1 bar pressure (Reproduced from Eh-pH Diagrams for Geochemistry, Brookins, D.G., © 1988/Springer-Verlag)

strongly correlated with concentrations of Zn, As and Sb (FOREGS, 2005). Anomalously high Pb levels reflect atmospheric precipitation, historical mining and smelting processes and industrial activities.

Air. Atmospheric concentrations of Pb in remote rural areas of the USA are in the range 0.9–1.2 ng/m^3, with average levels in urban air of around 4.5–9 ng/m^3 (Hutzell and Luecken, 2008); the highest concentrations are in the most densely populated areas. The input of Pb from atmospheric deposition is higher than that from P fertilisers on European agricultural soils (Nziguheba and Smolders, 2008).

4.4.5 Thallium

The major pathways of Tl in the environment are summarised in Figure 4.12.

Soil. Thallium tends to persist in soils and is leached in acid conditions. The greatest retention occurs in soils containing large amounts of clay, organic matter and Fe and Mn oxides. The concentrations of Tl in soils over Europe are in the range 0.05–24.0 mg/kg for topsoils and 0.01–21.3 mg/kg for subsoils (FOREGS, 2005). In China, Tl concentration of 0.011 mg/kg has been reported in garden soil (Zhou and Liu, 1985).

Water. Thallium exists in the monovalent (Tl^+), trivalent (Tl^{3+}) and tetravalent (Tl^{4+}) forms in the aqueous environment (Brookins, 1988). Under oxidising conditions, Tl oxides, including Tl_2O, Tl_2O_3 and Tl_2O_4, dominate in the aquatic environment (Figure 4.13). The monovalent Tl is more stable in inorganic complexes than the Tl^{3+}, which is more stable in organic complexes (WHO, 1996a). The hydroxide, carbonate and sulphate of Tl^+ are highly soluble in waters, like the corresponding K compounds, whereas the solubility of Tl^+ oxide, sulphide and halides is low, more like that of comparable compounds of Ag, Hg and Pb. The concentrations of dissolved Tl in Great Lakes waters has been found to be in the range 0.001–0.026 µg/l (Cheam *et al.*, 1995), while Tl levels in sea waters ranged from 0.012 to 0.016 µg/l (Flegal and Patterson, 1985). Henshaw *et al.* (1989) reported concentrations of up to 0.41 µg/l Tl in filtered water from fresh-water lakes in the eastern USA. In Europe, Tl concentration in stream waters is in the range $<$0.002–0.22 µg/l, with a mean of 0.009 µg/l (FOREGS, 2005).

Sediments. Thallium concentrations in stream sediments are in the range 0.15–23.1 mg/kg, depending on the local geology and geochemistry, but uncontaminated sediments can have Tl concentrations as low as 0.01 mg/kg (Bowen, 1966; Mathis and Kevern, 1975; Waidmann *et al.*, 1992). The level of Tl in stream sediments over Europe ranges from $<$0.02 to 7.90 mg/kg, with a mean of 0.477 mg/kg (FOREGS, 2005).

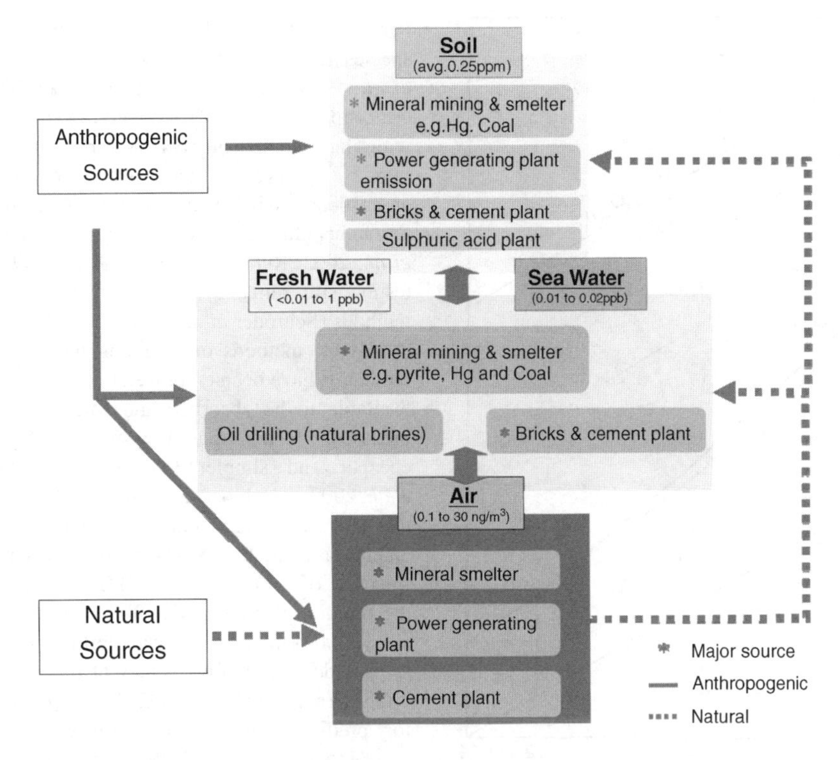

Figure 4.12 Pathways of Tl

Air. Mean values of 0.06 ng/m^3 of particulate Tl in air have been reported for Europe and of 0.22 ng/m^3 for North America (Bowen, 1979; Gaudry *et al.*, 2008). In Nebraska, USA, yearly mean values for Tl of 0.22 ng/m^3 $\pm$ 0.08 (range 0.07–0.48 ng/m^3) and of 0.15 $\pm$ 0.04 ng/m^3 were reported during the summers of 1973 and 1974, respectively (Struempler, 1975). In industrial and urban areas of Genoa, Italy, mean concentrations of 15 and 14 ng/m^3 Tl in air have been reported, with maximum values of 58 ng/m^3, although values were often below 1 ng/m^3 (Valerio *et al.*, 1988, 1989). In London, levels of 0.07–6.0 mg/kg Tl in dust have been reported (Bowen, 1979).

4.5 Effects on human receptors

4.5.1 Human exposure pathways

Human uptake of TTEs occurs through inhalation (respiratory exposure), ingestion of food and drink (oral exposure) and skin contact (dermal exposure) (Paustenbach, 2000). Among these pathways, exposure to TTEs in humans is primarily through oral exposure (Wragg and Cave, 2002). Exposure to airborne TTEs can be important in occupational settings, especially in industries that produce, refine, use or dispose of TTEs and their compounds or to people living near such sources (WHO, 1995). The main human exposure pathways of As, Cd, Pb, Hg and Tl are summarised in Figure 4.14.

4.5.2 Bioaccessibility

A proportion of TTEs taken through respiratory, oral and dermal exposure is absorbed in human organs or enters the human bloodstream, depending on their bioavailable fractions. There is limited information on the bioavailability of TTEs (Figure 4.15), and more research on the absorption and metabolism of TTEs during transit through the digestive tract is required before an assessment can be made of the risk to human health. In-vivo tests are methods for estimating the bioavailability of TTEs, using animals with similar gastrointestinal-tract characteristics to humans (Wragg and Cave, 2002), and juvenile swine, rabbits and rats have been used to predict the bioavailability of TTEs in humans (Ruby *et al.*, 1999). Unfortunately, the digestive physiology of rats and rabbits differs from that of humans, and rats also have a pre-stomach compartment with a very specific physiology. Although these models have a long history of use in toxicological studies, the aim for bioaccessibility and bioavailability studies with respect to soil is to mimic human physiology and model the interactions occurring between the soil contaminants and the digestive fluids. The animals considered to be most like humans physiologically are other primates, but few experiments have been conducted with these animals (Roberts *et al.*, 2007), because of financial and ethical considerations. Juvenile swine are considered to be a useful anatomical proxy for the human neonatal digestive tract (Miller and Ullrey, 1987; Moughan *et al.*, 1992) and have been successfully used to assess the bioavailability of Pb

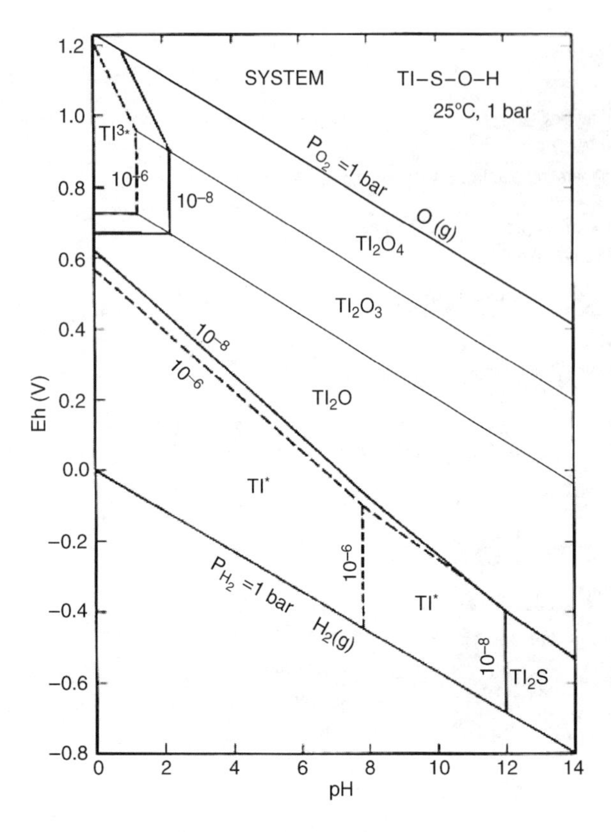

Figure 4.13 Eh-pH diagram for Tl in the system Tl-S-O-H at 25 °C and 1 bar pressure (Reproduced from Eh-pH Diagrams for Geochemistry, Brookins, D.G., © 1988/Springer-Verlag)

(Casteel *et al.*, 1996, 2006). These animals are now the preferred species for such experiments (Moughan *et al.*, 1994; Rowan *et al.*, 1994). Animal studies are expensive, they are open to criticism on the grounds of cruelty, and they cannot be conducted with a large enough number and variety of contaminated soils. In-vitro tests (tests performed in a laboratory dish or test tube, i.e. in an artificial environment ('in vitro' Latin for 'in glass')) allow researchers to overcome the limitations of animal tests (Oomen *et al.*, 2003; Ruby, 2004; Ruby *et al.*, 1999). However, animal testing still remains a necessary means to validate in-vitro methods (Schroder *et al.*, 2004).

In-vitro methods of assessing the bioaccessibility of contaminants have been employed to estimate the fraction of TTEs available to be absorbed and enter the human body. These methods, which are based on simulated human physiology, are cheaper and simpler than in-vivo animal tests (Oomen *et al.*, 2002). The bioaccessibility of TTEs represents the fraction that exists as a soluble phase in the gastrointestinal environment and that is available for absorption (Paustenbach, 2000). As the fraction of bioaccessible TTEs is generally higher than the fraction of bioavailable TTEs (Ruby *et al.*, 1996; Semple *et al.*, 2004), in-vitro bioaccessibility measurements are likely to provide conservative values of the bioavailable fraction of TTEs (Wragg and Cave, 2002). There are many in-vitro methods for predicting the bioaccessibility of TTEs in simulated gastrointestinal tracts. Of these, the Physiologically Based Extraction Test (PBET) has been recommended as a pragmatic method of assessing the bioaccessibility of contaminants (Cave *et al.*, 2003; Ruby, 2004). The PBET is designed to estimate the bioaccessible fraction of contaminants in the gastrointestinal tract of a 2- to 3-year-old child exposed to accidental soil ingestion (Ruby *et al.*, 1993). Leaching of contaminants from solid media in the gastrointestinal tract is predicted, based on a

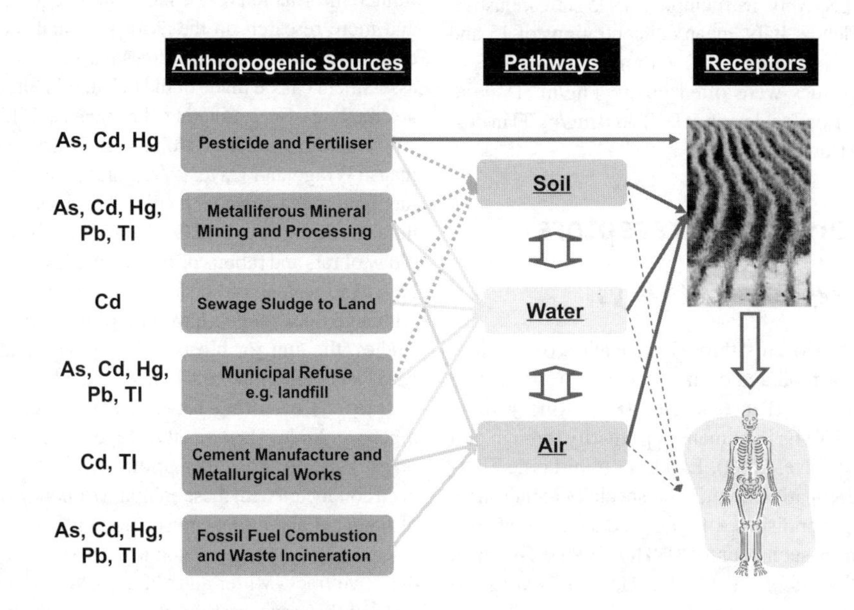

Figure 4.14 Anthropogenic sources, pathways and receptors of TTEs

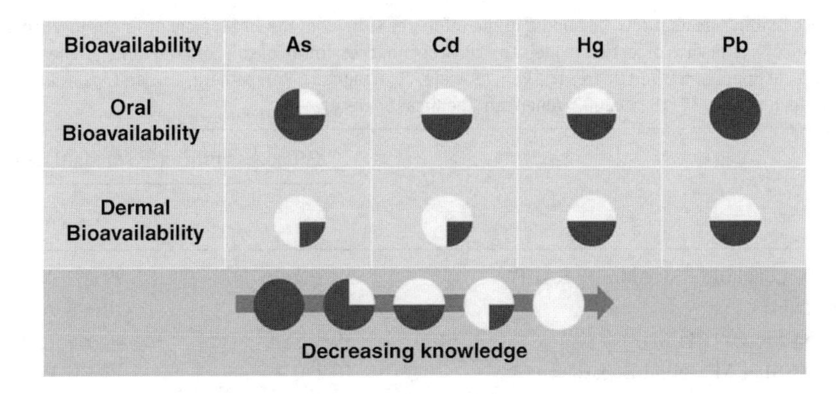

Figure 4.15 Extent of knowledge regarding the bioavailability of metals in soil (Modified, with permission, from Bioavailability of soil-borne chemicals: Abiotic assessment tools, Ruby, M.V., Human and Ecological Risk Assessment/10/4, © 2004/Taylor & Francis)

two-stage sequential extraction that uses various acids and enzymes to simulate both the stomach and the small intestine of humans (Wragg and Cave, 2002). The bioaccessibility of TTEs in soils and plants estimated by PBET is presented in Table 4.4. The bioaccessible fractions of TTEs (per cent) shown in the table are calculated by taking the concentrations of TTEs dissolved from a soil (or plant) matrix in simulated gastrointestinal solution and dividing them by the total concentrations of TTEs present in the soils (or plant) used.

The Relative Bioaccessibility Leaching Procedure (RBALP), previously called the Simplified Bioaccessibility Extraction Test (SBET), is a simplified bioaccessibility procedure developed for Pb bioaccessibility measurements. It is not physiologically based (it uses glyceine as the acid buffered medium) but has been validated against a swine model and is accepted for Pb bioaccessibility in soil by the US EPA (Drexler and Brattin, 2007). It is designed to estimate the bioaccessible fraction of Pb in the gastric solution alone (Wragg and Cave, 2002) and is therefore carried out in low-pH conditions. Hence, it indicates higher bioaccessibility of Cd and Pb than other in-vitro tests (Oomen *et al.*, 2002). A lung-fluid test has been used to examine the bioaccessibility of heavy metals, but only for As and platinum (Pt) (Colombo *et al.*, 2008). The bioaccessible fractions of As, Cd and Pb in the reference soil, Montana 2711, from different in-vitro assays are presented in Table 4.5 (Oomen *et al.*, 2002). The results varied widely, depending on the procedures used even in the same soil; hence they need to be further validated for site-specific human-health-risk assessments (Zagury *et al.*, 2009). Recent collaborative research lead by BARGE (Bioaccessibility Research Group of Europe), a network of European institutes and research groups formed in 1999/2000 (www.bgs.ac.uk/barge), has been aimed at producing a standardised procedure to ensure comparability between studies. This work has resulted in the development of a physiologically relevant bioaccessibility test (the Unified BARGE Method, UBM), which has been validated against a swine model for As, Cd and Pb in soils (Caboche, 2009) and undergone preliminary inter-laboratory trials (Wragg *et al.*, 2009). The UBM is being put forward as a standardised bioaccessibility for the assessment of human-health risk.

4.5.3 Human health effects

The effects on human health of adsorption of TTEs as a consequence of acute or chronic exposure are indicated in Table 4.6.

Table 4.4 Examples of the bioaccessibility of TTEs from some soils and plants, estimated by PBET

TTEs	Bioaccessibility (per cent)					
	Soils*		Plants**			
	Minimum	Maximum	Lettuce	Spinach	Carrot	Radish
As	29	55	–	–	–	–
Cd	–	–	42.41	55.73	45.23	52.61
Pb	0.2	83	49.58	30.18	43.52	60.69

*The range of bioaccessibility of TTEs was determined from PBET results of 7 samples for Pb and 3 samples for As (Ruby *et al.*, 1996).
**The higher bioaccessibility of TTEs in plants grown in non-contaminated (control) soil, between the gastric and the intestinal compartments (Intawongse and Dean, 2008).

Table 4.5 Results of bioaccessibility tests for As, Cd and Pb in the reference soil, Montana 2711, from different in-vitro assays (Modified, with permission, from Comparison of five in vitro digestion models to study the bioaccessibility of soil contaminants, Oomen, A.G., Hack, A., Minekus, M., Zeijdner, E., Cornelis, C., Schoeters, G., Verstraete, W., Van de Wiele, T., Wragg, J., Rompelberg, C.J.M., Sips, A.J.A.M. & Van Wijnen, J.H., Environmental Science & Technology/36/15, © 2002/American Chemical Society)

In-vitro assay	Bioaccessibility (± standard deviation) (per cent)		
	As	Cd	Pb
Relative Bioaccessibility Leaching Procedure (RBALP)	59 ± 2	99 ± 4	90 ± 2
German DIN Method (DIN)	50 ± 1	79 ± 8	68 ± 2
RIVM In-vitro Digestion Model (RIVM)	59 ± 1	40 ± 2	11 ± 2
Simulator of Human Intestinal Microbial Ecosystem of Infants (SHIME)	10 ± 0.4	6 ± 0.3	3 ± 0.3
TNO Gastrointestinal Model (TIM)	50 ± 1	58 ± 1	17 ± 3

4.5.3.1 Arsenic

Arsenic is essential for some organisms such as rats, sheep and chickens, but it is toxic to humans. Following ingestion, between 60 and 90 per cent of soluble As is absorbed from the gastrointestinal tract (ATSDR, 1990) and is subsequently stored in the heart, kidneys, liver, lungs, and to a lesser extent muscles and neural tissues. In addition to its accumulation in such organs, As is detected from hair and nails of people chronically exposed to As (Bencko *et al.*, 1971; Mandal *et al.*, 2003). There are significant differences in the toxicity of different As compounds. Organic As compounds such as arsenobetaine and arsenocholine are generally much less toxic than inorganic As species.

Table 4.6 Summary of human health effects of TTEs (Archer, 2006)

TTEs	Acute Exposure	Chronic Exposure
As	Vomiting; oesophageal and abdominal pain; bloody 'rice water' diarrhoea; facial and cerebral oedema; anorexia; insomnia; abnormal heart rhythm	Blood-vessel damage; black-foot disease leading to gangrene; skin pigmentation changes; hyperkeratosis; ulceration; small corns or warts on palms, soles and torso; Mee's lines; Raynaud's syndrome; myocardial ischemia; haematological changes; anaemia; hepatomegaly; damage to central or peripheral nervous system (pins and needles in hands and feet); skin epithelioma and other cancers; coma; cardiovascular failure and death; inhibition of cellular respiration, causing organ failure
Cd	Nausea; vomiting; increased salivation; choking; abdominal pains; metabolic acidosis; tenesmus; diarrhoea; headaches	Zn deficiency; nephrotoxicity; increased protein in urine; hypertension; infertility; anaemia; cardiovascular diseases; renal failure; cardiopulmonary collapse; itai-itai disease (a form of osteomalacia) and bone pseudofractures; cancer
Hg	Appetite loss; headaches; fever; irritability; fatigue; vomiting	Neuropathies causing paralysis, memory loss, tremor and impaired physical coordination; chromosome damage; teratogenesis; mutagenesis; cancer; death
Pb	Anorexia; dyspepsia; weakness in fingers, wrists or ankles; constipation	Neuropathies causing memory loss, reduced IQ, anaemia, haematopathies, nephropathies, ataxia, seizures, reduced male fertility, stunted growth in infants, coma and death
Tl	Vomiting; diarrhoea; headaches; constipation; alopecia	Nervous system damage, including paresthesia; muscular pain and weakness; mental confusion or delirium; convulsions; heart, liver and kidney diseases; death.

Acute exposure to As can cause gastrointestinal effects, nausea, vomiting, anuria (inability to urinate), neurological effects and, in serious cases, coma and death due to fluid loss and circulatory collapse (Cullen *et al.*, 1995). Acute exposure to As has also been shown to result in memory problems, difficulties in concentration, anxiety and mental confusion (Hall, 2002; Rodriguez *et al.*, 2003).

Chronic exposure to As increases the risk of gangrene of the externals, cancer of the skin and kidneys, and liver failure (WHO, 1996b). Chronic As poisoning causes arsenicosis and it is considered to be a modifying factor in atherosclerosis and related cardiovascular diseases, depending on other factors such as genetic background, diet and co-exposure (Simeonova and Luster, 2004). Arsenic and its compounds are known to be carcinogens and are classified as Group 1, 'carcinogenic to humans' (IARC, 1987). It has been demonstrated that As interacts with the nervous system in humans at several levels, but further studies are needed to understand the biological/biochemical processes involved. Chronic exposure to As is associated with peripheral neuropathy and demyelination (Heaven *et al.*, 1994; Steven and Greenberg, 1996; Vahidnia *et al.*, 2007). Encephalopathy and impairment of superior neurological functions, which are normally observed in patients with acute As poisoning, have also been identified in individuals with chronic occupational exposure to As (Rodriguez *et al.*, 2003).

4.5.3.2 Cadmium

Cadmium species are absorbed in human bodies by respiratory, oral and dermal exposure, though dermal absorption of Cd is much less than other exposure routes (ATSDR, 2008). Cadmium, especially inorganic Cd, is highly toxic to humans (Gleason *et al.*, 1969), and it has a long biological half-life of 10–35 years (Elinder *et al.*, 1976). The majority of the total body burden of Cd is in the liver and kidneys, mostly in the kidneys, while Cd is slowly excreted in urine and faeces (ATSDR, 2008). Most people have a kidney concentration below 30 mg/kg Cd, although many cigarette smokers have concentrations twice this level. The mode of action of Cd in the human body depends on the dose, duration and the route of exposure.

Many cases of acute and chronic Cd poisoning have been reported from industrialised countries such as the USA, Japan and Europe, mainly related to battery manufacture. Acute Cd toxicity is associated with severe abdominal pain, nausea, vomiting, diarrhoea, headache and vertigo, and it can lead to death within 24 hours, or 1–2 weeks later, following liver and kidney damage (Tsuchiya, 1980; Yasumura *et al.*, 1980).

Symptoms of chronic Cd toxicity include respiratory problems, renal dysfunction, disorders of Ca metabolism and bone disease such as osteoporosis, including spontaneous bone fracture. These are the symptoms of itai-itai disease in Japan (Tsuchiya, 1980). It was the Japanese experience that raised awareness and concern about the toxicity of Cd, especially in heavily industrialised regions of the world. One study on the effect of Cd on bone formation shows that it inhibits osteoblasts which rebuild bone (Chen *et al.*, 2009). Other symptoms of chronic Cd poisoning include tubular damage in the kidneys, anaemia, tooth discoloration (from the formation of cadmium sulphide) and loss of sense of smell. In an IARC monograph, Cd and its compounds are classified as a Group 1 carcinogen, 'carcinogenic to humans' (IARC, 1993). Cadmium is also listed as an endocrine-disrupting substance (Darbre, 2006) because of its ability to bind to cellular oestrogen receptors and to mimic the actions of oestrogens. There is some evidence that exposure to Cd may be a factor in the development of prostate cancer (Waalkes, 2003).

4.5.3.3 Mercury

Mercuric species are highly toxic, but of even greater concern is the ability of micro-organisms to create fat-soluble species such as monomethylmercury (CH_3Hg^+) and dimethylmercury ($(CH_3)_2Hg$) (Westcott and Kalff, 1996). These compounds are readily taken up by aquatic organisms, and may be concentrated to several mg/kg by higher organisms in the food chain, such as fish and shellfish (Wolfe *et al.*, 1998). General human exposure to Hg is primarily associated with eating fish and marine mammals that accumulate methylmercury in their tissues and with the use of dental amalgam (ATSDR, 1999). Most organic Hg, up to 90–100 per cent, and inorganic Hg, up to 10–15 per cent, is absorbed in the gastrointestinal tract (Archer, 2006); the kidneys accumulate higher levels of Hg than the brain or liver (Hussain *et al.*, 1998).

One of the effects of Hg in the central nervous system (CNS) is to damage glial cells known as astrocytes which absorb the excitatory neurotransmitter glutamate and convert it to the non-excitatory substance glutamine (Fonfria *et al.*, 2005). Classic symptoms of Hg poisoning include tremors, emotional excitability, insomnia, dementia and hallucinations – known as the Mad Hatter's syndrome.

The ingestion of acute toxic doses of any form of Hg results in shock, cardiovascular collapse, acute renal failure and severe gastritis and colitis. Clinical symptoms of acute toxicity include pharyngitis, dysphagia (difficulty in swallowing), abdominal pain, nausea, vomiting, bloody diarrhoea and shock. Later, swelling of the salivary glands, stomatitis, loosening of the teeth, nephritis, anuria and hepatitis occur (Stockinger, 1981).

Chronic exposure to either inorganic or organic Hg can permanently damage the brain, kidneys and developing fetus (ATSDR, 1999). With regard to carcinogenicity, methylmercury compounds are 'possibly carcinogenic to humans' (Group 2B), while metallic Hg and inorganic Hg compounds are 'not classifiable as to their carcinogenicity to humans' (Group 3) (IARC, 1993). Chronic exposure to dietary methylmercury or mercury vapour leads to typical neurotoxic symptoms including disruption of the nervous system, damage to brain functions, DNA and chromosomes and negative reproductive effects such as sperm damage, birth defects and miscarriages (Zahir *et al.*, 2005).

Some scientists and doctors have attributed autism to the past use of thimerosal in the triple vaccine against mumps, measles and rubella (MMR) (Clarkson, 2002; Parker *et al.*, 2004), but links between autism and the MMR vaccines have since been discredited (CDC, 2009). Thimerosal is no longer used, or has been reduced to trace amounts, in most vaccines, especially those routinely used for children 6 years of age or younger (US FDA, 2010), but it is still used as a preservative in some influenza vaccines (CDC, 2009).

4.5.3.4 Lead

Lead can affect all of the organs of the body, and Pb accumulation in tissues is independent of the route of exposure (DEFRA and Environment Agency, 2002). Inorganic Pb can be absorbed in humans by respiratory and oral exposure and, to much less extent, by dermal exposure. Organic Pb is found to be well absorbed through the skin in tests on animals. The severity of Pb toxicity is affected by its speciation, the duration of exposure and the pathway into the body (ATSDR, 2007).

One of the principal concerns about Pb is its harmful effect on developing the nervous system of children and its potential to affect intelligence (Bellinger *et al.*, 1992). Some of the effects of Pb on adult and child health are shown in Figure 4.16. In blood, Pb is usually bound to proteins in red blood cells, while approximately 40–75 per cent of Pb in plasma is bound to plasma proteins, mostly on albumin with some bound on γ-globulins (Ong and Lee, 1980). Lead that is not bound to proteins in plasma forms complexes with low-molecular-weight sulphydryl compounds such as the amino acid cysteine (Al-Modhefer *et al.*, 1991). Human exposure to Pb is therefore often monitored by analysing blood samples, although long-term cumulative Pb concentrations may be better determined on bone samples (ATSDR, 2007).

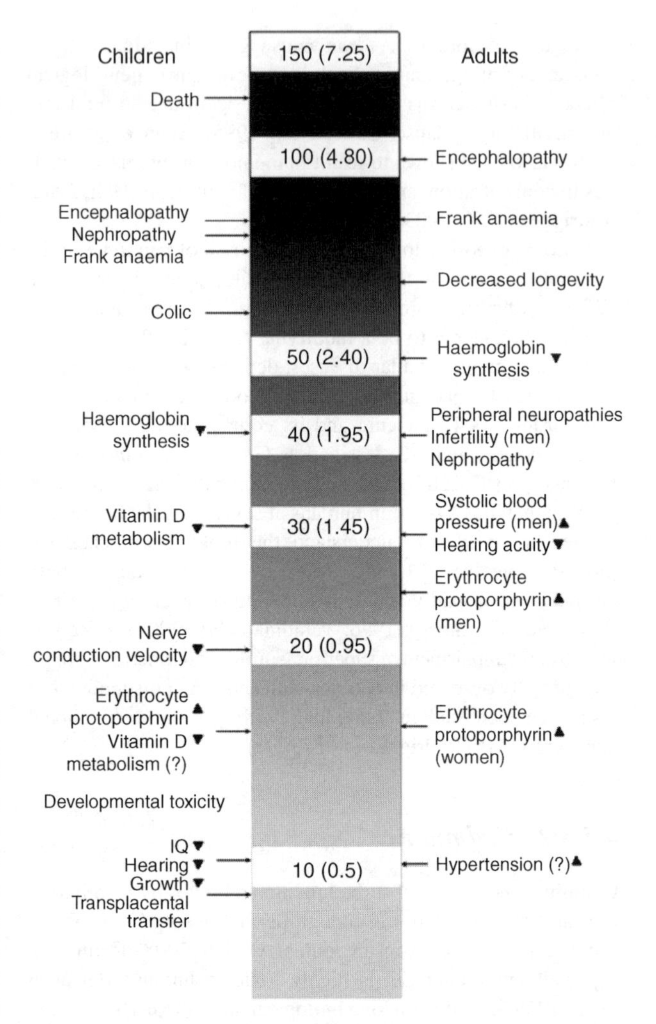

Figure 4.16 Effects of inorganic Pb on children and adults (Lowest Observed Adverse Effect Level, LOAEL). Blood Pb concentrations are given as g/dl (values in brackets are in µmol/l). The effects shown on the left side apply to children, those on the right side to adults

Typical symptoms of acute Pb toxicity include abdominal pains, constipation, cramps, nausea and vomiting, which typically occur at blood Pb concentrations of 100–200 µg/dl although colic has been reported in workers with blood concentrations of only 40–60 µg/dl (ATSDR, 2007; Green *et al.*, 1976). In children, brain and kidney damage occur at blood Pb concentrations of around 80 µg/dl (US EPA, 2000).

Chronic exposure to Pb generally leads to its accumulation in the bones. Around 94 per cent of the total body burden of Pb is present in bones in human adults, while in children this is 75 per cent (ATSDR, 2007; Lamadrid-Figueroa *et al.*, 2006). This reservoir for Pb can maintain blood Pb levels even after exposure to Pb has ceased, and the Pb can be transferred to the fetus during pregnancy as the maternal bones are resorbed for fetal skeleton production. Chronic Pb poisoning can have adverse neurological, cardiovascular, renal, hepatic, gastrointestinal, developmental, reproductive, endocrine and haematological effects (ATSDR, 2007). The most important diseases attributed to chronic exposure to Pb in adults are hypertension, chronic kidney disease and Pb-induced anaemia (Carmignani *et al.*, 1988).

Studies on the carcinogenicity of Pb and its compounds are inconclusive. A study suggested that there is a significant correlation between exposure to TEL and the incidence of colorectal cancer in workers (Fayerweather *et al.*, 1997). Lead is a suspected human carcinogen, and the US EPA (2000) classifies Pb as a Group B2 carcinogen, 'probable human carcinogen'. The IARC classification places organic Pb compounds in Group 3, 'not classifiable as to their carcinogenicity to humans', because of inadequate evidence in humans or experimental animals, while inorganic Pb compounds are in Group 2A, 'probably carcinogenic to humans' (IARC, 2006).

4.5.3.5 *Thallium*

Compared with other TTEs, few studies have been performed on the mechanism of Tl toxicity (Galván-Arzate and Santamaría, 1998). Only a few cases of Tl poisoning as a result of industrial exposure have been reported, and these are mainly the result of skin contact or inhalation. Increased Tl levels have been found in the urine of workers in a Zn smelter in eastern Germany, with levels as high as 28.6 μg/l in the part of the plant concerned with Tl production (WHO, 1996a). In Italy, slightly but significantly higher Tl levels were found in the urine of cement workers (0.4 μg/l) and cast iron workers (0.3 μg/l) than in a non-exposed group (0.2 μg/l) (Apostoli *et al.*, 1988).

Almost 100 per cent of ingested soluble Tl species is adsorbed in the gastrointestinal tract (Archer, 2006; ATSDR, 1992). Concentrations are significantly different in vertebrae (12.7 mg/kg), sternum (7.0 mg/kg), femur (16.4 mg/kg) and tibia (9.0 mg/kg) (Arnold, 1986). The distribution of Tl differs considerably from that reported for K in humans (Davis *et al.*, 1981). The endocrine glands, kidneys, liver and intestine (without content) showed the highest concentrations. With respect to the total amount per organ, liver and lung were found to contain 2 to 6 times as much Tl as the kidneys, and the brain about 1.5 to 2 as much (Arnold, 1986; Curry *et al.*, 1969).

The process of Tl poisoning in animals and humans is similar to the enzyme-inhibiting effects indicated in exposure to Pb and Hg (Kemper and Bertram, 1991). Symptoms of acute Tl toxicity depend on the age of the person, the route of administration and the dose (WHO, 1996a). Doses which have proved lethal in adults vary between 6 and 40 mg/kg body weight, with an average of 10–15 mg/kg body weight (Schoer, 1984). Acute exposure to Tl results in vomiting, diarrhoea, temporary hair loss, palmar erythema, acne, anhydrosis and dry scaly skin (Mulkey and Oehme, 1993; Peter and Viraraghavan, 2005). The triad of symptoms of gastroenteritis, polyneuropathy and alopecia is regarded as the classic Tl poisoning syndrome (Gastel, 1978), but in some cases gastroenteritis and alopecia are not observed.

Symptoms of chronic Tl poisoning are similar to those of acute intoxication, but they are generally milder (Goldblatt, 1989; Schoer, 1984). Chronic Tl poisoning causes anorexia, headache and pains in abdomen, upper arms and thighs and even all over the body (Peter and Viraraghavan, 2005). Alopecia, blindness and in extreme cases death can occur in chronic exposure to Tl (Zhang *et al.*, 1998). Limited studies reported chronic Tl effects on human reproduction, including menstrual cycle, libido, male potency and sperm production (Peter and Viraraghavan, 2005; WHO, 1996a).

4.6 Risk reduction

The effects of TTEs on the environment and human health have been known since ancient times. The contribution of natural sources has remained steady, but the scale and impact of anthropogenic sources have increased dramatically with industrialisation and globalisation. Mining, smelting and refining of most base metals have been spread across several countries (often in different continents), causing the accumulation of TTEs and their compounds in most environmental compartments. Industrial processes and wastes generated have increased the TTE exposure of populations living or working around such settings, especially through inhalation of contaminated air or ingestion of polluted water.

Municipal waste water containing domestic and commercial effluents is another significant source of TTEs to humans (Chon *et al.*, 2010). The contribution of TTEs from municipal waste water to receiving water bodies requires attention, as does catchment-based water management. The development of urban areas has increased the extent of hard surfaces and therefore affects the volume of surface waters with high concentrations of TTEs emitted by traffic and the abrasion of building materials (Chon *et al.*, 2010). As emissions of TTEs from human actions have increased, their impact on the environment has been extended from the local to the global scale.

Clearly, the most effective way of minimising environmental and human-health risks from TTEs is to reduce the probability of exposure, and risk-management methods are required based on understanding of the sources and different emission patterns of TTEs (Figure 4.17). Since most emissions are difficult to control, especially those from natural sources, risk-reduction strategies should reflect ways of reducing exposure to anthropogenic sources of TTEs via air, soils, waters and sediments. Considering the non-biodegradable properties of TTEs, humans, as final receptors in the food chain, may accumulate or absorb higher quantities of TTEs than other receptors.

Because it is difficult and expensive to remediate TTE pollution of environmental media, management of anthropogenic sources can be the most effective method of risk reduction. In the West, mining companies now generally follow a code of practice known as the global industry initiative. In addition to source management, there is increased regulation aimed at mitigating emissions of toxic chemicals internationally. The Registration, Evaluation, Authorisation and Restriction of Chemicals (REACH) is, for example, European legislation implemented in 2007 with the aim of improving risk management and controlling the volume of industrial chemicals in Europe (see Chapter 2). Similar actions to improve regulation and manage the risks due to TTEs are required elsewhere – especially in China and India because of the extent of fossil-fuel burning in these countries. International cooperation employing the same standards or guidelines to recover the chemical and ecological status of the environment may reduce emissions of TTEs and prevent a serious and long-lasting effect of these elements on the biosphere.

Recent advances in mass spectrometry are allowing stable-isotope variation in metallic elements, including Cd, Hg and Tl, to be determined more precisely than hitherto. An increasing number of such studies are showing that the stable-isotope composition of several metals varies significantly in nature, and this variability can be used to increase our understanding of their low-temperature biogeochemistry (Bullen and

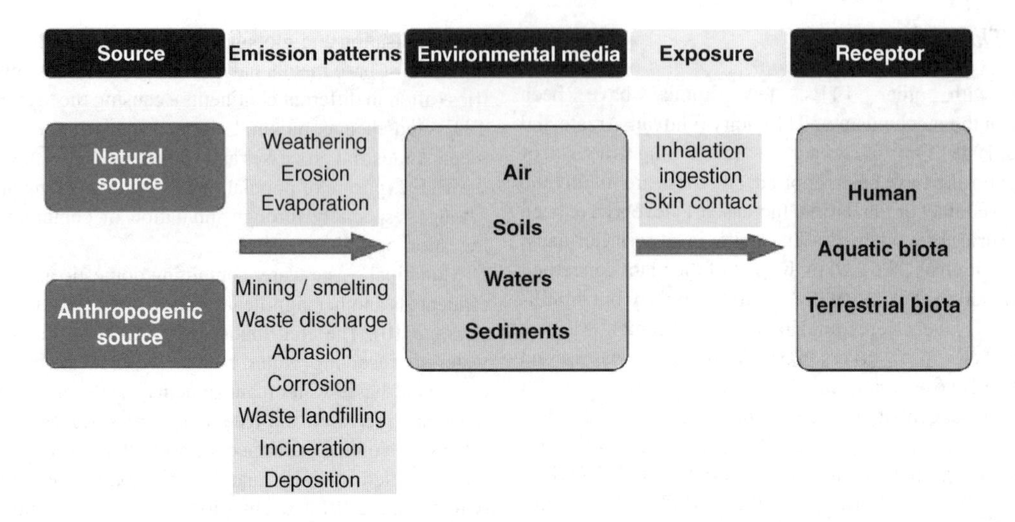

Figure 4.17 A source–pathway–receptor model for TTEs in the environment

Eisenhauer, 2009). Such research should great improve the risk management of TTEs such as Cd, Hg and Tl.

References

Al-Khashman, O.A. and Shawabkeh, R.A. (2006) Metals distribution in soils around the cement factory in southern Jordan. *Environmental Pollution*, **140**, 387–394.

Al-Modhefer, A.J.A., Bradbury, M.W.B. and Simmons, T.J.B. (1991) Observations on the chemical nature of lead in human blood serum. *Clinical Science*, **81**, 823–829.

Ansorena, J., Marino, N. and Legorburu, I. (1995) Agricultural use of metal polluted soils near an old lead-zinc mine in Oiartzun (Basque Country, Spain). *Environmental Technology*, **16**, 213–222.

Aoyama, H., Yoshida, M. and Yamamura, Y. (1988) Induction of lipid peroxidation in tissues of thallous malonate-treated hamster. *Toxicology*, **53**, 11–18.

Apostoli, P., Maranelli, G., Minoia, C., Massola, A., Baldi, C. and Marchiori, L. (1988) Urinary thallium: Critical problems, reference values and preliminary results of an investigation in workers with suspected industrial exposure. *Science of the Total Environment*, **71**, 513–518.

Archer, J.K. (2006) Aqueous exposure and uptake of heavy metals and metalloids by communities living alongside the mining contaminated Rio Pilcomayo, Bolivia. Ph.D. Thesis, Birkbeck, University of London.

Arnold, W. (1986) Excretion and distribution of thallium in human poisoning. Proceedings of 5th Symposium of Trace Elements. Jena, Germany, July 14th-17th, 1986 (in German).

ATSDR (1990) *Arsenic toxicity. Case studies in environmental medicine*. Agency for Toxic Substances and Disease Registry, US Department of Health and Human Services, Atlanta, Georgia.

ATSDR (1992) *Toxicological profile for thallium*. Agency for Toxic Substances and Disease Registry, US Department of Health and Human Services, Atlanta, Georgia.

ATSDR (1999) *Toxicological profile for mercury*. Agency for Toxic Substances and Disease Registry, US Department of Health and Human Services, Atlanta, Georgia.

ATSDR (2007) *Toxicological profile for lead*. Agency for Toxic Substances and Disease Registry, US Department of Health and Human Services, Atlanta, Georgia.

ATSDR (2008) *Draft toxicological profile for cadmium*. Agency for Toxic Substances and Disease Registry, US Department of Health and Human Services, Atlanta, Georgia.

Baur, W.H. and Onishi, H. (1978) Arsenic. In Wedepohl, K.H. (Ed.) *Handbook of geochemistry*, Springer-Verlag, Berlin, Heidelberg, New York.

Bellinger, D.C., Stiles, K.M. and Needleman, H.L. (1992) Low-level lead exposure, intelligence and academic achievement: A long-term following-up study. *Pediatrics*, **90**, 855–861.

Bencko, V., Dobisova, A. and Macaj, M. (1971) Arsenic in hair of a nonoccupationally exposed population. *Atmospheric Environment*, **5**, 275–279.

Bentley, R. and Chasteen, T.G. (2002) Arsenic curiosa and humanity. *The Chemical Educator*, **7**, 51–60.

BGS (1992) *Regional geochemistry of the lake district and adjacent areas*. British Geological Survey, Keyworth, Nottingham.

Biester, H., Gosar, M. and Müller, G. (1999) Mercury speciation in tailings of the Idrija mercury mine. *Journal of Geochemical Exploration*, **65**, 195–204.

Block, C. and Dams, R. (1975) Lead contents of coal, coal ash and fly ash. *Water, Air, and Soil Pollution*, **5**, 207–211.

Bowen, H.J.M. (1966) *Trace Elements in Biochemistry*. Academic, New York.

Bowen, H.J.M. (1979) *Environmental chemistry of the elements*. Academic Press, New York, London.

Bowen, H.J.M. (Ed.) (1982) *Environmental chemistry*. Specialist Periodical Report, Royal Society of Chemistry, London.

Boyle, R.W. and Jonasson, I.R. (1973) The geochemistry of arsenic and its use as an indicator element in geochemical prospecting. *Journal of Geochemical Exploration*, **2**, 251–296.

Breward, N., Williams, T.M. and Cummins, C. (1994) *Environmental geochemistry of the Mengapur and Sungai Luit mining areas, near Kuantan, Pahang, Malaysia*. WC/94/26R, British Geological Survey, Keyworth, Nottingham.

Brewster, U.C. and Perazzella, M.A. (2004) A review of chronic lead intoxication: an unrecognised cause of chronic kidney disease. *American Journal of Medical Science*, **327**, 341–347.

Brookins, D.G. (1988) *Eh-pH diagrams for geochemistry*. Springer-Verlag, Berlin, Heidelberg, New York, London, Paris, Tokyo.

Brumsack, H.J. (1977) Potential metal pollution in grass and soil samples around brickworks. *Environmental Geology*, **2**, 33–41.

Bullen, T.D. and Eisenhauer, A. (2009) *Metal stable isotopes in low-temperature systems: A primer*. *Elements*, **5**(6), Mineralogical Society of America, pp. 349–352.

Caboche, J. (2009) Validation d'un test de mesure de bioaccessibilité: Application à quatre éléments traces métallique dans les sols: As, Cd, Pb et Sb. Ph.D. Thesis, L'Institut National Polytechnique de Lorraine, Nancy.

Carmignani, M., Boscolo, P. and Preziosi, P. (1988) Cardiovascular actions of lead in rats as related to the level of chronic exposure. *Archives of Toxicology Supplement*, **12**, 326–329.

Casteel, S.W., Cowart, R.P., Weis, C.P., Henningsen, G.M., Hoffman, E., Brattin, W.J., Starost, M.F., Payne, J.T., Stockham, S.L., Becker, S.V. and Turk, J.R. (1996) A swine model for determining the bioavailability of lead from contaminated media. In Tumbleson, M.E. and Schook, L.B. (Eds), *Advances in swine in biomedical research*, **Vols 1** and **2**, Plenum Press, New York, pp. 637–646.

Casteel, S.W., Weis, C.P., Henningsen, G.M. and Brattin, W.J. (2006) Estimation of relative bioavailability of lead in soil and soil-like materials using young swine. *Environmental Health Perspectives*, **114**, 1162–1171.

Cave, M.R., Wragg, J., Palumbo, B. and Klinck, B.A. (2003) Measurement of the bioaccessibility of arsenic in UK soils. R&D Technical Report P5-062/TR2, Environment Agency.

CDC (2009) Thimerosal. Centers for Disease Control and Prevention. Available via. http://www.cdc.gov/vaccinesafety/Concerns/Thimerosal/Index.html. Accessed 22 January 2010.

Chandler, H.A. and Scott, M. (1986) A review of thallium toxicology. *Journal of the Royal Naval Medical Service*, **72**, 75–79.

Chattopadhyay, A. and Jervis, R.E. (1974) Multielement determination in market-garden soils by instrumental photon activation analysis. *Analytical Chemistry*, **46**, 1630–1639.

Cheam, V., Lechner, J., Desrosiers, R., Sekerka, I., Lawson, G. and Mudroch, A. (1995) Dissolved and total thallium in Great Lakes waters. *Journal of Great Lakes Research*, **21**, 384–394.

Chen, X., Zhu, G.Y., Gu, S.Z., Jin, T.Y. and Shao, C.L. (2009) Effects of cadmium on osteoblasts and osteoclasts in vitro. *Environmental Toxicology and Pharmacology*, **28**, 232–236.

Cheng, H., Hu, Y., Luo, J., Xu, B. and Zhao, J. (2009) Geochemical processes controlling fate and transport of arsenic in acid mine drainage (AMD) and natural systems. *Journal of Hazardous Materials*, **165**, 13–26.

Chernyak, Y.B. and Nussinov, M.D. (1976) Volatilization from solid particles of the regolith. *Nature*, **264**, 241.

Chon, H.S., Ohandja, D.G. and Voulvoulis, N. (2010) Implementation of EU Water Framework Directive: source assessment of metallic substances at catchment levels. *Journal of Environmental Monitoring*, **12**, 36–47.

Clarkson, T.W. (2002) The three modern faces of mercury. *Environmental Health Perspectives*, **110**, 11–23.

Clifton, J.C. (2007) Mercury exposure and public health. *Pediatric Clinics of North America*, **54**, 237–269.

Colombo, C., Monhemius, A.J. and Plant, J.A. (2008) Platinum, palladium and rhodium release from vehicle exhaust catalysts and road dust exposed to simulated lung fluids. *Ecotoxicology and Environmental Safety*, **71**, 722–730.

Cullen, N.M., Wolf, L.R. and St Clair, D. (1995) Pediatric arsenic ingestion. *The American Journal of Emergency Medicine*, **13**, 432–435.

Cummings, D.E., Caccavo, F., Fendorf, S. and Rosenzweig, R.F. (1999) Arsenic mobilization by the dissimilation Fe (III)-reducing bacterium shewanella alga BrY. *Environmental Science and Technology*, **33**, 723–729.

Curry, A.S., Grech, J.L., Spiteri, L. and Vassallo, L. (1969) Death from thallium poisoning: A case report. *European Journal of Toxicology*, **2**, 260–269.

Curtis, C.D. (2003) The aqueous geochemistry of metals in the weathering environment: Strengths and weaknesses in our understanding of speciation and process. *Mineralogical Magazine*, **67**, 235–246.

Darbre, P.D. (2006) Metalloestrogens: an emerging class of inorganic xenoestrogens with potential to add to the oestrogenic burden of the human breast. *Journal of Applied Toxicology*, **26**, 191–197.

Davis, L.E., Standefer, J.C., Kornfeld, M., Abercrombie, D.M. and Butler, C. (1981) Acute thallium poisoning: Toxicological and morphological studies of the nervous system. *Annals of Neurology*, **10**, 38–44.

DEFRA and Environment Agency (2002) Contaminants in soil: collation of toxicological data and intake values for humans. Department for Environment, Food and Rural Affairs and the Environment Agency.

Drasch, G., Horvat, M. and Stoeppler, M. (2004) Mercury. In Merian, E., Anke, M., Ihnat, M. and Stoeppler, M. (Eds.) *Elements and their compounds in the environment: Occurrence, analysis and biological relevance*, 2nd edn, Volume 2: Metals and their compounds, Wiley-VCH, pp. 931–1005.

Drexler, J.W. and Brattin, W.J. (2007) An in vitro procedure for estimation of lead relative bioavailability: with validation. *Human and Ecological Risk Assessment*, **13**, 383–401.

Dudas, M.J. (1987) Accumulation of native arsenic in acid sulfate soils in Alberta. *Canadian Journal of Soil Science*, **67**, 317–331.

Dudka, S., Piotrowska, M., Chlopecka, A. and Witek, T. (1995) Trace metal contamination of soils and crop plants by the mining and smelting industry in upper Silesia, South Poland. *Journal of Geochemical Exploration*, **52**, 237–250.

Elinder, C.G., Lind, B., Kjellstrom, T., Linnman, L. and Friberg, L. (1976) Cadmium in kidney cortex, liver, and pancreas from Swedish autopsies: Estimation of biological half-time in kidney cortex, considering calorie intake and smoking habits. *Archives of Environmental Health*, **31**, 292–302.

European Commission (2001) Ambient air pollution by As, Cd and Ni compounds. Position paper. European Communities.

European Parliament (2008) Export-ban of mercury and mercury compounds from the EU by 2011. European Parliament (Press release). Available via. http://www.europarl.europa.eu/sides/getDoc.do?pubRef=-//EP//TEXT + IM-PRESS + 20080520IPR29477 + 0 + DOC + XML + V0//EN. Accessed 11 January 2011.

Ewers, U. and Schlipköter, H.W. (1991) Lead. In Merian, E., Clarkson, T.W., Fishbein, L., Mallinckrodt, M.G., Piscator, M., Schlipköter, H.W., Stoeppler, M., Stumm, W. and SundermanJr., F.W. (Eds.) *Metals and their compounds in the envrionment: Occurrence,*

Analysis, and biological relevance, VCH, Weinheim, New York, Basel, Cambridge, pp. 971–1014.

Farmer, J.G., Eades, L.J., Graham, M.C. and Bacon, J.R. (2000) The phasing out of leaded petrol and changes in the $^{206}Pb/^{207}Pb$ isotopic ratio of atmospheric lead in Scotland (Contribution No. 1273). Proceedings of 11th International Conference on Heavy Metals in the Environment. University of Michigan, School of Public Health, Ann Arbor, MI, USA.

Fayerweather, W.E., Karns, M.E., Nuwayhid, I.A. and Nelson, T.J. (1997) Case-control study of cancer risk in tetraethyl lead manufacturing. *American Journal of Industrial Medicine*, **31**, 28–35.

Fergusson, J.E. (1990) *The heavy elements: chemistry, environmental impact and health effects*. Pergamon, Oxford.

Ferrara, R., Maserti, B.E., Andersson, M., Edner, H., Ragnarson, P., Svanberg, S. and Hernandez, A. (1998) Atmospheric mercury concentrations and fluxes in the Almadén District (Spain). *Atmospheric Environment*, **32**, 3897–3904.

Fitzgerald, W.F. and Lamborg, C.H. (2004) Geochemistry of mercury in the environment. In Lollar, B.S. (Ed.) *Environmental geochemistry: treaties on geochemistry*, **Volume 9**, Elsevier, Amsterdam.

Flegal, A.R. and Patterson, C.C. (1985) Thallium concentrations in seawater. *Marine Chemistry*, **15**, 327–331.

Fleischer, M., Sarofim, A.F., Fassett, D.W., Hammond, P., Shacklette, H.T., Nisbet, I.C.T. and Epstein, S. (1974) Environmental impact of cadmium: A review by the panel on hazardous trace substances. *Environmental Health Perspectives*, **5**, 253–323.

Fonfria, E., Vilaro, M.T., Babot, Z., Rodriguez-Farre, E. and Sunol, C. (2005) Mercury compounds disrupt neuronal glutamate transport in cultured mouse cerebellar granule cells. *Journal of Neuroscience Research*, **79**, 545–553.

FOREGS (2005) Geochemical Atlas of Europe. EuroGeoSurveys. Available via. http://www.gsf.fi/publ/foregsatlas/. Accessed 22 January 2010.

Forstner, U. (1980) Cadmium in polluted sediments. In Nriagu, J.O. (Ed.), *Cadmium in the environment, Part I: Ecological cycling*, John Wiley & Sons, New York, pp. 305–364.

Galván-Arzate, S. and Santamaría, A. (1998) Thallium toxicity. *Toxicology Letters*, **99**, 1–13.

Gastel, B. (1978) Clinical conferences at the Johns Hopkins Hospital: Thallium poisoning. *Johns Hopkins Medical Journal*, **142**, 27–31.

Gaudry, A., Moskura, M., Mariet, C., Ayrault, S., Denayer, F. and Bernard, N. (2008) Inorganic pollution in PM10 particles collected over three French sites under various influences: Rural conditions, traffic and industry. *Water, Air, and Soil Pollution*, **193**, 91–106.

Gerhardsson, L. (2004) Lead. In Merian, E., Anke, M., Ihnat, M. and Stoeppler, M. (Eds.) *Elements and their compounds in the environment: Occurrence, analysis and biological relevance*, 2nd edn. **Volume 2**: Metals and their compounds, Wiley-VCH, pp. 879–900.

Gleason, M.N., Gosselin, R.S., Hodge, H.C. and Smith, R.P. (1969) *Clinical toxicology of commercial products: acute poisoning (home and farm)* Williams and Wilkins, Baltimore.

Goldberg, E.D. (1976) Rock volatility and aerosol composition. *Nature*, **260**, 128–129.

Goldblatt, D. (1989) Pollutants and industrial hazards. In Rowland, L. P. (Ed.) *Merritt's textbook of neurology*, 8th edn, Lea and Febinger, Philadelphia, Pennsylvania, pp. 919–928.

Gosar, M., Pirc, S. and Bidovec, M. (1997) Mercury in the Idrijca River sediments as a reflection of mining and smelting activities of the Idrija mercury mine. *Journal of Geochemical Exploration*, **58**, 125–131.

Govett, G.J.S. (Ed.) (1983) *Handbook of exploration geochemistry*, **Volume 3**: Rock geochemistry in mineral exploration, Elsevier, Amsterdam, Oxford, New York.

Green, V.A., Wise, G.W. and Callenbach, J. (1976) Lead Poisoning. *Clinical Toxicology*, **9**, 33–51.

Gustin, M.S., Lindberg, S., Marsik, F., Casimir, A., Ebinghaus, R., Edwards, G., Hubble-Fitzgerald, C., Kemp, R., Kock, H., Leonard, J., Majewski, M., Montecinos, C., Owens, J., Pilote, M., Poissant, L., Rasmussen, P., Schaedlich, F., Schneeberger, D., Schroeder, W., Sommar, J., Turner, R., Vette, A., Wallschlaeger, D., Xiao, Z. and Zhang, H. (1999) Nevada STORMS project: Measurement of mercury emissions from naturally enriched surfaces. *Journal of Geophysical Research*, **104**, 21831–21844.

Hall, A.G. (2002) Chronic arsenic poisoning. *Toxicology Letters*, **128**, 69–72.

Heaven, R., Duncan, M. and Vukelja, S.J. (1994) Arsenic intoxication presenting with macrocytosis and peripheral neuropathy, without anemia. *Acta Haematologica*, **92**, 142–143.

Hem, J.D. (1992) Study and interpretation of the chemical characteristics of natural water. 3rd edn. U.S. Geological Survey Water Supply Paper.

Henshaw, J.M., Heithmar, E.M. and Hinners, T.A. (1989) Inductively coupled plasma mass spectrometric determination of trace elements in surface waters subject to acidic deposition. *Analytical Chemistry*, **61**, 335–342.

Herber, R.F.M. (2004) Cadmium. In Merian, E., Anke, M., Ihnat, M. and Stoeppler, M. (Eds.) *Elements and their compounds in the environment: Occurrence, analysis and biological relevance*, 2nd edn. **Volume 2**: Metals and their compounds, Wiley-VCH, pp. 689–708.

Hudson, R.J.M., Gherini, S.A., Watras, C.J. and Porcella, D.B. (1994) Modeling the biogeochemical cycle of mercury in lakes: The mercury cycling model (MCM) and its application to the MTL study lakes. In Watras, C.J. and Huckabee, J.W. (Eds.) *Mercury pollution: Integration and synthesis*, Lewis Publishers, pp. 473–526.

Hughes, M.F. (2002) Arsenic toxicity and potential mechanisms of action. *Toxicology Letters*, **133**, 1–16.

Hussain, S., Rodgers, D.A., Duhart, H.M. and Ali, S.F. (1998) Mercuric chloride-induced reactive oxygen species and its effect on anti-oxidant enzymes in different regions of rat brain. *Journal of Environmental Science and Health, B*, **32**, 395–409.

Hutzell, W.T. and Luecken, D.J. (2008) Fate and transport of emissions for several trace metals over the United States. *Science of the Total Environment*, **396**, 164–179.

IARC (1987) *IARC monographs on the evaluation of carcinogenic risks to humans: Overall evaluations of carcinogenicity (an updating of IARC monographs Volumes 1 to 42) Supplement 7*. International Agency for Research on Cancer, Lyon.

IARC (1993) *IARC monographs on the evaluation of carcinogenic risks to humans: Volume 58 beryllium, cadmium, mercury and exposures in the glass manufacturing industry*. International Agency for Research on Cancer, Lyon.

IARC (2006) *IARC monographs on the evaluation of carcinogenic risks to humans: Volume 87 Inorganic and organic lead compounds*. International Agency for Research on Cancer, Lyon.

Intawongse, M. and Dean, J.R. (2008) Use of the physiologically-based extraction test to assess the oral bioaccessibility of metals in vegetable plants grown in contaminated soil. *Environmental Pollution*, **152**, 60–72.

Irgolic, K.T., Greschonig, H. and Howard, A.G. (1995) Arsenic. In Townsend, A. (Ed.) *Encyclopaedia of analytical science*, Academic Press, London, pp. 168–184.

Jennings, S. and King, A.R. (2002) Geology, exploration history and future discoveries in the Red Dog district, western Brooks Range, Alaska. In Cooke, D.R. and Pongratz, J. (Eds.) *Giant ore deposits: characteristics, genesis and exploration*. Special Publication 4, CODES, pp. 151–158.

Jenny, H. (1941) *Factors of soil formation: A system of quantitative pedology*. McGraw-Hill, New York.

Johnson, M.S. and Eaton, J.W. (1980) Environmental contamination through residual trace metal dispersal from derelict lead-zinc mine. *Journal of Environmental Quality*, **9**, 175–179.

Kabata-Pendias, A. and Pendias, H. (1992) *Trace elements in soils and plants*. 2nd edn. CRC Press, Boca Raton, Ann Arbor, London.

Kabata-Pendias, A. (2001) *Trace elements in soils and plants*. CRC Press, Boca Raton, Florida.

Kabata-Pendias, A. and Sadurski, W. (2004) Trace elements and compounds in soil. In Merian, E., Anke, M., Ihnat, M. and Stoeppler, M. (Eds.) *Elements and their compounds in the environment: Occurrence, analysis and biological relevance*. **Volume 1**: General aspects, Wiley-VCH, pp. 79–99.

Kazantzis, G. (2000) Thallium in the environment and health effects. *Environmental Geochemistry and Health*, **22**, 275–280.

Kelley, K.D. and Hudson, T. (2007) Natural versus anthropogenic dispersion of metals to the environment in the Wulik River area, western Brooks Range, northern Alaska. *Geochemistry: Exploration, Environment, Analysis*, **7**, 87–96.

Kemper, F.H. and Bertram, H.P. (1991) Thallium. In Merian, E., Clarkson, T.W., Fishbein, L., Mallinckrodt, M.G., Piscator, M., Schlipköter, H.W., Stoeppler, M., Stumm, W. and SundermanJr., F. W. (Eds.) *Metals and their compounds in the environment: Occurrence, analysis, and biological relevance*, VCH, Weinheim, New York, Basel, Cambridge, pp. 1227–1241.

Lamadrid-Figueroa, H., Tellez-Rojo, M.M., Hernandez-Cadena, L., Mercado-Garcia, A., Smith, D., Solano-Gonzalez, M., Hernandez-Avila, M. and Hu, H. (2006) Biological markers of fetal lead exposure at each stage of pregnancy. *Journal of Toxicology and Environmental Health, Part A*, **69**, 1781–1796.

Leblanc, M., Achard, B., Othman, D.B. and Luck, J.M. (1996) Accumulation of arsenic from acidic mine waters by ferruginous bacterial accretions (stromatolites). *Applied Geochemistry*, **11**, 541–554.

Levi, P.E. (1997) Target organ toxicity. In Hodgson, E. and Levi, P.E. (Eds.) *A textbook of modern toxicology*, Appleton and Lange, Connecticut, pp. 199–228.

Lewis, J. (1985) Lead poisoning: A historical perspective. US Environmental Protection Agency. Available via. http://www.epa.gov/history/topics/perspect/lead.htm. Accessed 23 January 2010.

Lindh, U., Hudecek, R., Danersund, A., Eriksson, S. and Lindvall, A. (2002) Removal of dental amalgam and other metal alloys supported by antioxidant therapy alleviates symptoms and improves quality of life in patients with amalgam-associated ill health. *Neuroendocrinology Letters*, **23**, 459–482.

Lorscheider, F.L. and Vimy, M.J. (1993) Evaluation of the safety issue of mercury release from dental fillings. *Journal of the Federation of American Societies for Experimental Biology*, **7**, 1432–1433.

Mailman, R.B. (1980) Heavy metals. In Guthrie, F.E. and Perry, J.J. (Eds.) *Introduction to environmental toxicology*, Elsevier, North Holland, pp. 34–43.

Mandal, B.K. and Suzuki, K.T. (2002) Arsenic round the world: A review. *Talanta*, **58**, 201–235.

Mandal, B.K., Ogra, Y. and Suzuki, K.T. (2003) Speciation of arsenic in human nail and hair from arsenic-affected area by HPLC-inductively coupled argon plasma mass spectrometry. *Toxicology and Applied Pharmacology*, **189**, 73–83.

Manzo, L. and Sabbioni, E. (1988) Thallium toxicity and the nervous system. In Bondy, S.C. and Prasad, K.N. (Eds.) *Metal neurotoxicity*, CRC Press, Boca Raton, Florida, pp. 35–54.

Markovich, D. and James, K.M. (1999) Heavy metals mercury, cadmium and chromium inhibit the activity of the mammalian liver and kidney sulphate transporter sat-1. *Toxicology and Applied Pharmacology*, **152**, 181–187.

Martín-Doimeadios, R.C.R., Wasserman, J.C., Bermejo, L.F.G., Amouroux, D., Nevado, J.J.B. and Donard, O.F.X. (2000) Chemical availability of mercury in stream sediments from the Almadén area, Spain. *Journal of Environmental Monitoring*, **2**, 360–366.

Mathieson, P.W. (1995) Mercury: god of TH2 cells. *Clinical and Experimental Immunology*, **102**, 229–230.

Mathis, B.J. and Kevern, N.R. (1975) Distribution of mercury, cadmium, lead and thallium in a eutrophic lake. *Hydrobiology*, **46**, 207–222.

Mathys, W. (1981) Thallium in river sediments of North-West Germany: Its sources of input and its relation to other heavy metals. Proceedings of International Conference on Heavy Metals in the Environment. Amsterdam, September, 1981.

Mattinson, J.M. (2005) Zircon U-Pb chemical abrasion ('CA-TIMS') method: Combined annealing and multi-step partial dissolution analysis for improved prevision and accuracy of zircon ages. *Chemical Geology*, **220**, 47–66.

McBride, M.B. (1994) *Environmental chemistry of soils*. Oxford University Press, Oxford.

Merrington, G. and Alloway, B.J. (1994) The flux of Cd, Cu, Pb and Zn in mining polluted soils. *Water, Air, and Soil Pollution*, **73**, 333–334.

Miller, E.R. and Ullrey, D.E. (1987) The pig as a model for human-nutrition. *Annual Review of Nutrition*, **7**, 361–382.

Moeschlin, S. (1980) Thallium poisoning. *Clinical Toxicology*, **17**, 133–146.

Molina, J.A., Oyarzun, R., Esbri, J.M. and Higueras, P. (2006) Mercury accumulation in soils and plants in the Almaden mining district, Spain: one of the most contaminated sites on Earth. *Environmental Geochemistry and Health*, **28**, 487–498.

Moore, J.W. and Ramamoorthy, S. (1984) *Heavy metals in natural waters*. Springer-Verlag, New York.

Morillo, J., Usero, J. and Gracia, I. (2004) Heavy metal distribution in marine sediments from the southwest coast of Spain. *Chemosphere*, **55**, 431–442.

Moughan, P.J., Birtles, M.J., Cranwell, P.D., Smith, W.C. and Pedraza, M. (1992) The piglet as a model animal for studying aspects of digestion and absorption in milk-fed human infants. *World Review of Nutrition and Dietetics*, **67**, 40–113.

Moughan, P.J., Cranwell, P.D., Darragy, A.J. and Rowan, A.M. (1994) The domestic pig as a model animal for studying digestion in humans. Proceedings of VIth International Symposium on Digestive Physiology in Pigs. Bad Doberan, Germany, October 4th-6th, 1994.

Muezzinoglu, A. and Cizmecioglu, S.C. (2006) Deposition of heavy metals in a Mediterranean climate area. *Atmospheric Research*, **81**, 1–16.

Mulkey, J.P. and Oehme, F.W. (1993) A review of thallium toxicity. *Veterinary and Human Toxicology*, **35**, 445–453.

Ng, J.C., Wang, J. and Shraim, A. (2003) A global health problem caused by arsenic from natural sources. *Chemosphere*, **52**, 1353–1359.

Nriagu, J.O. (1980a) Cadmium in the atmosphere and in precipitation. In Nriagu, J.O. (Ed.) *Cadmium in the environment, Part I: Ecological cycling*, John Wiley & Sons, New York, Chichester, Brisbane, Toronto, pp. 71–114.

Nriagu, J.O. (1980b) Global cadmium cycle. In Nriagu, J.O. (Ed.) *Cadmium in the environment, Part I: Ecological cycling*, John Wiley & Sons, New York, Chichester, Brisbane, Toronto, pp. 1–12.

Nriagu, J.O. and Pacyna, J.M. (1988) Quantitative assessment of worldwide contamination of air, water and soils by trace metals. *Nature*, **333**, 134–139.

Nziguheba, G. and Smolders, E. (2008) Inputs of trace elements in agricultural soils via phosphate fertilizers in European countries. *Science of the Total Environment*, **390**, 53–57.

Ong, C.N. and Lee, W.R. (1980) Distribution of lead-203 in human peripheral blood in vitro. *British Journal of Industrial Medicine*, **37**, 78–84.

Oomen, A.G., Hack, A., Minekus, M., Zeijdner, E., Cornelis, C., Schoeters, G., Verstraete, W., Van de Wiele, T., Wragg, J., Rompelberg, C.J.M., Sips, A.J.A.M. and Van Wijnen, J.H. (2002) Comparison of five in vitro digestion models to study the bioaccessibility of soil contaminants. *Environmental Science and Technology*, **36**, 3326–3334.

Oomen, A.G., Rompelberg, C.J.M., Bruil, M.A., Dobbe, C.J.G., Pereboom, D.P.K.H. and Sips, A.J.A.M. (2003) Development of an in vitro digestion model for estimating the bioaccessibility of soil contaminants. *Archives of Environmental Contamination and Toxicology*, **44**, 281–287.

Pacifico, R., Adamo, P., Cremisini, C., Spaziani, F. and Ferrara, L. (2007) A geochemical analytical approach for the evaluation of heavy metal distribution in lagoon sediments. *Journal of Soils and Sediments*, **7**, 313–325.

Pacyna, E.G., Pacyna, J.M., Fudala, J., Strzelecka-Jastrzab, E., Hlawiczka, S., Panasiuk, D., Nitter, S., Pregger, T., Pfeiffer, H. and Friedrich, R. (2007) Current and future emissions of selected heavy metals to the atmosphere from anthropogenic sources in Europe. *Atmospheric Environment*, **41**, 8557–8566.

Page, A.L. and Bingham, F.T. (1973) Cadmium. *Residue Review*, **48**, 1–43.

Pan, J., Plant, J.A., Voulvoulis, N., Oates, C.J. and Ihlenfeld, C. (2010) Cadmium levels in Europe: Implications for human health. *Environmental Geochemistry and Health*, **32**, 1–12.

Parker, S.K., Schwartz, B., Todd, J. and Pickering, L.K. (2004) Thimerosal-containing vaccines and autistic spectrum disorder: A critical review of published original data. *Pediatrics*, **114**, 793–804.

Parsons, M.B. and Percival, J.B. (Eds) (2005) Mercury: Sources, measurements, cycles and effects. Volume 34 of short course series, Mineralogical Association of Canada.

Patterson, C. (1965) Contaminated and natural lead environments of man. *Archive of Environmental Health*, **11**, 344–360.

Paustenbach, D.J. (2000) The practice of exposure assessment: A state-of-the-art review. *Journal of Toxicology and Environmental Health, Part B*, **3**, 179–291.

Peter, A.L.J. and Viraraghavan, T. (2005) Thallium: A review of public health and environmental concerns. *Environment International*, **31**, 493–501.

Peterson, U. (1972) Geochemical and tectonic implications of south American metallogenic provinces. *Annals of the New York Academy of Sciences*, **196**, 3–38.

Plant, J.A., Gunn, A.G., Rollin, K.E., Stone, P., Morissey, C.J., Norton, G.E., Simpson, P.R. and Wiggans, G.N. (1998) The MIDAS Project (Multidataset analysis for gold in Europe)–Evidence from the British Caledonides. *Transactions of the Institute Mining and Metallurgy*, **107**, B77–B88.

Plant, J.A., Kinniburgh, D.G., Smedley, P.L., Fordyce, F.M. and Klinck, B.A. (2004) Arsenic and selenium: In treatise on geochemistry. *Environmental Geochemistry*, **9**, 17–66.

Preciado, H.F., Li, L.Y. and Weis, D. (2007) Investigation of past and present multi-metal input along two highways of British Columbia, Canada, using lead isotopic signatures. *Water, Air, and Soil Pollution*, **184**, 127–139.

Ravichandran, M. (2004) Interactions between mercury and dissolved organic matter–A review. *Chemosphere*, **55**, 319–331.

Reimann, C. and de Caritat, P. (1998) *Chemical elements in the environment: Factsheets for the geochemist and environmental scientist*. Springer Verlag, Berlin, Heidelberg, New York.

Reuter, J.H. and Perdue, E.M. (1977) Importance of heavy metal-organic matter interactions in natural waters. *Geochimica et Cosmochimica Acta*, **4**, 325–334.

Roberts, S.M., Munson, J.W., Lowney, Y.W. and Ruby, M.V. (2007) Relative oral bioavailability of arsenic from contaminated soils measured in the cynomolgus monkey. *Toxicological Sciences*, **95**, 281–288.

Robinson, G., Reynolds, J. and Gates, P. (1994) *365 days of nature and discovery*. Harry N. Adams, Inc., New York.

Rodriguez, V.M., Jimenez-Capdevile, M.E. and Giordano, M. (2003) The effects of arsenic exposure on the nervous system. *Toxicology Letters*, **145**, 1–18.

Romer, R.L. (2003) Alpha-recoil in U-Pb geochronology: effective sample size matters. *Contributions to Mineralogy and Petrology*, **145**, 481–491.

Rowan, A.M., Moughan, P.J., Wilson, M.N., Maher, K. and Tasman-Jones, C. (1994) Comparison of the ileal and faecal digestibility of dietary amino acids in adult humans and evaluation of the pig as a model animal for digestion studies in man. *British Journal of Nutrition*, **71**, 29–42.

Ruby, M.V., Davis, A., Link, T.E., Schoof, R., Chaney, R.L., Freeman, G.B. and Bergstrom, P. (1993) Development of an in-vitro screening-test to evaluate the in-vivo bioaccessibility of ingested mine-waste lead. *Environmental Science and Technology*, **27**, 2870–2877.

Ruby, M.V., Davis, A., Schoof, R., Eberle, S. and Sellstone, C.M. (1996) Estimation of lead and arsenic bioavailability using a physiologically based extraction test. *Environmental Science and Technology*, **30**, 422–430.

Ruby, M.V., Schoof, R., Brattin, W., Goldade, M., Post, G., Harnois, M., Mosby, D.E., Casteel, S.W., Berti, W., Carpenter, M., Edwards, D., Cragin, D. and Chappell, W. (1999) Advances in evaluating the oral bioavailability of inorganics in soil for use in human health risk assessment. *Environmental Science and Technology*, **33**, 3697–3705.

Ruby, M.V. (2004) Bioavailability of soil-borne chemicals: Abiotic assessment tools. *Human and Ecological Risk Assessment*, **10**, 647–656.

Ryu, Y., Lee, T.S., Lubguban, J.A., White, H.W., Kim, B.J., Park, Y.S. and Youn, C.J. (2006) Next generation of oxide photonic devices: ZnO-based ultraviolet light emitting diodes. *Applied Physics Letters*, **88**, 241108/1–3.

Sadiq, M. (1997) Arsenic chemistry in soils: An overview of thermodynamic predictions and field observations. *Water, Air, and Soil Pollution*, **93**, 117–136.

Sahl, K., de Albuquerque, C.A.R. and Shaw, D.M. (1978) Thallium. In Wedepohl, K.H. (Ed.) *Handbook of geochemistry*, Springer-Verlag, Berlin, Heidelberg, New York.

Schoer, J. (1984) Thallium. In Hutzinger, O. (Ed.) *The handbook of environmental chemistry, Part C: Anthropogenic compounds. Volume 3* Springer Verlag, Berlin, pp. 143–214.

Schroder, J.L., Basta, N.T., Casteel, S.W., Evans, T.J., Payton, M.E. and Si, J. (2004) Validation of the in vitro gastrointestinal (IVG) method to estimate relative bioavailable lead in contaminated soils. *Journal of Environmental Quality*, **33**, 513–521.

Schuster, E. (1991) The behavior of mercury in the soil with special emphasis on complexation and adsorption processes–a review of the literature. *Water, Air, and Soil Pollution*, **56**, 667–680.

Schwab, P., Zhu, D. and Banks, M.K. (2007) Heavy metal leaching from mine tailings as affected by organic amendments. *Bioresource Technology*, **98**, 2935–2941.

Semple, K.T., Doick, K.J., Jones, K.C., Burauel, P., Craven, A. and Harms, H. (2004) Defining bioavailability and bioaccessibility of contaminated soil and sediment is complicated. *Environmental Science and Technology*, **38**, 228A–231A.

Simeonova, P.P. and Luster, M.I. (2004) Arsenic and atherosclerosis. *Toxicology and Applied Pharmacology*, **198**, 444–449.

Simon, M., Jönk, P., Wühl-Couturier, G. and Halbach, S. (2006) Ullmann's Encyclopedia of Industrial Chemistry: mercury, mercury alloys, and mercury compounds. Wiley-VCH. Available via. http://www.mrw.interscience.wiley.com/emrw/9783527306732/ueic/article/a16_269/current/pdf. Accessed 4 March 2010.

Singh, R.P. and Agrawal, M. (2007) Effects of sewage sludge amendment on heavy metal accumulation and consequent responses of Beta vulgaris plants. *Chemosphere*, **67**, 2229–2240.

Smedley, P.L. and Kinniburgh, D.G. (2002) A review of the source, behaviour and distribution of arsenic in natural waters. *Applied Geochemistry*, **17**, 517–568.

Smith, A.H., Lingas, E.O. and Rahman, M. (2000) Contamination of drinking-water by arsenic in Bangladesh: A public health emergency. *Bulletin of the World Health Organization*, **78**, 1093–1103.

Smith, I.C. and Carson, B.L. (1977) *Trace metals in the environment: Thallium*. Ann Arbor Science Publishers, Inc., Ann Arbor, Michigan.

Song, Y., Wilson, M.J., Moon, H.S., Bacon, J.R. and Bain, D.C. (1999) Chemical and mineralogical forms of lead, zinc and cadmium in particle sise fractions of some wastes, sediments and soils in Korea. *Applied Geochemistry*, **14**, 621–633.

Sörme, L. and Lagerkvist, R. (2002) Sources of heavy metals in urban wastewater in Stockholm. *Science of the Total Environment*, **298**, 131–145.

Staudinger, K.C. and Roth, V.S. (1998) Occupational lead poisoning. American Family Physician. Available via. http://www.aafp.org/afp/980215ap/stauding.html. Accessed 22 January 2010.

Steven, A. and Greenberg, M.D. (1996) Acute demyelinating polyneuropathy with arsenic ingestion. *Muscle and Nerve*, **19**, 1611–1613.

Stockinger, H.E. (1981) The metals. In Clayton, G.D. and Clayton, F. E. (Eds.) *Patty's industrial hygeine and toxicology*, John Wiley & Sons, New York, pp. 1769–1792.

Struempler, A.W. (1975) Trace element composition in atmospheric particulates during 1973 and the summer of 1974 at Chadron, Neb. *Environmental Science and Technology*, **9**, 1164–1168.

Sun, G. (2004) Arsenic contamination and arsenicosis in China. *Toxicology and Applied Pharmacology*, **198**, 268–271.

Thévenot, D.R., Moilleron, R., Lestel, L., Gromaire, M.C., Rocher, V., Cambier, P., Bonté, P., Colin, J.L., de Pontevès, C. and Meybeck, M. (2007) Critical budget of metal sources and pathways in the Seine River basin (1994–2003) for Cd, Cr, Cu, Hg, Ni, Pb and Zn. *Science of the Total Environment*, **375**, 180–203.

Thornton, I. (1996) Sources and pathways of arsenic in the geochemical environment: Health implications. In Appleton, J.D., Fuge, R. and McCall, G.J.H. (Eds.) *Environmental geochemistry and health, 113*, Geological Society Special Publication pp. 153–161.

Tseng, C.H., Chong, C.K. and Heng, L.T. (2002) Epidemiologic evidence of diabetogenic effect of arsenic. *Toxicology Letters*, **133**, 69–76.

Tsuchiya, K. (1980) Clinical signs, symptons and prognosis of cadmium poisoning. In Nriagu, J.O. (Ed.) *Cadmium in the environment, Part II: Health effects*, John Wiley & Sons, Canada, pp. 40–52.

Uetani, M. (2007) Investigation of renal damage among residents in the cadmium-polluted Jinzu River basin, based on health examinations in 1967 and 1968. *International Journal of Environmental Health Research*, **17**, 231–242.

US EPA (1997) Mercury study report to congress: Volume III fate and transport of mercury in the environment. EPA-452/R-97-005, US Environmental Protection Agency.

US EPA (2000) Lead compounds. US Environmental Protection Agency. Available via http://www.epa.gov/ttnatw01/hlthef/lead.html. Accessed 23 January 2010.

US EPA (2005) Partition coefficients for metals in surface water, soil and waste. EPA/600/R-05/074, US Environmental Protection Agency, Washington D.C.

US EPA (2006) Revised reregistration eligibility decision for MSMA, DSMA, CAMA and cacodylic acid. US Environmental Protection Agency.

US FDA (2010) Thimerosal in vaccines. US Food and Drug Administration. Available via http://www.fda.gov/biologicsbloodvaccines/safetyavailability/vaccinesafety/ucm096228.htm. Accessed 11 January 2011.

USGS (1995) Preliminary compilation of descriptive geoenvironmental mineral deposit models. Open-File Report 95-831, US Geological Survey.

Vahidnia, A., Van der Voet, G.B. and de Wolf, F.A. (2007) Arsenic neurotoxicity – A review. *Human and Experimental Toxicology*, **26**, 823–832.

Valerio, F., Brescianini, C., Mazzucotelli, A. and Frache, R. (1988) Seasonal variation of thallium, lead, and chromium concentrations in airborne particulate matter collected in an urban area. *Science of the Total Environment*, **71**, 501–509.

Valerio, F., Brescianini, C. and Lastraioli, S. (1989) Airborne metals in urban areas. *International Journal of Environmental Analytical Chemistry*, **35**, 101–110.

Van der Veer, G. (2006) Geochemical soil survey of the Netherlands. Atlas of major and trace elements in topsoil and parent material; Assessment of natural an anthropogenic enrichment factors. Ph.D. Thesis, Utrecht University.

Waalkes, M.P. (2003) Cadmium carcinogenesis. *Mutation Research*, **533**, 107–120.

Waidmann, E., Stoeppler, M. and Heininger, P. (1992) Determination of thallium in sediments of the river elbe using isotope dilution mass spectrometry with thermal ionization. *Analytical Chemistry*, **338**, 572–574.

Wainipee, W., Weiss, D.J., Sephton, M.A., Coles, B.J., Unsworth, C. and Court, R. (2010) The effect of crude oil on arsenate adsorption on goethite. Water Research, doi:10.1016/j.watres.2010.05.056.

Walker, C.H., Hopkin, S.P., Sibly, R.M. and Peakall, D.B. (2006) *Principles of ecotoxicology*, 3rd edn. CRC Press.

Wang, D., Shi, X. and Wei, S. (2003) Accumulation and transformation of atmospheric mercury in soil. *Science of the Total Environment*, **304**, 209–214.

Wang, Q., Kim, D.K., Dionysiou, D.D., Sorial, G.A. and Timberlake, D. (2004) Sources and remediation for mercury contamination in aquatic systems – A literature review. *Environmental Pollution*, **131**, 323–336.

Wedepohl, K.H. (1991) The composition of the upper earth's crust and the natural cycles of selected metals. Metals in natural raw materials. Natural resources. In Merian, E., Clarkson, T.W., Fishbein, L., Mallinckrodt, M.G., Piscator, M., Schlipköter, H.W., Stoeppler, M., Stumm, W. and SundermanJr., F.W. (Eds.) *Metals and their compounds in the envrionment: Occurrence, analysis, and biological relevance*, VCH, Weinheim, New York, Basel, Cambridge, pp. 3–17.

Weidenhamer, J.D. (2009) Lead contamination of inexpensive seasonal and holiday products. *Science of the Total Environment*, **407**, 2447–2450.

Westcott, K. and Kalff, J. (1996) Environmental factors affecting methylmercury accumulation in zooplankton. *Canadian Journal of Fish and Aquatic Science*, **53**, 2221–2228.

White, D.E. (1967) Mercury and base-metal deposits with associated thermal and mineral waters. In Barnes, H.L. (Ed.) *Geochemistry of hydrothermal ore deposits*, Holt, Rinehart and Winston, New York, pp. 575–631.

WHO (1977) Environmental health criteria 3: Lead. World Health Organisation. Available via http://www.inchem.org/documents/ehc/ehc/ehc003.htm. Accessed 8 February 2010.

WHO (1995) Environmental health criteria 165: Inorganic lead. World Health Organisation. Available via http://www.inchem.org/documents/ehc/ehc/ehc165.htm. Accessed 8 February 2010.

WHO (1996a) *Environmental health criteria 182: Thallium*. World Health Organisation. Available via http://www.inchem.org/documents/ehc/ehc/ehc182.htm. Accessed 8 February 2010.

WHO (1996b) *Trace elements in human nutrition and health*. World Health Organisaiton, Geneva.

WHO (2005) *Mercury in drinking water*. World Health Organization, Geneva.

Wilhelm, S.M. (2001) Mercury in petroleum and natural gas: Estimation of emissions from production, processing, and combustion. EPA-600/R-01-066, US Environmental Protection Agency.

Windham, B. (2010) The results of removal of amalgam fillings: 60,000 documented cases. The National Recovery Plan. Available via http://www.thenaturalrecoveryplan.com/docs/research_docs/2010.07.28.03.24_Results_removal.pdf. Accessed 11 January 2011.

Wixon, B.G. and Davies, B.E. (1993) *Lead in soil: recommended guidelines*. Science Reviews, Northwood, pp. 25–38.

Wolfe, M.F., Schwarzbach, S. and Sulaiman, R.A. (1998) Effects of mercury on wildlife: A comprehensive review. *Environmental Toxicology and Chemistry*, **17**, 146–160.

Wong, S.C., Li, X.D., Zhang, G., Qi, S.H. and Min, Y.S. (2002) Heavy metals in agricultural soils of the Pear River Delta, *South China. Environmental Pollution*, **119**, 33–34.

Wragg, J. and Cave, M.R. (2002) In-vitro methods for the measurement of the oral bioaccessibility of selected metals and metalloids in soils: A critical review. RandD Technical Report P5-062/TR/01, Environment Agency of England and Wales.

Wragg, J., Cave, M., Taylor, H., Basta, N., Brandon, E., Casteel, S., Denys, S., Gron, C., Oomen, A., Reimer, K., Tack, K. and Van de Wiele, T. (2009) Interlaboratory trial of a unified bioaccessibility procedure. OR/07/027, British Geological Survey.

Yarlagadda, P.S., Matsumoto, M.R., Vanbenschoten, J.E. and Kathuria, A. (1995) Characteristics of Heavy Metals in Contaminated Soils. *Journal of Environmental Engineering*, **121**, 276–286.

Yasumura, S., Vartsky, D., Ellis, K.J. and Cohn, S.H. (1980) Cadmium in human beings. In Nriagu, J.O. (Ed.) *Cadmium in the environment, Part I: Ecological cycling*, John Wiley & Sons, pp. 12–28.

Zagury, G.J., Bedeaux, C. and Welfringer, B. (2009) Influence of mercury speciation and fractionation on bioaccessibility in soils. *Archives of Environmental Contamination and Toxicology*, **56**, 371–379.

Zahir, F., Rizwi, S.J., Haq, S.K. and Khan, R.H. (2005) Low dose mercury toxicity and human health. *Environmental Toxicology and Pharmacology*, **20**, 351–360.

Zhang, Z., Zhang, B., Long, J., Zhang, X. and Chen, G. (1998) Thallium pollution associated with mining of thallium deposits. *Science in China Series D: Earth Sciences*, **41**, 75–81.

Zhou, D.X. and Liu, D.N. (1985) Chronic thallium poisoning in a rural area of Guizhou province, *China. Journal of Environmental Health*, **48**, 14–18.

5

Radioactivity and radioelements

Jane A. Plant[1*], Barry Smith[2], Xiyu Phoon[3] and K. Vala Ragnarsdottir[4]

[1]Centre for Environmental Policy and Department of Earth Science and Engineering, Imperial College London, Prince Consort Road, London SW7 2AZ
[2]Intelliscience Ltd, 38A Station Rd, Nottingham, NG4 3DB
[3]Centre for Environmental Policy, Imperial College London, Prince Consort Road, London SW7 2AZ
[4]Faculty of Earth Sciences, School of Engineering and Natural Sciences, Askja, University of Iceland, Reykjavik 101, Iceland
**Corresponding author, email jane.plant@imperial.ac.uk*

5.1 Introduction

Radioelements (also called radioactive isotopes or radionuclides) are a class of unstable elements that undergo radioactive decay, emitting radiation, most commonly as alpha or beta particles and/or gamma rays (UNSCEAR, 2000; WHO, 1983). Acute radiation syndrome (ARS) or radiation sickness, usually from a large dose of radiation over a short period of time, can result in nausea, vomiting, headache and fatigue, and, more seriously, can lead to dizziness and low blood pressure. High levels of radiation especially when received in an acute dose can damage cells directly, resulting in changes in blood count and leading, in serious cases, to death. Chronic exposure leads to the production of free radicals and carcinogenesis; and it is reprotoxic. The extent of damage from acute or chronic exposure depends on the ionising potential of the radiation, the quantity of radiation (its duration and magnitude) and the sensitivity of the exposed tissue (WHO, 1983). Some radioelements are both chemically and radiologically toxic. For example, uranium (U) ingestion can cause poisoning and kidney disease (ATSDR, 1999) as well as potentially being radiologically toxic. On the other hand, radon (Rn), which is a member of the ^{238}U decay series and a noble gas, is chemically inert, but it and its decay products cause radiological harm including lung cancer (ATSDR, 2008).

In this chapter we discuss three types of radioactivity:

(a) Natural radioactivity, which includes cosmic and terrestrial radioactivity. The latter mainly reflects the decay of isotopes of U (which produces hazardous decay products such as Rn), thorium (Th) and potassium (K).

(b) Radioactivity from altered or refined naturally occurring radioelements, such as enriched U, in which the proportion of the fissile isotope ^{235}U has been enriched for nuclear fuel or weaponry, and depleted U, the residual material, enriched in ^{238}U after extraction of ^{235}U.

(c) Fission products such as strontium-90 (^{90}Sr) and caesium-137 (^{137}Cs) and heavy synthetic actinides such as americium (^{241}Am, ^{242}Am and ^{243}Am) and plutonium (^{239}Pu and ^{241}Pu) which form when a large nucleus, usually ^{235}U, undergoes fission in a nuclear reactor or in an atomic bomb explosion. Many of these isotopes are extracted and used in industry and nuclear medicine.

The number of radioactive isotopes is vast and they outnumber stable elements in the periodic table. Here, we focus on U, radium (Ra), Rn, ^{90}Sr and ^{241}Am as examples of the risk that radioelements pose to human health and the environment.

5.1.1 History

Radiation and radioactivity were first discovered in U salts by Henri Becquerel in 1896, although the harmful properties of radioactive substances were known in Europe from the fifteenth century. Miners in the silver (Ag) and later U mining district of the Erzgebirge at the present border of the Czech Republic and

Pollutants, Human Health and the Environment: A Risk Based Approach, First Edition. Edited by Jane A. Plant, Nikolaos Voulvoulis and K. Vala Ragnarsdottir.

Germany were affected by a disease known as 'Bergsucht' or miner's disease, now known to be lung cancer, caused by inhaling Rn gas and its decay products. The problem was not remedied until mine ventilation was improved by the founder of modern geology, George Bauer (Agricola), early in the sixteenth century. Radon continues to be of concern for human health. According to the US Environmental Protection Agency (USEPA, 2007), Rn-induced lung cancer is the sixth leading cause of cancer death there, causing 21 000 lung-cancer deaths per year. Studies of human exposure to U and its impact on health date back to 1824. Later in the same century, studies on humans suggested that U could be used to treat diabetes (Hodge *et al.*, 1973).

The health effects of radioelements remained poorly understood long after Becquerel's discovery of radioactivity. Radioactive spas were established to treat a variety of conditions, including arthritis. Examples include Jáchymov in the Czech Republic and Bad Gastein in Austria, and the 'health mines' established later at Boulder, Montana, USA. Indeed, therapy involving bathing in and drinking Rn-enriched water, and inhalation of Rn continues to be used in spas around the world.

In 1934, Marie Skłodowska-Curie, whose achievements included the theory of radioactivity and the discovery of the elements polonium (Po) and, with Pierre Curie, Ra (both of which are decay products of U), and who pioneered the use of ^{226}Ra to treat cancer, died aged 67. She obtained two Nobel prizes for her work. Her death from leukaemia was almost certainly the result of exposure to radiation.

Radium was looked upon initially as a source of health and healing, and this led to its widespread misuse. The first harmful effects of radiation identified, mainly from acute exposure to Ra, were damage to skin and blood formation. The main concern about chronic exposure to radioactivity now is carcinogenicity and mutagenicity. Evidence available for many years indicates that exposure to radiation damages cellular tissues and DNA, although at low doses such damage may be repaired by in-situ cellular processes. Radiation is used to destroy cancer cells, which have no repair mechanism and therefore cannot recover after they have been damaged by radiation. Techniques include external-beam radiotherapy, brachytherapy and radioisotope therapy (Gerber and Chan, 2008).

Radiation from the first sources, such as ^{226}Ra, mostly had poor penetration and was dangerous mainly at short distances and by direct contact with tissue (i.e. via ingestion and inhalation). Therefore the earliest controls on radiation exposure were aimed principally at minimising internal exposure and preventing burns and skin ulceration, which in serious cases led to hand amputation.

Prior to 1942, U was used mainly for colouring glass green or red and ceramic glazes yellow. In that year, controlled nuclear fission of ^{235}U was demonstrated, which led to the development of its use in nuclear weaponry and, subsequently, to generate electricity. Nuclear fission resulted in the production and discovery of a wide range of fission products such as ^{137}Cs (caesium 137) and new radioactive elements such as Pu, californium (Cf) and americium (Am). Following the atomic bombing of Hiroshima and Nagasaki in 1945 near the end of the second world war the significance of the delayed effects of radioactivity in causing cancer (leukaemias and solid tumours) was recognised (Lindell, 1996). The destructive power of the bombs and the human misery caused by them have influenced public attitudes to ionising radiation ever since. Concern continues to be driven by the increased understanding of the biological effects of radioactivity and by changing public attitudes on what are acceptable risks and benefits. Issues that have focused public concern include the atmospheric testing of nuclear weapons between 1944 and 1963, and high-profile nuclear accidents such as those at the Three Mile Island Reactor in the USA in 1979 (Hull, 1980) and at Chernobyl in the Ukraine in 1986 (IAEA, 1986). More recently, public concern about radioactivity has focused mainly on the proliferation of nuclear weapons, nuclear-waste repositories and nuclear-power generation.

Common types of radiation emitted during radioactive decay

Alpha particles consist of a helium nucleus (two protons and two neutrons; He^{2+} or $^{4}_{2}He^{2+}$) and have an energy normally in the range 3–9 MeV. They are the most highly ionising form of radiation but have low penetration in tissue. Chromosome damage from alpha particles is estimated to be about 100 times greater than that caused by an equivalent amount of other types of radiation (BSS, 1996). The alpha emitter ^{210}Po is suspected of playing a role in lung cancer and bladder cancer associated with tobacco smoke, and was also the agent used to induce acute radiation poisoning in Alexander Litvinenko (Cobain *et al.*, 2006; Hansard, 2007). Ionisation is also caused by recoiling nuclei following alpha emissions, which can also cause genetic damage, since the cations of many alpha-emitting soluble transuranic elements are chemically attracted to the net negative charge of DNA.

Beta particles. There are two forms of beta decay, β^- (nuclear electrons) and β^+ (nuclear positrons). Beta radiation has penetrating and ionising power intermediate between that of alpha and gamma radiation. In medicine, beta particles have been used to treat conditions such as eye and bone cancer, and beta emitters are also used as tracers in medical diagnostics. Beta particles can penetrate living matter and change the structure of molecules, including DNA, with the potential to cause cancer and death. If the mutated DNA is in ova or sperm it may be passed to offspring.

Gamma rays are emitted as high-energy (commonly in the range of a few hundred KeV to 10 MeV) electromagnetic radiation. Together with neutrons they are the most dangerous form of ionising radiation emitted by nuclear fission and explosions because of their high energy and high degree of penetration. They have the shortest wavelength in the electromagnetic spectrum, and therefore the greatest ability to penetrate, even subatomically. The most biologically damaging forms of gamma radiation are in the energy range 3–10 MeV. Gamma rays are not stopped by the skin, and can induce DNA alteration by whole-body irradiation.

Table 5.1 Particles typically emitted during nuclear decay or in reactions with cosmic radiation

Emitted particle	Mass (atomic mass units)	Electrical charge	Potential to cause ionisation	Penetrating power in matter
Alpha (α)	4	2+	moderately high	very low
Beta (β- (electron) or β+ (positron))	0	1+ or 1−	moderate	low
Gamma (γ)	0	0	moderate	high
X-ray	0	0	low	moderate
Fission fragment	70–160	high	very high	low
Neutron (n)	1	0	low (direct), high (indirect)	high
Proton (p)	1	+1	high	low
Heavy ion (e.g. C ion)	>>1	> 1, + or −	very high	very low
Pion (π)	~0.1	+1, 0, −1	high	high
Muon (m)	~0.1	+1, 0, −1	low	very high

The toxicological and medical effects of radiation on humans and the control of exposure to radiation are generally studied by radiobiologists and health physicists. Over the past half century, radiation legislation has developed mostly independently of chemical legislation. For example, radiological risk is concerned mainly with human health and uses death as the end point, whereas chemicals legislation such as the REACH legislation of the European Union includes a wider range of impacts on human health and the environment.

5.1.2 Basics of radioactivity

5.1.2.1 Types of radioactive decay and associated nuclear reactions

Most decay processes involve the emission of alpha or beta particles and gamma rays (see box). Most decay events involve the emission of at least two of these; it is common, for example, for an atom undergoing nuclear decay to emit both a beta particle and a gamma ray.

Cosmic rays (ionising radiation composed of highly energetic charged extra-terrestrial particles) also produce ionising radiation by their interaction with the atmosphere and the Earth's surface to form cosmogenic radioelements such as ^{3}H (tritium, an isotope of hydrogen, H) and ^{14}C (carbon 14). The potential decay and reaction products for both natural and man-made radioactive sources and cosmic ray interactions are listed in Table 5.1.

5.1.2.2 Measurement of radioactivity

5.1.2.2.1 Units of measurement The quantity of a given radioactive material in a known matrix is generally referred to as its radioactivity or, more commonly, activity. The international unit of measurement of radioactivity is the becquerel (Bq), which is defined as the activity of a quantity of radioactive material in which one nucleus decays per second. It is commonly expressed in Bq/kg in solids, Bq/l in liquids, and Bq/m^3 in air.

5.1.2.2.2 The concept of radiological dose The radiological dose[1] is the amount of energy transferred to a given mass of material, usually tissue.

The international unit of absorbed radiological dose is the gray (1 Gy is equal to 1 joule per kg; SI Base, m^2/s^2). One Gy represents the deposition of approximately 6.2×10^{12} MeV in 1 kg mass. The international radiation community use the sievert (Sv) for equivalent and effective dose, however, because this allows the impact of different particles and their effect on different tissues to be enumerated.

The **equivalent dose** in Sv is the product of the adsorbed dose of ionising radiation and a weighting factor (Q) which depends on the type of ionising radiation (Table 5.2). It enables a uniform whole-body dose to be calculated.

The **effective dose** in Sv is the product of the equivalent dose of ionising radiation for each tissue and a tissue-weighting factor (Table 5.3) and sums to 1. It is higher for tissues which are particularly sensitive to cancer such as the lungs and bone marrow (0.12) and lower for muscle or skin (0.01). It enables the relative risks of radiation to different tissues and organs to be calculated.

Table 5.2 Typical values of Q for a whole body exposed to alpha, beta and gamma radiation (After BSS, 1996)

Type of disintegration	Relative Q
Alpha	20
Beta	1
Gamma	1

[1] In other chapters the term 'dose' refers to the amount of a chemical substance administered.

Table 5.3 Typical tissue-weighting factors (After BSS, 1996)

Organ or tissue	Tissue-weighting factor
Gonads	0.20
Colon	0.12
Bone marrow (red)	0.12
Lung	0.12
Stomach	0.12
Bladder	0.05
Chest	0.05
Liver	0.05
Thyroid gland	0.05
Oesophagus	0.05
Skin	0.01
Bone surface	0.01
Adrenals, brain, small intestine, kidney, muscle, pancreas, spleen, thymus, uterus	0.05
Whole body	1.0

The absorbed dose from cosmic rays and cosmogenic isotopes also contributes to the background dose received by humans and the environment.

The world average adult annual background radiation dose is approximately 2.4 mSv per annum (NCRP, 1987; BSS, 1996). Approximately 8 per cent is from cosmic radiation and 11 per cent from internal radiation. Internal radiation comes principally from the decay of ^{40}K in the human body. ^{40}K, makes up 0.012 per cent of total K and decays to ^{40}Ca (calcium 40) and ^{40}Ar (argon 40) gas with a half-life of 1.3 billion years (Kaye and Laby, 1986). Potassium is an essential element for animals and plants and it makes up 0.20 per cent of the human body by weight. The 140 g of K in a normal human male contains about 4400 Bq of ^{40}K, which delivers doses of about 0.018 mSv to soft tissue and 0.014 mSv to bone each year (ANL, 2005). The rest of the world average adult annual background radiation is from gamma rays from naturally occurring radioelements, medical procedures and Rn gas, derived from the decay of ^{238}U in the ground.

The magnitude of the background dose varies significantly with altitude (mainly from cosmic radiation) and geology. For example, according to the UNSCEAR (2000) report, Ramsar in northern Iran has some of the highest known natural background radiation levels in the world (260 mSv per annum). This is mainly due to the presence of hot springs containing exceptionally high levels of ^{226}Ra and its decay products (Ghiassi-nejad *et al.*, 2002).

The estimated average annual radiation dose to UK and US residents is shown in Figure 5.1. Note the high proportion (63 per cent) from naturally occurring Rn gas (including ^{220}Rn, which is a decay product of ^{232}Th and is known as thoron) and terrestrial (gamma) radiation in the UK. Medical diagnostics accounted for approximately 15 per cent of the average dose of in the UK in 2000 (NRPB, 2000) and 2001–2003 (Watson *et al.*, 2005). In the USA, estimates of the medical contribution have increased to almost 50 per cent, mainly as a result of the increasing use of computed-tomography (CT) scans and interventional fluoroscopy. It has been estimated that the number of CT scans in the USA increased from about 3 million in 1980 to 62 million in 2006 (NCRP, 2009). The dose from medical interventions is now more than a hundred times the average dose estimated to be due to products and discharges of the nuclear industry (<0.2 per cent). One CT scan typically delivers a dose of between 50 and 100 mSv, whereas a conventional chest X-ray delivers 0.1 mSv and dental X-ray 1.6 mSv (NCRP, 2009). Such data has led to a re-evaluation of the risks associated with the use of CT scans in medicine, especially paediatric medicine (Golding and Shrimpton, 2002; Brenner and Hall, 2007).

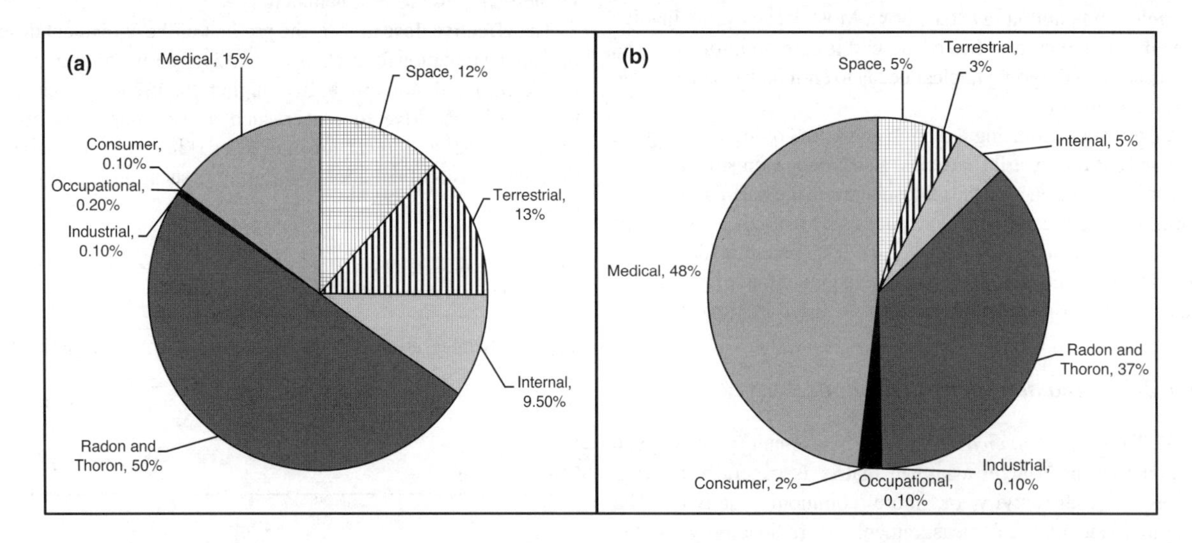

Figure 5.1 Estimated average annual radiation dose to (a) UK population (2.6 mSv in 2000) and (b) USA population (6.2 mSv in 2009) (NRPB, 2000; Watson *et al.*, 2005 and NCRP, 2009 respectively)

Table 5.4 Primary effective dose limits used in UK legislation (The Ionising Radiations Regulations, 1999)

Class of person	Limit of effective dose, mSv in any one calendar year
(1) Employees of 18 years old or above	6 or 20 if a designated worker (*100 mSv in any 5 years subject to a maximum of 50 mSv in any one year)
(2) Employees and trainees under 18 years	6
(3) Women of reproductive capacity	As per (1) and (2)
(4) Others (including children < 16)	1

*UK legislation allows for employees limits to be revised upwards should the primary limit prove impractical for the nature of work being undertaken.

5.1.2.3 Recommended dose limits

Dose limits for ionising radiation are recommended by international bodies such as the ICRP, and then set as legal limits within individual countries' legislation. For example, in the UK, dose limits for employees of industries that use radiation, members of the public, females of reproductive capacity, pregnant employees and others are set out in the Ionising Radiations Regulations, 1999 (Table 5.4). Whilst dose limits are used to set legal maximums in many countries, the practice of radiation protection requires that all doses are kept as low as reasonably practical (ALARP, as low as reasonably practicable; (or ALARA, as low as reasonably achievable, in the USA) as required by the concepts of exposure optimisation and limitation (BSS, 1996).

The use of such dose limits typically excludes: radiation exposures resulting from medical interventions, which must also take into account any health benefits that might be associated with the exposure and radiological dose; and situations such as nuclear emergencies where higher one-off doses may be considered acceptable by the appropriate authorities (BSS, 1996).

The relationship between effective radiological dose and risks attributable to a specific health-related endpoint continues to be the subject of debate, especially at low dose (UNSCEAR, 2001, 2006; ECRR-CERI, 2009; Wakeford, 2005; Allison, 2009). Currently accepted levels of risk associated with generic low-level exposure to radiation are around 0.005 per cent per mSv for fatal cancer (Wakeford, 2005), based on the linear no-threshold dose response model (LNT). Opponents of the LNT model consider this rate to be at least an order of magnitude too low (ECRR-CERI, 2009). Compilations of health risks associated with exposure to specific radionuclides have also been made (for example the Health Effects Assessment Summary Tables (HEAST) by the USEPA (www.epa.gov/radiation/heast) which include cancer slope factors for radionuclides in units of picocuries (pCi: 1 Bq = 27 pCi)).

5.1.3 Naturally occurring radioactive elements

There are naturally occurring radioisotopes throughout the periodic table (Figure 5.2), including those associated with the U and Th decay series, non-series radionuclides such as ^{40}K and cosmogenic radionuclides. (Friedlander *et al.*, 1981). The most

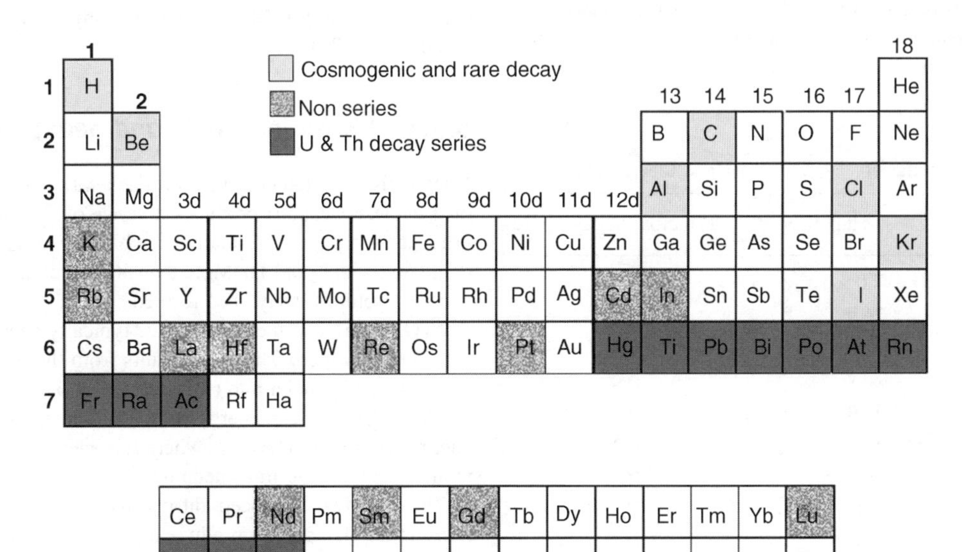

Figure 5.2 Periodic table, showing the occurrence of elements with naturally occurring radioactive isotopes

Table 5.5 Decay series of ^{238}U (From Kaye and Laby, 1986)

Isotope	Half-life	Principal decay modes
^{238}U	4.5×10^9 y	α.γ
^{234}Th	24.1 d	β.γ
^{234}Pa	6.75 h	β.γ
^{234}U	2.48×10^5 y	α.γ
^{230}Th	8.0×10^4 y	α.γ
^{226}Ra	1622 y	α.γ
^{222}Rn	3.82 d	α
^{218}Po	3.05 min	α
^{214}Pb	26.8 min	β
^{214}Bi	19.7 min	β.γ
^{214}Po	1.6×10^{-4} s	α
^{210}Pb	22.0 y	β.γ
^{210}Bi	5.01 d	β
^{210}Po	138.4 d	α
^{206}Pb	Stable	

Table 5.7 Decay series of ^{232}Th (Kaye and Laby, 1986)

Isotope	Half-life	Principal decay modes
^{232}Th	1.4×10^{10} y	α
^{228}Ra	5.8 y	β
^{228}Ac	6.1 h	β.γ
^{228}Th	1.9 y	α
^{224}Ra	3.7 d	α
^{220}Rn	56 s	α
^{216}Po	0.15 s	α
^{212}Pb	11 h	β.γ
^{212}Bi	61 min	β
^{212}Po	3.0×10^{-7} s	α
^{208}Tl	3.1 min	β.γ
^{208}Pb	Stable	

abundant members of these groups, their half-lives and principal decay modes are given in Tables 5.5, 5.6 and 5.7. Their abundance, composition and chemical form are frequently changed during extraction, processing and use, as in the mining and processing of U to prepare enriched or depleted U.

In the sections below we are concerned mainly with the U and Th decay series (Tables 5.5, 5.6 and 5.7) because of their abundance and the known health effects of some of their decay products (especially Ra and Rn), and the non-series radionuclide ^{40}K. Several of these radioelements, particularly those with long half-lives such as U, have been demonstrated to be chemically toxic as well as potentially radiotoxic (The Royal Society, 2002a, b).

Table 5.6 Decay series of ^{235}U (From Kaye and Laby, 1986)

Isotope	Half-life	Principal decay modes
^{235}U	7.0×10^8 y	α.γ
^{231}Th	26 h	β.γ
^{231}Pa	3.3×10^3 y	α.γ
^{227}Ac	22 y	β.α
^{227}Th	19 d	α.γ
^{223}Ra	11.4 d	α.γ
^{219}Rn	4.0 s	α.γ
^{215}Po	1.8×10^{-3} s	α
^{211}Pb	36.1 m	β.γ
^{211}Bi	2.2 m	α.γ
^{207}Tl	4.8 m	β
^{207}Pb	Stable	

In isolation, a decay series will in time reach a state of radiological equilibrium in which the activity (Bq) of each decay product is equal to the activity (Bq) of its precursor, but the different physicochemical properties of the decay products mean that in the environment they can behave differently from their parent isotopes and from each other, which can result in secular disequilibrium.

In particular, there are important differences in the chemical and physical properties of the decay products formed during the decay of ^{238}U to ^{206}Pb (Figure 5.3), So the different radioisotopes separate in environmental and biological systems, leading to disequilibrium. In decaying from ^{226}Ra to ^{222}Rn, for example, the atom changes from a metallic group II alkaline-earth element to a noble gas and then, after several days, to a metalloid (^{210}Po).

5.1.4 *Fission products and heavy actinides*

Nuclear fission (either during power generation or the detonation of a nuclear weapon) produces a series of radioactive fission products such as ^{137}Cs, many of which have short half-lives compared with U and Th. More than 300 different intermediate-weight isotopes may be formed in the fission process (NATO, 1996), and their abundance typically shows a bimodal distribution (Figure 5.4). This results from the breaking of the heavy unstable nuclei into two nuclei. The overall loss in mass results in the release of energy due to the conversion of mass into energy according to $E = mc^2$ where E is energy, m is mass and c is the speed of light in a vacuum.

The neutrons produced during fission react with U and transuranic elements to produce a series of heavy radioactive actinides such as Am and Pu isotopes and also react with stable elements such as sodium (Na), manganese (Mn), aluminium (Al) and silicon (Si) in the reactor core or natural environment to produce radioactive isotopes of these elements.

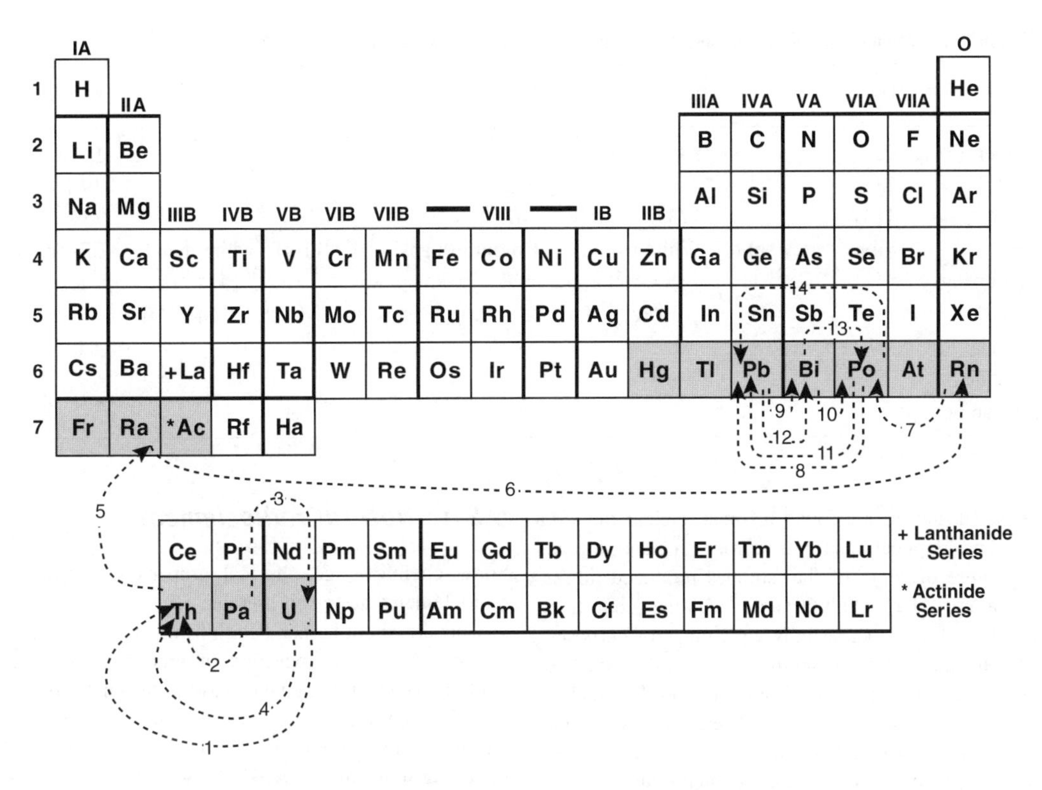

Figure 5.3 Periodic table, illustrating the decay of ^{238}U to ^{206}Pb and the chemical transitions involved

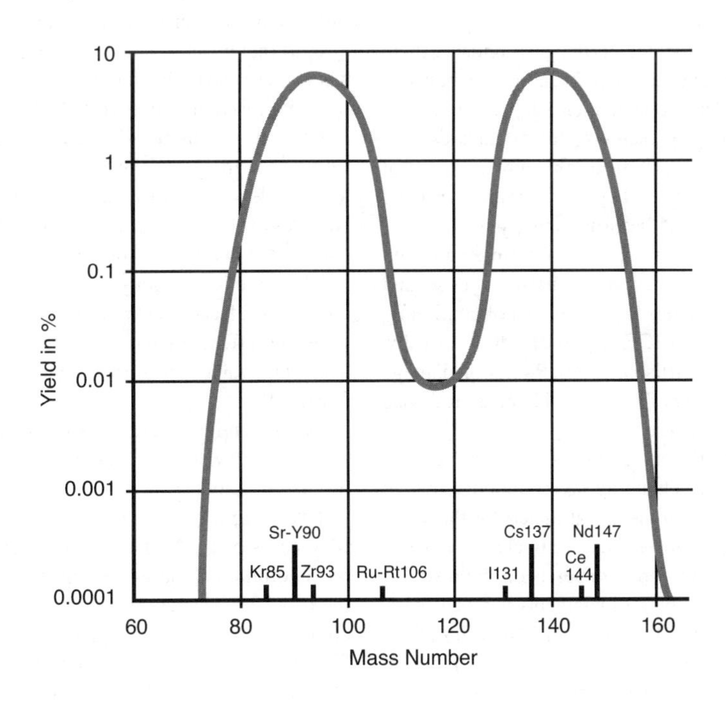

Figure 5.4 Typical fission-yield profile for neutron-induced fission of ^{235}U (Kaye and Laby, 1996)

Table 5.8 Physical and chemical properties of selected natural radioelements (Kaye and Laby, 1986)

	U	Ra	Rn	Th	K
Atomic number	92	88	86	90	19
Atomic weight	238	226	222	232	39.1
Boiling point (°C)	4131	1737	−61.9	4788	759
Appearance	Silvery-grey solid	White solid	Colourless gas	Silvery-white metal	Silvery-white metal
Naturally occurring isotopes	^{238}U, ^{235}U, ^{234}U	^{223}Ra, ^{224}Ra, ^{226}Ra, ^{228}Ra	^{222}Rn, ^{220}Rn, ^{219}Rn	^{227}Th, ^{228}Th, ^{230}Th, ^{231}Th, ^{232}Th, ^{234}Th	^{39}K, ^{40}K, ^{41}K
Most important oxidation states in the environment	+4, +6	+2	Noble	+4	+1

All of these substances are found in spent nuclear fuel and waste, and this has implications for its reprocessing and storage. The unique nature of some of the isotopes produced during fission, for example ^{60}Co (cobalt 60) and ^{241}Am has led to their extraction and use for an increasingly wide range of purposes. These include medical imaging and treatment, food production and processing, energy generation (both as fuel for nuclear reactors and as a source of heat in nuclear fuel cells) and industrial processes (IAEA, 2003a, b).

The radiotoxicity and toxicity of the fission product ^{90}Sr and the synthetic transuranic element ^{241}Am are discussed later.

5.2 Hazardous properties

The hazardous property of radioelements that distinguishes them from all other chemicals is their radioactivity, although they may also be chemically toxic. The same element, depending on the mode and rate of its decay, presents different hazards. For example, the radiological hazard from the alpha decay of ^{227}Th is greater than that from the beta and gamma decay of ^{234}Th, although the toxicological hazard is the same.

The detailed mechanisms by which radiation interacts with biological materials and the relationship between dose and biological effect at low dose levels are the subject of continuing research and debate (e.g. UNSCEAR, 2001, 2006; Busby, 1995, 1998; Wakeford, 2008; Allison, 2009). Radiation damage can initiate cancer if it damages the cell-cycle genes; it is also mutagenic and reprotoxic (CMR).

One of the ways in which the energy deposited by radiation is thought to damage cells is by changing deoxyribonucleic acid (DNA). Many radioelement cations are attracted by the weak negative charge of DNA. The type and degree of change to DNA and the likelihood of error-free repair depends on the dose.

The degree of hazard posed by radioelements varies according to their half-lives relative to the length of human life, their chemical species and physical form (for example if they are gases) and the extent to which they emit alpha, beta and/or gamma radiation and the energy of the radiation.

5.2.1 *Natural radioelements*

Some chemical and physical properties of selected natural radioelements are given in Table 5.8.

Concern about human health from natural radioactivity centres mainly on the ingestion or inhalation of alpha or beta emitters which, although they have little penetrating power, can, over time, damage tissues around them. If they are deposited in the lungs, for example, they can cause adverse health effects, including lung cancer (ATSDR, 1990, 1999).

Uranium is pyrophoric and U-metal powder and turnings can ignite spontaneously at room temperature in air and, in extreme conditions, can generate enough energy to cause a chemical explosion creating a range of particulate oxides which may be inhaled (ATSDR, 1999). Thus when depleted U is used in military armour or projectiles, or as counterweights in a large aircraft that crashes, it can form a fine dust that may be inhaled. Thorium is also highly flammable, and fine particles in the air can form explosive mixtures (IPCS, 2005). The main concern about inhalation of natural radioelements, however, is with the naturally occurring gas ^{222}Rn which has a half-life of only 3.8 days and the deposition of its alpha-emitting decay products on the surfaces of the lung. Damage can lead to the initiation of cancer and inheritable genetic defects.

The chemical toxicity of radioelements depends on the biochemistry of each species and may cause threshold and non-threshold effects including carcinogenesis and/or mutagenesis. The LD_{50} values for some U and Th compounds are shown in Table 5.9.

The potential hazards of some of the radioisotopes in the U and Th decay chains (Tables 5.5, 5.6 and 5.7) and ^{40}K are discussed below.

Uranium is a dense metal (density 19 g/cm^3). It has oxidation states of IV, V and VI, of which the IV and VI states are most common in nature. In solution, U exists predominantly as the free U(IV) species (in low-oxygen environments) or as the complex ion UO_2^{2+} (in high-oxygen environments), the latter forming particularly strong neutral or negatively charged complexes with CO_3 and HCO_3. Aqueous U species also have a strong affinity for phosphate, and while soluble U is excreted

Table 5.9 LD_{50} values of some uranium and thorium compounds (ATSDR, 1990, 1999)

Substance	Species	Route of exposure	LD_{50} (mg/kg)	Source
^{230}Th	Rat	Injection	42.7	ATSDR, 1990
Th nitrate	Rat	Intraperitoneal	60	ChemCAS*
Th nitrate	Mouse	Oral	1760	ChemCAS
Th nitrate	Mouse	Intravenous	45	ChemCAS
Th chloride	Rat	Intraperitoneal	1900	ChemCAS
Th chloride	Mouse	Intraperitoneal	332	ChemCAS
U	Rat	Oral	114	ATSDR, 1999
U	Mouse	Oral	136	ATSDR, 1999
Uranyl nitrate	Rabbit	Dermal	28	ATSDR, 1999
Uranyl nitrate	Guinea pig	Dermal	1190	ATSDR, 1999
Uranyl nitrate	Mouse	Dermal	4286	ATSDR, 1999
U fluoride	Rat	Intraperitoneal	20	ChemCAS
U fluoride	Rat	Intravenous	15	ChemCAS
U fluoride	Rabbit	Intravenous	5	ChemCAS

*Chemical Abstracts Service (CAS), a division of the American Chemical Society, is the most authoritative and comprehensive source for chemical information.

rapidly from the body it can concentrate on the growth surfaces of bone (Leggett, 1992).

All known U isotopes are radioactive and they decay mainly by alpha emission; those that occur naturally are given in Table 5.10. Uranium has two main decay series, ^{238}U and ^{235}U. These two isotopes occur at the present time in the proportion 99.3 per cent ^{238}U to 0.7 per cent ^{235}U.

The activity of ^{238}U in natural systems is approximately 20 times that of ^{235}U. It also has a greater number of decay products, including ^{226}Ra and ^{222}Rn, several of which have half-lives of significance for human health and/or are highly radiotoxic. The environmental significance of ^{235}U is small, therefore, but it is important commercially, as it is the isotope used for nuclear energy and nuclear weapons.

In their pure form, natural U, enriched U and depleted U, although they have different isotopic compositions, are almost identical chemically, undergoing the same chemical reactions in the environment and having the same chemical, biochemical and biological effects on the body. The radioactivity of depleted U, however, is 60 per cent that of natural U (WHO, 2001b; UNEP, 2001). The environmental behaviour of U and its mineralogy are summarised from the detailed account in Burns and Finch (1999).

The external dose from pure U, in the absence of its decay products, is small. However, pure U does not remain pure for long, as several short-lived decay products which emit beta and gamma radiation, such as ^{234}Th (β^-, γ), ^{234}Pa (β^-, γ) and ^{231}Th (β^-, γ) are soon formed (< 1 year). This is of potential concern in areas of conflict where depleted U has been used.

Uranium compounds differ in their toxicological effects due to their differing bioavailability and chemical behaviour. There is no universally accepted evidence that inhaled or ingested U causes cancer in humans (The Royal Society, 2001, 2002b). Studies of large cohorts of miners are confounded by the presence of other carcinogens such as arsenic (As) and Rn in the mining environment, and 14 epidemiological studies on workers in the U industry (exposed mainly to pure U) failed to provide conclusive evidence. Such studies have consistently failed to show excess deaths from cancer or kidney disease that can be attributed to inhaling or ingesting U. Such findings continue to be debated, and there is a need for more research (Balman, 2009).

While much is understood about the radiological and chemical toxicity of U, more needs to be done to understand its toxicological properties at a cellular level. At a population level, however, there is little, if any, evidence of clinical effects from naturally occurring U, even where exposures are high.

Table 5.10 Naturally occurring isotopes of U: decay modes, energies of emitted radiations and half-lives (Kaye and Laby, 1986)

Isotope	Decay series	Decay modes	α energies (MeV)	β^- energies (MeV)	γ energies (MeV)	Half-life
^{238}U	^{238}U series	α	4.20; 4.15	–	–	4.5×10^9 a
^{235}U	^{235}U series	α, γ	4.60; 4.40; 4.37	–	0.186; 0.144; 0.163	7×10^8 a
^{234}U	^{238}U series	α	4.77; 4.72	–	–	2.4×10^5 a

Table 5.11 Naturally occurring isotopes of Th: decay modes, energies of emitted radiations and half lives (Kaye and Laby, 1986)

Isotope	Decay series	Decay modes	α energies (MeV)	β^- energies (MeV)	γ energies (MeV)	Half-life
^{234}Th	^{238}U series	β^-, γ	–	0.18	0.093; 0.063	24 d
^{232}Th	^{232}Th series	α	4.02; 3.96	–	–	1.4×10^{10} a
^{231}Th	^{235}U series	β^-, γ	–	0.31; 0.29	0.026; 0.084; 0.081; 0.090	26 h
^{230}Th	^{238}U series	α	4.69; 4.62	–	-	7.3×10^3 a
^{227}Th	^{235}U series	α, γ	6.04; 5.98; 5.76	–	0.236; 0.050; 0.256	19 d
^{228}Th	^{232}Th series	α, γ	5.42; 5.34	–	0.084	1.9 a

Thorium has a number of naturally occurring isotopes, the longest lived being ^{232}Th (Table 5.11), which has a relatively simple decay chain (Table 5.7) but it is approximately four times as abundant as U in rocks and its specific toxicity in air and water is many times that of U; it even exceeds that of ^{239}Pu.

Radium is a group-2 (alkaline earth) element with several natural radioactive isotopes (Table 5.12) which emit energetic alpha particles with high Q values, and which, if ingested or inhaled, can deliver a significant local radiological dose. Radium has similar chemistry and geochemistry to the other group-2 essential elements Ca and magnesium (Mg), but, as the heaviest member of the group, it readily forms sparingly soluble salts (chlorides > carbonates > sulphates) and is more reactive than barium (Ba). Radium tends to accumulate in bone (ATSDR, 1990) and irradiates sensitive tissues such as bone marrow (Table 5.3). Some Ra isotopes are also gamma emitters and can consequently cause damage from outside the body.

^{226}Ra is the most abundant isotope of Ra (by mass), because it has a significantly longer half-life than the other isotopes and is associated with the more abundant ^{238}U decay chain.

Group-2 elements generally have low toxicity and are essential to living organisms, though soluble Ba ($BaCl_2$) is toxic to mammals (ATSDR, 2007). Radium might also have toxic chemical effects if present in significant concentrations, but these will be limited by its short half-life.

Radon is a heavy noble gas that is invisible, odourless and tasteless. It decays with the emission of beta and alpha particles and photons through a series of isotopes of Po and Bi, which, like ^{222}Rn, have short half-lives (in the range minutes to milliseconds; Tables 5.5, 5.6, 5.7 and 5.13), and finally Pb. One of the most hazardous properties of Rn and its decay products is the emission of energetic alpha particles capable of causing significant ionisation and tissue damage. Moreover, its decay products, also alpha and beta emitters, tend to attach to ambient aerosols that

Table 5.12 Naturally occurring isotopes of Ra: decay modes, energies of emitted radiations and half-lives (Kaye and Laby, 1986)

Isotope	Decay series	Decay modes	α energies (MeV)	β^- energies (MeV)	γ energies (MeV)	Half-life
^{228}Ra	^{232}Th series	β^-	–	0.039; 0.015	–	5.8 a
^{226}Ra	^{238}U series	α, γ	4.78; 4.60	–	0.186	1.6×10^3 a
^{224}Ra	^{232}Th series	α, γ	5.69; 5.45	–	0.241	3.7 d
^{223}Ra	^{235}U series	α, γ	5.75; 5.72; 5.61	–	0.269; 0.154; 0.324	11.4 d

Table 5.13 Naturally occurring isotopes of Rn: decay modes, energies of emitted radiations and half-lives (Kaye and Laby, 1986)

Isotope	Decay series	Decay modes	α energies (MeV)	β^- energies (MeV)	γ energies (MeV)	Half-life
^{222}Rn	^{238}U series	α	5.49	–	–	3.8 d
^{220}Rn	^{232}Th series	α	6.29	–	–	56 s
^{219}Rn	^{235}U series	α, γ	6.82; 6.55; 6.42	–	0.271; 0.402	4.0 s
^{218}Rn	^{238}U series	α	7.26	–	–	35 ms

can be deposited in the bronchi. Alpha particles emitted by radon and its decay products have a range of 50–70 μm in tissue, so they can irradiate all sensitive cells in lung tissue and, over time, initiate lung cancer because of DNA cell damage. Moreover the decay products of ^{222}Rn are heavy metals (Po, Pb, Bi) that are also toxic (Scott, 2007; ATSDR, 2007; Slikkerveer and de Wolff, 1989).

5.2.2 *Artificial radionuclides*

Strontium-90 (^{90}Sr) is a beta-emitting fission product which is an important component of nuclear waste and nuclear-weapons fallout. It is of concern because of its high fission yield (between 1 and 7.5 per cent, depending on the target nuclide: ^{235}U, ^{238}U, ^{241}Pu, ^{239}Pu, or ^{232}Th) and the energy of the reacting neutron, and its half-life (28.9 years). During its decay to stable zirconium (Zr) via yttrium-90 (^{90}Y, half-life 2.67 days), two beta particles are emitted (Kaye and Laby, 1986), one of which is of particularly high energy (E_{max} 2.28 MeV) with a large penetration range (from 0.1 to 1 cm). Hence ^{90}Sr sources are useful in industrial thickness gauges, as a radioactive tracer in medicine and agriculture, and in radiotherapy, including the treatment of bone and prostate cancer (Lewington, 2005; Stewart and Jones, 2005). It is a Group-2 alkaline-earth element with similar behaviour to Ca, forming sparingly soluble carbonates and sulphates and being susceptible to cation-exchange reactions, becoming strongly sorbed onto clays and organic matter. These latter factors reduce its overall environmental mobility and result in its concentration in topsoil, from which it may be ingested and recycled by grazing animals. The environmental behaviour of Sr is described in detail by Ragnarsdottir *et al.* (2001).

The chemical properties of ^{90}Sr are identical to those of other Sr isotopes and similar to other Group-2 alkaline-earth elements, such as Ca. Hence, if ingested or inhaled, ^{90}Sr can be deposited in areas of active bone growth close to blood-forming tissues. This, combined with the energy of beta particles from its decay product (^{90}Y) and its half-life, make it one of the most hazardous radioisotopes.

Americium (atomic number 95) is a synthetic transuranic element (Seaborg, 1946) produced by the successive neutron activation of lighter actinide elements such as U and the decay of intermediate Pu isotopes in a nuclear reactor. Radioactive isotopes of Am can also be produced by the direct decay of Pu isotopes (^{241}Pu, ^{242}Pu, ^{243}Pu, and ^{245}Pu) dispersed as a result of nuclear explosions and other accidental emissions involving irradiated reactor fuel such as the reactor explosion at Chernobyl and discharges from nuclear-fuel reprocessing facilities in the USA, Soviet Union, France and the UK. Whilst 16 radioactive isotopes of Am have been identified, the most stable are ^{241}Am (half-life 432.2 years), ^{242m}Am (a metastable state of ^{242}Am with a half-life 150 years) and ^{243}Am (half-life 7370 years). The remaining isotopes all have half-lives of less than 51 hours (Kaye and Laby, 1986).

^{241}Am is formed from the following sequential neutron-induced nuclear reactions:

(1) ^{238}U (n,γ) ^{239}Pu

(2) ^{239}Pu (n,γ) ^{240}Pu

(3) ^{240}Pu (n,γ) ^{241}Pu

(4) ^{241}Pu (β^- decay) ^{241}Am

Whilst the decay products of ^{242m}Am and ^{243}Am enter the natural ^{238}U and ^{235}U decay chains, the decay of ^{241}Am occurs via the non-natural neptunium (Np) decay series and results in the production of stable ^{209}Bi.

All three isotopes are present in nuclear waste from the production and processing of Pu. However, studies at US DOE sites such as Hanford indicate that ^{241}Am is typically the most abundant isotope, accounting for $>$ 90 per cent of the total Am inventory (Peterson *et al.*, 2007).

The comparatively short half-life of ^{241}Am (432.2 years) and the emission of alpha particles (5.49 and 5.44 MeV) and gamma rays (0.060 MeV) make it a useful source for a wide range of industrial and medical applications. Its most common use is in smoke detectors, where alpha particles from a small quantity (typically 0.3 μg) of ^{241}Am cause a constant current to flow through an ionisation chamber. The presence of smoke particles reduces ionisation, and interrupts electrical current flow, triggering the alarm.

Alpha particles from ^{241}Am are also used to generate neutrons via their reaction with beryllium (Be) in portable medical and industrial equipment: for moisture and density gauges, for example.

$$^{9}Be(\alpha, n)^{12}C + n$$

The low-energy gamma rays from the decay of ^{241}Am are used as a portable source for gamma and X-ray radiography in industry and medicine.

5.3 Sources

5.3.1 *Natural radioactivity*

The isotopes ^{238}U, ^{235}U, ^{232}Th and ^{40}K decay with half-lives so long that significant amounts of these elements remain in the Earth. Uranium, Th and K are lithophile elements with Pauling electronegativities of 1.7, 1.3 and 0.8 respectively, and they are strongly partitioned into the Earth's continental crust. Natural materials containing significant quantities of these radioactive isotopes have come to be referred to as Naturally Occurring Radioactive Material (NORM). The exact definition of the levels of radioactivity required for a natural mineral or material to be classified as a NORM varies internationally (IAEA, 2005).

Uranium, Th and K are present in trace amounts in all crustal rocks, and they are highly concentrated in certain types of mineral deposits (Plant *et al.*, 1999). The average crustal abundance of U is 2–3 mg/kg (Plant *et al.*, 1985, 1999; Ragnarsdottir and Charlet, 2000; WHO, 2001a) and that of Th 10 mg/kg (Levinson, 1974), but the U and Th contents of rocks can vary by factors of as much as 60 000 and 10 000, respectively (Bowie and Plant, 1983). The large ionic radii and valences of U and Th mean that in igneous rocks they become highly concentrated in the most fractionated granites and rhyolites (Plant *et al.*, 1999), which can contain up to 22 mg/kg U and 100 mg/kg Th, and in some alkaline igneous complexes. The lowest concentrations of U and Th are in ultrabasic and basic rocks such as ophiolites, which range from 0.1 mg/kg to 0.4 mg/kg if they are hydrothermally altered (Valsami-Jones and Ragnarsdottir, 1997). Lower crustal tonalites also generally have low concentrations of U and Th, especially where they have been subjected to granulite-facies metamorphism (Weaver and Tarney, 1980). In felsic rocks, U and Th are both enriched in accessory minerals such as zircon (U), allanite (Th) and monazite (Th), while K is enriched in rock-forming minerals such as feldspar and mica. Uranium occurs in ore minerals such as uraninite (UO_2) and other oxides in mineralised rocks.

The behaviour of U and Th differ in sedimentary rocks. Thorium has only one oxidation state, Th^{4+}, and it forms solid solutions with Ce and Zr in resistate minerals such as monazite ($(Ce,La)PO_4$) and thorite ($ThSiO_4$) which become concentrated in heavy-mineral sands and placers. Uranium as U^{4+} behaves like Th, and U can also be concentrated with heavy minerals. In oxidising environments, however, U can also occur as the U^{6+} uranyl ion (UO_2^{2+}). This oxyanion can have relatively high concentrations in the surface environment under oxidising conditions but is easily reduced and fixed in reducing environments, for example by decomposing organic matter (e.g. Higgo *et al.*, 1993; Ragnarsdottir and Charlet, 2000). The average U content of sedimentary rocks is 1.2–1.3 mg/kg and of Th 5.8 mg/kg (Langmuir, 1978; Kaye and Laby, 1986). Ocean sediments range from 0.3 to 3.8 mg/kg U (Church, 1973; Ben Othman *et al.*, 1989), though concentrations as high as 7.7 mg/kg U have been reported in Mn nodules (Ben Othman *et al.*, 1989). Sedimentary rocks such as sandstone and greywacke can contain up to 10 mg/kg Th, while shale and mudstone can average 10–13 mg/kg Th. Shale deposited in anoxic basins such as the Upper Cambrian shale of Southern Sweden contain up to 1000 mg/kg U.

Uranium concentrations in phosphate rocks are frequently in the range 20–120 mg/kg U (Langmuir, 1978), and phosphorites can contain 2500 mg/kg U or more. Vast tonnages occur in a belt of Late Cretaceous to Eocene age phosphorite that extends from Morocco into Egypt, Israel and Jordan (Boujo and Jiddou, 1989; Wininger, 1954). Similar deposits occur in Florida. The radioelements in such rocks not only emit Rn locally but, since rock phosphate is commonly used as fertiliser, they can also add to the body burden of U and its decay products because of their uptake into food crops, especially where rock phosphate is used in intensive agriculture (e.g. Zielinski *et al.*, 1997; WHO, 2001b).

The distribution of U over Europe has been described by Plant *et al.* (2003), based on geochemical data prepared by the Forum of European Geological Surveys (FOREGS). The highest background levels (5–50 mg/kg) occur over crystalline Variscan blocks (e.g. the Massif Central in France and the Meseta in Spain) associated with post-orogenic high-heat-production (HHP) granites. High U values also occur over parts of the Alpine terrain, especially in Slovenia, where there are historical U workings, and southern Italy, associated with contemporary volcanism. In contrast to these areas, much of the Caledonides of north-west Europe and the Precambrian of the Baltic Shield and East European craton and its overlying sedimentary cover have low uranium values (< 4 mg/kg). The distribution of Th is broadly similar to that of K, although high values of the latter occur over southern Norway, much of Sweden and south Finland, which have been attributed to the presence of potassic schists, gneiss, granite and Old Red Sandstone rocks (FOREGS, 2005).

Uranium ore deposits can be important sources of U and its decay products into the environment. Several classifications of ore deposits have been proposed. The most recent is in the OECD (2005), which classifies U deposits according to their economic importance. Here we follow the classification of U ore deposits of Plant *et al.* (1999) which uses the deposit types of OECD (1998) but orders them according to geological setting rather than economic importance (Table 5.14).

Generally, the ore deposits that are of most concern environmentally are (1) of vein type, associated with granites such as those of the Massif Central of France, and (2) lignite and phosphorite deposits (Smith *et al.*, 1998), most of which are no longer economic. The giant unconformity-type deposits that now represent much of the world's known U reserves, such as the deposits of the Athabasca basin in Canada, are difficult to detect at the surface using either radiometric or chemical methods (Allan and Richardson, 1974; Hoffman, 1983) and are therefore less of a threat to the environment.

Radium concentrations in rocks and soils reflect those of U, provided there is secular equilibrium. Radon is formed by the decay of ^{226}Ra and is a significant contaminant affecting indoor air quality, especially in areas of mineralised granites, uraniferous shales or phosphatic rocks including phosphorites. There are no Rn sinks, due to its short radioactive half-life (3.8 days) (ATSDR, 2008). The outer 10 km of the Earth's crust contains 10^{24} Bq Rn. It is emitted from the ground, especially in areas with soils containing uraniferous rocks. Sources of Rn are given in Table 5.15. The level of Rn and its decay particles in the air depends mainly on its source and the extent of dilution. The gas may accumulate in stagnant air, such as in mines, caves or homes where there is inadequate ventilation (WHO, 1983). Radon in houses in the UK varies 100 fold between different parts of the country, the highest levels overlapping those in mines although the mean radon level (20 Bq) in the UK is slightly less than half that in USA (46 Bq). The principle source of Rn in the home is the underlying ground and associated soil gas (Ball and Miles, 1993).

The component of the average annual radiation dose to the UK population from rocks, and radon and thoron gas, is approximately 63 per cent (Figure 5.1).

Table 5.14 Classification of U deposits (Plant *et al.*, 1999)

A IGNEOUS PLUTONIC AND VOLCANIC ASSOCIATION

1 Igneous Plutonic Association

1.1 *Magmatic U deposits – formed by differentiation of evolved uraniferous magmas*
- 1.1.1 Alkaline complex deposits

1.2 *Formed as a result of high-to-low temperature hydrothermal activity associated with high-level granite magmatism*
- 1.2.1 Granite-associated deposits, including vein-type deposits
- 1.2.2 Perigranitic vein deposits
- 1.2.3 Metasomatite deposits

2 Igneous Plutonic and Volcanic Association

2.1 *Deposits associated with granite magmatism and acid volcanic and volcaniclastic sequences in anorogenic settings*
- 2.1.1 Breccia complex deposits

3 Igneous Volcanic Association

3.1 *Formed as a result of high-to-low temperature hydrothermal activity associated with high-level mainly felsic volcanics*
- 3.1.1 Volcanic deposits

B METAMORPHIC ASSOCIATION

1 Formed by metamorphic fluids probably derived from igneous or sedimentary rocks previously enriched in U

1.1 Synmetamorphic deposits

1.2 Vein deposits in metamorphic rocks

C SEDIMENT OR SEDIMENTARY-BASIN ASSOCIATION

1 Continental

1.1 *Associated with large post-orogenic sedimentary basins having mainly clastic fill – formed or modified in some cases by intra-basinal fluid flow*
- 1.1.1 Quartz-pebble conglomerate deposits
- 1.1.2 Unconformity-related deposits
- 1.1.3 Sandstone deposits
- 1.1.4 Sedimentary-hosted vein deposits
- 1.1.5 Collapse-breccia deposits
- 1.1.6 Lignite deposits

1.2 *Penecontemporaneous with sedimentation or formed by surface weathering*
- 1.2.1 Superficial deposits

2 Marine

2.1 *Oceanic*
- 2.1.1 Phosphorite deposits

2.2 *Epicontinental*
- 2.2.1 Black-shale deposits

Main types are based on classes used by the OECD/NEA (1998).

5.3.2 Anthropogenic sources

The hazards from radioactivity and radioelements which are of the greatest public concern are associated with the nuclear industry. The average dose to people in Western democracies such as the UK is extremely small (Siegel and Bryan, 2004), but many people are afraid of the consequences of a serious accidental release of radioactivity, however unlikely that may be. Anthropogenically derived materials containing only naturally occurring radioactive isotopes have come to be referred to as technologically enhanced naturally occurring radioactive material (TENORM). Typical sources of TENORM and associated levels of activity are given in Table 5.16.

5.3.2.1 Nuclear power and weaponry

Nuclear power and weaponry use energy from naturally occurring radioactive substances. They are now based mainly on U as

Table 5.15 Sources of Rn (WHO, 1983)

Source	Rn production per year (Bq)
Soil	9×10^{19}
Plants and ground water	$< 2 \times 10^{19}$
Oceans	9×10^{17}
Houses	3×10^{16}
Natural gas	3×10^{14}
Coal	2×10^{13}

feedstock, but it is likely that Th-based reactors will be developed in the future. Here we focus on U sources that have been enriched or depleted by humans. Uranium and its decay products can be released into the environment from industrial processes associated with the nuclear fuel cycle, which includes mining, milling, and processing U ores or U products (ATSDR, 1999).

5.3.2.2 Uranium mining and processing

Uranium mine tailings can provide a source of U and its decay products long after mine closure. Uranium mining contamination has been reported from the Navajo Nation in the south-west US (Abdelouas *et al.*, 1999; Colon *et al.*, 2001), where large-volume tailings contain about 85 per cent of the radioactivity of the unprocessed ore, mainly as U and isotopes of Ra and Rn. A study of six mining sites in the USA reported U concentrations of up to 10 mg/l at pH ~7 in leachate waters below the tailings dam of the Midnite mine in eastern Washington, which was in operation from 1956 to 1982 (Landa and Gray, 1995). A pale yellow precipitate containing 6 per cent $U_3O_8.nH_2O$ (schoepite)

Table 5.16 Typical sources and levels of TENORM in Bq/g or Bq/l. NA indicates data not available. Adapted from http://www.epa.gov/radiation/tenorm/sources.html

Source		Radiation level [Bq/g]		
		low	average	high
Geothermal energy production wastes		0.370	4.89	9.41
Oil and gas production wastes	Produced water	0.004	NA	333
	Pipe/tank scale	<0.009	<7.40	>3700
Drinking water treatment wastes	Treatment sludge	0.048	0.407	433
	Treatment plant filters	NA	1480	NA
Waste water treatment wastes	Treatment sludge	0.000	0.074	1.741
	Treatment plant ash	0.000	0.074	0.815
Al production wastes	Ore (bauxite)	0.163	NA	0.274
	Product			
	Production wastes	NA	0.144–0.207	NA
Coal ash	Bottom ash	0.059	0.130–0.170	0.285
	Fly ash	0.074	0.215	0.359
Cu mining and production wastes		0.026	0.444	3.06
Fertiliser and fertiliser production wastes	Ore (Florida)	0.259	0.064–1.46	0.230–2.00
	Phosphogypsum	0.270	0.433–0.900	1.36
	Phosphate fertiliser	0.019	0.211	0.778
Gold and silver mining wastes				
Rare earth (monazite, xenotime, bastnasite) extraction wastes		0.211	NA	119
Titanium production wastes			0.296	0.907
	Rutile		0.730	NA
	Ilmenite	NA	0.211	
	Wastes	0.144	0.444	1.67
U mining wastes	Uranium mining overburden	<7.50	<7.50	<7.50
	U in-situ leachate evaporation pond	0.111	1.11	111
	Solids	11.1		
Zircon mining wastes	Wastes	3.22	2.52	48.1

was reported for a distance of 0.8 km downstream of the site in 1978. Water in the adjoining surface waters have been reported to contain up to 5.7 mg/l U. Concentrations of up to 20 mg/l U have also been reported in shallow aquifers where the pH is low (Langmuir, 1997) near to the Canyon City ore processing site in Colorado, which operated from 1958 to 1989.

Uranium tailings and waste rock have been reported to cover 1518 ha with a tailings volume of 161×10^6 m^3 in the south of the former German Democratic Republic (Hagen, 2007). After the end of the second world war, the Soviet Wismut company developed the third-largest U mining province in the world (after the US and Canada) in Saxony and Thuringia to provide a source of U for Soviet nuclear weaponry (Beleites, 1988).

The global U-tailing inventory at mine and mill sites has been estimated to be accumulating at a rate of 25–75 million tonnes per year (NEA, 1984). Emissions of Rn from uncovered U tailings range from 0.5 to 10 Bq/m^2/s. Tailings engineering can help reduce the rate of Rn emission by adjusting the thickness and area of the tailings per unit mass of U. Covering tailings with a few metres of soil can reduce Rn emission by several orders of magnitude, to as low as 0.01–1.0 Bq/m^2/s (WHO, 1983). In the US, the DOE is responsible for remediating mill tailings and associated groundwater under the Uranium Mill Tailings Radiation Control Act (UMTRCA, 1978, 1988). By 1999, 24 of the main UMTRCA sites of concern had been remediated (Brady *et al.*, 2002).

Other anthropogenic sources of U include P mining and milling, tailings for rock-phosphate fertiliser and detergent production, and geothermal power stations (Fisher, 1998). Some coals contain high levels of U and its decay products, and in some situations thermal power stations can emit as much or more radioactivity than nuclear power plants (Ewing, 1999; McBride *et al.*, 1978). Production wells in oil and gas fields can also contain high levels of Ra in brine. This is deposited as Ra sulphate along with Ba sulphate as scale on pipe work and plant during extraction and requires treatment as radioactive waste (Fisher, 1998; Botezatu and Grecea, 2004).

5.3.2.3 *Nuclear-power generation*

Most civil nuclear reactors use comparatively lowly enriched U (up to 6 per cent ^{235}U), whilst military reactors used for powering ships and submarines use more highly enriched U due to the need for a smaller reactor core. In both cases, U may be isotopically enriched through the use of a variety of techniques including gaseous diffusion and centrifugation. An example of the change in composition of U fuel during three years of power generation is shown in Table 5.17.

No U should be released from nuclear power plants if the design of the fuel assembly is adequate (ATSDR, 1999), although the nuclear reactor accident which occurred at Chernobyl in the Ukraine in 1986 did release significant quantities of U and its fission products into the environment and the fallout is still affecting parts of Europe. Earlier nuclear accidents included a fire in a prototype reactor at Windscale (now called Sellafield) in

Table 5.17 Heavy-metal composition of nuclear fuel enriched to 4.2 per cent, before and after use for about 3 years (40 000 megawatt days per tonne). Minor actinides include Np, Am and curium (Cm). Secondary radioactivity produced from the irradiation of structural material such as zirconium and stainless steel is not included (www.whatisnuclear.com)

	Percentage in initial fuel	Percentage in fuel after 3 years
Uranium	100	93.49
Enrichment (i.e. per cent ^{235}U)	4.2	0.71
Plutonium	0.00	1.27
Minor actinides	0.00	0.14
Fission products	0.00	5.15

1957 in the UK. The accident at the Three Mile Island Unit nuclear power plant near Middletown, Pennsylvania, in 1979, was the most serious in the operating history of the US commercial nuclear power industry, although it led to no deaths or injuries to workers or people in nearby communities. The main concern with such accidents is generally not with U itself but with its fission products such as ^{90}Sr and ^{131}I (iodine 131), which are relatively short lived and can cause tissue damage and, in some cases, cancer if ingested. In the West, authorities have usually acted promptly, for example destroying contaminated milk and issuing potassium iodide tablets to saturate the thyroid glands of exposed individuals with stable iodine.

5.3.2.3.1 Fukushima On 11 March 2011 an under-sea, magnitude-9.1 earthquake occurred off the north-east coast of Japan. The earthquake was the most powerful ever to have hit Japan, and was one of the five most powerful in the world since modern record-keeping began in 1900. The Fukushima Daiichi nuclear power station and all the other nuclear power plants along the north-east coast of Japan withstood the earthquake. However, the subsequent tsunami, with waves of up to 15 metres, disabled the Fukushima Daiichi nuclear power plant, which had been designed to withstand 6-metre waves.

The Fukushima Daiichi plant was first commissioned in 1971, and consisted of six reactor units. Three units had been shut down prior to the earthquake, for planned maintenance. The other three were shut down automatically after the earthquake and the remaining decay heat of the fuel was being cooled with power from emergency generators. The tsunami disabled these emergency generators. Over the following weeks there was evidence of partial nuclear meltdowns, and explosions, and a possible uncovering of spent fuel pools in some units. Releases of isotopes such as ^{137}Cs and ^{131}I caused large-scale evacuation and concern about the contamination of local food and water supplies.

The events at units 1, 2 and 3 have been rated at Level 7 (major release of radioactive material with widespread health and environmental effects requiring implementation of planned and extended countermeasures) on the International Nuclear Event Scale.

The bodies of two nuclear workers were discovered in the plant, and there is evidence that they were killed by the tsunami.

It will be a considerable time before it will be possible to estimate the number of cancers caused by this event – in workers at the plant who were directly exposed to radiation, and in the wider public who may have been exposed to the radioactive iodine and caesium. However, the number is likely to be very much smaller than the number of people killed by the earthquake and tsunami, which, according to the Japanese National Police Agency, exceeded 10 000 with more than 17 000 people still listed as missing two weeks after the earthquake and tsunami (The Telegraph on Line, 25 March, 2011).

The accident at the Fukushima Daiichi nuclear plant clearly indicates the need for continued vigilance and updating of risk assessments and mitigation measures for the effects of natural disasters on nuclear plants and their environs in the light of increasing scientific knowledge and awareness of potential hazards.

Whilst our understanding of the long-term consequences and contributing factors to the events at Fukushima Daiichi are still in the early stages, it is clear that excellent communication and action by scientists, planners, politicians and the general public are needed to ensure that the risks from such events are minimised in the future.

5.3.2.4 Nuclear weaponry

Fallout from atmospheric testing of nuclear weapons continued until the atmospheric test ban treaty of 1963 between the USA, UK and USSR. This followed the detonation by the USSR of the then biggest-ever nuclear device at Novaya Zemla which caused radioactive-fallout levels to peak all around the world, causing great public concern (Lindell, 1996).

Contamination from the production of nuclear weapons is particularly important in the US and the former Soviet Union. In the US, contamination is located at 64 environmental management sites in 25 states, for example at the Hanford site in Washington State, where nine Pu production reactors were built. The total inventory at the Hanford site was found to be $1.33–1.37 \times 10^{19}$ Bq, of which between 8.1×10^{15} Bq and 2.4×10^{17} Bq were released to the ground (National Research Council, 2001). Contamination in the former Soviet Union and former Warsaw Pact counties is more extensive because of fewer controls during the communist era; there have been a series of serious accidents exposing the local population to increased radioactivity, such as those at Chelyabinsk near Kyshtym in the 1950s and 60s which released radioactivity into the nearby Techa river causing large radiation doses to exposed populations. One of these accidents, the Kyshtym disaster of 1957, was an explosion in a high-level nuclear storage tank.

The Soviet Union conducted 456 nuclear tests (340 underground and 116 above ground) at Semipalatinsk-21, renamed Kurchatov, in eastern Kazakhstan from 1949 until 1989, increasing the exposure of the local population and the environment to radioactivity. The actual extent of damage inflicted by nuclear testing at Semipalatinsk continues to be the subject of much research and debate. One study (Akanov *et al.*, 2009) noted that the population in Semipalatinsk and areas adjacent to the test site received more than 500 millisieverts of radiation in one exposure. In the most heavily contaminated areas, people received an estimated effective equivalent dose of 2000 millisieverts during the years of testing. The rate of cancer in those living in the area most exposed to the radiation, remains 25–30 per cent higher than elsewhere in Kazakhstan.

A total of 1054 nuclear tests were conducted by the US; over 100 of them at sites in the Pacific Ocean, more than 900 of them at the Nevada test site, and ten on miscellaneous sites elsewhere in the US. Until 1962, most were atmospheric. Several groups of people who lived downwind of the Nevada test site and US military workers at various tests have won compensation for their exposure, especially after the Radiation Exposure Compensation Act of 1990 allowed for the systematic filing of compensation claims.

5.3.2.5 Depleted uranium and conventional weaponry

Nuclear fuel is produced by the enrichment of natural ^{235}U from about 0.7 per cent in uranium ore to about 3.5 per cent; weapons-grade U results from further enrichment (to >80 per cent). The residue from the enrichment process, mainly ^{238}U with about 0.2 per cent ^{235}U, is used chiefly in armour-piercing shells because of its pyrophoric properties on impact and its high density (19 g cm^3). This depleted U is also used as a counterweight in ships and aircraft (WHO, 2001b; UNEP, 2001; The Royal Society 2001, 2002a).

In nature, U occurs with its decay products (Tables 5.5 and 5.6). Depleted U, has been available only since about 1940, when enrichment of ^{235}U began. Depleted U therefore contains only low levels of many of the naturally occurring decay products of U and has only about 60 per cent of its radioactivity. Both U and depleted U, and their immediate decay products (e.g. ^{234}Th, ^{234}Pa and ^{231}Th), emit alpha and beta particles with a small amount of gamma radiation. Some depleted U is also contaminated with transuranic elements and fission products such as ^{238}Pu, ^{239}Pu, ^{240}Pu, ^{241}Am, ^{237}Np and ^{99}Tc (technetium 99) (Rich, 1988; CHPPM, 2000) because the same enrichment plants are used for processing both natural U and spent U nuclear-fuel rods (UNEP, 2001; The Royal Society, 2001; 2002a, b).

A recent survey of depleted U in Kosovo provided an analysis of penetrators found in conflict areas (UNEP, 2001). There was up to 12 Bq/kg of Pu isotopes and up to 61 kBq/kg of ^{236}U, together with a much higher activity concentration of ^{238}U – 12700 kBq/kg. Depleted U used in warfare has the potential to contaminate the environment, although few military personnel appear to have been contaminated (Smith, 2007). The potential long-term risk to the local population from exposure to depleted

U used in military conflict remains uncertain, since its subsequent fate has not been identified and its dispersal in the environment is likely to vary (The Royal Society, 2002b).

5.3.2.6 *Radioactive waste*

Radioactive waste includes waste from a wide range of materials and processes. While the term is most commonly used to refer to wastes from nuclear-power generation it also includes a wide range of wastes from other industries including the oil and gas industry as well as nuclear medicine. The classification of nuclear waste differs between countries. For example, in the UK materials such as U, Pu and spent nuclear fuel are not declared as waste (HSE, 2007) because they are reprocessed and may be reused.

While each country is free to classify and control radioactive wastes in its own way, at an international level this is overseen by the International Atomic Energy Agency's Joint Convention on the Safety of Spent Fuel Management and on the Safety of Radioactive Waste Management and its Basic Safety Standards (BSS, 1996). These categorise radioactive waste into different types according to the level and nature of radioactivity present (Table 5.18).

Hence there are many different types of wastes with different management requirements. The greatest volumes comprise the exempt, very low-level, and low-level wastes. The volume of high-level waste is relatively small but continues to increase as nuclear reactors across the world come to the end of their useful life. For example, it is estimated that the inventory of reactor waste in the US will reach 1.3×10^{21} Bq by 2020 (Ewing, 1999). The decay of radionuclides from a reference inventory is shown in Figure 5.5 on the next page.

Initially the bulk of the radioactivity is due to short-lived isotopes such as ^{137}Cs and ^{90}Sr. After 1000 years the main hazard is from the decay of ^{241}Am, ^{243}Am, ^{237}Np and Pu isotopes. Up to 10^6 years, ^{99}Tc, ^{210}Pb and ^{226}Ra dominate. The US Department of Energy (DOE) has shown that after 10^5 years the radiological dose associated with ingesting a given weight of spent nuclear fuel is about the same as ingesting the same weight of 0.2 per cent U ore (Siegel and Bryan, 2004).

Holmes *et al.* (2002) compared the characteristics of a large unconformity-type U ore deposit in the Athabasca basin with those of proposed nuclear-waste repositories and concluded that the Earth is able to store large volumes of radioactive materials without any chemical or radioactive contamination of the environment (Figure 5.6). The latest EPA standard for nuclear-waste repositories seeks to limit maximum exposure from all pathways for individuals living up to 18 km away to 0.15 mSv/yr, less than that on the shores of the contaminated Lake Karachay near Chelyabinsk where the dose is about 0.2 mSv/hr (Cochran *et al.*, 1993; Siegel and Bryan, 2004).

5.3.2.7 *Other accidents involving radioactive materials*

There have been other accidental exposure to humans as a result of the use of substances such as ^{137}Cs in medical devices. For example, in the Goiania in Brazil in 1987 a stolen capsule caused widespread contamination and killed four people. There have been similar accidents with ^{60}Co sources in Mexico city in 1962

Table 5.18 Broad categorisation of radioactive waste materials based on IAEA (2003a, b, c, 2005, 2006)

Category	Material	Management	Isolation timescale
Exempt and very low-level wastes	Uranium mine and other tailings; some coal and wood ash; phosphate wastes; depleted U	No radiological restrictions, disposal as for other mine wastes	
Low-level wastes	Most nuclear maintenance wastes contaminated with fission nuclides Some hospital and medical wastes	Isolation and storage at an enclosed surface or near-surface facility	Tens of years
Intermediate-level wastes	U-mine and other tailings. Some coal ash. Some wood ash. Some medical sources and wastes	Isolation and storage at an enclosed surface or geological disposal facility	Hundreds of years
High-level and transuranic wastes	Separated fission products (i.e. Cs-137 and Sr-90 are the significant nuclides). Retired medical, industrial and research devices. Spent fuel if not reprocessed. Retired weapon-grade material	Isolation and storage within a deep geological disposal facility	Thousands of years

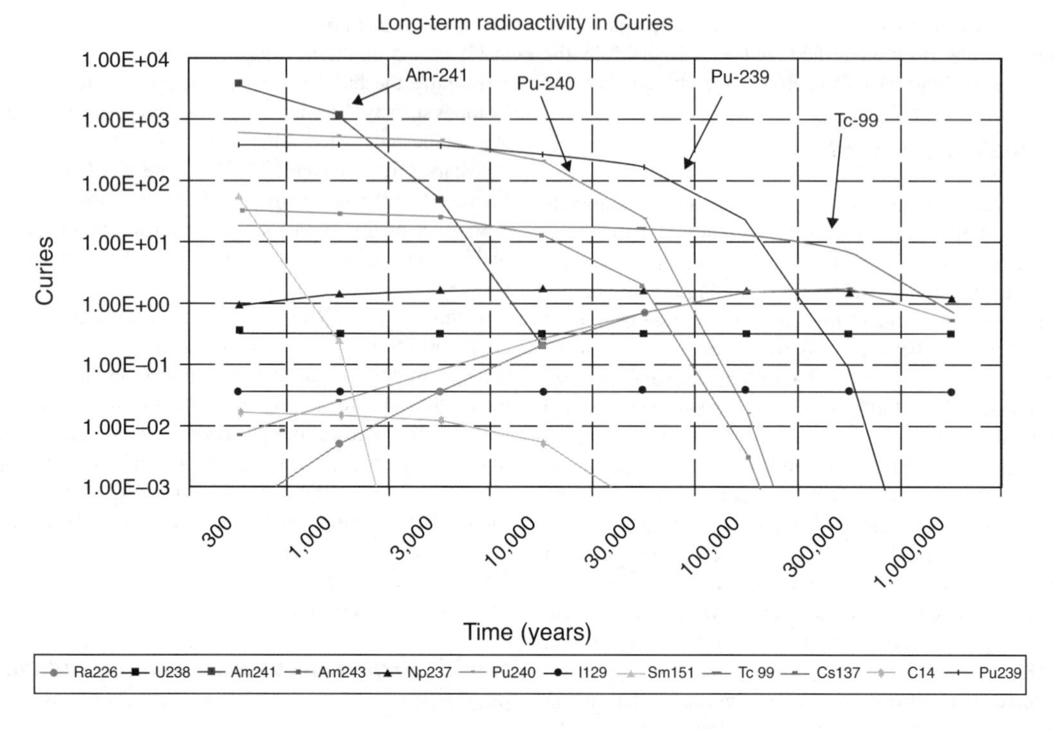

Figure 5.5 The activity of some radioactive nuclides as a function of time up to 1 million years from 1 million tons of reactor-fuel wastes, burned to 45 megawatt days/kg (http://www.whatisnuclear.com/)

and in Italy in 1975, but most of these accidents have caused lethal injury to only one or two people (Nenot, 1996).

5.4 Environmental pathways

Uranium compounds exhibit a variety of different chemical properties. For example, U trioxide (UO_3), uranyl chloride, uranyl nitrate and uranyl ethanoate are relatively soluble, whereas U dioxide (uraninite, UO_2) is relatively insoluble (e.g. ATSDR, 1999; Burns and Finch, 1999; Ragnarsdottir and Charlet, 2000), providing that the environment is low in oxygen.

5.4.1 Soils

Worldwide, there are few systematic high-resolution surveys of radioactivity in soil. Data available from international agencies are mostly low-resolution temporal monitoring data collected to identify atmospheric contamination from nuclear testing, for example. The increasing availability of high-sensitivity, high-resolution airborne gamma spectrometry, however, has enabled U, as equivalent U-(^{214}Bi), other isotopes in the U and Th series, and K to be determined in soils routinely over large areas (Jones and Scheib, 2007).

The concentration of naturally occurring radioactivity in soil reflects the levels of U, Th and K in the soil parent material and chemical processes in the soil. Uranium, for example, can become concentrated in organic-rich horizons (Porcelli *et al.*, 1997). In addition, concentrations may reflect the addition of natural-series radionuclides from fertiliser (U, Th and K), military conflicts and training, and mining (UNSCEAR, 1982; Zielinski *et al.*, 1997; The Royal Society, 2002b; Ledvina *et al.*, 1996; McConnell *et al.*, 1998).

Levels of U in soils away from sources of contamination or mineralisation have median values of 1 to 2 mg/kg. Values of 4 mg/kg are considered to represent the upper baseline level for U (WHO, 2001b). The median for U in subsoils over Europe (reported by FOREGS, 2005) is 2.03 mg/kg, and 2.0 mg/kg in topsoil. For Th, the values were 8 mg/kg in subsoils and 7.24 mg/kg in topsoils, and for K_2O 2.02 per cent in subsoil and 1.92 per cent in topsoil. Much higher concentrations of U occur in soils over U mineralisation. Concentrations >200 mg/kg are not unusual over mine sites. Levels of up to 53 mg/kg occur in soil over mineralised Variscan granites in the Massif Central of France and northern Portugal, for example (FOREGS, 2005). Levels of U in surface soils over phosphorite in north Africa and the Middle East, can be as high as 200 mg/kg. Uranium levels can also be high near U-processing plants (British Geological Survey, 1992, Chenery *et al.*, 2002), mine tailings and process waste streams (Ledvina *et al.*, 1996; McConnell *et al.*, 1998), and agricultural environments in which U-rich phosphate fertilisers have been used (e.g. Zielinski *et al.*, 1997).

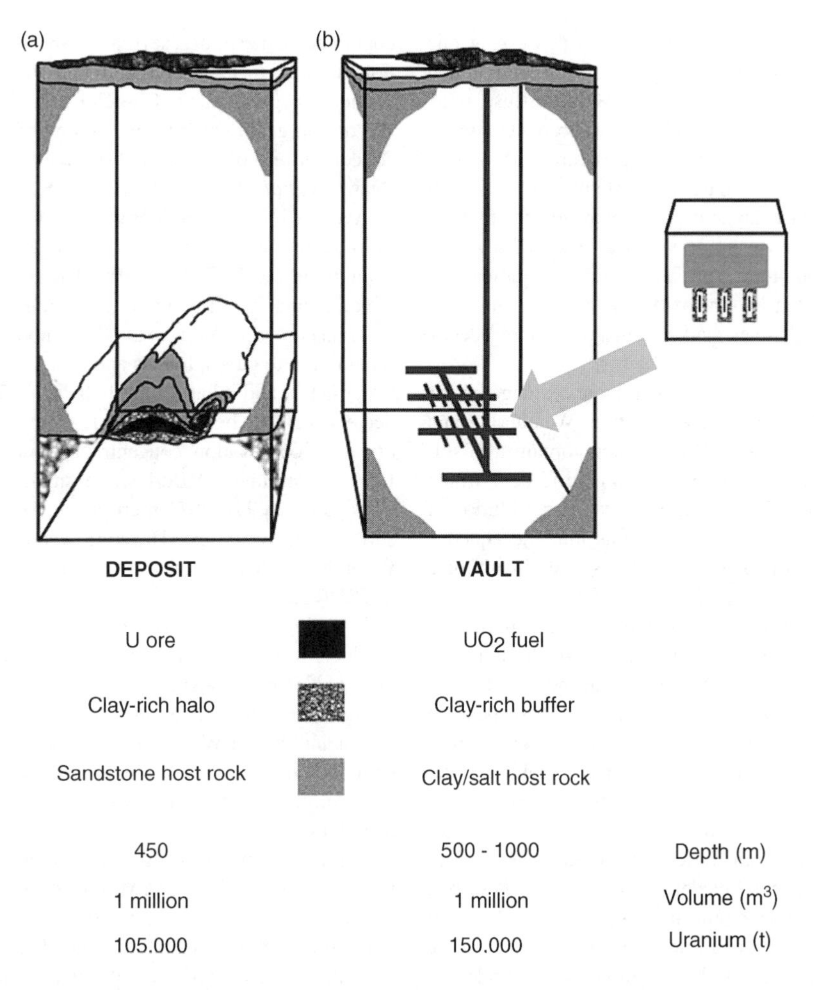

Figure 5.6 Comparison between an unconformity-type U deposit and a geological setting chosen for a radioactive waste repository. (a) High-heat-production (HHP) granites capable of regenerating convective flow are present in the basement; (b) HHP granites are not present in the basement (Reprinted from Goodwin, B.W., Cramer, J.J. and McConnell, D.B. (1988). The Cigar Lake uranium deposit: an analogue for nuclear fuel waste disposal, CEC natural analogue working group, Third Report. Report of the Commission of the European Communities, pp. 181–190)

Concentrations of other U and Th series isotopes in soils are rarely in secular equilibrium due to their different chemical behaviour. Soils rich in iron and/or manganese can have high levels of Ra isotopes due to the long-term sorption of dissolved Ra in reducing groundwater or springs, for example.

The presence of artificial radionuclides in soils frequently reflects atmospheric deposition as a result of nuclear-weapons testing, small discharges from nuclear and thermal power stations and research reactors, and from radioactive substances used in medicine, research and technology. There have also been several minor nuclear incidents involving the release of radionuclides into the atmosphere and, subsequently, soil.

The grave nuclear accident at the Chernobyl nuclear power plant in the Ukraine in 1986 caused serious radioactive contamination of soil by Pu, Am, and Cs isotopes from atmospheric fallout. Contamination across Europe extended thousands of miles from the accident site (Hohenemser *et al.*, 1986; INSAG, 1986).

5.4.2 Water

Concentrations of radioelements such as U, K, Ra and Rn are detected in many natural waters, depending on the pH and redox conditions and the activity of ligands such as HCO_3^- and Cl^-. Levels of Th are generally low. Under reducing conditions at low pH, U is present as the aqua U^{4+} species, which is only sparingly soluble. As the pH increases, U hydroxides such as UOH_3^+, $U(OH)_3^+$, $U(OH)_4^o$, $U(OH)_5^o$ are stable. Ryan and Rai (1983), Parks and Pohl (1988) and Rai *et al.* (1990) showed that the solubility of $UO_2(s)$ is independent of pH up to 300 °C and that $U(OH)_5^-$ is only of minor importance.

Radiolysis of water can occur when an α particle from U decay loses its energy by ionising water. One α particle with the energy of 1 MeV can ionise 10^5 water molecules (Dubessy *et al.*, 1988), producing the stable secondary products H_2 and O_2. Other species produced include H_2O_2 and HO_2, which are powerful oxidising agents. If organic matter is present, organic free radicals (organic compounds or moieties possessing one or more unpaired electrons) are also produced (Rigali and Nagy, 1997). In the surface environment, H can diffuse away easily, but this is not the case in U ores whereby U can be mobilised readily. Hence uraninite surfaces rarely show true $UO_{2.00}$ stoichiometry in natural systems (Cathelineau *et al.*, 1982), the surface layer of UO_2 being closer to UO_3 in composition. Uranium in nuclear fuel consists of >95 per cent UO_2 so that long-term risk assessment of radioactive waste sites is dominated by the oxidation of ^{238}U and ^{235}U oxides and by the half-lives and chemistry of their decay products.

The U concentration of river waters in large drainage basins may be determined by carbonate dissolution. Weathering and oxidation of uraniferous black shales, which contain iron sulphides, can produce large quantities of sulphuric acid which reduces the pH of water. This in turn leads to the dissolution of carbonates and the input of carbonate ions into the aqueous environment, leading to high levels of U as uranyl-carbonate complexes in rivers (Palmer and Edmond, 1993).

A significant proportion of U in rivers is transported as suspended particles (Martin and Whitfield, 1983). Uranium and other radioelements such as Th can be transported through natural waters as colloids, which can remain in stable suspension because of mutual charge repulsion (Stumm and Morgan, 1996). Rivers, especially turbulent ones, can transport colloids over long distances, but when they enter estuaries colloidal suspensions are destabilised by increases in ionic strength (Mayer, 1982). Uranium transported by the Kalix River into the Baltic Sea from northern Sweden is dominated by U sorbed onto Fe-oxyhydroxide and biogenic phases greater than 0.45 μm in size. The colloid fraction carries 30–90 per cent of the transported U, transported mainly by humic acids (Anderson *et al.*, 1998).

The distribution of U in stream water over Europe varies by four orders of magnitude, from <0.002 to 11.1 μg/l (FOREGS, 2005). The highest values occur mainly over sediments and sedimentary basins in south and central Europe and reflect the distribution of major dissolved ions and hence the residence time of the water rather than bedrock geology. The distribution of Th is quite different, with high values over south Sweden and Finland, for example, reflecting the low pH conditions of the waters and their high levels of organic matter (FOREGS, 2005).

The average concentration of U in surface water in the USA is 1.6 μg/l whilst that in groundwater is estimated to be 4.7 μg/l (Drury, 1981). Contaminated surface and groundwaters such as those associated with U mining, nuclear sites, and mining of rocks and materials such as coal and phosphate ore which may also contain trace quantities of U do however occur throughout the United States (ATSDR, 1999). Uranium concentrations as high as 226 μg/l have been observed in surface waters outside a radioactive-waste disposal site in Massachusetts (Cottrell *et al.*, 1981) and as high as 23 000 μg/l in groundwaters and surface waters at the Urvan site in Colorado (ATSDR, 1999) from which most of the U for the Manhattan Project was mined. Although much of the U is contained on these sites, several notable incidents have occurred. For example, in 1979 a break in the tailings pond at the United Nuclear Corp mill site in Church Rock, New Mexico resulted in the discharge of 93 million gallons of tailings fluid with U concentrations of up to 5700 μg/l into the Rio Puerco river (Dawson *et al.*, 1997; ATSDR, 1999).

Average U concentrations in groundwater range from 0.1 μg/l in reducing conditions to 100 μg/l in oxidising conditions (Langmuir, 1997; WHO, 2001b). The average Th concentration, which is unaffected by redox conditions, is 1 μg/l.

Transport via water is also an important pathway for Rn, including into caves, homes and mines. The solubility of Rn in water is low and it has a short half-life of 3.8 days, so most of it decays before being released from groundwater (O'Neil *et al.*, 2006). Radon concentrations in large aquifers in the USA are around 8.8 Bq/l while smaller aquifers contain an average of 28.9 Bq/l (Cothern *et al.*, 1986), depending on the surrounding rock type. Drinking water sourced from groundwater has an estimated average concentration of 20 Bq/l Rn (ATSDR, 2008).

Leaching can lead to the dispersion of U and its decay products from mine and tailings into groundwater. Concentrations of U in pore waters in mill tailings in North America can be as high as 0.5 mg/l (Abdelouas *et al.*, 1999). In the former Soviet Union and former Warsaw Pact countries, extensive radioactive contamination of groundwater has occurred because of the use of solution mining of U ore deposits using aggressive acid-leach techniques. The Straz U deposit in the Czech Republic, for example, was mined using sulphuric and nitric acids and ammonia. Approximately 266 Mm^3 of the north Bohemian aquifer extending over 24 km^2 is estimated to be contaminated, threatening the catchment of the Plucnice River (Slezak, 1997).

In the US, the DOE estimates that nuclear-weapons production has led to the radioactive contamination of approximately 63 Mm^3 of soil and 1310 Mm^3 of groundwater at 25 sites such as the Hanford Reservation in Washington State. A complex mixture of radioactive waste may have entered the vadose zone. (Siegel and Bryan, 2004). The Mayak chemical combine, the main weapons production facility in Russia, discharged large quantities of medium-level waste directly into the Techa river between 1950 and 1951, exposing 1.24×10^5 people to excess radioactivity. Medium-level waste was also discharged into the nearby Lake Karachay, creating a radioactive pollution plume that extends up to 5 km from the lake. In 1967, dust from the lake-shore sediments was blown over an area of 2700 km^2, causing ^{137}Cs and ^{90}Sr contamination. In 1957 a high-level-waste storage tank exploded during the Kystym disaster, releasing 7.4×10^{17} Bq of radioactivity into the atmosphere, about 90 per cent of which fell out near the tank. The rest contaminated an area of about 23 000 km^2 with high levels of ^{90}Sr (Medvedev, 1976; Trabalka *et al.*, 1980).

Effluents from Th mining, milling and recovery plants, as well as water from P and phosphate fertiliser production and tin-ore processing can contribute some Th as ^{230}Th and ^{232}Th into surface and groundwater (ATSDR, 1990). Leachate from U tailings can contaminate groundwater with ^{230}Th from the decay of ^{238}U.

Food and drinking water can also be important pathways for natural radioelements and contribute to internal exposure of radioactivity. The level of radioelements ingested depends on the soil, the use of rock-phosphate fertiliser and the type of food and its preparation prior to consumption (Arogunjo *et al.*, 2005). The level of U tends to be higher in root vegetables than other types of food and there is generally more U in water than food (ATSDR, 1999). Levels of radioelements can be reduced if foodstuffs that are normally eaten raw are washed prior to consumption (Arogunjo *et al.*, 2005). Peeling removes most of the radioactivity in potatoes (Carvalho *et al.*, 2009).

A study by Adriano and co-workers (2001) showed that concentration ratios (ratio of mean radioelement concentration in plant tissue (Bq/kg dry weight) to mean radioelement concentration (Bq/kg dry weight) in soil) declined with more vigorous cleaning of root vegetables. A light wash of carrots had a concentration ratio of 3.7×10^{-3} while scrubbing reduced this to 1.1×10^{-3} (Adriano, 2001).

In the UK, cereals and bread are the main sources of ^{238}U, accounting for up to 65 per cent of total U intake in children aged 1.5–2.5 years. In adults, the total dietary exposure is 0.028 mg/kg of body weight per day (FSIS, 2004).

The amounts of Rn and Th obtained from food are negligible, although drinking water is important for Rn exposure. It has been estimated that between 3.7 and 22.2 Bq per day is consumed, assuming that 2 litres of water is imbibed (Cothern *et al.*, 1986). The intake of ^{230}Th, ^{232}Th and U through drinking water is estimated to be 0.00222, 0.00074 and 0.03 Bg/l, respectively (ATSDR, 1990, 1999). Intake of radioactivity and radioelements from mineral water has been investigated by several authors. Bomben *et al.* (1996) showed U concentrations in bottled mineral waters from Argentina to range from 0.04 to 11 µg/l, with a mean of 1.3 µg/l; while DeCamargo and Mazzilli (1996) showed the concentrations of U in mineral waters from Brazil to be 0.08 to 2.0 µg/l. Kovács *et al.* (2004) showed that the radiochemical dose from drinking bottled mineral waters in Hungary was considerably higher than the limit of 0.1 mSv suggested by WHO, especially for children. The issue of the ingestion of U by infants in food and water has been further highlighted as an area of particular concern by the European Food Safety Agency (Castoldi *et al.*, 2009). The UK Food Standards Agency undertook a survey of natural radioactivity in 100 commonly available brands of bottled waters (FSA, 2004) and reported that none breached any legal limits.

The radiological dose has been determined for Italian red and white wines (Desideri *et al.*, 2010). Mean concentrations of ^{238}U, ^{234}U and ^{210}Po were $3.63 \times 10^{-3} \pm 2.19 \times 10^{-3}$, $4.41 \times 10^{-3} \pm 3.10 \times 10^{-3}$ and $6.85 \times 10^{-2} \pm 3.79 \times 10^{-2}$ Bq/l for red wine and $5.08 \times 10^{-3} \pm 4.20 \times 10^{-3}$, $5.59 \times 10^{-3} \pm 8.49 \times 10^{-3}$ and $3.92 \times 10^{-2} \pm 2.63 \times 10^{-2}$ Bq/l for white wine. For most of the samples, the ^{210}Po concentration is 10 times higher than that of ^{238}U, probably due to fallout of atmospheric Po. Uranium concentrations in wine were lower than in water, but the Po concentration was higher. The effective radiation dose from U and Po ingested in wine (0.5 l per day) ranged from 2.68 to 4.91×10^{-2} mSv per year.

Table 5.19 Levels of natural radioelements in some foods, mBq/kg (Santos *et al.*, 2002)

Food type	^{238}U	^{232}Th
Cabbage	0.17	0.05
Lettuce	4.10	13.0
Water cress	15.00	2.50
Spinach	14.00	16.00
Tomato	1.30	1.10
Pumpkin	0.70	0.10
Potato	0.11	0.98
Onion	0.22	0.02
Carrot	1.30	1.90
Rice	2.60	3.20
Banana	0.03	<0.02
Apple	0.18	0.08
Pasta	26.00	2.80
Sugar	0.58	3.20
Coffee	25.00	23.0
Milk	48.00	-

The content of ^{238}U and ^{232}Th in some foodstuffs is shown in Table 5.19.

A recent study of soil to plant (potato) radionuclide transfer in the vicinity of an old U mine in Portugal has shown significant uptake of Ra, reaching 12.9 Bq/kg dry weight. The potatoes were irrigated with water containing 13.7 Bq/l U (Carvalho *et al.*, 2009).

Mineral waters have often been shown to be rich in U, Ra and Rn, although levels will vary considerably depending on the local geology (ASTDR, 1999; WHO, 2001b; ASTDR, 2008; Reimann and Birke, 2010). This is typically because such waters are associated with high degrees of water-rock interaction and areas of high geothermal gradient.

In the past, legislation on the chemical content of mineral and table waters was often less stringent than for tap water, as they were commonly considered under separate regulations (food safety and drinking-water quality). However, European and national regulations are now in place which have harmonised regulated acceptable levels of potentially harmful substances in mineral waters and drinking waters (EU, 2003, 2009). The presence of U, Ra and Rn in medicinal waters lies outside the scope of these regulations. At present the World Health Organisation has estimated that U levels in drinking water of up to 15 µg/l are considered tolerable for adults, on the basis of its nephrotoxicity, whilst concentrations of Ra and Rn are regulated by their radiological impact (WHO, 2009).

5.4.3 Air

The amount of Rn released from water into air is influenced by water use: boiling and splashing increase its release, so most of the Rn release from water in homes is in bathrooms and kitchens (WHO, 1983). Whilst ATSDR (2008) has estimated that around 1/1000th of the Rn in water can enter air in this way the amount is highly variable and can be much higher. Radon concentrations are also dependent on the volume of water used, the volume of the house and the effectiveness of ventilation (WHO, 1983).

Most Rn is released into the atmosphere by the decay of ^{226}Ra. The average estimated global soil emission rate is 59.3 Bq/cm^2 per year, equivalent to a global release of 7.4×10^{19} Bq per year. Outdoor air concentrations of Rn in the USA average 14.8 Bq/m^3, depending on factors such as soil type and bedrock, while average indoor air concentrations are higher, at around 46.25 Bq/m^3 (ATSDR, 2008).

Ambient levels of Rn in air and water are influenced by the composition of the soil and bedrock. The highest concentrations are usually related to silica-rich igneous rocks such as rhyolite and granite (ATSDR, 2008), especially where these are fractured and/or mineralised (Plant *et al.*, 1999).

Thoron in air can be from both natural and anthropogenic sources. Volcanic ash can contain 1.06 µg/g ^{232}Th, and coal fly ash contains 4.5–37 µg/g. ^{230}Th has been detected in dust from U ore processing and milling and uraniferous tailings, and its levels are especially high where U ore is being crushed. It is estimated that 7.4 GBq of ^{230}Th is released into the atmosphere annually from U milling, coal-fired power plants, phosphate-rock processing and other mineral extraction and processing facilities (ATSDR, 1990).

5.5 Bioaccessibility and bioavailability

The terminology for bioavailability and bioaccessibility for radioelements differs from that used for other chemicals. Transfer rates within an organism are critical in biokinetic modelling because of the relationship between residence time and radiological dose, and the need to enable organ-specific doses to be calculated. Moreover, radiation biology puts more emphasis on intake via inhalation, especially for particulate matter entering the lungs, and exchanges between lung fluid and blood.

5.5.1 *Inhalation*

Biokinetic models used to calculate radiological dose from inhalation (ICRP-1994) consider the size and shape of particles to estimate the fraction reaching sites where irradiation of lung tissue or transport into the lymphatic systems can occur. The mineralogy of particulates is used to determine the solubility of the particle (and any associated radionuclides) in lung fluids (for example Ansoborlo *et al.*, 2002). From these data, the residence time of radionuclides in the lungs, the rates of clearance and the efficiency of transfer to other systems in the body can all be determined.

Table 5.20 Default absorption/lung clearance parameter values

ICRP Publication 66 absorption type	F (fast)	M (moderate)	S (slow)
Model parameters:			
Fraction dissolved rapidly, *fr*	1	0.1	0.001
Dissolution rate:			
Rapid (per day), *sr*	100	100	100
Slow (per day), *ss*	–	0.005	0.0001

F (fast) – materials that are readily absorbed into blood. The default for such materials is *sr*=100 per day (t $^1/_2$ approximately 10 minutes).
M (moderate) – materials with intermediate rates of absorption, where *fr* = 0.1 per cent; *sr* = 100 per day; and *ss* = 0.005 per day.
S (slow) – relatively insoluble materials. It is assumed that for most of the material the rate of absorption to blood is 0.0001 per day and that 0.1 per cent of the deposited material is rapidly absorbed. This is represented by *fr* = 0.001; *sr* = 100 per day; and *ss* = 0.0001 per day.

Data have typically been obtained from:

- biokinetic studies on animals and humans in which inhalation exposure to a particular particle-size distribution and chemistry is monitored.
- Experimental studies of the solubility of particles in synthetic lung fluids.

Methods have been standardised (ICRP-1994; ICRP-68, 1994) to enable radiological doses to be calculated and to classify inhaled particulates (Table 5.20) and particle loadings (Table 5.21) for typical workers and members of the public (ICRP-2002).

Table 5.21 Deposition of inhaled airborne particles after occupational and public exposure

Region[a]	Occupational[b] (per cent)	Public[c] (per cent)
ET1	33.9	14.2
ET2	39.9	17.9
BB (bronchial)	1.8	1.1
bb (bronchiolar)	1.1	2.1
AI	5.3	11.9
Total deposit	82	47.3

[a]The extrathoracic airways consist of the anterior nasal passages (ET1) and the posterior nasal and oral passages, pharynx and larynx (ET2). The thoracic regions are bronchial and bronchiolar (BB and bb) and alveolar-interstitial (AI). For the purposes of external monitoring, the retention in the chest would be the activity retained in the thoracic region.
[b]Occupational exposure. Worker, 5-µm AMAD (sg = 2.5), 3.5-µm AMTD, density 3.0 g/cm^3, shape factor 1.5; fraction breathed through nose is 1.31 per cent sitting and 69 per cent light exercise; mean ventilation rate is 1.2 m^3/h.
[c]Environmental exposure (indoors at home). Adult male, 1-µm AMAD (sg = 2.47), 0.69-µm AMTD, density 3.0 g/cm^3, shape factor 1.5; fraction breathed through nose is 1.55 per cent 33.3 per cent sleeping, 25 per cent sitting, 40.6 per cent light exercise and 1.0 per cent heavy exercise; ventilation rate is 0.78 m^3/h.

Table 5.22 Dissolution kinetics

Type	Americium	Uranium	Thorium	Radium	Strontium
F		UO_4			Sr
M	Am	Mixed oxides, UF_4, UO_3	Th-rich black sands	Ra	$SrTiO_3$
S		UO_2 and U_3O_8	Th oxides and hydroxides		

Examples of absorption and clearance type for various elements and some associated chemical species discussed in this chapter are given in Table 5.22.

Inhaled particles and aerosols may be removed from the lung through direct ingestion during the early stages of inhalation, the mucociliary elevator or by the lymphatic system. A tabulation of age-dependent doses to members of the public resulting from the inhalation of a wide range of radionuclides has been compiled in ICRP-71 (1995). However, where material specific data on bioavailability is available, it is recommended that additional calculations based on this data should be undertaken unless the calculations in ICRP-71 can be shown to be conservative.

5.5.2 Ingestion

In radiobiology and biokinetic modelling, the terms 'gut uptake factor' and 'fractional uptake from the small intestine (f_1)' are used in the discussion of bioavailability. Both terms correspond to the proportion of ingested substance entering the hepatic circulation from the small intestine.

More recently the term 'alimentary tract transfer factor (f_A)' (ICRP-2006) has been introduced to:

- allow the total amount of radioactive material entering the blood from the whole of the alimentary canal (including the mouth, stomach and upper and lower intestine) to be calculated; $f_A >$ or $= f_1$,
- facilitate research into the variability and complexity of the uptake of radioactive materials into the alimentary tract (ICRP-100, 2006; Dublineau *et al.*, 2006).

ICRP has not yet published lists of f_A values.

While material specific data should be used for modelling to reflect different solubilities in the gut and the ability to transfer across the gut wall, generic f_1 values for a particular element and age of receptor are often used (Table 5.23). Uncertainties in f_1 values and their variation within populations and between age groups is a potential source of error in biokinetic modelling and the estimation of dose coefficients (ICRP-100,2006). For example, reported variations in f_1 for U species range from 0.002 for U (IV) compounds, 0.02 for more soluble U(VI) compounds to as high as 0.20 for unspecified compounds in food stuffs (Wrenn *et al.*, 1985).

Variations in f_1 with age show the increased uptake in the young and especially newborns (for example group-2 elements as a result of rapid bone growth).

Table 5.23 Typical default element-specific f_1 values (ICRP, 2001)

Element	f_1 (adult)	f_1 (10-year- old)	f_1 (newborns)
Am	0.0005	0.0005	0.005
U	0.02	0.02	0.04
Th	0.0005	0.0005	0.005
Ra	0.20	0.30	0.6
Sr	0.30	0.40	0.6

5.5.3 Effects on receptors

The effects of radiation on human health have been determined from studying those exposed to different levels of radiation in nuclear explosions or occupational settings and from patients treated with X-rays and other radiation. For example, Court Brown and Doll (1958 and 1965) determined leukaemia risk not only from observations on Japanese atomic-bomb survivors, but also patients who have been treated with X-rays for ankylosing spondylitis. The most extreme exposures were suffered by survivors of atomic-bomb blasts in Hiroshima and Nagasaki at the end of the second world war.

No increase in cancer frequency has been observed below an external dose of 0.2 Gy (20 rad). Workers involved in cleaning up after the Chernobyl nuclear accident received 100 mSv (Ewing, 1999), approximately twice the legal annual dose for radiological workers in the US. Radiation-protection standards are usually based on a zero threshold dose–response curve, with public exposure limited to 1 mSv per year (0.1 rem per year).

Acute doses of 0.1–5 Gy damage organs at the cellular level through single-strand and double-strand DNA breaks, leading to mutations and or cell death after a few divisions of the irradiated cell. Cells that reproduce rapidly, such as those in bone marrow or the gastrointestinal tract, are more sensitive than longer-lived cells such as nerve cells.

Cancer is the main effect of low-radiation doses. Laboratory studies have shown that alpha, beta and gamma radiation can produce cancer in almost every tissue and organ in the animals studied (ATSDR, 2001). Besides cancer there is little evidence of other human health problems from low-level radiation (ATSDR, 2001).

The main routes of exposure to natural radioelements, depending on their properties, are through food, water and air. Uranium exposure is mainly through food and water, and Rn exposure is primarily through inhalation. In order for a radioelement to be considered as a risk, a source, pathway and receptor must be in place. Some of the effects of exposure to natural radioelements include bone cancer, lung cancer, female breast cancer, thyroid cancer, skin cancer, pancreatic cancer, leukaemias, liver disease and kidney disease. These effects can occur many years after initial exposure to Th and/or U and/or their decay products (ATSDR, 1990, 1999). Different cancers have different latencies; leukaemia can appear 2 years after exposure, while solid tumours can arise 20 years after exposure. Much of the evidence on the adverse health effects and types of cancer caused by Th and ^{226}Ra are from their historical use in medicine and are relatively unimportant for the general population. The evidence relating to risks from ^{222}Rn, however, is of immediate practical importance to millions of people worldwide. The evidence that it can cause lung cancer has been confirmed by many studies of miners in Canada, China, the Czech Republic, France, Sweden, the UK and the USA (Doll, 1996).

5.5.3.1 *Oral exposure*

The main chronic toxicological effect of U in exposed humans was considered to be degradation of renal functioning (The Royal Society, 2002a; WHO, 2001a, b). However, whilst renal effects have been demonstrated in animal studies, acute poisoning incidents and epidemiological studies indicate that the clinical significance of these findings is still uncertain. For example, in epidemiological studies of human populations with chronic long-term exposure to U in drinking water (usually the main source of dietary U; ATSDR, 1999), reports of clinical and sub-clinical effects on the human renal system were inconsistent (Kurttio *et al.*, 2002, 2006; Magdo *et al.*, 2007; Seldén, *et al.*, 2009).

Other toxicological studies of oral U exposure, using data from exposed humans and animals and/or in-vitro models, indicate a diverse range of potential effects, including development, reproduction, increased oxidative stress, diminished bone growth, mental health effects and DNA damage (WHO, 2001b; Brugge *et al.*, 2005).

For example, recent animal-based research has provided evidence that U produces an oestrogenic response in female mice (Raymond-Whish, *et al.*, 2007) and disturbs their fertility (Arnault *et al.*, 2008), as well as promoting leukaemic transformation of haematopoietic cells (Miller *et al.*, 2005, 2009). Such research also confirms the induction of genotoxicity in certain fish (Lourenco *et al.*, 2010), the proximal cells of rat kidney (Thiebault *et al.*, 2007) and the lung cells of rats (Monleau *et al.*, 2006).

Human exposure has been linked to: genomic instability in human osteoblast cells and weak genotoxic effects in a cohort of Gulf war veterans (Miller *et al.*, 2003, 2009; Smirnova *et al.*, 2005); cytotoxicty to human lung cells and oxidative stress in lung epithelial cells (Wise *et al.*, 2007; Periyakaruppan *et al.*, 2007); and chromosomal aberrations and alterations to gene expression in populations and cell cultures from individuals exposed to U (Wolf, 2004; Prat *et al.*, 2010; Milacic, 2008; Milacic and Simic, 2009).

Knowledge of the adverse effects of ^{226}Ra is derived mainly from historical reports (Macklis, 1993). Between 1917 and 1926, the US Radium Corporation was engaged in the extraction and purification of Ra as a major supplier of luminescent watches to the military. Their plant in New Jersey employed over 100 workers, mainly women, to paint watch faces and instruments. Unfortunately, many of the women licked their brushes to make them more pointed and in doing so ingested Ra – in some cases accumulating large body burdens, leading to the development of bone sarcomas and carcinomas of the paranasal sinuses and mastoids. Between 1925 and 1930 a quack medicine, 'Radithor', was marketed as a cure for a range of conditions including impotence and high blood pressure. One individual, a millionaire Eben Byers, died from severe radiation damage after consuming more than 1000 bottles of this 'medicine' (Lindell, 1996). Information obtained as a result of these experiences helped to establish intake limits.

5.5.3.2 *Inhalation exposure*

Inhalation is the main route of exposure for Rn. The daily intake of Rn in the USA is an estimated 926 Bq per day based on an average indoor air concentration of Rn of 46 Bq/m^3 (USEPA, 2003) and a breathing rate of 20 m^3 per day. In Europe, the geometric mean Rn concentration in indoor air ranges from $<$30 Bq/m^3 (UK, Holland Denmark, Lithuania, Bulgaria) to as high as 105 Bq/m^3 (Finland, Estonia, Czech Republic, Hungary and Albania) (Appleton and Miles, 2010). Occupational exposures to Rn, especially those in U or phosphate mining, can be as high as 410 kBq/m^3 (Somlai *et al.*, 2006).

In the United States, the average intake of ^{230}Th through inhalation of air and ingestion of drinking water are 0.0259 and less than 2.2 mBq per day respectively whilst corresponding intakes for ^{232}Th were estimated to be 0.0259 and less than 0.74 mBq per day (ATSDR, 1990). Levels vary between countries and regions, but average daily intakes of U from inhalation are estimated to be between 0.259 and 0.0259 mBq per day (ATSDR, 1999).

Both epidemiological studies and animal studies indicate that exposure to Rn gases can increase the risk of lung cancer. This can be exacerbated by smoking tobacco, since Rn particles adhere to particles in tobacco smoke (ATSDR, 2008). It has been suggested that lung cancers caused by Rn exposure may be indistinguishable from those attributed to cigarette smoke (Chaffey and Bowie, 1994). The first epidemiological study of radon and lung cancer was made among miners at Jachymov, in the Czech Republic just before the second world war. After the war, the Soviets mined massive amounts of uranium from this region with little regard for health and as a result many miners contracted lung cancer (Clarke, 1996).

Lung-cancer mortality in a cohort of 4320 miners, first employed during 1948–1959 at the Jáchymov and Horní Slavkov U mines in west Bohemia and monitored until 1 January 1991, showed that the highest risk was for miners who had spent more than 20 per cent of their employment at Jáchymov, where they were exposed to much higher levels of As in dust. Arsenic and U are frequently associated with mineralised leucogranites containing tin (Sn) and U mineralisation (Tomasek *et al.*, 1994).

If all the studies of miners are pooled, the overall excess of 1900 lung cancers due to Rn gas exceeds the overall excess of 400 total cancers in the survivors of the atomic bombings of Japan in 1945 (Clarke, 1996). Such studies also show that risk varies in proportion to exposure at levels below those suffered by miners. Radon can also cause carcinoma of the nasal sinuses.

Following inhalation, Rn delivers doses to the liver, bone marrow and kidney between 0.2 and 1 per cent of that to the lungs, but there is no epidemiological evidence of an excess of associated cancers (Doll, 1996). The use of Rn therapy continues to be used in spas around the world, but there is no evidence of an excess of associated cancers caused by these low doses.

5.5.3.3 Dermal exposure

Dermal absorption of U has not been quantified in humans, but animal studies show that water-soluble U compounds are the most easily absorbed (ATSDR, 1999).

Thorium X (^{226}Ra) was commonly used in topical dermatological preparations in the first half of the twentieth century, but since the 1950s it has been linked to cases of basal-cell carcinoma. For example, a 59-year-old patient had been given multiple applications of a preparation containing Th X to his scalp when he was 7 years of age as a treatment for alopecia. Several decades later, malignant papules appeared on his scalp. It is has been suggested that the age of the patient, part of the body treated and the total dose of radiation would affect the type of neoplasm produced (Rajaratnam *et al.*, 2007).

Radiological effects such as ulceration and dermal atrophy have not been observed from dermal exposure to Rn, because target cells are deeper than the range of the alpha particles emitted (Charles, 2007).

5.5.3.4 Injection

Some compounds containing natural radioelements have been used in radiography. Thorotrast, a colloid containing 25 per cent ^{232}Th dioxide, was used as a radiographic contrast medium for angiography between 1928 and 1955. Thorium has a long half-life and thorotrast had a biological half-life of 700 years in the body. Typically 1–75 ml of the solution was injected, resulting in 129–962 Bq/kg body weight, of which 59 per cent was accumulated by the liver, 29 per cent by the spleen and 9 per cent by the bone marrow (Kaul and Muth, 1978). Granulomas were diagnosed at the injection site 4–6 years after administration. Other effects include haematological problems, which became apparent 20 years after exposure (Harrist *et al.*, 1979; Frank, 1980). Studies showed that those treated had increased their risk of developing liver cancer by more than 100 times and of developing leukaemia by more than 10 times. The risk of other cancers, including myeloma, was more than doubled (Andersson and Storm, 1992).

A more recent study of Portuguese patients given Thorotrast between 1928 and 1959 showed a significantly increased risk of death from liver cancer, chronic liver disease, diseases of the digestive and respiratory systems, and neoplastic and non-neoplastic haematological problems (dos Santos Silva *et al.*, 1999). According to Trott (1988) 10 times more people may have died from Thorotrast-induced cancer than from radiation-induced cancer following the bombing of Nagasaki and Hiroshima in 1945 (Quinlan and Scopa, 1976).

After 5 minutes of intravenous administration of U as uranyl nitrates for tracer studies into humans, 25 per cent of U is present in the blood. This drops to 5 per cent after 5 hours, 1 per cent after 20 hours and less than 0.5 per cent after 100 hours. U has been found to be mainly in the skeleton (10 per cent), kidneys (14 per cent) and other soft tissues (6 per cent) 2.5 hours after injection (Basset *et al.*, 1948).

A short-lived isotope of Ra, ^{224}Ra, was injected into patients in Germany shortly after the end of the second world war to treat bone tuberculosis and ankylosing spondylitis. An increase in bone sarcomas occurred, with an incidence that reflected dose and duration of treatment (Quinlan and Scopa, 1976).

5.6 Risk reduction

Radiation has been and continues to be an emotive issue (Fisk, 1996). Psychometric studies of public perception of risk have shown that dangers associated with radioactivity are the most dreaded and least understood hazards (Slovic, 1987). For this reason, the US stopped building civil nuclear power plants more than 30 years ago, following the Three Mile Island accident, although this accident did not kill or injure anyone – a fact that can be compared with the significant loss of life in the oil and coal industries over the same period of time. The fossil-fuel industries have also caused serious environmental damage associated with accidental spills of oil, as well as global warming and ecosystem damage. The moratorium (until recently) on new nuclear build in the US and UK failed to recognise the important contribution that the nuclear industry can make to energy security in the West and reductions in greenhouse gas emissions globally. The need for nuclear power is increasingly recognised including by environmentalists such as James Lovelock. In his view, nuclear energy is the only realistic alternative to fossil fuels that has the capacity to fulfil the large-scale energy needs of society while also reducing greenhouse emissions (Lovelock, 2006).

One of the problems with the public perception of the nuclear industry has been the secrecy surrounding it from the early years of the development of nuclear weaponry. As Fisk has suggested in relation to the nuclear industry generally, ‘We need to shed the philosophy of secrecy with which it is often surrounded. We can hardly complain about public perception of risk if our body language in public is one of secrecy and confidentiality. One of the ways of overcoming people's fears is through public information. But that is not enough. People do not believe what they are being told, because they do not believe they are being told the full story. We need openness as well' (Fisk, 1996). In the UK, public access to information has been pioneered by the Royal Commission on Environmental Pollution. There has also been strong pressure for openness in relation to the development of nuclear-waste repositories, for example from The Royal Society.

Unlike many other environmental hazards, the risk from radiation has been recognised almost since its discovery and radiological protection has developed alongside the development of the nuclear industry. Indeed, it has been suggested that Agricola, who invented methods to ventilate mines affected by Rn in the Erzgebirge in the fifteenth century, was the first radiation protection officer (Lindell, 1996). Late in 1896 following Rontgen's discovery of X-rays in 1895 and Becquerel's discovery of radioactivity in 1896, a paper by Fuchs (Lindell, 1996) gave simple recommendations on avoiding skin burns, and by 1921 the British X-ray and Radium Protection Committee had proposed ambitious radiation-protection measures, followed soon afterwards by other European countries and the US. In 1928 the International Commission on Radiological Protection (ICRP) was established. There were no dose limits, but there were recommendations on working-hours restrictions, the use of protective barriers and shielding and remote handling methods for radium. Also, in 1931, the League of Nations commissioned a report on radiological protection.

Following the discoveries of nuclear fission in the late 1930s, radiological protection was focused on nuclear weaponry. Large Pu-producing reactors and isotope-separation plants were built at what is now known as the Oakridge site in the US, to produce weapons-grade ^{235}U as part of the Manhattan Project. This had a well-established Health Physics department, which carried out physical measurements and assessment. However, the most reliable information on the cancer risks of radiation was obtained from studies of survivors of the atomic bombing of Japan in 1945. In 1949 experts from the US, UK and Canada and, for the first time, members of the public met to discuss dose limits and in 1950 a reconstituted ICRP was established. In 1951, the International Atomic Energy Agency (IAEA) was established in Vienna with a division of nuclear safety concerned with applying many ICRP recommendations for risk reduction. Considerable work to limit the risks from radiation took place throughout the period of atmospheric testing, especially after the 1954 test series on Bikini Atoll. These tests were carried out by the US and released more fission products, by orders of magnitude, than any previous explosions. In response, the UN created its scientific committee UNSCEAR (United Nations Scientific Committee on the Effects of Atomic Radiation).

Today, radiological protection is carried out by international agreements. These are based on the basic safety standards (BSS, 1996) of the FAO, IAEA, OECD (NEA), WHO and the European Commission Councils directive for the protection of the general public and workers against the risks from ionising radiation. Hence efforts to contain and minimise the risks of radiation are international and use sophisticated methods unlike the situation with many other hazardous substances.

Until March 2011, the risks to individuals from the nuclear industry had been negligible. The most serious accidents and contamination episodes had been in the Czech Republic, the Ukraine, Kazakhstan and other countries in the Soviet bloc. The most notorious of these in the West was the Chernobyl accident in 1986. In March 2011, however, the Fukushima Daiichi nuclear plant was overwhelmed by a tsunami much greater than the defences that the reactors had been designed to withstand. This led some other countries such as Germany and Italy to cancel their nuclear programmes. It is worth noting, however, that the power plant was relatively old and perhaps not best suited for its setting. Nevertheless the reactor and all the other nuclear plants along the north-east coast of Japan withstood the worst known earthquake ever to have hit that country. Few countries are as tectonically unstable as Japan, and it is now clear that nuclear plants can be designed to withstand massive earthquakes. The risks of tsunamis and other hazards must be taken into account when planning the siting, security and maintenance of any nuclear installations, both in Japan and in all other countries.

Risks from nuclear accidents to workers must be set against the risks of mining coal, which claims thousands of lives each year, and those of recovering oil and gas. The pollution risks from accidents such as Fukushima Daiichi are likely to be small compared to the damage to ecosystems and human health caused by the combustion of fossil fuels. Burning these fuels releases large quantities of sulphur dioxide, carbon dioxide and nitrogen oxides, which are causing widespread acidification, especially of the oceans, as well as releasing large quantities of toxic trace elements such as arsenic and mercury into the environment. The risk of social unrest in countries which lack adequate power-generating capacity in the future must also be considered.

Much has been learnt about radiation risks from studies of potential nuclear waste repositories. Analysis of the risks through potential exposure pathways requires consideration of the rates of release and dispersion of different radionuclides. Much is now known that can help predict the risk of their release and dispersion from nuclear-waste sites and contaminated areas. This is due to the improved understanding of the processes involved. These processes include: contact of the waste with groundwater; degradation of the waste and release of radioactive aqueous species and particulate matter; transport of aqueous species and colloids through the saturated and vadose zones; and controls on the uptake of radionuclides by exposed populations or ecosystems. Also, measurements and thermodynamic calculations, for example for nuclear-waste repositories, are increasingly site specific.

The risk of radiation exposure as a result of nuclear conflict remains, but this is a socio-political rather than a scientific issue. The radiation risk to individuals from radiation is now mainly from natural radioactivity (gamma rays) and Rn gas, medical diagnostics and treatments, especially in the US, and agriculture where uraniferous rock-phosphate fertiliser is used. Many developed countries now have publicly available gamma and Rn potential maps, which can be used to determine the need for radiation-protection measures. These include building-standard legislation to ensure adequate ventilation and isolation of buildings from potential Rn sources by the use of impermeable membranes. Some risks, such as those from the use of depleted uranium in conflict zones, remain poorly understood and unquantified.

Medical diagnostics and treatment is the main area where efforts to reduce risk would now seem to be required. It has been calculated that Thorotrast, the colloid containing 25 per cent ^{232}Th dioxide, used as a radiographic contrast medium for angiography between 1928 and 1955 may have caused the deaths of 10 times more people from cancer than the atomic bombing of Nagasaki and Hiroshima in 1945 (Trott, 1988; Grebe, 1954). The frequent use of CT scans, especially in private medical health systems such as those of the US, must be a cause for concern. It is worth noting that the medical use of radiation presently contributes about 75 times more radiation to the average UK citizen than the entire nuclear industry. In the US the equivalent figure is 500 times!

References

Abdelouas, A., Lutze, W. & Nuttall, H. E. (1999). Uranium contamination in the subsurface: characterization and remediation. In P. C. Burns & R. Finch (Eds), Uranium: Mineralogy, Geochemistry and the Environment (pp. 433–473). Washington, DC: Mineralogical Society of America.

Adriano, D.C. (2001) Trace elements in terrestrial environments: Biogeochemistry, Bioavailability, and risk of metals, 2nd edn. Springer-Verlag, New York.

Akanov, A., Meirmanov, S., Indershiev, A., Musahanova, A., and Yamashita, S. (2009) Nuclear explosions and public health development. Radiation Health Risk Sciences **11**: 323–333

Allan, R.J. and Richardson, K.A. (1974) Uranium and potassium distribution by lake sediment geochemistry and airborne gamma-ray spectrometry, A comparison of reconnaissance techniques, CIM Bull. **67**, pp. 109–120.

Anderson, R.T. Rooney-Varga, J. N. Gaw, C. V. & Lovley, D.R. (1998). Anaerobic benzene oxidation in the Fe (III) reduction zone of petroleum-contarninated aquifers. Environmental Science and Technology **32**: 1222–1229.

Appleton, J.D. and Miles, J.C.H. (2010) Soil uranium, soil gas radon and indoor radon empirical relationships in the UK and other European countries. In: 10th International Workshop on the Geological Aspects of Radon Risk Mapping, Prague, Czech Republic, 20–21 Sept 2010.

Allison, W (2009) Radiation and Reason – The impact of Science on a Culture of Fear, Crown Octavo, York, UK. 216 pages.

Andersson, M. and Storm, H.H. (1992) Cancer incidence among Danish thorotrast-exposed patients. J Natl Cancer Inst **84**: 1318–1325.

ANL (2005) Radiological and Chemical Fact Sheets to Support Health Risk Analyses for Contaminated Areas. Argonne National Laboratory Health Fact Sheets.

Ansoborlo, E.; Chazel, V.; Hengé-Napoli, M. H.; Pihet, P.; Rannou, A.; Bailey, M. R.; Stradling, N. (2002) Determination of the Physical and Chemical Properties, Biokinetics, and Dose Coefficients of Uranium Compounds Handled During Nuclear Fuel Fabrication in France. Health Physics, **82**, 279–289.

Arnault E, Doussau M, Pesty A, Gouget B, Van der Meeren A, Fouchet P, Lefèvre B. (2008) Natural uranium disturbs mouse folliculogenesis in vivo and oocyte meiosis in vitro. Toxicology; **247**, 80–87.

Arogunjo, A. M., Ofuga, E. E. and Afolabi, M. A. (2005) Levels of natural radionuclides in some Nigerian cereals and tubers. Journal of Environmental Radioactivity **82**: 1–6.

ATSDR (1990) Toxicological Profile for thorium. Agency for Toxic Substances and Disease Registry, Atlanta, Georgia. 174p.

ATSDR (1999) Toxicological Profile for uranium. Agency for Toxic Substances and Disease Registry Atlanta, Georgia. 422 p.

ATSDR (2001) Cancer factsheet. Avaliable online [http://www.atsdr.cdc.gov/com/cancer-fs.html] Accessed 14 January 2011.

ATSDR (2007) Toxicological Profile for barium. Agency for Toxic Substances and Disease Registry, Atlanta, Georgia. 184p.

ATSDR (2008) Toxicological Profile for radium. Agency for Toxic Substances and Disease Registry, Atlanta, Georgia. 132p.

Busby, C. C., (1995), *Wings of Death: Nuclear Pollution and Human Health*. (Aberystwyth: Green Audit).

Balman, R.J. (2009) Coroner's inquest into the death of Stuart Raymond Dyson; September 10th 2009, Smethwick Council Chambers, Smethwick West Midlands, UK. [http://www.nonuclear.se/files/dyson_coronor_rule43_20090918.pdf] Accessed 19 January 2009.

Ball, T.K. and Miles, J.C.H. (1993) Geological and geochemical factors affecting the radon concentration in homes in Cornwall and Devon, UK. Environ. Geochem. Health **15**: 27–36.

Bassett SH, Frenkel A, Cedars N, *et al.* (1948). The excretion of hexavalent uranium following intravenous administration. II. Studies on human subjects. USAEC Report UR-37.

Beleites, M. (1988) Pitchblend – Uranium Mining in the GDR and its Impacts.

Ben Othman, D., White, W.M. and Patchett, J. (1989) The geochemistry and age of Timiskaming alkali volcanic and the Otto syenite stock, Abiti, Ontario. Earth Planet Sci, Lett. **94**: 1–21.

British Geological Survey (1992) Geochemical Atlas North Wales and Liverpool Bay, BGS, Keyworth, UK.

Botezatu, E and Grecea, C (2004) Radilogical impact assessment on behalf of the oil and gas industry. The Journal of Preventative Medicine **12**(1–2) 16:21.

Bomben A. M., Equillor H. E., and Oliveira A. A. (1996) Ra–226 and natural uranium in Argentinean bottled mineral waters. Radiation Protection Dosimetry **67**(3), 221–224.

Bowie, S.H.U. and Plant, J. A., (1983) Radioactivity in the environment. Thornton (Editor), Applied Environmental geochemistry. Academic Press, London.

Boujo A and El Houssein Ould Jiddou (1989). The Eocene phosphorite deposits of Bofal and Loubboira, Mauritania. In: Notholt AJG, Sheldon RP and DF Davidson (eds.) Phosphate deposits of the world. Vol. 2. Phosphate rock resources. Cambridge Univ. Press, Cambridge, UK: 207–213.

Brady P. V., Jove-Colon C., Carr G., and Huang F. (2002) Soil Radionuclide Plumes. In Geochemistry of Soil Radionuclides (ed. P. Zhang and P. Brady), pp. 165–190. Soil Science Society of America, Madison, WI. Brown, W.M.C. and Doll, R. (1961) Leukaemia in Childhood and Young Adult Life. British Medical Journal 1:981.

Brugge D, de Lemos JL, Oldmixon B (2005) Exposure pathways and health effects associated with chemical and radiological toxicity of natural uranium: a review. Rev Environ Health, **20**, 177–193.

Brenner DJ, Hall EJ. (2007) Computed tomography – an increasing source of radiation exposure. New England Journal Medicine; **357**: 2277–84.

BSS (1996) Food and Agricultural Organization of the United Nations; International Atomic Energy Agency; International Labour Organisation; OECD Nuclear Energy Agency; Pan American Health Organization; World Health Organisation: International Basic Safety Standards for Protection Against Ionizing Radiation and for the Safety of Radiation Sources. Safety Series No 115, IAEA, Vienna.

Busby, C. C. (1995), Wings of Death: Nuclear Pollution and Human Health (Aberystwyth: Green Audit).

Busby, C. C., 1998, Recalculating the Second Event Error. http://www.llrc.org/secevnew.htm.

Burns, P.C. and Finch, R. (1999) Uranium: Mineralogy, geochemistry adn the environment. Reviews in Mineralogy, **38**: 679p.

Carvalho, F. P., Oliveira, J M., M. Orquidia Neves, Abreu, M. M., and Vicente, E. M. (2009) Soil to plant (Solanum tuberosum L.) radionuclide transfer in the vicinity of an old uranium mine. Geochemistry: Exploration, Environment Analysis, **9**, 275–278.

Castoldi, A.F., Ferrari, P., Bordajandi, L.R., Curtui, V., Arcella, D., Fabiansson, S and Heppner C (2009) Toxicology Letters, **189**, S230–S231.

Cathelineau, M., Cuney, M., Leroy, J., Lhote, F., Nguyen Trung, C., Pagel, M. and Poty, B. (1982) Caractères minéralogiques des pechblendes de la province hercynienne d'Europe—Comparaison avec les oxydes d'uranium du Protérozoique de différents gisements d'Amérique du Nord, d'Afrique et d'Australie, I.A.E.A. Symp. on The Geology of Vein and Similar Type Uranium Deposits Lisbon, 1979, Proc. Int. At. Energy Agency (I.A.E.A.), Vienna (1982), pp. 159–177.

Chenery, S.R.N., Ander, E. L., Perkins, K. M. and Smith, B. (2002). Uranium anomalies identified using G-Base data—Natural or anthropogenic? A uranium isotope pilot study, British Geological Survey: Internal Report No: IR/02/001.

Chaffey, C.M. and Bowie, C. (1994) Radon and health – an update. Journal of Public Health **16**(4): 465–470.

Charles, M.W. (2007) Radon exposure of the skin: I. Biological effects. Journal of Radiological Protection **27**: 231–252.

CHPPM (2000). Depleted uranium – Human Exposure Assessment and Health Risk Characterisation in Support of the Environmental Exposure Report "Depleted Uranium in the Gulf" of the Office of the Special Assistant for Gulf War Illnesses (OSAGWI). Health Risk Assessment Consultation No. 26-MF-7555–00D. National Technical Information Service, Springfield VA. Available at: http://www.gulflink.osd.mil/chppm_du_rpt_index.html.

Church, S.E. (1973) Limits of sediment involvement in the genesis of orogenetic volcanic rocks. Contrib. Mineral. Petrol. **39**: 17–32.

Clarke, R.H. (1996) Natural Sources. Radiation Protection Dosimetry **68**: 37–42.

Cobain, I., Vasagar, J and Glendinning, L (2006) Poisoned former KGB man dies in Hospital. The Guardian, 24th November 2006.

Cochran T.B. Norris, R.S. and Suokko, K.L. (1993) Radioactive Contamination at Chelyabinsk-65, Russia, Annual Review of Energy and the Environment: 507–528.

Colon, C., Jove P., Brady, M., Siegel, M., and Lindgren, E. (2001) Historical Case Analysis of Uranium Plume Attenuation. NUREG/CR-6705 SAND2000–2254. Washington, DC: U.S. Nuclear Regulatory Commission, Office of Nuclear Regulatory Research, February, 2001.

Cothern, C.R., Lappenbusch, W.L., Michel, J. (1986) Drinking-water contribution to natural background radiation. Health Phys **50**: 33–47.

Cottrell WD, Haywood FF and Witt DA, (1981). Radiological survey of the Shpack Landfill, Norton, Massachusetts. Report: ISS DOE/EV-0005/31. DE82004939.

Court Brown WM, Doll R. (1958) Expectation of life and mortality from cancer among British radiologists. Br Med J; **ii**: 181–7.

Court Brown W.M and Doll R (1965) Mortality from cancer and other causes after radiotherapy for ankylosing spondylitis. Br Med J; **ii**: 1327–1332.

Dawson, E., Madsen, G.E and Spykerman, B.R (1997) Public Health Issues Concerning American Indian and Non-Indian Uranium Millworkers. Journal of Health and Social Policy, Vol **8**(3) 41–56.

DeCamargo I.M.C. and Mazzilli B. (1996) Determination of uranium and thorium isotopes in mineral spring waters. Journal of Radioanalytical and Nuclear Chemistry–Letters **212**(4), 251–258.

Desideri, D., Roselli, C and Meli, M.A (2010) Intake of ^{210}Po, ^{234}U and radionuclides with wine in Italy. Food and Chemical Toxicology, **48**, 650–657.

Doll, R. (1996) Epidemiological Evidence of Hazard. Radiation Protection Dosimetry **68**: 97–103.

dos Santos Silva, I., Jones, M.E., Malveiro, F., Swerdlow, A.J. (1999) Mortality in the Portuguese Thorotrast study. Radiation Research **152**: S88–92.

Drury JS. (1981). Uranium in U. S. surface, ground, and domestic waters. Volume 1. U. S Department of Commerce. National Technical Information Services.

Dublineau, I. Grison, S. Grandcolas, L. Baudelin, C. Tessier, C. Suhard, D. Frelon, S. Cossonnet, C. Claraz, M. Ritt, J. Paquet, P. Voisin, P and Gourmelon, P (2006). Absorption, accumulation and biological effects of depleted uranium in Peyer's patches of rats. Toxicology. **227**, 227–239.

Dubessy, J., Pagel, M., Beny, J-M., Christensen, H., Hickel, B., Kosztolanyi, C. and Poty, B. (1988) Radiolysis evidenced by H_2-O_2 and H_2-bearing fluid inclusions in three uranium deposits. Geochimica et Cosmochimica Acta **52**: 1155–1167.

Ewing, R. C. 1999. Nuclear waste forms for actinides. Proceedings of the National Academy of Sciences of the USA **96**: 3432–3439.

ECRR-CERI (2009) The Lesvos Declaration, 6th May 2009—International Conference Criticisms and Developments in the Assessment of Radiation Risk 5th and 6th May 2009. Molyvos (Mithymna) Island of Lesvos, Greece.

EU (2009) The exploitation and marketing of natural mineral waters. European Commission Directive 2009/54/EC of 18 June 2009.

EU (2003). Establishing the list, concentration limits and labelling requirements for the constituents of natural mineral waters and the conditions for using ozone-enriched air for the treatment of natural mineral waters and spring waters. European Commission Directive 2003/40/EC of 16 May 2003.

Fisk, D.J. (1996) Opening address on behalf of the secretary of state for the environment. Radiation Protection Dosimetry **68**: 1–2.

Fisher, R.S., 1998, Geologic and geochemical controls on naturally occurring radioactive materials (NORM) in produced water from oil, gas, and geothermal operations. Environmental Geosciences, v. **5**, no. 3, p. 139–150.

Frank, A.L. (1980) Diseases associated with exposure to metals-thorium. In: Last JMM-R, ed. Public health and preventive medicine. 11th ed. New York, NY: Appleton-Century-Crofts, pp. 681–682.

FOREGS (2005) Geochemical Atlas of Europe. Forum of the European Geological Surveys Directors. Available online: [http://www.gsf.fi/publ/foregsatlas/] Accessed 1 February 2008.

FSA (2004) Analysis of the natural radioactivity content of bottled waters. UK Food Standards Agency, Report 67/04 September 2004. (See http://www.food.gov.uk/multimedia/webpage/bottledwatersamples.).

FSIS (2004) Uranium-238 in the 2001 Total Diet Study. Food Survey Information Sheet 56/04.

Friedlander, GJ.W. Kennedy, E.S. Macias and Miller J.M. (1981) Nuclear and Radiochemistry, John Wiley and Sons.

Gerber, D. E and Chan, T.A (2008) Recent advances in radiation therapy American Family Physician, **78**(11), 1263–4.

Ghiassi-nejad, M., Mortazavi, S.M.J., Cameron, J.R., Niroomand-rad, A. and Karam, P. A. (2002) Very high background radiation areas of Ramsar, Iran: preliminary biological studies. Health Physics **82**: 1.

Grebe, S.F., (1954) Beitrag zur Frage der Thorotrastspatschadigung eine myeloische Leukamie nach diagnostischer Thorotrastapplikation, Strahlentherapie **94**, pp. 311–319.

Goodwin, B.W., Cramer, J.J. and McConnell, D.B. (1988) The Cigar Lake uranium deposit: an analogue for nuclear fuel waste disposal, CEC natural analogue working group, Third Report. Report of the Commission of the European Communities, pp. 181–190.

Golding SJ, Shrimpton PC. (2002) Radiation dose in CT: are we meeting the challenge? British Journal Radiology; **75**: 1–4.

Hagen, M. (2007). The Wismut Uranium Mining Rehabilitation Project Running for 15 Years – Lessons Learned and Achievements [http://www.scientific.net/AMR.20–21.243] Accessed 6 July 2010.

Hansard (2007). Daily Hansard—Written Ministerial Statements: Alexander Litvinenko 22 May 2007: Column 74WS.

Harrist, T.J., Schiller, A.L., Trelstad, R.L., Mankin, H.J. and Mays, C. W. (1979). Thorotrast-associated sarcoma of bone. A case report and review of the literature. Cancer **44**: 2049–2058.

Higgo, J.J.W., Williams, G.M., Harrison, I., Warwick, P., Gardiner, M. and Longworth, G., (1993) Colloid transport in a glacial sand aquifer: Laboratory and field studies. Colloids Surfaces A: Physiochem. Eng. Aspects **73**: 179–200.

Hodge H. C., Stannard J. N., and Hursh J. B. (1973) Handbook of Experimental Pharmacology. Springer–Verlag, Berlin, Germany.

Hoffman, S.J. (1983) Geochemical exploration for unconformity-type uranium deposits in permafrost terrain, Hornby bay basin, Northwest territories, Canada. Journal of Geochemical Exploration **19**: 11–32.

Holmes, D.C., Plant, J.A. and Shaw, R., (2002a). Radioactive waste management. Proc. Environmental Science Forum & Workshop, Seoul, Korea.

Hohenemser, C., Deicher, M., Hofsäss, H., Lindner, G., Recknagel, E and Budnik, J.I (1986) Agricultural Impact of Chernobyl: A Warning. Nature **321**, 817.

Hull, A. P. (1980). A critique of source term and environmental measurement at Three Mile Island. *IEEE Trans. Nucl. Sci.*, NS-27.

HSE (2007) Fundamentals of the management of radioactive waste. Guidance from the UK Health and Safety Executive, the Environment Agency and the Scottish Environment Protection Agency to nuclear licensees, December 2007, NRW02, 20pp. Crown Copyright.

IAEA (1986) Summary Report on the Post-Accident Review Meeting on the Chernobyl Accident. Safety Series No. 75-INSAG-1. IAEA, Vienna.

IAEA (2003a) Categorization of radioactive sources. IAEA-TEC-DOC-1344, International Atomic Energy Agency, Vienna, Austria.

IAEA (2003b) Predisposal Management of Low and Intermediate Level Radioactive Waste, IAEA Safety Standards Series No. WS-G-2.5, IAEA, Vienna.

IAEA (2003c) Predisposal Management of High Level Radioactive Waste, IAEA Safety Standards Series No. WS-G-2.6, IAEA, Vienna.

IAEA (2005) IAEA (2005) Management of Waste from the Use of Radioactive Material in Medicine, Industry, Agriculture, Research and Education, IAEA Safety Standards Series No. WS-G-2.7, IAEA, Vienna.

IAEA (2006) Storage of Radioactive Waste IAEA Safety Standard Series Guide No WS-G-6.1. Pub1254, IAEA Vienna.

ICRP-66 (1994) International Commission on Radiological Protection. Human respiratory tract model for radiological protection, ICRP Publication 66, Annals of the ICRP, Vol24(1–3).

ICRP-68 (1994) International Commission on Radiological Protection. Dose Coefficients for Intakes of Radionuclides by Workers, ICRP Publication 68, Annals of the ICRP 24(4).

ICRP-71 (1995) Age-dependent doses to members of the public from intake of radionuclides: part 4. Inhalation dose coefficients, ICRP Publication 71. Annals of the ICRP 25/3–4, Pergamon Press, Oxford, UK.

ICRP-89 (2002). Basic Anatomical and Physiological Data for Use in Radiological Protection: Reference Values. ICRP Publication 89. International Commission on Radiological Protection, Pergamon Press, New York.

ICRP (2001). The ICRP Database of Dose Coefficients: Workers and Members of the Public. CD 1 Ver. 2. 01. International Commission on Radiological Protection, Elsevier Science, New York.

ICRP-100 (2006). Human Alimentary Tract Model for Radiological Protection. ICRP Publication 100. Annals of the ICRP 2006:36/1–2, Pergamon Press, Oxford, UK.

INSAG (1986). Report on post-Chernobyl accident review meeting, IAEA, Vienna, 24 Sept., pp. 140.

IPCS (2005) Thorium. International Programme on Chemical Safety and the Commission of the European Communities.

Jones, D, and Scheib, C. (2007). A preliminary interpretation of the Tellus airborne radiometric data. British Geological Survey Commission Report for the Geological Survey of Northern Ireland CR/07/061. 70 pp.

Kaye, G.W.C. and Laby, T.H. (1986) Tables of Physical and Chemical Constants, Longman, 477 pp.

Kaul A., Muth H. (1978) Thorotrast kinetics and radiation dose. Results from studies in Thorotrast patients and from anumal experiments. Radiat Environ Biophys **15**: 241–259.

Kovács, T., Bodrogil E., Dombovári, P., Somlai, J, Németh, Cs, Capote, A and Tarján, S (2004) ^{238}U, ^{226}Ra, ^{210}Po concentrations of bottled mineral waters in Hungary and their committed effective dose. Radiation Protection Dosimetry, **108**(2) 175–181.

Kurttio, P., Auvinen, A., Salonen, L., Saha, H., Pekkanen, J., Mäkeläinen, I., Väisänen, S.B., Penttilä, I.M and Komulainen, H. (2002) Renal Effects of Uranium in Drinking Water. Environmental Health Perspectives, **110**, 337–342.

Kurttio P, Harmoinen A, Saha H, Salonen L, Karpas Z, Komulainen H, Auvinen A. (2006) Kidney toxicity of ingested uranium from drinking water. Am J Kidney Dis. **47**, 972–982.

Landa, E.R. and Gray, J.R. (1995) US Geological Survey research on the environmental fate of uranium mining and milling wastes. Environmental Geology **26**: 19–31.

Langmuir, D. (1978) Uranium solution-mineral equilibria at low temperatures with applications to sedimentary ore deposits. Geochimica Cosmochimica Acta **42**: 547–597.

Langmuir, D. (1997) Aqueous Environmental Geochemistry. Prentice Hall, New Jersey.

Ledvina R., Kolar L., and Frana J. (1996) Uranium, thorium and some other elements in topsoils of the Trebon region from the aspect of production contamination. Rostlinna Vyroba **42**(2), 73–78.

Lindell, B. (1996) The history of radiation protection pp. 83–95 from Radiation Protection Dosimetry. Proceedings of a conference vol. **68** Nos. 1/ 2.

Leggett R. W. (1992) A generic age-specific biokinetic marker for calcium like elements, Radiation Protection Dosimetry **41**(2–4), 183–198.

Levinson, A.A. (1974) Introduction to Exploration Geochemistry. Applied Publishing Limited, Calgary, Canada.

Lewington, V.J (2005) Bone-Seeking Radionuclides for Therapy. Journal of Nuclear Medicine, **46**, 38S–47S.

Lourenço J, Castro BB, Machado R, Nunes B, Mendo S, Gonçalves F, Pereira R. (2010) Genetic, Biochemical, and Individual Responses of the Teleost Fish Carassius auratus to Uranium. Arch Environ Contam Toxicol in press.

Lovelock, J (2006) *The Revenge of Gaia: Why the Earth is Fighting Back—and How we Can Still Save Humanity* Penguin 2006.

Macklis, R.M. (1993) The great radium scandal. Scientific American Magazine. August 1993 pp 78–83.

Magdo HS, Forman J, Graber N, Newman B, Klein K, Satlin L, Amler RW, Winston JA, Landrigan PJ. (2007) Grand rounds: nephrotoxicity in a young child exposed to uranium from contaminated well water. Environmental Health Perspectives. **115**, 1237–1241.

Martin, J.M. and Whitfield, M. (1983) The significance of the river input of chemical elements to the ocean, p. 265. In C. S. Wong, E. Boyle, K.W. Bruland,] D. Burton,and E. D. Goldberg (eds.), Trace Metals in Seawater. NATO Conference Series, Plenum Press, New York.

Mayer, M. (1982) Aggregation of colloidal iron during estuarine mixing: Kinetics, mechanism and seasonally. Geochimica et Cosmochimica Acta **46**: 2527–2535.

McBride J. P.R. E. Moore, R.E., Witherspoon, J.P. and Blanco, R.E. (1978) Radiological Impact of Airborne Effluents of Coal and Nuclear Plants. Science **202**: 1045–1050.

McConnell M. A., Ramanujam V.M.S., Alcock N. W., Gabehart G. J., and Au W.W. (1998) Distribution of uranium–238 in environmental samples from a residential area impacted by mining and milling activities. Environmental Toxicology and Chemistry **17**(5), 841–850.

Medvedev, Z. (1976) Two decades of dissidence. New Scientist **72**: 264.

Milacic S. (2008) Health investigations of depleted-uranium clean-up workers. Med Lav **99**,(5): 366–370.

Milacic S., Simic J. (2009) Identification of health risks in workers staying and working on the terrains of contaminated deplted uranium. J Radiat Res (Tokyo) **50**,(3): 213–222.

Miller AC, Brooks K, Stewart M, Anderson B, Shi L, McClain D, Page N. (2003) Genomic instability in human osteoblast cells after exposure to depleted uranium: delayed lethality and micronuclei formation. Journal Env. Radioact. **64**,(2–3), 247–59.

Miller, A.C., Bonait-Pellie, C., Merlot, R.F., Stewart, M and Lison P. D. (2005) Leukemic transformation of hematopoietic cells in mice internally exposed to depleted uranium. Mol Cell Biochem, **279**, (1–2), 97–104.

Miller, A.C., Stewart, M and Rivas, R (2009) DNA methylation during depleted uranium induced leukemia. Biochimie, **91**,(10), 1328–30.

Monleau M, De Méo M, Paquet F, Chazel V, Duménil G, Donnadieu-Claraz M (2006) Genotoxic and inflammatory effects of depleted uranium particles inhaled by rats. Toxicol Sci. **89**(1), 287–95.

NATO (1996) NATO handbook on the medical aspects of NBC defensive operations. North Atlantic Treaty Organisation.

NCRP (1987) NCRP Report No. 94 Exposure of the Population in the United States and Canada from Natural Background Radiation. National Council on Radiation Protection.

NCRP (2009) NCRP Report No. 160 on increased average radiation exposure of the US population. National Council on Radiation Protection.

NEA (1984) Long term radiological aspects of management wastes from uranium mining and milling. OECD, Paris.

Nenot, J.C. (1996) Radiation accidents, Radiation Protection Dosimetry **68**: 111–118.

NRPB (1998) Living with radioactivity. HPA.

NRPB (2000) Radiation Exposure of the UK Population from Medical and Dental X-ray Examinations. National Radiological Protection Board.

National Research Council (2001) Science and Technology for Environmental Cleanup at Hanford. National Academy Press.

OECD Nuclear Energy Agency (2005) Uranium Resources, Production and Demand, 2004, OECD, Paris.

OECD Nuclear Energy Agency (1998) Uranium Resources, Production and Demand, 1997, OECD, Paris.

O'Neil, M.J., Heckelman, P.E., Koch, C.B., *et al.,* (2006) eds. The Merck index. 14th ed. Whitehouse Station, NJ: Merck & Co., Inc., pp 1393–1394.

Palmer, M.R. and Edmond, J.M. (1993) Uranium in river water. Geochimica et Cosmochimica Acta **57**: 4947–4955.

Parks, G.A. and Pohl, D.C. (1988) Hydrothermal solubility of uraninite. Geochimica et Cosmochimica Acta **52**: 863–875.

Periyakaruppan A, Kumar F, Sarkar S, Sharma CS, Ramesh GT. (2007) Uranium induces oxidative stress in lung epithelial cells. Arch Toxicol, **81**,(6), 389–95.

Peterson, J., MacDonell, M., Haroun, L., Monette, F., Hildebrand, R.D. and Taboas, A. (2007). Radiological and chemical factsheets to suppor health risk analyses for contaminated areas. Argonne National Laboratory.

Plant, J.A., O'Brien, C., Tarney, J. and Hurdley, J. (1985) Geochemical criteria for the recognition of high heat production granites. pp. 263–285 in High Heat Production (HHP) Granites, Hydrothermal Circulation and Ore Genesis. Institution of Mining and Metallurgy, London.

Plant, J.A., Simpson, P.R., Smith, B. and Windley, B.F. (1999) Uranium ore deposits – products of the radioactive earth. pp 321 – 432 in Uranium: Mineralogy, Geochemistry and the Environment (P.C. Burns and R. Finch eds) Reviews in Mineralogy 38: Mineralogical Society of America.

Plant, J.A., Reeder, S., Salminen, R., Smith, D.B., Tarvainen, T., de Vivo, B. and Petterson, M.G. (2003) The distribution of uranium over Europe: geological and environmental significance. Applied Earth Science: IMM Transactions section B **112**: 221–238.

Porcelli D., Andersson P. S., Wasserburg G. J., Ingri J., and Baskaran M. (1997) The importance of colloids and mires for the transport of uranium isotopes through the Kalix river watershed and Baltic Sea. Geochimica Et Cosmochimica Acta **61**,(19), 4095–4113.

Prat O., Bérenguer F., Steinmetz G., Ruat S., Sage N., Quéméneur E. (2010) Alterations in gene expression in cultured human cells after acute exposure to uranium salt: involvement of a mineralization regulator. Toxicol In Vitro **24**: 160–168.

Quinlan MF, Scopa J. (1976) Thorotrast-induced haemangioendothelial sarcoma—a lesson from the past. Aust N Z J Med. **6**(4): 329–335.

Rai, D., Felmy, A.R. and Ryan, J.L. (1990) Uranium(IV) hydrolysis constants and solubility product of UO2.cntdot.xH2O(am). Inorganic chemistry **29**: 260–264.

Ragnarsdottir, K.V. and Charlet, L. (2000) Uranium behaviour in natural environments. Environmental Mineralogy—Microbial Interactions, Anthropogenic Influences, Contaminated Land and Waste Management, Mineralogical Society, London: pp. 245–289.

Ragnarsdottir, K.V., Fournier, P., Oelkers, E.H. and Harrichoury, J-C. (2001) Experimental determination of the complexation of strontium and cesium with acetate in high-temperature aqueous solutions. Geochimica et Cosmochimica Acta **65**: 3955–3964.

Rajaratnam, R., Balasubramaniam, P. and Marsden, J. R. (2007) Thorium X and skin cancer: still a problem in the 21st century. Clinical and Experimental Dermatology **32**: 125–126.

Raymond-Whish, S., Mayer, L.P., O'Neal, T., Martinez, A., Sellers, M. A, Christian, P.J. Marion, S.L., Begay, C., Propper, C.R., Hoyer, P.B and Dyer C.A. (2007) Drinking Water with Uranium below the U.S. EPA Water Standard Causes Estrogen Receptor–Dependent Responses in Female Mice. Environmental Health Perspectives; **115**, 1711–1716.

Reimann, C and Birke, M (2010) Geochemistry of European Bottled Mineral Water. Borntraeger Science Publishers, Stuttgart, Germany. 280 p.

Rich B.L. (1988). Health physics manual of good practices at uranium facilities, 2–17–2–24. US Department of Energy, EGG–2530 UC–41. DOE: Washington, DC, USA.

Rigali, M.J. and Nagy, B. (1997) Organic free radicals and micropores in solid graphitic carbonaceus matter at the Oklo natural fission reactors, Gabon. Geochimica et Cosmochimica Acta **61**: 357–368.

Ryan, J.L. and Rai, D. (1983) The solubility of uranium(IV) hydrous oxide in sodium hydroxide solutions under reducing conditions. Polyhedron **2**: 947–952.

Santos, E.E., Lauria, D.C., Amaral, E.C.S. and Rochedo, E.R. (2002) Daily ingestion of ^{232}Th, ^{238}U, ^{226}Ra, ^{228}Ra and ^{210}Pb in vegetables by inhabitants of Rio de Janeiro City. Journal of Environmental Radioactivity **62**: 75–86.

Scott, B.R. (2007) Biological basis for radiation hormesis in mammalian cellular communities. Int. J. Low Radiation **4**: 1–16.

Seaborg, G.T. (1946). The Transuranium Elements. Science **104** (2704): 379–386.

Siegel, M. D. and Bryan, C. R. (2004) Environmental geochemistry of radioactive contamination, 205–262. In: Lollar, B. S. (Ed), Environmental Geochemistry, vol. 9. In: Holland, H. D. and Turekian, K. K. (Eds) Treatise on Geochemistry, Elsevier-Pergamon, Oxford.

Seldén A,I, Lundholm C, Edlund B, Högdahl C, Ek BM, Bergström BE, Ohlson CG (2009) Nephrotoxicity of uranium in drinking water from private drilled wells. Environ Res. **109**, 486–494.

Slezak, J. (1997) National Experience on Groundwater Contamination Associated with Uranium Mining and Milling in the Czech Republic, DIAMO s.p. Straz pod Ralskem, Czech Republic.

Slikkerveer, A. and de Wolff, F.A. (1989) Pharmacokinetics and toxicity of bismuth compounds. Med Toxicol Adverse Drug Experience **4**: 303–303.

Smith, B. (2007) The MOD depleted uranium program independent review board: closure report. Nottingham, UK, British Geological Survey, 28pp.

Smith, B., Powell, A.E., Milodowski.,A.E, Hards,V.L., Hutchins, M. G., Amro., A., Gedeon, R., Kilani, K, Scrivens. S.M. and Galt, V. (1998) Identification, investigation and remediation of groundwater containing elevated levels of uranium-series radionuclides: a case study from the Eastern Mediterranean. In:Panayides, I., Xenophontos, C. and Malpas, J. (Eds). Proceedings of the Third International Conference on Geology of the Eastern Mediterranean, Geological Survey of Cyprus, p355–363.

Slovic, P. (1987) Perception of risk. Science **236**: 280–285.

Smirnova V.S., Gudkov S.V., Shtarkman I.N., Chernikov A.V., Bruskov V.I. (2005) Yhe genotoxic action of uranyl ions on DNA invitro caused by the genereatiion of reactive oxygen species. Biofizika %0 (3): 456–463.

Somlai J, Gorjánácz Z, Várhegyi A, *et al.* (2006) Radon concentration in houses over a closed Hungarian uranium mine. Sci Total Environ **367**(2–3): 653–665.

Stewart, A.J. and Jones, B (2005) Radiobiological Concepts for Brachytherapy. In: Devlin P. Brachytherapy. Applications and Techniques. Lippincott Williams and Wilkins, Philadelphia, 448 pp.

The Royal Society, Depleted Uranium Working Group (2002a). The health hazards of depleted uranium munitions Part II, The Royal Society. London, 134 pp.

The Royal Society, Depleted Uranium Working Group The health hazards of depleted uranium munitions: a summary. Journal of Radiological Protection **22**. 131–139.

Thiébault C, Carrière M, Milgram S, Simon A, Avoscan L, Gouget B (2007) Uranium induces apoptosis and is genotoxic to normal rat kidney (NRK-52E) proximal cells. Toxicol Sci. **98**,(2), 479–87.

Tomásek L, Darby SC, Fearn T, Swerdlow AJ, Placek V, and Kunz E. (1994) Patterns of lung cancer mortality among uranium miners in West Bohemia with varying rates of exposure to radon and its progeny. Radiation Research **137**: 251–261.

Trabalka, J.R., Eyman, L.D. and Auerback, S.I. (1980). Analysis of the 1957–1958 Soviet nuclear accident. Science **209**: 345–353.

Trott, N.G. (1988) Radionuclides in Brachytherapy: Radium and after. Br J Radiol Suppl. **21**: 1–54.

Uranium Mill Tailings Radiation Control Act (UMTRCA), (1978) US office of Environmental Management, Department of Energy, Pub L P.L. 95–604, 92 Stat. 3021; 42 USC 2014, 2021, 2022, 2111, 2113, 2114, 2201, 7901, 7911–7925, 7941, 7942.

Uranium Mill Tailings Radiation Control Act (Amendments UMTRCA), (1988) US office of Environmental Management, Department of Energy P.L. 100–616, 102 Stat. 3192; 42 USC 7901 note, 7916, 7922.

UNSCEAR (1982). Ionizing radiation: Sources and biological effects.

UNSCEAR (2000) Sources and effects of ionizing radiation. Report to the General Assembly. New York: United Nations Scientific Committee on the Effects of Atomic Radiation.

UNSCEAR, (2001) Hereditary effects of radiation. Report to the general assemble of the United Nations, UNSCEAR 2001, United Nations, New York, 2001.

UNSCEAR, (2006), Effects of Ionizing Radiation. Report to the general assemble of the United Nations, UNSCEAR 2006, United Nations, New York, 2008.

UNEP_(2001) "Depleted Uranium in Kosovo, Post-Conflict Environmental Assessment. UNEP Scientific Team Mission to Kosovo" (5th–19th November 2000). United Nations Environment Programme, Geneva, March 2001.

USEPA (2003) Assessment of risks from radon in homes. Environment Protection Agency. Avaliable online [http://www.epa.gov/radon/risk_assessment.html] Accessed 8 December 2010.

USEPA (2007). "A Citizen's Guide to Radon". 2007–11–26. [http://www.epa.gov/radon/pubs/citguide.html.] Retrieved 26 June 2008.

Valsami-Jones, E. and Ragnarsdottir, K.V. (1997) Controls on uranium and thorium behaviour in ocean-floor hydrothermal systems: examples from the Pindos ophiolite, Greece. Chem. Geol. **135**: 263–274.

Wakeford, R (2005) Cancer Risk among nuclear workers. Journal of Radiological Protection, **25**, 255.

Wakeford, R (2008) What to believe and what not to believe. Journal of Radiological Protection, **28**, 5–7.

Watson, S.J., Jones, A.L., Oatway, W.B and Hughes, J.S (2005) Ionising Radiation Exposure of the UK Population: 2005 Review. HPA-RPD-001, HPA Chilton, Oxfordshire, UK.

Weaver, B.L. & Tarney, J. (1980) Rare earth geo chemistry of Lewisian granulite-facies gneisses, northwest Scotland: implications for the petrogenesis of the Archaean lower continental crust. Earth and Planetary Sci. Letters **51**: 279–296.

WHO (1983) Selected Radionuclides. Environmental Health Criteria 25. World Health Organization, Geneva.

WHO (2001a) WHO Guidance on Exposure to Depleted Uranium. For Medical Officers and Programme Administrators. World Health Organization, Geneva.

WHO (2001b) Depleted Uranium, Sources, Exposure and Health Effects. World Health Organization, Geneva.

WHO (2009) Guidelines for drinking-water quality. World Health Organization, Geneva.

Wininger, R. D. (1954) Minerals for Atomic Energy. Van Nostrand, Toronto.

Wise SS, Thompson WD, Aboueissa AM, Mason MD, Wise JP Sr., (2007) Particulate depleted uranium is cytotoxic and clastogenic to human lung cells. Chem Res Toxicol **20**(5), 815–20.

Wolf G., Arndt D., Kotschy-Lang N., Obe G. (2004) Chromosomal aberrations in uranium and coal miners. Int J Radiat Biol **80**,(2): 147–153.

Wrenn M.E., Durbin P., and Willis D. (1985) Metabolism of ingested uranium and radium. Health Physics **48**, 601–633.

Zielinski R. A., AsherBolinder S., Meier A. L., Johnson C. A., and Szabo B. J. (1997) Natural or fertilizer-derived uranium in irrigation drainage: A case study in southeastern Colorado, USA. Applied Geochemistry **12**(1), 9–21.

6

Industrial chemicals

Danelle Dhaniram, Alexandra Collins, Khareen Singh and Nikolaos Voulvoulis*

Centre for Environmental Policy, Imperial College London, Prince Consort Road, London SW7 2AZ

**Corresponding author, email n.voulvoulis@imperial.ac.uk*

6.1 Introduction

The use of industrial chemicals dates back to as early as 7000 BC when alkalis and limestone were refined by Middle Eastern artisans for glass production and the dyeing of linen and leather was carried out in temple workshops in Mesopotamia, northern Syria and Egypt, where the sacred vestments for gods and priests were dyed. The production of soap from boiling animal fat or olive oil to dryness with ashes from wood fires (containing potassium hydroxide) in the sixth century BC by the Phoenicians and Celts marked another milestone in the chemical industry. Gunpowder or black powder, a primitive explosive, was developed by the Chinese using sodium and/or potassium nitrate with sulphur by the tenth century AD and chemicals used for tanning were manufactured by the sixteenth century (Taylor and Sudnick, 1984; Morehouse and Subramaniam, 1986; Aftalion, 1991; Heaton, 1994).

Large-scale chemical production only began in the nineteenth century, during the second industrial revolution, and was based on the emergence of communication links which signalled that production and transport of goods were no longer dependent on the power and speed of animals and the vagaries of wind and water (Chandler, 2005). In 1823, the British entrepreneur, James Muspratt, began mass-producing the soda ash needed for soap and glass using a process developed by Nicolas Leblanc in 1790, and by the 1850s advances in organic chemistry had enabled companies to produce synthetic dyes for the textile industry from coal tar. In the 1890s, German companies began mass-producing sulphuric acid and, at about the same time, companies began using electrolysis to prepare caustic soda and chlorine. Synthetic fibres changed the textile industry when rayon (made from wood fibres) was introduced in 1914, while advances in the manufacture of plastics led to the synthesis of celluloid and nylon by Du Pont in 1928. By the 1920s and 30s organic chemicals including benzene, toluene, naphthalene, phenol, pyridine and anthracene were being used to synthesise a wide range of other industrial chemicals (Williams, 1982).

The current production and use of industrial chemicals has increased immensely over the years. According to the American Chemistry Council the production of the largest-volume top 100 chemicals increased from 397 million tonnes in 1990 to 502 million tonnes in 2000, up approximately 20 per cent. Between 2002 and 2006, world exports of industrial chemicals also grew at an average annual rate of 17 per cent (Global-production.com, 2007). Industrial chemicals are now central to the global economy and the contribution of the chemicals industry to improvements in life expectancy, human health and living standards for most people in Western-style civilisations is widely acknowledged (Royal Commission on Environmental Pollution, 2003). Approximately 80 per cent of the industrial chemicals used worldwide for the production of polymers are obtained from bulk petrochemicals (natural gas and crude oil), synthetic fibres (polyester, nylon and acrylics) and plastics (polyethylene and polyvinylchloride), the remainder being used to make construction materials, agrochemicals and a wide variety of consumer goods.

Humans and the environment have therefore been continuously exposed to these chemicals, and in recent decades scientific studies have revealed that exposure to these chemicals have been linked to human health effects and environmental concerns. Industrial chemicals such as polychlorinated dibenzodioxins and polychlorinated biphenyls (PCBs) have shown unexpected effects, despite being banned in the Western world since the 1970s. PCBs, for example, have been linked to cancer of the liver and other health impacts, as this chemical is fairly resistant to degradation and its stability and lipophilicity have resulted in

Pollutants, Human Health and the Environment: A Risk Based Approach, First Edition. Edited by Jane A. Plant, Nikolaos Voulvoulis and K. Vala Ragnarsdottir.

accumulation in the environment and in organisms. Humans therefore have been and will continue to be exposed to them, particularly in industrialised countries where PCBs can be inhaled in small amounts through the air or ingested through food (Kimbrough, 1987). These chemicals have been suggested to exert carcinogenic effects by interfering directly with the cell cycle, or, in the case of hormone-dependant cancers such as breast cancer, by interfering with processes controlled by steroid hormones such as oestrogen.

Many of these chemicals are currently ubiquitous in water, air, soil and sediment. There is a growing awareness that synthetic chemicals are widespread in the biosphere, including in human tissues (Royal Commission on Environmental Pollution, 2003) and over the past 50 years new highly sensitive methods of chemical analysis have been developed and links between industrial chemicals and human exposure and health have been proposed. As well as potential chronic effects on human health and the environment, concerns have been heightened by accidents at industrial chemical plants such as Flixborough in the UK in 1974, Seveso in Italy in 1976, Bhopal in India in 1984 and Toulouse in France in 2001. Examples of the chronic effects due to human exposure of industrial chemicals and consequences of industrial chemical accidents are illustrated in Table 6.1.

In this chapter, four industrial chemicals are discussed in relation to their health and environmental impacts: (1) brominated flame retardants (BFRs) such as polybrominated diphenyl ethers (PBDEs), (2) polycyclic aromatic hydrocarbons (PAHs), (3) the aromatic hydrocarbon benzene and (4) trichloroethylene (TCE), a chlorinated hydrocarbon. BFRs are comparatively new and their exposure and fate are still uncertain, as are their physiological effects. Benzene and TCE are well established and their impacts are better known. By comparing these four chemicals at different levels of scientific understanding we have sought to illustrate the diversity of chemicals in the industrial sector and their potential harm to human health and the environment. We shall consider their hazardous properties, their natural and anthropogenic sources and their behaviour in the environment. Their key human health impacts, including their acute and chronic effects, are also reviewed, as well as issues of risk reduction.

6.2 Hazardous properties

Industrial chemicals have become an integral part of daily life in modern societies and were developed to improve the quality of human life. These chemicals are produced and used by industry for various purposes, but their chemical, physical or biological properties can pose a potential risk for life, health and the environment. The adverse effects on humans and the environment from industrial chemicals due to their hazardous properties are increasingly recognised (Gebel *et al.*, 2009). These chemicals have shown persistence, bioaccumulation and toxicity (PBT) as well as CMR characteristics (carcinogenic, mutagenic and reproductive disorders). These properties mean that many of them have the potential to have a significant impact on human health and the environment and can be classified as persistent organic pollutants (POPs) as listed under the Stockholm Convention. They are resistant to environmental degradation (chemical, biological and photolytic processes) and clearly illustrate the 'grasshopper' effect, enabling them to travel across the globe frequently accumulating in the polar regions (Jones and de Voogt, 1999).

Internationally, the principles of PBT as indicators of hazard came to form part of various initiatives in the late 1990s. Organisations such as United Nations Environment Programme (UNEP), POPs Convention, EU, OSPAR, US EPA and the UK Chemicals Stakeholder Forum have undertaken prioritisation exercises to identify such substances (Wildey *et al.*, 2004). Some of these substances have the potential to build up in environmental organisms, leading to long-term subtle effects (such as reduced fertility). Table 6.2 gives an overview of the main PBT criteria used by different organisations. The industrial chemicals chosen for further analysis in this chapter all have typical hazardous characteristics and are ubiquitous in the environment.

6.2.1 *Brominated flame retardants (BFRs)*

Brominated flame retardants have similar physiochemical properties to many persistent organic pollutants (POPs) in the environment. They are persistent and bioaccumulate in higher trophic-level species in the food chain, including marine mammals, some birds and fish and humans (Yang, 2008; Gevao *et al.*, 2008). Recent studies have also illustrated the presence of this chemical group in dairy products as well as human breast milk (LaKind, Berlin *et al.*, 2008; Fernandes, Tlustos *et al.*, 2009; Gensler, Schwind *et al.*, 2009). Their similar molecular structure to PCBs and dioxins and their resistance to degradation have raised concerns that they may cause similar problems (Ejarrat, 2008). They are easily leached over product lifetimes because they are not covalently bonded to the polymers to which they are added (Almqvist, 2006; Bawden, 2004).

Toxicological endpoints of concern for environmental levels of PBDEs in the environment have been suggested to be thyroid hormone disruption, neurodevelopment deficits, immunotoxicity, cancer and reproductive effects (Janssen, 2005; Brouwer, Morse *et al.*, 1998; Mercado-Feliciano and Bigsby, 2008). The available toxicology on PBDEs is surprisingly limited, given their widespread use, bioaccumulative potential and structural similarity to thyroid hormones and PCBs (McDonald, 2002). The thermal breakdown of products such as PBDEs produce brominated/chlorinated dibenzo-p-dioxins and dibenzofurans (PBDDs/DFs and mixed PXDDs/DF) (Figure 6.1) with further serious environmental and human health implications (Leung, 2007).

Table 6.1 Examples of some industrial chemicals and related health impacts

Industrial chemical	Use	Case study	Exposure period	Health impact
n-hexane	Main component in petroleum distillates	Hisanaga *et al.*, 2002; in 1962 workers manufacturing vinyl sandals were exposed to 500–2000 mg/l n-hexane in air. 296 workers were examined and 93 were found to have polyneuropathy. The results suggested that about 100 mg/l n-hexane could cause sensory polyneuropathy, and about 500 mg/l n-hexane could cause sensory-motor polyneuropathy in workers	The average exposure period for workers was 6 months	Chronic effect: Polyneuropathy (neurological disorder that occurs when many peripheral nerves throughout the body malfunction simultaneously)
Vinyl chloride	Chemical intermediate	Heldaas *et al.*, 1984; in 1974 a cohort of 454 male workers producing vinyl chloride and polyvinyl chloride were studied. It was assumed that the vinyl chloride concentration in the air was about 2000 mg/l from 1950 to 1954, about 1000 mg/l from 1955 to 1959, about 500 mg/l from 1960 to 1967, and about 100 mg/l from 1968 to 1974. During autoclave cleaning the exposure level may have been as high as 3000 mg/l	Workers with more than one year's work experience in the study plant between 1950 and 1969	Chronic effects: 23 new cases of cancer were observed and 1 case of liver angiosarcoma was found. There were 5 cases of lung cancer, and 4 cases of malignant melanoma of the skin were observed
2-bromo-propane	Solvent	Hisanaga, *et al.*, 2002; in 1994 at a tactile switch assembling process, parts of the switch were dipped in cleaning baths with a solvent containing 2-bromo-propane and polytetrafluoroethylene to prevent flux and fume from spreading after soldering	The average exposure period for workers was 9.5±3.6 months	Chronic effects: reproductive disorders amenorrhoea observed in the female workers and azoospermia or oligospermia in the male workers Thirty three workers were exposed in the process of assembling tactile switches at the Korean electric company. Females (64%) had amenorrhoea, and 75% of males had azoospermia, oligozoospermia or reduced sperm motility. The mean 2-bromopropane concentrations in 14 area samples was 12.4 mg/l, ranging from 9.2–19.6 mg/l. Short-term exposure levels were monitored and ranged from 106 to 4360 mg/l 2-bromopropane

(continued)

Table 6.1 (*Continued*)

Industrial chemical	Use	Case study	Exposure period	Health impact
2,3,7,8-tetrachlorodibenzo-p-dioxin (impurity of hexachlorophene)	Disinfectant	Mocarelli, 2009; Consonni *et al.*, 2008; in 1976 a vessel used to produce hexachlorophene exploded in Seveso, Italy. The chemical 2,4,5-trichloro-phenol was being produced from 1,2,4,5-tetrachlorobenzene by the nucleophilic aromatic substitution reaction with sodium hydroxide. 2,4,5-trichlorophenol was intended as an intermediate for hexachlorophene. This batch process was interrupted due to an Italian law requiring routine shutdown of the plant operations. The residual heat in the jacket then heated the upper layer of the mixture to the critical temperature of 180 °C, starting a slow runaway decomposition, and after seven hours a rapid runaway reaction ensued when the temperature reached 230 °C. The relief valve eventually opened and 6 tonnes of material were distributed over an 18 km^2 area, including 1 kg of 2,3,7,8-tetrachlorodibenzodioxin (TCDD), which is normally seen only in trace amounts of less than 1 mg/l. However, in the higher-temperature conditions associated with the runaway reaction, TCDD production exceeded 100 mg/l	Immediately after explosion and 30 years onwards	Acute effects: 300 individuals developed chloracne (an acne-like eruption of blackheads, cysts, and pustules) Chronic effects: children were diagnosed with high blood levels of thyroid-secreting hormone (TSH), which is associated with failing thyroid and the potential for permanent damage to a baby's developing body and brain Excesses of lymphatic and haematopoietic tissue (neoplasms) 34 deaths from circulatory diseases, chronic obstructive pulmonary disease and diabetes mellitus

Polychlorinated biphenyl (PCB)	Coolant and insulating fluid. It is also used as a stabilising additive in products	Koppe and Keys, 2001; in 1979 the Yucheng disaster led to 2000 people being poisoned in Taiwan by ingesting contaminated rice oil. PCBs were used in heating pipes due to their insulating properties and resistance to high temperatures The rice oil was found to contain a large amount of PCB, which was believed to have leaked from a heating pipe	Immediately after consumption and onwards	Acute effects: adults showed increased skin allergies, chloracne and spine and joint diseases. Chronic effects: one quarter of the children born to poisoned mothers died before the age of 4 as a result of respiratory infections. At the age of 8 years, children still had nail deformities, chronic otitis media (middle-ear inflammation) together with bronchitis
Methyl isocyanate (MIC)	Intermediate chemical used in the production of pesticides, rubbers and adhesives	Willey *et al.*, 2007; Mishra *et al.*, 2009; Dhara and Dhara, 2002; in 1984 30–40 tonnes of MIC gas was leaked from the Union Carbide plant in Bhopal, India and released over approximately 30 square miles. This was caused by large amounts of water entering the tank containing MIC. The resulting exothermic reaction increased the temperature inside the tank to over 200 °C, raising the pressure to a level the tank was not designed to withstand. This forced the emergency venting of pressure from the MIC holding tank, releasing a large volume of toxic gases into the atmosphere	Immediately after accident and 24 years onwards	Acute effects: coughing, vomiting and irritant effects on the eyes and respiratory tract leading to ocular lesions and respiratory impairment Chronic effects: differences in various immune responsive parameters of the exposed individuals, including increased cytokine secretion by stimulated lymphocytes in vitro and higher immunoglobulin levels Reproductive difficulties and birth defects among children born to affected women Estimation of 16 000 deaths immediately after the accident and from gas-related diseases

Table 6.2 Overview of PBT criteria

Organisation	Bioaccumulation	Long-range transport potential	Toxicity (L(E) C50; NOEC – no observed effect concentration; CMR)	Persistence
UNEP POPs Convention	Half-life in water >2 months or in sediment >6 months or in soils >6 months	BCF or BAF >5000 or log K_{ow} >5 or monitoring data in biota	Measured levels far from source or monitoring data in remote area or multi-media modelling evidence and half-life in air >2 days	Evidence of adverse effect on human health or the environment or toxicity characteristics indicating potential damage to human health or environment
OSPAR PBT Criteria	Not readily biodegradable or half-life in water >50 days	log K_{ow}>= 4 or BCF >= 500	Not applicable	Acute aquatic toxicity L(E) C50≤1 mg/l or long-term NOEC≤0.1 Chronic NOEC < 0.01 mg/l or CMR cat 1&2 or endocrine-disrupting effects mg/l or mammalian toxicity: CMR or chronic toxicity
EU vPvB-criteria	Half-life >60 d in marine- or freshwater or >180 d in marine or freshwater sediment	BCF >5000 or log K_{ow} >5	Not applicable	Not applicable
EU PBT criteria	Half-life >60 days in marine water or >40 days in fresh water) or >180 days in marine sediment or >120 days in fresh-water sediment	BCF >2000 or log K_{ow} >4.5	Not applicable	Chronic NOEC < 0.01 mg/l or CMR cat 1&2 or endocrine-disrupting effects
US EPA Control Action	Transformation half-life >2 months	BCF >1000	Not applicable	Toxicity data based on level of risk concern
US EPA Ban Pending	Transformation half-life >6 months	BCF ≥ 5000	Not applicable	Toxicity data based on level of risk concern

BCF = bioconcentration factor; BAF = bioaccumulation factor; Log K_{ow} = logarithm of octanol-water partition coefficient; CMR = carcinogenic, mutagenic, reprotoxic.

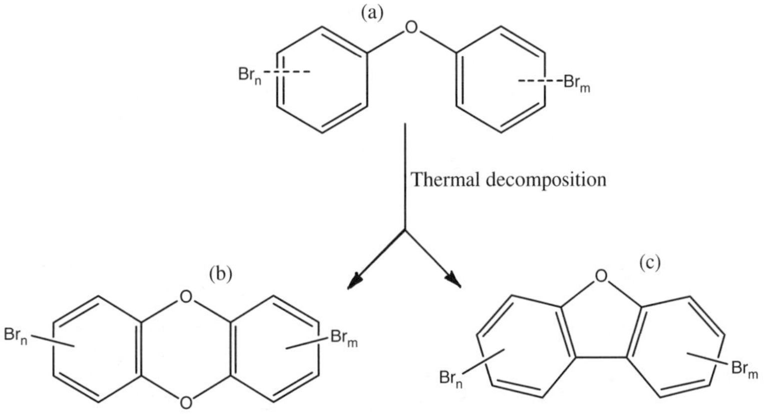

Figure 6.1 Example of thermal decomposition of polybrominated diphenyl ethers. (a) polybrominated diphenyl ether (general structure) (Leung, 2007); (b) polybrominated dibenzo-p-dioxins; (c) polybrominated dibenzofurans. (n and m represent the number of bromine atoms, which can vary; the same process occurs for the chlorinated molecules)

6.2.2 Trichloroethylene (TCE)

Trichloroethylene evaporates rapidly when released into the atmosphere. It has an atmospheric half-life of around 7 days and is degraded primarily by reaction with hydroxyl radicals. Therefore it is not considered to be a persistent atmospheric contaminant. It also volatilises easily from soil and surface water, but its high mobility within soils may result in high rates of percolation to subsurface regions where degradation is much slower. As a consequence, TCE is frequently a significant and persistent groundwater contaminant (Williams-Johnson *et al.*, 1997).

TCE has a low to moderate ability to bioaccumulate in aquatic organisms, with a Log K_{ow} of 2.53 (Pearson *et al.*, 1975; Lide, 1993) and some plants (Schroll *et al.*, 1994). Little is known about its bioaccumulation in terrestrial organisms, but it has been detected in foodstuffs such as dairy products (IARC, 1995). The toxicity of TCE mainly reflects its depressing effect on the central nervous system. Acute exposure results in drowsiness, mucus-membrane irritation, headaches and fatigue (Nomiyama and Nomiyama, 1977; Stewart *et al.*, 1974). Chronic exposure can affect the Central Nervous System (CNS), with reduced word association, short-term memory loss, sleep disturbances, vertigo, loss of appetite, headaches and ataxia. TCE is also suspected to be hepatotoxic and to cause several skin diseases which collectively are known as TCE-related generalised skin disorders.

The International Agency for Research on Cancer (IARC) considers TCE to be 'probably carcinogenic to humans' (IARC, 1995) and the USA's National Toxicology Programme states that it is 'reasonably anticipated to be a human carcinogen', based on meta-analysis of animal, cohort, case studies and community studies (Wartenberg *et al.*, 2000).

Exposure to TCE results in the production of a complex mix of the parent compound and its metabolites, many of which have carcinogenic and toxicological properties in the body (Chiu *et al.*, 2006). It undergoes metabolism via two competing metabolic pathways: oxidation by cytochrome P450 and by conjugation with glutathione by glutathione–S-transferase (GST). Oxidation is quantitatively the major pathway (Figure 6.2).

Oxidation by cytochrome P450 occurs in the liver, initially forming an unstable intermediate (b) 2,2,3 trichlorooxirane (Miller and Guengerich, 1982), which then forms (c) chloral and leads to trichloroethanol which is responsible for most of the CNS effects of TCE (Bolt, 1981).

The minor but important metabolites (i) dichloroacetic acid, (f) oxalic acid and (e) *N*-(hydroxyacetyl)-aminoethanol are formed from the oxidative pathway (Brüning and Bolt, 2000). Dichloroacetic acid (DCA) and trichloroacetic acid (TCA) induce liver tumours in experimental rodents (Chiu *et al.*, 2006) and increase the risk of cancer. In humans, for example, DCA suppresses insulin production and increases glycogenosis (Bull, 2004; Bull *et al.*, 2004; Lingohr *et al.*, 2001), which in turn increases the incidence of liver cancer in humans (Adami *et al.*, 1996; LaVecchia *et al.*, 1994; Rake *et al.*, 2002; Wideroff *et al.*, 1997).

Although the glutathione –S-transferase pathway is a minor route, it produces metabolites such as (dichlorvinyl) glutathione and (m) dichlorovinyl-cysteine (Figure 6.2) which are chemically reactive and unstable, and even a small amount of these metabolites can alter cell functions. Both metabolites have been linked to the genotoxic and carcinogenic effects of TCE (Brüning and Bolt, 2000; Chiu *et al.*, 2006).

6.2.3 Benzene

Benzene has a short half-life in environmental media, the actual length depending on the medium (Table 6.3). It is degraded in the atmosphere by photochemically produced hydroxyl radicals, with a half-life estimated to be 13 days (Atkinson, 2000) calculated from its rate constant of 1.23×10^{-12} cm^3/molecule-s at

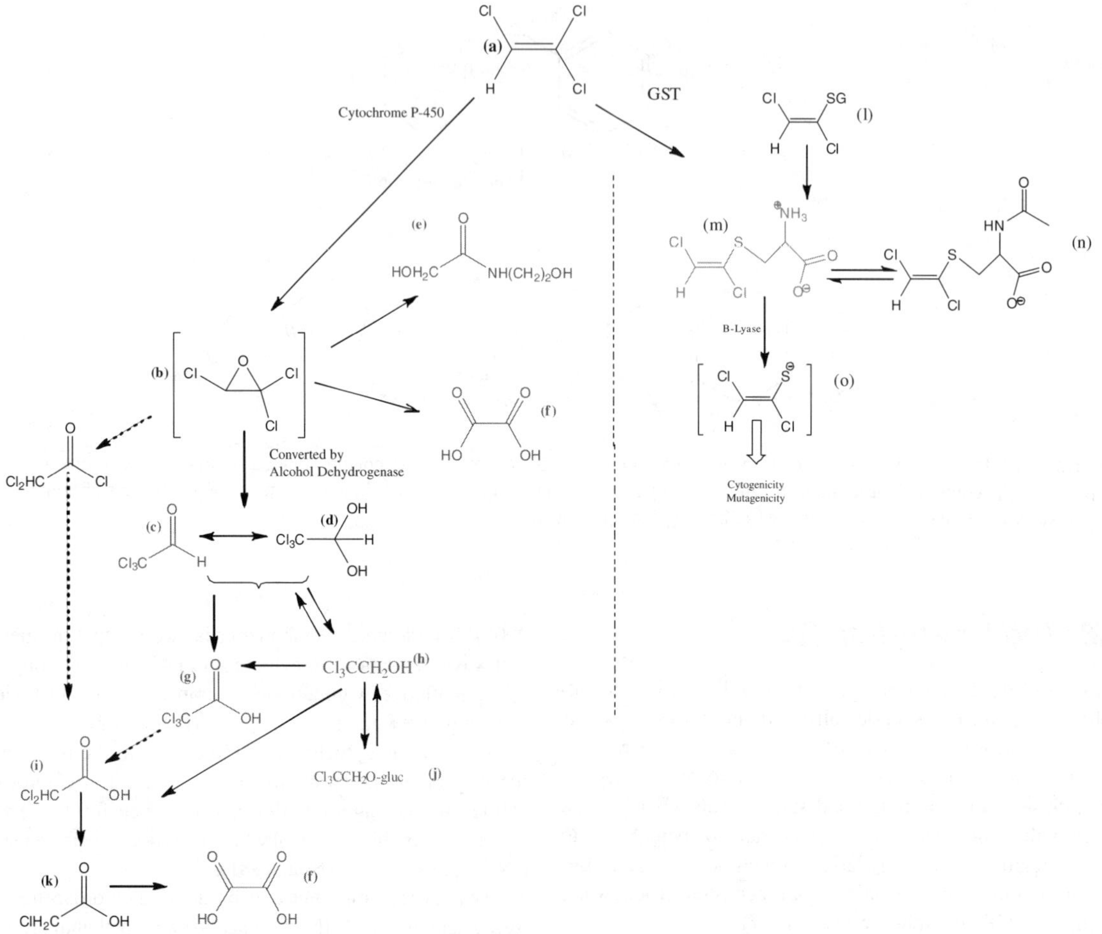

Figure 6.2 The major metabolites of the cytochrome metabolic pathway. h, trichloroethanol; j, trichloroethanol glucuronide; g, trichloroacetic acid; c, trichloroacetylaldehyde (chloral)

25 °C. In estuarine waters, one source reported the half-life to be 6 days (Lee and Ryan, 1979). In sea water, however, the half-life has been reported to be significantly lower, just 5 hours (Neff and Sauer, 1996). This may be due to the presence of micro-organisms, or possibly the effect of sodium chloride on benzene's susceptibility to degradation.

On the whole, therefore, benzene does not persist for a long time in the environment and is usually degraded to phenol in the atmosphere (Lay *et al*., 1996).

Benzene damages the blood-making system of bone marrow and is a risk factor for leukaemia and multiple myeloma. Benzene is first metabolised in the liver and forms a range of hydroxylated and open-ringed products which are transported to the bone marrow. The phenol produced can be hydroxylated to toxic metabolites such as hydroquinone or catechol (Snyderand Hedli, 1996). Hydroquinone is thought to inhibit cell replication in the bone marrow, by covalently bonding to the spindle fibre proteins in cells (Snyder and Hedli, 1996), damaging DNA and leading to bone-marrow depression and eventually aplastic anaemia. Survivors of aplastic anaemia can contract myeloid leukaemia, which suggests that benzene is carcinogenic and mutagenic (Snyder, 2004). Benzene metabolites can also oxidise DNA, leading to similar mutagenic effects. Oxidation of the original metabolites can form highly reactive products that bind to cellular macromolecules. Snyder *et al.* (1989) reported that when benzene was administered to mice, it was less potent than a combination of its metabolites.

The first stage of phenol formation can occur via two possible routes. In the first route, hydroxyl radicals are initially formed from hydrogen peroxide in a reaction involving Cytochrome P450 (Synder and Hedli, 1996):

$$(1)\ \mathrm{CYP\,450} + \mathrm{NADPH} + \mathrm{H}^{+} + \mathrm{O_2} \rightarrow \mathrm{CYP\,450} + \mathrm{NADP} + \mathrm{HOOH}$$

$$(2)\ \mathrm{HOOH} \rightarrow 2\dot{O}H$$

These hydroxyl radicals then hydroxylate benzene, forming phenol (Figure 6.3) (Balakrishnan and Reddy, 1970).

Table 6.3 Half-lives of benzene in different environmental compartments

Condition	Estimated half-life
Vapour phase	13 days (Atkinson, 2000)
Aqueous solution	103 days (Buxton *et al.*, 1988)
Sea water	5 hours (Neff and Sauer, 1996)
Estuarine waters	6 days (Lee and Ryan, 1979)
Atmosphere (ozone radicals)	466 years (Verschueren, 1985)

The second route involves Cytochrome reacting directly with benzene to form benzene oxide, then rearranging to form phenol (Figure 6.3).

6.2.4 Polyaromatic hydrocarbons (PAHs)

Polyaromatic hydrocarbons (PAHs) are a group of chemicals that are formed during the incomplete combustion of organic substances (ATSDR, 1995b). There are more than 100 PAHs and they usually occur as a complex mixture rather than a single compound. There is a group of 16 PAHs that is usually regularly referred to as the US-EPA 16 PAHs, as a result of their hazardous nature, widespread exposure and general availability of data (ATSDR, 1995b) Fifteen from this group have been classified as 'reasonably anticipated to be human carcinogens' based on sufficient evidence of carcinogenicity in experimental animals (IARC 1973, 1983, 1987); they are discussed in Section 6.5.2.3.1 on acute effects.

Polycyclic aromatic compounds have bioconcentration factors (BCF) that range from 10 to 10 000 (ATSDR, 1995b) (Table 6.4); values above 1000 are characterised as high and therefore bioaccumulative (Franke *et al.*, 1994). In general, higher molecular weight PAHs are more bioaccumulative (ATSDR, 1995b). The logarithm of the octanol/water partition coefficient (Log K_{ow}) is usually estimated to determine the bioaccumulation potential, as it is easily available and comparable. The general rule is that if the Log K_{ow} is greater than 3, the chemical has a bioaccumaltive potential. The selected PAHs all have values above 3 and can therefore be classified as bioaccumulative. It is believed that Log K_{ow} values that are above 6 and have a molecular weight above 600 would not be expected to bioaccumulate, but in the cases below, both the Benzo[a]pyrene and Benzo[b]fluoranthene have molecular weights below 300 and can be deemed as potentially bioaccumulative.

In an aerobic aquatic environment, low molecular weight PAHs degrade more rapidly than PAHs with more than three rings (and therefore with a higher molecular weight). The compounds with more than three rings are believed to be extremely stable to biodegradation and thus persistent (Santodonato *et al.*, 1981). It has been estimated that it takes weeks to months for PAHs with three rings to break down in soil, and PAHs with four or more rings have been classified as resistant to biodegradation (ATSDR, 1995b). Table 6.4 shows the estimated half-life of PAHs in soil and the number of rings in the PAH structure. The three-ring PAHs have significantly shorter half-lives than the four and five-ring structures. Chrysene is estimated to be particularly persistent.

Benzo(a)pyrene has been reported to cause developmental problems in the fetuses of experimental animals, but the other PAHs have not been investigated sufficiently to identify similar effects (Wisconsin Department of Health Services, 2000). The carcinogenicity of the PAHs reflects their transformation to reactive electrophilic intermediates that readily undergo hydrolysis and covalently bind to DNA proteins and other cellular molecules, leading to mutations and tumour initiation (Jerina *et al.*, 1980; Harvey, 1991). The products of the PAH metabolism include epoxide intermediates, dihydrodiols, phenols and quinones (ATSDR, 1995b). The bay region of an arene is the indentation formed where three rings fuse together (Figure 6.4) and the diol epoxide that forms at this region is considered to be the ultimate carcinogen for alternant polyarenes (Jerina *et al.*, 1980). Alternant PAHs have an equally distributed electron density, whereas non-alternant PAHs have an uneven electron density from one portion of the molecules to another, which can affect the toxicological effects (ASTDR, 1995b).

Non-alternant polyarenes do not have a bay region and their carcinogenicity reflects a different process (Amin *et al.*, 1985).

Table 6.4 Log K_{ow} values of selected PAH compounds

PAH	Anthracene	Benz[a] anthracene	Benzo[a] pyrene	Benzo[b] fluoranthene	Chrysene	Phenanthrene	Reference
Log K_{ow}	4.45	5.61	6.06	6.04	5.16	–	ATSDR, 1995b
Estimated half-life of PAH in soil (days)	50–134	162–261	229–309	211–294	371–387	50–134	Park *et al.*, 1990
Number of rings	3	4	5	5	4	3	

Figure 6.3 Formation of phenol from benzene

BFRs are emitted during production and enhanced levels are found close to BFR-producing factories (Lassen and Søersen, 1999). Sources include effluent and flue gases from factories and other facilities processing BFRs, as well as grey water from washing textiles (Almqvist, 2006; Leung, 2007).

Some of the major sources of BFRs are in the use and disposal of consumer products such as textiles, furniture, plastics, circuit boards and building materials. Table 6.6 shows the major uses of penta, octa and deca-BDE in consumer products. Research conducted by Johnson *et al.* (2010) illustrated that house dust is a primary exposure pathway of PBDEs and supports the use of dust PBDE concentrations as a marker for its exposure. Treatment of these products at facilities such as incineration, recycling plants and wastewater treatment plants are major sources of these chemicals to the environment (Watanabe, 2003; North, 2004; Osako *et al.*, 2004; Hale *et al.*, 2002).

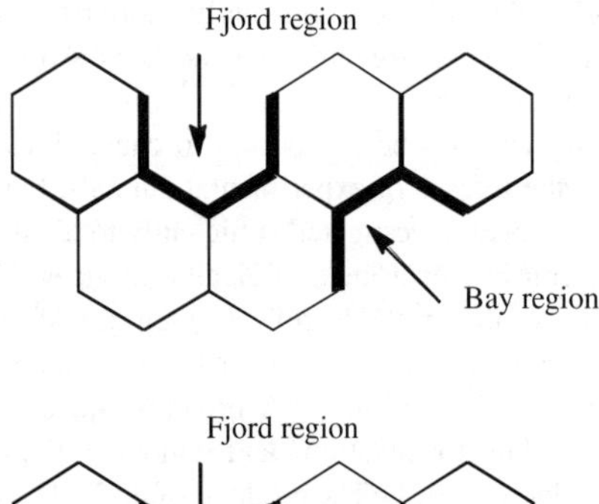

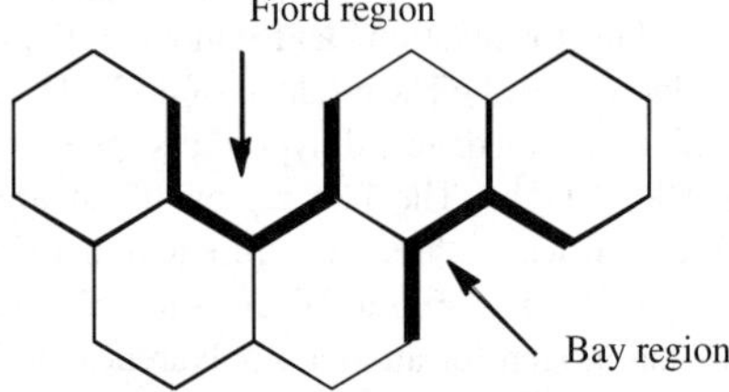

Figure 6.4 Molecular regions of polyarenes

It has been suggested that they can be converted into active bay-region epoxides in the presence of cytochrome P450 and other enzymes (ATSDR, 1995b).

A summary of the hazardous properties of the industrial chemicals discussed here is given in Table 6.5.

6.3 Sources

Emissions and releases of industrial chemicals can occur during every stage of their life cycles, from production to processing, through manufacturing, to their use in downstream production and products to disposal (Figure 6.5).

6.3.1 Brominated flame retardants (BFRs)

Brominated flame retardants have no known natural sources, but PBDEs with higher levels of oxygen are produced by marine organisms (Palm *et al.*, 2002).

6.3.2 Benzene

Benzene is a natural constituent of crude oil (Brief *et al.*, 1980) and it has been detected in air, water and human biological samples (US EPA, 2010). It is a by-product of many combustion processes, including forest fires, wood burning and organic wastes and can be released from volcanoes. Benzene is also one of the world's major commodity chemicals, most of it being obtained from petrochemical and petroleum refining industries (US EPA, 2010a).

The main sources of benzene in the environment include vehicle exhausts, automobile refuelling, hazardous waste sites, waste water from industries that use benzene and chemical manufacturing sites. Benzene has been detected in soil following the release of industrial discharges, the disposal to land of wastes containing benzene and following the underground injection of such wastes (ATSDR, 2007). Smoking is the main source of benzene exposure to the public. The estimated daily intake from smoking is between 5.9 and 73.0 μg benzene per day (Brunnemann *et al.,*1990).

Table 6.5 Hazardous properties of the BFRs, Benzene, PAHs and TCE based on based on Annex III of the Hazardous Waste Regulations (England and Wales, 2005)

Hazardous properties	BFRs	Benzene	PAHs (e.g. benzo(a)pyrene)	TCE
Flammability	Used as flame retardants	Easily ignited under all normal temperature conditions	Highly flammable	Not flammable
Irritant	Some indication of skin-sensitising potential may be considered	Eyes and skin	Eyes and skin	Eyes and skin
Harmful	Harmful if swallowed	May cause lung damage if swallowed	May cause lung damage if swallowed	Harmful to aquatic organisms
Toxicity	The US National Toxicology Program conducted 2-year feeding studies with BFRs and showed that doses at 2500 g/kg/day resulted in neoplastic nodules in the liver in rats. There are also concerns about developmental neurotoxicity in mice.	Danger of serious damage to health by prolonged exposure through inhalation	LD_{50} Subcutaneous rat 50 mg/kg	LD_{50} Oral, rat, 4920 mg/kg
		LD_{50} Oral, rat 2990 mg/kg	LDL_o (oral, rat): 2250 mg/kg	LC_{50} Inhalation, mouse, 4 h, 8450 mg/kg
		LC_{50} Inhalation, rat, female, 4 h, 44 700 mg/m^3		LD_{50} Dermal, rabbit >20 000 mg/kg
		LD_{50} Dermal, rabbit, 8263 mg/kg		
Ecotoxicity	$LC(EC)_{50}$ values in crustaceans are below 1 mg/l for PBDEs, which classifies these chemicals as very toxic to aquatic organisms	LC_{50} *Pimephales promelas* (fathead minnow), 15.00–32.00 mg/l, 96 h	Very toxic to aquatic organisms; may cause long-term adverse effects in the aquatic environment	May cause long-term effects in the aquatic environment
			LC_{50} (*P. lucida*): 1.2–3.7 mg/l/ 24h	Toxicity to fish LC_{50}, Pimephales promelas (fathead minnow), 41.00 mg/l, 96 h
Carcinogenicity	Not classified by the IARC due to lack of information	May cause cancer. May cause heritable genetic damage	Should be treated as a carcinogen	May cause cancer. Anticipated by the IARC to be carcinogenic
		IARC rated as known to be carcinogenic	Classified by the IARC as probably carcinogenic	
Mutagenicity	Negative	Laboratory experiments have shown mutagenic effects	Evidence of mutagenic effects	Experiments have shown mutagenic effects
Bioaccumulation	High bioaccumulation potential Log K_{OW} = 5.9–6.265	*Leuciscus idus* (Golden orfe), 3 d Log K_{OW} =2.13	High bioaccumulation potential Log K_{OW}= 6.06	Medium potential Log K_{OW} = 2.53
Reference	de Wit, 2002 Wang *et al.*, 2006 IARC, 1986	Sigma-Aldrich, 2010a VWR, 2004 IARC 1982 Sangster, 1989	Sigma-Aldrich, 2010b IARC, 1995 Pinsuwan and Yalkowsky, 1995	Sigma-Aldrich, 2010c VWR, 2007 IARC, 2010 Mackay *et al.*, 1993

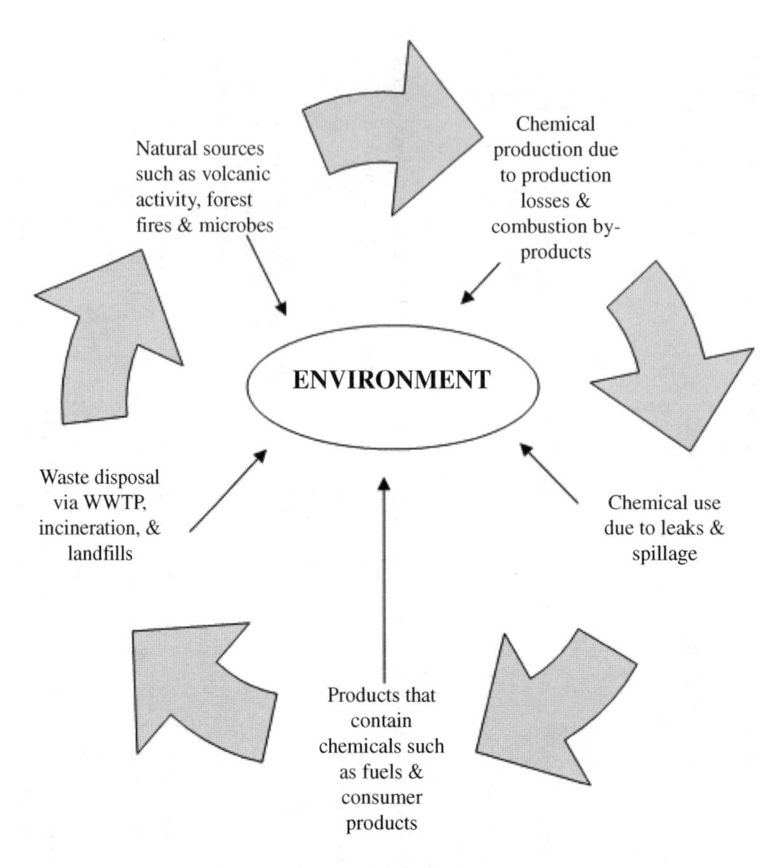

Figure 6.5 Sources of industrial chemicals to the environment

Oil spills from tanker washings, exploration, transport and handling and wastes are other significant sources of benzene (Cheremisinoff and Morresi, 1979).

6.3.3 Polyaromatic hydrocarbons (PAHs)

Natural sources of PAHs include volcanic activity and forest fires. Benzo [a] pyrene can also be formed through biosynthesis, by bacteria, algae and various higher plants (Verschueren, 1996). The major source of PAHs in soils uncontaminated by anthropogenic sources is from natural fires (Wild and Jones, 1995).

The major anthropogenic source of PAHs to the environment is through the incomplete combustion of organic materials, principally coal, oil, petrol and wood (Wild and Jones, 1995). Wild and Jones (1995) estimated the sources of PAHs to the UK environment (Table 6.7); the largest source of PAHs is domestic coal burning, followed by vehicle emissions.

The Annual Report for 2007 on the UK PAH Monitoring and Analysis Network (AEA, 2009) indicates that the dominant source of the US-EPA 16 PAHs, in the UK, is from transport with an estimated 62 per cent (Figure 6.6). The total emission in terms of the US-EPA 16 is dominated by the emission of naphthalene (AEA, 2009).

It has been found that the levels of PAHs associated with cigarette smoke are 1.5–4.0 times higher than other indoor combustion sources (Sheldon *et al.*, 1993). Up to 17 different PAHs have been identified in the mainstream smoke from cigarettes (Gmeiner *et al.*, 1997) and all of the EPA's priority PAHs are present in cigarette smoke, 14 of which are known to have carcinogenic properties (IARC, 1983). The levels of PAHs in cigarettes have been found to vary between brands and with tobacco blend, with total PAH ranging from 1.0 to 1.6 μg per cigarette (Ding *et al.*, 2005). Individual PAH levels have also been found to vary widely: levels of benzo[k]fluoranthene are below 10 ng/cigarette and levels of naphthalene around 500 ng/cigarette (Ding *et al.*, 2005).

6.3.4 Trichloroethylene (TCE)

There are few natural sources of TCE and the amounts released from them are negligible. They include production by one species of red microalga and some temperate, subtropical and tropical algae (Abrahamsson *et al.*, 1995).

The main release of TCE into the environment now occurs during its manufacture, use and disposal (Figure 6.7). In the past, it was used mainly in dry-cleaning processes (Wartenberg *et al.*, 2000; Doherty, 2000) due to its relatively low cost, rapid evaporation rate, low flammability and efficiency in dissolving a wide range of organic substances (Doherty, 2000), but it has now been replaced for this purpose mainly by tetrachlorethylene or

Table 6.6 Uses of PBDEs in consumer products (Schecter *et al.*, 2005; Rahman *et al.*, 2001)

PBDEs	Contents	Applications	Finished products
Deca-BDE	>97% deca-BDE	Circuit boards, protective coatings	Computers, ship interiors, electronic parts
	<3% nona-BDE	Panels, electrical components	Lighting panels, housing of electrical appliances
	Trace octa-BDE	Cross-linked wire cable, foam tubing, weather protection moisture barriers	Power cables, insulation of heating tubes, marine appliances, building control instruments
		Electrical components, conduits, electronics devices	Television and electronic devices, electro-mechanical parts, underground junction boxes
		Cable sheets	Wire and cables, floor mats
		Transportation	Conveyor belts, foamed pipes for insulation
Octa-BDE	37–35% octa-BDE	Model parts	High impact polystyrene
	44% hepta-BDE		Plastic casing for computers and electronic equipment
	10–12% hexta-BDE		Polyethylene: plastic electronic wires and cables, minor use in textiles and upholstery
	10–11% nona-BDE		Acrylonitrile-butadiene-styrene: plastics computer monitors, television housings, circuit boards
	<1% deca-BDE		Polyurethane: plastic foams for mattresses, upholstered furniture
			Television, computer casings, hairdryers, automotive parts
		Television cabinets and back covers, electrical housing	Smoke detectors, office machines, housing of electrical appliances
		Electrical connectors, automotive interior parts	Computers, connectors, automotive industry
Penta-BDE	50–60% penta-BDE	Printed circuit boards	Paper laminates for printed circuit boards
	24–30% tetra-BDE	Electrical components, connectors	Switches, fuses, stereos
	4–8% hexa BDE	Cushioning/packaging materials	Furniture, sound insulation, wood imitation
		Coatings	Carpets, automotive seating, furniture, tents, military safety clothing
		Circuit boards, coatings	Electrical equipment, military/marine applications, construction panels

Table 6.7 Summary of annual PAH emissions to the UK atmosphere (tonnes) (From Wild and Jones, 1995)

Compound	Vehicles	Coal-fired power stations	Coke-manufacture, smokeless fuel production and industrial coal use	Domestic coal combustion	Oil-free power stations and industrial oil users	Domestic wood combustion	Oil-free power stations and industrial oil users	Domestic wood combustion	Waste incineration etc	Stubble burning	Industrial processing	Total
Naphthalene	7.4	–	–	–	–	–	–	–	0.0014	–	–	7.4
Acenaphthene/ fluorene	12.8	0.13	0.32	0.7	0.32	0.7	0.32	–	0.0043	–	1.39	15
Phenanthrene	27.4	0.95	0.24	40	1.5	40	1.5	–	0.00014	–	7.74	78
Anthracene	8.7	–	–	14	–	14	–	–	0.00014	–	–	23
Fluoranthene	7.3	0.56	0.144	180	0.24	180	0.24	0.92	0.0082	1.6	2.5	193
Pyrene	7.8	1.20	0.3	160	0.31	160	0.31	1.1	0.022	1.8	4.77	177
Benz[a] anthracene/ chrysene	3.4	0.3	0.078	110	0.28	110	0.28	1.1	0.010	1.8	1.83	119
Benzo[b] fluoranthene	1.4	0.0054	0.014	43	–	43	–	0.32	0.0045	0.5	0.017	45
Benzo[a]pyrene	2.1	0.0058	0.0015	28	0.015	28	0.0015	0.29	0.0003	0.5	0.066	31
Benzo [ghi]perylene	1.7	–	–	22		22	–	0.061	0.0060	0.1	–	24
ΣPAH	80.2	3.1	0.8	600	2.7	600	2.7	3.8	0.056	6.3	18.3	712

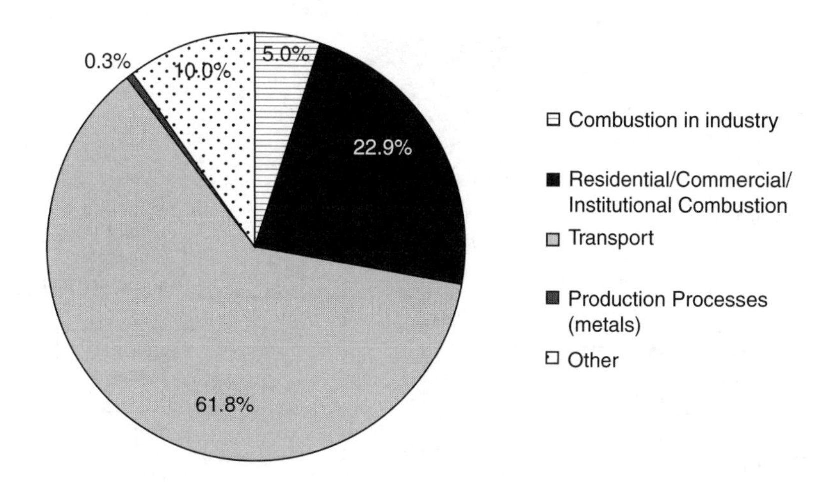

Figure 6.6 Estimated emission of the US-EPA 16 PAHs in the UK (AEA, 2009)

1-1-1-trichloroethane (Williams-Johnson *et al.*, 1997). One of its principal uses is as a degreaser in metal cleaning (Corden *et al.*, 2005) and to a lesser extent other materials, including inorganic fibres, asbestos, glass, carbon fibres, silicon carbide and plastics.

TCE is found in several consumer products, including paint strippers, adhesive solvents, paints and varnishes (Williams-Johnson *et al.*, 1997). It is used in the manufacture of organic chemicals, including polyvinyl chloride, pentachloroethane, other polychlorinated aliphatic hydrocarbons, flame retardants and insecticides and to produce CFC substitutes, particularly hydrochlorofluorocarbons and hydrofluorocarbons. The demand for TCE increased after ozone-damaging CFCs were phased out under the Montreal Protocol.

6.4 Environmental pathways

There are significant differences in the way industrial chemicals enter the environment (Figure 6.8).

6.4.1 Brominated flame retardants (BFRs)

6.4.1.1 Water

Brominated flame retardants have high log k_{ow} and hence are poorly soluble in water. The small amounts of BFRs that occur are bound to particles which settle with sediments beneath lakes and rivers. These sediments act as reservoirs for BFRs, sometimes for years (Qiu *et al.*, 2007). However, numerous studies

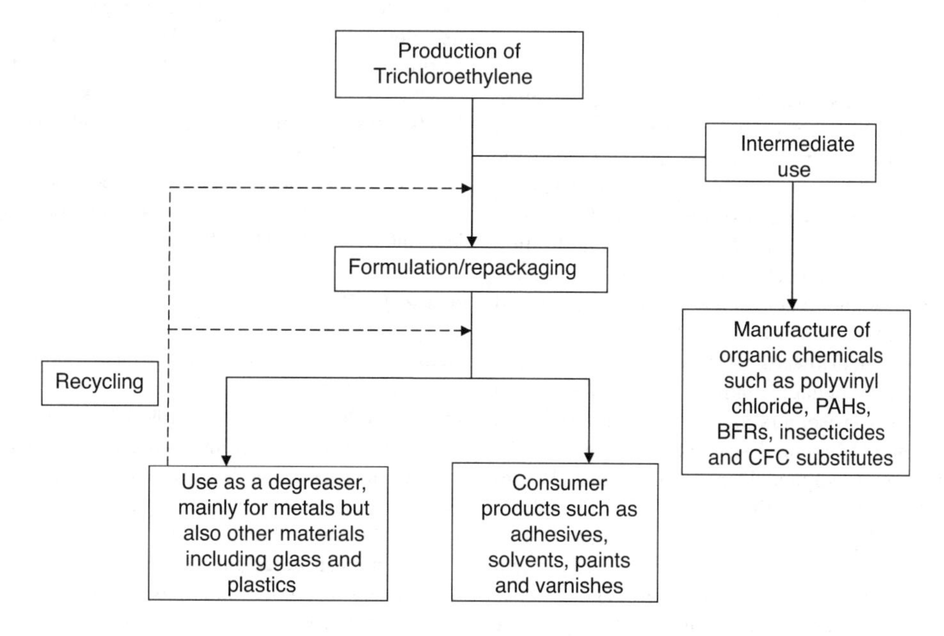

Figure 6.7 The major uses of TCE (Adapted from Corden *et al.*, 2005 and Williams-Johnson *et al.*, 1997)

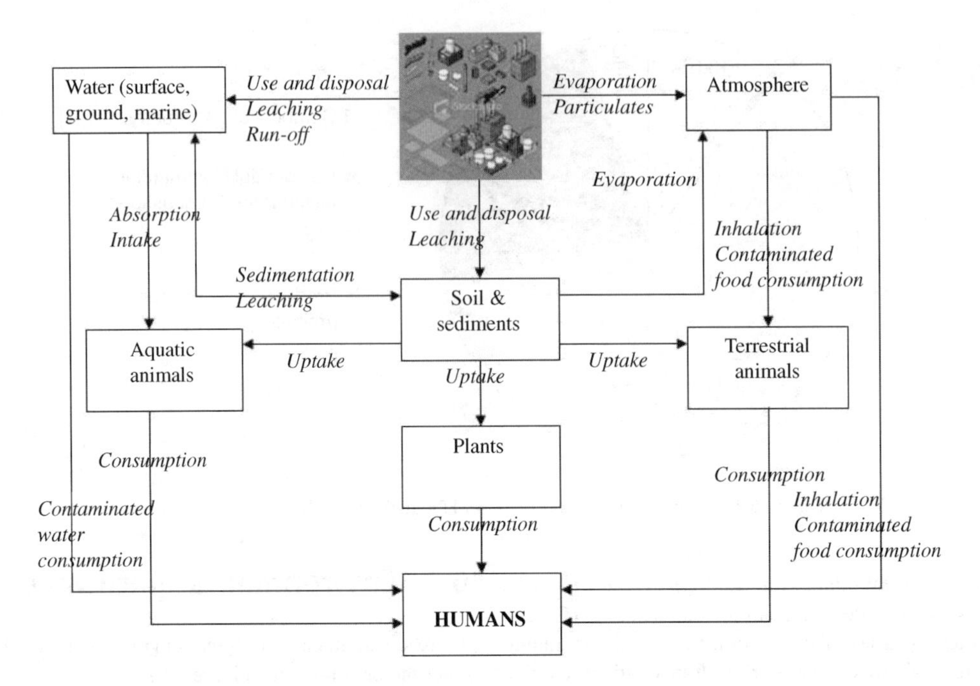

Figure 6.8 Pathways of industrial chemicals in the environment

show that flame retardants such as PBDE can bioaccumulate in marine and freshwater fish, water being an exposure pathway (Zennegg *et al.*, 2003; Xiang-Zhou *et al.*, 2008). Some lower-molecular-weight BFRs (e.g. tetra- and penta-congeners of PBDE) may accumulate in fish to concentrations of 10×10^{-9} to 1×10^{-6} g of PBDE per gram of fresh fish (Yu *et al.*, 2009).

6.4.1.2 Soils

Soils are important in the distribution and biogeochemical cycling of BFRs because of their large sorption capacity and they are a major reservoir for these compounds (Zou, 2007). Polybrominated diphenyl ethers such as the tetra- to octa- isomers are predicted to have low mobility, based on their range of estimated log K_{ow} values from 4.56 to 6.23 (van Esch, 1994). Volatilisation from moist soil surfaces is expected to be an important process for tetra- and penta-bromodiphenyl ethers based on estimated Henry's Law constants of 8.5×10^{-6} and 2.2×10^{-5} atm-m^3/mol respectively (Meylan and Howard, 1991). However, adsorption to soil is expected to attenuate volatilisation. If released into water, the tetra- to octa-isomers are predicted to adsorb to suspended solids and sediment. Two biodegradation tests for a commercial pentabromodiphenyl ether mixture and a hexabromodiphenyl ether, illustrated that pentabromodiphenyl did not degrade after 29 days in a ready biodegradation test. Hexabromodiphenyl ether, present at 100 mg/l, reached 15 per cent of its theoretical biological oxygen demand (BOD) in 4 weeks using an activated sludge inoculum at 30 mg/l and the Japanese MITI test (European Union Joint Research Centre *et al.*, 2001).

6.4.2 Benzene

6.4.2.1 Air

Benzene is widespread in the environment, but the main exposure pathway is via the atmosphere, usually from its manufacture or from the petrol industry and vehicle exhausts. Benzene levels measured in ambient outdoor air average 6 µg/m^3 (range 2–9 µg/m^3) globally (Franke *et al.*, 1994). Benzene levels inside houses or offices are usually higher than levels outside, with homes attached to garages or occupied by smokers having the highest levels. Seasonal variations occur, with higher levels in less well-ventilated buildings. People living around some hazardous waste sites, petroleum-refining operations, petrochemical manufacturing sites, or petrol stations may be exposed to higher levels of benzene in air (ATSDR, 1995a).

6.4.2.2 Water

Benzene is a light non-aqueous-phase liquid (LNAPL) and because of this it migrates through soil and tends to collect in the water table (Morgan *et al.*, 1993). Benzene contamination of groundwater can result from leaking underground petrochemical storage tanks, petrol stations, landfills and the improper disposal of hazardous waste. Benzene may also be released to water via industrial effluent (Franke *et al.*, 1994; US EPA, 2010a).

Benzene can be degraded by microbial activity under both aerobic and anaerobic conditions, depending on such factors as sunlight penetration, the type of microbe present and temperature and oxygen content. Benzene in water has been shown to be degraded by an enriched aerobic bacterial culture. The compound began to degrade after 12 hours of incubation, half had degraded after 60 hours and it was almost completely degraded after 90 hours (Zhang and Bouwer, 1997). Hydrolysis is not expected to be important in degradation because the stable aromatic ring does not have any hydrolysable functional groups (Howard, 1997).

6.4.2.3 Soil

Benzene can contaminate land via industrial discharges, fuel leaks or spillages (ATSDR, 2007). Historically, leaks or spillages of petroleum and solvents have resulted in contamination over wide areas. Benzene's mobility in soil means that it normally migrates through the soil to water sources. Although it has a moderate tendency to adsorb to organic matter, this depends on the water content and porosity of the soil and environmental conditions such as temperature (Morgan *et al.*, 1993).

Benzene volatilises readily from soil, with increasing rates at shallower depths, especially from soils with a high air-filled porosity such as sands and gravels (Environment Agency, 2010). The volatility of benzene means that microbial degradation is not as important as for contaminated waters, although it does still occur (Morgan *et al.*, 1993).

The behaviour of benzene in soil is also influenced by the presence of other compounds; its solubility and volatility, for example, are reported to decrease in the presence of other hydrocarbons.

6.4.2.4 Food

Food contaminated with benzene is a potential exposure pathway, although benzene has a low to moderate bioconcentration potential in aquatic organisms. (M.M. Miller *et al.*, 1985). The contribution of diet to total benzene exposure in urban, rural and city locations is only a small fraction of total exposure (Concawe, 1999).

6.4.2.5 Consumer products

Cigarette smoke is an important exposure pathway for benzene. The Concawe report (Concawe, 1999) estimated that smoking 20 cigarettes a day contributed an estimated 400 μg benzene per day, which is greater than the total amount absorbed by persons exposed to benzene emitted from motor vehicles for a prolonged period, such as traffic wardens.

6.4.3 Polyaromatic hydrocarbons (PAHs)

6.4.3.1 Air

When PAHs are released to air, they can undergo photolysis, though this is retarded if they adsorb to airborne particulates in water. They can also be transformed and removed from air by reaction with ozone and nitrous oxides (US EPA, 2010b). PAHs are stable in the atmosphere and can travel long distances, enabling them to be detected far from their sources.

6.4.3.2 Water

The low solubility and high stability of PAHs prevent them from undergoing hydrolysis and instead they adsorb to particulates in water. They are generally stable to biodegradation, but have been shown to be metabolised in micro-organisms in some natural waters (US EPA, 2010b). Adsorption to sediments makes them less susceptible to other forms of degradation.

PAHs are detected in some groundwater and studies reveal that biodegradation may be important (Martens *et al.*, 1997). They can enter groundwater from ash, coal tar or creosote, disposed to landfills, which may lead to contamination of drinking water. They are, however, likely to adsorb to particulates and be removed before reaching domestic supplies, so exposure to humans via this pathway is unlikely to be significant.

6.4.3.3 Soil

PAHs tend to adsorb strongly to soil, which helps to prevent leaching to groundwater.

6.4.4 Trichloroethylene (TCE)

6.4.4.1 Air

Trichloroethylene is released to the air primarily from metal cleaning and degreasing operations but also at sewage treatment works and from landfills. It is highly volatile and therefore air is its most important exposure pathway. In the UK, TCE is classified as one of the 50 most significant volatile organic compounds (VOC) released to the environment (Environment Agency, 2010). Global estimations of TCE emission have been reported for different regions using audited production and sales data (Figure 6.9).

6.4.4.2 Water

Trichloroethylene is moderately soluble and can migrate through soil into groundwater and water bodies (Wu and Schaum, 2001). TCE is the most frequently reported contaminant of groundwater there, with 9–34 per cent of drinking-water sources in the US contaminated by it (Bourg *et al.*, 1992).

6.4.4.3 Soil

Trichloroethylene is a common soil contaminant (Chiu *et al.*, 2006) and it has been estimated that there are at least 15000 TCE hazardous sites in the US (US EPA, 2011). Its volatilisation from soil is slower than from water, although it is more rapid than for most other VOCs. One study showed that on average 37 per cent of the TCE applied to soil was volatilised 168 hours after treatment at 12 °C and 45 per cent was volatilised at 21 °C (Park *et al.*, 1988).

The sorption of organic compounds to soils has been found to be related to the organic-carbon content of the soil (Kenaga, 1980; Urano and Murata, 1985). Experimentally measured sorption

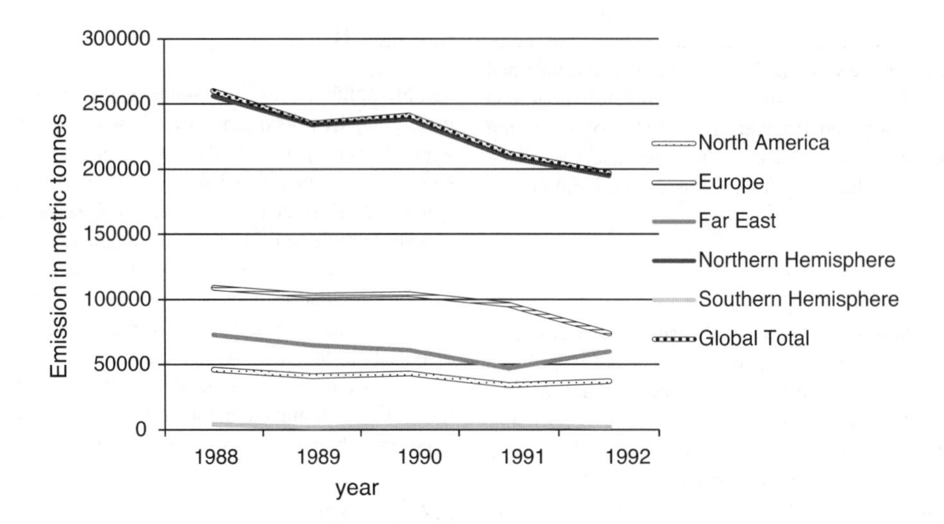

Figure 6.9 Global estimates of TCE emissions (McCulloch and Midgley, 1995)

coefficients for soil organic carbon (K_{oc} values) for TCE indicate medium to high mobility in soils. Moreover, the absorption of TCE has been shown to be related to the organic carbon matter components, for example soils with high fat contents have high TCE absorption rates (Kenaga, 1980; Swarm *et al.*, 1983).

6.4.4.4 Food

Trichloroethylene has been reported in food in the UK and the USA. The food groups with the highest concentrations are butter and margarine (73.6 pg/kg), cheese (3.8 pg/kg) and ready-to-eat cereals 3.0 pg/kg (McConnell *et al.*, 1975). The USA Food and Drug Administration and EPA recommend that bottled water for human consumption should not contain more than 0.005 mg/l of TCE (Williams-Johnson *et al.*, 1997). TCE may be present in food if contaminated water is used or if equipment cleaned with solvents is used in its preparation; some food-packaging materials also containing TCE.

Small amounts of TCE have also been detected in fruit and vegetables, possibly as a result of bioconcentration in plants or contamination after harvesting. It has been demonstrated that over time TCE can be absorbed from the atmosphere and concentrated in food products such as rape-seed oil, butter and cream. Therefore, undesirable levels in food could result despite ambient room levels not exceeding acceptable levels (Grob *et al.*, 1990).

6.5 Human health

6.5.1 Exposure routes

This section further describes the effects of the industrial chemicals investigated on human health.

6.5.1.1 Brominated flame retardants (BFRs)

The major routes of entry for BFRs into the body are through inhalation and ingestion of contaminated food (especially of the lower brominated congeners; Betts, 2004) and water. Inhalation of contaminated particles or direct contact with materials that have been treated with BFRs can increase human exposure (Vos *et al.*, 2003). Karlsson *et al.* (2007) estimated the maximum daily human exposure via inhalation for BFRs was 1.9 ng per day for females (4.1 per cent of overall daily intake) and 2.0 ng per day for males (4.4 per cent overall daily intake). They reported the presence of PBDEs in air, dust and human plasma in five randomly selected households in Sweden.

A major route of intake is food of animal origin, especially fish. Schecter *et al.* (2004) reported levels of PBDEs in foods in the US using a market-based survey of 30 food types from three major supermarket chains in Dallas, Texas. The highest levels of PBDEs were found in fish, followed by meat; the lowest levels were in dairy products. In Sweden, the major human exposure route is via fish-related food, and the levels of tetrabromodiphenyl ether (TeBDE) and pentabromodiphenyl ether (PeBDE) in fish can be used to estimate this route of exposure. Normal fish intake in Sweden is about 30 g per day, and, if herring is used as a model fish, the estimated intake is approximately 0.3 μg TeBDE per person per day and about 0.1 ug PeBDE per person per day (van Esch, 1994).

6.5.1.2 Benzene

Benzene can enter the body via inhalation, ingestion and skin absorption. Inhalation is the most important human exposure pathway. Inhalation may occur via the volatilisation of benzene from contaminated shower water or during cooking and laundering. Benzene may also be absorbed into the body via ingestion. It has been detected in drinking water, bottled water, alcohol and food (Franke *et al.*, 1994).

Table 6.8 Local concentrations of TCE air, water and food and the calculated human intake (ECB, 2000)

Concentration in air, water and biota	Daily human consumption or intake	Calculated human intake via indirect exposure (mg/kg per day) based on 70 kg person and intake of 100 per cent
Air: 65 μg/m^3 (PEClocal$_{air}$ measured near a production plant)	20 m^3	0.019
Drinking water: 21 μg/l	0.002 m^3	6×10^{-4}
Fish: 479 μg/kg	0.115 kg	7.9×10^{-4}
Leaf/stem crop: 6 μg/kg	1.2 kg	1.0×10^{-4}
Root crop: 6 μg/kg	0.384 kg	3.3×10^{-5}
Meat: 192 μg/kg	3.01 kg	8.3×10^{-4}
Milk/diary products: 20 μg/kg	0.561 kg	1.6×10^{-4}
		Total = 0.22

6.5.1.3 Polyaromatic hydrocarbons (PAHs)

Humans can be exposed to PAHs via inhalation, ingestion and dermal contact. The main exposure pathway is through inhalation, especially from cigarette smoke, automobile emissions and exhausts (Wisconsin Department of Health Services, 2000).

Ingestion is an important pathway if PAHs are present in drinking water. Shellfish that live in contaminated water are also a pathway for exposure via ingestion, as well as charcoal-smoked foods. Dermal absorption of PAHs can also occur if there is direct contact with contaminated soil or water.

6.5.1.4 Trichloroethylene (TCE)

The main exposure route of TCE is by inhalation. The Aerometric Information Retrieval System (AIRS) monitored outdoor air quality using 115 monitors across 14 states in the USA. The mean level of TCE in air during 1998 was 0.88 μg/l, which, assuming an inhalation rate of 20 m^3 air per day, gives an estimated exposure rate of 18 μg per day per person (Wu and Schaum, 2001). Data collected from Hamburg, Germany, showed ambient air concentrations of TCE ranging from 0.8 to 18.5 μg/m^3 (Bruckmann and Kersten, 1988). A study in Finland reported levels of 0.27 and 36 μg/m^3 in ambient air in a suburban and an industrialised area, respectively (Kroneld, 1989). It has been demonstrated that workers from urban areas had higher blood levels (0.763 ng/l) of TCE than workers in rural areas (0.180 ng/l), showing that exposure varies greatly with location (Brugnone *et al.*, 1994). TCE has been reported in samples of mothers' milk from urban areas in the USA (Erickson *et al.*, 1982).

Exposure can also occur through the contamination of potable water sources. TCE can be a persistent contaminant of subsurface waters. The average ingestion of TCE from water in the USA has been calculated by ATSDR as 2–20 μg per day (Wu and Schaum, 2001). Brown *et al.* (1984) demonstrated that dermal contact provides a significant proportion of the total body burden of VOCs and this can occur from contact with contaminated water during bathing and washing. The use of contaminated water in the home, for example for cooking, bathing and laundering, will also increase indoor air levels because of the volatility of TCE. The transfer of TCE from shower water to air has been found to have a mean efficiency of 61 per cent, independent of water temperature. This suggests that showering in TCE-contaminated water for 10 minutes could result in a daily exposure by inhalation comparable to that from drinking contaminated tap water (McKone and Knezovich, 1991).

Occupational exposure to TCE can occur from its role as an intermediate in polyvinyl chloride production. Increased incidence of TCE-related diseases has been reported in workers in the aerospace, electronics, metal and iron, printing, dry-cleaning, painting and chemicals industries (Zhao *et al.*, 2005; Chang *et al.*, 2005; Raaschou-Nielsen *et al.*, 2003).

An EU risk-assessment for TCE (ECB, 2000) aimed at estimating the daily intake of TCE regionally, based on typical human consumption and inhalation rates (Table 6.8) showed that inhalation is responsible for around 90 per cent of the total intake of TCE.

6.5.2 Physiological effects

6.5.2.1 Brominated flame retardants (BFRs)

6.5.2.1.1 Acute effects Recent animal studies have indicated the ability of PBDEs to enhance the growth of oestrogen-dependent breast-tumour cells, (Darnerud *et al.*, 2001; She *et al.*, 2002; McDonald, 2002; Wenning, 2002; Hooper and She, 2003). They have also been shown to delay puberty in male rats and decrease the growth of androgen-dependent tissues (Stoker *et al.*, 2005). Animal studies consistently show the neurobehavioural toxicity of PBDEs in rodents, and pronounced effects on thyroid homeostasis have also been reported (Hallgren and Darnerud, 2002; Hallgren *et al.*, 2001). When administered to mice for 14 consecutive days, in olive oil, hepatic lesions were induced in the liver (Tada *et al.*, 2006).

6.5.2.1.2 Chronic effects Other researchers have reported permanent neurodevelopmental disturbances in rodents exposed to PBDEs pre and/or post-natally (Branchi *et al.*, 2002, 2003; Eriksson *et al.*, 2001; Eriksson *et al.*, 2002; Viberg *et al.*, 2003a, 2003b), at doses below those which cause maternal toxicity. Eriksson *et al.* (2006) reported that PBDEs and PCBs can interact and enhance the development of neurobehavioural defects if exposure occurs at a critical stage of brain development. In their study, a single oral dose of PCB or PBDE, or a combined dose of PCB and PBDE, was administered to 10-day-old mice. Mice exposed to the combined dose or the high dose of PBDE exhibited significant impairment in spontaneous motor behaviour at the ages of four and six months. In vivo and in vitro animal studies show nervous system, reproductive, developmental and endocrine effects, as well as cancer at high-doses of BFRs (Schecter *et al.*, 2005b).

6.5.2.2 Benzene

6.5.2.2.1 Acute effects Benzene vapours > 20 000 mg/kg have been reported to be fatal via inhalation within minutes because of brain damage (Cheremisinoff and Morresi, 1979). Benzene concentrates in brain lipids and affects the circulation, causing convulsions and/or paralysis, followed by loss of consciousness and eventually death (Cheremisinoff and Morresi, 1979). Accidental ingestion of benzene is treated as poisoning. In comparison, the dermal absorption of benzene is believed to be minor, but can result in erythema (Figure 6.10) and may also lead to blisters.

Acute workplace exposure to benzene is associated with early reversible haematoxicity or irreversible bone-marrow damage (if there is more prolonged exposure in high doses) (Snyder and Hedli, 1996). Workers in a benzene factory were reported to show haematoxic effects in the form of: anaemia, leucopoenia and thrombocytopenia. A decrease in number in more than one type of cell has been reported to lead to irreversible bone marrow aplasia.

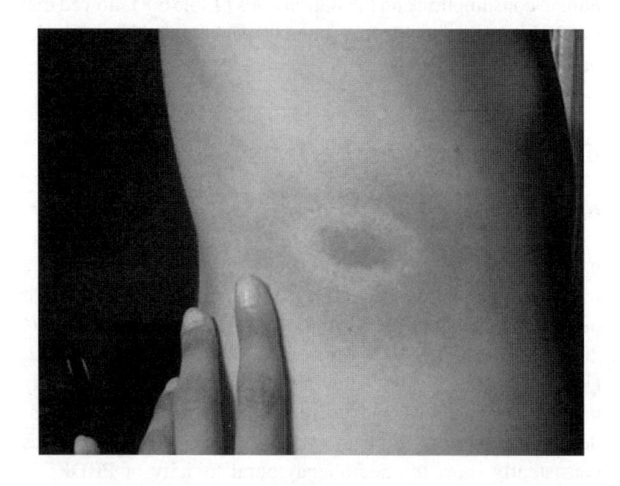

Figure 6.10 An example of the Erythematous rash (Source: Wikimedia Commons, 2007)

6.5.2.2.2 Chronic exposure Animal carcinogenicity data support the evidence that exposure to benzene increases the risk of cancer in multiple organ systems (US EPA, 2010a). A 2-year carcinogenicity study conducted by the National Toxicology Program, carried out on rats and mice of both sexes, used almost pure benzene. Doses ranging from 0 to 200 mg/kg body weight were force-fed to the rats. After 2 years both the male and female rats and mice had clear indications of carcinogenicity. Exposure resulted in increased incidences of Zymbal gland carcinomas and carcinomas of the oral cavity, as well as increased incidences of skin carcinomas and bronchiolar carcinomas. Malignant tumours of the ovary cells and mammary glands in female mice were also reported (National Toxicology Program, 1986).

Benzene has also been seen to exhibit mutagenic effects through studies of occupational exposure. Workers exposed to a concentration of benzene below 31 ppm for 10 to 26 years had significantly more chromosome breaks and gaps in peripheral lymphocytes than were found in the controls, which led to the conclusion that exposure resulted in mutagenicity (Sasiadek *et al.*, 1989).

Benzene toxicity causes bone-marrow depression and leukaemogenesis, which is caused by damage to the haematopoietic cells and a variety of their functions (Snyder and Hedli, 1996). Epidemiologic studies have revealed that the relative risk of contracting leukaemia after being exposed to benzene is 6.97 times the risk in an unexposed group (Yin *et al.*, 1987). In one study, a group of 784 workers who produced rubber hydrochloride and were exposed to benzene concentrations of 10 to 100 ppm for up to 9 years had a relative risk of 10 for myelogenous and acute monocytic leukaemia (Infante *et al.*, 1977). In another study, 680 workers who were exposed to benzene concentrations exceeding 6.88 mg/m^3 for 30 years had a relative risk of 3.93 for leukaemia and other lymphopoietic cancers (Wong, 1987).

6.5.2.3 Polyaromatic hydrocarbons (PAHs)

6.5.2.3.1 Acute effects The EPA has reported that PAHs could lead to red-blood-cell damage and anaemia (US EPA, 2010b). Benzo[a]pyrene (BaP), in particular, is a known immunomodulator (Booker and White, 2005) and has been linked to autoimmune diseases including rheumatoid arthritis. An experiment conducted on mice exposed to BaP dermally, resulted in the mice developing anaemia (Booker and White, 2005).

6.5.2.3.2 Chronic exposure Chronic exposure to PAHs has been linked to cancer and reproductive defects. The IARC has classified the following 15 PAHs as being *reasonably anticipated to be human carcinogens* based on sufficient evidence of carcinogenicity in experimental animals (IARC, 1973, 1983, 1987):

- benz[*a*]anthracene (56–55–3);
- benzo[*b*]fluoranthene (205–99–2);
- benzo[*j*]fluoranthene (205–82–3);
- benzo[*k*]fluoranthene (207–08–9);
- benzo[*a*]pyrene (50–32–8);
- dibenz[*a,h*]acridine (226–36–8);
- dibenz[*a,j*]acridine (224–42–0);
- dibenz[*a,h*]anthracene (53–70–3);
- 7*H*-dibenzo[*c,g*]carbazole (194–59–2);
- dibenzo[*a,e*]pyrene (192–65–4);
- dibenzo[*a,h*]pyrene (189–64–0);
- dibenzo[*a,i*]pyrene (189–55–9);
- dibenzo[*a,l*]pyrene (191–30–0);
- indeno[1,2,3-*cd*]pyrene (193–39–5);
- 5-methylchrysene (3697–24–3).

Evidence for their carcinogenic effects includes the development of papilomas in the fore-stomach of mice, as well as lung adenomas and hepatomas, when benz[*a*]anthracene was administered by gavage. Skin papilomas were also induced when the benz[*a*]anthracene was applied topically to mice (IARC, 1973). Similar tumours were observed when mice were exposed to benzo[*a*]pyrene. An increased incidence of tumours was observed in the forestomach, skin and trachea, when benzo[*a*] pyrene was administered by gavage, topically and by inhalation respectively (IARC, 1973). The skin carcinomas and papillomas were visible in mice as well as rats, guinea pigs and rabbits. The PAHs deemed to be carcinogenic have shown similar effects as those discussed above. There is, however, inadequate evidence for carcinogenicity in humans (IARC, 1973, 1983), although mortality studies have shown that exposure to PAHs from coke-oven emissions has led to increased incidences of lung and genito-urinary cancers in oven workers (Redmond *et al.*, 1972). Exposure to other chemical mixtures that contain PAHs, such as cigarette smoke, coal tar and bitumens, have also been reported to increase the incidence of lung cancer in humans (IARC, 1985).

Benzo(a)pyrene has been reported to cause developmental problems in laboratory animal fetuses. Other PAHs have not been investigated adequately (Wisconsin Department of Health Services, 2000). Some polyaromatic hydrocarbons become carcinogenic only after undergoing metabolism.

Lutz *et al.* (1978) reported studies in mice that showed that benzo(a)pyrene is a skin carcinogen, but not a liver carcinogen because the enzyme glutathione-*S*-transferase inhibits the binding of diol expoxides to DNA (Hesse *et al.*, 1982), and more of this enzyme is found in rat liver than in skin and mammary glands.

6.5.2.4 Trichloroethylene (TCE)

6.5.2.4.1 Acute effects The acute effects of TCE are well established and are related primarily to its suppression of the central nervous system. In fact, it was once widely used as an anaesthetic. Other acute effects include dizziness, drowsiness, nausea and headaches (Chiu *et al.*, 2006).

Human exposure to 27 mg/kg TCE for 1–4 hours causes fatigue and drowsiness, together with mucous membrane irritation (Nomiyama and Nomiyama, 1977). Exposure to 110 mg/l for two 4-hour periods decreased performance in perception, memory, reaction time and dexterity tests (Salvini *et al.*, 1971).

Such acute effects on the nervous system typically resolve within a few hours, although in one case long-term residual oculomotor and ciliary reflex dysfunction occurred, as well as impaired neuropsychological performance (Feldman *et al.*, 2007).

The inhalation of TCE, along with the ingestion of ethanol, results in a syndrome called degreaser's flush (Figure 6.11), which typically causes erythema (vasodilatation of blood vessels) within 30 minutes of exposure (Stewart *et al.*, 1974). This normally subsides within one hour.

6.5.2.4.2 Chronic effects The USA's National Toxicology Programme states that TCE is 'reasonably anticipated to be a human carcinogen', based on a meta-analysis of different types of studies (Wartenberg *et al.*, 2000). Statistically higher risks of liver and kidney cancer were found in individuals with increased TCE exposure across all study types except community studies (Wartenberg *et al.*, 2000). These studies also support an association between TCE and non-Hodgkin's lymphoma, cervical cancer, Hodgkin's disease and multiple myeloma. Since 2000, there have been more studies of the link between TCE and cancer using more rigorous exposure assessments (Scott and Chiu, 2006; Wartenberg *et al.*, 2000; Diot *et al.*, 2002; Garabrant *et al.*, 2003; Nietert *et al.*, 1998). Two large-cohort studies (Raaschou-Nielsen *et al.*, 2003; Zhao *et al.*, 2005) and three case-control studies (Brüning *et al.*, 2003; Pesch *et al.*, 2000; Charbotel *et al.*, 2006) found statistical evidence of a link between TCE exposure and kidney cancer. Most additional cohort studies also supported a link between TCE and liver and biliary cancers, either as combined or separate categories (Hansen, 2004; Raaschou-Nielsen *et al.*, 2003; Lee *et al.*, 2003; Morgan and Cassady, 2002). Additionally, statistically significant associations were found between non-Hodgkin's lymphoma and increased exposure to TCE (Hansen, 2004; Raaschou-Nielsen *et al.*, 2003). Associations between TCE exposure and cancers of the bladder, breast and oesophagus were also highlighted in the review by Scott and Chiu (2006).

Other health concerns associated with TCE exposure relate to the endocrine and reproductive systems and fetal development (Chiu *et al.*, 2006).

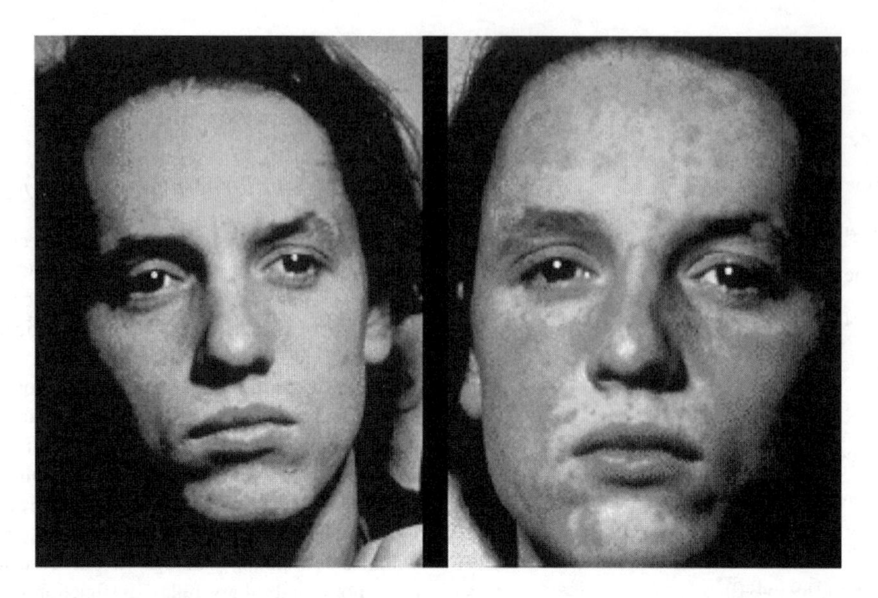

Figure 6.11 An example of degreaser's flush from The National Institute for Occupational Safety and Health (NIOSH) (www.cdc.gov/niosh/topics/skin/occderm-slides/ocderm18.html)

Several experimental studies on rats, chicks and rabbits suggest that prenatal exposure to TCE decreases the viability of fetuses and causes birth defects (Healy *et al.*, 1982; Schwetz *et al.*, 1975; Elovaara *et al.*, 1979; Johnson *et al.*, 2003).

Evidence of developmental problems in humans associated with TCE comes mainly from studies of communities exposed to contaminated drinking-water supplies. Goldberg *et al.* (1990) showed that the distribution of patients with congenital heart disease in Arizona, USA correlated closely with an area supplied with TCE-contaminated water. Moreover, once the contaminated wells were closed in 1981 the incidence of heart disease fell dramatically. TCE in the concentration range 50–250 mg/kg in chicks affects several elements of epithelial-mesenchymal cell transformation, which is essential for the development of the septa and valves of the human heart (Boyer *et al.*, 2000).

Other developmental effects observed in communities exposed to TCE-contaminated drinking water include oral clefts, neural-tube defects (NTDs), eye defects, choanal atresia (blockage of the nasal passage by abnormal bone or soft-tissue development), hypospadias (abnormal opening of the urethra in males) and small size for gestation age (Bove *et al.*, 1995, 2002; Massachusetts Department of Public Health, 1996).

Both animal and human studies have reported altered autoimmune responses from TCE exposure (Griffin *et al.*, 2000; Cai *et al.*, 2007). For example, a study of metal degreasers, with high TCE exposure, found different Interleukin-4 (IL-4) and IFN-y interferon levels when compared to non-exposed office workers at the same factory (Iavicoli *et al.*, 2005). IL-4 is a cytokine that is responsible for the differentiation of naïve T helper cells into type two T helper cells, which are essential for a humoral immune response.

It is now widely accepted that industrial exposure to TCE or its metabolites can cause serious skin disorders (Kamijima *et al.*, 2007). Several human studies have documented skin disorders related to industrial TCE exposure (Dai *et al.*, 2004; Xia *et al.*, 2004; Phoon *et al.*, 1984). These are often known as TCE-related generalised skin disorders and include symptoms similar to hypersensitivity syndrome and Stevens-Jonson syndrome and toxic epidermal necrolysis (Figure 6.12). Both Stevens-Jonson syndrome and toxic epidermal necrolysis cause the separation of the epidermis from the dermis (Kamijima *et al.*, 2007).

A review of TCE-related generalised skin disorders highlighted the increased occurrence in Asian countries since the 1990s, which has been attributed to rapid industrialisation (Hisanaga *et al.*, 2002).

6.5.3 Organs of accumulation

6.5.3.1 Brominated flame retardants (BFRs)

Concentrations of PBDEs in trout from Lake Ontario have increased from 0.54 ng/g in 1978 to 190 ng/g wet weight in whole fish samples in 2002 (Luross *et al.*, 2002). Dusts contaminated with PBDEs can be inhaled and have been found to accumulate in tissues of various marine animals. Hexabromocyclodecane (HBCD, related to PBDEs) has been detected in liver and blubber samples from harbour seals and harbour porpoises from the North Sea (Law *et al.*, 2006).

Human studies have documented the ability of PBDEs to disrupt the normal functioning of thyroid hormone (Patrick, 2009; Zoeller, 2005). Kuriyama *et al.* (2005) reported that developmental exposure to low-dose PBDE-99 not only caused persistent neurobehavioural effects but also permanently affected adult male reproductive health. Johnson-Restrepo *et al.* (2005) conducted research on the impact of PBDEs in human adipose tissue in individuals in New York. Fifty-two samples of

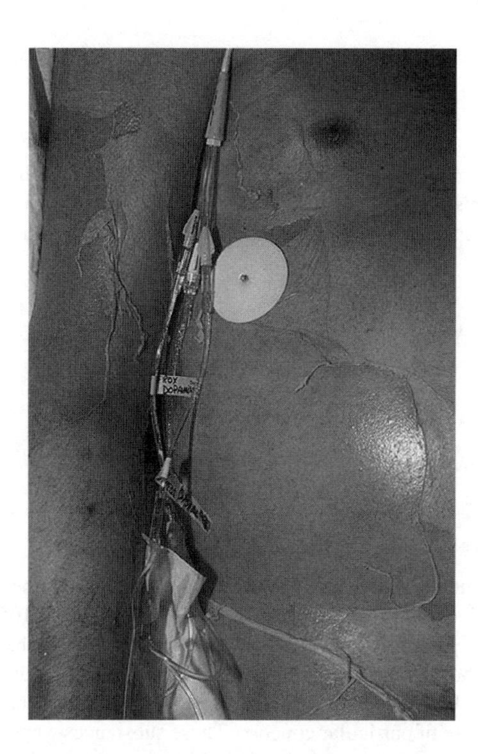

Figure 6.12 An example of toxic epidermal-necrolysis (Source: Wikimedia Commons 2011)

human adipose tissue were collected and analysed between 2003 and 2004. Average levels were 10 to 100 times greater than those reported in Europe, with no difference between genders. The levels of PBDEs in US mothers' milk were also the highest in the world, 10–100 times higher than European levels (Schecter *et al.*, 2003).

These findings point to the potential health risks PBDEs pose for nursing infants. Schecter *et al.* (2007) recently completed the first peer-reviewed study to examine human tissue prior to and immediately after birth. Liver tissues were obtained from four stillborn fetuses and seven live-born infants (gestational age ranging from 20.5 to 39 weeks). Only two lived longer than 4 hours and none were fed formula or nursed, so tissue levels reflect uterine PBDE intake only. All samples were contaminated with between 4 and 98 ng/g lipid, indicating that PBDEs are transferred from maternal to fetal tissue (Schecter *et al.*, 2007).

6.5.3.2 Benzene

Benzene accumulates in the eyes, blood, female reproductive system and bone marrow (CCOHS, 1997).

Benzene is easily absorbed into the body by inhalation and ingestion. After this it is swiftly dispersed around the body, particularly in the fatty tissues. Metabolism occurs primarily in the liver and to a lesser extent in bone marrow (CCOHS, 1997). High concentrations of metabolites of benzene have been found in the bone marrow of mice given large doses of benzene. The occurrence of the metabolites in the bone marrow correlated with their disappearance from the liver, providing evidence that bone marrow is a target for the metabolites of benzene (Andrews *et al.*, 1977).

The half-life of benzene in the body is 1 to 2 days; most of it is exhaled through the lungs unchanged or excreted as metabolites in urine (CCOHS, 1997).

Evidence of benzene's teratogenicity was obtained from studies of 14 children of women working in chemical laboratories and the printing industry. The women were exposed to a mixture of benzene and other solvents during pregnancy and this resulted in the children having an increased frequency of chromatic and isochromatid breaks (Funes-Cravioto *et al.*, 1977), resulting from the benzene crossing the placenta and concentrating in the cord blood in quantities equal to or greater than the maternal blood (Dowty *et al.*, 1976).

6.5.3.3 Polyaromatic hydrocarbons (PAHs)

The distribution patterns and transport mechanisms differ between PAHs, reflecting their different lipophilic and hydrophilic properties. For example, benzo(a)pyrene in the body is absorbed via intestinal lymphatic drainage and transported to the vascular circulation where it is sequestered with lipoproteins in thoracic-duct lymph (ATSDR, 2009).

Most PAHs are metabolised by cytochrome P450-dependent oxidation. The group of enzymes that facilitate this reaction come from the large family of enzymes known as cytochrome P450 enzymes, all of which contain an active site consisting of a heme iron protein (ATSDR, 2009). The hepatic enzyme CYP1A is thought to be responsible for most PAH metabolism (Anzenbacher and Anzenbacherová, 2001). In addition to the liver, PAHs may also be stored and metabolised in the kidneys and adipose tissue and to some extent in the adrenal glands, testes, thyroid, lungs, skin, sebaceous glands and small intestine (ATSDR, 2009).

Metabolism results initially in the formation of epoxides, which are subsequently converted to dihydrodiol derivatives and phenols. Conjugates or additional metabolites of these chemicals are excreted in the bile and urine. Urinary excretion of PAHs is thought to be relatively efficient, due to the widespread distribution of metabolising enzymes in the body (ATSDR, 2009).

6.5.3.4 Trichloroethylene (TCE)

Trichloroethylene is readily absorbed across biological membranes, due to its volatility and lipophilicity. After inhalation it is absorbed rapidly through the alveolar endothelium in the lungs as a result of its high blood/gas partition coefficient. Gastrointestinal absorption is also efficient, via passive diffusion (Lash *et al.*, 2000).

Studies of the distribution of TCE in the body demonstrate that TCE has an adipose tissue: blood partition coefficient

Table 6.9 Environmental quality standards (EQS) for the priority substances listed in WFD (European Parliament and Council, 2008)

Name of substance	Annual average EQS for inland waters (μg/l)	Annual average EQS for other surface waters (μg/l)	Maximum allowable concentration EQS for inland waters (μg/l)	Maximum allowable concentration EQS for other surface waters (μg/l)
Benzene	10	8	50	50
Benzo(a)pyrene	0.05	0.05	0.1	0.1
Benzo(b)fluor-anthene Benzo(k)fluor-anthene	$\Sigma = 0.03$	0.03	–	–
Benzo(g,h,i)-perylene Indeno(1,2,3-cd)-pyrene	$\Sigma = 0.02$	0.02	–	–
Trichloro-ethylene	10	10	–	–
Brominated diphenylether	0.0005	0.0002	–	–

ranging from 25 to 45 and a half-life estimated at 3.5–5 hours, higher than other tissues, where TCE has a half-life of 2–4 minutes (Lash *et al.*, 2000). Due to this, organs high in adipose tissue will be likely targets for TCE accumulation. For example, exposure of rats to TCE results in the accumulation of TCE in the fat of brains for 17 hours after exposure; this has been found to affect motor and emotional behaviour of exposed rats (Savolainen, Pfäffli *et al.*, 1977).

Partition coefficients for richly perfused tissues are also high, approximately 1.2 in male rats (Lash *et al.*, 2000). Therefore, tissues such as the liver, kidneys and lungs are likely to accumulate TCE. The muscle/blood partition coefficient is only 0.5 in adult rats, suggesting that muscle tissue is neither a major storage site nor a site of metabolism for TCE.

6.6 Risk reduction and future trends

6.6.1 Risk reduction

An estimated one thousand new chemicals enter the market every year and about 100 000 chemical substances are used in industry, but only limited data on their chronic toxicity to humans are available (McKone and Knezovich, 1991; RCEP, 2003). Each chemical has its own distinctive properties and hazards, and the whole life cycle of each chemical should be considered when assessing its risks and benefits.

Within the European Union, the REACH regulation (2006/1907/EC) is concerned with the Registration, Evaluation, Authorisation and Restriction of Chemical substances. It aims to improve the protection of human health and the environment through the better and earlier identification of the intrinsic properties of chemical substances. It came into force in June 2007.

Additionally, the Water Framework Directive (WFD) (Directive 2000/60/EC) has aimed to protect the environment from chemical damage. The Daughter Directive of the WFD, Directive on Priority Substances (Directive 2008/105/EC), lists 33 priority substances, or group of substances, that have been shown to be of major concern for European Waters. Of these, 11 have been identified as priority hazardous substances which are of particular concern. These substances were prioritised based on their individual bioaccumulation potential, toxicity, persistence and risk to human health, and in relation to the monitored and modelled concentration of each substance in the aquatic environment. The regulations requires that all member states of the European Union progressively reduce pollution from priority substances and cease or phase out emissions, discharges and losses of priority hazardous substances. Benzene is listed as a priority substance, and brominated diphenylethers and PAHs as priority hazardous substances. The regulation has introduced legally binding environmental quality standards for these substances in surface waters (Table 6.9). Trichloroethylene (TCE) has not been identified as a priority substance, but as it was included in the Directive 86/280/EEC (which will be repealed in 2012), which limits discharges and sets quality objectives for certain dangerous substances, environmental quality standards for it were included in the WFD in order to maintain its regulation throughout Europe (Table 6.9).

In the UK the use of chemicals in the workplace is regulated by Control of Substances Hazardous to Health (COSHH) legislation. This requires employers to control substances that are hazardous to health by conducting risk assessments, implementing control measures and providing relevant training to employees (HSE, 2010). The Control of Substances Hazardous to Health (COSHH) legislation also imposes workplace exposure limits for chemicals that could harm health (HSE, 2010).

The impact of legislation such as REACH, WFD and COSHH on the reduction of risk of chemicals will be discussed in this section, along with other strategies that have helped to reduce risk.

6.6.1.1 Brominated flame retardants (BFRs)

The potential of these compounds to be PBT and to break down into more harmful congeners in the environment (Knoth *et al.*, 2007; Morf *et al.*, 2005; Webster *et al.*, 2008) means that there are several national and international risk-reduction strategies.

Production of BFRs has been greatly reduced due to restrictions on pentaBDE and octaBDE in Europe and several US states (Song *et al.*, 2006; Leung, 2007; Liang *et al.*, 2008; Morf *et al.*, 2005; Xian *et al.*, 2008). PBDE congeners were included in the 1998 list of chemicals for priority action by the Oslo and Paris Convention (OSPAR) and they were also included as priority hazardous substances under the EC Water Framework Directive (Mercado-Feliciano and Bigsby, 2008) and were listed as persistent organic pollutants (POPs) under the Stockholm Convention (Leung, 2007). The environmental levels of the pentaBDE congeners are expected to persist for decades, due to their large inventory and release from the technosphere, soils and sediments (Song *et al.*, 2006).

Within Europe, the 'Restriction of the use of certain Hazardous Substances in electrical and electronic equipment directive' (RoHs) prevents the sale of electrical and electronic equipment with more that 0.1 per cent by weight of polybrominated biphenyl (PBBs) and polybrominated diphenyl ether (PBDEs) flame retardants.

Under the REACH legislation (2006/1907/EC), producers and importers of BFRs must register BFRs with the European Chemicals Agency (ECHA). The commonest commercial BFRs (Deca-BDE, TBBPA and HBCD) are currently in the pre-registration phases under REACH.

Deca-BDE has already been subject to prior risk assessments, which concluded that no restrictions were necessary. However, concerns remain regarding its persistence and toxicity in the environment and uncertainty with respect to human health. For these reasons, Sweden and Norway have both banned Deca-BDE and called for an EU-wide ban (Kemmlein *et al.*, 2009).

A risk assessment for TBBPA concluded that it posed no risk to human health but acknowledged the risk to the aquatic environment. As a result, the EU approved a Risk Reduction Strategy which addresses this risk through the reduction of emissions from industrial sites, and therefore restrictions to the marketing and use of this chemical do not need to be applied (Kemmlein *et al.*, 2009).

Risk assessments have found that HBCD has PBT properties (Kemmlein *et al.*, 2009), so a few studies are currently in progress to assess the need to regulate it more tightly.

6.6.1.2 Benzene

Restrictions and concentration limits for benzene (in Annex 17 (XVII) of the REACH regulation), do not permit benzene levels of more than 5 mg/kg in toys or part of toys. Moreover, with a few exceptions, products on the market must not contain more than 0.1 per cent by mass of benzene (2006/1907/EC).

In order to reduce the risk of exposure of benzene to humans and the environment, international and national regulations have set guideline exposure limits. IARC has classified benzene as a category 1 carcinogen, indicating that there is evidence of carcinogenicity in humans (IARC, 1987). This classification communicates the hazardous nature to users of benzene. In terms of occupational exposure, the Occupational Safety and Health Administration (OSHA), which is the main federal agency that deals with the enforcement of health and safety legislation in the United States, has set a Permissible Exposure Limit (PEL) (for an 8-hour time-weighted average, TWA) to 3.44 mg/m^3 for general industries, the construction industry as well as the shipyard industry (OSHA, 2005). This regulation stipulates that respirators must be used during the time that is needed to implement the necessary engineering controls, as well as in the cases that work practice controls or engineering are not sufficient to comply with the PEL (OSHA, 2005). This ensures that exposure to the workers via inhalation is reduced. In the UK, the Health and Safety Executive (HSE) also set the EH40 Occupational Exposure Limit (OEL) to 3.44mg/m^3 for benzene (HSE, 2005).

The World Health Organisation (WHO) has set a guideline limit of 0.01 mg/l for drinking-water supplies (WHO, 2004). The EPA has set a maximum contaminant level of 0.005 mg/l for benzene in drinking water (US EPA, 2002), which is the same stringent value set by the US Food and Drug Administration (FDA, 2004). The European Union has set an even more conservative limit for benzene in drinking water, of 0.001 mg/l under Council Directive 98/83/EC on the quality of water intended for human consumption (European Union, 1998).

In order to control the release of benzene to the environment, the Superfund Amendments and Reauthorisation Act (SARA) of 1986, requires that owners and operators of certain facilities that manufacture, import, process or use benzene should report any releases of benzene to environmental media, over the legal threshold (US Congress, 1986).

6.6.1.3 Polyaromatic hydrocarbons (PAHs)

In order to reduce the risk of exposure of PAHs to humans and the environment, international and national regulators have set guideline exposure limits. The Occupational Safety and Health Administration (OSHA) regulates the benzene-soluble fraction of the coal-tar pitch volatiles and mineral-oil mist, which contains several PAHs. The permissible exposure limit (PEL) (with an 8-hour time-weighted average) is 0.2 mg/m^3 for coal-tar pitch and 5 mg/m^3 for mineral-oil mist (OSHA, 2005). This indicates that the more hazardous PAHs are found in the coal-tar pitch volatiles.

PAHs are also regulated by the Clean Water Effluent Guidelines, in the Code of Federal Regulations (US EPA, 1981). The PAHs are regulated as a group of chemicals controlled as Total Toxic Organics. The point-source categories for which the PAHs have specific regulatory limitations include organic chemicals, plastics and synthetic fibres, coke-making and the manufacture of non-ferrous metals (ATSDR, 1995b).

PAHs are also regulated as hazardous waste, under the Resource Conservation and Recovery Act, when the PAHs are discarded as commercial chemical products, off-speciation species, container residues as well as spill residues (US EPA, 1980). The European Commission has set maximum levels for certain contaminants in foodstuffs, including PAHs. The Scientific Committee on Food (SCF) suggested that benzo(a)pyrene can be used for a marker of occurrence and effect for carcinogenic PAHs in food, based on examinations of the PAH profiles in food and on the evaluation of carcinogenicity studies of coal tar in mice (European Commission, 2002). The SCF advises that the estimated maximum daily intake of benzo[a]pyrenefrom food is approximately 420 ng benzo[a]pyrene per person, which is equivalent to approximately 6 ng/kg bw/day for a person weighing 70 kg (European Commission, 2002). The parameters for benzo[a]pyrene in food range from the very conservative 1 μg/kg wet weight for baby food and infant formulae to 10 μg/kg wet weight for bivalve molluscs (Commission of the European Communities, 2006).

6.6.1.4 *Trichloroethylene (TCE)*

The discontinuation of the use of TCE in medicine, some consumer products and dry cleaning has helped to reduce human exposure.

Under the REACH regulation (2006/1907/EC), producers and importers of TCE must register TCE with the European Chemicals Agency (ECHA). The ECHA states that where TCE is essential to meet quality and safety standards, it can continue to be used provided that strict control measures are in place to guarantee compliance with the EU Solvent Emission Directive 1999/13/EC (SED). The directive requires that sectors which consume more than 1 tonne of TCE per annum reduce their TCE emissions to 20 mg/m^3 if the flow rate is above 100 g/h.

Users consuming less than 1 tonne of TCE per year fall outside the SED legislation. Hence in May 2008 recommendations on risk-reduction measures for TCE, along with benzene and 2-methoxy-2-methylbutane (TAME) were published in the Official Journal of the European Union. Section 1 of these recommendations is concerned with the risk reduction for workers exposed to TCE and recommended that an agreement be made with the European Chlorinated Solvents Association to restrict the sale of TCE to purchasers who comply with the charter for the safe use of trichloroethylene in metal cleaning. This requires that there is no direct connection between any volume containing halogenated solvents such as TCE and the outside environment during normal operation. Other methods to reduce the risk posed by TCE include remediating contaminated land, protecting and monitoring groundwater resources and improved advice to users of consumer products containing TCE.

Within the UK, TCE is also covered by COSHH workplace exposure limits. Exposure over 8 hours should not exceed 100 ppm and not exceed 115 ppm over 15 minutes (Hansen, 2004).

6.6.2 *Future trends and further reading*

This chapter has demonstrated the concerns surrounding the unintentional consequences of the production and use of selected industrial chemicals on humans and the environment. It has also highlighted the diverse sources and pathways of industrial chemicals, along with the time and weight of evidence required in the past to prove whether chemicals are dangerous to human health. The source–pathway–target relationships are essential in understanding the interactions between the physiochemical properties of these chemicals and the potential impacts to assess potential risks. With an estimated 1000 new chemicals entering the market yearly, there is increasing need for frameworks to assess the risks associated with manufacturing, use and disposal of these chemicals. The implementation of the EU REACH legislation is expected to be the main driver for this chemical regulation within Europe. This has the objective of providing a high level of protection to human health and the environment as well as increasing producer responsibilities for ensuring that chemicals that are placed on the market are safe to use. However, REACH will not be fully implemented until 2018, so the full extent of its impacts cannot be fairly assessed to date. Constant vigilance is therefore essential to prevent the replacement of old, known hazards with new, unknown hazards and to ensure the sustainable management of new chemicals throughout their development and use. This will be required especially as REACH only covers chemicals manufactured or imported within the EU, whilst finished products manufactured outside the EU remain unregulated. This highlights the need for international legislative frameworks and holistic management of the production and use of industrial chemicals.

References

Abrahamsson, K., Ekdahl, A., Collen, J. and Pedersen, M. (1995) Marine algae – a source of trichloroethylene and perchloroethylene. *Limnology and Oceanography*, **40**(7), 1321.

Adami, H.O., Chow, W.H., Nyren, O., Berne, C., Linet, M.S., Ekbom, A., Wolk, A., McLaughlin, J.K. and Fraumeni J. F. (1996) Excess risk of primary liver cancer in patients with diabetes mellitus. *Journal of the National Cancer Institute*. **88**(20), 1472.

AEA Group (2009) Annual Report for 2007 on the UK PAH Monitoring and Analysis Network. AEA Group. Report number: AEA/R2686/Issue 1.

Aftalion, F. (1991) *A History of the International Chemical Industry*, 2nd Revised edition. Chemical Heritage Foundation, USA.

Almqvist, H. (2006) Organic hazardous substances in graywater from Swedish households. *Journal of Environmental Engineering*, **132** (8),901–908.

Amin, S., Huie, K. and Hecht, S.S. (1985) Mutagenicity and tumor initiating activity of methylated benzo[b]fluoranthenes. *Carcinogenesis*, **6**(7), 1023.

Andrews, L. S., Lee, E.W., Witmer, C.M., Kocsis, J.J. and Snyder, R. (1977) Effects of toluene on the metabolism, disposition and hemopoietic toxicity of [3H] benzene. *Biochemical Pharmacology*, **26**(4), 293–300.

Anzenbacher, P. and Anzenbacherova, E. (2001) Cytochromes P450 and metabolism of xenobiotics. *Cellular and Molecular Life Sciences*, **58**(5), 737–747.

Atkinson, R. (2000) Atmospheric chemistry of VOCs and NOx. *Atmospheric Environment*, **34**(12–14), 2063–2101.

ATSDR (1995a) National exposure registry benzene subregistry baseline technical report. Report number: PB95255766Atlanta, Georgia, U.S. Department of Health and Human Services. Agency for Toxic Substances and Disease Registry.

ATSDR (1995b) Toxicological profile for polycyclic aromatic hydrocarbons. Atlanta, Georgia, U.S. Department of Health and Human Services. Agency for Toxic Substances and Disease Registry.

ATSDR (1995c) *Toxicological profile for trichloroethylene*. Update Draft for Public Comments. Atlanta, GA, Agency for Toxic Substances and Disease Registry,.

ATSDR (2007) *Benzene*. Atlanta, Georgia, U.S. Department of Health and Human Services. Agency for Toxic Substances and Disease Registry.

ATSDR (2009) *Case Studies in Environmental Medicine Toxicity of Polycyclic Aromatic Hydrocarbons (PAHs)*. U.S. Department of Health and Human Services, Agency for Toxic Substances and Disease Registry. Atlanta, Georgia.

Balakrishnan, I. and Reddy, M. P. (1970) Mechanism of reaction of hydroxyl radicals with benzene in the. gamma. Radiolysis of the aerated aqueous benzene system. *Journal of Physical Chemistry*, **74** (4),850–855.

Bawden, J. (2004) Emerging Contaminants: Identifying chemical compounds of concern and addressing adverse effects. School of Natural Resources, University of Arizona, p 22.

Betts, K. (2004) Another brominated flame retardant in the environment. *Environmental Science and Technology*, **38**(12), 214A–215A.

Bolt, H.M. (1981) Biologische Arbeitsstoff-Toleranzwerte: BAT-Wert für Trichloräthylen. *Arbeitsmed.Sozialmed. Präventivmed*, **16**, 116–121.

Booker, C.D. and White, K.L. (2005) Benzo(a)pyrene-induced anemia and splenomegaly in NZB/WF1 mice. *Food and Chemical Toxicology*, **43**(9), 1423–1431.

Bourg, A., Mouvet, C. and Lerner, D.N. (1992) A review of the attenuation of trichloroethylene in soils and aquifers. *Quarterly Journal of Engineering Geology and Hydrogeology*, **25**(4), 359.

Bove, F., Shim, Y. and Zeitz, P. (2002) Drinking water contaminants and adverse pregnancy outcomes: a review. *Environmental Health Perspectives*, **110**(1), 61.

Bove, F.J., Fulcomer, M.C., Klotz, J.B., Esmart, J., Dufficy, E.M. and Savrin, J.E. (1995) Public drinking water contamination and birth outcomes. *American Journal of Epidemiology*, **141**(9), 850.

Boyer, A., Finch, W. and Runyan, R. (2000) Trichloroethylene inhibits development of embryonic heart valve precursors in vitro. *Toxicological Sciences*, **53**, 109–117.

Branchi, I., Capone, F., Alleva, E. and Costa, L.G. (2003) Polybrominated diphenyl ethers: Neurobehavioral effects following developmental exposure. *Neurotoxicology*, **24**(3), 449–462.

Branchi, I., Alleva, E. and Costa, L.G. (2002) Effects of perinatal exposure to a polybrominated diphenyl ether (PBDE 99) on mouse neurobehavioural development. *Neurotoxicology*, **23**(3), 375–384.

Brief, R. S., Lyncha, J., Bernath, T. and Scala, R. A. (1980) Benzene in the workplace. *American Industrial Hygiene Association Journal*, **41**(9), 616–623.

Brouwer, A., Morse, D.C., Lans, M.C., Schuur, A.G., Murk, A.J., Klasson-Wehler, E., Bergman, A. and Visser, T.J. (1998) Interactions of persistent environmental organohalogens with the thyroid hormone system: Mechanisms and possible consequences for animal and human health. *Toxicology and Industrial Health*, **14**(1–2), 59–84.

Brown, H. S, Bishop, D.R. and Rowan, C.A. (1984) The role of skin absorption as a route of exposure for volatile organic compounds (VOCs) in drinking water. *American Journal of Public Health*, **74** (5),479.

Bruckmann, P. and Kersten, W. (1988) The occurrence of chlorinated and other organic trace compounds in urban air. *Chemosphere*, **17** (12),2363–2380.

Brugnone, F., Perbellini, L., Giuliari, C., Cerpelloni, M. and Soave, M. (1994) Blood and urine concentrations of chemical pollutants in the general population. *La Medicina del lavoro*, **85**(5), 370.

Brüning, T. and Bolt, H.M. (2000) Renal toxicity and carcinogenicity of trichloroethylene: key results, mechanisms, and controversies. *CRC Critical Reviews in Toxicology*, **30**(3), 253–285.

Brüning, T., Pesch, B., Wiesenhutter, B., Rabstein, S., Lammert, M., Baumüller, A. and Bolt, H.M. (2003) Renal cell cancer risk and occupational exposure to trichloroethylene: results of a consecutive case-control study in Arnsberg, Germany. *American Journal of Industrial Medicine*, **43**(3), 274–285.

Brunnemann, K.D., Kagan, M.R., Cox, J.E. and Hoffmann, D. (1990) Analysis of 1, 3-butadiene and other selected gas-phase components in cigarette mainstream and sidestream smoke by gas chromatography–mass selective detection. *Carcinogenesis*, **11** (10),1863.

Bull, R.J. (2004) Trichloroethylene and liver tumors in mice. In: Proceedings of New Scientific Research Related to the Health Effects of Trichlorothylene, 26-27 February 2004, Washington DC. Washington, DC:U.S. Environmental Protection Agency.

Bull, R. J, Sasser, L.B. and Lei, X.C. (2004) Interactions in the tumor-promoting activity of carbon tetrachloride, trichloroacetate, and dichloroacetate in the liver of male B6C3F1 mice. *Toxicology*, **199**, 169–183.

Buxton, G.V., Greenstock, C.L., Helman, W.P. and Ross, A.B. (1988) Critical review of rate constants for reactions of hydrated electrons, hydrogen atoms and hydroxyl radicals. *Physics and Chemistry Reference Data*, **17**, 513–886.

Cai, P., König, R., Khan, M.F., Kaphalia, B.S. and Ansari, G.A.S. (2007) Differential immune responses to albumin adducts of reactive intermediates of trichloroethene in MRL + / + mice. *Toxicology and Applied Pharmacology*, **220**(3), 278.

CCOHS (Canadian Centre for Occupational Health and Safety) (1997) *Health Effects of Benzen*. Available from: http://www.ccohs.ca/oshanswers/chemicals/chem_profiles/benzene/health_ben.html. Accessed 9 April 2010.

Chandler, A.D. (2005) Shaping the industrial century: the remarkable story of the modern chemical and pharmaceutical industries. President and Fellows of Harvard College, Harvard studies in business history, USA.

Chang, Y.M., Tai, C.F., Yang, S.C., Lin, R.S., Sung, F.C., Shih, T.S. and Liou, S.H. (2005) Cancer incidence among work ers potentially exposed to chlorinated solvents in an electronics factory. *Journal of Occupational Health*, **47**(2), 171.

Charbotel, B., Fevotte, J., Hours, M., Martin, J.L., Bergeret, A. (2006) Case–control study on renal cell cancer and occupational exposure to trichloroethylene. *Part II: Epidemiological aspects. Annals of Occupational Hygiene*, **50**(8), 777–787.

Cheremisinoff, P.N. and Morresi, A.C. (1979) *Benzene: basic and hazardous properties*. M. Dekker, New York.

Chiu, W.A., Caldwell, J.C., Keshava, N. and Siegel Scott, C. (2006) Key scientific issues in the health risk assessment of trichloroethylene. *Environmental Health Perspectives*, **114**, 9.

Commission of the European Communities (2006) Commission Regulation (EC) No 1881/2006 of 19 December 2006 setting maximum levels for certain contaminants in foodstuffs. *Official Journal of the European Union* (1881).

CONCAWE (1999) Environmental exposure to benzene. Report number: 2/99. Concawe, Brussels.

Consonni, D., Pesatori, A.C., Zocchetti, C., Sindaco, R., D'Oro, L.C., Rubagotti, M. and Bertazzi, P.A., (2008) Mortality in a population exposed to dioxin after the Seveso, Italy, accident in 1976: 25 years of follow-up. *American Journal of Epidemiology* **167**(7), 847–858.

Corden, C., Lawton, K., Stavrakaki, A. and Warwick, O. (2005) Risk reduction strategy and analysis of advantages and drawbacks for trichloroethylene, a report for Defra. Report number 060203. Available from http://www.defra.gov.uk/environment/quality/chemicals/documents/report060203.pdf. Accessed 5 July 2010.

Dai, Y., Leng, S., Li, L., Niu, Y., Huang, H., Cheng, J. and Zheng, Y. (2004) Genetic polymorphisms of cytokine genes and risk for trichloroethylene-induced severe generalized dermatitis: A case-control study. *Biomarkers*, **9**(6): 470–478.

Darnerud, P.O., Eriksen, G.S., Johannesson, T., Larsen, P.B., Viluksela, M. (2001) Polybrominated diphenyl ethers: occurrence, dietary exposure, and toxicology. *Environmental Health Perspectives* **109** (Suppl 1), 49–68.

De Wit, C. A. (2002) An overview of brominated flame retardants in the environment. *Chemosphere*, **46**, 583–624.

Dhara, V.R. and Dhara, R., (2002) The Union Carbide disaster in Bhopal: A review of health effects. *Archives of Environmental Health*, **57**(5), 391–404.

Ding, Y.S., Trommel, J.S., Yan, X.J., Ashley, D. and Watson, C. H. (2005) Determination of 14 polycyclic aromatic hydrocarbons in mainstream smoke from domestic cigarettes. *Environmental Science and Technology*, **39**(2), 471–478.

Diot, E., Lesire, V., Guilmot, J.L, Metzger, M.D, Pilore, R, Rogier, S., Stadler, M., Diot, P., Lemarie, E. and Lasfargues, G. (2002) Systemic sclerosis and occupational risk factors: a case–control study. *Occupational and Environmental Medicine*, **59**(8), 545.

Doherty, R. (2000). A history of the production and use of carbon tetrachloride, tetrachloroethylene, trichloroethylene and 1 1, 1-trichloroethane in the United States: Part 1 – Historical background; carbon tetrachloride and tetrachloroethylene. *Environmental Forensics*, **1**(2), 69–81.

Dowty, B.J., Laseter, J.L. and Storer, J. (1976) The transplacental migration and accumulation in blood of volatile organic constituents. *Pediatric Research*, **10**(7), 696–701.

ECB (2000) IUCLID dataset for trichloroethylene, European Commission – European Chemicals Bureau.

Ejarrat, E. (2008) Effect of sewage sludges contaminated with polybrominated diphenylethers on agricultural soils. *Chemosphere*, **71**(6),1079–1086.

Elovaara, E., Hemminki, K. and Vainio, H. (1979) Effects of methylene chloride, trichloroethane, trichloroethylene, tetrachloroethylene and toluene on the development of chick embryos. *Toxicology*, **12**(2), 111–119.

Environment Agency (2010) http://www.environment-agency.gov.uk/cy/busnes/pynciau/llygredd/39281.aspx. Accessed 7 April 2010.

Erickson, M.D., Whitaker, D.A., Waddell, R.D., Harris, B.S.H., Hartwell, TD. and Pellizzari, E.D. (1982) Purgeable organic compounds in mother's milk. *Bulletin of Environmental Contamination and Toxicology*, **28**(3), 322–328.

Eriksson, P., Ankarberg, E., Viberg, H. and Fredriksson, A. (2001) The developing cholinergic system as target for environmental toxicants, nicotine and polychlorinated biphenyls (PCBs): implications for neurotoxicological processes in mice. *Neurotoxicity Research*, **3**(1), 37–51.

Eriksson, P., Viberg, H., Jakobsson, E., Orn, U. and Fredriksson, A. (2002) A brominated flame retardant, 2, 2′,4,4′,5-pentabromodiphenyl ether: Uptake, retention, and induction of neurobehavioral alterations in mice during a critical phase of neonatal brain development. *Toxicological Sciences*, **67**(1), 98–103.

Eriksson, P., Fischer, C. and Fredriksson, A. (2006) Polybrominated diphenyl ethers, a group of brominated flame retardants, can interact with polychlorinated biphenyls in enhancing developmental neurobehavioral defects. *Toxicological Sciences*, **94**(2), 302–309.

European Commission (2002) European Commission Health and Consumer Protection Directorate-General. *Opinion of the Scientific Committee on Food on the risks to human health of Polycyclic Aromatic Hydrocarbons in food.* Brussels—Belgium, European Commission. Report number: SCF/CS/CNTM/PAH/29 Final.

European Parliament and Council (2008) environmental quality standards in the field of water policy, amending and subsequently repealing Council Directives 82/176/EEC, 83/513/EEC, 84/156/EEC, 84/491/EEC, 86/280/EEC and amending Directive 2000/60/EC of the European Parliament and of the Council. *Official Journal of the European Community*, L348.

European Union (1998) Council Directive 98/83/EC of 3 November 1998 on the quality of water intended for human consumption. *Official Journal of the European Communities* (OJL330, 5.12.98, p.32).

European Union Joint Research Centre et al. (2001) European Union Risk Assessment Report. Diphenyl ether, pentabromo derivative (pentabromodiphenyl ether). Luxembourg: Office for Official Publications of the European Committees, pp. 1-124.

FDA (Food and Drug Administration) (2004) Beverages. *Bottled water. Report number: Code of Federal Regulations. 21 CFR 165. 110.* Food and Drug Administration, Washington, DC.

Feldman, R., White, R., Currie, J.N., Travers, P.H. and Simmons, L. (2007) Long-term follow-up after single toxic exposure to trichloroethylene. *American Journal of Industrial Medicine*, **8**(2), 119–126.

Fernandes, A.R., Tlustos, C. et al. (2009) Polybrominated diphenylethers (PBDEs) and brominated dioxins (PBDD/Fs) in Irish food of animal origin. *Food Additives and Contaminants Part B – Surveillance*, **2**(1), 86–94.

Franke, C., Studinger, G., Berger, G., Böhling, S., Bruckmann, U., Cohors-Fresenborg, D. and Jöhncke, U. (1994) The assessment of bioaccumulation. *Chemosphere*, **29**(7), 1501–1514.

Funes-Cravioto, F., Kolmodin-Hedman, B., Lindsten, J., Nordenskjöld, M., Zapata-Gayon, C., Lambert, B., Norberg, E., Olin, R. and Swensson, Å. (1977) Chromosome aberrations and sister-chromatid exchange in workers in chemical laboratories and a rotoprinting factory and in children of women laboratory workers. *The Lancet*, **310**(8033), 322–325.

Garabrant, D., Lacey, J., Laing T.J., Gillespie, B.W., Mayes, M.D., Cooper, BC. and Schottenfeld, D. (2003) Scleroderma and solvent exposure among women. *American Journal of Epidemiology*, **157** (6),493.

Gebel, T., Lechtenberg-Auffarth, E. and Guhe, C. (2009) About hazard and risk assessment: Regulatory approaches in assessing safety in the European Union chemicals legislation. 37th Annual Conference of the European-Teratology-Society, Arles France. Reproductive Toxicology 28 (2), 188–195.

Gensler, M., Schwind, K.H. et al. (2009) Polybrominated diphenylether (PBDE) in animal food. *Fleischwirtschaft*, **89**(5), 105–110.

Gevao, B., Muzaini, S. and Helaleh, M. (2008) Occurrence and concentrations of polybrominated diphenyl ethers in sewage sludge from three wastewater treatment plants in Kuwait. *Chemosphere*, **71** (2),242–247.

Global-production.com. (2007). Trends in Global Production and Trade. http://www.global-production.com/chemicals/trendstudy/index.htm. Assessed 12/03/2010.

Gmeiner, G., Stehlik, G. and Tausch, H. (1997) Determination of seventeen polycyclic aromatic hydrocarbons in tobacco smoke condensate. *Journal of Chromatography A*, **767**(1–2), 163–169.

Goldberg, S.J., Lebowitz, M.D., Graver, E.J. and Hicks, S. (1990) An association of human congenital cardiac malformations and drinking water contaminants. *Journal of the American College of Cardiology*, **16**, 155.

Griffin, J.M., Blossom, S.J., Jackson, S.K., Gilbert, K.M. and Pumford, N.R. (2000) Trichloroethylene accelerates an autoimmune response by Th1 T cell activation in MRL +/+. *Immunopharmacology*, **46** (2),123.

Grob, K., Frauenfelder, C. and Artho, A. (1990) Uptake by foods of tetrachloroethylene, trichloroethylene, toluene, and benzene from air. *Zeitschrift für Lebensmittel-Untersuchung und -Forschung. A, Food research and technology*, **191**(6), 435.

Hale, R.C., La Guardia, M.J., Harvey, E. and Mainor, T.M. (2002) Potential role of fire retardant-treated polyurethane foam as a source of brominated diphenyl ethers to the US environment. *Chemosphere*, **46**(5), 729–735.

Hallgren, S., Sinjari, T., Hakansson, H. and Darnerud, P.O. (2001) Effects of polybrominated diphenyl ethers (PBDEs) and polychlorinated biphenyls (PCBs) on thyroid hormone and vitamin A levels in rats and mice. *Archives of Toxicology*, **75**(4), 200–208.

Hallgren, S. and Darnerud, P.O. (2002) Polybrominated diphenyl ethers (PBDEs), polychlorinated biphenyls (PCBs) and chlorinated paraffins (CPs) in rats – testing interactions and mechanisms for thyroid hormone effects. *Toxicology*, **177**(2–3), 227–243.

Hansen, J. (2004) Cohort studies of cancer risk among Danish workers exposed to TCE. In: Proceedings of Symposium on New Scientific Research Related to the Health Effects of Trichloroethylene, 26-27 February 2004, Washington, DC. U.S. Environmental Protection Agency, Washington, DC.

Harvey, R.G. (1991) *Polycyclic Aromatic Hydrocarbons: Chemistry and Carcinogenicity*. Cambridge University Press, Cambridge.

Healy. T.E.J., Poole, T.R. and Hopper, A. (1982) Rat fetal development and maternal exposure to trichloroethylene 100 ppm. *British Journal of Anaesthesia*, **54**(3), 337–341.

Heaton, C.A. (1993) *The Chemical Industry*. Kluwer Academic Publishers.

Heldaas, S., Langard, S. L. and Andersen, A. (1984) Incidence of cancer among vinyl chloride and polyvinyl chloride workers. *British Journal of Industrial Medicine*, **41**, 25–30.

Hesse, S., Jernstrom, B., Martinez, M., Moldeus, P., Christodoulides, L. and Ketterer, B. (1982) Inactivation of DNA-binding metabolites of benzo[a]pyrene and benzo[a]pyrene-7, 8-dihydrodiol by glutathione and glutathione S-transferases. *Carcinogenesis*, **3**(7), 757.

Hisanaga, N, Jonai, H., Yu, X., Ogawa, Y., Mori, I., Kamijima, M., Ichihara, G., Shibata, E., Takeuchi, Y. (2002) Stevens–Johnson syndrome accompanied by acute hepatitis in workers exposed to trichloroethylene or tetrachloroethylene. *National Institute of Industrial Health*, **6**(21), 1.

Hooper, K. and She, J.W. (2003) Lessons from the polybrominated diphenyl ethers (PBDEs): Precautionary principle, primary prevention, and the value of community-based body-burden monitoring using breast milk. *Environmental Health Perspectives*, **111**(1), 109–114.

Howard, P. (1997) *Handbook of Environmental Fate and Exposure Data for Organic Chemicals*. CRC Press, Florida.

HSE (Health and Safety Executive) (2010) Control of Substances Hazardous to Health (COSHH). [Online] Available from: http://www.hse.gov.uk/coshh/ Accessed 27 September 2010.

IARC (1973) Certain polycyclic aromatic hydrocarbons and heterocyclic compounds. *Monographs on the evaluation of carcinogenic risk of the chemical to man. World Health Organization*, International Agency for Research on Cancer, Lyon, France.

IARC (1982) *IARC Monographs on the evaluation of carcinogenic risks to humans. Volume 29 Some Industrial Chemicals and Dyestuffs*, International Agency for Research on Cancer, Lyons, France.

IARC (1995) Monographs on the evaluation of carcinogenic risks to humans. Vo1ume 63. Drycleaning, some chlorinated solvents, and other industrial chemicals. International Agency for Research on Cancer, Lyon, France.

IARC (1983) Monographs on the evaluation of the carcinogenic risk of chemicals to humans. Polynuclear aromatic compounds: Part 1. Chemical, environmental and experimental data. *World Health Organization*, International Agency for Research on Cancer, Lyon, France.

IARC (1985) Monographs on the evaluation of the carcinogenic risk of chemicals to man. Polynuclear aromatic compounds: Part 4. Bitumens, coal-tars and derived products, shale oils and soots. *World Health Organization*, International Agency for Research on Cancer, Lyon, France.

IARC (1986) IARC Monographs on the evaluation of carcinogenic risks to humans. *Volume 41. Some halogenated hydrocarbons and pesticide exposures*, International Agency for Research on Cancer, Lyons, France.

IARC (1987) *Monographs on the evaluation of carcinogenic risks to humans: An updating of IARC monographs. World Health Organization*, International Agency for Research on Cancer, Lyon, France.

IARC (2010) IARC Monographs on the Evaluation of Carcinogenic Risks to Humans. *Volume 92. Some non-heterocyclic polycyclic aromatic hydrocarbons and some related exposures*. International Agency for Research on Cancer, Lyons, France.

Iavicoli, I., Marinaccio, A. and Carelli, G. (2005) Effects of occupational trichloroethylene exposure on cytokine levels in workers. *Journal of Occupational and Environmental Medicine*, **47**(5), 453–457.

Infante, P.F., Rinsky, R.A., Wagoner, J.K. and Young, R.J. (1977) Leukaemia in benzene workers. *Lancet*, **2**(8028), 76–78.

Janssen, R. (2005) Decision support for integrated wetland management. *Environmental Modelling and Software*, **20**(2), 215–29.

Jerina, D.M, Sayer, J.M., Thakker, D.R., Yagi, H., Levin, W., Wood, A. W. and Conney, A.H. (1980) Carcinogenicity of polycyclic aromatic hydrocarbons: the bay-region theory. Carcinogenesis, fundamental mechanisms and environmental effects. Proceedings of the thirteenth Jerusalem Symposium on Quantum Chemistry and Biochemistry, held in Jerusalem, Israel, April 28–May 2, 1980, Springer.

Johnson, P.D., Goldberg, S.J., Mays, M.Z. and Dawson, B.V. (2003) Threshold of trichloroethylene contamination in maternal drinking waters affecting foetal heart development in the rat. *Environmental Health Perspectives*, **111**(3), 289–292.

Johnson, P.I., Stapleton, H.M., Slodin, A. and Meeker, J.D. (2010) Relationships between polybrominated diphenyl ether concentrations in house dust and serum. *Environmental Science and Technology*, **44**, 5627–5632.

Johnson-Restrepo, B., Kannan, K, Rapaport, D.P. and Rodan, B.D. (2005) Polybrominated diphenyl ethers and polychlorinated biphenyls in a marine food web of coastal Florida. *Environmental Science and Technology*, **39**(21), 8243–8250.

Jones, K.C. and de Voogt, P. (1999) Persistent organic pollutants (POPs): state of the science. *Environmental Pollution*, **100**(1–3), 209–221.

Kamijima, M., Hisanaga, N., Wang, H. and Nakajima, T. (2007) Occupational trichloroethylene exposure as a cause of idiosyncratic generalized skin disorders and accompanying hepatitis similar to drug hypersensitivities. *International Archives of Occupational and Environmental Health*, **80**(5), 357.

Karlsson, M., Julander, A., van Bavel, B. and Hardell, L. (2007) Levels of brominated flame retardants in blood in relation to levels in household air and dust. *Environment International*, **33**(1),62–69.

Kemmlein, S., Herzke, D. and Law, R.J. (2009) Brominated flame retardants in the European chemicals policy of REACH – Regulation and determination in materials. *Journal of Chromatography A*, **1216**(3),320–333.

Kenaga, E. (1980) Predicted bioconcentration factors and soil sorption coefficients of pesticides and other chemicals. *Ecotoxicology and Environmental Safety*, **4**(1), 26–38.

Kimbrough, R.D. (1987) Human health-effects of polychlorinated-biphenyls (PCBS) and polybrominated biphenyls (PBBS). *Annual Review of Pharmacology and Toxicology*, **27**, 87–111.

Knoth, W., Mann, W., Meyer, R. and Nebhuth, J. (2007) Polybrominated diphenyl ether in sewage sludge in Germany. *Chemosphere*, **67**(9), 1831–1837.

Koppe, J. G. and Keys, J. (2001) Late lessons from early warnings: the precautionary principle 1896-2000: PCBs and the precautionary principle. Environmental Issue report No 22. European Environmental Agency.

Kroneld, R. (1989) Volatile pollutants in the environment and human tissues. *Bulletin of Environmental Contamination and Toxicology*, **42**(6), 873–877.

Kuriyama, S.N., Talsness, C.E., Grote, K. and Chahoud, I. (2005) Developmental exposure to low-dose PBDE-99: Effects on male fertility and neurobehavior in rat offspring. *Environmental Health Perspectives* **113**(2), 149–154.

LaKind, J. S., Berlin, Jr.C. M., et al. (2008) Lifestyle and polybrominated diphenyl ethers in human milk in the United States: a pilot study. *Toxicological and Environmental Chemistry* **90**(6): 1047–1054.

Lash L.H., Fisher J.W., Lipscomb J.C. and Parker J.C. (2000) Metabolism of trichloroethylene. *Environmental Health Perspectives*, **108**(2), 177–200.

Lassen, C. L. and Søersen, L.I. (1999) Brominated flame retardants, substance flow analysis and assessment of alternatives. Danish Environmental Protection Agency, Denmark.

LaVecchia, C., Negri, E., Franceschi, S., D'Avanzo, B. and Boyle, P. (1994) A case-control study of diabetes mellitus and cancer risk. *British Journal of Cancer*, **70**, 950.

Law, R.J., Bersuder, P., Allchin, C.R. and Barry, J. (2006) Levels of the flame retardants hexabromocyclododecane and tetrabromobisphenol A in the blubber of harbor porpoises (*Phocoena phocoena*) stranded or bycaught in the UK, with evidence for an increase in HBCD concentrations in recent years. *Environmental Science and Technology*, **40**(7), 2177–2183.

Lay, T. H., Bozzelli, J. W. and Seinfeld, J. H. (1996) Atmospheric photochemical oxidation of benzene: benzene OH and the benzene–OH adduct (hydroxyl-2, 4-cyclohexadienyl) O2. *Journal of Physical Chemistry*, **100**(16), 6543–6554.

Lee, L.J.H., Chung, C.W., Ma, Y.C., Wang, G.S., Chen, P.C., Hwang, Y.H. and Wang, J.D. (2003) Increased mortality odds ratio of male liver cancer in a community contaminated by chlorinated hydrocarbons in groundwater. *Occupational and Environmental Medicine*, **60**(5), 364–369.

Lee, R. F. and Ryan, C. (1979) Microbial degradation of pollutants in marine environments. Report number: USEPA-600/9-72-012. US Environment Protection Agency.

Leung, A. (2007) Spatial distribution of polybrominated diphenyl ethers and polychlorinated dibenzo-p-dioxins and dibenzofurans in soil and combusted residue at Guiyu, an electronic waste recycling site in southeast China. *Environmental Science and Technology*, **41**(8),2730–2737.

Liang, S.X., Zhao, Q., Qin, Z.F., Zhao, X.R., Yang, Z.Z. and Xu, X.B. (2008) Levels and distribution of polybrominated diphenyl ethers in various tissues of foraging hens from an electronic waste recycling area in South China. *Environmental Toxicology and Chemistry*, **27**(6),1279–1283.

Lide, D. R. (1993) *CRC Handbook of Chemistry and Physics*. CRC Press, Boca Raton.

Lingohr, M.K., Thrall, B.D. and Bull, R.J. (2001) Effects of dichloroacetate (DCA) on serum insulin levels and insulin-controlled signalling proteins in livers of male B6C3F1 mice. *Toxicological Sciences*, **59**(1), 178–184.

Luross, J.M., Alaee, M., Sergeant, D.B., Cannon, C.M, Whittle, D.M, Solomon, K.R. and Muir, D.C.G. (2002) Spatial distribution of polybrominated diphenyl ethers and polybrominated biphenyls in lake trout from the Laurentian Great Lakes. Dioxin 2000 Conference, Monterey California. Chemosphere, 46 (5), 665–672.

Lutz, W. K., Viviani, A. and Schlatter, C. (1978) Nonlinear dose–response relationship for the binding of the carcinogen benzo(a) pyrene to rat liver DNA in vivo. *Cancer Research*, **38**(3), 575.

Mackay, D., Shiu, W. Y. and Ma, K. C. (1993) *Illustrated Handbook of Physical-Chemical Properties and Environmental Fate for Organic Chemicals*, Vols 1 and 2. Lewis Publishers, Chelsea, Michigan.

Martens, D., Maguhn, J., Spitzauer, P. and Kettrup, A. (1997) Occurrence and distribution of polycyclic aromatic hydrocarbons (PAHs) in an agricultural ecosystem. *Fresenius' Journal of Analytical Chemistry*, **359**(7), 546–554.

Massachusetts Department of Public Health, Centers for Disease Control and Prevention, Massachusetts Health Research Institute (1996). *Final report of the Woburn environmental and birth study*. Massachusetts Department of Public Health, Boston, Massachusetts.

McConnell, G., Ferguson, D.M. and Pearson, C.R. (1975) Chlorinated hydrocarbons and the environment. *Endeavour* **34**, 13–8.

McCulloch, A. and Midgley, P. (1995) The production and global distribution of emissions of trichloroethene, tetrachloroethene and dichloromethane over the period 1988-1992. *Atmospheric Environment*, **30**(4): 601.

McDonald, T.A. (2002) A perspective on the potential health risks of PBDEs Dioxin 2000 Conference, Monterey California. Chemosphere (46) 5 745–755.

McKone, T.E., Knezovich, J.P. (1991) The transfer of trichloroethylene (TCE) from a shower to indoor air: experimental measurements and their implications. *Journal of the Air and Waste Management Association*. **41**(3): 282 -?

Mercado-Feliciano, M. (2008) The polybrominated diphenyl ether mixture DE-71 is mildly estrogenic. *Environmental health perspectives* **116**(5): 605–611.

Meylan, W.M. and Howard, P.H. (1991) Bond contribution method for estimating Henrys law constants. *Environmental Toxicology and Chemistry*, **10**(10), 1283–1293.

Miller, M.M., Wasik, S.P., Huang, G.L., Shiu, W.Y. and Mackay, D. (1985) Relationships between octanol-water partition coefficient and aqueous solubility. *Environmental Science and Technology*, **19**(6),522–529.

Miller, R. E. and Guengerich, F. P. (1982) Oxidation of trichloroethylene by liver microsomal cytochrome P-450: evidence for chlorine migration in a transition state not involving trichloroethylene oxide. *Biochemistry*, **21**(5), 1090–1097.

Mishra, PK., Dabadghao, S., Modi, GK., Desikan, P., Jain, A., Mittra, I., Gupta, D., Chauhan, C., Jain, S. K. and Maudar, K.K. (2009) In utero exposure to methyl isocyanate in the Bhopal gas disaster: evidence of persisting hyperactivation of immune system two decades later. *Occupational and Environmental Medicine*, **66**(4), 279–279.

Mocarelli, P. (2009) Seveso, the dioxin disaster. What we have learnt 30 years later. Birth Defects Research. Part A, *Clinical and Molecular Teratology*, **85**(3), 236–236.

Morehouse, W. and Subramaniam, A. (1986) *The Bhopal Tragedy: What Really Happened and What It Means for American Workers and Communities at Risk*. Learning Resources in International Studies 190. The Apex Press.

Morf, L.S., Tremp, J., Gloor, R., Huber, Y., Stengele, M. and Zennegg, M. (2005) Brominated flame retardants in waste electrical and electronic equipment: substance flows in a recycling plant. *Environmental Science and Technology*, **39**(22), 8691–8699.

Morgan, P., Lewis, S.T. and Watkinson, R.J. (1993) Biodegradation of benzene, toluene, ethylbenzene and xylenes in gas-condensate-contaminated ground-water. *Environmental Pollution*, **82**(2), 181–190.

Morgan, J.W. and Cassady, R.E. (2002) Community cancer assessment in response to long-time exposure to perchlorate and trichloroethylene in drinking water. *Journal of Occupational and Environmental Medicine*, **44**(7), 616–621.

National Toxicology Program (1986) Toxicology and carcinogenesis studies of benzene in f344/n rats and b6c3f1 mice. Report number: 289. U.S. Department of Health and Human Services.

Neff, J. M. and Sauer, T. C. (1996) *Environmental Science and Research* 163–75.

Nietert, P.J., Sutherland, S.E., Silver, R.M., Pandey, J.P., Knapp, R.G., Hoel, D.G. and Dosemeci, M. (1998) Is occupational organic solvent exposure a risk factor for scleroderma? *Arthritis and Rheumatism*, **41**(6), 111.

Nomiyama, K. and Nomiyama, H. (1977) Dose–response relationship for trichloroethylene in man. *International Archives of Occupational and Environmental Health*, **39**(4), 237–248.

North, K.D. (2004) Tracking polybrominated diphenyl ether releases in a wastewater treatment plant effluent, Palo Alto, California. *Environmental Science and Technology*, **38**(17), 4484–4488.

Osako, M., Kim, Y.J. and Sakai, S.I. (2004) Leaching of brominated flame retardants in leachate from landfills in Japan. *Chemosphere*, **57**(10), 1571–1579.

OSHA. (2005) Air contaminants. Occupational safety and health standards for shipyard employment. Code of Federal Regulations. 29 CFR 1915.1028. U.S. Department of Labor, Occupational Safety and Health Administration, Washington, DC.

Palm, A., Cousins, I. T., Mackay, D., Tysklind, M., Metcalfe, and Alaee, M. (2002) Assessing the environmental fate of chemicals of emerging concern: a case study of the polybrominated diphenyl ethers. *Environmental Pollution*, **117**(2), 195–213.

Park, K.S., Soresen, D.L., Sims, J.L. and Adams, V.D. (1988) Volatilization of wastewater trace organics in slow rate land treatment systems. *Hazardous Waste and Hazardous Materials*, **5**(3), 219–229.

Park, K. S., Sims, R. C., Dupont, R. R., Doucette, W. J. and Matthews, J. E. (1990) Fate of PAH compounds in two soil types: influence of volatilization, abiotic loss and biological activity. *Environmental Toxicology and Chemistry*, **9**(2), 187–195.

Patrick, L. (2009) Thyroid disruption: mechanisms and clinical implications in human. *Health Alternative Medicine Review*, **14**(4), 326–346.

Pearson, C.R., McConnell, G. and Cain, R.B. (1975) Chlorinated C1 and C2 hydrocarbons in the marine environment [and Discussion]. Proceedings of the Royal Society of London. Series B, Biological Sciences, 189 (1096), 305–332.

Pesch, B., Haerting, J., Ranft, U., Klimpel, A., Oelschlägel, B. and Schill, W. (2000) Occupational risk factors for renal cell carcinoma: agent-specific results from a case-control study in Germany. *International Journal of Epidemiology*, **29**(6), 1014–1024.

Phoon, W.H., Chan, M.O.Y., Rajan, V.S., Tan, K.J., and Thirumoorty, T. (1984) Stevens–Johnson syndrome associated with occupational exposure to trichloroethylene. *Contact Dermatitis*, **10**(5): 270.

Pinsuwan, S., Li, A. and Yalkowsky, S. H. (1995) Correlation of octanol/water solubility ratios and partition coefficients. *Journal of Chemical and Engineering Data*, **40**(3), 623–626.

Qiu, X., Marvin, C.H. and Hites, R. A. (2007) Dechlorane Plus and other flame retardants in a sediment core from Lake Ontario *Environmental Science andTechnology*, **41**(17), 6014–6019.

Raaschou-Nielsen. O., Hansen, J., McLaughlin, J.K., Kolstad, H., Christensen, J.M., Tarone, R.E. and Olsen, J.H. (2003) Cancer risk among workers at Danish companies using trichloroethylene: a cohort study. *American Journal of Epidemiology*, **158**, 1182–1192.

Rake, J.P., Visser, G., Labrune, P., Leonard, J.V., Ullrich, K. and Smit, G.P. (2002) Glycogen storage disease type I: diagnosis, management, clinical course and outcome. Results of the European Study on Glycogen Storage Disease Type I (ESGSD I). *European Journal of Paediatrics*, **161**(1), S20.

Rahman, F., Langford, K. H., Scrimshaw, M. D. and Lester, J. N. (2001) Polybrominated diphenyl ether (PBDE) flame retardants. *Science of the Total Environment*, **275**, 1–17.

Redmond, C. K., Ciocco, A., Lloyd, W. J. and Rush, H. W. (1972) Long-term mortality study of steelworkers: VI – Mortality from malignant neoplasms among coke oven workers. *Journal of Occupational and Environmental Medicine*, **14**(8), 621.

RCEP (Royal Commision on Environmental Pollution) (2003) Chemicals in products: safeguarding the environment and human health, 24th Report Royal Commission on Environmental Pollution, London.

Salvini, M., Binaschi, S. and Riva, M. (1971) Evaluation of the psychophysiological functions in humans exposed to trichloroethylene. *British Journal of Industrial Medicine*, **28**(3), 293.

Sangster, J. (1989) Octanol–water partition coefficients of simple organic compounds. *Journal of Physical and Chemical Reference Data*, **18**(3), 1111–1229.

Santodonato, J., Howard, P. and Basu, D. (1981) Health and ecological assessment of polynuclear aromatic hydrocarbons. *Journal of Environmental Pathology and Toxicology*, **5**(1), 364.

Sasiadek, M., Jagielski, J. and Smolik, R. (1989) Localization of breakpoints in the karyotype of workers professionally exposed to benzene. *Mutation Research*, **224**(2), 235–240.

Savolainen, H., Pfäffli, P. et al. (1977) Trichloroethylene and 1,1, 1-trichloroethane: effects on brain and liver after five days intermittent inhalation. *Archives of Toxicology*, **38**(3), 229–237.

Schecter, A., Johnson-Welch, S., Tung, K.C., Harris, T. R., Paepke, O. L. and Rosen, R. (2007) Polybrominated diphenyl ether (PBDE) levels in livers of US human fetuses and newborns. *Journal of Toxicology and Environmental Health. Part A, Current Issues* **70**(1),1–6.

Schecter, A., Papke, O., Joseph, J. E. and Tung, K. C. (2005) Polybrominated diphenyl ethers (PBDEs) in US computers and domestic carpet vacuuming: Possible sources of human exposure. *Journal of Toxicology and Environmental Health. Part A, Current Issues*, **68**, 501–513.

Schecter, A., Papke, O., Tung, K.C., Joseph, J., Harris, T.R. and Dahlgren, J. (2005) Polybrominated diphenyl ether flame retardants in the US population: Current levels, temporal trends, and comparison with dioxins, dibenzofurans, and polychlorinated biphenyls. *Journal of Occupational and Environmental Medicine*, **47**(3), 199–211.

Schecter, A., Papke, O., Tung, K.C., Staskal, D. and Birnbaum, L. (2004) Polybrominated diphenyl ethers contamination of United States food. *Environmental Science and Technology*, **38**(20), 5306–5311.

Schecter, A., Pavuk, M., Papke, O., Ryan, J.J., Birnbaum, L. and Rosen, R. (2003) Polybrominated diphenyl ethers (PBDEs) in US mothers' milk. *Environmental Health Perspectives*, **111**(14), 1723–1729.

Schroll, R., Bierling, B., Cao, G., Doerfler, U., Lahaniati, M., Langenbach, T., Scheunert, I. and Winkler, R. (1994) Uptake pathways of organic chemicals from soil by agricultural plants. *Chemosphere*, **28**(2), 297–303.

Schwetz, B.A., Leong, K.J. and Gehring, P.J. (1975) The effect of maternally inhaled trichloroethylene, perchloroethylene, methyl chloroform. *Toxicology and Applied Pharmacology*, **32**(1), 84.

Scott, C.S. and Chiu, W.A. (2006) Trichloroethylene cancer epidemiology: a consideration of select issues. *Environmental Health Perspectives*, **114**(9), 1471.

She, J.W., Petreas, M., Winkler, J., Visita, P., Mckinney, M. and Kopec, D. (2002) PBDEs in the San Francisco Bay area: measurements in harbor seal blubber and human breast adipose tissue. DIOXIN 2000 Conference, Monterey California. Chemosphere, **46**(5), 697–707.

Sheldon, L., Clayton, A., Keever, J., Perritt, R. and Whitaker, D. (1993) *Indoor concentrations of polycyclic aromatic hydrocarbons in California residences*. Air Resource Board, Sacremento.

Sigma-Aldrich (2010a) Benzene Safety Data Sheet. http://www .sigmaaldrich.com/united-kingdom.html. Accessed 12 November 2010.

Sigma-Aldrich (2010b) Benzo[a]pyrene Safety Data Sheet. http://www .sigmaaldrich.com/united-kingdom.html. Accessed 12 November 2010.

Sigma-Aldrich. (2010c) Trichloroethylene Safety Data Sheet. http:// www.sigmaaldrich.com/united-kingdom.html. Accessed 12 November 2010.

Snyder, R. (2004) Xenobiotic metabolism and the mechanism (s) of benzene toxicity. *Drug Metabolism Reviews*, **36**(3), 531–547.

Snyder, R., Dimitriadis, E., Guy, R., Hu, P., Cooper, K., Bauer, H., Witz, G. and Goldstein, B.D. (1989) Studies on the mechanism of benzene toxicity. *Environmental Health Perspectives*, **82**, 31.

Snyder, R. and Hedli, C.C. (1996) An overview of benzene metabolism. *Environmental Health Perspectives*, **104** (Suppl 6), 1165- ?.

Song, M., Chu, S., Letcher, R.J. and Seth, R. (2006) Fate, partitioning, and mass loading of polybrominated diphenyl ethers (PBDEs) during the treatment processing of municipal sewage. *Environmental Science and Technology*, **40**(20), 6241–6246.

Stewart, R., Hake, C., Lebrum, A.J. and Peterson, J.E. (1974) Effects of trichloroethylene on behavioral performance capabilities. In *Behavioral Toxicology, Early Detection of Occupational Hazards* (C. Xintaras, B.L. Johnson,and I. deGroot,eds) National Institute for Occupational Safety and Health, Cincinnati, OH.

Stoker, T.E., Cooper, R.L., Lambright, C.S., Wilson, V.S., Furr, J. and Gray, L.E. (2005) In vivo and in vitro anti-androgenic effects of DE-71, a commercial polybrominated diphenyl ether (PBDE) mixture. *Toxicology And Applied Pharmacology*, **207**(1), 78–88.

Swarm, R.L., Laskowski, D.A., McCall, P.J. et al. 1983. A rapid method for the estimation of the environmental parameters octanol/ water partition coefficient, soil sorption constant, water to air ratio, and water solubility. *Research Reviews* **85**, 17–28.

Tada, Y., Fujitani, T., Yano, N., Takahashi, H., Yuzawa, K., Ando, H., Kubo, Y., Nagasawa, A., Ogata, A. and Kamimura, H. (2006) Effects of tetrabromobisphenol A, brominated flame retardant, in ICR mice after prenatal and postnatal exposure. *Food and Chemical Toxicology*, **44**(8), 1408–1413.

Taylor, G.D. and Sudnick, P.E. (1984) *Du Pont and the International Chemical Industry*. Twayne Publishers, Boston.

Urano, K. and Murata, C. (1985) Adsorption of principal chlorinated organic compounds on soil. *Chemosphere*, **14**(3–4), 293–299.

US Congress (1986) Superfund Amendments and Reauthorization Act of 1986. 99th Congress, 2nd Session, October, 17, 1986.

US EPA (1980) U.S. Environmental Protection Agency. Code of Federal Regulations. Report number: 40 CFR 261.33. Where *Washington, DC*.

US EPA (1981) Toxic pollutants/effluent standards. U.S. Environmental Protection Agency. Code of Federal Regulations. Report number: 40 CFR 401.15. Washington, DC.

US EPA (2010a) Integrated Risk Information System Benzene. [Online] Available from: http://www.epa.gov/NCEA/iris/subst/ 0276.htm Accessed 7 April 2010.

US EPA (2010b) Technical Factsheet on: Polycyclic Aromatic Hydrocarbons (PAHs). Available from: http://www.epa.gov/ogwdw000/ pdfs/factsheets/soc/tech/pahs.pdf Accessed 12 September 2010.

US EPA (2002) Office of Ground Water and Drinking Water. National primary drinking water regulations. U.S.EPA Report number: EPA816F02013*Washington, DC*.

US EPA (2011) http://cfpub.epa.gov/ncea/cfm/recordisplay.cfm? deid=215006 Accessed 16 June 2011.

van Esch G.J. (1994) Brominated Diphenyl Ethers. Environmental Health Criteria 162. United Nations Environment Programme, the International Labour Organisation and the World Health Organization. Published where?.

Verschueren, K. (1985) Handbook of environmental data on organic chemicals. *Soil Science*, **139**(4), 376.

Verschueren, K. (1996) *Handbook of Environmental Data on Organic Chemicals. In*: 3rd edition. Von Nostrand Reinhold, New York.

Viberg, H., Fredriksson, A., Jakobsson, E., Orn, U. and Eriksson, P., (2003a) Neurobehavioral derangements in adult mice receiving decabrominated diphenyl ether (PBDE 209) during a defined period of neonatal brain development. *Toxicological Sciences*, **76**(1), 112–120.

Viberg, H., Fredriksson, A. and Eriksson, P. (2003b) Neonatal exposure to polybrominated diphenyl ether (PBDE 153) disrupts spontaneous behaviour, impairs learning and memory, and decreases hippocampal cholinergic receptors in adult mice. *Toxicology and Applied Pharmacology*, **192**(2), 95–106.

Vos, J.G., Becher, G., van den Berg, M., de Boer, J. and Leonards, P. E. G. (2003) Brominated flame retardants and endocrine disruption. *Pure and Applied Chemistry*, **75**(11–12), 2039–2046.

VWR International (2007) Trichloroethylene Safety Data Sheet. http://www.jencons.co.uk/app/GenericPage?page=/search/msds.jsp?en_GB_msds. Accessed 25 September 2010.

VWR International (2004) Benzene Safety Data Sheet. http://www.jencons.co.uk/app/GenericPage?page=/search/msds.jsp?en_GB_msds. Accessed 25 September 2010.

Wang, Y.W., Zhao, C.Y., Ma, W.P., Liu, H.X., Wang, T. and Jiang, G.B. (2006) Quantitative structure–activity relationship for prediction of the toxicity of polybrominated diphenyl ether (PBDE) congeners. *Chemosphere*, **64**, 515–524.

Wartenberg, D., Reyner, D. and Scott, C.S. (2000) Trichloroethylene and cancer: epidemiologic evidence. *Environmental Health Perspectives*, **108**(2) 161.

Watanabe, I. (2003) Environmental release and behavior of brominated flame retardants. *Environment International* **29**(6), 665–682.

Webster, L., Russell, M., Adefehinti, F., Dalgarno, E.J. and Moffat, C.F. (2008) Preliminary assessment of polybrominated diphenyl ethers (PBDEs) in the Scottish aquatic environment, including the Firth of Clyde. *Journal of Environmental Monitoring*, **10**(4), 463–473.

Wenning, R.J. (2002) Uncertainties and data needs in risk assessment of three commercial polybrominated diphenyl ethers: probabilistic exposure analysis and comparison with European Commission results. DIOXIN 2000 Conference, California. Chemosphere, **46** (5), 779–796.

WHO (2004) *Guidelines for drinking-water quality. 3rd ed. Geneva, Switzerland.* Available from: http://www.who.int/water_sanitation_health/dwq/gdwq3/en/ Accessed 9 April 2010.

Wideroff, L., Gridley, G., Mellemkjaer, L., Chow, W.H., Linet, M. and, Keehn, S. (1997) Cancer incidence in a population-based cohort of patients hospitalized with diabetes mellitus in Denmark. *Journal of the National Cancer Institute*, **89**, 1360.

Wikimedia Commons (2007) Bulls Eye Lyme Disease Rash. Hannah Garrison. http://commons.wikimedia.org/wiki/File:Bullseye_Lyme_Disease_Rash.jpg

Wikimedia Commons (2011) toxic epidermal-necrolysis. Thomas Habif. http://commons.wikimedia.org/wiki/File:Toxic-epidermal-necrolysis.jpg

Wild, S.R. and Jones, K.C. (1995) Polynuclear aromatic hydrocarbons in the United Kingdom environment: a preliminary source inventory and budget. *Environmental Pollution*, **88**(1), 91–108.

Wildey, R.J., Barnes, H. and Girling, A.E. (2004) *Persistence*, Bioaccumulation Potential And Toxicity – 'PBT'. Peter Fisk Associates, Herne Bay.

Willey, R.J., Hendershot, D.C. and Berger, S. (2007) *The Accident in Bhopal: Observations 20 Years Later*. Wiley Inter Science industry publication, Chemical and Engineering News (April 19).

Williams, T.I. (1982) *A short history of twentieth-century technology c. 1900–c. 1950*. Clarendon Press, Oxford, 405 pp

Williams-Johnson, M., Eisenmann, C.J. and Donkin, S.G. (1997) Toxicological profile for trichloroethylene. U.S. Department of Health and Human Services. Agency for Toxic Substances and Disease Registry.

Wisconsin Department of Health Services (2000) Information on Toxic Chemicals: Polyaromatic Hydrocarbons (PAHs) Report number: 4606. Wisconsin Department of Health Services, Madison, WI.

Wong, O. (1987) An industry wide mortality study of chemical workers occupationally exposed to benzene. I. General results. *British Medical Journal*, **44**(6), 365–381.

Wu, C. and Schaum, J. (2001) Sources, emissions and exposures for trichloroethylene (TCE) and related chemicals. EPA/600/R-00/099. U.S. Environmental Protection Agency, Washington, DC.

Xia, L.H., Huang, H.L., Kuang, S.R., Liu, H.F. and Kong, L.Z. (2004) [A clinical analysis of 50 cases of medicament-like dermatitis due to trichloroethylene.] In *Chinese. Zhonghua Lao Dong Wei Sheng Zhi Ye Bing Za Zhi*, **22**(3), 207–210.

Xian, Q.M., Ramu, K., Isobe, T., Sudaryanto, A., Liu, X., Gao, Z., Takahashi, S., Yu, H. and Tanabe, S. (2008) Levels and body distribution of polybrominated diphenyl ethers (PBDEs) and hexabromocyclododecanes (HBCDs) in freshwater fishes from the Yangtze River, China. *Chemosphere*, **71**(2), 268–276.

Xiang-Zhou, M., Liping, Y., Ying, G., Bi-Xian, M. and Eddy, Y.Z. (2008) Congener-specific distribution of polybrominated diphenyl ethers in fish of China: Implication for input sources. *Environmental Toxicology and Chemistry*, **27**(1), 67–72.

Yang, F. (2008) Detection of polybrominated diphenyl ethers in tilapia (Oreochromis mossambicus) from O'ahu, Hawaii. *Journal of Environmental Monitoring*, **10**(4), 432–434.

Yin, S.N., Li, G.L., Tain, F.D., Fu, Z.I., Jin, C., Chen, Y.J., Luo, S.J., Ye, P.Z., Zhang, J.Z. and Wang, G.C. (1987) Leukaemia in benzene workers: a retrospective cohort study. *British Journal of Industrial Medicine*, **44**(2), 124–128.

Yu, M., Luo, X.J., Wu, J.P., Chen, S.J. and Mai, B.X. (2009) Bioaccumulation and trophic transfer of polybrominated diphenyl ethers (PBDEs) in biota from the Pearl River Estuary, South China. *Environment International*, **35**(7), 1090–1095.

Zennegg, M., Kohler, M., Gerecke, A.C., Schmid, P. (2003) Polybrominated diphenyl ethers in whitefish from Swiss lakes and farmed rainbow trout. *Chemosphere*, **51**(7), 545–553.

Zhang, W. and Bouwer, E. J. (1997) Biodegradation of benzene, toluene and naphthalene in soil-water slurry microcosms. *Biodegradation*, **8**(3), 167–175.

Zhao, Y., Krishnadasan, A., Kennedy, N., Morgenstern, H. and Ritz, B. (2005) Estimated effects of solvents and mineral oils on cancer incidence and mortality in a cohort of aerospace workers. *American Journal of Industrial Medicine*, **48**(4), 249–258.

Zoeller, R.T. (2005) Environmental chemicals as thyroid hormone analogues: New studies indicate that thyroid hormone receptors are targets of industrial chemicals. *Molecular and Cellular Endocrinology*, **242**(1–2), 10–15.

Zou, M. (2007) Polybrominated diphenyl ethers in watershed soils of the Pearl River Delta, China. Occurrence, inventory, and fate. *Environmental Science and Technology*, **41**, 8262–8267.

7

Agricultural pesticides and chemical fertilisers

Rebecca McKinlay[1], Jason Dassyne[2], Mustafa B. A. Djamgoz[3], Jane A. Plant[4] and Nikolaos Voulvoulis [2]*

[1]Centre for Toxicology, University of London School of Pharmacy, 29–39 Brunswick Square, London
[2]Centre for Environmental Policy, Imperial College London, Prince Consort Road, London SW7 2AZ
[3]Department of Life Sciences, Imperial College London, Prince Consort Road, London SW7 2AZ
[4]Centre for Environmental Policy and Department of Earth Science and Engineering, Imperial College London, Prince Consort Road, London SW7 2AZ
**Corresponding author, email n.voulvoulis@imperial.ac.uk*

7.1 Introduction

Before the middle of the nineteenth century in Europe and North America and until much later in Asian countries such as China, food production involved the cycling of human, animal and marine wastes back to soil to maintain fertility based on the presence of humus. There were no chemical fertilisers, artificial hormones or genetically modified crops. Pesticides were mainly simple toxic salts of elements such as copper or arsenic, and insecticidal derivatives of plants such as chrysanthemums. Chemical agriculture dates from the publication in 1813 by Sir Humphrey Davy and in 1840 by the German chemist J. von Liebig of works arguing that inorganic fertilisers could replace manure because it was not the humus that was important to plant growth, but the chemicals it contained (Davy, 1813; Liebig, 1847). In 1843 the Rothamsted Experimental Station was founded in the UK and in the same year one of its founders, J. B. Lawes established a highly profitable superphosphate fertiliser factory in London.

The use of artificial chemicals in agriculture was minimal until after the second world war, when dichlorodiphenyltrichloroethane (DDT) and other synthetic pesticides were introduced, along with subsidies to increase agricultural production. Much agriculture is now based on chemical fertilisers manufactured from feedstocks such as rock phosphate, or from ammonia manufactured using the Haber process. The Food and Agriculture Organisation (FAO) estimated in 2008 that 136 million tonnes of synthetic nitrogen, 39 million tonnes of phosphate and 35 million tonnes of potash would be used globally in 2010.

Currently more than 507 different pesticides are used in food production in the USA (US EPA, 2010). Moreover, many chemicals are added during the storage and transport of food as well as in food processing and manufacture; and more than 3800 additives are now used as colourants, emulsifiers, flavourings and preservatives in processed food and beverages (Humphrys, 2002). It is not possible to discuss here all of the chemicals used in modern food production. In this chapter, we are concerned mainly with pesticides and artificial chemical fertilisers used in agriculture.

Chemical pesticides (the suffix 'cide' meaning to kill) are designed to poison one or several forms of life. A pesticide is defined by the FAO as any substance or class of substances intended for preventing, destroying or controlling the life cycle of any pest. These include vectors of human and animal disease, unwanted species of plants and animals which cause harm or interfere with the production, processing, storage, transport or marketing of food, agricultural commodities, wood and wood products or animal feedstuffs (FAO, 2010a). Chemical pesticides are also used for many other purposes, including treating ectoparasite infestations on companion animals and on recreational facilities such as golf courses (McKinlay *et al.*, 2008a).

Pollutants, Human Health and the Environment: A Risk Based Approach, First Edition. Edited by Jane A. Plant, Nikolaos Voulvoulis and K. Vala Ragnarsdottir.

Throughout history, pests and crop diseases have plagued humanity, from the lice and locusts described in ancient Egypt by the writers of Exodus to the Irish famines in 1845 and 1852 when potato blight caused mass starvation, disease and emigration. Chemicals have had a place in crop protection for thousands of years. Elemental sulphur was used for fumigation in pre-classical Greece, as mentioned in the *Odyssey* (Homer, *Odyssey* 22, 480–495) and it remains an important fungicide, especially in organic agriculture. Pyrethrum-containing powder from chrysanthemums was used as an insecticide by the Chinese from around 1000 BC, and toxic derivatives of arsenic, mercury and lead were in use in Europe from around the fifteenth century AD. Derris, a preparation containing the plant-derived insecticide rotenone, was first formulated in the mid-1800s and continues to be used by organic farmers on a restricted basis (Ecobichon and Joy, 1993).

Many natural plant-derived insecticides such as pyrethrum break down rapidly in the environment, making them difficult to store and reducing their potency. Their overall effectiveness is low, and most are broad spectrum, rather than targeting only harmful species. Compounds of arsenic, barium, boron, copper, fluorine, mercury, selenium, sulphur, thallium, zinc, elemental phosphorus and sulphur, which were the main elements used in inorganic insecticides (Carter, 1952), are also broad-spectrum pesticides, and most are highly persistent and toxic to humans. Bordeaux mixture, the first pesticide to be used on an industrial scale, is prepared by reacting dilute copper sulphate with sodium hydroxide (Carter, 1952), and is used to treat mildew and fungal infections (Mukerjee and Srivastava, 1957; WHO, 1998; Moolenaar and Beltrami, 1998). It was used in the nineteenth century in traditional European wine-growing regions to treat vine diseases spread by pests carried on botanical vine specimens from America, such as the aphid *Phylloxera vastatrix* which caused the Great French Wine Blight.

The first organochlorine pesticide was gamma hexachlorocylohexane (HCH), now known as lindane. It was first prepared by Faraday in 1825, but its insecticidal properties were not recognised until after the discovery of DDT in 1939 (Stenersen, 2004). DDT, a simple, easily manufactured chlorinated hydrocarbon which kills insects and was thought initially not to pose any significant risk to humans, animals or plants, revolutionised pesticide use. Others followed, such as herbicides based on chlorophenoxyacetic acid, which selectively remove broad-leaved weeds from grain fields, together with other technological innovations such as high-yielding crop varieties, improved irrigation techniques and farm machinery. Throughout Asia, Europe and North and South America, food production soared. World grain production increased on average 2.6 per cent a year between 1950 and 1990, to 1700 million tons per annum (Brown *et al.*, 1997). Yields increased so spectacularly that it became known as the 'green revolution', in spite of the fact that the chemical fertilisers used were mainly nitrogen, potassium and phosphate with no consideration of other trace elements essential to the health of humans and animals (Chapter 3). Globally, pesticide use has continued to increase dramatically, with an estimated 2.4 million tonnes, worth over 32 billion US dollars, used worldwide in 2000 and 2001 (US EPA, 2009).

Opposition to chemical agriculture began in the UK, US and Europe in the 1920s. It was based on the studies of distinguished scientists such as those conducted by Sir Robert McCarrison on the Hunza tribesmen in what is now Pakistan, Sir Albert Howard in India, and F. H. King in China, Japan and Korea. The work of others such as the Austrian philosopher Rudolf Steiner were also influential. They argued in favour of traditional biologically based agriculture (now referred to as 'organic farming') on the basis that a healthy humus-rich soil would produce vigorous, disease-resistant crops which would, in turn, promote the well-being of animals and humans (Williamson and Pearse, 1951). King (1944) argued strongly that human waste should be returned to land on the basis of the large populations sustained by agricultural systems in the East.

McCarrison spent seven years with the Hunza and reported that during that time he saw no cases of heart disease, cancer, appendicitis, peptic ulcer, diabetes or multiple sclerosis. He attributed the Hunza's 'remarkable health', longevity and cheerful temperament to their diet and traditional cultivation methods (McCarrison, 1961). At the same time the nutrionist J. B. Orr, later the first director of the United Nations FAO, began research on diet and demonstrated the crucial importance of essential trace elements to the health of humans and animals (Orr, 1936, 1948; see also Chapter 3). The Soil Association was established in 1946. Its first president, Lady Eve Balfour, was one of the most influential advocates of traditional biological farming in the UK. Nevertheless, by the early 1940s authoritative commentators such as Griggs were claiming that pesticides can control disease, artificial fertilisers could replace manure and herbicides could dispose of weeds, so the problems of hunger and famine had been overcome (Griggs, 1986).

The most important turning point in the way that pesticides were perceived was the publication of the book *Silent Spring* by Rachel Carson (Carson, 1962). This book caused widespread public concern and led to the banning of DDT in the United States in 1972. Despite many attacks on Carson's work, such as those coordinated by the US pesticide industry trade group and the National Agricultural Chemical Association (Orlando, 2002), there is increasing recognition that pesticides are poisonous to humans, animals and plants. The insecticides described in *Silent Spring,* mainly organochlorines such as aldrin, dieldrin, lindane and DDT, are now banned in most developed countries because of their broad-spectrum toxicity, persistence and bioaccumulation.

By the 1970s and 1980s, organochlorine pesticides had been replaced extensively by less persistent organophosphate and carbamate compounds. Organophosphates were initially developed from the neurotoxic agents prepared as nerve gas by the Nazis for chemical warfare during the second world war. The replacement of organochlorines was not without hazard. In areas where organophosphates were phased in to control mosquitos, acute organophosphate poisoning, especially amongst applicators and small children, gained prominence as a public health issue since their lack of persistence meant that these pesticides needed to be applied more frequently than organochlorines, and proper safety procedures were often ignored (Moffett, 2006)

There have been thousands of serious poisonings and fatalities caused by pesticides worldwide and they are acknowledged to be a threat to illiterate rural populations, especially in hot countries where wearing protective clothing is uncomfortable and where more insecticides, which tend to be more acutely toxic than herbicides, are used. Organophosphate pesticides, especially parathion, are responsible for the most deaths (Satoh, 2006). Another cause of greater exposure to pesticides in the tropics is the prophylactic spraying of DDT and other persistent compounds to control malarial mosquitoes (Beard, 2006). Stenersen (2004) reports that, according to the Chinese National Statistics Bureau, there were >48 000 pesticide poisonings in 1995, including 3000 fatalities, while another Chinese government estimate placed farm-worker fatalities due to pesticides at 7000–10 000 annually. Even in the USA it has been estimated that approximately 45 000 farmers, crop dusters and workers involved in pesticide manufacture and use are poisoned each year, resulting in 200 deaths (Schubel and Linss, 1971). Suicide as a result of intentional pesticide poisoning, especially in developing countries, is now also a major concern (WHO, 2008a).

There have also been serious accidents caused by pesticides, such as mass poisoning from the consumption of methylmercury-dressed grain in Iraq in 1956, 1960 and 1971 (Bakir *et al.*, 1973) and the poisoning of thousands of people in Turkey between 1955 and 1959 from the ingestion of hexachlorobenzene-treated grain (Hayes, 1982). Catastrophic incidents killing thousands of people have struck chemical factories involved in pesticide manufacture, for example the 1984 incident at Bhopal (India) in a plant manufacturing methyl isocyanate, and that in Seveso (Italy) in 1976 during the production of the pesticide and pesticide intermediary pentachlorophenol. The use of Agent Orange, a mixture of two phenoxyl herbicides, by the US as a defoliant during the Vietnam war in the 1960s had an immediate detrimental effect on human health and it continues to affect both human health and the vegetation there (Westing, 1975).

There is also increasing concern about the chronic effects of low levels of pesticides in food, water, the environment generally and human tissues and body fluids, including breast milk (Gliden *et al.*, 2010), which partly reflects improved analytical capability. There is mounting evidence of potential damage to humans and animals from long-term exposure, especially during fetal development. Concerns have also been raised over impacts on non-target insect populations, especially pollinators such as bees (Gross, 2008). There is also concern about pesticides used on genetically modified (GM) crops, since farmers tend to increase the quantities of herbicides used as resistant weeds proliferate (Benbrook, 2004).

Recently, the Royal Commission on Environmental Pollution (RCEP) investigated the effects on the health of residents and bystanders from crop spraying and concluded that a link between resident and bystander pesticide exposure and chronic ill health is plausible (RCEP, 2005). The Commission recommended a more precautionary approach to passive exposure to pesticides and indicated an urgent need for research to investigate the size and nature of the problem.

Most countries now have regulations and legislation to control pesticide use. In the EU, for example, pesticide residues in food and water are monitored, and maximum residue levels (MRLs) put in place for all active ingredients. In most cases, MRLs are decided on the basis of a risk assessment by the European Food Safety Authority (EC, 2009). For a pesticide to be approved for use in the EU, the manufacturers must produce a dossier containing human and environmental toxicology data. There is also legislation on acceptable operator exposure levels (EC, 2006). Similar legislation exists in the United States (US EPA, 2008; US FDA, 2007).

Pesticide manufacturers have responded to public concerns by trying to create compounds, particularly insecticides, which have lower mammalian toxicity and greater specificity, to enable problem organisms to be targeted without destroying beneficial ones or endangering human health. They have also increased the purity and selectivity of their products, ensuring that formulations and toxic metabolites are broken down rapidly. Nevertheless, concerns remain. There are also new and emerging problems from the use of pesticide formulations, most notably the problem of endocrine disruption which is defined as the capability of certain pollutants and naturally occurring substances to interfere with the synthesis, transport and action of hormones and other chemical messengers in the body (Diamanti-Kandarakis *et al.*, 2009).

In this chapter, we first discuss the hazardous properties, sources, environmental pathways and effects on receptors of pesticides, and then go on to discuss the same issues for fertilisers.

7.2 Pesticides

7.2.1 Hazardous properties

The most hazardous property of pesticides is their toxicity. Many insecticides which are neurotoxic were developed on the assumption that the nervous system of insects was very different from that of humans. Modern molecular biology has shown, however, that there are many similarities, even at the genetic level (RCEP, 2005). The most toxic xenobiotics almost always have bonds between carbon and halogens, direct bonds between carbon atoms and phosphorus or between carbon atoms and metals. These are easy to manufacture but rare in naturally occurring biomolecules. Moreover, different stereoisomers can have different biological properties. Natural biological systems usually use one stereoisomer of a substance, whereas many pesticides are mixtures of stereoisomers. New manufacturing methods, however, aim to prepare purer products of only the most active stereoisomers (Stenersen, 2004).

Pesticides have been classified into several categories according to their biochemical mode(s) of action (Ecobichon, 2001;

Table 7.1 Mode of action of some common classes of pesticides

Targeted system or biological process	Class of pesticide	Main mode of action	References
Nervous system (excitation and conduction of impulses)	Avermectin	Chloride channel modulator	Wolstenholme and Rogers, 2005
	Organophosphate	Cholinesterase inhibitor	O'Brien, 1963
	Neonicotinoid	Cholinergic agonist (mimic)	Tomizawa and Casida, 2005
	Carbamate	Cholinesterase inhibitor	Casida, 1963
	Pyrethroid	Voltage-gated sodium channel modulator	Vijerberg and Bercken, 1990
	Phenylpyrazole	GABA-gated chloride channel modulator	Caboni *et al.*, 2003
	Chlorinated hydrocarbon – cyclodiene organochlorine	GABA-gated chloride channel antagonist (blocker)	Beeman, 1982
	Benzenehexachloride isomers	Voltage-gated sodium channel modulator	Casida, 1993
	Oxadiazine	Voltage-gated sodium channel blocker	Wing *et al.*, 2000
	Spinosyn	Nicotinic acetylcholine receptor agonist (mimic)	Salgado, 1998
	Pyridine azomethine	Selective feeding blocker	Harrewijn and Kayser, 1997
Cellular metabolism	Petroleum-based products	Membrane disruption leading to blockage of breathing apparatus by mechanical suffocation	Don-Pedro, 1989
	Fatty acids	Membrane disruption	Don-Pedro, 1990
	Microbial derivatives	Membrane disruption in insect midgut	Tanaka and Omura, 1993
	Neem oil extract	Prothoracicotropic hormone inhibitor; phagostimulant disruptor	Schmutterer, 1990
	Amidinohydrazone	Electron transport inhibitor (site II)	Hollingshaus, 1987
	Pyridazinone	Electron transport inhibitor (site I)	Büchel, 1972
	Pyrazole	Electron transport inhibitor (site I)	Lummen, 1998
	Pyrrole	Disruption (uncoupling) of oxidative phosphorylation	Black *et al.*, 1994
	Organotin	Disruption (uncoupling) of oxidative phosphorylation	Snoeij *et al.*, 1987
Growth and development	Azadirachtin (botanical from neem oil)	Inhibitor of prothoracicotropic hormone; disruption of phagostimulation	Schmutterer, 1990

Table 7.1 (*Continued*)

Targeted system or biological process	Class of pesticide	Main mode of action	References
Growth and development	Carboxamide mite growth inhibitors	Unknown or non-specific mode of action	Dekeyser, 2004
	Diacylhydrazine insect growth regulators	Ecdysone (hormone) agonist/ disruptor	Dhadialla *et al.*, 1998
	Benzoylurea insect growth regulators	Chitin synthesis inhibitor	Xu *et al.*, 2003
	Substituted melamine insect growth regulators	Chitin synthesis inhibitor	Koehler and Patterson, 1989
	Tetrazine mite growth inhibitors	Chitin synthesis inhibitor	Eberle *et al.*, 2003
	Triazines	Photosynthesis inhibitor	Shimabukuro and Swanson, 1969

Gregus and Klaasen, 2001; Stenersen, 2004; Brown, 2006) (Table 7.1).

7.2.1.1 Enzyme inhibitors

These chemicals react with an enzyme or transport protein and inhibit its normal function. For example, carbamate and organophosphorus insecticides are acetylcholinesterase inhibitors (AchEIs); they inhibit the enzyme that breaks down the excitatory neurotransmitter, acetylcholine, increasing the level and duration of its action and hence the excitatory state (O'Brien, 1963; Matsumara, 1975). Although cholinesterase inhibition by carbamates is partially reversible, organophosphate poisoning is not, because the insecticide remains bound to cholinesterase. Organophosphates work in the same way as nerve gas and belong to a group of chemicals classified as weapons of mass destruction by the United Nations according to UN Resolution 687. Some enzyme inhibitors are effective herbicides. Glyphosate, for example, inhibits enzymes that are important in amino-acid synthesis. Enzyme inhibitors show selectivity, depending on the prevalence of the enzymes they inhibit in non-target organisms. For example, AchEIs generally do not affect plants which lack nervous systems, while inhibitors of amino-acid synthesis are less toxic to animals which do not synthesise essential amino acids.

7.2.1.2 Pesticides that disturb biochemical signalling systems

These types of pesticides are usually extremely potent and often more selective than other classes. They frequently act as agonists, imitating endogenous biochemical substances thereby causing the signaling to be too strong or long lasting, and/or to be transmitted at the wrong time. Nicotine and the neonicotinoids, for example, are acetylcholine agonists which are not degraded by acetylcholinesterase after signal transmission has ended. Hence, they produce a similar effect to that caused by AchEIs. The neonicotinoids stimulate insect acetylcholine receptors more effectively than human receptors, so they are considered to be more toxic to insects than humans (Brown, 2006).

The herbicide 2,4-dichlorophenoxy acetic acid (2,4-D) mimics plant auxin hormones, which are essential in coordinating many growth and behavioural processes in the plant life cycle. In contrast, some pesticides are auxin antagonists and block receptor sites for signaling substances including, in some cases, intracellular sites.

Organochlorine insecticides alter the ability of excitable tissues, central nervous system (CNS) and peripheral nervous system (PNS), including the selective permeability to Na^+ and K^+ ions, to generate and conduct impulses. Some insecticides also affect the transport of Ca^{2+} in sarcoplasmic reticulum, where Ca^{2+} calcium normally is released during muscle contraction and absorbed during muscle relaxation (Doherty, 1979). For example, parathion and organophosphate insecticides depress reticular Ca^{2+} uptake. The most important action of organochlorine insecticides at the neuromuscular junction is their inhibitory action on ATP-promoted uptake of Ca^{2+} by the sarcoplasmic reticulum (Huddart *et al.*, 1974). Such processes are also important in the initiation and propagation of some types of cancer cells (Doherty, 1979).

Cyclodiene-type organochlorine insecticides affect Cl^- channels by inhibiting the gamma-amino-butyric-acid (GABA) receptor, which has a general inhibitory effect on the nervous system. In contrast, bifenazate is a GABA receptor agonist (mimic), stimulating the receptor and disrupting nerve-impulse transmission. Avermectins, which are derived from soil organisms, also bind to Cl^- channels, causing an inhibitory effect which can lead to insect death (Brown, 2006).

The cyclodiene organochlorine pesticides tend to accumulate through food chains because of their persistence and high lipid solubility. Concerns have been raised about their carcinogenic, mutagenic and reprotoxic (CMR) behaviour, which may not be apparent until after several years of chronic exposure (Coats, 1990). Pyrethroids are synthetic versions of pyrethrins. They act primarily on voltage-gated Na^+ channels, resulting in prolonged nerve excitation and transmission, tremors, and death in insects. Many insect-growth regulators, such as thiazdiazine-type insecticides, mimic a protein called 'juvenile hormone' and prevent metamorphosis, moulting and reproduction, thereby causing the insect to die. Such insecticides are considered to have low toxicity to mammals, including humans.

7.2.1.3 Pesticides that generate free-radical cascades

The classic example is the herbicide paraquat, which captures an electron from the electron-transport chain in mitochondria. This then reacts with molecular oxygen to produce a superoxide anion, beginning an aggressive free-radical cascade whereby many biomolecules can be damaged or destroyed by one paraquat molecule. Copper acts in a similar manner because Cu^{2+} takes up an electron to become Cu^+, donating the spare electron to oxygen to produce the superoxide anion.

7.2.1.4 Pesticides that degrade pH gradients across membranes

These substances, such as ammonia phenols and acetic acid, work by carrying H^+ ions from outside mitochondria or chloroplasts into their more alkaline interiors. This disrupts cell function, since the pH gradient across membranes is particularly important for energy production by mitochondria in animals and plants and for photosynthesis by chloroplasts in plants. In addition, since many intracellular enzymes are sensitive to cell pH, normal cell metabolism can be severely affected.

7.2.1.5 Electron-transport inhibitors

Electron transport disruptors, such as pyrroles, inhibit the production of the energy-rich compound ATP by phosphorylation, thereby effectively closing down an organism's ability to produce energy from food.

7.2.1.6 Pesticides that dissolve in lipophilic membranes and disturb their physical structure

Alcohols, petrol, chlorinated hydrocarbons and organic solvents such as toluene have this type of general cellular toxicity.

7.2.1.7 Substances that disturb the electrolytic or osmotic balance

Sodium chloride (NaCl) is the best-known example of this group. Since the trans-membrane Na^+-concentration gradient is crucial for cells' excitability, communication with the outside medium, transport of many biochemicals in and out of cells, and osmotic balance, adding NaCl to cells, would have a general deleterious effect on cellular functioning and well-being.

7.2.1.8 Strong acids, alkalis, oxidants or reductants that destroy tissue proteins and DNA

These chemicals destroy organisms by dissolving tissue proteins and, because of their potential damage to DNA, may initiate cancer. Sulphuric acid is one of the commonest. It is widely used to kill potato haulms prior to harvest, in order to speed harvest times and to force the plants to direct extra resources to their tubers. However, its use risks the contamination and acidification of water courses and severe irritation of the eyes and respiratory tract if humans are exposed. Accordingly, strict regulations are in place to limit the quantities sprayed and prevent spray drift travelling beyond the area of application (NAAC, 2010).

7.2.1.9 Pesticides as endocrine-disrupting chemicals

The growth and development of fetuses and infants are regulated by a complex array of chemical signals, which are vulnerable to disruption. The adverse effects of their disruption may not become manifest until later in life and can be passed onto subsequent generations via genetic or epigenetic changes. Endpoints include reproductive malformations, increased risk of certain cancers, neurodevelopmental defects, and increased risks of diabetes and obesity (Diamanti-Kandarakis *et al.*, 2009).

Many pesticides have been identified as known or suspected endocrine-disrupting chemicals (EDCs) (McKinlay *et al.*, 2008a). Their modes of action depend on their individual or group chemical nature (Table 7.2). The best-studied are those that mimic or otherwise affect the action of steroid hormones such as oestrogen. Compounds that bind to the oestrogen receptor can bind to both subtypes, α and β (Kuiper *et al.*, 1997), have agonist or antagonistic effects and vary in their potencies (Celik *et al.*, 2008). However, most are weaker than endogenous oestrogens and oestrogen-receptor-targeting drugs such as tamoxifen and diethylstilbestrol (Bolger *et al.*, 1998).

Production of the enzyme aromatase, which transforms testosterone into oestradiol, is inhibited by some compounds such as tributyltin (Saitoh *et al.*, 2001). Similarly, androgens and their

Table 7.2 EDC pesticides and their effects

Mode of action	Examples and effects	References
Disruptors of steroid synthesis, metabolism and transport	Atrazine, benomyl and carbendazim: induce aromatase activity, decreasing androgen and elevating oestrogen levels. Endosulfan: weak aromatase inhibitor; also interferes with steroid hormone storage and release. Fenoxycarb: interferes with metabolism of testosterone. Heptachlor: increases testosterone andandrostenedione production. Lindane: increases metabolism of oestrogens. Prochloraz: inhibits aromatase activity and significantly decreases steroidogenesis. Vinclozolin: interacts with the pregnane X cellular receptor, interfering with the manufacture of enzymes responsible for steroid hormone metabolism	Sanderson, 2000; Morinaga, 2004; Haake *et al.*, 1987; Verslycke, 2004; Eriko *et al.*, 2003; Robert *et al.*, 1988; Cocco, 2002; Mason *et al.*, 1987
Steroid hormone receptor agonists and antagonists	2,4D: synergistic androgenic effects when combined with testosterone. Aldicarb: inhibits 17 beta-estradiol and progesterone activity; shows weak oestrogenic effects. Cypermethrin: mimics the action of oestrogen, metabolites also have oestrogenic effects. DDT and metabolites: mimics the action of oestrogens indirectly by stimulating the production of their receptors; antagonises the action of androgens via binding competitively to their receptors. Dieldrin: antagonises the action of androgens via binding competitively to their receptors and inhibiting the genetic transcription they induce. Endosulfan: mimics the actions of oestrogens indirectly by stimulating the production of their receptors; antagonises the action of androgens via binding competitively to their receptors. Lindane: binds to oestrogen receptors without activating them. Prochloraz: antagonises the cellular androgen and oestrogen receptors. Vinclozolin: potent androgen receptor antagonist; competitively inhibits the binding of androgen to its receptor; inhibits androgen-inducing gene expression	Kim, 2005; Klotz, 1997; Cocco, 2002; Chen *et al.*, 2002; McCarthy, 2006; Tapiero, 2002; Andersen, 2002; Bulayeva, 2004; Lemaire, 2004; Vonier, 1996; Fang *et al.*, 2003; Grunfeld, 2004; Cooper, 1989
Disruptors of thyroid hormone synthesis, metabolism and transport	Ziram, fenbuconazole, mancozeb, maneb and metriam: inhibit the production of thyroid hormones. Cyhalothrin: decreases their secretion. Lindane: decreases blood thyroxine concentration	Marinovich *et al.*, 1997; Cocco, 2002; Akhtar *et al.*, 1996; Beard, 1999
Thyroid hormone receptor agonists and antagonists	Dimethoate: disrupts the action of the thyroid hormones. Ioxynil: antagonises the action of thyroid hormones and the expression of the genes coding for their cellular receptors. Malathion: binds to thyroid hormone receptors	Mahjoubi-Samet *et al.*, 2005; Ishihara, 2003; Sugiyama, 2005
Aryl hydrocarbon receptor agonists and antagonists	Prochloraz: aryl hydrocarbon receptor agonist	Vinggaard, 2006

(*continued*)

Table 7.2 (*Continued*)

Mode of action	Examples and effects	References
Disruption of neuroendocrine control of hormone synthesis	Atrazine: disrupts the hypothalamic control of lutenising hormone and prolactin levels; induces aromatase activity, increasing oestrogen production. Malathion: inhibits catecholamine secretion. Parathion: inhibits catecholamine secretion, increases nocturnal synthesis of melatonin and causes gonadotrophic hormone inhibition. Thiram: reduces the conversion of dopamine to noradrenalin in the adrenal glands	Cooper *et al.*, 2000; Caroldi and De Paris, 1995; Cocco, 2002

receptors are targets for EDC pesticides (Lemaire *et al.*, 2004; Andersen, 2002). Gene expression triggered by hormones can be blocked as well as initiated by such EDC-receptor interactions, as in the blockage of androgen-mediated gene expression by the fungicide vinclozolin (Gray, 1998).

Circulating thyroid hormone levels are decreased by some pesticides (Buckler-Davis, 1998; Howdeshall, 2002). This can have a variety of disruptive effects, including hastening the excretion of thyroid hormones, preventing the transport of iodine into the thyroid, preventing thyroxine manufacture by interfering with the action of thyroid peroxidase, binding to or otherwise interfering with the transport proteins carrying iodine through the body, preventing its conversion to triiodothyronine and interfering with the thyroid receptors (Köhrle, 2008).

Outside the laboratory, EDC-mediated effects are usually the result of a cocktail of substances which may act together with each other and with endogenous hormones in an additive or synergistic manner. Consequently, damaging effects can be seen from lower than expected concentrations of compounds in these mixtures, particularly when similarly acting EDCs are combined (e.g. xeno-oestrogens and thyroid-disrupting chemicals) (Kortenkamp, 2007).

7.2.1.10 Carcinogenic, mutagenic and reprotoxic properties

Since resistance to cancer and successful reproduction depend on genetic stability and many pesticides affect this, pesticides may have CMR properties. As discussed in Chapter 2, REACH legislation covers all chemicals produced in or imported into the EU in quantities greater than 1 tonne per year. Substances which are found to have CMR properties are classified alongside persistent, bioaccumulative and toxic substances and confirmed EDCs as substances of very high concern. Their sale and use are strictly regulated (ECHA, 2007). Since pesticides are already covered by specific legislation, they are exempt from the chemical safety assessment because the chemical assessment process will have been completed (ECHA, 2010).

7.2.2 Sources

Accurate data on the use of agricultural pesticides and areas treated have been collected by the Pesticide Safety Directorate in the UK since 1974. These show that total pesticide use increased steadily, reaching a peak in the mid to late 1990s before decreasing slightly (Figure 7.1). Overall, Europe followed a similar trend, with a steady application rate of 2–2.5 kg per hectare of farmland per year after the early 1990s.

The types and quantities of pesticides used vary according to the types of farm, crops and livestock managed. Extensive grazing, set-aside and most fodder crops use little, whereas high-value crops, including greenhouse crops and most fruit and vegetables, use much more (Snoo *et al.*, 1997). Fruit and vegetables use the most insecticides and fungicides, and conventionally farmed cereal crops (i.e. not organically farmed) are sprayed with herbicides (Matthews, 1999). Pesticide formulations containing the same active ingredients but intended for different crops can have very different toxicological properties, due to use of different adjuvants.

Besides agriculture, there are several other important sources of pesticides in the environment. For example, they are used as a means of maintaining pavements, parks and sports facilities such as golf courses in urban or peri-urban areas (Denham and White, 1998). The pesticides used on golf courses include chlorpyrifos, carbendazim, chlorthalonil and iprodione, which are more toxic than those permitted for home use or in parks. Pesticides are also used for domestic, medicinal and veterinary purposes, to control weeds and household pests and to treat ectoparasites and fungal infections in humans and animals (Grey *et al.*, 2006; Menegaux *et al.*, 2006). These types of use have received little attention in the UK as a pathway of human exposure; they are outside the scope of this chapter but are discussed in McKinlay *et al.* (2008b). Data on the use of pesticides for agricultural purposes are closely monitored and put into the public domain, but there is no such requirement for pesticides used municipally or in private homes.

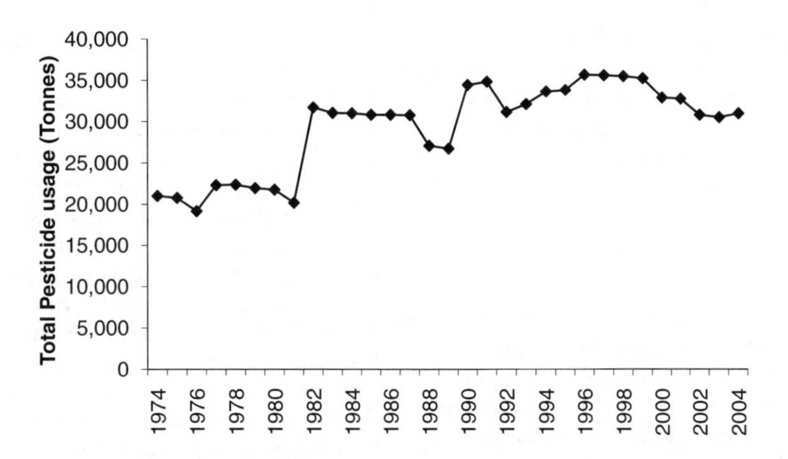

Figure 7.1 UK pesticide use 1974–2004 (Reproduced, with permission, from Calculating human exposure to endocrine disrupting pesticides via agricultural and non-agricultural exposure routes, McKinlay, R.; Plant, J.A.; Bell, J.N.B.; Voulvoulis, N. Science of the Total Environment/398/1-2/1-12 © 2008 Elsevier)

7.2.3 Environmental pathways

An outline of the pathways and fate of pesticides used in agriculture following their release into the environment is shown in Figure 7.2.

7.2.3.1 Soil

Most pesticides can form hydrogen bonds with organic carbon, binding them to soil particles (Borisover and Graber, 1996). They can also form ionic bonds with clays, becoming trapped between layers in the clay lattice (Sheng, 2001; Xu, 2001). Compounds which are strongly adsorbed to soil organic matter (SOM) can persist for long periods of time. If laboratory tests show that more than 70 per cent of a substance forms non-extractable residues after 100 days in soil and the substance has a mineralisation rate of less than 5 per cent in 100 days, it is deemed to be persistent. In field trials, if 50 per cent or more remains in the soil after 3 months and it takes more than a year for 90 per cent or more to disappear, the pesticide is confirmed as

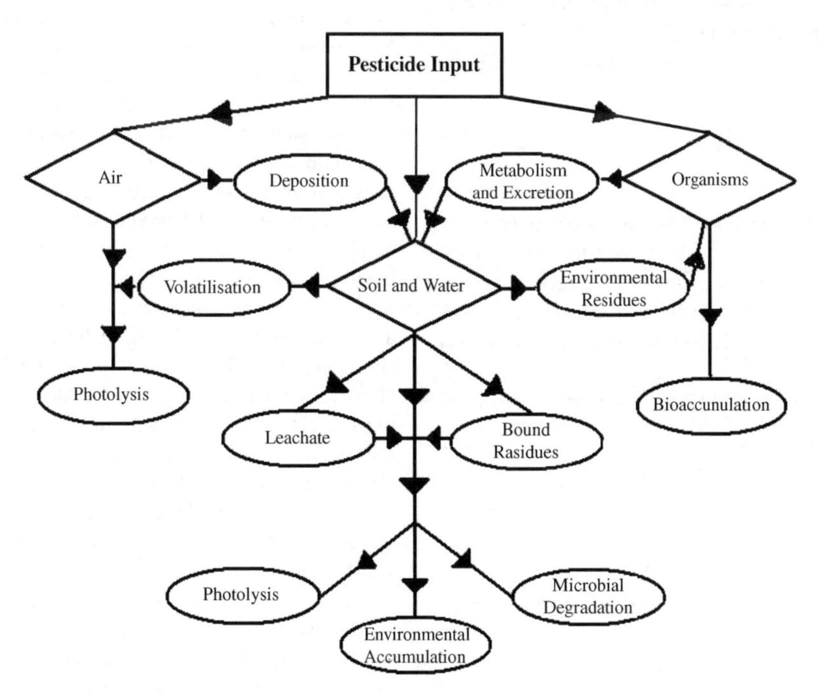

Figure 7.2 Pathways and fate of agricultural pesticides in the environment (Reproduced, with permission, from Calculating human exposure to endocrine disrupting pesticides via agricultural and non-agricultural exposure routes, McKinlay, R.; Plant, J.A.; Bell, J.N.B.; Voulvoulis, N. Science of the Total Environment/398/1-2/1-12 © 2008 Elsevier)

being persistent (EC, 2000). Persistence varies with soil type. When applied to SOM-poor sandy soil, nitrofen, for example, degraded far more quickly than the same pesticide applied to SOM-rich soils: after 16 weeks, 2 per cent of the original application remained in the sandy soil compared with 15 per cent in the SOM-rich soil (Murty, 1982).

7.2.3.2 Water

Compounds can travel in water, either sorbed to soil particles, or as colloids or as dissolved species, depending on their hydrophilicity. Pesticide contamination of groundwater is a problem in arable areas. Between 1990 and 2000, the UK water industry spent £1bn on infrastructure to deal with pesticide contamination and a further £100 million on running costs to remove pesticide residue during water treatment (EUREAU, 2001). In Europe, triazine herbicides are the most problematic and frequently detected pesticides in groundwater, but a wider variety are detected in rivers (Table 7.3). The water-borne residues of 38 pesticides are listed as being of global health significance by the WHO. These include the highly persistent organochlorine pesticides (such as aldrin and dieldrin), triazine herbicides and a variety of organophosphates and pyrethroids (WHO, 2008b).

Pesticides can be removed or broken down using methods such as filtration through activated carbon, chemical coagulation, advanced oxidation and ozonation. This may reduce the concentration of the parent compound, but the breakdown product itself may be toxic. Some endocrine-disrupting pyrethroids can produce persistent breakdown products which are also EDCs (McCarthy, 2006). People in rural areas who use private wells for their water supplies are particularly at risk from water-borne pesticides.

7.2.3.3 Air

In air, pesticides can be transported as vapour or bound to dust particles, depending on factors such as volatility. Direct exposure is a particularly important pathway for people employed in agriculture or pest control (Van Tongeren *et al.*, 2002; Ambroise, 2005; Bouvier *et al.*, 2006). Living or working near areas treated with pesticides also increases exposure (Arya, 2005; RCEP, 2005; Vlacke *et al.*, 2006). Rural residents are at particularly high risk, not only from agricultural spray drift but also if they use sprayed land for walking and exercising their dogs (RCEP, 2005; Valcke *et al.*, 2006). Although there is no legal requirement for farmers applying most pesticides to notify the public of their use or to provide information on the compounds used, many follow voluntary recommendations (DEFRA, 2006) which advise a distance of 2 m between sprayed crops and sensitive areas such as homes or footpaths (BCPC, 2005). These recommendations fail, however, to take into account field topography, vehicle speed and spraying in unfavourable weather conditions – and the fact that very fine droplets and vapour can travel far beyond 2 m (RCEP, 2005).

Pesticides can also reach humans and other environmental receptors indirectly from exposure through food, beverages, water, soil and air. Possible pathways of human exposure are summarised in Figure 7.3.

Table 7.3 Problematic pesticides frequently detected in European groundwater and rivers (EURAU, 2001)

Groundwater	River water
Atrazine and related products	Diuron
Simazine	Isoproturon
Mecoprop	Atrazine and related products
Bentazone	Simazine
	Mecoprop
	MCPA
	Chlortoluron
	Glyphosate (AMPA)

7.2.3.4 Food

Pesticides used in food production, including EDCs, can reach receptors via numerous pathways from both point and diffuse sources. Such pesticides have varying degrees of persistence and may breakdown or, in the case of salts of toxic trace elements, be changed into different species (Chapter 4) and moved through different environmental compartments (air, soil, sediment, water, and organisms). Compounds may accumulate in one or more of these compartments, travel between them, or break down to toxic or non-toxic metabolites (Figure 7.4).

Contamination of food by agricultural pesticides is an established exposure pathway. Pesticide residues are present in conventionally produced foods ranging from fresh fruit and vegetables (Newsome *et al.*, 2000; Andersen and Poulsen, 2001; Jong and Snoo, 2001) to processed baby foods (Stepan, 2005). Meat, eggs and milk products frequently contain traces of persistent pesticides such as organochlorines, despite their use having been illegal for many years (Herrera *et al.*, 1996; Jong and Snoo, 2001). The concentrations detected are usually low, and are often dismissed as being of little concern to human health (Herrera *et al.*, 1996; Kieszak *et al.*, 2002; Rawn *et al.*, 2004). However, variations in individual food preferences, residues in individual food items and variations in the amount of a particular food consumed can increase individual pesticide exposure (Hamilton *et al.*, 2004). Over time, some of these pollutants may accumulate in human tissues to levels comparable to those shown to cause abnormalities in humans and wildlife (Dewailly *et al.*, 1999; Nakagawa *et al.*, 1999; Olea *et al.*, 1999; Charlier, 2002). The consumption of sport fish contaminated with bioaccumulated pollutants can also be a source of exposure to some particularly hazardous pesticides, such as organochlorine compounds (Anderson *et al.*, 1998).

Pesticide residues in food are the subject of much legislation, nationally and internationally (PSD, 2006), and stringent monitoring programmes are in place (PSD, 2007). People who

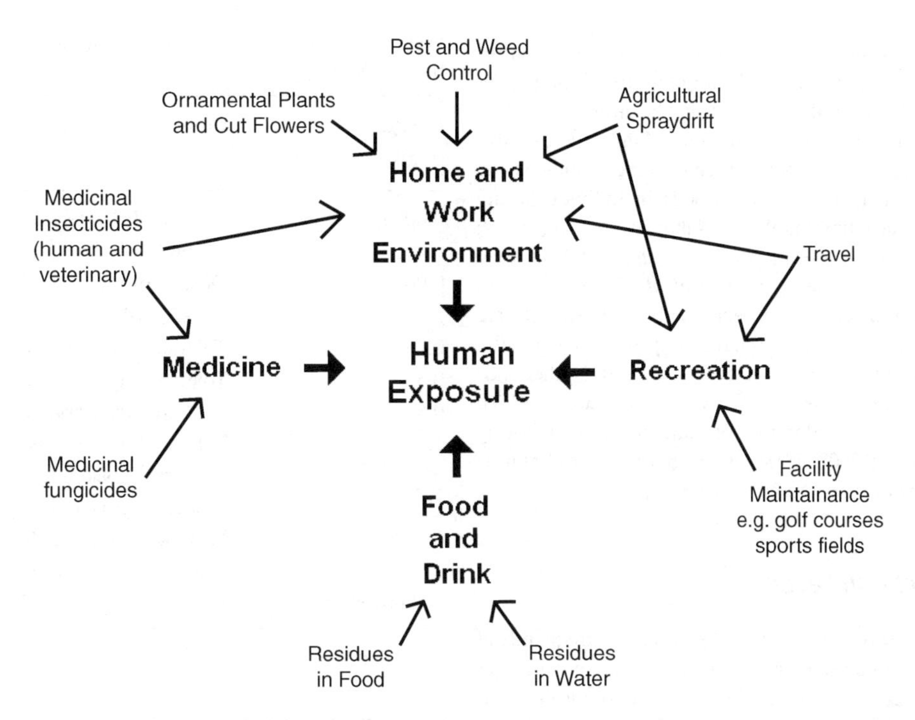

Figure 7.3 Sources and pathways for human exposure to pesticides (Reproduced, with permission, from Geochemical Atlas of Europe. Part 1: Background Information, Methodology and Maps, Salminen, R., Batista, M.J. Bidovec, M., Demetriades, A., De Vivo, B., De Vos, W., Duris, M., Gilucis, A., Gregorauskiene, V., Halamic, J., Heitzmann, P., Lima, A., Jordan, G., Klave,r G., Klein, P., Lis, J., Locutura, J., Marsina, K., Mazreku, A., O'Connor, P.J., Olsson, S.Å., Ottesen, R.-T., Petersell, V., Plant, J.A., Reeder, S., Salpeteur, I., Sandström, H, Siewers, U., Steenfelt, A., & T. Tarvainen © 2005/Geological Society of Finland)

consume oily fish, meat and dairy products are likely to receive a higher dose of persistent compounds (such as organochlorines, which bioaccumulate up the food chain because of their lipophilicity) than people whose diets are vegetable based (Darnerud *et al.*, 2006; Bro-Rasmussen, 1996). Moreover, people whose diet contains a high proportion of organic food have less pesticide exposure than those who do not (Lu *et al.*, 2006; Fenske, 2005). Organophosphates, carbamates and other less persistent compounds can reach humans via their diets. Crops requiring frequent pesticide applications, such as fruit, potatoes and some brassicas contribute more to overall exposure than those that require less, such as grains, some field vegetables (Beaton, 2006) and root vegetables (Givens *et al.*, 2006). Infants and children have disproportionately high exposures to these pesticides from dietary sources, as they eat more per unit body-weight and their diets tend to contain more fruit and vegetable products (Givens *et al.*, 2006). Measuring human exposure from any pesticide source is difficult. Short-term exposure to pesticides that do not accumulate in fatty tissues can be calculated using the quantities of their metabolites in urine, and this method has been used to determine the occupational exposure of agricultural workers (Tuomainen *et al.*, 2002; Hardt and Angerer, 2003; Coronado *et al.*, 2004; Jagt *et al.*, 2004), the effectiveness of an organic diet in reducing childhood pesticide exposure (Lu *et al.*, 2006) and as an exposure indicator in national surveys (Murphy *et al.*, 1983; Kieszak *et al.*, 2002; Barr *et al.*, 2005). Sampling to determine the presence of pesticides in the environment can also give an indication of human environmental exposure, particularly when used in combination with urine-metabolite monitoring (Curwin *et al.*, 2005; Ott, 2005). Unfortunately, such tests cannot differentiate between exposure to the pesticides and exposure to metabolites.

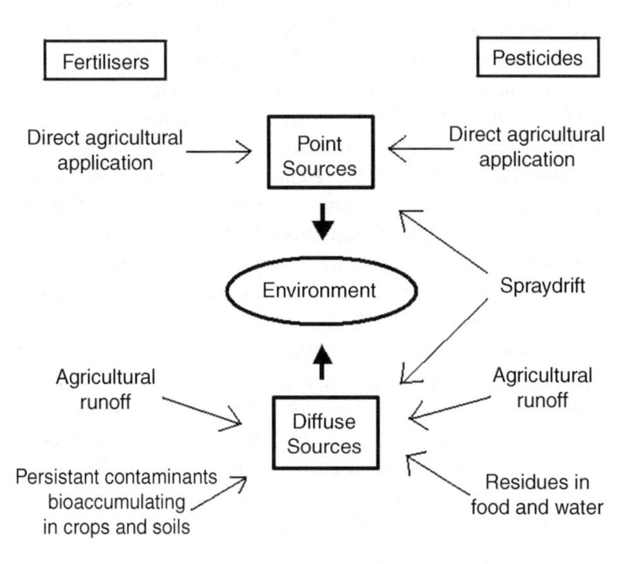

Figure 7.4 Primary pathways of environmental exposure to fertilisers and pesticides

Measuring body burdens of lipophilic pesticides requires invasive procedures such as blood sampling, tissue biopsy, or the collection of breast milk; the latter may be difficult to obtain and excludes all but lactating women. Many studies have attempted to determine the body burden of some pesticides and xeno-oestrogens (Fernandez *et al.*, 2004) in individuals and groups such as lactating mothers (Nakagawa *et al.*, 1999; Sun, 2005), pesticide appliers (Arbuckle *et al.*, 2005), farming families and cancer patients (Rusiecki *et al.*, 2005). Unfortunately, different protocols have been used, so the results are difficult to compare. Some studies have sampled human populations nationally, for example in the USA and Netherlands (Murphy *et al.*, 1983; WWF, 2003; Barr *et al.*, 2005). WWF has recommended the adoption of national biomonitoring programmes (WWF, 2003, 2004, 2005), but so far Belgium is the only European country to comply (WWF, 2003).

7.2.4 Effects on receptors

Pesticides and toxic trace elements in fertilisers are frequently at such low concentrations, typically in the range of parts per million or billion, that their risk to human health or the environment has been questioned (Wade *et al.*, 2002). However, chemical mixtures can act additively or synergistically to cause biological effects, even when the levels of the individual chemicals are well below non-observed/non-adverse-effect levels (NOAELs) (Kortenkamp, 2008). Low-dose mixtures may not cause acute or chronic toxicity, but they can interfere with biochemical signalling involved in growth and development, and with the maintenance of homeostasis (Diamanti-Kandarakis *et al.*, 2009; Sharpe, 2004). Organisms may be more affected during fetal development, particularly during organogenesis (between the third and eighth weeks in humans), and future generations can be affected if gametogenesis is perturbed (Sultan, 2001; Hardell *et al.*, 2006; Sharpe, 2006).

Many pesticides affect the nervous system causing subtle or overt damage following acute or chronic administration; some, such as organochlorines, organophosphates, carbamates and bipyridines can cause severe damage.

7.2.4.1 Organochlorine pesticides

The insecticide DDT and its analogues, the pyrethroids, the cylcodiene compounds, lindane and many other pesticides have neurotoxic properties (Doherty, 1979; Ecobichon and Joy, 1993). The impact of this group of pesticides on organisms is determined primarily by their concentration in tissues, especially the nervous system (Ecobichon and Joy, 1993). Dose–response data for DDT in humans are sparse, but estimates of its toxicity are available (Table 7.4).

Table 7.4 Dosage-response of humans to DDT (Adapted from Hayes, 1963)

Dosage (mg/kg per day)	Consequence
0.0004	Dosage of the general US population in 1968
0.0025	Dosage of the general US population in 1953–1954
0.25	Tolerated by workers for 19 years
0.5	Tolerated by volunteers for 21 months
0.5	Tolerated by workers for 6.5 years
1.5	Tolerated by volunteers for 6 months
6[a]	Moderate poisoning in one man
10[a]	Moderate poisoning in some men
16–286[a]	Prompt vomiting at higher doses, all poisoned, convulsions in some
Unknown[a]	Fatal

[a]Single dose only.

DDT toxicity following acute ingestion usually lasts 1–2 days (Ecobichon and Joy, 1993). Studies on volunteers indicated that amounts below 10 mg/kg usually produce few symptoms; levels between 10 and 16 mg/kg produce moderate to severe symptoms, and exposures of up to 20 mg/kg can induce convulsions and death (Ecobichon and Joy, 1993). In similar experiments, amounts up to 500 mg were reported to produce no effects except hypersthesias of the mouth and lips (Hayes, 1959). At 750 mg, sensory disturbances involving the face developed with some motor impairment (Hayes, 1959). Following the ingestion of 1500 mg (approximately 20 mg/kg), severe paresthesia of the face and mouth developed (Hayes, 1959). Equilibrium was impaired for a few hours and the subjects experienced dizziness, confusion and tremor of the extremities, accompanied by general malaise, headache and profound fatigue (Hayes, 1959). However, according to Ecobichon and Joy (1993), ingestion of DDT has proved fatal although the effects of chronic toxicity of DDT may emerge slowly.

Cyclodiene and hexachlorocyclohexane derivatives (HCH-CYCs) such as lindane and dieldrin are toxic at lower levels than DDT. Symptoms reflect CNS effects non-neurological sequalae (Ecobichon and Joy, 1993). Acute exposure to 20–40 mg/kg causes convulsions, and higher exposures can be fatal (Ecobichon and Joy, 1993). Symptoms of acute poisoning include exaggerated motor reflexes, myoclonic jerking and convulsions, while headache, nausea, vomiting and dizziness may occur (Ecobichon and Joy, 1993).

The most typical response to ingestion of toxic amounts of the HCH-CYCs is convulsive seizures, which can cause neuronal cell death and persisting neurological symptoms, especially in children (Ecobichon and Joy, 1993). Jager (1970) identifies two types of chronic exposure syndromes from HCH-CYCs. The first occurs when constant exposure results in the slow accumulation of the insecticide with progressive symptomology (Jager, 1970). The second occurs when the intake of the insecticide remains below that required for overt symptoms but the subject is rendered sensitive to acute exposure (Jager, 1970).

Chronic exposure to levels greater than 0.03 mg/kg/day may cause intoxication and convulsions in some people, and repeated chronic exposure may induce histopathological changes in the liver and kidney (Ecobichon and Joy, 1993).

Persistent neurological sequelae may result from acute or chronic exposure to HCH-CYCs, and changes in electroencephalogram (EEG) occur; some degree of abnormality may persist for weeks or months (Ecobichon and Joy, 1993). Symptoms may be more persistent from chronic exposures. Other studies suggest that chronic exposure to HCH-CYCs causes alterations in neurological and psychological functions. Selective hearing loss in cases of acute and chronic poisoning involving organochlorine or organophosphate agents has been reported (Muminov, 1972).

7.2.4.2 *Organophosphate pesticides*

Organophosphate insecticides such as malathion, methyl parathion, fenitrothion, diazinon and guthion are not as acutely toxic and persistent as organochlorine pesticides (Ecobichon and Joy, 1993). Acute toxicity is associated with accumulation of the neurotransmitter acetylcholine at nerve terminals (Ecobichon and Joy, 1993). The signs and symptoms of organophosphate insecticide poisoning are summarised in Table 7.5.

Mild organophosphate poisoning results in symptoms similar to the common cold (Ecobichon and Joy, 1993). In cases of severe poisoning, symptoms include marked muscle weakness and twitching (Ecobichon and Joy, 1993). Effects on the CNS from severe poisoning include tension, anxiety, restlessness, nervousness, giddiness, impairment of memory and the ability to concentrate, and insomnia with excessive dreaming and nightmares (Ecobichon and Joy, 1993). Sometimes, fatigue, lethargy, apathy, withdrawal and depression can occur (Durham *et al.*, 1965). The sudden development of respiratory insufficiency within 1–4 days of exposure can cause paralysis, coma and death if untreated (Senanayake and Karalliedde, 1987).

Organophosphates can produce delayed neurological lesions in exposed individuals, but the chronic effects are difficult to detect (Ecobichon and Joy, 1993) and are therefore not well documented. In a study of previously poisoned individuals and

Table 7.5 Signs and symptoms of poisoning from organophosphate insecticides (Ecobichon and Joy, 1993)

Nervous tissue and receptors affected	Site affected	Manifestations
Parasympathetic automatic (muscarinic receptors) post-ganglionic nerve fibres	Exocrine glands	Increases salivation, lachrymation, perspiration
	Eyes	Miosis (pinpoint and nonreactive), ptosis, blurring of vision, conjunctival injection, bloody tears
	Gastrointestinal tract	Nausea, vomiting, abdominal tightness, swelling and cramps, diarrhoea, tenesmus, faecal incontinence
	Respiratory tract	Excessive bronchial secretions, rhinorrhea, wheezing, oedema, tightness in chest, bronchospasms, bronchoconstriction, cough, bradpynea, dysnea
	Cardiovascular system	Bradycardia, decrease in blood pressure
	Bladder	Urinary frequency and incontinence
Parasympathetic and sympathetic autonomic fibres (nicotinic receptors)	Cardiovascular system	Tachycardia, pallor, increase in blood pressure
Somatic motor nerve fibres (nicotinic receptors)	Skeletal muscles	Muscle fasciulations (eyelids, fine facial muscles), cramps, diminished tendon reflexes, generalised muscle weakness in peripheral and respiratory muscles, paralysis, flaccid or rigid tone Restlessness, generalised motor activity, reaction to acoustic stimuli, tremulousness, emotional lability, ataxia
Brain (acetylcholine receptors)	Central nervous system	Drowsiness, lethargy, fatigue, mental confusion, inability to concentrate, headache, pressure in head, generalised weakness
		Coma with absence of reflexes, tremors, Cheyne-Stokes respiration, dyspnoea, convulsions, depression of respiratory centres, cyanosis

controls, differences involving the CNS were apparent, including in cognitive function, academic skills, abstraction, flexibility of thinking and simple motor skills (Savage *et al.*, 1988). Twice as many cases had test scores characteristic of individuals with cerebral damage or dysfunction (Savage *et al.*, 1988). In patients poisoned by parathion, tension, depression, crying spells, anxiety, mild anorexia and insomnia lasted up to 3 weeks after exposure (Grob *et al.*, 1950). Studies on agricultural pilots suggest that mental alertness and reflex responses could be impaired by continuous exposure to organophosphates, but they recovered when removed from the source for a period of time (Durham *et al.*, 1965). Several behavioural sequelae such as impaired vigilance, reduced concentration and information-processing speed, memory deficit, linguistic disturbances, depression, anxiety and irritability have been linked to poisoning by organophosphate insecticides (Levin and Rodnitzky, 1976).

7.2.4.3 Chronic diseases

Chronic human diseases such as Alzheimer's disease (AD) and Parkinson's disease (PD) have been linked to pesticide exposure (Fleming *et al.*, 1994; Barbeau *et al.*, 1986; Semchuk *et al.*, 1991; Butterfeld *et al.*, 1993; Gauthier *et al.*, 2001). Alzheimer's disease is defined by specific neuropathological and neurochemical features such as senile plaques (Duyckaerts *et al.*, 1986) and neurofibrillary tangles (NFT) (Ball, 1977). The aetiology of AD remains poorly understood, but it is suggested that it could be multifactorial, involving genetic predisposition, exposure to environmental factors modulated by biological aging (Gauvreau, 1987) and, more recently, a lack of B vitamins in the diet (de Lau *et al.*, 1970; Politis *et al.*, 2010). The toxic effects of pesticides on humans such as the generation of free radicals by DDT, dieldrin and paraquat or inhibition of acetylcholinesterase (AchE) by parathion and fenitrothion may contribute to the cholinergic system deficiency and production of free radicals observed in AD (Gauthier *et al.*, 2001). In one epidemiological study, a long-lasting residue of DDT (pp-DDT) was found in the majority of AD as well as PD cases, suggesting a link with pesticides (Fleming *et al.*, 1994).

Parkinson's disease is a late-onset, progressive motor disease marked by selective degradation of dopaminergic neurons of the *substantia nigra* and the formation of fibrillar cytoplasmic inclusions (Baba *et al.*, 1998). The cause is unknown, but epidemiological studies suggest an association with pesticides and other environmental toxins. Cancer has been reported at high rates among agricultural workers and farmers (Costello *et al.*, 2009). Chronic exposure to a common pesticide such as rotenone can reproduce the anatomical, neurochemical and neuropathological features of PD (Steenland, 1996). In a study by Richardson *et al.* (2009), elevated serum levels of the organochlorine pesticide beta-hexachlorocyclohexane (beta-HCH) were linked to PD.

Many pesticides have been found to be abundant in the nervous system of exposed individuals. Pesticides targeting ion channels (Hendy and Djamgoz, 1988) and neurotransmitter receptors (Lang and Bastian, 2007) have been found to be associated with cancer initiation and progression (Djamgoz, 2011). In addition, in a recent study on mice using 'activity-based protein profiling', Nomura and Casida (2011) tested 29 widely used organophosphorus and thiocarbamate pesticides and showed that these can affect a multitude of unintended targets in the brain, including several neurologically significant enzymes.

Voltage-gated sodium channels (VGSCs), are the main targets of DDT and some pyrethroid insecticides (Bloomquist, 1996). The main effect of these pesticides is to prolong channel opening and kill insects by hyper-excitation. Importantly, VGSC over-expression and activity have also been associated with metastatic progression of several carcinomas, including those of lung, ovary, colon and cervix (Djamgoz, 2011). Exposure to DDT, especially in early life, has been shown to increase the risk of several cancers (Cohn *et al.*, 2007, 2010; Rogan and Chen, 2005). Accordingly, the International Agency for Research on Cancer has classified DDT as a possible human carcinogen.

Several neurotransmitter receptors and ligand-gated ion channels have been found to be involved in cancer (e.g. Lang and Bastian, 2007). GABA receptors are targeted by organochlorines, so organochlorine residues may be an important aetiological factor in breast cancer and this could involve oestrogenic activity (Wolff *et al.*, 1993; Hoyer *et al.*, 1998). However, some studies (e.g. Hunter *et al.*, 1997) failed to associate organochlorines with breast cancer. These inconsistent findings could reflect the fact that GABA produces mixed effects on cancer cells. For example, Azuma *et al.* (2003) and Takehara *et al.* (2007) reported its promotory effects on prostate and pancreatic cancer, whilst Ortega (2003) proposed a generally inhibitory role of GABA on tumour-cell migration. Several pesticides can mimic GABA or block GABA receptors. Further work is required to understand the role of neurotransmitter systems in the cancer process before the potential carcinogenic action of pesticides can be evaluated fully. Links between exposure to pesticides and endocrine disruption in humans were suggested as early as 1949, when low sperm counts were observed in men involved in the aerial application of DDT (Singer, 1949). More recently, exposure to EDC pesticides has been implicated in the aetiologies of various cancers (Garry, 2004; Mathur *et al.*, 1998), miscarriage and other reproductive disorders (Garry, 2004; Nicolopoulou-Stamati and Pitsos, 2001), genital deformities (Baskin *et al.*, 2001), other birth defects (Schreinemachers, 2003), behavioural abnormalities (Zala and Penn, 2004) and skewed offspring-sex ratios (Garry, 2004; Mackenzie *et al.*, 2005). Elevated rates of disease are reported in populations living in areas with high exposure to EDCs (including pesticides), such as Windsor, Canada (Gilbertson and Brophy, 2001), densely populated agricultural areas in Gaza, Palestinian Territories (Safi, 2002), children exposed to pesticides in their homes (Menegaux *et al.*, 2006) and female agricultural workers in Jaipur, India (Mathur *et al.*, 1998). Two meta-analyses of studies examining links between pesticide exposure and prostate cancer reported a positive correlation, with the strongest identified by the meta-analysis covering the greatest number of studies and the longest time (Van Maele-Fabry and Willems, 2004).

Some pesticides are inhibitors of aromatase, the enzyme which converts testosterone to oestradiol. *In vitro* testing has shown that several commonly used medicinal azole fungicides, including bifonazole, miconazole, and clotrimazole, are as potent inhibitors of aromatase as anti-oestrogen drugs (Scholz *et al.,* 2004; Trösken *et al.,* 2006). Although some aromatase inhibitors can be used safely as anti-cancer agents under medical supervision, exposing healthy individuals to such substances, particularly during sensitive periods of development such as the differentiation and development of the sex organs in embryogenesis, is of concern (Dickerson and Gore, 2007).

7.3 Fertilisers

7.3.1 Hazardous properties

Both inorganic and organic fertilisers have hazardous properties. Anhydrous ammonia is one of the most dangerous substances used in agriculture. It has a boiling point of –33°C and must be stored and applied under pressure. This requires specialist equipment and protective clothing (which must be properly maintained) and training to ensure safe use. Protective clothing must be worn while handling or applying it, since it is extremely caustic. Upon release, it boils and reacts immediately with any available source of water, including that found in biological tissues. Ammonia gas is an irritant and can damage the respiratory tract (Pritchard, 2007). Anhydrous ammonia can corrode copper, zinc, aluminium and their alloys, and reacts explosively with strong oxidising agents, hydrocarbons, ethanol and some halogen-containing compounds. In combination with chlorine, it forms the irritant chloramine gas (Yost, 2007).

Ammonium nitrate, a powerful oxidising agent, can irritate the eyes, skin and respiratory tract. Chronic exposure can cause muscular weakness, depression, headache and mental impairment. It acts as an oxidising agent, presenting a fire risk in places where it is stored in large quantities, and can be induced to decompose explosively (HSE, 1986). Numerous accidents have been caused by its accidental detonation: the Oppau explosion in Germany in 1921, for example, which involved ammonium nitrate and ammonium sulphate, killed 561 people and injured almost 2000. To reduce this risk, the nitrogen content and combustible content of ammonium-nitrate fertilisers is tightly regulated and only a limited amount can be stored on farms (HSE, 1986; US EPA, 1997). Urea is an irritant to the eyes and the respiratory and gastrointestinal tracts, and can decompose into toxic products. When in a solid or highly concentrated form, it can also react explosively with substances such as strong oxidising agents and calcium or sodium hypochlorate (Young, 2007).

Other fertilisers which are potentially harmful during handling and application include potassium chloride and calcium oxide (quicklime). Calcium oxide reacts with water to form calcium hydroxide. It can cause severe irritation if the water is from damp skin, the eyes or mucous membranes (Lide, 2004). Although potassium and phosphate fertilisers themselves are not thought to pose any direct threat to the health of humans and other animals, they may be contaminated with toxic trace elements and radioisotopes if these elements are present in the source rock. Repeated application can cause these elements to accumulate in soil to unsafe levels (Chapters 4 and 5). Some compounds, such as potassium chloride and sulphate, can make toxic trace elements such as cadmium more bioavaliable by reacting to form compounds which are absorbed preferentially by plants (Zhao *et al.*, 2004).

Manures and human sewage can contain toxic trace elements and other pollutants, including EDCs. Although these do not bioaccumulate in grazing animals, the *in utero* development of their offspring can be disrupted, including impaired ovarian development (Fowler *et al.*, 2008; Rhind *et al.*, 2010). Improperly composted manure can contain parasites and other pathogens transmissible to humans such as pig roundworm (*Ascaris suum*), and pathenogenic *Salmonella*, *Campylobacter* and *E. coli* species (Strauch, 1991). Allowing large amounts to ferment in a confined space such as a slurry pit can create a great deal of methane, a potential explosion hazard. Ammonia and hydrogen sulphide are also produced and can reach levels hazardous to health (Groves and Ellwood, 1991).

As well as being used as fertilisers, animal manures can be used as cheap feedstuffs, rich in crude protein. Guano can be used to increase the nitrogen-to-carbon ratio of ruminant feed (FAO, 2010b), but this can pose health risks both to the animals themselves and to humans eating their products, from the high levels of TTEs and transmission of parasites and prions (McLoughlin *et al.*, 1988; Tokarnia *et al.*, 2000; Henry, 2004).

7.3.2 Sources

Fertiliser use depends on its cost, soil nutrient status, and the crop grown. In most developed countries, soil-nutrient analyses are used to determine the nutrients that should be added. Arable crops have much higher requirements than pasture. Overall, an estimated 102.4 million tonnes of nitrogen, 37.2 million tonnes of phosphorus and 22.9 million tonnes of potassium-containing fertilisers were applied to crops worldwide in 2009–2010. The adoption of controlled-release fertilisers, which better target nutrient release, is a factor in falling demand for fertilisers (Trenkel, 1997).

Increased demand for food, especially meat, fish and dairy products, and the adoption of biofuels are now increasing the global demand for fertilisers, particularly in developing countries, despite the global economic downturn (FAO, 2008) (Table 7.6).

7.3.3 Pathways and environmental fate

Nitrogen- and phosphate-containing fertilisers are used to replace and augment natural nitrogen and phosphate. In nature, both move through the environment following cycles. Nitrogen moves between soil, water and biomass as a result of

Table 7.6 Forecast regional and sub-regional fertiliser consumption, 2007/2008–2011/2012 (FAO, 2008)

Regions and sub regions	Nitrogen		Phosphorus		Potassium	
	Share of world consumption (%)	Annual growth (%)	Share of world consumption (%)	Annual growth (%)	Share of world consumption (%)	Annual growth (%)
World		1.4		2.0		2.4
Africa	3.4	2.9	2.5	1.0	1.6	2.0
North America	13.5	0.3	12.0	0.5	17.1	0.7
Latin America	6.3	2.4	13.0	2.8	17.5	2.9
West Asia	3.5	1.7	3.3	1.0	1.4	2.4
South Asia	19.6	2.2	20.5	3.5	10.9	4.2
East Asia	38.3	1.3	36.1	1.9	35.2	3.3
Central Europe	2.7	1.8	1.5	1.2	2.4	1.0
West Europe	8.4	−0.3	5.6	−0.7	9.5	0.0
East Europe and Central Asia	3.0	2.4	2.0	4.5	3.1	1.6
Oceania	1.4	4.9	3.5	1.7	1.3	2.1

nitrogen-fixing bacteria and lightning, which turns atmospheric nitrogen into ammonia; ammonia is also released by the decomposition of plant and animal matter by micro-organisms (Delwiche, 1970). Nitrogen fertilisers are applied to crops in both slow and fast-release forms. Slow-release forms, such as urea formaldehyde compounds and sulphur or polymer-coated fertiliser particles provide a steady supply of nitrogen over several weeks. Fast-release forms, such as ammonium nitrate and urea, have the advantage of being available to growing plants immediately but pose a greater pollution threat. Ammonia and nitrous oxide released from the breakdown of nitrogen-containing compounds are greenhouse gases (Matson *et al.*, 1998).

Phosphate is weathered from rock and soil, and can become immobilised on clay particles or leach into water. It can then be absorbed by plants and enter the food chain, be deposited in sediments in lakes and rivers, or travel into the sea where it eventually ends up deposited on the deep sea bed. Dead organisms release phosphate as they decompose back into the soil and water. Eventually, phosphates may end up on the sea floor, from where they may re-enter the cycle if tectonic activity raises the Earth's crust, and rocks formed from phosphate-bearing sediments are presented to the atmosphere for weathering. As with nitrogen, slow-release and fast-release phosphate-containing fertilisers can be used to tailor soil phosphate availability to plant growth. Fast-release fertilisers contain water-soluble forms of phosphate, such as superphosphate or ammonium phosphate, whilst slow-release ones such as rock phosphate contain insoluble forms which become available over time (Shaviv and Mikkelsen, 1993).

Both inorganic and organic nitrogen and phosphorus-containing fertilisers (including human and animal waste) can pollute water courses and cause eutrophication. Intensive agriculture increases nitrate levels in stream water significantly (Shaviv and Mikkelsen, 1993). Areas of Europe such as south-east England, northern and central France and parts of Italy and central Europe have high nitrate levels, reflecting intensive agriculture (Figure 7.5).

Soils treated with inorganic fertilisers can undergo chemical changes which affect their quality, and in some cases render soil unusable (Scherr, 1999). Superphosphate, for example, is made by treating rock phosphate with sulphuric or phosphoric acid, and releases these acids as it degrades. Urea, which degrades to ammonia, releases ammonium hydroxide, and when coated with sulphur or combined with formaldehyde (urea formaldehyde) releases sulphuric acid or formaldehyde. Sulphuric acid is also released by ammonium sulphate fertiliser and, in soils with high calcium contents, reacts to form calcium sulphate (gypsum), which accumulates in dry conditions making it difficult for plants and micro-organisms to absorb water. In contrast, in waterlogged soils, sulphur-containing compounds react anaerobically to form hydrogen sulphide. Ammonia compounds destroy organic matter via oxidation, making soil more prone to erosion and releasing nitrates. Also, persistent pollutants such as trace elements in fertilisers can accumulate in soils over time (Chapter 4).

7.3.4 Effects on the environment and human health

When phosphate and nitrate exceed the absorption capacity of a terrestrial or aquatic ecosystem, the plant species balance changes in favour of plants such as algae (some toxic) and coarse grasses which grow well in nutrient-rich environments. This leads to eutrophication. In aquatic environments, sunlight cannot penetrate and growth of bottom-dwelling organisms is inhibited. This, coupled with the oxygen demands of rotting vegetation, reduces oxygen levels, destroying fish and other aquatic animals. Eutrophication can be rapid, but can take a long time to reverse. For example, sediment core data indicate that Lake Seebergsee in the Swiss Alps took only 8 years to undergo eutrophication from intensive cattle grazing in the thirteenth century, but 88 years to recover after it was aban-

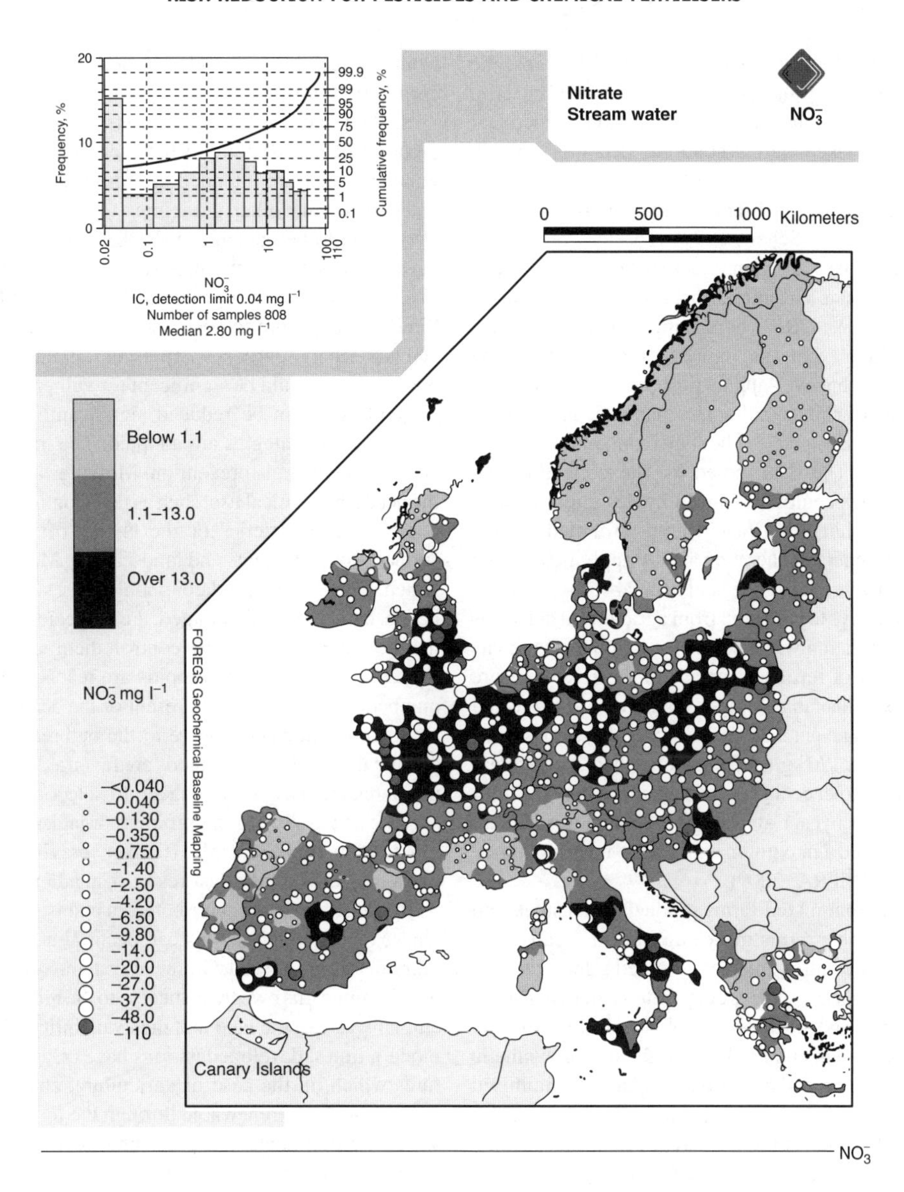

Figure 7.5 Nitrate concentrations in European streams (Salminen *et al.*, 2005)

doned in the 1600s (Hausmann *et al.*, 2002). In terrestrial ecosystems the diversity of plant species is reduced as tall, nutrient-loving species out-compete smaller species (Walker *et al.*, 2009). In humans, the consumption of very large quantities of nitrates has been implicated in the aetiology of methaemoglobinaemia, a condition where the ferric iron found in haemoglobin is oxidised by nitrates present in the blood and inhibits the transport of oxygen, leading to hypoxia and cyanosis. Young infants are most susceptible. Chronic exposure to lower but excessive quantities of nitrates has been implicated in increasing rates of gastrointestinal cancers, although this only applies to the nitrate-containing compounds absorbed via the consumption of processed meats (Milkowski *et al.*, 2010).

Nitrate concentrations in food and drinking water are a cause for concern. Levels in foods produced using synthetic fertilisers are often found to be raised and some, such as leafy salad vegetables, can reach concentrations exceeding those deemed legally acceptable (Worthington, 2001; Pussemier *et al.*, 2006).

7.4 Risk reduction for pesticides and chemical fertilisers

Risks can be reduced by reducing or eliminating the hazard or by taking measures to prevent it reaching sensitive receptors such as vulnerable organisms or ecosystems. Although organic agriculture may help eliminate many agricultural hazards, the use of artificial fertilisers and pesticides has increased crop yields, so the costs and benefits of conventional chemical agriculture must be taken into account in deciding how to reduce risk.

Reliance on artificial fertilisers causes many problems other than eutrophication (Scherr, 1999). The loss of organic matter from soils makes them prone to erosion, and trace elements vital to human and animal health such as selenium and iodine can be depleted. Returning animal and human waste to land could help prevent the loss of organic matter and depletion of nutrients, but over-application can lead to eutrophication via nutrient leaching, and some manures, particularly those derived from sewage, are heavily contaminated with persistent pollutants. The risks posed by the collection and application of manures could be reduced by designing processing facilities such as slurry pits to avoid build-up of methane and heat, for example by energy generation and by applying no more of the slurry or manure than the soil can absorb. In many countries, areas sensitive to pollution from agricultural activities are protected by legislation which limits the number of animals kept and the amount of fertiliser spread. In the UK these are known as 'Nitrate Vulnerable Zones'. Manures are protected from the elements to prevent leachate escaping. Livestock must produce no more than 170 kg of nitrogen per hectare per year (unless the farm is >80 per cent grassland, in which case the limit is 250 kg per hectare per year), and fertiliser application from all sources may not exceed a maximum application rate designated for each crop type (DEFRA, 2009).

The move away from mixed farms producing livestock and crops to specialist arable farms makes the use of locally produced animal manure unfeasible, so synthetic fertilisers or processed organic fertilisers such as dehydrated chicken manure or sewage sludge are used. Slow or controlled-release fertilisers could reduce the environmental risks from the use of synthetic fertiliser. Fertiliser losses to the environment can be minimised by timing the application to provide peak nutrient availability when the nutrient demands of the crop are highest. The production of organic fertilisers is more involved generally than that of synthetic fertilisers, making them more expensive. They therefore command only a very small share of the fertiliser market (Chen *et al.*, 2008). Some fertilisers are in granules coated with organic thermoplastic or resin polymers, which may release potentially harmful compounds. The problem of persistent pollutants in sewage sludge is another issue which is difficult to resolve. Risk-reduction options currently include avoiding the use of sewage sludge, monitoring levels of pollutants so only acceptable accumulation limits are reached, or targeting it to crops less likely to suffer from or transmit the pollutants. Sludge application may cause reproductive and other abnormalities in the offspring of pregnant animals grazed on treated fields, although most of the pollutants do not accumulate in their tissues (Fowler *et al.*, 2008; Rhind *et al.*, 2010). Hence animals destined for slaughter could be fed fodder grown on treated land, but this would remain off limits to breeding stock or crops grown for human consumption. Methods of removing persistent pollutants from sewage sludge are clearly needed to allow the safe reincorporation of the nutrients it contains into the food chain.

In 2005, the Royal Commission for Environmental Pollution reported that a 2 m boundary between sprayed areas and homes, a voluntary measure intended to minimise human exposure, was inadequate, and other guidelines to minimise spraydrift were also found to be inadequate (RCEP, 2005). This demonstrated the urgent need for the exposure of bystanders such as rural residents to be properly assessed and reduced in many areas. Similarly, exposure to pesticides used for municipal, domestic and veterinary purposes represent an unquantified route of exposure that could be significant for vulnerable people. Overall pesticide use can be reduced significantly if integrated pest-management strategies are adopted. The first line in integrated pest management is prevention. Measures are taken that reduce the need for chemical use, such as the adoption of pest or disease-resistant crop varieties or the choice of weed-resistant hard surfaces for pavements and landscaping. Management practices which increase pest-predator populations, such as the creation of beetle banks, are encouraged. Pest species are identified and monitored, and action to control them is taken only when predetermined action thresholds are reached. These can be the numbers of a pest species present or the occurrence of environmental conditions conducive to the outbreak of the pest. Non-chemical pest-control methods are favoured, and when chemical measures are used, care is taken to select a chemical that is as specific as possible to the problem organism and poses as little risk to humans as possible. Blanket spraying with non-specific pesticides is used as a last resort (Chandler, 2008).

In nature, nutrients cycle between plants, animals and the soil. Under most conditions this can occur almost indefinitely since nutrients are conserved within the system. Systems of agriculture which mimic this owe their success to the return of all human and animal waste to the land and the availability of cheap labour. In modern times, developed nations have created linear infrastructures which, in the case of agriculture, channel raw materials that are often not renewable through the food chain and into the wider environment, often causing pollution or causing the nutrients to be lost to the deep oceans. Many obstacles need to be overcome before this system can be transformed into one which is sustainable and poses less risk to human and environmental health. Safely recycling human waste, for example, is difficult at present, since it arrives at the sewage works mixed with industrial waste, household cleaning products and run-off water, all of which contain persistent pollutants. Upgrading the sewage systems to carry sewage and other waste separately would be a costly project which would be difficult to incentivise or implement in the near future. Developing countries, particularly those in water-stressed regions, could, however, develop their nascent systems in such a way that human waste was collected separately from waste water. There is also scope to develop new integrated processes to recover phosphate, urea, other nutrients and valuable metals and trace elements after sewage treatment and to use methane gas for energy. Novel techniques such as ozonation and microwave are being investigated to enable phosphate recovery (Saktaywin *et al.*, 2005; Liao *et al.*, 2005). The need to adopt cradle-to-cradle processes as advocated by Braungart is discussed further in the final chapter of this book.

References

Akhtar, N., S. Kayani, S.A., Ahmad, M.M.and M. Shahab (1996) Insecticide-induced changes in secretory activity of the thyroid gland in rats. *Journal of Applied Toxicology* **16**(5), 397–400.

Ambroise, D. (2005) Cancer mortality among municipal pest-control workers. *International Archives of Occupational and Environmental Health* **78**(5), 387–93.

Andersen, H. R. (2002) Effects of currently used pesticides in assays for estrogenicity, androgenicity, and aromatase activity *in vitro*. *Toxicology and Applied Pharmacology* **179**(1), 1–12.

Andersen, J. H. and M. E. Poulsen (2001) Results from the monitoring of pesticide residues in fruit and vegetables on the Danish market, 1998–99. *Food Additives and Contaminants* **18**(10), 906–31.

Anderson, H. A., Falk, C., Hanrahan, L., Olson, J., Burse, V. W., Needham, L., Paschal, D., Patterson, D. Jr., and Hill, R. H. Jr (1998) Profiles of Great Lakes Critical Pollutants: A Sentinel Analysis of Human Blood and Urine. *Environmental Health Perspectives* **106**(5),279–89.

Arbuckle, T. E., Cole, D. C., Ritter, L. and Ripley B. D. (2005) Biomonitoring of herbicides in Ontario farm applicators. *Scandinavian Journal of Work, Environment and Health* **31**, 90–7.

Arya, N. (2005) Pesticides and human health: why public health officials should support a ban on non essential residential use. *Canadian Journal of Public Health* **96**, 89–92.

Azuma, H., Inamoto, T., Sakamoto, T., Kiyama, S., Ubai, T., Shinohara, Y., Maemura, K., Tsuji, M., Segawa, N., Masuda, H., Takahara, K., Katsuoka, Y. and Watanabe, M. (2003) Gamma-aminobutyric acid as a promoting factor of cancer metastasis; induction of matrix metalloproteinase production is potentially its underlying mechanism. *Cancer Research* **63**, 8090–6.

Baba, M., Nakajo, S., Tu, P.H., Tomita, T., Nakaya, K., Lee, V.M., Trojanowski, J.Q. and Iwatsubo, T. (1998) Aggregation of alpha-synuclein in Lewy bodies of sporadic Parkinson's disease and dementia with Lewy bodies. *American Journal of Pathology* **152**, 879–84.

Bakir, F., Damluji, S.F., Amin-Zaki, L., Murtadha, M., Khalidi, A., al-Rawi, N.Y., Tikriti, S., Dhahir, H.I., Clarkson, T.W., Smith, J.C. and Doherty, R.A. (1973) Methylmercury poisoning in Iraq. *Science* **181**(4096), 230–41.

Ball, M.J. (1977) Neuronal loss, neurofbrillary tangles and granulo-vacuolar degeneration in hippocampus with ageing and dementia. *Acta Neuropathoogy* **37**, 111–18.

Barbeau, A., Roy, M., Cloutier, T., Plasse, L. and Paris, S. (1986) Environmental and genetic factors in the etiology of Parkinson's disease. *Advances in Neurology* **45**, 299–306.

Barr, D.B., Allen, R., Olsson, A.O., Bravo, R., Caltabiano, L.M., Montesano, A., Nguyen, J., Udunka, S., Walden, D., Walker, R.D., Weerasekera, G., Whitehead, R.D. Jr, Schober, S.E. and Needham, L.L. (2005) Concentrations of selective metabolites of organophosphorus pesticides in the United States population. *Environmental Research* **99**(3), 314–26.

Baskin, L.S., Himes, K. and Colborn, T. (2001) Hypospadias and endocrine disruption: Is there a connection? *Environmental Health Perspectives* **109**(11), 1175–83.

BCPC (2005) During Spraying: Avoiding Drift. *Crop Protection Agency Best Practice Guides*. Crop Protection Agency: 1.

Beard, A.A.P. (1999) Thyroid function and effects on reproduction in ewes exposed to the organochlorine pesticides lindane or pentachlorophenol (PCP) from conception. *Journal of Toxicology and Environmental Health. Part A* **58**(8), 509–30.

Beard, J. (2006) DDT and human health. *Science of the Total Environment* **355**(1–3) 78–89.

Beaton, C. (ed.) (2006) *The Farm Management Handbook 2006/7*. Edinburgh: Scottish Agricultural College.

Beeman, R. W. (1982) Recent Advances in Mode of Action of Insecticides. *Annual Review of Entomology* **27**, 253–81.

Benbrook, C.M. (2004) Genetically engineered crops and pesticide use in the United States: The first nine years. *BioTech InfoNet Technical Paper Number* **7**, 1–53.

Black, B.C., Hollingworth, R.M., Ahammadsahib K.I., Kukel C.D. and Donovan, S. (1994) Insecticidal action and mitochondrial uncoupling activity of AC-303, 630 and related halogenated pyrroles. *Pesticide Biochemistry and Physiology* **50**(2), 115–28.

Bloomquist, J.R. (1996) Ion channels as targets for insecticides. *Annual Review of Entomology* **41**, 163–90.

Bolger, R., Weise, T.E., Ervin, K., Nestich, S. and Checovich, W. (1998) Rapid screening of environmental chemicals for estrogen receptor binding capacity. *Environmental Health Perspectives* **106**(9),551–7.

Borisover, M.D. and Graber, E.D. (1996) Specific interactions of organic compounds with soil organic carbon. *Chemosphere* **34**(8),1761–6.

Bouvier, G., Blanchard, O., Momas, I. and Seta, N. (2006) Environmental and biological monitoring of exposure to organophosphorus pesticides: application to occupationally and non-occupationally exposed adult populations. *Journal of Exposure Science Environmental Epidemiology* **16**, 417–26.

Bro-Rasmussen, F. (1996) Contamination by persistent chemicals in food chain and human health. *Science of the Total Environment* **188**(Suppl 1), S45–60.

Brown, A.E. (2006) *Mode of action of pesticides and related pest control chemicals for production agriculture, ornamentals and turf*. Maryland Cooperative Extension. Pesticide Information Leaflet No. 43.

Brown, L.R., Renner, M. and Flavin, C. (1997) *Vital signs 1997: The environmental trends that are shaping our future*. W.W. Norton for Worldwatch Institute, New York.

Büchel, K.H. (1972) Mechanisms of action and structure activity relations of herbicides that inhibit photosynthesis. *Pesticide Science* **3**, 89–110.

Buckler-Davis, F. (1998) Effects of environmental synthetic chemicals on thyroid function. *Thyroid* **8**(9), 827–56.

Bulayeva, N. (2004) Xenoestrogen-induced ERK-1 and ERK-2 activation via multiple membrane-initiated signaling pathways. *Environmental Health Perspectives* **112**(15), 1481–7.

Butterfeld, P.G., Valanis, B.G., Spencer, P.S., Lindeman, C.A. and Nutt, J.G. (1993) Environmental antecedents of young-onset Parkinson's disease. *Neurology* **43**, 1150–8.

Caboni, P., Sammelson, R.E. and Casida, J.E. (2003) phenylpyrazole insecticide photochemistry, metabolism, and GABAergic action: ethiprole compared with fipronil. *Journal of Agricultural and Food Chemistry* **51**(24), 7055–61.

Caroldi, S. and De Paris, P. (1995) Comparative effects of two dithiocarbamates disulfiram and thiram, on adrenal catecholamine content and on plasma dopamine-beta-hydroxylase activity. *Archives of Toxicology* **69**(10), 690–3.

Carson, R. (1962) *Silent Spring*. The Riverside Press, Boston.

Carter, R.H. (1952) The inorganic insecticides. In: Stefferud, A. (ed.) *The Yearbook of Agriculture 1952*. Washington, D. C., United States Government Printing Office. pp. 218–222.

Casida, J.E. (1963) Mode of action of carbamates. *Annual Review of Entomology* **8**, 39–58.

Casida, J.E. (1993) Insecticide action at the GABA-gated chloride channel: Recognition, progress, and prospects. *Archives of Insect Biochemistry and Physiology* **22**(1–2) 13–23.

Celik, L., Lund, J.D.D. and Schiøtt, B. (2008) Exploring Interactions of endocrine-disrupting compounds with different conformations of the human estrogen receptor α ligand binding domain: a molecular docking study. *Chemical Research in Toxicology* **21**(11), 2195–206.

Chandler, D. (2008) The consequences of the "cut off" criteria for pesticides: alternative methods of cultivation. *European Parliament Policy Department B, Structural and Cohesion Policies* 1–45.

Charlier, C.J. (2002) Determination of organochlorine pesticide residues in the blood of healthy individuals. *Clinical Chemistry and Laboratory Medicine* **40**(4), 361–4.

Chen, D., Suter, H., Islam, A., Edis, R., Freney, J. R. and Walker, C.R. (2008) Prospects of improving efficiency of fertiliser nitrogen in Australian agriculture: a review of enhanced efficiency fertilisers. *Australian Journal of Soil Research* **46**(4), 289–301.

Chen, H., Xiao, J., Hu, G., Zhou, J., Xiao, H., and Wang, X. (2002) Estrogenicity of organophosphorus and pyrethroid pesticides. *Journal of Toxicology and Environmental Health. Part A* **65**(19),1419–35.

Coats, J.R. (1990) Mechanisms of toxic action and structure–activity relationships for organochlorine and synthetic pyrethroid insecticides. *Environmental Health Perspectives* **87**, 255–62.

Cocco, P. (2002) On the rumors about the silent spring. Review of the scientific evidence linking occupational and environmental pesticide exposure to endocrine disruption health effects *Cadernos de Saúde Pública* **18**(2), 379–402.

Cohn, B.A., Cirillo, P.M. and Christianson, R.E. (2010) Prenatal DDT exposure and testicular cancer: A nested case-control study. *Archives of Environmental and Occupational Health* **65**, 127–34.

Cohn, B.A., Wolff, M.S., Cirillo, P.M. and Sholtz, R.I. (2007) DDT and breast cancer in young women: new data on the significance of age at exposure. *Environmental Health Perspectives* **115**(10), 1406–14.

Cooper, R.L. (1989) Effect of lindane on hormonal control of reproductive function in the female rat. *Toxicology and Applied Pharmacology* **99**(3), 384.

Cooper, R.L. and Stoker, T.E., Tyrey, L., Goldman, J.M. and McElroy, W.K. (2000) Atrazine disrupts the hypothalamic control of pituitary-ovarian function. *Toxicological sciences* **53**(2), 297–307.

Coronado, G.D., Thompson, B., Strong, L., Griffith, W.C. and Islas, I. (2004) Agricultural task and exposure to organophosphate pesticides among farmworkers. [see comment] *Environmental Health Perspectives* **112**(2), 142–7.

Costello, S., Cockburn, M., Bronstein, J., Zhang, X. and Ritz, B. (2009) Parkinson's Disease and Residential Exposure to Maneb and Paraquat From Agricultural Applications in the Central Valley of California. *American Journal of Epidemiology* **169**(8), 919.

Curwin, B.D., Hein, M.J., Sanderson, W.T., Nishioka, M.G., Reynolds, S.J. Ward, E.M. and Alavanja, M.C. (2005) Pesticide contamination inside farm and nonfarm homes. *Journal of Occupational and Environmental Hygiene* **2**(7), 357–67.

Darnerud, P.O., Atuma, S., Aune, M., Bjerselius, R., Glynn, A., Grawé, K.P. and Becker, W. (2006) Dietary intake estimations of organohalogen contaminants (dioxins, PCB, PBDE and chlorinated pesticides, e.g. DDT) based on Swedish market basket data. *Food and Chemistry Toxicology* **44**(9), 1597–606.

Davy, H. (1813) *Elements of Agricultural Chemistry. In a Course of Lectures for the Board of Agriculture*. London, Longman, Hurst Rees Orme and Brown.

DEFRA (2009) Guidance for Farmers in Nitrate Vulnerable Zones. Leaflet 1, Summary of the Guidance for Farmers in NVZs. http://www.defra.gov.uk/environment/quality/water/waterquality/diffuse/nitrate/documents/leaflet1.pdf. Last accessed 3 September 2010.

DEFRA (2006) Pesticides: Code of Practice for Using Plant Protection Products. Department for Environment, Food and Rural Affairs, http://www.pesticides.gov.uk/safe_use.asp?id=64. Last accessed 28 December 2010.

Dekeyser, M.A. (2004) Acaricide mode of action. *Pest Management Science* **61**(2), 103–110.

de Lau, L.M.L. Koudstaal, P.J., Witteman, J.C.M., Hofman, A., Breteler, M.M.B. (2006) Dietary folate, vitamin B_{12}, and vitamin B_6 and the risk of Parkinson disease. *Neurology* **67**, 315–18.

Delwiche, C.C. (1970) The nitrogen cycle. *Scientific American* **223**(3), 137–46.

Denham, C. and White, I. (1998) Differences in urban and rural Britain. Office of National Statistics. http://www.statistics.gov.uk/articles/population_trends/urbrurdif_pt91.pdf. Accessed 5 September 2010.

Dewailly, E., Mulvad, G., Pedersen, H.S., Ayotte, P., Demers, A., Weber, J.P., and Hansen, J.C. (1999) Concentration of organochlorines in human brain, liver, and adipose tissue autopsy samples from Greenland. [erratum appears in *Environ Health Perspect* 2000 Mar;108(3):A112]. *Environmental Health Perspectives* **107**(10), 823–8.

Dhadialla, T.S., Carlson, G.R. and Le, D.P. 1998. New insecticides with ecdysteroidal and juvenile hormone activity. *Annual Review of Entomology* **43**, 545–69.

Diamanti-Kandarakis, E., Bourguignon, J.P., Giudice, L.C., Hauser, R., Prins, G.S., Soto, A.M., Zoeller, R.T., and Gore, A. C. (2009) Endocrine-disrupting chemicals: An Endocrine Society Scientific Statement. *Endocrine Reviews* **30**(4), 293–342.

Dickerson, S.M. and Gore, A.C. (2007) Estrogenic environmental endocrine-disrupting chemical effects on reproductive neuroendocrine function and dysfunction across the life cycle. *Reviews in Endocrine and Metabolic Disorders* **8**(2), 143–59.

Djamgoz, M.B.A. (2011) Bioelectricity of cancer: Voltage-gated ion channels and direct-current electric fields. In C.E. Pullar (ed.), *The Physiology of Bioelectricity in Development, Tissue Regeneration and Cancer*. Boca Raton, CRC Press.

Doherty, J.D. (1979) Insecticides affecting ion transport. *Pharmacology and Therapeutics* **7**(1), 123–51.

Don-Pedro, K.N. (1989), Mode of action of fixed oils against eggs of *Callosobruchus maculatus* (F.). *Pesticide Science* **26**, 107–15.

Don-Pedro, K.N. (1990) Insecticidal activity of fatty acid constituents of fixed vegetable oils against *Callosobruchus maculatus* (F.) on cowpea. *Pesticide Science* **30**, 295–302.

Durham, W.F., Wolfe, H.R. and Quinby, G.E. (1965) Organophosphorus insecticides and mental alertness. *Archives of Environmental Health* **10**, 55–66.

Duyckaerts, C., Hauw, J.J., Bastenaire, F. and Piette, F. (1986) Laminar distribution of neocartical senile plaques in senile dementia of Alzheimer type. *Acta Neuropathologia*. **70**, 249–56.

Eberle, M., Farooq, S., Jeanguenat, A., Mousset, D., Steiger, A., Trah, S., Zambach, W. and Rindlisbacher, A. (2003) Azine derivatives as a new generation of insect growth regulators. *CHIMIA International Journal for Chemistry* **57**(11), 705–9.

EC (2000) *Guidance Document on Persistence in Soils. Working Document*. European Commission Directorate General for Agriculture pp 1–17.

EC (2006) *Guidance for the Setting and Application of Acceptable Operator Exposure Levels (AOELs)*. European Commission Health and Consumer Protection Directorate-General. Directorate E – Safety of the food chain, E3 – Chemicals, Contaminants, Pesticides.

EC (2009) *Guidance Document on the Procedures Relating to the Authorisation of Plant Protection Products Following Inclusion of an Existing Active Substance in Annex I of Council Directive 91/414/EEC*. European Commission Health and Consumer Protection Directorate-General. Directorate E – Food Safety: plant health, animal health and welfare, international questions, E1 – Plant health.

ECHA (2007) *Guidance for the Preparation of an Annex XV Dossier on the Identification of Substances of Very High Concern*. European Chemicals Agency. Guidance for the Implementation of REACH pp. 58.

ECHA (2010) *Chemical Safety and Report*. REACH and CLP Guidance. European Chemicals Agency. http://guidance.echa.europa.eu/chemical_safety_en.htm. Last Accessed 2 August 2010.

Ecobichon, D.J. (2001) Toxic effects of pesticides. In: Klaassen, C. (ed.) *Cassarett and Doull's Toxicology. The Basic Science of Poisons*. New York, McGraw-Hill, pp. 763–810.

Ecobichon, D.J. and Joy, R.M. (1993) *Pesticides and Neurological Diseases*, 2nd edition. Boca Raton, CRC Press.

Eriko, M., Harada, S., Nishikawa, J. I. and Nishihara, T. (2003) Endocrine disruptors induce cytochrome P450 by affecting transcriptional regulation via pregnane X receptor. *Toxicology and Applied Pharmacology* **193**(1), 66.

EUREAU (2001) *Keeping Raw Drinking Water Safe From Pesticides*. EUREAU Position Paper http://www.water.org.uk/static/files_archive/1Full_Report.pdf. Last accessed 5 September 2010.

Fang, H., Tong, W., Branham, W.S., Moland, C.L., Dial, S.L., Hong, H., Xie, Q., Perkins, R., Owens, W. and Sheehan, D.M. (2003) Study of 202 natural, synthetic, and environmental chemicals for binding to the androgen receptor. *Chemical Research in Toxicology* **16**(10), 1338–58.

FAO (2008) *Current World Fertilizer Trends and Outlook to 2011/12*. Rome, Food and Agriculture Organisation of the United States: 57.

FAO (2010a) *AGP – International Code of Conduct on the Distribution and Use of Pesticides*. http://www.fao.org/agriculture/crops/core-themes/theme/pests/pm/code/en/ Last accessed 17 June 2010.

FAO (2010b) Manure. *Animal Feed Resources Information System*. Food and Agriculture Organisation of the United Nations http://www.fao.org/ag/aga/agap/frg/AFRIS/Data/476.htm. Accessed 17 August 2010.

Fenske R.A. (2005) State-of-the-art measurement of agricultural pesticide exposures. *Scandinavian Journal of Work, Environment and Health* **31** (Suppl. 1), 67–73.

Fernandez, M., Rivas, A., Olea-Serrano, F., Cerrillo, I., Molina-Molina, J.M., Araque, P., Martínez-Vidal, J.L., and Olea, N. (2004) Assessment of total effective xenoestrogen burden in adipose tissue and identification of chemicals responsible for the combined estrogenic effect. *Analytical and Bioanalytical Chemistry* **379**(1), 163–70.

Fleming, L., Mann, J.B., Bean, J., Thomas, B. and Sanchez-Ramos, J. R. (1994) Parkinson's disease and brain levels of organochlorine pesticides. *Annals of Neurology* **36**, 100–3.

Fowler, P.A., Dorà, N.J., McFerran, H., Amezaga, M.R., Miller, D.W., 3, Lea, R.G., Cash, P., McNeilly, A.S., Evans, N.P., Cotinot, C., Sharpe, R.M., and Rhind, S.M. (2008) *In utero* exposure to low doses of environmental pollutants disrupts fetal ovarian development in sheep. *Molecular Human Reproduction* **14**(5), 269–80.

Gauthier, E., Fortier, I, Courchesne, F., Pepin, P., Mortimer, J. and Gauvreau, D. (2001) Environmental pesticide exposure as a risk factor for Alzheimer's disease: A case-control study. *Environmental Research* **86**(1), 37–45.

Gauvreau, D. (1987) Le paradigme de la maladie d'Alzheimer. *Interface* **8**, 16–21.

Garry, V.F. (2004) Pesticides and children. *Toxicology and Applied Pharmacology* **198**(2), 152–63.

Gilbertson, M. and Brophy, J. (2001) Community health profile of Windsor, Ontario, Canada: anatomy of a Great Lakes area of concern. *Environmental Health Perspectives* **109**, 827–43.

Givens, M.L., Lu, C., Bartell, S.M. and Pearson, M.A. (2006) Estimating dietary consumption patterns among children: a comparison between cross-sectional and longitudinal study designs. *Environmental Research* **99**(3), 1065–79.

Gliden, R.C., Huffling, K. and Sattler, B. (2010) Pesticides and health risks. *Journal of Obstetric, Gynecologic, and Neonatal Nursing* **39**(1),103–10.

Gray, L.E. Jr (1998) Xenoendocrine disrupters: laboratory studies on male reproductive effects. *Toxicology Letters* **102–103**, 331–5.

Gregus, Z.K. and Klaasen, C.D. (2001) Mechanisms of toxicity. In: Klaassen, C. (ed.) *Cassarett and Doull's Toxicology. The Basic Science of Poisons*. New York, McGraw-Hill, pp. 35–83.

Grey, C., Nieuwenhuijsen, M.J., Golding, J.,ALSPAC Team (2006) Use and storage of domestic pesticides in the UK. *Science of the Total Environment* **3682**–3465–70.

Griggs, B. (1986) *The Food Factor: Why We Are What We Eat*. USA, Penguin Group.

Grunfeld, H. (2004) Effect of *in vitro* estrogenic pesticides on human oestrogen receptor alpha and beta mRNA levels. *Toxicology Letters* **151**(3), 467–80.

Grob, D., Garlick, W.L. and Harvey, A.M. (1950) The toxic effects in man of the anticholinesterase insecticide parathion (p-nitro-phenly diethyl thionpphosphate). *Bulletin Johns Hopkins Hospital* **87**, 106–29.

Gross, M. (2008) Pesticides linked to bee deaths. *Current Biology* **18**(16),R684.

Groves, J.A. and Ellwood, P.A. (1991) Gases in agricultural slurry stores. *Annals of Occupational Hygiene* **35**(2), 139–51.

Haake, J., Kelley, M., Keys, B. and Safe, S. (1987) The effects of organochlorine pesticides as inducers of testosterone and benzo(a) pyrene hydroxylases. *General Pharmacology* **18**(2), 165–9.

Hamilton, D., Ambrus, Á., Dieterle, R., Felsot A., Harris, C. Petersen, B., Racke, K., Wong, S.-S., Gonzalez, R., Tanaka, K., Earl, M., Roberts, G. and Bhula, R. (2004) Pesticide residues in food – acute dietary exposure. *Pest Management Science* **83**(4), 311–39.

Hardell, L., Bavel, B.V., Lindstrom, G., Eriksson, M., and Carlberg, M. (2006) In utero exposure to persistent organic pollutants in relation to testicular cancer risk. *International Journal of Andrology* **29**(1), 228–34.

Hardt, J.J. and Angerer, J. (2003) Biological monitoring of workers after the application of insecticidal pyrethroids. *International Archives of Occupational and Environmental Health* **76**(7), 492–8.

Harrewijn, P. and Kayser, H. (1997) Pymetrozine, a fast-acting and selective inhibitor of aphid feeding. *In-situ* studies with electronic monitoring of feeding behaviour. *Pesticide Science* **49**(2),130–40.

Hausmann, S., Lotter, A.F., van Leeuwen, J.N.F., Ohlendorf, Ch., Lemke, G., Gronlund, E. and Sturm, M. (2002) Interactions of climate and land use documented in the varved sediments of Seebergsee in the Swiss Alps. *The Holocene* **12**, 279–89.

Hayes, W.J. (1959) Pharmacology and toxicology of DDT. In: Müller, P. (ed.) *The Insecticide Dichlorodiphenyltrichloroethane and its Significance*, **Vol 2** Basel, Birkhäuser Verlag, pp. 11–119.

Hayes, W.J. (1963) *Clinical Handbook on Economic Poisons: Emergency Information for Treating Poisonings*. U.S. Public Health Service Publication No. 476. Washington D.C., U.S. Government Printing Office.

Hayes, W.J. Jr, (1982) *Pesticides Studies in Man*. Baltimore, Williams and Wilkins Co.

Hendy, C.H. and Djamgoz, M.BA. (1988) Deltamethrin raises potassium activity in the microenvironment of the central nervous system of the cockroach: An assessment of the potential role of the blood–brain barrier in insecticide action. *Pesticide Science* **24**, 289–98.

Henry, C. (2004) New BSE rule will change re-feeding of poultry litter to ruminants. *Manure Matters*. UNL's Livestock Environmental Issues Committee, Nebraska. http://www.p2pays.org/ref/15/14481.pdf. Accessed on 17 August 2010.

Herrera, A., Ariño, A., Conchello, P., Lázaro, R., Bayarri, S., Pérez-Arquillué, C., Garrido, M. D., Jodral M. and Pozo, R. (1996) Estimates of mean daily intakes of persistent organochlorine pesticides from Spanish fatty foodstuffs. *Bulletin of Environmental Contamination and Toxicology* **56**(2), 173–7.

Hollinshaus, J.G. (1987) Inhibition of mitochondrial electron transport by hydramethylnon: A new amidinohydrazone insecticide. *Pesticide Biochemistry and Physiology* **27**(1), 61–70.

Howdeshall, K. L. (2002) A model of the development of the brain as a construct of the thyroid system. *Environmental Health Perspecitves* **110** (Suppl 3), 337–48.

Høyer, A.P., Grandjean, P., Jørgensen, T., Brock, J.W. and Hartvig, H. B. (1998) Organochlorine exposure and risk of breast cancer. *Lancet* **352**, 1816–20.

HSE (1986) *Storing and Handling Ammonium Nitrate*. Health and Safety Executive, pp 1–12.

Huddart, H., Greenwood, M. and Williams, M.J. (1974) The effect of some orgophosphorus and organochlorine compounds on calcium uptake by sarcoplasmic reticulum isolated from insect and crustacean skeletal muscle. *Journal of Comparative Physiology B: Biochemical, Systemic, and Environmental Physiology*. **98**(2), 139–50.

Humphrys, J. (2002) *The Great Food Gamble*. Great Britain, Coronet Books.

Hunter, D.J., Hankinson, S.E., Laden, F., Colditz, G.A., Manson, J.E., Willett, W.C., Speizer, F.E., and Wolff, M.S. (1997) Plasma organochlorine levels and the risk of breast cancer. *The New England Journal of Medicine* **337**, 1253–8.

Ishihara, A. (2003) The effect of endocrine disrupting chemicals on thyroid hormone binding to Japanese quail transthyretin and thyroid hormone receptor. *General and Comparative Endocrinology* **134**(1),36–43.

Jager, K.W. (1970) *Aldrin, Dieldrin, Endrin and Telodrin – An Epidemiological and Toxicological Study of Long-term Occupational Exposure*. New York, Elsevier.

Jagt, K., Tielemans, E., Links, I., Brouwer, D. and van Hemmen, J. (2004) Effectiveness of personal protective equipment: relevance of dermal and inhalation exposure to chlorpyrifos among pest control operators. *Journal of Occupational and Environmental Hygiene* **1**(6), 355–62.

Jong, F.M. and Snoo, G.R. (2001) Pesticide residues in human food and wildlife in The Netherlands. *Mededelingen* **66**(2b), 815–22.

Kieszak, S.M., Naeher, L.P., Rubin, C.S., Needham, L.L., Backer, L., Barr, D. and McGeehin, M. (2002) Investigation of the relation between self-reported food consumption and household chemical exposures with urinary levels of selected nonpersistent pesticides. *Journal of Exposure Analysis and Environmental Epidemiology* **12**(6),404–8.

Kim, H. (2005) Effects of 2,4-D and DCP on the DHT-induced androgenic action in human prostate cancer cells. *Toxicological Sciences* **88**(1), 52–9.

King, F.C. (1944) *Gardening with Compost*. London, Faber and Faber.

Klotz, D. (1997) Inhibition of 17 beta-estradiol and progesterone activity in human breast and endometrial cancer cells by carbamate insecticides. *Life Sciences* **60**(17), 1467–75.

Koehler, P.G. and Patterson, R.S. (1989) Effects of chitin synthesis inhibitors on German cockroach (Orthoptera: Blattellidae) mortality and reproduction. *Journal of Economic Entomology* **82**(1), 143–8.

Köhrle, J. (2008) Environment and endocrinology: The case of thyroidology. *Annales d'Endocrinologie* **69**(2), 116–22.

Kortenkamp, A. (2007) Ten years of mixing cocktails: A review of combination effects of endocrine-disrupting chemicals. *Environmental Health Perspectives* **115** (Suppl 1), 98–105.

Kortenkamp, A. (2008) Low dose mixture effects of endocrine disrupters: implications for risk assessment and epidemiology. *International Journal of Andrology* **31**(2), 233–40.

Kuiper, G.G., Carlsson, B., Grandien, K., Enmark, E., Haggblad, J., Nilsson, S. and Gustafsson, J.A. (1997) Comparison of the ligand binding specificity and transcript tissue distribution of estrogen receptors alpha and beta. *Endocrinology* **138**(3), 863–70.

Liao, P.H., Wong, W.T. and Lo, K.V. (2005) Release of phosphorus from sewage sludge using microwave technology *Journal of Environmental Engineering and Science* **4**(1), 77–81.

Lang, K. and Bastian, P. (2007) Neurotransmitter effects on tumor cells and leukocytes. *Progress in Experimental Tumor Research* **39**, 99–121.

Lemaire, G. (2004) Effect of organochlorine pesticides on human androgen receptor activation *in vitro*. *Toxicology and Applied Pharmacology* **196**(2), 235–46.

Lemaire, G., Terouanne, B., Mauvais, P., Michel, S. and Rhamani, R. (2004) Effect of organochlorine pesticides on human androgen receptor activation *in vitro*. *Toxicology and Applied Pharmacology* **196**(2), 235–46.

Levin, H.S. and Rodnitzky, R.L. (1976) Behavioural aspects of organophosphate pesticides in man. *Clinical Toxicology* **9**, 391.

Lide, D.R. (ed.) (2004) *CRC Handbook of Chemistry and Physics*. London, Taylor and Francis Ltd.

Liebig, J.V. (1847) *Chemistry in its Application to Agriculture and Physiology*, 3rd edition. Philadelphia, Campbell.

Lu, C., Toepel, K., Irish, R., Fenske, R.A., Barr, D.B., and Bravo, R. (2006) Organic diets significantly lower children's dietary exposure to organophosphorus pesticides. *Environmental Health Perspectives* **114**(2), 260–3.

Lümmen, P. (1998) Complex I inhibitors as insecticides and acaricides. *Biochimica et Biophysica Acta (BBA) – Bioenergetics* **1364**(2)287–96.

Mackenzie, C.A., Lockridge, A. and Keith, M. (2005) Declining sex ratio in a first nation community. *Environmental Health Perspectives* **113**(10), 1295–8.

Mahjoubi-Samet, A.A., Hamadi, F., Soussia, L., Fadhel, G. and Zegha, N. (2005) Dimethoate effects on thyroid function in suckling rats. *Annales d'Endocrinologie* **66**(2), 96–104.

Marinovich, M.M., Viviani, B., Corsini, E., Guizzetti, M., Ghilardi, F. and Galli, C.L. (1997) Thyroid peroxidase as toxicity target for dithiocarbamates. *Archives of Toxicology* **71**(8), 508–12.

Mason, J.I., Carr, B.R. and Murry, B.A. (1987) Imidazole antimycotics: selective inhibitors of steroid aromatization and progesterone hydroxylation *Steroids* **50**1–3179–89.

Mathur, V., Bhatnagar, P., Sharma, R.G., Acharya, V. and Sexana, R. (2002) Breast cancer incidence and exposure to pesticides among women originating from Jaipur. *Environment International* **28**(5), 331–6.

Matson, P. A., Naylor, R., and Oritz-Monasterio, I. (1998) Integration of environmental, agronomic, and economic aspects of fertiliser management. *Science* **280**(5360), 112–5.

Matsumara, F. (1975) *Toxicology of Insecticides*. New York, Plenum Press.

Matthews, G.A. (1999) *Application of Pesticides to Crops*. London, Imperial College Press.

McCarrison, R. (1961) *Nutrition and Health*. London, Faber and Faber.

McCarthy, J.C. (2006) Estrogenicity of pyrethroid insecticide metabolites. *Journal of Environmental Monitoring* **8**(1), 197–202.

McKinlay, R., Plant, J.A., Bell, J.N.B. and Voulvoulis, N. (2008a) Calculating human exposure to endocrine disrupting pesticides via agricultural and non-agricultural exposure routes. *Science of the Total Environment* **398**(1–3) 1–12.

McKinlay, R., Plant, J.A., Bell, J.N.B. and Voulvoulis, N. (2008b) Endocrine disrupting pesticides: Implications for risk assessment. *Environment International*. **34**, 168–83.

McLoughlin, M.F., McIlroy, S.G. and Neill, S.D. (1988) A major outbreak of botulism in cattle being fed ensiled poultry litter. *The Verterinary Record* **122**(24), 579–81.

Menegaux, F.F., Baruchel, A., Bertrand, Y., Lescoer, B., Leverger, G., Nelken, B., Sommelet, D., Hemon, D. and Clavel, J. (2006) Household exposure to pesticides and risk of childhood acute leukaemia. *Occupational and Environmental Medicine* **63**(2), 131–4.

Milkowski, A., Garg, H.K., Coughlin, J.R. and Bryan, N.S. (2010) Nutritional epidemiology in the context of nitric oxide biology: A risk–benefit evaluation for dietary nitrite and nitrate. *Nitric Oxide* **22**(2), 110–19.

Moffett, D.B. (2006) *Public health impacts of organophosphates and carbamates.*, In Gupta, C. (ed.), *Toxicology of Organophosphate and Carbamate Compounds*. London, Academic Press, Chapter 40, pp. 599–606.

Moolenaar, S.W., and Beltrami, P. (1998) Heavy metal balances of an Italian soil as affected by sewage sludge and bordeaux mixture applications *Journal of Environmental Quality* **27** 828–35.

Morinaga, H. (2004) A benzimidazole fungicide, benomyl, and its metabolite, carbendazim, induce aromatase activity in a human ovarian granulose-like tumor cell line (KGN). *Endocrinology* **145**(4), 1860–9.

Mukerjee, L.N. and Srivastava, S.N. (1957) Bordeaux mixture and related compounds as emulsifiers. *Colloid and Polymer Science* **150**(2), 148–51.

Muminov, A.I. (1972) Functional state of the organ of hearing in persons with pesticide poisoning. (Translation). *Vestnik Otorinolaringol* **34**(6), 33–5.

Murphy, R., Kutz, F.W. and Strassman, S.C. (1983) Selected pesticide residues or metabolites in blood and urine specimens from a general population survey. *Environmental Health Perspectives* **48**, 81–6.

Murty, A.S. (1982) Persistence and mobility of nitrofen (niclofen, TOK) in mineral and organic soils. *Journal of Environmental Science and Health. Part B, Pesticides, Food Contaminants, and Agricultural Wastes* **17**(2), 143–52.

NAAC (2010), Sulphuric acid code of conduct. *National Association of Agricultural Contractors* http://www.naac.co.uk/Codes/AcidCode.aspx. Last accessed 29 December 2010.

Nakagawa, R., Hirakawa, H., Lida, T., Matsueda, T. and Nagayama, J. (1999) Maternal body burden of organochlorine pesticides and dioxins. *Journal of AOAC International* **82**(3), 716–24.

Newsome, W.H., Doucet, J., Davies, D. and Sun, W.F. (2000) Pesticide residues in the Canadian Market Basket Survey – 1992 to 1996. *Food Additives and Contaminants* **17**(10), 847–54.

Nicolau, G. (1983) Circadian rhythms of RNA, DNA and protein in the rat thyroid, adrenal and testis in chronic pesticide exposure. III. Effects of the insecticides (dichlorvos and trichlorphon). *Physiologie* **20**(2), 93–101.

Nicolopoulou-Stamati, P. and Pitsos, M.A. (2001) The impact of endocrine disrupters on the female reproductive system. *Human Reproduction Update* **7**(3), 323–30.

Nomura, D.K. and Casida, J.E. (2011) Activity-based protein profiling of organophosphorus and thiocarbamate pesticides reveals multiple serine hydrolase targets in mouse brain. *Journal of Agricultural and Food Chemistry* **59**(7), 2808–15.

O'Brien, R.D. (1963) Mode of action of insecticides, binding of organophosphates to cholinesterases. *Journal of Agricultural and Food Chemistry* **11**(2), 163–6.

Olea, N., Olea-Serrano, F., Lardelli-Claret, P., Rivas, A. and Barba-Navarro, A. (1999) Inadvertent exposure to xenoestrogens in children. *Toxicology and Industrial Health* **15**1–2151–8.

Orlando, L. (2002) *Industry Attacks on Dissent: From Rachel Carson to Oprah*. Dollars and Sense http://www.dollarsandsense.org/archives/2002/0302orlando.html. Last accessed 20 July 2010.

Orr, J.B. (1936) *Food, Health and Income: Report on a Survey of Adequacy of Diet in Relation to Income*. London, Macmillan.

Orr, J.B. (1948) *Soil Fertility: The Wasting Basis of Human Society*. London, Pilot Press.

Ortega, A. (2003) A new role for GABA: inhibition of tumor cell migration. *Trends in Pharmacological Science* **24**(4), 151–4.

Ott, M.G. (2005) Exposure assessment as a component of observational health studies and environmental risk assessment. *Scandinavian journal of work, environment and health* **31**(Suppl1) 110–115.

Politis, A., Olgiati, P., Malitas, P., Albani, D., Signorini, A. *et al.* (2010) Vitamin B12 levels in Alzheimer's disease: association with clinical features and cytokine production. *Journal of Alzheimers Disease* **19**(2), 481–8.

Pritchard, J.D. (2007) *Ammonia – Toxicological Overview*. Health Protection Agency, pp. 1–13.

PSD (2006) UK Pesticide Law, England and Wales, Pesticide Safety Directorate http://www.pesticides.gov.uk/approvals.asp?id=869 Last accessed 29 December 2010.

PSD (2007) Keeping our food safe: measuring, monitoring and assessing residues, Pesticide Safety Directorate. http://www.pesticides.gov.uk/food_safety.asp?id=632. Last accessed 29 December 2010.

Pussemier, L., Larondelle, Y., Van Peteghem, C. and Huyghebaert, A. (2006) Chemical safety of conventionally and organically produced foodstuffs: a tentative comparison under Belgian conditions. *Food Control* **17**, 14–21.

Rawn, D.F.K., Roscoe, V., Krakalovich, T. and Hanson, C. (2004) N-methyl carbamate concentrations and dietary intake estimates for apple and grape juices available on the retail market in Canada. *Food Additives and Contaminants* **21**(6) 555–63.

RCEP (2005) *Crop Spraying and the Health of Residents and Bystanders*. T. Blundell. London, Royal Commission on Environmental Pollution: 176.

Rhind, S.M., Kyle, C.E., Mackie, C., McDonald, L., Zhang, Z., Duff, E.I., Bellingham, M., Amazega, M.R. and Mandon-Pepin, B. (2010) Maternal and fetal tissue accumulation of selected endocrine disrupting compounds (EDCs) following exposure to sewage sludge-treated pastures before or after conception. *Journal of Environmental Monitoring* **12**, 1582–93.

Richardson, J.R., Shalat, S.L., Buckley, B., Winnik, B., O'Suilleabhain, P., Diaz-Arrastia, R., Reisch, J. and German, D. C. (2009) Elevated serum pesticide levels and risk of Parkinson disease. *Archives of Neurology* **66**(7), 870–5.

Robert, W.C., Ralph, L.C., Georgia, L.R., McElroy, W.K. and Chang, J. (1988) Possible antiestrogenic activity of lindane in female rats. *Journal of Biochemical Toxicology* **3**(3), 147–58.

Rogan, W.J. and Chen, A. (2005) Health risks and benefits of bis(4-chlorophenyl)-1,1,1-trichloroethane (DDT) *Lancet* **366**, 763–73.

Rusiecki, J.A., Matthews, A., Sturgeon, S., Sinha, R., Pellizzari, E., Zheng, T. and Baris, D. (2005) A correlation study of organochlorine levels in serum, breast adipose tissue, and gluteal adipose tissue among breast cancer cases in India. *Cancer Epidemiology Biomarkers and Prevention* **14**(5), 1113–24.

Safi, J.M. (2002) Association between chronic exposure to pesticides and recorded cases of human malignancy in Gaza Governorates (1990–1999) *Science of the Total Environment* **284**(1-3), 75–84.

Saitoh, M., Yanase, T., Morinaga, H., Tanabe, M., Mu, Y.M. et al. (2001) Tributyltin or triphenyltin inhibits aromatase activity in the human granulosa-like tumor cell line KGN. *Biochemical and Biophysical Research Communications* **289**(1), 198–204.

Saktaywin, W., Tsuno, H., Nagare, H., Soyama, T. and Weerapakkaroon, J. (2005) Advanced sewage treatment process with excess sludge reduction and phosphorus recovery. *Water Research* **39**(5), 902–10.

Salgado, V.L. (1998) Studies on the mode of action of spinosad: insect symptoms and physiological correlates. *Pesticide Biochemistry and Physiology* **60**(2), 91–102.

Salminen, R., Batista, M.J., Bidovec, M., Demetriades, A., De Vivo, B. *et al.* (2005) *Geochemical Atlas of Europe. Part 1: Background Information, Methodology and Maps*. Brussels, Belgium, EuroGeoSurveys. http://www.gtk.fi/publ/foregsatlas/ Last Accessed 3 September 2010.

Sanderson, J.T. (2000) 2-Chloro-s-triazine herbicides induce aromatase (CYP19) activity in H295R human adrenocortical carcinoma cells: a novel mechanism for estrogenicity? *Toxicological Sciences* **54**(1), 121–7.

Satoh, T. (2006) Global epidemiology of organophosphate and carbamate poisonings. In C. Gupta (ed.), *Toxicology of Organophosphate and Carbamate Compounds*. Academic Press, pp. 89–100.

Savage, E.P., Keefe, T.J., Mounce, L.M., Heaton, R.K., Lewis, J.A. and Burcar, P.J. (1988) Chronic neurological sequelae of acute organophosphate pesticide poisoning. *Archives of Environmental Health*, **43**(1), 38.

Scherr, S. J. (1999) Soil degradation: a threat to developing-country food security by 2020? *International Food Policy Research Institute* pp. 1–65.

Schmutterer, H. (1990) Properties and potential of natural pesticides from the neem tree, *Azadirachta indica. Annual Review of Entomology* **35**, 271–97.

Scholz, K., Lutz, R.W. *et al.* (2004) Comparative assessment of the inhibition of recombinant human CYP19 (aromatase) by azoles used in agriculture and as drugs for humans. *Endocrine Research* **30**(3), 387–94.

Schreinemachers, D.M. (2003) Birth malformations and other adverse perinatal outcomes in four US wheat-producing states. *Environmental Health Perspectives* **111**(9), 1259–64.

Schubel, F. and Linss, G. (1971) Inhalation allergy induced by the fungicide Zineb 80. *Das Deutsche Gesundheitswesen* **26**(25), 1187–9.

Semchuk, K.M., Love, E.J. and Lee, R. (1991) Parkinson's disease and exposure to rural environmental factors: A population based case-control study. *Canadian Journal of Neurological Sciences* **18**, 279–86.

Senanayake, N. and Karalliedde, L. (1987) Neurotoxic effects of organophosphorus insecticides: An intermediate syndrome. *New England Journal of Medicine* **216**, 761–3.

Sharpe, R.M. (2004) How strong is the evidence of a link between environmental chemicals and adverse effects on human reproductive health? *British Medical Journal* **328**(7437), 447–51.

Sharpe, R.M. (2006) Pathways of endocrine disruption during male sexual differentiation and masculinisation. *Bailliere's Best Practice and Research in Clinical Endocrinology and Metabolism* **20**(1), 91–110.

Shaviv, A., and Mikkelsen, R. L. (1993) Controlled-release fertilizers to increase efficiency of nutrient use and minimize environmental degradation - A review. *Nutrient cycling in agroecosystems* **35**(1-2), 1–12.

Sheng, G.G. (2001) Potential contributions of smectite clays and organic matter to pesticide retention in soils. *Journal of Agricultural and Food Chemistry* **49**(6), 2899–907.

Shimabukuro, S.H. and Swanson, H.R. (1969) Atrazine metabolism, selectivity, and mode of action. *Journal of Agricultural and Food Chemistry* **17**(2), 199–205.

Singer, P.L. (1949) Occupational oligiospermia. *Journal of American Medical Directors Association* **140**, 1249.

Snoeij, N.J. Penninks, A.H. and Seinen, W. (1987) Biological activity of organotin compounds – An overview *Environmental Research* **44**(2), 335–53.

Snoo, G.R.de, de Jong, F.M.W., van der Poll, R.J., Janzen, S.E., van der Veen, L.J. and Schuemie, M.P. (1997) Variation of pesticide use among farmers in Drenthe: A starting point for environmental protection. *Med. Fac. Landbouww, University of Gent.* **62/2a**: 199–212.

Steenland, K. (1996) Chronic neurological effects of organophosphate pesticides. *British Medical Journal*, **312**(7042), 1312–13.

Stenersen, J. (2004) *Chemical Pesticides: Mode of Action and Toxicology*. Florida, CRC Press.

Stepan, R.R. (2005) Baby food production chain: pesticide residues in fresh apples and products. *Food Additives and Contaminants* **22**(12), 1231–42.

Strauch (1991) Survival of pathogenic micro-organisms and parasites in excreta, manure and sewage sludge. *Revue Scientifique et Technique* **10**(3), 813–46.

Sugiyama, S. (2005) Detection of thyroid system-disrupting chemicals using *in vitro* and in vivo screening assays in *Xenopus laevis. Toxicological sciences* **88**(2), 367–74.

Sultan, C.C. (2001) Environmental xenoestrogens, antiandrogens and disorders of male sexual differentiation. *Molecular and Cellular Endocrinology* **178**(1–2) 99–105.

Sun, S. S. (2005) Persistent organic pollutants in human milk in women from urban and rural areas in northern China. *Environmental research* **99**(3), 285–93.

Takehara, A., Hosokawa, M., Eguchi, H., Ohigashi, H., Ishikawa, O., Nakamura, Y. and Nakagawa, H. (2007) γ-aminobutyric acid (GABA) stimulates pancreatic cancer growth through overexpressing $GABA_A$ receptor π subunit. *Cancer Research* **67**, 9704–12.

Tanaka, Y. and Omura, S. (1993) Agroactive compounds of microbial origin. *Annual Review of Microbiology* **47**, 57–87.

Tapiero, H.T. (2002) Estrogens and environmental estrogens. *Biomedicine and Pharmacotherapy* **56**(1), 36.

Tokarnia, C.H., Döbereiner, J., Peixoto, P.V. and Moraes, S.S. (2000) Outbreak of copper poisoning in cattle fed poultry litter. *Veterinary and Human Toxicology* **42**(2), 92–5.

Tomizawa, M. and Casida, J.E. (2005) Neonicotinoid insecticide toxicology: mechanisms of selective action. *Annual Review of Pharmacology and Toxicology* **45**, 247–68.

Trenkel, M.E. (1997) *Controlled Release and Stabilised Fertilisers in Agriculture*. United Nations Food and Agriculture Organisation, pp 1–156.

Trösken, E.R., Adamska, M., Arand, M., Zarn, J.A., Patten, C., Völkel, W. and Lutz, W.K. (2006) Comparison of lanosterol-14 (alpha)-demethylase (CYP51) of human and *Candida albicans* for inhibition by different antifungal azoles. *Toxicology* **228**(1), 24–32.

Tuomainen, A.A., Kangas, J.A. *et al.* (2002) Monitoring of pesticide applicators for potential dermal exposure to malathion and biomarkers in urine. *Toxicology Letters* **134**1–3125–32.

US EPA (1997) Explosion hazard from ammonium nitrate. United States Environmental Protection Agency, Office of Solid Waste and Emergency Response, pp 1–5.

US EPA (2008) Federal Insecticide, Fungicide, and Rodenticide Act http://www.epa.gov/lawsregs/laws/fifra.html. Last accessed 20 July 2010.

US EPA (2009) 2000–2001 Pesticide Market Estimates. http://www.epa.gov/oppbead1/pestsales/01pestsales/table_of_contents2001.htm. Last Accessed 12 June 2010.

US EPA (2010) Pesticide reregistration status. http://www.epa.gov/pesticides/reregistration/status_page_z.htm. Last accessed 17 July 2010.

US FDA (2007) Federal Food, Drug, and Cosmetic Act. http://www.fda.gov/RegulatoryInformation/Legislation/FederalFoodDrugandCosmeticActFDCAct/default.htm. Last accessed 20 July 2010.

Van Maele-Fabry, G. and Willems, J.L. (2004) Prostate cancer among pesticide applicators: a meta-analysis. *International Archives of Occupational and Environmental Health* **77**(8), 559–70.

Van Tongeren, M., Nieuwenhuijsen, M.J. *et al.* (2002) A job-exposure matrix for potential endocrine-disrupting chemicals developed for a study into the association between maternal occupational exposure and hypospadias. *Annals of Occupational Hygiene* **46**(5), 465–77.

Verslycke, T. (2004) Testosterone and energy metabolism in the estuarine mysid *Neomysis integer* (Crustacea: Mysidacea) following exposure to endocrine disruptors. *Environmental Toxicology and Chemistry* **23**(5), 1289–96.

Vijerberg, H.P.M. and Bercken, J.V. (1990) Neurotoxicological effects and the mode of action of pyrethroid insecticides. *Critical Reviews in Toxicology* **21**(2), 105–26.

Vinggaard, A.A. (2006) Prochloraz: an imidazole fungicide with multiple mechanisms of action. *International Journal of Andrology* **29**(1), 186–92.

Vlacke, M., Samuel, O., Bouchard, M., Dumas, P., Belleville D. and Tremblay, C. (2006) Biological monitoring of exposure to organophosphate pesticides in children living in peri-urban areas of the Province of Quebec, Canada. *International Archives of Occupational and Environmental Health* **79**, 568–77.

Vonier, P.M. (1996) Interaction of environmental chemicals with the estrogen and progesterone receptors from the oviduct of the American alligator. *Environmental Health Perspectives* **104**(12), 1318.

Wade, M.G., Foster, W.G., Younglai, E.V., McMahon, A., Leingartner, K., Yagminas, A., Blakey, D., Fournier, M., Desaulniers, D. and Hughs, C.L. (2002) Effects of subchronic exposure to a complex mixture of persistent contaminants in male rats: systemic, immune, and reproductive effects. *Toxicological Sciences* **67**, 131–43.

Walker, K.J., Preston, C.D. and Boon, C.R. (2009) Fifty years of change in an area of intensive agriculture: plant trait responses to habitat modification and conservation, Bedfordshire, *England. Biodiversity and Conservation* **18**(13), 3597–613.

Westing, A.H. (1975) Environmental consequences of the second Indochina war: a case study. *Ambio* **4**(5/6) 216–22.

Wing, K.D., Sacher, M., Kagaya, Y., Tsurubuchi, Y., Mulderig, L., Connair, M. and Schnee, M. (2000) Bioactivation and mode of action of the oxadiazine indoxacarb in insects. *Crop Protection* **19**(8–10) 537–545.

WHO (1998) *Copper: Environmental health criteria; 200*. Geneva, World Health Organisation. http://www.inchem.org/documents/ehc/ehc/ehc200.htm. Last accessed 29 December 2010.

WHO (2008a) Clinical management of acute pesticide intoxication: prevention of suicidal behaviours. Geneva, World Health Organisation Press. http://www.who.int/mental_health/prevention/suicide/pesticides_intoxication.pdf. Last accessed 29 December 2010.

WHO (2008b) *Guidelines for Drinking Water Quality: Third Edition, Incorporating the First And Second Addenda. Volume 1: Recommendations*. World Health Organisation, Geneva pp. 1–668.

Williamson, G.S. and Pearse, I. (1951) *The Passing of Peckham, 1951*. London, Peckham Health Center.

Wolff, W.S., Toniolo, P.G., Lee, E.W., Rivera, M. and Dubin, N. (1993) Blood levels of organochlorine residues and risk of breast cancer. *Journal of the National Cancer Institute* **85**(8) 648–52.

Wolstenholme, A.J. and Rogers, A.T. (2005) Glutamate-gated chloride channels and the mode of action of the avermectin/milbemycin anthelmintics. *Parasitology* **131**, S85–S95.

Worthington, V. (2001) Nutritional quality of organic versus conventional fruits, vegetables, and grains. *Journal of Alt0ernative and Complementary Medicine* **7**, 161–73.

WWF (2003) WWF-UK National Biomonitoring Survey 2003, World Wildlife Fund: 107.

WWF (2004) Chemical check up: an analysis of chemicals in the blood of members of the European Parliament, World Wildlife Fund. http://www.pops.int/documents/meetings/poprc/submissions/Comments_2006/wwf/checkupmain.pdf. Last accessed 29 December 2010.

WWF (2005) Generations: Results of the WWF's European Family Biomonitoring Survey. G. Watson. Brussels, Belgium, World Wildlife Fund.

Xu, J.C. (2001) Fate of atrazine and alachlor in redox-treated ferruginous smectite. *Environmental Toxicology and Chemistry* **20**(12), 2717–24.

Xu, X., Qian, X., Li, Z., Huang, Q. and Chen, G. (2003) Synthesis and insecticidal activity of new substituted *N*-aryl-*N*′-benzoylthiourea compounds. *Journal of Fluorine Chemistry* **121**(1), 51–4.

Yost, D.M. (2007) Ammonia and liquid ammonia solutions. *Systemic Inorganic Chemistry* Read Books, pp 132–54.

Young, J.A. (2007) Urea. *Journal of Chemical Education* **84**(9), 1421.

Zala, S. and Penn, D.J. (2004) Abnormal behaviours induce0d by chemical pollution: a review of the evidence and new challenges. *Animal Behaviour* **68**, 649–64.

Zhao, Z.Q., Zhu, Y.G., Li, H.Y., Smith, S.E. and Smith, F.A. (2004) Effects of forms and rates of potassium fertilizers on cadmium uptake by two cultivars of spring wheat (*Triticum aestivum,* L.). *Environment International* **29**(7), 973–8.

8

Pharmaceuticals and personal-care products

James Treadgold[1], Qin-Tao Liu[2*], Jane A. Plant[3] and Nikolaos Voulvoulis[1]

[1]Centre for Environmental Policy, Imperial College London, Prince Consort Road, London SW7 2AZ
[2]Current address: Dow Corning (China) Holding Co., Ltd.
[3]Centre for Environmental Policy and Department of Earth Science and Engineering, Imperial College London, Prince Consort Road, London SW7 2AZ
**Corresponding author, email qliu56@btinternet.com*

8.1 Introduction

8.1.1 Pharmaceuticals

The earliest pharmaceuticals seem to have been plants, which palaeopharmacological studies indicate have been used to treat illness since prehistoric times (Ellis, 2000). The earliest compilation to describe the medicinal properties of plant species is thought to be the *Sushruta Samhita*, an Indian Ayurvedic treatise attributed to Sushruta, the father of surgery, in the sixth century BC (Dwivedi and Dwivedi, 2007). Further descriptions of the therapeutic effects of plant extracts, animal parts and minerals are given by Pedanius Dioscorides in the book *De Materia Medica*, published in the first century AD. *The Divine Farmer's Materia Medica*, which is thought to have been compiled around 960–1280 AD, includes hundreds of plant and animal medicines discovered and researched by Shen Nong (also known as the Yan emperor), the legendary ruler of China six thousand years ago (Yang, 1998). Other contributions to the *Materia Medica* were made by Islamic physicians, and the book remains one of the most influential texts on herbal medicine (Rashed, 1996).

Ancient Chinese medicine used various plants and minerals to treat illnesses, including low mood, fevers and back pain. For example, *Dichroa febrifuga*, an evergreen shrub that grows in Nepal and China, is one of the fifty most important plants in traditional Chinese herbalism. The powerful antimalarial alkaloids contained in its roots and leaves have been used to treat fevers since at least the first century AD (Manandhar, 2002). *Aloe vera*, recorded in Dioscorides' *De Materia Medica*, is used today for the treatment of burns and wounds (Volger and Ernst, 1999). The North American Plains Indians used species of *Echinacea* for its general medical properties (Wishart, 2007).

The era of modern Western pharmacology probably dates from the early nineteenth century, when small molecules and a series of alkaloids, including morphine, quinine, caffeine and later cocaine, were isolated and purified for medicinal use. By 1829, scientists had identified the compound salicin in willow, and by the end of the nineteenth century, acetylsalicylic acid had been patented by Bayer as aspirin. The discovery of the potent antibiotic, penicillin, by Fleming in 1928 and its development by Chain, Florey and Heatley in the 1940s marked another important milestone in the development of the modern pharmaceutical industry. More recently, drug molecules, often known as designer drugs, have been developed, such as ondansetron (an antinausea drug), ibuprofen (a non-steroidal anti-inflammatory drug, NSAID) and many selective serotonin re-uptake inhibitors (SSRIs) for the treatment of depression.

Pharmaceuticals are used for both human and veterinary purposes. Most pharmaceuticals have been of great value in the treatment of illness and the alleviation of pain and distress. However, since the 1990s, there have been increasing concerns about their presence and pseudo-persistence in the

Pollutants, Human Health and the Environment: A Risk Based Approach, First Edition. Edited by Jane A. Plant, Nikolaos Voulvoulis and K. Vala Ragnarsdottir.

environment, and their potential effects on wildlife and human health (Halling-Sørensen *et al.*, 1998; Daughton and Ternes, 1999; Heberer, 2002a). For example, the feminisation and masculinisation of fish in many rivers downstream of sewage treatment plants (STPs) has been attributed to the presence of natural and synthetic steroid oestrogens, including ethinyl estradiol (EE_2), and possibly the interaction of these drugs with other endocrine-active compounds (EACs) such as polychlorinated biphenyls (PCBs), pharmaceuticals and surfactants (Vos *et al.*, 2000; Jobling *et al.*, 2004; Hinck *et al.*, 2009).

Pharmacologically active substances enter the environment from a variety of anthropogenic sources and through different pathways. Human pharmaceuticals that are excreted or flushed into lavatories are released into the aquatic environment continuously by STPs (Ternes, 1998). Active pharmaceutical ingredients (APIs) in veterinary pharmaceuticals deposited on land by treated farm animals can be found in soils and can enter surface waters through run-off or leach into groundwater (Boxall *et al.*, 2003). Hundreds of different compounds from a variety of different therapeutic classes have been detected in soils, lakes, rivers, groundwaters and estuaries in countries across the globe (Kümmerer, 2008). Although their measured concentrations are only in the nanogram to low microgram range, their biological effects and their continuous release into surface waters from STPs means that aquatic life is chronically exposed to a mixture of biologically potent chemicals.

Excreted and incorrectly disposed compounds can be detected in the environment as parent compounds, metabolites or conjugates (Ternes, 2000; Kümmerer, 2004a). Some conjugates can be converted back to the active compounds by bacterial action in STPs (Jones *et al.*, 2001a). Furthermore, depending on the properties of the APIs and the nature of the receiving environment, APIs can also undergo biotic and abiotic transformation processes both in the environment (Liu and Williams, 2007; Liu *et al.*, 2009a) and during waste-water treatment (Escher *et al.*, 2010). For example, biodegradation and photodegradation can produce transformation products that coexist with the parent APIs in STPs and in the environment, so it is important that the ecological effects of the reaction mixtures should be understood (Liu *et al.*, 2009b). The ecotoxicology of many APIs is poorly understood, but data for the chronic effects of individual pharmaceuticals and their mixtures is mounting (Haeba *et al.*, 2008).

Pharmaceuticals for veterinary use in the EU have been regulated since the 1990s and their assessment and authorisation is similar to that of agrochemicals. Regulation or approval of human pharmaceuticals is based on efficacy, safety, residues and quality control of the engineering and manufacturing processes; it varies between different countries and regions. New EU guidelines for testing the environmental impacts of human pharmaceuticals have been published by the European Medicines Agency (EMEA, 2006). They are based on principles similar to those used for testing other chemicals under the new EU Registration, Evaluation, Authorisation and restriction of CHemicals legislation (REACH, 2008). Hence both hazard assessment, i.e. PBT (persistence, bioaccumulation and toxicity), CMR (carcinogenicity, mutagenicity and reproductive toxicity) and risk assessment are used. However, the general perception is that this legislation is implemented less strictly for human pharmaceuticals than for industrial chemicals. Environmental data alone would not be sufficient to have a pharmaceutical restricted or banned, because the benefits of pharmaceuticals to humans are considered to outweigh their potential risks to the environment. Furthermore, most pharmaceuticals assessed so far by the Swedish Association of the Pharmaceutical Industry show predicted environmental concentration/predicted no-effect concentration (PEC/PNEC) ratios less than one (FASS, 2008).

8.1.2 Personal-care products

Personal care products (PCPs) include a diverse group of chemicals, such as additives, fragrances, preservatives and surfactants, contained in cosmetics, toiletries and other household cleaning products. They are considered together with human pharmaceuticals mainly because of the similarity of their release from humans and their ubiquitous presence in surface waters (Boyd *et al.*, 2003). Furthermore, some PCPs, such as surfactants, have similar physical and chemical (but not biological) properties to pharmaceuticals, i.e. they are ionisable compounds with one or more p*K*a values. Phthalates, for example, have a wide range of uses in PCPs depending on their chain lengths and degree of branching. It is important to consider whether exposure to PCPs is sufficiently significant to cause harm to wildlife or human health, and a risk-based approach is needed to assess their safety.

EU and US regulations currently have different approaches to the treatment of PCP ingredients. The components of PCPs are covered by the new REACH regulation for chemicals in the EU (REACH, 2008), which includes hazard and environmental risk assessment. However, the US law does not presently require the disclosure of chemical ingredients in PCPs. In one survey, nearly 100 volatile organic compounds (VOCs), of which ten are regulated in the US as toxic or hazardous chemicals, were found in six samples of fresheners and laundry products (Steinemann, 2009). The problem is compounded by the frequent lack of material safety data sheets (Barrett, 2005).

In this chapter, we examine the potential impacts of pharmaceuticals and personal care products (PPCPs) on the environment and human health. The examples used include antibiotics, NSAIDs, cardiovascular drugs, antidepressants and phthalates. The selection criteria are based on their difference in molecular structure, mode of action and therapeutic groups. Data on individual compounds for the specific therapeutic classes are used to demonstrate the potential hazards and risks of PPCPs in the environment generally.

8.2 Hazardous properties

Pharmaceutical substances are designed to have a biological effect when administered to humans and animals. One key drug design consideration is to have the appropriate pharmacokinetics, such as a half-life of hours in the body (at stomach pH $\sim$2) in order to have the effect required. This means that APIs are normally resistant to biodegradation at a pH less than four. Drug safety mainly refers to ensuring, as much as possible, that the only effects of the drug are those for 'curing' or suppressing the symptoms of diseases. However, sometimes drugs may have undesirable side effects. Conventional STP techniques are designed for removing organic molecules, nutrients and heavy metals but may not be effective in removing micro-pollutants, such as pharmaceuticals. Since patients are continually releasing pharmaceuticals into the sewage system, APIs are often detected in surface waters and there is increasing concern about their potential chronic toxicological effects on aquatic species. This section outlines some of the hazardous side effects that PPCPs can have on humans; their effects on other species in the environment are discussed later in physiological effects.

8.2.1 Antibiotics

Antibiotics kill or inhibit the growth of bacteria. They are a hugely diverse group of chemicals that can be divided into subgroups such as β-lactams, tetracyclines, macrolides, quinolones and sulphonamides; some occur naturally in the environment. These complex molecules are used for the prevention and treatment of diseases in humans, farmed animals and aquaculture (Sarmah *et al.*, 2006).

Depending on the therapeutic class of antibiotic administered, general gastrointestinal side effects, such as diarrhoea, nausea, vomiting and abdominal pain may occur, while headache, dizziness and restlessness are common central nervous system (CNS) effects. Moreover, failure to complete a course of prescribed antibiotics can lead to the build-up of resistance, while overuse can reduce healthy bacteria in the gastrointestinal system, leaving users prone to further infections.

For example, two patients treated with ciprofloxacin have been reported to have developed acquired transitory von Willebrand syndrome, which causes difficulty in blood clotting (Castaman *et al.*, 1995) and in addition to the known gastrointestinal upsets associated with clarithromycin, an elderly patient also developed thrombocytopenic purpura, causing the blood not to clot properly (Oteo *et al.*, 1994).

8.2.2 Non-steroidal anti-inflammatory drugs

NSAIDs are a group of unrelated organic acids that have analgesic, anti-inflammatory and antipyretic properties. Most NSAIDs act by inhibiting both isomers of the cyclo-oxygenase enzymes, which results in the direct inhibition of the biosynthesis of prostaglandins and thromboxanes from arachidonic acid (Vane and Botting, 1998). Inhibition of COX-2, the enzyme responsible for inflammation, is thought to be responsible for delivering some of the therapeutic effects of NSAIDs, whereas inhibition of COX-1 is thought to produce some of their toxic effects.

The most common side effects of NSAIDs are associated with gastrointestinal disturbances such as nausea and diarrhoea, and CNS-related side effects including headache, tinnitus, depression and insomnia. Anaemias and thrombocytopenia are also associated with use of the drugs. Hughes and Sudell (1983) reported a rare case of a patient developing haemolytic anaemia after a two-week course of naproxen, and Roderick *et al.* (1993) showed that aspirin can cause a multitude of symptoms including haematemesis, melaena, bloody stools and ulcers, albeit at low frequencies.

8.2.3 Antidepressants

Antidepressants are used to treat mood disorders such as depression or dysthymia. They are classified into different groups, depending on their structure or the central neurotransmitters they act upon. Before the 1950s, opiates and amphetamines were used as antidepressants (Weber and Emrich, 1988), but they were superseded by monoamine oxidase inhibitors (MAOIs) and, more recently, by selective serotonin reuptake inhibitors (SSRIs) and serotonin-noradrenalin reuptake inhibitors (SNRIs).

Antimuscarinic side effects, including dry mouth and constipation, are associated with taking antidepressants. Drowsiness is also a common side effect, and in some cases insomnia may occur. Adverse neurological effects include headache, peripheral neuropathy, tremors and tinnitus; while gastrointestinal side effects include stomatitis and gastric irritation with nausea and vomiting.

Abnormal platelet aggregation has been noted as a side effect of fluoxetine given to a severely underweight patient (Alderman *et al.*, 1992). Hyponatraemia and the syndrome of inappropriate secretion of an antidiuretic hormone (SIADH) have been reported in over 700 cases, yet over 10 million patients are exposed to SSRIs worldwide, suggesting that side effects are negligible (Liu *et al.*, 1996).

8.2.4 Cardiovascular drugs

Cardiovascular drugs are a diverse group of chemicals that are used for treating disorders of the cardiovascular system. Calcium-channel blockers are used primarily for the dilation of coronary and peripheral arteries and arterioles. Beta-blockers act by competitively inhibiting $beta_1$ and $beta_2$ receptor subtypes and are used for hypertension and the prevention and treatment of heart attacks.

Depending on the specific beta-blocker drug, side effects occur because of the selective or non-selective inhibition of $beta_2$ receptors, which are found mainly in non-cardiac tissue, including bronchial tissue, peripheral blood vessels, the uterus and the pancreas. The most serious adverse effects are heart failure, heart block and bronchospasm. Adverse effects of calcium-channel blockers include effects on the vasodilatory system, such as dizziness, flushing, headache, hypotension and palpitations.

Treatment of patients with nifedipine, a dihydropyridine calcium-channel blocker, significantly reduces the ability of platelets to aggregate (Ośmiałowska *et al.*, 1990) and four patients who underwent routine coronary bypass surgery while receiving nifedipine suffered sudden circulatory collapse

(Goiti, 1985). A patient taking the beta-blocker atenolol for coronary thromboses developed retroperitoneal fibrosis (Johnson and McFarland, 1980), and atrial fibrillation was induced in six out of twelve predisposed patients after intravenous injection with 2.5 mg atenolol (Rassmussen *et al.*, 1982).

8.2.5 Phthalates

Phthalates are used to increase the flexibility and durability of plastics in the enteric coatings of pharmaceutical pills and in time-release mechanisms of pharmaceutical capsules. They are used in a range of cosmetics and as solvents in PCPs (Barrett, 2005; Rudel and Perovich, 2009). The main phthalates in PCPs are dibutyl phthalate in nail polish, diethyl phthalate in perfumes and lotions, and dimethyl phthalate in hair spray (Barrett, 2005), while dibutyl phthalate and diethyl phthalate are used in pharmaceutical formulations (Hernández-Díaz *et al.*, 2009).

Patients using the ulcerative colitis drug asacol showed levels of monobutyl phthalate, a metabolite of dibutyl phthalate (DBP), 50 times higher than the mean for non-users in the urinary system (Hernández-Díaz *et al.*, 2009). It has been found that adult men with average amounts of phthalates in their urine had lower levels of testosterone and oestrogen in their blood (Meeker *et al.*, 2008), and research by Swan *et al.* (2008) indicates the antiandrogenic properties of phthalates. This has been suggested to be a factor in testicular dusgenesis syndrome, which, in the worst cases, is linked to testicular cancer.

8.3 Anthropogenic sources

A detailed overview of human and veterinary pharmaceutical sources is given by Ruhoy and Daughton (2008). The principal sources of human pharmaceuticals entering the environment are from patient use, and from industries and services that are connected to the sewerage system such as manufacturing plants, hospitals, care homes, prisons and residential areas. The importance of secondary sources from over-prescription and unwanted medicines has also been suggested (Daughton and Ruhoy, 2009).

Veterinary pharmaceuticals enter the environment from farm animals and aquaculture. Human pharmaceuticals may be introduced into agricultural land via the application of biosolids to fields, while both human and veterinary pharmaceuticals can be disposed to landfill sites (Figure 8.1).

Because of advances in analytical chemistry, concentrations of pharmaceuticals in the low ng/l to high μg/l levels can now be determined readily in waters. There is disagreement about whether such low concentrations of pharmaceuticals are harmful to human health, and the current thinking is that this will depend on the modes of action of individual pharmaceuticals and on species sensitivity. However, it is suggested here that basic physical chemistry may shed some light on this. The Avogadro constant (N_A), i.e. the number of molecules in one mole of chemical, is large: the 2006 CODATA recommended value of N_A

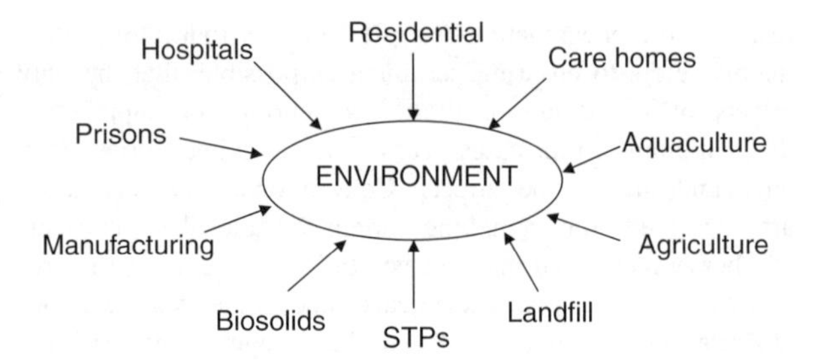

Figure 8.1 Primary sources of human and veterinary pharmaceuticals in the environment

is $6.0221417930 \times 10^{23}$ per mol. Using N_A and the molecular weight of a pharmaceutical (Mw), the numbers of molecules per litre (*N*) corresponding to measured concentrations (C) can be calculated (Equation 8.1).

$$\mathrm{N\ (molecules\ per\ } l) = [\mathrm{C\ (ng/l)} \times \mathrm{NA\ (mol - 1)}]/[109\ (\mathrm{ng/g}) \times \mathrm{Mw\ (g/mol)}] \quad (8.1)$$

It is not surprising that the number of molecules of a pharmaceutical can be very large even when it occurs at very low concentrations (e.g. ng/l) (Table 8.1). Moreover, this suggests that even below the detection limit the numbers of molecules can still be more than 10^{10} for the relatively small molecules that are typical of modern drugs. Together with the consideration of dose-response curves that could demonstrate U-shaped responses for some PPCPs, this approach can be particularly complementary for assessing drinking water quality or food intake risks at the molecular level. For example, calculated or estimated numbers of molecules can give some indications of PPCP levels in drinking water especially if concentrations cannot be measured because they are lower than the analytical detection limit.

8.3.1 Antibiotics

Worldwide antibiotic consumption has been estimated to be between 100 000 and 200 000 tonnes per annum (Wise, 2002). Several hundred different antibiotic substances are used extensively in human and veterinary medicine and aquaculture, with considerable potential for contamination of the environment. Measured concentrations of antibiotics in source effluent are detailed in Table 8.2.

In Western countries, manufacturing plants are not considered a major source of antibiotics in the environment because of the on-site treatment of production waste water. In contrast, the environmental standards for manufacturing pharmaceuticals in developing countries are often not regulated (Larsson and Fick, 2009), and many classes of antibiotics have been detected in effluent at high mg/l concentrations (Larsson *et al.*, 2007; Li *et al.*, 2008a, b; Lin *et al.*, 2008; Lin and Tsai, 2009).

Table 8.1 Calculated numbers of molecules (per l) based on measured concentration cited in the references

Therapeutic class	Compound name	Concentration (ng/l)	Number of molecules (per l)	Reference
Antibiotic	Nalidixic acid	0.94 (low)	2.44×10^{12}	Fisher and Scott, 2008
Antidepressant	Fluoxetine	18	3.50×10^{13}	Barnes *et al.*, 2004
Cardiovascular	Propranolol	42	9.75×10^{13}	Lin *et al.*, 2008
NSAID	Ibuprofen	417 (med)	1.22×10^{15}	Thomas *et al.*, 2007

Table 8.2 Measured concentrations (μg/l) of antibiotics in effluent

Source	Therapeutic class	Compound	Concentration	Reference
Manufacturing plant (Croatia)	Sulphonamide	Sulphaguanidine	>1100	Babić *et al.*, 2007
Manufacturing plant (Croatia)	Sulphonamide	Sulphamethazine	>400	Babić *et al.*, 2007
Manufacturing plant (India)	Quinolone	Ciprofloxacin	28 000–31 000	Larsson *et al.*, 2007
Manufacturing plant (India)	Quinolone	Enrofloxacin	780–900	Larsson *et al.*, 2007
Manufacturing plant (Taiwan)	Cephalosporin	Cephalexin	0.027 (median)	Lin *et al.*, 2008
Manufacturing plant (Taiwan)	Cephalosporin	Cephradine	0.001 (median)	Lin *et al.*, 2008
Manufacturing plant (Taiwan)	Tetracycline	Oxytetracycline	0.023 (median), 7.44 (max)	Lin and Tsai, 2009
Manufacturing plant (Taiwan)	Tetracycline	Tetracycline	0.025 (median), 9.66 (max)	Lin and Tsai, 2009
Hospitals (New Mexico)	Quinolone	Ofloxacin	4.9–35.5	Brown *et al.*, 2006
Rikshospitalet hospital (Norway)	Quinolone	Ciprofloxacin	14.0 (median), 39.8 (max)	Thomas *et al.*, 2007
Ullevål hospital (Norway)	Quinolone	Ciprofloxacin	24.0 (median), 54.0 (max)	Thomas *et al.*, 2007
Hospitals (Spain)	Macrolide	Erythromycin	0.025 (mean), 0.01–0.03	Gómez *et al.*, 2006
Hospitals (Switzerland)	Quinolone	Ciprofloxacin	2–83	Hartmann *et al.*, 1998
Hospitals (Taiwan)	Imidazole	Metronidazole	1.59 (median)	Lin *et al.*, 2008
Hospitals (Taiwan)	Tetracycline	Tetracycline	0.089 (median), 0.455 (max)	Lin and Tsai, 2009
Regional discharges (Taiwan)	Penicillin	Ampicillin	0.042 (median)	Lin *et al.*, 2008
Regional discharges (Taiwan)	Cephalosporin	Cefazolin	5.89 (median)	Lin *et al.*, 2008
Regional discharges (Taiwan)	Imidazole	Metronidazole	0.314 (median)	Lin *et al.*, 2008
Regional discharges (Taiwan)	Quinolone	Nalidixic acid	0.178 (median)	Lin *et al.*, 2008
Animal husbandries (Taiwan)	Cephalosporin	Cefazolin	0.053 (median)	Lin *et al.*, 2008
Animal husbandries (Taiwan)	Lincosamide	Lincomycin	56.8 (median)	Lin *et al.*, 2008
Dairy farm (Australia)	Penicillin	Nalidixic acid	0.00094–0.173	Fisher and Scott, 2008
Dairy farm (Australia)	Sulphonamide	Sulphasalazine	0.076–0.321	Fisher and Scott, 2008
Dairy (New Mexico)	Lincosamide	Lincomycin	0.7–6.6	Brown *et al.*, 2006
Swine farm (Malaysia)	Sulphonamide	Sulphameth-oxypyridazine	0.00512–0.0950	Malintan and Mohd, 2006

Antibiotics that are used to treat humans are mostly dispensed as prescriptions from pharmacies or as a treatment in hospitals. However, in some countries (e.g. China) antibiotics can be purchased over the counter. It is reported that community (i.e. not hospital) use of antibiotics in the UK is about 70 per cent (House of Lords, 1998) and 75 per cent in the US (Wise, 2002). In Germany, about 75 per cent of antibiotics are used in the community while 25 per cent are used in hospitals (Kümmerer and Henninger, 2003). Thomas *et al.* (2007) found that only 10 per cent of selected antibiotics detected in a local STP in Oslo came from hospitals. The measured concentrations of antibiotics in hospital waste water is often in the ng/l range (Lin *et al.*, 2008; Lin and Tsai, 2009), but some APIs have been measured in mg/l concentrations (Hartmann *et al.*, 1998; Brown *et al.*, 2006; Gómez *et al.*, 2006; Thomas *et al.*, 2007). Measured concentrations from residential drains are more scarce, but Lin *et al.* (2008), found various antibiotics at concentrations in the low ng/l to low μg/l range, and Brown *et al.* (2006) reported measured μg/l concentrations of ofloxacin in residential facilities in New Mexico.

Agricultural contributions come from the use of antibiotics in veterinary medicine and plant agriculture. Antibiotics are administered for the treatment of infections in domesticated animals in veterinary surgeries, and treatment is often continued on the farm. Streptomycin and oxytetracycline are primarily used for fruit crops, but in the USA antibiotics applied to plants account for less than 0.5 per cent of total antibiotic use (McManus *et al.*, 2002).

The livestock industry has intensified over the last few decades and operates concentrated animal feeding operations (CAFOs) for the production of human food from beef and dairy cattle, pigs, sheep and poultry. Pharmaceutical compounds including antibacterial and antimicrobial agents are administered at therapeutic doses for disease treatment and at non-therapeutic doses for growth promotion and increased food efficiency (Bloom, 2001). Depending on the volume of pharmaceutical compounds used, CAFOs can generate large volumes of wastes, containing compounds that can pose risks to ecosystems and human health (Lee *et al.*, 2007). Various antibiotic classes, including sulphonamides and lincosamides have been detected in the environment in the USA and Australia, derived from dairy farms (Brown *et al.*, 2006; Fisher and Scott, 2008).

The use of antibiotics for therapeutic purposes and as prophylactic agents in aquaculture for the production of molluscs, crustaceans, fish and aquatic plants is the most direct release of antibiotics into the aquatic environment. Only a small number of compounds are approved for the treatment of fish, including amoxicillin, flumequine, oxytetracycline, sulphamerazine and thiamphenicol (Lalumera *et al.*, 2004), which are often administered as feed additives or by injection (Bloom, 2001). These substances are most commonly detected in the sediment below fish-farming structures (Jacobsen and Berglind, 1998; Björklund *et al.*, 1990, 1991; Coyne *et al.*, 1994).

Municipal landfill sites are often used for the disposal of household and industrial wastes. Although modern landfill sites are designed to collect and reduce the leachate produced from the decomposition of waste and rainwater, the toxicity and treatment of the leachate produced is of concern (Visvanathan *et al.*, 2007). Moreover, older landfill sites that do not recover leachate can leach pollutants directly to soil and surrounding watercourses. Antibiotics have been detected in leachate plumes, often at ng/l to μg/l concentrations; mg/l concentrations have also been detected (Barnes *et al.*, 2004; Holm *et al.*, 1995).

8.3.2 Antidepressants

Between 1975 and 1988, antidepressant prescriptions more than doubled in the UK, with a total of 23.4 million prescriptions issued by GPs (Middleton *et al.*, 2001). Sertraline and fluoxetine were the two most prescribed generic antidepressants in 2007 [1]. As they are mostly prescribed for human use, they enter into the environment through STPs. However, some antidepressants are also used for the treatment of animals and may enter into the environment through leaching and run-off (Vogel *et al.*, 1986).

Fluoxetine, a selective serotonin re-uptake inhibitor, has been detected in ng/l concentrations from various sources including drug production facilities, landfill waste water and effluent from animal husbandries (Barnes *et al.*, 2004; Lin *et al.*, 2008). Gómez *et al.* (2006) and Lin *et al.* (2008) detected carbamazepine at ng/l concentrations in a variety of sources including hospital waste waters, and high concentrations of citalopram were measured (770–840 μg/l) from drug-production facilities in India (Larsson *et al.*, 2007) (Table 8.3).

8.3.3 Cardiovascular drugs

The most likely sources for cardiovascular drugs entering into the environment are hospitals, care homes and residential areas (Table 8.4). The concentrations measured in hospital waste waters are usually in the ng/l range, with a few reported cases of μg/l concentrations (Gómez *et al.*, 2006; Larsson *et al.*, 2007; Thomas *et al.*, 2007; Lin *et al.*, 2008; Lin and Tsai, 2009). Measured concentrations of atenolol in residential areas in Taiwan were 1.03 μg/l (Lin *et al.*, 2008). The highest recorded source concentrations were from drug-production facilities in India, where losartan was measured at 2400–2500 μg/l and metoprolol concentrations were 800–950 μg/l (Larsson *et al.*, 2007). In comparison, atenolol and acebutolol were present in only ng/l concentrations from drug production facilities in Taiwan (Lin *et al.*, 2008).

8.3.4 Non-steroidal anti-inflammatory drugs

NSAIDs are among the most-prescribed drugs in England (Jones *et al.*, 2002). They are also available over the counter,

[1] http://drugtopics.modernmedicine.com/drugtopics/data/articlestandard//drugtopics/072008/491181/article.pdf

Table 8.3 Measured concentrations (μg/l) of antidepressants in effluent

Source	Therapeutic class	Compound	Concentration	Reference
Manufacturing plant (India)	Antidepressants	Citalopram	770–840	Larsson *et al.*, 2007
Manufacturing plant (Taiwan)	Antidepressants	Carbamazepine	7.81 (median)	Lin *et al.*, 2008
Manufacturing plant (Taiwan)	Antidepressants	Fluoxetine	0.154 (median)	Lin *et al.*, 2008
Manufacturing plant (Taiwan)	Antidepressants	Paroxetine	0.003 (median)	Lin *et al.*, 2008
Hospitals (Spain)	Antidepressants	Carbamazepine	0.03–0.07, 0.04 (mean)	Gómez *et al.*, 2006
Hospitals (Taiwan)	Antidepressants	Carbamazepine	0.163 (median)	Lin *et al.*, 2008
Regional discharges (Taiwan)	Antidepressants	Carbamazepine	0.138 (median)	Lin *et al.*, 2008
Animal husbandries (Taiwan)	Antidepressants	Carbamazepine	0.003 (median)	Lin *et al.*, 2008
Animal husbandries (Taiwan)	Antidepressants	Fluoxetine	0.013 (median)	Lin *et al.*, 2008
Landfill (Oklahoma)	Antidepressants	Fluoxetine	0.018	Barnes *et al.*, 2004

so they are detected in a wide range of environments. Most NSAIDs are generally detected at ng/l concentrations, but higher concentrations have been found in source effluents (Table 8.5). Lin *et al.* (2008) measured a median concentration of diclofenac at 20.7 μg/l from drug-production facilities in Taiwan and a median concentration of acetaminophen at 37.0 μg/l from a Taiwanese hospital. Further research into these sources showed an extremely high maximum concentration of 1.5 mg/l ibuprofen measured from a pharmaceutical production facility (Lin and Tsai, 2009). Median concentrations of 46.9 μg/l and 197 μg/l of

Table 8.4 Measured concentrations (μg/l) of cardiovascular drugs in effluent

Source	Therapeutic class	Compound	Concentration	Reference
Manufacturing plant (India)	Cardiovascular drug	Losartan	2400–2500	Larsson *et al.*, 2007
Manufacturing plant (India)	Cardiovascular drug	Metoprolol	800–950	Larsson *et al.*, 2007
Manufacturing plant (Taiwan)	Cardiovascular drug	Acebutolol	0.006 (median)	Lin *et al.*, 2008
Manufacturing plant (Taiwan)	Cardiovascular drug	Atenolol	0.016 (median)	Lin *et al.*, 2008
Manufacturing plant (Taiwan)	Cardiovascular drug	Propranolol	63.9 (max)	Lin and Tsai, 2009
Manufacturing plant (Taiwan)	Cardiovascular drug	Salbutamol	0.001 (median)	Lin *et al.*, 2008
Manufacturing plant (Taiwan)	Cardiovascular drug	Tulobuterol	0.001 (median)	Lin *et al.*, 2008
Rikshospitalet hospital (Norway)	Cardiovascular drug	Metoprolol	3.41 (median), 25.097 (max)	Thomas *et al.*, 2007
Ullevål hospital (Norway)	Cardiovascular drug	Metoprolol	0.591 (median), 2.232 (max)	Thomas *et al.*, 2007
Hospital (Spain)	Cardiovascular drug	Atenolol	0.1–122, 3.4 (mean)	Gómez *et al.*, 2006
Hospital (Spain)	Cardiovascular drug	Propranolol	0.2–6.5, 1.35 (mean)	Gómez *et al.*, 2006
Hospitals (Taiwan)	Cardiovascular drug	Acebutolol	0.185 (median)	Lin *et al.*, 2008
Hospitals (Taiwan)	Cardiovascular drug	Atenolol	1.61 (median)	Lin *et al.*, 2008
Hospitals (Taiwan)	Cardiovascular drug	Metoprolol	0.145 (median)	Lin *et al.*, 2008
Hospital (Taiwan)	Cardiovascular drug	Propranolol	0.054 (median), 0.225 (max)	Lin and Tsai, 2009
Hospitals (Taiwan)	Cardiovascular drug	Propranolol	0.042 (median)	Lin *et al.*, 2008
Hospitals (Taiwan)	Cardiovascular drug	Salbutamol	0.022 (median)	Lin *et al.*, 2008
Hospitals (Taiwan)	Cardiovascular drug	Terbutaline	0.038 (median)	Lin *et al.*, 2008
Regional discharges (Taiwan)	Cardiovascular drug	Acebutolol	0.223 (median)	Lin *et al.*, 2008
Regional discharges (Taiwan)	Cardiovascular drug	Atenolol	1.03 (median)	Lin *et al.*, 2008
Regional discharges (Taiwan)	Cardiovascular drug	Salbutamol	0.009 (median)	Lin *et al.*, 2008
Animal husbandries (Taiwan)	Cardiovascular drug	Atenolol	0.052 (median)	Lin *et al.*, 2008

Table 8.5 Measured concentrations (μg/l) of NSAIDs in effluent

Source	Therapeutic class	Compound	Concentration	Reference
Manufacturing plant (Taiwan)	NSAID	Acetaminophen	0.009 (median)	Lin *et al.*, 2008
Manufacturing plant (Taiwan)	NSAID	Acetaminophen	0.124 (median), 418 (max)	Lin and Tsai, 2009
Manufacturing plant (Taiwan)	NSAID	Diclofenac	20.7 (median)	Lin *et al.*, 2008
Manufacturing plant (Taiwan)	NSAID	Diclofenac	0.053 (median), 229 (max)	Lin and Tsai, 2009
Manufacturing plant (Taiwan)	NSAID	Famotidine	0.025 (median)	Lin *et al.*, 2008
Manufacturing plant (Taiwan)	NSAID	Fenbufen	0.031 (median)	Lin *et al.*, 2008
Manufacturing plant (Taiwan)	NSAID	Ibuprofen	0.101 (median)	Lin *et al.*, 2008
Manufacturing plant (Taiwan)	NSAID	Ibuprofen	45.9 (median), 1500 (max)	Lin and Tsai, 2009
Manufacturing plant (Taiwan)	NSAID	Naproxen	1.05 (max)	Lin and Tsai, 2009
Rikshospitalet hospital (Norway)	NSAID	Diclofenac	1.55 (median), 14.9 (max)	Thomas *et al.*, 2007
Rikshospitalet hospital (Norway)	NSAID	Ibuprofen	1.22 (median), 8.96 (max)	Thomas *et al.*, 2007
Rikshospitalet hospital (Norway)	NSAID	Paracetamol	197 (median), 1368 (max)	Thomas *et al.*, 2007
Ullevål hospital (Norway)	NSAID	Diclofenac	0.784 (median), 1.629 (max)	Thomas *et al.*, 2007
Ullevål hospital (Norway)	NSAID	Ibuprofen	0.417 (median), 0.987 (max)	Thomas *et al.*, 2007
Ullevål hospital (Norway)	NSAID	Paracetamol	46.9 (median), 177.674 (max)	Thomas *et al.*, 2007
Hospitals (Spain)	NSAID	Acetaminophen	0.5–29, 16.02 (mean)	Gómez *et al.*, 2006
Hospitals (Spain)	NSAID	Diclofenac	0.06–1.9, 1.4 (mean)	Gómez *et al.*, 2006
Hospitals (Spain)	NSAID	Ibuprofen	1.5–151, 19.77 (mean)	Gómez *et al.*, 2006
Hospitals (Spain)	NSAID	Ketorolac	0.2–59.5, 4.2 (mean)	Gómez *et al.*, 2006
Hospitals (Taiwan)	NSAID	Acetaminophen	37.0 (median)	Lin *et al.*, 2008
Hospitals (Taiwan)	NSAID	Acetaminophen	62.3 (median), 186.500 (max)	Lin and Tsai, 2009
Hospitals (Taiwan)	NSAID	Diclofenac	0.328 (median), 70 (max)	Lin and Tsai, 2009
Hospitals (Taiwan)	NSAID	Diclofenac	0.286 (median)	Lin *et al.*, 2008
Hospitals (Taiwan)	NSAID	Famotidine	0.094 (median)	Lin *et al.*, 2008
Hospitals (Taiwan)	NSAID	Fenbufen	0.015 (median)	Lin *et al.*, 2008
Hospitals (Taiwan)	NSAID	Ibuprofen	0.282 (median)	Lin *et al.*, 2008
Hospitals (Taiwan)	NSAID	Ibuprofen	0.119 (median), 0.300 (max)	Lin and Tsai, 2009
Hospitals (Taiwan)	NSAID	Ketoprofen	0.0096 (median), 0.231 (max)	Lin and Tsai, 2009
Hospitals (Taiwan)	NSAID	Naproxen	0.47 (median)	Lin *et al.*, 2008
Hospitals (Taiwan)	NSAID	Naproxen	0.760 (median), 1.110 (max)	Lin and Tsai, 2009
Regional discharges (Taiwan)	NSAID	Acetaminophen	8.06 (median)	Lin *et al.*, 2008
Regional discharges (Taiwan)	NSAID	Diclofenac	0.184 (median)	Lin *et al.*, 2008
Regional discharges (Taiwan)	NSAID	Famotidine	0.014 (median)	Lin *et al.*, 2008
Regional discharges (Taiwan)	NSAID	Ibuprofen	0.747 (median)	Lin *et al.*, 2008
Regional discharges (Taiwan)	NSAID	Naproxen	0.278 (median)	Lin *et al.*, 2008
Animal husbandries (Taiwan)	NSAID	Acetaminophen	0.012 (median)	Lin *et al.*, 2008
Animal husbandries (Taiwan)	NSAID	Diclofenac	0.004 (median)	Lin *et al.*, 2008
Animal husbandries (Taiwan)	NSAID	Fenoprofen	0.008 (median)	Lin *et al.*, 2008
Animal husbandries (Taiwan)	NSAID	Ibuprofen	0.863 (median)	Lin *et al.*, 2008
Animal husbandries (Taiwan)	NSAID	Ketoprofen	0.164 (median)	Lin *et al.*, 2008
Animal husbandries (Taiwan)	NSAID	Naproxen	1.77 (median)	Lin *et al.*, 2008
Aquacultures (Taiwan)	NSAID	Acetaminophen	0.021 (median)	Lin *et al.*, 2008
Aquacultures (Taiwan)	NSAID	Diclofenac	0.004 (median)	Lin *et al.*, 2008
Aquacultures (Taiwan)	NSAID	Ibuprofen	0.05 (median)	Lin *et al.*, 2008
Landfill (Oklahoma)	NSAID	Acetaminophen	0.009	Barnes *et al.*, 2004
Landfill (Oklahoma)	NSAID	Codeine	0.24	Barnes *et al.*, 2004
Landfill (Oklahoma)	NSAID	Ibuprofen	0.018	Barnes *et al.*, 2004

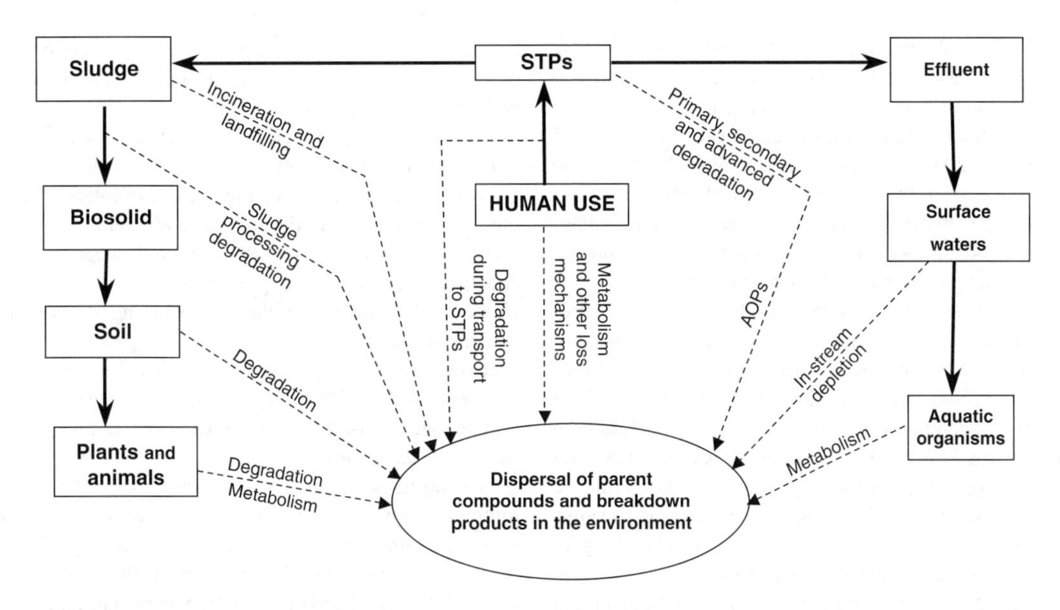

Figure 8.2 Pathways and fate of pharmaceuticals after human use. AOPs refer to Advanced Oxidation Processes (Reproduced by kind permission of Virginia L. Cunningham)

acetaminophen were measured in effluent from two Norwegian hospitals (Thomas *et al.*, 2007), while Gómez *et al.* (2006) measured a mean concentration of acetaminophen at 16.0 μg/l. Acetaminophen and ibuprofen were measured in ng/l concentrations in water samples from a municipal landfill in Oklahoma, USA (Brown *et al.*, 2004).

8.3.5 Phthalates

Due to the many uses of phthalates and their ubiquitous nature, it is difficult to link measured concentrations to specific sources (Fromme *et al.*, 2004). Concentrations of phthalates in indoor air are generally higher than outdoor concentrations (Rakkestad *et al.*, 2007; Rudel *et al.*, 2003) and urban and suburban concentrations are higher than rural and remote locations (Rudel and Perovich, 2009).

Phthalate metabolites have also been detected in human urine samples, where diet is the most probable source (Tsumura *et al.*, 2001; Fromme *et al.*, 2007). However, the plastic enteric coatings of tablets may also lead to the detection of phthalates in urine. A urinary sample collected three months after starting taking asacol for the treatment of ulcerative colitis, was measured at 16.9 μg/l, more than two orders of magnitude higher than the 95th percentile for males reported in the 1999–2000 National Health and Nutrition Examination Survey (Hernández-Díaz *et al.*, 2009). After this discovery, six other asacol users were identified and the mean urinary concentration of monobutyl phthalate was found to be 50 times higher than the mean for non-users (2257 versus 46 μg/l, $p < 0.0001$) (Hernández-Díaz *et al.*, 2009).

8.4 Pathways and environmental fate

Potential pathways of pharmaceuticals to the environment after human use are shown in Figure 8.2. Once medicines are released from the source, pathways into the environment depend on the physical and chemical properties of the APIs and the properties of the environmental compartments. This section will outline the pathways and fate of human and veterinary drugs in relevant environmental compartments.

Regardless of their route of entry or the aqueous compartment, pharmaceutical concentration and persistence are governed by similar physical, chemical and biological processes (Gurr and Reinhard, 2006; Liu *et al.*, 2009a). Pharmaceuticals may be sorbed to the sediment of a river (Liu *et al.*, 2004) or transformed by photodegradation (Lam and Mabury, 2005; Liu and Williams, 2007), biodegradation (Kim *et al.*, 2005; Perez *et al.*, 2005; Quintana *et al.*, 2005) and/or hydrolysis (Waterman and Adami, 2005; El-Gindy *et al.*, 2007). Liu *et al.* (2009a) emphasised that both biotic and abiotic transformation processes may occur in natural surface waters, and they developed a test strategy for measuring the multiple kinetics of photodegradation, biodegradation and hydrolysis in river waters simultaneously. However, in spite of various removal processes in surface waters, there are still concerns about the potential of pharmaceuticals to reach drinking water through groundwater.

8.4.1 Antibiotics

After an API acts in the body, various reactions, many of them enzyme catalysed, can cause the drug to be excreted as the parent compound, metabolites or conjugates (Cunningham, 2004).

Ciprofloxacin is eliminated from the body principally by urinary excretion and to a lesser extent by faecal excretion. About 40–50 per cent of an oral dose is excreted unchanged in the urine and about 15 per cent as metabolites, while faecal excretion over 5 days accounts for 20–35 per cent of an oral dose (Vance-Bryan *et al.*, 1990). Amoxicillin is excreted as 80–90 per cent parent compound and 10–20 per cent metabolites, while chloramphenicol leaves the body as 5–10 per cent unchanged compound and 70–90 per cent as glucuronides (Hirsch *et al.*, 1999).

After elimination from the human body or from improper disposal of unused medication, parent compound and metabolites enter the sewerage system. Influent concentrations of antibiotics to waste-water-treatment facilities are generally higher than effluent concentrations (Göbel *et al.*, 2005a; Lindberg *et al.*, 2005; Gros *et al.*, 2006), and antibiotics are frequently detected in sewage sludge. Lindberg *et al.* (2006) measured a maximum concentration of 7.7 mg/kg ciprofloxacin in Swedish STPs, and Göbel *et al.* (2005b) measured a 0.012–0.063 mg/kg range of clarithromycin from German and Swiss STPs. In 2005, 995 000 tonnes of sewage sludge was applied to English and Welsh agricultural fields as organic fertilisers,[2] providing a pathway for human pharmaceuticals to enter the agricultural environment (Kinney *et al.*, 2006a). Pharmaceuticals can also be leached by precipitation from biosolids applied to land and enter surface and groundwater (Ternes *et al.*, 2004a).

Within STPs, pharmaceuticals with an octanol/water partition coefficient of less than one are likely to partition to the aqueous phase. Several studies have been carried out to investigate the occurrence of antibacterial drugs in STP effluents across Europe and the US (Ternes, 1998; Hirsch *et al.*, 1999; Andreozzi *et al.*, 2003; Ternes *et al.*, 2003). Hirsch *et al.* (1999) detected six antibiotics in a German STP, at maximum concentrations ranging from 0.24 µg/l to the highest concentration of 6 µg/l, measured for erythromycin-H_2O. Further studies showed that various therapeutic classes of antibiotics were measured in French, Greek, Italian and Swedish STPs, at concentrations in the low µg/l range (Andreozzi *et al.*, 2003).

Antibiotics have been detected in several surface waters across Europe and the US. Hirsch *et al.* (1999) measured 1.7 µg/l erythromycin-H_2O and ng/l concentrations of five other antibiotics in German river waters, and Kolpin *et al.* (2002) measured a maximum concentration 1.9 µg/l sulphamethoxazole. Hirsch *et al.* (1999) also measured sulphamethoxazole and sulphamethazine in groundwater samples at concentrations of 0.47 µg/l and 0.16 µg/l respectively, while Lindsey *et al.* (2001) measured 0.22 µg/l of sulphamethoxazole in the US groundwater and Sacher *et al.* (2001) measured 0.41 µg/l of sulphamethoxazole in German groundwater sites. Kinney *et al.* (2006b) detected a concentration range of 0.15–0.61 ng/l for erythromycin in drinking water, but the frequency of detection was generally very low.

Veterinary pharmaceuticals enter the environment either from direct excretion of medicated animal's faeces or from the application of animal manure to land (Boxall *et al.*, 2003; Sarmah *et al.*, 2006). Irrespective of the source, veterinary antibiotics enter the environment via soil, and their behaviour is determined by their physical and chemical properties, including water solubility, lipophilicity, volatility and partition potential. Depending on the partition coefficients into soil and the soil organic carbon content (K_d and K_{oc}), antibiotics may be either mobile or persistent in the soil. Sulphonamide antibiotics have low K_{oc} values and are mobile in the soil, whereas tetracycline and macrolide antibiotics are less mobile (Kay *et al.*, 2005). However, the properties of the soil, including pH, organic carbon content, ionic strength and cation exchange capacity, can influence the sorption behaviour of antibiotics (Ter Laak *et al.*, 2006; Sassman and Lee, 2005). Antibiotics can persist in soil, leach to groundwater, run-off to surface waters or be taken up by biota (Boxall *et al.*, 2006).

Antibiotics such as oxytetracycline and oxolinic acid are routinely administered to aquaculture sites as a preventative measure against microbial pathogens and as prophylactic agents (Björklund *et al.*, 1991; Hirsch *et al.*, 1999). Halling-Sørensen *et al.* (1998) calculated that 70–80 per cent of drugs administered in aquaculture remain in the environment. Antibiotics residues can also be transported into fresh-water and marine sediments, where they have been shown to accumulate (Richardson and Bowron, 1985; Halling-Sørensen *et al.*, 1998).

8.4.2 Antidepressants

Fluoxetine is extensively metabolised by demethylation in the liver to its primary active metabolite norfluoxetine, and excretion is mainly via urine. The half-life for fluoxetine in the human body is about 1–3 days while the half-life for norfluoxetine is approximately 4–16 days (Altamura *et al.*, 1994). Calisto and Esteves (2009) reported the metabolites and excretion rates of psychiatric drugs and their presence in the environment.

In Norwegian STPs, measured effluent concentrations of SSRIs are lower than influent concentrations, indicating some removal during treatment. Sertraline and fluoxetine reduced from 2.0 to 0.9 ng/l and from 2.4 to 1.3 ng/l, respectively, and citalopram reduced from 612 to 382 ng/l (Vasskog *et al.*, 2006). High quantities of fluoxetine have been found in biosolids produced by a STP, ranging from 0.1 to 4.7 mg/l (Kinney *et al.*, 2006b). Many psychiatric drugs including diazepam, nordiazepam, oxazepam, fluoxetine and amitriptyline have been detected in high ng/l to low µg/l concentrations from STP effluents across Europe (Ternes *et al.*, 2001; Heberer, 2002b; Metcalfe *et al.*, 2003; Togola and Budzinski, 2008).

As conventional STPs were not specifically designed to remove pharmaceutical compounds, antidepressants could have entered the environment from the application of biosolids to land and from the release of effluent into receiving waters. Fluoxetine, diazepam and nordiazepam have been detected in surface waters at concentrations ranging from 2.4 to 88 ng/l (Ternes, 2001; Kolpin *et al.*, 2002; Togola and Budzinski, 2008), but the highest recorded measurement was for venlafaxine at 1000 ± 400 ng/l in samples downstream of the Pecan Creek Water Reclamation Plant in the USA (Schultz and Furlong, 2008). Even after

[2] http://www.defra.gov.uk/environment/statistics/waste/download/xls/wrtb11-12.xls

degradation in STPs and after biotic and abiotic processes in surface waters, some antidepressants have still been detected in finished drinking-water samples, albeit at low ng/l concentrations (Halling-Sørensen *et al.*, 1998; Zuccato *et al.*, 2000; Jones *et al.*, 2001b; Togola and Budzinski, 2008).

8.4.3 Cardiovascular drugs

Losartan is excreted in urine and in the faeces via bile as parent drug and metabolites. About 35 per cent of an oral dose is excreted in the urine and about 60 per cent in the faeces. The half-life of losartan in human bodies is about 1.5 to 3 hours, while the half-life of one of its metabolites, EXP3174 is approximately 3 to 9 hours (Lo *et al.*, 1995).

Propranolol has been measured in both the influent and effluent of STPs, and in some cases measured concentrations were reduced during STP treatment (Bendz *et al.*, 2005). However, most research indicates that effluent concentrations are higher than measured influent concentrations (Fono and Sedlack, 2005; Gros *et al.*, 2006; Roberts and Thomas, 2006). Higher effluent concentrations have also been reported for atenolol (Bendz *et al.*, 2005; Gros *et al.*, 2006). This may be due to the cleavage of conjugates to produce parent APIs in STPs (Heberer, 2002b; Miao *et al.*, 2002). Although atenolol has been reported to have higher concentrations in STP effluents, nanogram concentrations are still removed to sludge during sewage treatment and are present in biosolids applied to agricultural fields (Lapen *et al.*, 2008; Edwards *et al.*, 2009).

Albuterol, atenolol, metoprolol, propranolol and sotalol have been found in surface waters at low concentrations ranging from 1 to 107 ng/l (Castiglioni *et al.*, 2004; Bendz *et al.*, 2005; Fono and Sedlack, 2005; Zuccato *et al.*, 2005; Bound and Voulvoulis, 2006; Gros *et al.*, 2006; Roberts and Thomas, 2006). This reduction in concentration could be due to abiotic and biotic degradation processes in surface waters. For example, propranolol, metoprolol and atenolol have been found to undergo relatively fast direct and indirect photolysis in river waters (Liu and Williams, 2007; Liu *et al.*, 2009a). Metoprolol and atenolol underwent biodegradation in river waters under light conditions (Liu *et al.*, 2009a). Data are much more limited in groundwater. Sacher *et al.* (2001) reported sotalol at maximum concentrations of 560 ng/l in groundwater samples, and <5 ng/l concentrations of atenolol, metropolol and propranolol have been detected in drinking-water supplies in Germany (Webb *et al.*, 2003).

8.4.4 Non-steroidal anti-inflammatory drugs

Diclofenac is metabolised to 4′-hydroxydiclofenac, 5-hydroxydiclofenac, 3′-hydroxydiclofenac and 4′,5-dihydroxydiclofenac in the human body. It is then excreted in the form of glucuronide and sulphate conjugates, mainly in urine (about 65 per cent) and also in bile (about 35 per cent) (Davies and Anderson, 1997).

During waste-water treatment in STPs, NSAIDs generally decrease in concentration between influent and effluent concentrations. Acetaminophen has been found to decrease in concentration from 0.13 μg/l to below the limit of detection and from 26.1 to 5.99 μg/l (Gros *et al.*, 2006). Diclofenac has been found to decrease from an average concentration of 2.33 to 1.56 ng/l (Quintana and Reemtsma, 2004) and ibuprofen from 7.74 to 1.98 μg/l and from 33.8 to 4.24 μg/l (Roberts and Thomas, 2006).

After discharge into surface water, concentrations are usually detected at low ng/l concentrations (Castiglioni *et al.*, 2004; Alvarez *et al.*, 2005; Bendz *et al.*, 2005; Roberts and Thomas, 2006; Zhang *et al.*, 2007), with the occasional measurement in the low μg/l concentration range (Ashton *et al.*, 2004; Comoretto and Chiron, 2005; Bound and Voulvoulis, 2006). Many NSAIDs undergo photodegradation in surface waters. Buser *et al.* (1998) showed that there was significant elimination of diclofenac in a Swiss lake, concluding that photodegradation was the possible cause. In contrast, ibuprofen is relatively resistant to photodegradation in surface waters (Lin and Reinhard, 2005).

Diclofenac and ibuprofen have been detected in sludge and biosolids at 0.31 to 7.02 mg/kg and 0.12 mg/kg (Ternes *et al.*, 2004b). Diclofenac can enter surface waters due to run-off after periods of heavy rainfall or leach into groundwater from terrestrial compartments. Several pharmaceuticals, including acetaminophen, diclofenac and ibuprofen, have been detected in groundwater samples, originating from the application of biosolids to land or from landfill leachate (Heberer *et al.*, 2004; Kreuzinger *et al.*, 2004; Scheytt *et al.*, 2004; Verstraeten *et al.*, 2005). Some NSAIDs, such as ibuprofen, naproxen and ketoprofen have also been detected in low ng/l concentrations in drinking water (Vieno *et al.*, 2005; Kinney *et al.*, 2006b; Loraine and Pettigrove, 2006; Mompelat *et al.*, 2009).

8.4.5 Phthalates

Phthalates that are used in PPCPs are most likely to enter into the environment through the washing off of cosmetics and the excretion of phthalates into waste-water systems (Barrett, 2005). Clara *et al.* (2010) showed that a number of phthalates have been detected in STPs ($n = 15$) at ng/l concentrations and that influent concentrations are higher than effluent concentrations. Phthalates are readily sorbed to sewage solids and are thus removed from the aqueous phase (Marttinen *et al.*, 2003; Oliver *et al.*, 2005) but Fromme *et al.* (2002) showed that phthalates are present in surface waters and sediments at μg/l concentrations.

In addition, phthalates that are used in hair sprays and fragrances may enter into the atmospheric environment and could be deposited in house dust (Abb *et al.*, 2009; Bornehag *et al.*, 2005). Becker *et al.* (2004) measured levels of di(2-ethylhexyl)phthalate (DEHP) in house dust and also the levels of DEHP in urinary metabolites of 254 children, though correlations suggested that house dust was not a major contributor to total DEHP exposure. However, Adibi *et al.* (2003) suggested that inhalation may be an important pathway for exposure to the lower-molecular-weight phthalates diethyl phthalate (DEP), dibutyl phthalate (DBP) and butyl benzyl phthalate (BBP).

8.5 Physiological effects

Pharmaceutical compounds are manufactured to have a specific biological effect on humans and animals. Many of these compounds can enter the aquatic environment, both as the parent chemical and as metabolites, and there are concerns that they may have adverse effects on non-target species. This section will examine the harmful effects that antibiotics, antidepressants, cardiovascular drugs, NSAIDs and phthalates may be having on fauna in the terrestrial and aquatic environment.

8.5.1 Antibiotics

It is well known that bacteria have become resistant to a number of antimicrobial compounds that have been used to treat bacterial infections in humans. *Staphylococcus* spp. developed resistance to penicillin soon after the mass introduction of the antibiotic in 1947 (Gould, 1957), and methicillin-resistant *Staphylococcus aureus* (MRSA) is now a major problem (Enright *et al.*, 2002).

As a result, there is a concern for the development of new resistant strains in environmental bacteria biofilms (Schwartz *et al.*, 2006). Resistant genes and resistant bacteria have been detected in many environmental compartments (Zhang *et al.*, 2009), including sewage effluents and sewage sludge (Kim and Carlson, 2007), manure and soils (Thiele-Bruhn, 2003) and aquatic environments (Alexy and Kümmerer, 2006). However, it is still debatable whether these resistant bacteria have developed from environmental concentrations of antimicrobials or from excretion from humans and other animals (Kümmerer, 2004b, 2009a, b).

Schwartz *et al.* (2003) investigated the resistance of bacteria in a number of environmental compartments including hospital waste water, surface water and drinking water. Resistant bacteria, including *enterococci*, *staphylococci* and *enterobacteriaceae*, were detected, showing the highest resistant levels in hospital waste water, and some resistant heterotrophic bacteria were found in drinking-water samples.

Farming practice and farm animals serve as a reservoir for antibiotic resistance in the environment. Livestock supplied with feed containing 240 g tylosin per tonne resulted in a 2.1 per cent resistance level in field soils and a 25.8 per cent resistance level in cattle manure (Onan and LaPara, 2003). Chen *et al.* (2007) found macrolide, lincosamide and streptogramin B resistance in bacteria in a number of matrices including bovine manure, swine manure, compost of swine manure and swine-waste lagoons; the highest levels of resistance were in swine manure.

8.5.2 Antidepressants

Fluoxetine is so far the most acutely toxic human pharmaceutical to aquatic life (Fent *et al.*, 2006), with reported acute toxicity ranging from EC_{50} (48 h, alga) = 0.024 mg/l (Brooks *et al.*, 2003) to LC_{50} (48 h) = 2 mg/l (Kümmerer, 2004a), and it is possible that these effects may be carried over into aquatic ecosystems. In chronic toxicity studies, Flaherty and Dodson (2005) found the reproduction of *Daphnia magna* to be enhanced when exposed to a concentration of 36 μg/l fluoxetine. However, Péry *et al.* (2008) found that reproduction was significantly reduced at exposure concentrations of 31 μg/l fluoxetine, and there was 40 per cent mortality at day 21 at a concentration of 241 μg/l. Fong (1998) also showed fluoxetine to induce mussel spawning. Chronic toxicity studies using the SSRI sertraline hydrochloride showed that 100 per cent mortality of *Daphnia magna* was achieved when they were exposed to a concentration of 0.32 mg/l for 21 days and the number of days to reproduction was increased when 100 per cent mortality was not achieved (Minagh *et al.*, 2009).

The antidepressants fluoxetine and sertraline and their respective metabolites, norfluoxetine and desmethylsertraline, were found in brain, liver and muscle tissues in fish species bluegill (*Lepomis macrochirus*), channel catfish (*Ictalurus punctatus*) and black crappie (*Pomoxis nigromaculatus*) in an effluent-dominated stream in North Texas (Brooks *et al.*, 2005). In addition, the discharge of sewage into Fourmile Creek in the USA resulted in the accumulation of low ng/l concentrations of fluoxetine and sertraline and their respective metabolites in the brain tissue of White Sucker fish (Schultz *et al.*, 2010). Antidepressants can bioaccumulate in the tissue of Japanese medaka (*Oryzias latipes*), but a period of depuration has been found to result in the reduction of fluoxetine and norfluoxetine (Paterson and Metcalfe, 2008). Moreover, it has been found that four weeks of fluoxetine exposure to Japanese medaka (*Oryzias latipes*) at concentrations ranging from 0.1 to 5 μg/l does not result in any changes in adult reproductive parameters, though abnormalities, including oedema, curved spine, incomplete development and non-responsiveness, were observed in developing medaka embryos (Foran *et al.*, 2004).

8.5.3 Cardiovascular drugs

β_2-adrenoceptors are found in the heart and liver of fish (Reid *et al.*, 1992; Gamperl *et al.*, 1994) and also in reproductive tissues (Haider and Baqri, 2000); hence aquatic invertebrates may be adversely impacted by some beta-blockers. Exposure of Japanese medaka to propranolol resulted in a 48 h LC_{50} value of 24.3 mg/l, but increased mortality was not observed for metoprolol and nadolol. Egg production was not affected in two-week exposure studies to propranolol but growth of medaka was significantly reduced at concentrations of 0.5 mg/l. Male and female plasma steroid levels were significantly decreased at all concentrations tested, and male testosterone levels were significantly decreased and female medaka plasma estradiol were significantly increased at propranolol concentrations > 0.1 mg/l (Huggett *et al.*, 2002).

In aquatic toxicity tests with drugs including NSAIDs, anti-epileptics and cardiovascular drugs, propranolol was found to be the most toxic out of the ten prescription drugs tested against *Daphnia magna* and *Desmodesmus subspicatus*, with EC_{50} values of 7.5 mg/l and 5.8 mg/l, respectively. The EC_{50} value for metoprolol (7.3 mg/l) was the second most lethal for

Desmodesmus subspicatus. In comparison, both of the cardiovascular drugs tested were the least toxic to the duckweed *Lemna minor* (Cleuvers, 2003).

Ferrari *et al.* (2004) investigated the ecotoxicity of six pharmaceuticals (carbamazepine, clofibric acid, diclofenac, ofloxacin, sulphamethoxazole and propranolol), and propranolol was found to be the most toxic in many of the acute and chronic studies. In 48-h mortality studies using the crustaceans *Ceriodaphnia dubia* and *Daphnia magna*, propranolol had the lowest EC_{50} values, 1510 μg/l and 2750 μg/l respectively. In chronic studies, the rotifer *Brachionus calyciflorus* and the crustacean *Ceriodaphnia dubia* had the lowest no-observed-effect concentrations, 180 μg/l and 9 μg/l, in 48-h and 7-day reproduction studies.

8.5.4 Non-steroidal anti-inflammatory drugs

Diclofenac is responsible for the largest ecological disaster involving pharmaceutical compounds in recent times. Between 1991 and 2000, Prakash *et al.* (2003) showed a greater than 90 per cent decline in two species of *Gyps* vulture populations in northern India. Even though the drop in numbers was first observed in the 1990s, it was not until 2004 that this was linked to the use of diclofenac in the treatment of livestock (Oaks *et al.*, 2004). Dead cattle containing high diclofenac residues were allowed, for cultural reasons, to rot in the open air. As vultures fed on the rotting carcases they received a fatal dose. Diclofenac is considered to be safe for the treatment of cattle, but the drug proved to be one of the most toxic to vultures, with an LD_{50} value in the range 0.098–0.225 mg/kg (Swan *et al.*, 2006), causing death from a combination of increased reactive oxygen species and interference with uric acid transport (Naidoo and Swan, 2009). As a result, three species (oriental white-backed vulture, *Gyps bengalensis*, long-billed vulture, *Gyps indicus*, and slender-billed vulture, *Gyps tenuirostris*) are at a high risk of global extinction and are IUCN red-listed as critically endangered population declining (IUCN, 2009).

Diclofenac has been shown to have effects on invertebrates and vertebrates across a number of trophic levels in the aquatic environment. In 30-min luminescence tests on the bacterium *Vibrio fisheri*, diclofenac had lower EC_{50} concentration than carbamazepine, clofibric acid, ofloxacin, propranolol or sulphamethoxazole (Ferrari *et al.*, 2004). In the crustacean, *Daphnia magna*, 48-h mortality studies have produced EC_{50} values of 39.9 mg/l and 44.7 mg/l (Haap *et al.*, 2008). Using the rainbow trout (*Oncorhynchus mykiss*) as a model for histopathological and bioaccumulation studies, Schwaiger *et al.* (2004) showed concentration-related accumulation of diclofenac in the liver, kidneys and gills. At 5 μg/l concentrations and above, individuals exposed to diclofenac showed significant renal changes, including severe hyaline-droplet degeneration accompanied by an accumulation of proteinaceous material within the tubular lumina. Accumulation of diclofenac in the gills resulted in degenerative and necrotic changes in pillar cells as well as dilation of the capillary walls at 100 μg/l.

8.5.5 Phthalates

A monitoring study in the Netherlands found two phthalate esters, DEHP and DBP, in freshwater, marine-water and sediment samples. Even though the concentrations measured in freshwater samples were in the low μg/l range, low μg/kg concentrations were measured in fish lipid (Peijnenburg and Struijs, 2006). Phthalates have also been shown to bioconcentrate in other aquatic organisms (Brown and Thompson, 1982b; Staples *et al.*, 1997), but Brown and Thompson (1982a) found that DEHP or di-isodecyl phthalate (DIDP) did not show any acute or chronic effects on *Daphnia magna* at concentrations up to 100 μg/l. Scholz (2003) reported that short-chain monoesters, such as mono-isononyl phthalate (MINP) and mono-*n*-hexyl/*n*-octyl/n-decyl-phthalate ($MC_{8/10}P$) have the greatest acute effects.

8.6 Risk assessment, communication and reduction

Risk is a function of both hazard and exposure. PPCPs cover a wide range of chemicals, some of which are bioactive compounds and others ionisable chemicals (Cunningham, 2004). Most of PCPs are developed in various formulations for effective product application during manufacturing processes, so the final products may contain volatile solvents (Barrett, 2005). Hazards associated with PPCPs can be related to their mechanisms of action (MoA), PBT and CMR properties with regard to both environmental and human health risks (Figure 8.3). These are linked to the physical, chemical and biological properties of PPCPs. On the other hand, exposure during production, transport and use, and in the environment, is relevant to the use pattern, pathways and fate of PPCPs to the workplace, communities, STPs and natural environmental compartments.

Environmental-risk assessment of PPCPs is currently undertaken using risk-characterisation ratios (RCR) in environmental compartments such as air, water and soil (EMEA, 2006; REACH, 2008). For example, the water-sediment compartment

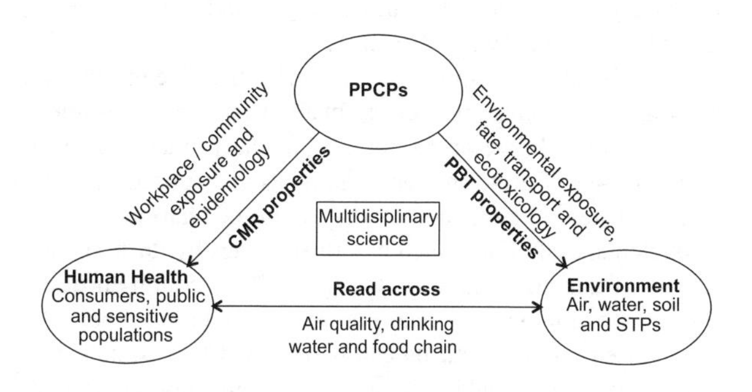

Figure 8.3 Links between environmental and human-health risks of PPCPs (Reproduced by kind permission of Qin-Tao Liu)

is most relevant to human pharmaceuticals, whilst the soil compartment can be important to veterinary drugs. RCR is the ratio of predicted environmental concentration (PEC) over predicted no-effect concentration (PNEC). If the RCRs are greater than 1, it is considered that there are potentially significant environmental risks (FASS, 2008).

At this stage, risk mitigation or reduction will be required unless the RCR can be further refined. The sciences underlying risk assessment are multidisciplinary and have made significant advances in recent decades, presenting great challenges to risk assessors and regulators. In reality, risk assessment can have uncertainties that may be small or large depending on information availability, i.e. uncertainty factors that have been included in the calculation of PEC and PNEC values. First, the PBT assessment methods for chemicals may not be suitable for PPCPs, which are usually ionic compounds with low vapour pressures. Secondly, some test methods that are recommended under REACH for chemical assessment, e.g. phototransformation, are still not included in the current EMEA guideline. Moreover, the EMEA guideline suggests the use of a market penetration factor (F_{pen}) for environmental-risk assessment. This factor varies depending on the type and stage the drugs have reached. For example, a cancer drug will have a much lower F_{pen} than a painkiller. Similarly a new drug is likely to have lower F_{pen} than a generic over-the-counter drug. The default number of F_{pen}, according to EMEA guidelines, is 1 per cent; however industries are allowed to refine the F_{pen} based on evidence and available data.

On the other hand, human-health risk assessment requires not only an understanding of CMR properties of PPCPs and their safety to individuals such as workers, consumers or patients, but also epidemiological studies of population responses and their effect on sensitive people such as children, pregnant women and the elderly (Figure 8.3). Furthermore, the human-health risks of PPCPs also include secondary routes from environmental compartments, through inhalation, skin sensitisation, drinking water or food-chain bioaccumulation.

There are different methods of risk reduction of the PPCPs over the product life cycles, related either to reduction of their hazardous properties or reduction of their release and exposure. Methods of risk reduction range from risk communication/education, green product design, reduction of use and waste, improving treatment techniques in STPs and substitution of hazardous products with safer ones.

Eco-labelling of pharmaceuticals can be effective in communicating information to doctors, pharmacists and consumers (FASS, 2008). However, in addition to the primary route of direct release of urine and faeces to STPs and subsequently to surface waters, there is a secondary route of pharmaceuticals through disposal of unwanted or leftover drugs by flushing into sewers (Daughton and Ruhoy, 2009). Risk communication should also aim to reduce over-prescription and inform patients that unused pharmaceuticals can be returned to hospitals and pharmacies. This could reduce the environmental impact of pharmaceuticals and unintentional risks to humans, and improve the quality and cost-effectiveness of health care.

Secondly, with high-throughput screening technologies coupled with combinatorial and synthetic chemistry, greener pharmaceuticals, such as solid acids, may be discovered, modified and developed (Clark, 2002, 2008). Green product and process design, together with a green sales force, has the potential to provide alternatives to environmentally hazardous pharmaceuticals and to minimise PPCPs wastes (Taylor, 2009). However, it is challenging for pharmaceutical companies to balance human versus environmental-safety issues, and a life-cycle-assessment approach will be needed to address the efficiency, efficacy, reliability and safety of pharmaceuticals (Tucker, 2006).

Thirdly, waste-water-treatment techniques may be improved for better removal of PPCPs from effluents. STPs were designed to remove large amount of organic carbon and nutrients from human wastes rather than PPCPs at low concentrations. Major removal processes in STPs are biodegradation and partitioning to sludge. However, recent development of potentially more effective treatment techniques for PPCPs include active carbon adsorption, membrane technology, nanofiltration and various advanced oxidation processes, such as ozone and UV oxidation. For example, dosing waste water with ozone at concentrations of 10 mg/l and 15 mg/l reduced the concentrations of five antibiotics, five beta-blockers, four antiphlogistics, two lipid-regulator metabolites, carbamazepine, estrone and two polycyclic musk fragrances below the limit of detection in the effluent of a biological STP (Ternes *et al.*, 2003).

8.7 Future trends

More and more PPCPs are likely to be developed and enter the global market as a result of demographic changes, improved health care and more affluent lifestyles. With population growth and ageing populations, sales of generic and prescription pharmaceuticals are likely to increase. Better health care will result in a longer average lifespan and further accelerate the global demand for PPCPs. Ultimately, the potential increase in consumption may lead to significant increase in continuous release of PPCPs into aquatic and terrestrial ecosystems. Although only EE2 and diclofenac have so far showed adverse environmental effects (to sex change in fish and death in vulture populations in India, respectively), there are increasing public concerns on the safety of other PPCPs on aquatic and soil biota and, subsequently, human health. A potentially drier climate in the future could increase concentrations in river water, putting even more pressure on water companies trying to provide clean, pure water. Climate change means that the temperatures of rivers and marine waters are likely to increase, and this could have an impact on population-level responses to PPCPs in the environment. In order to prepare for these problems, we suggest that future research should be prioritised in the following areas.

- A better understanding of the hazardous properties (PBT and CMR) of PPCPs through measurements, MoA, read-across and modelling approaches related to the environment and human health. The emerging issues in this area include antibacterial resistance and endocrine-disrupting properties of some pharmaceuticals (Sumpter, 2005; Sumpter and Johnson, 2005). Some extrapolation or read-across approaches may be useful for relating human-health endpoints to environmental species, and vice versa.

- A better understanding of exposure scenarios of PPCPs in the workplace, communities and the environment. This includes the investigation of exposure of PPCPs and their transformation products in mixtures as well as population responses through epidemiological studies.
- A better understanding of risk-assessment uncertainties for improving confidence in risk-characterisation ratios.
- The application of multidisciplinary science, life-cycle assessment and integrated test strategies for a better understanding of PPCP safety, both in the environment and for human health.
- Better treatment techniques for removing PPCPs at STPs, which requires not only the reduction of PPCP concentrations in effluents, but also reduction of their toxicity to wildlife species.
- Green initiatives. Although few PPCPs have been proved to pose significant risks to wildlife species and human health, more PPCP companies are showing interests in green product and process design that will improve the industry's reputation through sustainability and provide an opportunity for product and process innovation. Therefore green initiatives have the potential to make a great contribution to the mitigation of possible risks and enhance corporate responsibility.

References

Abb, M., Heinrich, T., Sorkau, E., Lorenz, W. (2009) Phthalates in house dust. *Environment International*, **35**, 965–970.

Adibi, J. J., Perera, F. P., Jedrychowski, W., Camann, D. E., Barr, D., Jacek, R., Whyatt, R. M. (2003) Prenatal exposures to phthalates among women in New York city and Krakow, Poland. *Environmental Health Perspectives*, **111**, 1719–1722.

Alderman, C. P., Moritz, C. K., Ben-Tovim, D. I. (1992) Abnormal platelet aggregation associated with fluoxetine therapy. *The Annals of Pharmacotherapy*, **26**(12), 1517–1519.

Alexy, R., Kümmerer, K. (2006) Antibiotics for Human Use. In: Reemtsma, T. and Jekel, M. (eds) *Organic Pollutants in the Water Cycle*. Weinheim, Wiley.

Altamura, A. C., Moro, A. R., Percudani, M. (1994) Clinical pharmacokinetics of fluoxetine. *Clinical pharmacokinetics*, **26**(3), 201–214.

Alvarez, D. A., Stackelberg, P. E., Petty, J. D., Huckins, J. N., Furlong, E. T., Zaugg, S. D., Meyer, M. T. (2005) Comparison of a novel passive sampler to standard water-column sampling for organic contaminants associated with wastewater effluents entering a New Jersey stream. *Chemosphere*, **61**(5), 610–622.

Andreozzi, R., Raffaele, M., Nicklas, P. (2003) Pharmaceuticals in STP effluents and their solar photodegradation in aquatic environment. *Chemosphere*, **50**, 1319–1330.

Ashton, D., Hilton, M., Thomas, K. V. (2004). Investigating the environmental transport of human pharmaceuticals to streams in the United Kingdom. *Science of the Total Environment*, **333**(1–3), 167–184.

Babić, S., Mutavdžić, D., Ašperger, D., Horvat, A.J.M., Kaštelan-Macan, M. (2007) Determination of veterinary pharmaceuticals in production wastewater by HPTLC-videodensitometry. *Chromatographia*, **65**, 105–110.

Barnes, K. K., Christenson, S. C., Kolpin, D. W., Focazio, M. J., Furlong, E. T., Zaugg, S. D., Meyer, M. T., Barber, L. B. (2004) Pharmaceuticals and other organic wastewater contaminants within a leachate plume downgradient of a municipal landfill. *Groundwater Monitoring and Remediation*, **24**(2), 119–126.

Barrett, J. R. (2005) The ugly side of beauty products. *Environmental Health Perspectives*, **113**(1), A24.

Becker, K., Seiwert, M., Angerer, J., Heger, W., Koch, H. M., Nagorka, R., Roßkamp, E., Schlüter, C., Seifert, B., Ullrich, D. (2004) DEHP metabolites in urine of children and DEHP in house dust. *International Journal of Hygiene and Environmental Health*, **207**, 409–417.

Bendz, D., Paxeus, N. A., Ginn, T. R., Loge, F. J. (2005) Occurrence and fate of pharmaceutically active compounds in the environment, a case study: Hoje River in Sweden. *Journal of Hazardous Materials*, **122**(3), 195–204.

Björklund, H., Bondestam, J., Bylund, G. (1990) Residues of oxytetracycline in wild fish and sediments from fish farms. *Aquaculture*, **86**, 359–367.

Björklund, H. V., Råbergh, C.M.I., Bylund, G. (1991) Residues of oxolinic acid and oxytetracycline in fish and sediments from fish farms. *Aquaculture*, **97**, 85–96.

Bloom, R. A. (2001) Use of veterinary pharmaceuticals in the United States. In: Kummerer, K. (eds) *Pharmaceuticals in the Environment: Sources, Fate, Effects and Risks,* 2nd edition. Berlin, Springer pp 149–163.

Bornehag, C.-G., Lundgren, B., Weschler, C. J., Sigsgaard, T., Hagerhed-Engman, L., Sundell, J. (2005) Phthalates in indoor dust and their association with building characteristics. *Environmental Health Perspectives*, **113**(10), 1399–1404.

Bound, J. P. and Voulvoulis, N. (2006). Predicted and measured concentrations for selected pharmaceuticals in UK rivers: Implications for risk assessment. *Water Research*, **40**(15), 2885–2892.

Boxall, A. B. A., Kolpin, D. W., Halling-Sørensen, B., Tolls, H. (2003) Are veterinary medicines causing environmental risks? *Environmental Science and Technology*, **37**(15), 286A–294A.

Boxall, A. B. A., Johnson, P., Smith, E. J., Sinclair, C. J., Stutt, E., Levy, L. S. (2006) Uptake of veterinary medicines from soils into plants. *Journal of Agricultural and Food Chemistry*, **54**(6), 2288–2297.

Boyd, G. R., Reemtsma, H., Grimm, D. A., Mitra, S. (2003) Pharmaceuticals and personal care products (PPCPs) in surface and treated waters of Louisiana, USA and Ontario, Canada. *The Science of the Total Environment*, **311**, 135–149.

Brooks, B. W., Foran, C. M., Richards, S. M., Weston, J., Turner, P. K., Stanley, J. K., Solomon, K. R., Slattery, M., La Point, T. W. (2003) Aquatic ecotoxicology of fluoxetine. *Toxicology Letters*, **142**, 169–183.

Brooks, B. W., Chambliss, K., Stanley, J. K., Ramirez, A., Banks, K. E., Johnson, R. D., Lewis, R. J. (2005) Determination of selected antidepressants in fish from an effluent-dominated stream. *Environmental Toxicology and Chemistry*, **24**(2), 464–469.

Brown, D., Thompson, R. S., (1982a) Phthalates and the aquatic environment: Part I The effect of di-2-ethyl-hexyl phthalate (DEHP) and di-isodecyl phthalate (DIDP) on the reproduction of *Daphnia magna* and observations on their bioconcentration. *Chemosphere*, **11**(4), 417–426.

Brown, D., Thompson, R. S., (1982b) Phthalates and the aquatic environment: Part II The bioconcentration and depuration of di-2-ethyl-hexyl phthalate (DEHP) and di-isodecyl phthalate (DIDP) in mussels (*Mytilus edulis*). *Chemosphere*, **11**(4), 427–435.

Brown, K. D., Kulis, J., Thomson, B., Chapman, T. H., Mawhinney, D. B. (2006) Occurrence of antibiotics in hospital, residential, and dairy effluent, municipal wastewater, and the Rio Grande in New Mexico. *Science of the Total Environment*, **366**, 772–783.

Buser, H.-R., Poiger, T., Muller, D. (1998) Occurrence and fate of the pharmaceutical drug diclofenac in surface waters: Rapid photodegradation in a lake. *Environmental Science and Technology*, **32**(22), 3449–3456.

Calisto, V., Esteves, V. I. (2009) Psychiatric pharmaceuticals in the environment. *Chemosphere*, **77**(10), 1257–1274.

Castaman, G., Lattuada, A., Mannucci, P. M., Rodeghiero, F. (1995) Characterisation of two cases of acquired transitory von Willebrand syndrome with ciprofloxac in: Evidence for heightened proteolysis of von Willebrand factor. *American Journal of Haematology*, **49**, 83–86.

Castiglioni, S., Fanelli, R., Calamari, D., Bagnati, R., Zuccato, E. (2004) Methodological approaches for studying pharmaceuticals in the environmentby comparing predicted and measured concentrations in River Po, Italy. *Regulatory Toxicology and Pharmacology*, **39**(1), 25–32.

Chen, J., Yu, Z., Michel, F. C., Wittum, T., Morrison, M. (2007) Development and application of real-time PCR assays for quantification of *erm* genes conferring resistance to macrolide-lincosamides-streptogramin B in livestock manure and manure management systems. *Applied and Environmental Microbiology*, **73**(14), 4407–4416.

Clara, M., Windhofer, G., Hartl, W., Braun, K., Simon, M., Gans, O., Scheffknecht, C., Chovanec, A. (2010) Occurrence of phthalates in surface run-off, untreated and treated wastewater and fate during wastewater treatment. *Chemosphere*, **78**, 1078–1084.

Clark, J. H. (2002) Solid acids for green chemistry. *Accounts of Chemical Research*, **35**(9), 791–797.

Clark, J. H. (2008) Greener pharmaceuticals. *European Journal of Pharmaceutical Sciences*, **34**(1), S23–S24.

Cleuvers, M. (2003) Aquatic ecotoxicity of pharmaceuticals including the assessment of combination effects. *Toxicology Letters*, **142**, 185–194.

Comoretto, L., Chiron, S. (2005) Comparing pharmaceutical and pesticide loads into a small Mediterranean river. *Science of the Total Environment*, **349**(1–3), 201–210.

Coyne, R., Hiney, M., O'Conner, B., Kerry, J., Cazabon, D., Smith, P. (1994) Concentration and persistence of oxytetracycline in sediments under a marine salmon farm. *Aquaculture*, **123**, 31–42.

Cunningham, V. L. (2004) Special characteristics of pharmaceuticals related to environmental fate. In: Kümmerer, K. (ed.) *Pharmaceuticals in the Environment: Sources, Fate, Effects and Risks, 2nd edition*. Berlin, Springer, Germany, pp. 13–24.

Daughton, C. G., Ruhoy, L. S. (2009) Environmental footprint of pharmaceuticals: the significance of factors beyond direct excretion to sewers. *Environmental Toxicology and Chemistry*, **28**(12), 2495–2521.

Daughton, C. G., Ternes, T. A. (1999) Pharmaceuticals and personal care products in the environment: agents of subtle change? *Environmental Health Perspectives*, **107**(6), 907–938.

Davies, N. M., Anderson, K. E. (1997) Clinical pharmacokinetics of diclofenac. Therapeutic insights and pitfalls. *Clinical pharmacokinetics*, **33**(3), 184–213.

Dwivedi, G., Dwivedi, S. (2007) Sushruta – the Clinician-Teacher par Excellence. *The Indian Journal of Chest Diseases and Allied Sciences*, **49**(4), 243–244.

Edwards, M., Topp, E., Metcalfe, C. D., Li, H., Gottschall, N., Bolton, P., Curnoe, W., Payne, M., Beck, A., Kleywegt, S., Lapen, D. R. (2009) Pharmaceuticals and personal care products in tile drainage following surface spreading and injection of dewatered municipal biosolids to an agricultural field. *Science of the Total Environment*, **407**, 4220–4230.

El-Gindy, A., Hadad, G. M., Mahmoud, W. M. M. (2007) High performance liquid chromatographic determination of etofibrate and its hydrolysis products. *Journal of Pharmaceutical and Biomedical Analysis*, **43**(1), 196–203.

Ellis, L. (ed.) (2000) *Archaeological Method and Theory: An Encyclopaedia*. London, Taylor and Francis.

EMEA [European Medicines Agency] (2006) Guideline on the environmental risk assessment of medicinal products for human use. Document reference EMEA/CHMP/SWP/4447/00. Committee for Medicinal Products for Human Use, London, UK.

Enright, M. C., Robinson, D. A., Randle, G., Feil, E. J., Grundmann, H., Spratt, B. G. (2002) The evolutionary history of methicillin-resistant *Staphylococcus aureus* (MRSA). *Proceedings of the National Academy of Sciences of the United States of America*, **99**(11), 7687–7692.

Escher, B. I., Bramaz, N., Lienert, J., Neuwoehner, J., Straub, J. O. (2010) Mixture toxicity of the antiviral drug tamiflu (R) (oseltamivir ethylester) and its active metabolite oseltamivir acid. *Aquatic Toxicology*, **96**(3), 194–202.

FASS (2008) Swedish environmental classification of pharmaceuticals, *The Swedish Association of the Pharmaceutical Industry*, www.fass.se.

Fent, K., Weston, A. A., Caminada, D. (2006) Ecotoxicology of human pharmaceuticals. *Aquatic Toxicology*, **76**, 122–159.

Ferrari, B., Mons, R., Vollat, B., Fraysse, B., Paxéus, N., Giudice, R. L., Pollio, A., Garric, J. (2004) Environmental risk assessment of six human pharmaceuticals: Are the current environmental risk assessment procedures sufficient for the protection of the aquatic environment? *Environmental Toxicology and Chemistry*, **23**(5), 1344–1354.

Fisher, P. M. J., Scott, R. (2008) Evaluating and controlling pharmaceutical emissions from dairy farms: a critical first step in developing a preventative management approach. *Journal of Cleaner Production*, **16**, 1437–1446.

Flaherty, C. M., Dodson, S. I. (2005) Effects of pharmaceuticals on *Daphnia* survival, growth and reproduction. *Chemosphere*, **61**, 200–207.

Fong, P. P. (1998) Zebra mussel spawning is induced in low concentrations of putative serotonin reuptake inhibitors, *The Biological Bulletin*, **194**(2), 143–149.

Fono, L. J., Sedlack, D. L. (2005) Use of the chiral pharmaceutical propranolol to identify sewage discharges into surface waters. *Environmental Science and Technology*, **39**(23), 9244–9252.

Foran, C. M., Weston, J., Slattery, M., Brooks, B. W., Huggett, D. B. (2004) Reproductive assessment of Japanese medaka (Oryzias latipes) following a four week fluoxetine (SSRI) exposure. *Archives of Environmental Contamination and Toxicology*, **46**, 511–517.

Fromme, H., Küchler, T., Otto, T., Pilz, K., Müller, J., Wenzel, A. (2002) Occurrence of phthalates and bisphenol A and F in the environment. *Water Research*, **36**, 1429–1438.

Fromme, H., Lahrz, T., Piloty, M., Gebhart, H., Oddoy, A., Rüden, H. (2004) Occurrence of phthalates and musk fragrances in indoor air and dust from apartments and kindergartens in Berlin (Germany). *Indoor Air*, **14**, 188–195.

Fromme, H., Gruber, L., Schlummer, M., Wolz, G., Böhmer, S., Angerer, J., Mayer, R., Liebl, B., Bolte, G. (2007) Intake of phthalates and di(2-ethylhexyl)adipate: Results of the Integrated Exposure Assessment Survey based on duplicate diet samples and biomonitoring data. *Environment International*, **33**, 1012–1020.

Gamperl, A.K., Wilkinson, M., Boutilier, R. G. (1994) β-Adrenoreceptors in the Trout (*Oncorhynchus mykiss*) Heart: Characterisation, Quantification, and Effects of repeated Catecholamine Exposure. *General and Comparative Endocrinology*, **95**, 259–272.

Göbel, A., Thomsen, A., McArdell, C. S., Joss, A., Giger, W. (2005a) Occurrence and sorption behaviour of sulfonamides, macrolides, and trimethoprim in activated sludge treatment. *Environmental Science and Technology*, **39**(11), 3981–3989.

Göbel, A., Thomsen, A., McArdell, C. S., Alder, A. C., Giger, W., Theiss, N., Loffler, D., Ternes, T. A. (2005b) Extraction and determination of sulfonamides, macrolides, and trimethoprim in sewage sludge. *Journal of Chromatography A*, **1085**(2), 179–189.

Goiti, J. J. (1985) Calcium channel blocking agents and the heart. *British Medical Journal*, **291**, 1505.

Gómez, J. M., Petrović, M., Fernández-Alba, A. R., Barceló, D. (2006) Determination of pharmaceuticals of various therapeutic classes by solid-phase extraction and liquid chromatography-tandem mass spectrometry analysis in hospital effluent wastewaters. *Journal of Chromatography A*, **1114**, 224–233.

Gould, J. C. (1957) Origin of antibiotic resistant Staphylocci. *Nature*, **180**, 282–283.

Gros, M., Petrovic, M., Barcelo, D. (2006) Development of a multi-residue analytical methodology based on liquid chromatography-tandem mass spectrometry (LC-MS/MS) for screening and trace level determination of pharmaceuticals in surface and wastewaters. *Talanta*, **70**, 678–690.

Gurr, C. J., Reinhard, M. (2006) Harnessing natural attenuation of pharmaceuticals and hormones in rivers. *Environmental Science and Technology*, **40**(9), 2872–2876.

Haap, T., Triebskorn, Köhler, H. R. (2008) Acute effects of diclofenac and DMSO to *Daphnia magna*: Immobilisation and hsp70-induction. *Chemosphere*, **73**, 353–359.

Haeba, M. H., Hilscherová, K., Mazurová, E., Bláha, L. (2008). Selected endocrine disrupting compounds (vinclozolin, flutamide, ketoconazol and dicofol): effects on survival, occurrence of males, growth, molting and reproduction of *Daphnia magna*. *Environmental Science and Pollution Research*, **15**(3), 222–227.

Haider, S., Baqri, S. S. (2000) Beta-adrenoceptor antagonists reinitiate meiotic maturation in *Clarias batrachus* oocytes. *Comparative Biochemistry and Physiology. Part A, Molecular and Integrative Physiology*, **126**(4), 517–525.

Halling-Sørensen, B., Nielsen, S. N., Lanzky, P. F., Ingerslev, F., Holten Lützheft, H. C., Jørgensen, S. E. (1998) Occurrence, fate and effects of pharmaceutical substances in the environment – a review. *Chemosphere*, **36**(2), 357–393.

Hartmann, A., Alder, A. C., Koller, T., Widmer, R. M. (1998) Identification of fluoroquinolone antibiotics as the main source of *umu*C genotoxicity in native hospital wastewater. *Environmental Toxicology and Chemistry*, **17**(3), 377–382.

Heberer, T. (2002a) Occurrence, fate, and removal of pharmaceutical residues in the aquatic environment: a review of recent research data. *Toxicology Letters*, **131**(1–2), 5–17.

Heberer, T. (2002b) Tracking persistent pharmaceutical residues from municipal sewage to drinking water. *Journal of Hydrology*, **266**, 175–189.

Heberer, T., Mechlinski, A., Fanck, B., Knappe, A., Massmann, G., Pekdeger, A., Fritz, B. (2004) Field studies on the fate and transport of pharmaceutical residues in bank filtration. *Ground Water Monitoring and Remediation*, **24**(2), 70–77.

Hernández-Díaz, S., Mitchell, A. A., Kelley, K. E., Calafat, A. M., Hauser, R. (2009) Medications as a potential source of exposure to phthalates in the U.S. population. *Environmental Health Perspectives*, **117**(2), 185–189.

Hinck, J. E., Blazer, V. S., Schmitt, C. J., Papoulias, D. M., Tillitt, D. E. (2009) Widespread occurrence of intersex in black basses (*Micropterus* spp.) from U.S. rivers, 1995–2004. *Aquatic Toxicology*, **95**, 60–70.

Hirsch, R., Ternes, T., Haberer, K., Kratz, K. L. (1999) Occurrence of antibiotics in the aquatic environment. *The Science of the Total Environment*, **225**, 109–118.

Holm, J. V., Ruegge, K., Bjerg, P. L., Christensen, T. H. (1995) Occurrence and distribution of pharmaceutical organic compounds in the groundwater downgradient of a landfill (Grindsted, Denmark). *Environmental Science and Technology*, **29**(5), 1415–1420.

House of Lords, UK (1998) House of Lords Select Committee on Science and Technology. Seventh Report. The Stationary Office, London.

Huggett, D. B., Brooks, B. W., Peterson, B., Foran, C. M., Schlenk, D. (2002) Toxicity of select beta adrenergic receptor-blocking pharmaceuticals (β-blockers) on aquatic organisms. *Archives of Environmental Contamination and Toxicology*, **43**, 229–235.

Hughes, J. A., Sudell, W. (1983) Haemolytic anaemia associated with naproxen. *Arthritis and Rheumatism*, **26**(8), 1054.

IUCN (2009) IUCN Red List of Threatened Species. Version 2009.2. www.iucnredlist.org Downloaded on 09 November 2009.

Jacobsen, P., Berglind, L. (1998) Persistence of oxytetracycline in sediments from fish farms. *Aquaculture*, **70**, 365–370.

Jobling, S., Casey, D., Rodgers-Gray, T., Oehlmann, J., Schulte-Oehlmann, U., Pawlowski, S., Baunbeck, T., Turner, A. P., Tyler, C.R. (2004) Comparative responses of molluscs and fish to environmental estrogens and estrogenic effluent. *Aquatic Toxicology*, **66**, 207–222.

Johnson, J. N., McFarland, J. (1980) Retroperitoneal fibrosis associated with atenolol. *British Medical Journal*, **280**, 864.

Jones, O.A.H., Voulvoulis, N., Lester, J. N. (2001a) Humans pharmaceuticals in the aquatic environment: a review. *Environmental Technology*, **22**, 1383–1394.

Jones, O.A.H., Lester, J. N., Voulvoulis, N. (2001b) Pharmaceuticals: a threat to drinking water? *Trends in Biotechnology*, **23**, 163–167.

Jones, O.A.H., Voulvoulis, N., Lester, J. N. (2002) Aquatic environmental assessment of the top 25 English prescription pharmaceuticals. *Water Research*, **36**, 5013–5022.

Kay, P., Blackwell, P. A., Boxall, A.B.A. (2005) Column studies to investigate the fate of veterinary antibiotics in clay soils following slurry application to agricultural land. *Chemosphere*, **60**(4), 497–507.

Kim, S., Eichhorn, P., Jensen, J. N., Weber, A. S., Aga, D. S. (2005) Removal of antibiotics in wastewater: Effect of hydraulic and solid retention times on the fate of tetracycline in the activated sludge process. *Environmental Science and Technology*, **39**(15), 5816–5823.

Kim, S. C., Carlson, K. (2007) Temporal and spatial trends in the occurrence of human and veterinary antibiotics in aqueous and river sediment matrices. *Environmental Science and Technology*, **41**(1), 50–57.

Kinney, C. A., Furlong, E. T., Zaugg, S. D., Burkhardt, M. R., Werner, S. L., Cahill, J. D., Jorgensen, G. R. (2006a) Survey of organic wastewater contaminants in biosolids destined for land application. *Environmental Science and Technology*, **40**(23), 7207–7215.

Kinney, C. A., Furlong, E. T., Werner, S. L., Cahill, J. D. (2006b) Presence and distribution of wastewater-derived pharmaceuticals in soil irrigated with reclaimed water. *Environmental Toxicology and Chemistry*, **25**(2), 317–326.

Kolpin, D.W., Furlong, E. T., Meyer, M. T., Thurman, E. M., Zaugg, S. D., Barber, L. B., Buxton, H. T. (2002) Pharmaceuticals, hormones, and other organic wastewater contaminants in U.S. streams, 1999–2000: A national reconnaissance. *Environmental Science and Technology*, **36**(6), 1202–1211.

Kreuzinger, N., Clara, M., Strenn, B., Vogel, B. (2004) Investigation on the behaviour of selected pharmaceuticals in the groundwater after infiltration of treated wastewater. *Water Science and Technology*, **50**(2), 221–228.

Kümmerer, K., Henninger, A. (2003) Promoting resistance by the emission of antibiotics from hospitals and households into effluents. *Clinical Microbiology and Infection*, **9**, *(12)*, 1203–1214.

Kümmerer, K. (ed.) (2004a) *Pharmaceuticals in the Environment: Sources, Fate, Effects and Risks*, 2nd edition. Verlag, Springer.

Kümmerer, K. (2004b) Resistance in the environment. *Journal of Antimicrobial Chemotherapy*, **54**, 311–320.

Kümmerer, K. (ed.) (2008) *Pharmaceuticals in the Environment: Sources, Fate, Effects and Risks*, 3rd edition. Berlin, Springer Verlag.

Kümmerer, K. (2009a) Antibiotics in the aquatic environment – A review – Part I. *Chemosphere*, **75**, 417–434.

Kümmerer, K. (2009b) Antibiotics in the aquatic environment – A review – Part II. *Chemosphere*, **75**, 435–441.

Lalumera, G. M., Calamari, D., Galli, P., Castiglioni, S., Crosa, G., Fanelli, R. (2004) Preliminary investigation on the environmental occurrence and effects of antibiotics used in aquaculture in Italy. *Chemosphere*, **54**, 661–668.

Lam, M. W., Mabury, S. A. (2005) Photodegradation of the pharmaceuticals atorvastatin, carbamazepine, levofloxacin, and sulfamethoxazole in natural waters. *Aquatic Sciences*, **67**(2), 177–188.

Lapen, D. R., Topp, E., Metcalfe, C. D., Li, H., Edwards, M., Gottschall, N., Bolton, P., Curnoe, W., Payne, M., Beck, A (2008) Pharmaceuticals and personal care products in tile drainage following land application of municipal biosolids. *Science of the Total Environment*, **399**, 50–65.

Larsson, D. G. J., de Pedro, C., Paxeus, N. (2007) Effluent from drug manufactures contains extremely high levels of pharmaceuticals. *Journal of Hazardous Materials*, **148**(3), 751–755.

Larsson, D. G. J., Fick, J. (2009) Transparancy through the production chain-a way to reduce pollution from the manufacturing of pharmaceuticals? *Regulatory Toxicology and Pharmacology*, **53**, 161–163.

Lee, L. S., Carmosini, N., Sassman, S. A., Dion, H. M., Sepúlveda, M. S. (2007) Agricultural contributions of antimicrobials and hormones on soil and water quality. *Advances in Agronomy*, **93**, 1–68.

Li, D., Yang, M., Hu, J., Ren, L., Zhang, Y., Li, K. (2008a) Determination and fate of oxytetracycline and related compounds in oxytetracycline production wastewater and the receiving river. *Environmental Toxicology and Chemistry*, **27**(1), 80–86.

Li, D., Yang, M., Hu, J., Zhang, Y., Chang, H., Jin, F. (2008b) Determination of penicillin G and its degradation products in a penicillin production wastewater treatment plant and the receiving river. *Water Research*, **42**, 307–317.

Lin, A. Y.-C., Reinhard, M. (2005) Photodegradation of common environmental pharmaceuticals and estrogens in river water. *Environmental Toxicology and Chemistry*, **24**(6), 1303–1309.

Lin, A.Y.C., Tsai, Y. T. (2009) Occurrence of pharmaceuticals in Taiwan's surface water: Impact of waste streams from hospitals and pharmaceutical production facilities. *Science of the Total Environment*, **407**, 3793–3802.

Lin, A.Y.C., Yu, T. H., Lin, C. F. (2008) Pharmaceutical contamination in residential, industrial, and agricultural waste streams: Risk to aqueous environments in Taiwan. *Chemosphere*, **74**, 131–141.

Lindberg, R. H., Wennberg, P., Johansson, M. I., Tysklind, M., Andersson, B.A.V. (2005) Screening of human antibiotic substances and determination of weekly mass flows in five sewage treatment plants in Sweden. *Environmental Science and Technology*, **39**(10), 3421–3429.

Lindberg, R. H., Olofsson, U., Rendahl, P., Johansson, M. I., Tysklind, M., Andersson, B.A.V. (2006) Behaviour of fluoroquinolones and trimethoprim during mechanical, chemical and active sludge treatment of sewage water and digestion of sludge. *Environmental Science and Technology*, **40**(3), 1042–1048.

Lindsey, M. E., Meyer, M., Thurman, E. M. (2001) Analysis of trace levels of sulfonamide and tetracycline antimicrobials in groundwater and surface water using solid phase extraction and liquid chromatography mass spectrometry. *Analytical Chemistry*, **73**(19), 4640–4646.

Liu, B. A., Mittmann, N., Knowles, S. R., Shear, N. H. (1996) Hyponatraemia and the syndrome of inappropriate secretion of antidiuretic hormone associated with the use of selective serotonin reuptake inhibitors: a review of spontaneous reports. *Canadian Medical Association Journal*, **155**(5), 519–527.

Liu, Q.-T., Riddle, A.M., Robinson, P.F., Gray, N. (2004) Roles of partitioning and phototransformation in predicting the fate and movement of pharmaceuticals in UK and US rivers, in: Proceedings 4th International Conference on Pharmaceuticals and EDCs in Water, National Ground Water Association, Minneapolis, Minnesota, 13–15 October, pp. 48–62.

Liu, Q.-T., Williams, H. (2007) Kinetics and degradation products for direct photolysis of β-blockers in water. *Environmental Science and Technology*, **41**(3), 803–810.

Liu, Q.-T., Cumming, R.I., Sharpe, A.D. (2009a) Photo-induced environmental depletion processes of β-blockers in river waters, *Photochemical and Photobiological Sciences*, **8**, 768–777.

Liu, Q.-T., Williams, T.D., Cumming, R.I., Holm, G., Hetheridge, M. J., Murray-Smith, R.M. (2009b) Comparative aquatic toxicity of propranolol and its photodegraded mixtures: algae and rotifer screening, *Environmental Toxicology and Chemistry*, **28**(12), 2622–2631.

Lo, M-W., Goldberg, M. R., McCrea, J. B., Lu, H., Furtek, C.I., Bjornsson, T. D. (1995) Pharmacokinetics of losartan, an angiotensin II receptor antagonist, and its active metabolite

EXP3174 in humans. *Clinical Pharmacology and Therapeutics*, **58**, 641–649.

Loraine, G. A., Pettigrove, M. E. (2006) Seasonal variations in concentrations of pharmaceuticals and personal care products in drinking water and reclaimed wastewater in Southern California. *Environmental Science and Technology*, **40**(3), 687–695.

Malintan, N. T., Mohd, M. A. (2006) Determination of sulphonamides in selected Malaysian swine wastewater by high-performance liquid chromatography. *Journal of Chromatography A*, **1127**, 154–160.

Manandhar, N. P. (2002) *Plants and People of Nepal*. Oregon, Timber Press.

Marttinen, S. K., Kettunen, R. H., Sormunen, K. M., Rintala, J. A. (2003) Removal of bis(2-ethylhexyl) phthalate at a sewage treatment plant. *Water Research*, **37**, 1385–1393.

McManus, P. S., Stockwell, V. O., Sundin, G. W., Jones, A. L. (2002) Antibiotic use in plant agriculture. *Annual Review of Phytopathology*, **40**, 443–465.

Meeker, J. D., Calafat, A. M., Hauser, R. (2008). Urinary metabolites of di(2-ethylhexyl) phthalate are associated with decreased steroid hormone levels in adult men. *Journal of Andrology*, **30**(3), 287–297.

Metcalfe, C. D., Miao, X.-S., Koenig, B. G., Struger, J. (2003) Distribution of acidic and neutral drugs in surface waters near sewage treatment plants in the lower Great Lakes, Canada. *Environmental Toxicology and Chemistry*, **22**(12), 2881–2889.

Miao, X.-S., Koenig, B. G., Metcalfe, C. D. (2002) Analysis of acidic drugs in the effluents of sewage treatment plants using liquid chromatography–electrospray ionisation tandem mass spectrometry. *Journal of Chromatography A*, **952**, 139–147.

Middleton, N., Gunnell, D., Whitley, E., Dorling, D., Frankel, S. (2001) Secular trends in antidepressant prescribing in the UK, 1975–1998. *Journal of Public Health Medicine*, **23**(4), 262–267.

Minagh, E., Hernan, R., O'Rourke, K., Lyng, F. M., Davoren, M. (2009) Aquatic ecotoxicity of the selective serotonin reuptake inhibitor sertraline hydrochloride in a battery of freshwater test species. *Ecotoxicology and Environmental Safety*, **72**, 434–440.

Mompelat, S., Le Bot, B., Thomas, O. (2009) Occurrence and fate of pharmaceutical products and by-products, from resource to drinking water. *Environment International*, **35**, 803–814.

Naidoo, V., Swan, G. E. (2009) Diclofenac toxicity in *Gyps* vultures is associated with decreased uric acid excretion and not renal portal vasoconstriction. *Comparative Biochemistry and Physiology, Part C*, **149**, 269–274.

Oaks, J. L., Gilbert, M., Virani, M. Z., Watson, R. T., Meteyer, C. U., Rideout, B. A., Shivaprasad, H. L., Ahmed, S., Chaudhry, M. J. I., Arshad, M., Mahmood, S., Ali, A., Khan, A. A. (2004) Diclofenac residues as the cause of vulture population decline in Pakistan. *Nature*, **427**, 630–633.

Oliver, R., May, E., Williams, J. (2005) The occurrence and removal of phthalates in a trickle filter STW. *Water Research*, **39**, 4436–4444.

Onan, L. J., LaPara, T. M. (2003) Tylosin-resistant bacteria cultivated from agricultural soil. *FEMS Microbiology Letters*, **220**(1), 15–20.

Ośmiałowska, Z., Nartowicz-Słoniewska, M., Słomiński, J. M., Krupa-Wojciechowska, B. (1990) Effect of nifedipine monotherapy on platelet aggregation in patients with untreated essential hypertension. *European Journal of Clinical Pharmacology*, **39**, 403–404.

Oteo, J. A., Gómez-Cadiñanos, R. A., Rosel, L., Casas, J. M. (1994) Clarithromycin-induced thrombocytopenic purpura. *Clinical Infectious Diseases*, **19**, 1170–1171.

Paterson, G., Metcalfe, C. D. (2008) Uptake and depuration of the antidepressantfluoxetine by the Japanese medaka (*Oryzias latipes*). *Chemosphere*, **74**, 125–130.

Peijnenburg, W. and Struijs, J. (2006). Occurrence of phthalate esters in the environment of the Netherlands. *Ecotoxicology and Environmental Safety*, **63**(2), 204–215.

Perez, S., Eichhorn, P., Aga, D. S. (2005) Evaluating the biodegradability of sulfamethazine, sulfamethoxazole, sulfathiazole, and trimethoprim at different stages of sewage treatment. *Environmental Toxicology and Chemistry*, **24**(6), 1361–1367.

Péry, A. R. R., Gust, M., Vollat, B., Mons, R., Ramil, M., Fink, G., Ternes, T., Garric, J. (2008) Fluoxetine effects assessment on the life cycle of aquatic invertebrates. *Chemosphere*, **73**, 300–304.

Prakash, V., Pain, D. J., Cunningham, A. A., Donald, P. F., Prakash, N., Verma, A., Gargi, R., Sivakumar, S., Rahmani, A. R. (2003) Catastrophic collapse of Indian white-backed *Gyps bengalensis* and long-billed *Gyps indicus* vulture populations. *Biological Conservation*, **109**, 381–390.

Quintana, J. B., Reemtsma, T. (2004) Sensitive determination of acidic drugs and triclosan in surface water and wastewater by ion-pair reverse-phase liquid chromatography/tandem mass spectrometry. *Rapid Communications in Mass Spectrometry*, **18**(7), 765–774.

Quintana, J. B., Weiss, S., Reemtsma, T. (2005) Pathways and metabolites of microbial degradation of selected acidic pharmaceuticals and their occurrence in municipal wastewater treated by a membrane bioreactor. *Water Research*, **39**(12), 2654–2664.

Rakkestad, K. E., Dye, C. J., Yttri, K. E., Holme, J. A., Hongslo, J. K., Schwarze, P. E., Becher, R. (2007) Phthalate levels in Norwegian indoor air related to particle size fraction. *Journal of Environmental Monitoring*, **9**, 1419–1425.

Rashed, R. (ed.) (1996) *Encyclopaedia of the History of Arabic Science*. London, Routledge.

Rassmussen, K., Anderson, K., Wang, H. (1982) Atrial fibrillation induced by atenolol. *European Heart Journal*, **3**(3), 276–281.

REACH (2008). Regulation Evaluation and;1; Authorisation of Chemicals. European Commission. http://www.hse.gov.uk/reach

Reid, S. D., Moon, T. W., Perry, S. F. (1992) Rainbow trout hepatocyte β-adrenoceptors, catecholamine responsiveness, and effects of cortisol. *The American Journal of Physiology-Regulatory, Integrative, and Comparative Physiology*, **262**, 794–799.

Richardson, M. L., Bowron, J. M. (1985) The fate of pharmaceutical chemicals in the aquatic environment. *The Journal of Pharmacy and Pharmacology*, **37**(1), 1–12.

Roberts, P. H., Thomas, K. V. (2006) The occurrence of selected pharmaceuticals in wastewater effluent and surface waters of the lower Tyne catchment. *Science of the Total Environment*, **329**(1–3), 143–153.

Roderick, P. J., Wilkes, H. C., Meade, T. W. (1993) The gastrointestinal toxicity of aspirin: an overview of randomised controlled trials. *British Journal of Clinical Pharmacology*, **35**, 219–226.

Rudel, R. A., Camann, D. E., Spengler, J. D., Korn, L. R., Brody, J. G. (2003) Phthalates, alkylphenols, pesticides, polybrominated diphenyl ethers, and other endocrine-disrupting compounds in indoor air and dust. *Environmental Science and Technology*, **37**(20), 4543–4553.

Rudel, R. A., Perovich, L. J. (2009) Endocrine disrupting chemicals in indoor and outdoor air. *Atmospheric Environment*, **43**, 170–181.

Ruhoy, I. S., Daughton, C. G. (2008) Beyond the medicine cabinet: An analysis of where and why medicines accumulate. *Environment International*, **34**, 1157–1169.

Sacher, F., Lange, F. T., Brauch, H-J., Blankenhorn, I. (2001) Pharmaceuticals in groundwaters. Analytical methods and results of a monitoring programme in Baden-Württemberg, Germany. *Journal of Chromatography A*, **938**, 199–210.

Sarmah, A. K., Meyer, M.T., Boxall, A. B. A. (2006) A global perspective on the use, sales, exposure pathways, fate and effects of veterinary antibiotics (VAs) in the environment. *Chemosphere*, **65**(5), 725–759.

Sassman, S. A., Lee, L. S. (2005) Sorption of three tetracyclines by several soils: assessing the role of pH and cation exchange. *Environmental Science and Technology*, **39**(19), 7452–7459.

Scheytt, T., Mersmann, P., Leidig, M., Pekdeger, A., Heberer, T. (2004) Transport of pharmaceutically active compounds in saturated laboratory columns. *Ground Water*, **42**(5), 767–773.

Scholz, N. (2003) Ecotoxicity and biodegradation of phthalate monoesters. *Chemosphere*, **53**, 921–926.

Schultz, M. M., Furlong, E. T. (2008) Trace analysis of antidepressant pharmaceuticals and their select degradates in aquatic matrices by LC/ESI/MS/MS. *Analytical Chemistry*, **80**(5), 1756–1762.

Schultz, M. M., Furlong, E. T., Kolpin, D. W., Werner, S. L., Schoenfuss, H. L., Barber, L. B., Blazer, V. S., Norris, D. O., Vajda, A. M. (2010) Antidepressant pharmaceuticals in two U. S. effluent-impacted streams: Occurrence and fate in water and sediment, and selective uptake in fish neural tissue. *Environmental Science and Technology*, **44**, 1918–1925.

Schwaiger, J., Ferling, H., Mallow, U., Wintermayr, H., Negele, R. D. (2004) Toxic effects of the non-steroidal anti-inflammatory drug diclofenac. Part I: Histopathological alterations and bioaccumulation in rainbow trout. *Aquatic Toxicology*, **68**, 141–150.

Schwartz, T., Kohnen, W., Jansen, B., Obst, U. (2003) Detection of antibiotic-resistant bacteria and their resistance genes in wastewater, surface water, and drinking water biofilms. *FEMS Microbiology Ecology*, **43**(3), 325–335.

Schwartz, T., Volkmann, H., Kirchen, S., Kohnen, W., Schön-Hölz, K., Jansen, B., Obst, U. (2006) Real-time PCR detection of *Pseudomonas aeruginosa* in clinical and municipal wastewater and genotyping of the ciprofloxacin-resistant isolates. *FEMS microbiology ecology*, **57**(1), 158–167.

Staples, C. A., Peterson, D. R., Parkerton, T. F., Adams, W. J. (1997) The environmental fate of phthalate esters: A literature review. *Chemosphere*, **35**(4), 667–749.

Steinemann, A.C. (2009) Fragranced consumer products and undisclosed ingredients. *Environmental Impact Assessment Review*, **29**, 32–38.

Sumpter, J. P. (2005) Endocrine disrupters in the aquatic environment: an overview. *Acta Hydrochimica et Hydrobiologica*, **33**, 9–16.

Sumpter, J. P., Johnson, A. C. (2005) Lessons from endocrine disruption and their application to other issues concerning trace organics in the aquatic environment. *Environmental Science and Technology*, **39**, 4321–4332.

Swan, G. E., Cuthbert, R., Quevedo, M., Green, R. E., Pain, D. J. *et al.* (2006) Toxicity of diclofenac to *Gyps* vultures. *Biology Letters*, **2**, 279–282.

Swan, S. H., GuilletteJr, L. J., Myers, J. P., vom Saal, F. S. (2008) Epidemiological studies of reproductive effects in humans. In: S. E. Jorgensen and B. Fath,eds *Encyclopaedia of Ecology*. Oxford: Academic Press, pp. 1383–1388.

Taylor, D. (2009) Sustainable development and production of human pharmaceuticals. In: *A Healthy Future – Pharmaceuticals in a Sustainable Society*. Apoteket AB, MistraPharma & Stockholm County Council, pp 58–73.

Ter Laak, T. L., Gebbink W. A., Tolls, J. (2006) The effect of pH and ionic strength on the sorption of sulfachloropyridazine, tylosin and oxytetracycline to soil. *Environmental Toxicology and Chemistry*, **25**(4), 904–911.

Ternes, T. A. (1998) Occurrence of drugs in German sewage treatment plants and rivers. *Water Research*, **32**(11), 3245–3260.

Ternes, T. A. (2000) Pharmaceuticals and metabolites as contaminants of the aquatic environment: An overview. *Abstracts of Papers [American Chemical Society]*, **219**, 317–318.

Ternes, T. A. (2001) Analytical methods for the determination of pharmaceuticals in aqueous environmental samples. *Trends in Analytical Chemistry*, **20**(8), 419–434.

Ternes, T. A., Bonerz, M., Schmidt, T. (2001) Determination of neutral pharmaceuticals in wastewaters and rivers by liquid chromatography–electrospray tandem mass spectrometry. *Journal of Chromatography A*, **938**, 175–185.

Ternes, T. A., Stüber, J., Herrmann, N., McDowell, D., Ried, A., Kampmann, M., Teiser, B. (2003). Ozonation: a tool for removal of pharmaceuticals, contrast media and musk fragrances from wastewater? *Water Research*, **37**, 1976–1982.

Ternes, T. A., Joss, A., Siegrist, H. (2004a) Scrutinizing pharmaceuticals and personal care products in wastewater treatment. *Environmental Science and Technology*, **38**(20), 392A–399A.

Ternes, T. A., Herrmann, N., Bonerz, M., Knacker, T., Siegrist, H., Joss, A. (2004b) A rapid method to measure the solid-water distribution coefficient (K-d) for pharmaceuticals and musk fragrances in sewage sludge. *Water Research*, **38**(19), 4075–4084.

Thiele-Bruhn, S. (2003) Pharmaceutical antibiotic compounds in soils—a review. *Journal of Plant Nutrition and Soil Science*, **166** (2), 145–167.

Thomas, K. V., Dye, C., Schlabach, M., Langford, K. H. (2007) Source to sink tracking of selected human pharmaceuticals from two Oslo city hospitals and a wastewater treatment works. *Journal of Environmental Monitoring*, **9**, 1410–1418.

Togola, A., Budzinski, H. (2008) Multi-residue analysis of pharmaceutical compounds in aqueous samples. *Journal of Chromatography A*, **1177**, 150–158.

Tsumura, Y., Ishimitsu, S., Kaihara, A., Yoshii, K., Nakamura, Y., Tonogai, Y. (2001) Di(2-ethylhexyl) phthalate contamination of retail packed lunches caused by PVC gloves used in the preparation of foods. *Food Additives and Contaminants*, **18**(6), 569–579.

Tucker, J. L. (2006) Green chemistry, a pharmaceutical perspective. *Organic Process Research and Development*, **10**(2), 315–319.

Vance-Bryan, K., Guay, D. R., Rotschafer, J. C. (1990) Clinical pharmacokinetics of ciprofloxacin. *Clinical Pharmacokinetics*, **19**(6), 434–461.

Vane, J. R., Botting, R. M. (1998) Anti-inflammatory drugs and their mechanism of action. *Inflammation Research*, **47**(2), S78–S87.

Vasskog, T., Berger, U., Samuelsen, P-J., Kallenborn, R., Jensen, E. (2006) Selective serotonin reuptake inhibitors in sewage influents and effluents from Tromsø, Norway. *Journal of Chromatography A*, **1115**, 187–195.

Verstraeten, I. M., Fetterman, G. S., Meyer, M. T., Bullen, T., Sebree, S. K. (2005) Use of tracers and isotopes to evaluate vulnerability of water in domestic wells to septic waste. *Ground Water Monitoring and Remediation*, **25**(2), 107–117.

Vieno, N. M., Tuhkanen, T., Kronberg, L. (2005) Seasonal variation in the occurrence of pharmaceuticals in effluents from a sewage treatment plant and in the recipient water, *Environmental Science and Technology*, **39**(21), 8220–8226.

Visvanathan, C., Choudhary, M. K., Montalbo, M. T., Jegatheesan, V. (2007) Landfill leachate treatment using thermophilic membrane bioreactor. *Desalination*, **204**, 8–16.

Vogel, G. W., Minter, K., Woolwine, B. (1986) Effects of chronically administered antidepressant drugs on animal behaviour. *Physiology and Behaviour*, **36**, 659–666.

Volger, B. K., Ernst, E. (1999) Aloe vera: a systematic review of its clinical effectiveness. *The British Journal of General Practice*, **49** (447), 823–828.

Vos, J. G., Dybing, E., Greim, H. A., Ladefoged, O., Lambré, C., Tarazona, J. V., Brandt, I., Vethaak, A. D. (2000) Health effects of endocrine-disrupting chemicals on wildlife, with special reference to the European situation. *Critical Reviews in Toxicology*, **30**(1), 71–133.

Waterman, K. C., Adami, R. C. (2005) Accelerated aging: Prediction of chemical stability of pharmaceuticals. *International Journal of Pharmaceutics*, **293**,(1–2), 101–125.

Webb, S., Ternes, T., Gibert, M., Olejniczak, K. (2003) Indirect human exposure to pharmaceuticals via drinking water. *Toxicology Letters*, **142**, 157–167.

Weber, M. M., Emrich, H. M. (1988) Current and historical concepts of opiate treatment in psychiatric disorders. *International Clinical Psychopharmacology*, **3**(3), 255–266.

Wise, R. (2002) Antimicrobial resistance: priorities for action. *Journal of Antimicrobial Chemotherapy*, **49**, 585–586.

Wishart, D. J. (2007) *Encyclopaedia of the Great Plains Indians*. University of Nebraska Press.

Yang, S.-Z. (1998) *The Devine Farmer's Materia Medica – A Translation of Shen Nong Ben Cao Jing*, 1st Edition, Blue Poppy Press. ISBN 0-936185-96-1.

Zhang, S., Zhang, Q., Darisaw, S., Ehie, O., Wang, G., (2007) Simultaneous quantification of polycyclic aromatic hydrocarbons (PAHs), polychlorinated biphenyls (PCBs), and pharmaceuticals and personal care products (PPCPs) in Mississippi river water, in New Orleans, Louisiana, USA. *Chemosphere*, **66**, 1057–1069.

Zhang, X.-X., Zhang, T., Fang, H. H. P. (2009) Antibiotic resistant genes in water environment. *Applied Microbiology and Biotechnology*, **82**, 397–414.

Zuccato, E., Calamari, D., Natangelo, M., Fanelli, R. (2000) Presence of therapeutic drugs in the environment. *Lancet*, **355**, 1789–1790.

Zuccato, E., Castiglioni, S., Fanelli, R. (2005) Identification of the pharmaceuticals for human use contaminating the Italian aquatic environment. *Journal of Hazardous Materials*, **122**(3), 205–209.

9

Naturally occurring oestrogens

Olwenn V. Martin[1]* and Richard M. Evans[2]

[1]*Institute for the Environment, Brunel University, Kingston Lane, Uxbridge, Middlesex, UB8 3PH*
[2]*Centre for Toxicology, School of Pharmacy, University of London, 29–39 Brunswick Square, London*
**Corresponding author, email olwenn.martin@brunel.ac.uk*

9.1 Introduction

The realisation that synthetic chemicals can mimic the actions of endogenous hormones is relatively recent. Plant and animal products have been used throughout history, however, for their putative properties as contraceptives or to increase fertility. Pedianius Dioscorides, a Greek surgeon in the Roman army of the Emperor Nero, recommended the use of several herbs, sometimes with the added ingredient of the kidney of a mule, as contraceptives or abortifacients in his *De Materia Medica* (Dioscorides, 2000). The role of silphium, a member of the giant fennel family, as a means of birth control is thought to have been so widespread in antiquity that it became extinct (Riddle, 1991).

More recently, since the 1930s, natural and synthetic hormones have been used in aquaculture to change the sex of fish (Pandian and Sheela, 1995), long before the identification of intersex fish downstream of sewage effluent plants was recognised as an environmental issue (Jobling *et al.*, 1998). There are also modern examples of natural compounds present in the environment acting as endocrine disrupters. Potent oestrogenic effects were reported in gilts fed mouldy corn as early as 1928, and these effects were later attributed to a zearalenone, a mycotoxin produced by *Fusarium* species of fungus (Stob *et al.*, 1962). In the early 1940s, ewes grazing on red clover pastures in Western Australia suffered severe breeding problems that came to be known as 'clover disease' (Bennetts *et al.*, 1946). The clover was found to contain up to 5 per cent dry weight isoflavones, including genistein, formononetin and biochanin A (Adams, 1995). Endocrine disruption of fish exposed to phyto-oestrogens downstream of pulp and paper-mill effluents has also been reported (Larsson *et al.*, 2000).

Hence, even relatively common and generally benign natural compounds can have serious effects at unusual doses or in unusual places, particularly if they are hormonally active compounds. This chapter is concerned with naturally occurring oestrogens, including steroid, phyto- and myco-oestrogens as examples.

9.1.1 Steroid oestrogens

Steroid hormones are derived from cholesterol, with which they share their four-ringed cyclopentanophenanthrene structure. Oestrogens are produced mainly by the ovary in vertebrate species and they are generally regarded as female hormones, being primarily responsible for female differentiation and reproductive function. However, oestrogens are also produced in the male testis, and in the adrenal gland, brain and fatty tissues of both sexes, and they are involved in many other developmental and homeostatic processes, including skeletal maintenance and cardiovascular health.

Steroidogenesis is a complex process regulated by many interacting pathways (Stocco *et al.*, 2005). Cholesterol is first transformed into progestagens, such as pregnenolone and progesterone, which are in turn converted into androgens. These biosynthetic reactions are catalysed by a cascade of specific enzymes, many of which belong to the cytochrome P450 (CYP) superfamily. In particular, the CYP19 enzyme aromatase controls the final conversion of androstenedione, 16α-hydroxydehydroisoandrosterone sulphate and testosterone into oestrone, oestriol and oestradiol, respectively (Simpson *et al.*, 1997). Additionally, 17β-dehydroxysteroid dehydrogenases can catalyse the conversion of oestrone to oestradiol and vice-versa.

Pollutants, Human Health and the Environment: A Risk Based Approach, First Edition. Edited by Jane A. Plant, Nikolaos Voulvoulis and K. Vala Ragnarsdottir.

17β-oestradiol is the most abundant and potent natural oestrogen in all vertebrates. In women of reproductive age, it is produced primarily in the granulosa cells of the preovulatory follicle. In men and postmenopausal women, the principal oestrogen is **oestrone** and it is aromatised in peripheral tissues (adipose and skin) from circulating androstenedione produced by the adrenal gland. Testicular steroidogenesis is estimated to account for 15 per cent of the circulating levels of oestrogens in men. **Oestriol** is produced specifically during pregnancy in the fetal liver. Various brain-cell types, including neurons, express aromatase, the activity of which has been shown to be stimulated by testosterone (Nelson and Bulun, 2001).

Over 98 per cent of circulating 17β-oestradiol is tightly bound to proteins, either albumin or sex hormone binding globulin (SHBG). Free oestradiol is readily available to cross cell membranes and is believed to be largely responsible for its biological activity, although there is some evidence that albumin-bound oestradiol is also biologically active (Pardridge, 1981; Toniolo *et al.*, 1994). Steroid oestrogens are metabolised to less-active hydroxylated and/or conjugated water-soluble metabolites that can subsequently be excreted in urine or faeces. Although this conversion takes place mainly in the liver, some of the CYP enzymes involved in the oxidative conversion to hydroxylated metabolites are expressed at low levels in some target tissues (Zhu and Conney, 1998). Some of the water-soluble conjugates are excreted via the bile duct and partly reabsorbed from the intestinal tract by a process called enterohepatic circulation. The chemical structures of steroid oestrogens are shown in Figure 9.1.

9.1.2 Phyto-oestrogens

Phyto-oestrogens are chemicals that occur naturally in plants (phyto-, Greek prefix indicating plant origin) and that have structural or functional similarities to steroid oestrogens. A formal definition could be 'any plant substance or metabolite that induces biological responses in vertebrates and can mimic or modulate the actions of endogenous oestrogens, usually by binding to oestrogen receptors' (COT, 2003). Phyto-oestrogens

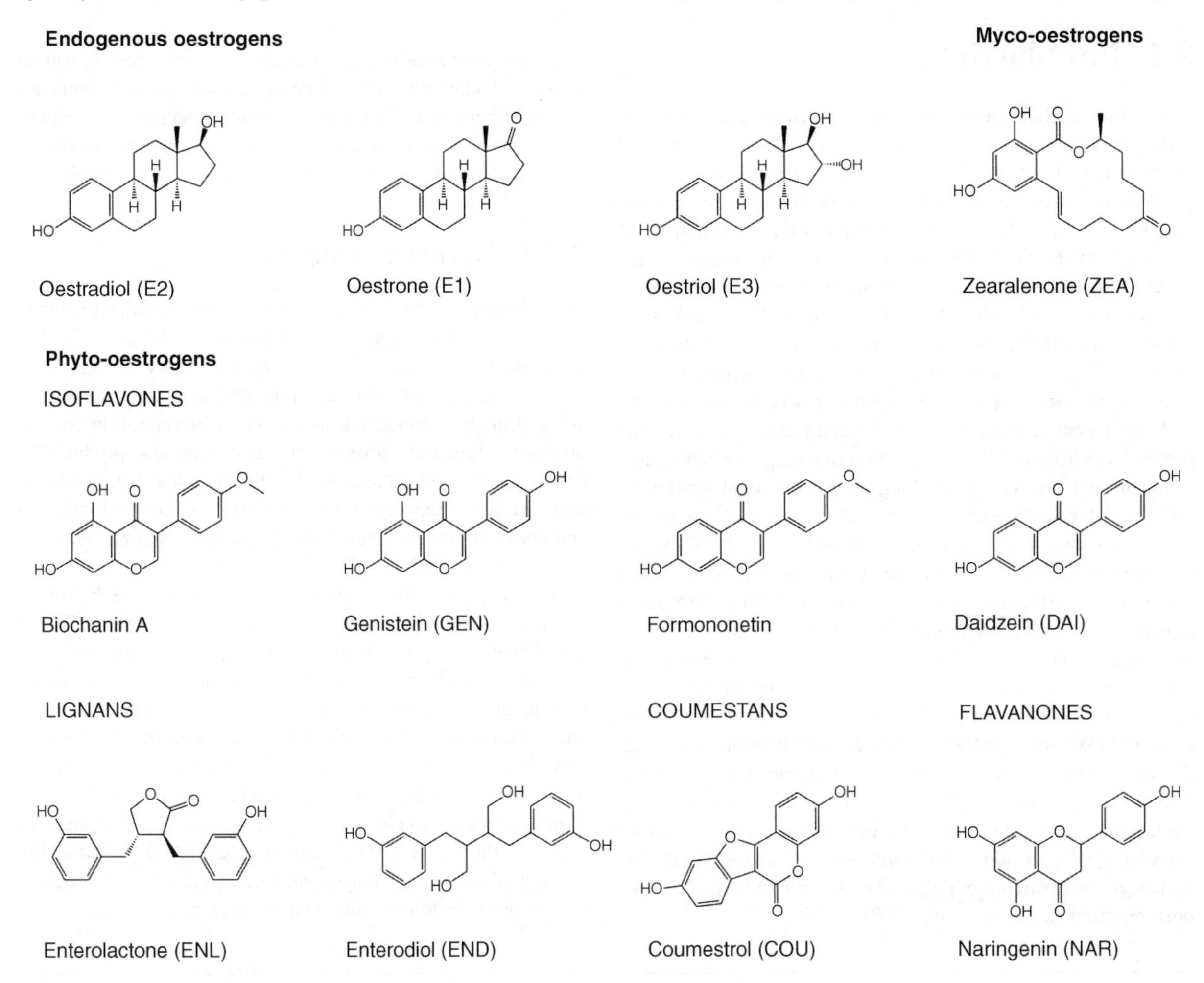

Figure 9.1 Chemical structures of steroid oestrogens, phyto-oestrogens and myco-oestrogens

are thus usually considered to act via the oestrogen receptor (ER – from the American spelling, estrogen).

The chemical structures of commonly studied phyto-oestrogens are shown in Figure 9.1.

There are three major groups of phyto-oestrogens:

- **Isoflavones**, including biochanin A, daidzein, formononetin and genistein. Isoflavones are particularly common in Eastern diets (for example in soya).
- **Lignans**, including enterolactone and enterodiol. Lignans are found in high concentrations in grains, seeds such as linseed, and other fibre-rich foods including broccoli and cauliflower; they are probably the most common phyto-oestrogens in the Western diet.
- **Coumestans**, including coumestrol, one of the more oestrogenic phyto-oestrogens, are found in various beans such as split peas, pinto beans, and lima beans, alfalfa and clover sprouts.

Other phyto-oestrogens include resveratrol in red grapes and wine (Gehm *et al.*, 1997), naringenin in hops and beer (Milligan *et al.*, 1999), liquiritigenin in liquorice (Mersereau *et al.*, 2008) and trigonelline in coffee (Allred *et al.*, 2009). Information on phyto-oestrogens and their associated foods is presented in Table 9.1. Phyto-oestrogens are found in most foods that contain plant material, and human exposure is frequent, widespread and substantial. Flavonoids, which include isoflavones, coumestans and prenylated flavonoids, can constitute up to 7 per cent of the dry weight of certain plants (COT, 2003). Phyto-oestrogens may also be added to foods or may replace other foods, for example the growing popularity of soya milk instead of cow's milk, in part due to the wish to avoid potential allergies or intolerance to the alternative animal product.

Phyto-oestrogens differ from the man-made 'xeno-oestrogens' such as the plasticiser bisphenol A and the persistent environmental pollutants polychlorinated biphenyls (PCBs), in that they occur naturally in plants, and phyto-oestrogens have generally, although perhaps speculatively, been thought overall to have beneficial effects. This may be part of the wider issue where there is public concern over perceived 'bad' synthetic chemicals instead of an appropriate recognition that any chemical, including naturally occurring compounds, may have beneficial and adverse effects. The competing claims of beneficial and adverse effects of phyto-oestrogens have been comprehensively reviewed (Patisaul and Jefferson, 2010), and will be described below in the section on human health effects.

9.1.3 Myco-oestrogens

Myco-oestrogens are naturally occurring, secondary metabolites produced by fungi, such as the *Fusarium* species, that can grow on cereals including corn, barley, wheat, hay and oats. The major group of mycoestrogens are the resorcyclic acid lactones (RALs), of which the most prominent member is zearalenone (Hartmann *et al.*, 2008). Zearalenone was identified in 1962 when it was found that its consumption resulted in lower reproductive capacity in farm animals, especially swine (Stob *et al.*, 1962). Zearalenone has several derivatives: α- and β-zearalenol, zeranol, taleranol and zearalanone. The activity of both the parent compound and its metabolites must be considered when assessing likely risks and effects (Minervini *et al.* 2001). Zearalenone is one of the most potent of the phyto/myco-oestrogens, but has oestrogenic activity of around 10 to 1000 fold less than that of the endogenous oestrogen, 17β-estradiol.

Myco-oestrogens are part of the larger group of mycotoxins, which includes ochratoxin A, a mycotoxin that has been classified as a possible human carcinogen (IARC group 2B), but these effects are outside the scope of this chapter.

9.2 Hazardous properties

Naturally occurring steroid hormones absorbed exogenously will have similar biological activity in the human body to those produced endogenously. Whether exogenous steroids have a significant impact on the biological processes in the body is thought to depend on the exogenous contribution in relation to endogenous production. To understand the hazardous properties of exogenous steroid oestrogens, it is therefore of interest to summarise first the main physiological functions and mode-of-action of endogenous oestrogens.

9.2.1 Physiological functions of endogenous oestrogens

Oestrogens have traditionally been thought of, and are indeed required, as the hormones specific to female sexual maturation and development, including that of secondary sex organs such as the breast. Together with progesterone and the gonadotropins, they also coordinate the female menstrual cycle during a woman's reproductive years. Furthermore, oestrogens and progesterone are necessary to maintain pregnancy.

A defined role for oestrogens in male reproductive function remains elusive, although oestradiol, produced by Sertoli cells in the testes, is thought to play a role in spermatogenesis (Couse and Korach, 1999). Similarly, the role of oestrogens in prostatic function remains to be fully elucidated.

More recently, it has been recognised that oestrogens are involved in a wide variety of other functions, such as bone and lipid metabolism and cardiovascular function. Osteoporosis in post-menopausal women is evidence of a role for oestrogens in maintaining bone mass. More recently, the importance of oestrogens in bone development and mineralisation has also been demonstrated in men (Bilezikian *et al.*, 1998).

Oestrogens have a protective effect on the cardiovascular system, as illustrated by the striking gender-related differences in cardiovascular risk. This is thought to be an indirect effect of oestrogens on lipoprotein metabolism (Couse and Korach, 1999).

Table 9.1 Phyto-oestrogens and myco-oestrogens: ER/AR activity, levels in food, levels in human tissues

Chemical name	Endocrine activity	Occurrence in foods	Human tissue levels
Coumestrol (COM)	Shows ER binding (Kuiper *et al.*, 1997); oestrogenic *in vitro* and *in vivo* (Whitten and Patisaul, 2001); shows ER binding and oestrogenic *in vitro* (mitogen) (Han *et al.*, 2002); shows ER binding, oestrogenic *in vitro* (gene, mitogen) (Matsumura *et al.*, 2005) Shows no AR binding (Fang *et al.*, 2003)	Alfalfa sprouts, various beans	
Genistein (GEN)	Oestrogenic *in vitro* (Ranhotra and Teng, 2005), oestrogenic *in vitro* and *in vivo* (Whitten and Patisaul, 2001); shows ER binding and oestrogenic *in vitro* (mitogen) (Han *et al.*, 2002); oestrogenic *in vitro* (gene) (Overk *et al.*, 2005); binds ER, oestrogenic *in vitro* (gene) (Overk *et al.*, 2005); oestrogen modulator *in vitro* against 17β-oestradiol and pesticides (Verma *et al.*, 1997); shows ER binding and oestrogenic *in vitro* (gene, mitogen) (Matsumura *et al.* 2005) Shows AR binding (Beck *et al.*, 2003; Fang *et al.*, 2003)	Vegetables: 1–3 mg/100g Soya products: 20–1100 mg/100 g Non-soya legumes: 0–80 mg/100 g *all reviewed in Whitten and Patisaul (2001)* GEN is the most oestrogenic component of red clover (Overk *et al.*, 2005) Miso: 3.2–394 μg/g Tofu: 1.4–9 μg/g Soya sauce: 0.1–2.6 μg/ml (Takamura-Enya *et al.*, 2003)	83.9 (9.2–303) nM, plasma (mean (range)), pregnant women, aged 20–30; Japan, 1985 (Adlercreutz *et al.*, 1999) 7.1 nM, Serum (mean); healthy women, aged 34–65; New York, 1985–91 (Zeleniuch-Jacquotte *et al.*, 1998) 2.05 (0–5.1) ng/ml, on Western diet; 12.1 (0–25.5) ng/ml, on vegetarian diet; 74.6 (0–319) ng/ml, on high soya diet; plasma (median (10–90%range)) (Safford *et al.*, 2003) 691 +/− 690 nM (EP), 806 +/− 1238 nM (NEP); plasma (mean +/− SD), after 10 wk high soya diet (Wiseman *et al.*, 2004) 0.5–276 nM mean range, plasma levels for different populations and diets summarised (Whitten and Patisaul, 2001)
Biochanin A (GEN precursor)	Oestrogenic *in vitro* and *in vivo* (Whitten and Patisaul, 2001); shows ER binding and oestrogenic *in vitro* (gene) (Overk *et al.*, 2005); shows ER binding and oestrogenic *in vitro* (mitogen) (Han *et al.*, 2002) Shows AR binding (Beck *et al.*, 2003)		

Daidzein (DAI)	Oestrogenic *in vitro* and *in vivo* (Whitten and Patisaul, 2001), shows ER binding, oestrogenic *in vitro* (gene) (Overk *et al.*, 2005); shows ER binding and oestrogenic *in vitro* (mitogen) (Han *et al.*, 2002); shows no ER binding and oestrogenic *in vitro* (gene, mitogen) (Matsumura *et al.*, 2005) Shows AR binding (Beck *et al.*, 2003)	Vegetables: 1–3 mg/100 g Soya products: 20–900 mg/100 g Non-soya legumes: 0–10 mg/100 g Fava beans: 100 mg/100 g *all reviewed in Whitten and Patisaul (2001)*	45.5 (2–243) nM, plasma (mean (range)), pregnant women, aged 20–30; Japan, 1985 (Adlercreutz *et al.*, 1999) 3.1 nM, Serum (mean), healthy women, aged 34–65; New York, 1985–91 (Zeleniuch-Jacquotte *et al.*, 1998) 8.2 (1.2–36) ng/ml, plasma (mean (range)), men; UK, 1997 (Morton *et al.*, 1997) 11.3 (u.d. – 62) ng/ml, prostatic fluid (mean (range)), men; UK, 1997 (Morton *et al.*, 1997) 3.2 (0–8.5) ng/ml, on Western diet; 12.7 (0.7–24.7) ng/ml, on vegetarian diet; 30.4 (0–143) ng/ml, on high soya diet; plasma (median (10–90% range)) (Safford *et al.*, 2003) 369 +/− 456 nM (EP), 310 +/− 244 nM (NEP), plasma (mean +/− SD), after 10 wk high soya diet (Wiseman *et al.*, 2004) 0.6–107 nM, plasma levels for different populations and diets summarised: mean range (Whitten and Patisaul, 2001)
Formononetin (DAI precursor)	Oestrogenic *in vitro* and *in vivo* (Whitten and Patisaul, 2001); binds ER, oestrogenic *in vitro* (gene) (Overk *et al.*, 2005); shows ER binding and oestrogenic *in vitro* (mitogen) (Han *et al.*, 2002) Shows AR binding (Beck *et al.*, 2003)		

(*continued*)

Table 9.1 (*Continued*)

Chemical name	Endocrine activity	Occurrence in foods	Human tissue levels
Equol (DAI metabolite)	Oestrogenic *in vitro* and *in vivo* (Whitten and Patisaul, 2001); shows ER binding and oestrogenic *in vitro* (mitogen) (Han *et al*., 2002); shows ER binding, oestrogenic *in vitro* (gene, mitogen) (Matsumura *et al*., 2005) Anti-androgenic *in vivo* (but by sequestering DHT and not by AR binding) (Lund *et al*., 2004); shows AR binding (Fang *et al*., 2003)		71.1 (0.6–404) nM, Plasma (mean (range)), pregnant women, aged 20–30; Japan, 1985 (Adlercreutz *et al*., 1999) 0.53 nM, serum (mean), healthy women, aged 34–65; New York, 1985–91 (Zeleniuch-Jacquotte *et al*., 1998) 0.57 (0.05–8.5) ng/ml, plasma (mean (range)), men; UK, 1997 (Morton *et al*., 1997) 0.5 (u.d. – 5.1) ng/ml, prostatic fluid (mean (range)), men; UK, 1997 (Morton *et al*., 1997) 0.4 (0–2.1) ng/ml, on Western diet; 0.4 (0.1–0.6) ng/ml, on vegetarian diet; 3.3 (0–3.3) ng/ml, on high soya diet; plasma (median (10–90% range)) (Safford *et al*., 2003) 364 +/− 396 nM (EP), 2 +/− 7 nM (NEP), plasma (mean +/− SD), after 10 wk high soya diet (Wiseman *et al*., 2004) Not all humans produce equol from daidzein, only around 30% do so (Cassidy *et al*., 2006) 0.1–5.5 nM, plasma levels for different populations and diets summarised: mean range (Whitten and Patisaul, 2001)
o-desmethylangolensin (DMA, a DAI metabolite)	Oestrogenic *in vitro* and *in vivo* (Whitten and Patisaul, 2001)		31.2 (1.3–194) nM, plasma (mean (range)); pregnant women, aged 20–30; Japan, 1985 (Adlercreutz *et al*., 1999) 0.47 nM, serum (mean), healthy women, aged 34–65; New York, 1985–91 (Zeleniuch-Jacquotte *et al*., 1998)

			82 +/− 92 nM (EP), 100 +/− 99 nM (NEP); plasma (mean +/− SD), after 10 wk high soya diet (Wiseman *et al.*, 2004) <0.1–3.3 nM, plasma levels for different populations and diets have been summarised: mean range (Whitten and Patisaul. 2001)
Enterolactone (ENL)	Oestrogenic *in vitro* (Whitten and Patisaul, 2001); oestrogen modulator *in vivo* (mice with MCF7 tumour) and *in vitro* (MCF7 production of VEGF) (Bergman Jungestrom *et al.*, 2007)	Data for lignans as a group: fruit: 60–200 mg/100 g vegetables: 100–400 mg/100 g cereals: 100–700 mg/100 g flaxseed: 68,000 mg/100 g soy products: 900 mg/100 g (Whitten and Patisaul, 2001) The plant lignans, matairesinol and secoisolariciresinol, are present in foods but are modified by gut microflora (to ENL and END)	12.9 (0.5–58) nM, plasma (mean (range)), pregnant women, aged 20–30; Japan, 1985 (Adlercreutz *et al.*, 1999) 3.9 (0.05–12.3) ng/ml, plasma (mean (range)), men; UK, 1997 (Morton *et al.*, 1997) 20.3 (u.d. – 156) ng/ml, prostatic fluid (mean (range)), men; UK, 1997 (Morton *et al.*, 1997) 21.23 nM, serum (mean), healthy women, aged 34–65; New York, 1985–91 (Zeleniuch-Jacquotte *et al.*, 1998) 4.6 (0.9–8.4) ng/ml, on Western diet; 75.4 (16.7–134) ng/ml, on vegetarian diet; 6.2 (0–36.6) ng/ml, on high soya diet: plasma (median (10–90% range)) (Safford *et al.*, 2003) 3.9–752 nM, plasma levels for different populations and diets summarised: mean range (Whitten and Patisaul, 2001)
Enterodiol (END)	Oestrogen modulator *in vivo* (mice with MCF7 tumour) and *in vitro* (MCF7 production of VEGF) (Bergman Jungestrom *et al.*, 2007)	See ENL	1.64 (0–9.6) nM, plasma (mean (range)), pregnant women, aged 20–30; Japan, 1985(Adlercreutz *et al.*, 1999) 2.6 (u.d. – 10.4) ng/ml, prostatic fluid (mean (range)), men; UK, 1997 (Morton *et al.*, 1997)

(*continued*)

Table 9.1 (*Continued*)

Chemical name	Endocrine activity	Occurrence in foods	Human tissue levels
			1.5 nM, serum (mean), healthy women, aged 34–65; New York, 1985–91 (Zeleniuch-Jacquotte *et al.*, 1998) 0.4 (0–0.9) ng/ml, on Western diet; 5.1 (0–16) ng/ml, on vegetarian diet; 1.7 (0–9.2) ng/ml, on high soya diet; plasma (median (10–90% range)) (Safford *et al.*, 2003) 0.4–65.6 nM, plasma levels for different populations and diets have been summarised: mean range (Whitten and Patisaul, 2001)
Zearalenone (ZEA)	Oestrogenic *in vitro* (Olsen *et al.*, 2005; Whitten and Patisaul, 2001; Le Guevel and Pakdel, 2001) Four metabolites show AR binding (Fang *et al.*, 2003)	ZEA and metabolites thereof are often found in cereal grains and derived foods. ZEA was detected in almost every one of 140 raw maize samples, and was >100 μg/kg in 42% of samples (Scudamore and Patel, 2000)	
Naringenin	Oestrogenic *in vitro* (Whitten and Patisaul, 2001); shows ER binding and oestrogenic *in vitro* (mitogen) (Han *et al.*, 2002); weakly oestrogenic *in vitro* (gene), but not oestrogenic *in vitro* (mitogen) (Ruh *et al.*, 1995); oestrogen modulator *in vivo* and *in vitro* (gene and mitogen) (Ruh *et al.*, 1995); shows slight AR binding (Fang *et al.*, 2003); not anti-androgenic *in vitro* (yeast or PC3(AR)2), abstract only (Zierau *et al.*, 2003)	Occurs particularly in hops (COT working group on phyto-oestrogens, 2000)	0.6 +/− 0.4 μM (mean +/− SD, orange juice) and 6 +/− 5.4 μM (grapefruit juice); plasma levels following 8 ml/kg juice (subjects weighed 73 +/− 15 kg); levels at t0 were essentially zero (Erlund *et al.*, 2001)

8-prenylnaringenin (8-PN)	Shows ER binding and oestrogenic *in vitro* (gene) (Milligan *et al.*, 1999; Overk *et al.*, 2005; Milligan *et al.*, 2002; Milligan *et al.*, 2000); oestrogenic *in vivo* (Schaefer *et al.*, 2003); shows ER binding, oestrogenic *in vitro* (gene, mitogen) (Matsumura *et al.*, 2005) Anti-androgenic in one *in vitro* assay, not in another, abstract only (Zierau *et al.*, 2003); not androgenic (Milligan *et al.*, 2000)	Most oestrogenic component in hops (Overk *et al.*, 2005) Human exposure may be from the metabolism of other phytochemicals into 8-PN rather than direct exposure to 8-PN in, e.g. beer (Possemiers *et al.*, 2005) Beer: 3 brands = 0.22, 0.52 and 4 ng/ml (Takamura-Enya *et al.*, 2003)	2.5 ng/ml (50 mg dose), 10 ng/ml (250 mg) and 34 ng/ml (750 mg dose); peak serum levels following single oral doses (Rad *et al.*, 2006) 8-PN itself has oestrogenic metabolites (2 out of 12 metabolites, human liver microsomes) (Zierau *et al.*, 2004)
Isoxanthohumol	Binds ER, oestrogenic *in vitro* (gene) (Overk *et al.*, 2005); oestrogenic *in vitro* (gene-ishikawa) but not in YES (Milligan *et al.*, 1999; Milligan *et al.*, 2000) Not androgenic (Milligan *et al.*, 2000)	Present in strong ales at up to 4 mg/L	May be metabolised to 8-prenylnaringen (Possemiers *et al.*, 2005)
Resveratrol	Shows no ER binding, oestrogenic *in vitro* (gene), not oestrogenic *in vitro* (mitogen) (Matsumura *et al.*, 2005)	Occurs in grapes and wine	
Ferutinine	Oestrogenic *in vitro* (gene) (Ikeda *et al.*, 2002)	Found in plants from the Umbelliferae family, parsley and carrot	
Tschimgine	Oestrogenic *in vitro* (gene) (Ikeda *et al.*, 2002)		
Tschimganidine	Oestrogenic *in vitro* (gene) (Ikeda *et al.*, 2002)		

Oestrogenic *in vitro*: oestrogen agonist, e.g. in ER-CALUX, MCF7 or similar assay; this is more than simply binding the receptor, and means there was activation/agonism as well. (gene) indicates oestrogenicity measured as a change in gene expression, e.g ER-CALUX; (mitogen), indicates oestrogenicity measured as increased proliferation of oestrogen-sensitive cell line, usually MCF7. **Oestrogen modulator:** describes reduction in oestrogen effect on co-application; includes ER antagonists if the mode of action was not **proven** to be ER antagonism. **ER antagonist:** reduces effect of oestrogen, has no agonist activity alone, antagonism of ER was demonstrated (for example binding of ER with no agonism). **u.d.:** undetectable in assay. **EP:** subject was an 'equol producer', based on urinary production >1 umol/24 hr after 10 wk high soya diet. **NEP:** subject was a 'non-equol producer', based on urinary production <1 μmol/24 h after 10 wk high soya diet

Finally, increasing attention has focused on the influence of oestrogens on brain and cognitive function, including learning and memory, as well as the pathology of neurodegenerative diseases such as Alzheimer's (Couse and Korach, 1999).

It is therefore clear that steroid oestrogens are implicated in many vital physiological functions. It is also important to stress the profound implications of the developmental stage at which exposure to exogenously elevated concentrations of steroid oestrogens occurs. Whilst exposure during adulthood may result in transient disruption of homeostatic processes, exposure during critical windows of development such as fetal life, neonatally, in childhood or at puberty could result in permanent organisational changes. This itself is consistent with an emergent medical hypothesis relating to the developmental origins of adult disease, also commonly referred to as the 'Barker hypothesis' (Barker, 2004).

9.2.2 Oestrogen signalling

Oestrogen signalling is part of the endocrine system, and is important for human health as well as being a target for adverse effects when humans are exposed to substances that mimic or interfere with the system, the so-termed 'endocrine disrupters'.

The ER is a hormone receptor whose physiological ligand is 17β-oestradiol. The ER acts mainly upon the genome as an intracellular transcription factor, and two receptor subtypes exist, ERα and ERβ, encoded by the genes ESR1 and ESR2, respectively (Dahlman-Wright *et al.*, 2006). Some ERs have also been found on cell membranes where they act via non-transcriptional pathways (Levin, 2002; Kelly and Levin, 2001).

ERα was cloned in 1986 and was considered to be the only ER until ERβ was discovered in 1996 (Kuiper *et al.*, 1996). Subsequently there has been substantial interest in the role of the two subtypes in health and disease, particularly in cancer (Pearce and Jordan, 2004) and specifically in breast cancer (Fox *et al.*, 2008; Rice and Whitehead, 2006; Shupnik, 2007). The cellular and tissue expression of ERα and ERβ are different (Kuiper *et al.*, 1997); *in-vitro* cell models suggest that the two subtypes counteract each other when co-expressed. For example, cells that proliferate if ERα is present alone remain quiescent when both subtypes are co-expressed (Koehler *et al.*, 2005). It is not clear what overall tissue response should be expected when the two subtypes are expressed in different cells within a given tissue.

The existence of two receptor subtypes, with potentially different effects, has complicated the definition of oestrogenicity. Prior to interest in ERβ, 'oestrogenic' was defined as 'stimulation of uterine growth and induction of the progesterone receptor in the uterus' (Dahlman-Wright *et al.*, 2006). However, these effects are a result of ERα activation and not ERβ. Use of this definition would mean that ERβ ligands would not qualify as oestrogens. Induction of a proliferative response may be the hallmark of ERα activation; however, such a clear apical response is harder to select for ERβ, and it may be that ERβ ligands either evoke a range of different effects through ERβ and/or that ERβ itself has a wider repertoire of effects (Koehler *et al.*, 2005).

In addition to the genomic effects of oestradiol through ERα and ERβ, oestradiol also has rapid, non-genomic effects on cells (Wendler *et al.*, 2010). These effects were identified when it was observed: (1) that application of oestradiol could have effects that occurred too rapidly to be explained by the classical, genomic mechanism, which is relatively slow because of the requirement for gene transcription and protein translation (Levin, 2008); (2) that oestradiol could evoke cellular events even when it was attached to large bulky molecules that could not traverse the cell membrane and activate cytoplasmic receptors (Zivadinovic *et al.*, 2005); and (3) that oestradiol could trigger events in cells without classical receptors or that could not be blocked by inhibitors of the classical receptors. Rapid effects include changes in cellular cAMP (cyclic adenosine monophosphate) and intracellular calcium and modulation of ion channels, kinases, phosphatases and other enzymes, for example the rapid effects of oestradiol on voltage-gated sodium channel may play a role in cell adhesion (Fraser *et al.*, 2010). Rapid effects may be mediated by the classical receptors relocated to the membrane (Levin, 2009) or through novel receptors such as GPR30, a G-protein coupled receptor now identified as a membrane receptor for oestradiol (Mizukami, 2010).

The hazard assessment of exogenous steroid oestrogens is therefore complicated by the existence of the endogenous ligand and a multiplicity of receptor and signalling mechanisms in the intact organism. This is related to the controversy over the existence of a threshold for adverse responses to low doses of endocrine disrupters. Hence, while some argue that the endocrine system is dynamic and able to correct minor disturbances through homeostatic control (at least in adults), others contend that the presence of hormones at physiologically active concentrations signify that the threshold has already been achieved and that endocrine control via feedback loops is likely to give rise to non-monotonic dose-response curves. With regard to the debate surrounding the effects of phyto-oestrogens, a greater understanding of the consequence of subtype selectivity could substantially alter our hazard assessment of other xeno-oestrogens, some of which may also show particular ER selectivity, similarly to the phyto-oestrogens.

9.2.3 Carcinogenicity of steroid oestrogens

Steroid oestrogens and some of their conjugates were evaluated for carcinogenicity by the International Agency for Research on Cancer (IARC, 1979). Relevant results and conclusions are summarised below.

In experimental animals, administration of oestradiol by subcutaneous injection or implantation has been found to result in an increased incidence of mammary, pituitary, uterine, cervical, vaginal and lymphoid tumours and interstitial-cell tumours of the testis in mice and an increased incidence of mammary and pituitary tumours in rats. Injections in neonatal mice resulted in precancerous and cancerous vaginal lesions in later life.

Similarly, administration of oestrone resulted in an increased incidence of mammary tumours in mice and in pituitary, adrenal and mammary tumours and bladder tumours in association with stones in rats. Oestriol increased the incidence and accelerated the appearance of mammary tumours in both male and female mice and produced kidney tumours in hamsters.

In humans, there is evidence that administration of conjugated oestrogens is causally related to an increased risk of developing endometrial carcinoma. IARC also found a possible increased risk of developing breast cancer following oestrogen therapy. As a result, steroid oestrogens are classified as Group 1 carcinogens (sufficient evidence). Additionally, steroid oestrogens administered prenatally or neonatally have been found to impair fertility and have teratogenic effects.

The mechanism by which oestrogens exert their carcinogenic effects remains unclear, as they lack mutagenic activity in bacterial and mammalian test systems. Several epigenetic mechanisms of tumour induction by oestradiol have been proposed. There has been some interest in the metabolic conversion of oestradiol to 4-hydroxyoestradiol by the CYP1B1 4-hydroxylase followed by further activation of this catechol to reactive semiquinone or quinine. Oestradiol also induces various chromosomal and genetic lesions in cell cultures. These data support a dual role of oestrogens in carcinogenesis as a weak carcinogen and weak mutagen and as a hormone-stimulating receptor-mediated cell proliferation of damaged cells (Liehr, 2000).

9.2.4 Hazards of phyto-oestrogens and myco-oestrogens

Unlike other pollutants, phyto-oestrogens have been studied as much for their beneficial properties as for their hazards. This creates an interesting situation, since the assumptions are often directly opposite in these two research areas. The pros and cons of phyto-oestrogens have been reviewed by Patisaul and Jefferson (2010). Phyto-oestrogens fulfil many of the criteria for classification as endocrine disrupters but are generally perceived to be beneficial to humans, and strong evidence of adverse effects appears to be lacking. Responses to this currently unresolved duality include referring to phyto-oestrogens as 'endocrine active' rather than 'endocrine disrupting', since the latter would indicate an adverse health effect.

9.2.4.1 Oestrogenicity

Hazard identification for phyto-oestrogens and myco-oestrogens has focused on their potential for adverse effect on oestrogen signalling in the endocrine system, principally by acting in an oestrogenic fashion and mimicking the effects of endogenous oestradiol, but also by acting as anti-oestrogens, for example in opposing the effects of oestradiol. In this context, phyto-oestrogens may be deemed 'endocrine disrupters'. Endocrine disrupters may act directly on a target receptor, for example as agonists or antagonist of the oestrogen, androgen or thyroid hormone receptors, or they may have less direct mechanisms, for example in altering the production or metabolism of an endogenous steroid.

Phyto-oestrogens (such as isoflavones and coumestans) and myco-oestrogens show oestrogenicity in a variety of in-vitro assays, including cell-proliferation assays, such as the ESCREEN which uses a cell line that shows a strong proliferative response to oestrogens, and reporter-gene systems (Shaw and McCully, 2002; Whitten and Patisaul, 2001). One of the most studied phyto-oestrogens is genistein, which has activity *in vitro* at concentrations in the nanomolar to low micromolar range (Whitten and Patisaul, 2001). For comparison, the endogenous oestrogen, 17β-oestradiol, typically has activity at concentrations below one nanomolar in similar assays.

At high doses, i.e. above those at which an oestrogenic response is seen, phyto-oestrogens may also shown anti-oestrogenic effects. For example, genistein evokes a maximal effect at around 1 μM in cell proliferation assays, but evokes sub-maximal effects at 10 μM (Matsumura *et al.*, 2005). However it has not been demonstrated whether this effect is due to the phyto-oestrogen acting as a receptor antagonist or by activating an additional pathway, or even if the decline in effect is secondary to toxicity. The finding that phyto-oestrogens may have both oestrogenic and anti-oestrogenic activities, has led to the proposal that they should be considered as selective oestrogen receptor modulators, SERMs (Setchell, 2001), although other authors consider that this conclusion is not yet supported by the experimental evidence (Matsumura *et al.*, 2005). SERMs are chemicals that can act to either mimic or antagonise the activities of oestradiol, depending on the tissue type, tissue status, receptor expression and presence of oestradiol, and receive substantial clinical interest, for example raloxifene (Howell, 2008).

The myco-oestrogen zearalenone was shown to act via the oestrogen receptor in the 1970s (Kiang *et al.*, 1978) and has since been shown to be active in many *in-vitro* assays for oestrogenicity, with endpoints ranging from binding to the isolated receptors, activation of a reporter gene under the control of the oestrogen receptor, and production of cell proliferation in cell lines that show oestrogen-sensitive proliferation (Le Guevel and Pakdel, 2001; Olsen *et al.*, 2005).

It has been proposed that the biological effects of phyto-oestrogens are mainly due to their preferential activation of the ERβ receptor subtype, rather than ERα (McCarty, 2006). Indeed, whilst the endogenous ligand, 17β-oestradiol, activates both subtypes equally, genistein shows a 30-fold selectivity for ERβ over ERα (Dahlman-Wright *et al.*, 2006) and a thorough comparison of the subtype selectivity of nine phyto-oestrogens found up to 100-fold selectivity for ERβ (Harris *et al.*, 2005). It is important to note, however, that the effects of an ER ligand will depend not only on subtype selectivity but also on the tissue expression of receptor subtypes, and the signalling pathways that are activated by each subtype in the tissue or cell of interest. It should also be noted that if some phyto-oestrogens such as coumestrol show similar binding affinity to the ERs as the endogenous ligand oestradiol, the relative potency of phyto-oestrogens derived from oestrogenicity assays *in vitro* is several orders of magnitude lower than oestradiol (Table 9.2).

Table 9.2 Relative binding affinities (RBA) for the ERs and relative *in vitro* oestrogenic potencies for some phyto-oestrogens (Adapted from COT, 2003)

Phyto-oestrogen	Range of relative binding affinity	Range of potency
Coumestrol	0.002–1.0	0.0003–0.09
Equol	0.001–0.003	0.00023–0.001
Genistein	0.0001–0.0088	0.000001–0.002
Biochanin A	0.0015–0.00005	0.0000001–0.0045
Daidzein	0.0001–0.0008	0.0000024–0.00014

9.2.4.2 Androgenicity

Although the term 'phytoandrogen' has been coined (Chen and Chang, 2007), it is not as frequently used as is 'phyto-oestrogen'; for example, phytoandrogen returns only one entry in PubMed (October 2010) whilst phyto-oestrogen returns almost eight thousand entries. Nonetheless, Fang *et al.* (2003) found that 14 out of the 17 phyto-oestrogens that they tested bound to the androgen receptor (AR). Phyto-oestrogens that showed AR binding included genistein, equol and zearalenol; non-binders included coumestrol. Phyto-oestrogens may have effects on the androgen system that are more complicated than simple receptor interactions. For example, equol has been reported to act as a novel anti-androgen by increasing sequestration of dihydrotestosterone, the physiological ligand for AR, rather than directly affecting receptor binding (Lund *et al.*, 2004). Genistein has been shown to decrease the level of AR protein by increasing its ubiquitination and thus increasing proteasomal degradation (Basak *et al.*, 2008). This latter observation was made on an androgen-sensitive prostate-cancer cell line, leading the authors to propose that there may be a therapeutic application for genistein in prostate cancer.

Table 9.1 includes reports of androgenicity or anti-androgenicity for phyto-oestrogens.

9.2.4.3 Other effects

In addition to their effects on the oestrogen or androgen receptors, phyto-oestrogens have also been reported to affect a wide spectrum of other receptor and enzyme targets.

Phyto-oestrogens may have indirect effects on oestrogen signalling though cross-talk between the ER and other transcription factors or by other mechanisms. For example, flavonoids, including genistein are positive in a reporter gene assay for the AP1 transcription factor (Gopalakrishnan *et al.*, 2006), and genistein has been shown to activate the mitogen-activated protein kinases (MAPK) signalling pathway (Liao *et al.*, 2007). Genistein is also an inhibitor of tyrosine kinases, intimately associated with growth factor signalling (Yan *et al.*, 2010). A further potential complexity is that steroid hormones and growth factors (e.g. epidermal growth factor) interact in their receptor-driven intracellular signalling cascades. In addition, phyto-oestrogens can inhibit DNA topoisomerase I and II and act as antioxidants. Phyto-oestrogens may also affect oestradiol metabolism, for example by inhibiting oestrogen sulphation (Harris *et al.*, 2004; Kirk *et al.*, 2001). Other effects on cytochrome p450 enzymes include competitive inhibition of aromatase at micromolar concentrations (Karkola and Wahala, 2009).

The aryl hydrocarbon receptor (AhR), which is activated by a wide range of xenobiotics, is also activated by isoflavone and flavanone phyto-oestrogens, including genistein, daidzein and naringenin, but not by other types, such as flavones, chalcones and coumestans (Amakura *et al.*, 2003). Potent AhR agonists are known to be endocrine disrupters, for example dioxin (TCDD), although most phyto-oestrogens are not as potent on the AhR as these endocrine-disrupting chemicals. However in *in vitro* gene expression assays, naringenin has been found to have no effect via the AhR when tested alone and to oppose the effect of TCDD when tested in combination (Kim *et al.*, 2004).

9.3 Sources

9.3.1 Natural sources

Steroid oestrogens are produced by all vertebrates and, as such, they can be detected in produce of animal origin such as meat, eggs and dairy, and their excretion is a major pathway of these substances into the environment.

Phyto-oestrogens and myco-oestrogens occur naturally in plants and moulds. Pasture plants that are reported to contain phyto-oestrogens include most of the common legumes, such as clover and soya, beans, peas and lentils. Some plants, such as alfalfa, usually contain low amounts of phyto-oestrogens but increase their content in response to aphids or fungal pathogens (Adams, 1995).

9.3.2 Anthropogenic sources

9.3.2.1 Steroid oestrogens

In addition to natural sources, steroid oestrogens have been synthesised principally as pharmaceutical active compounds used for human contraception, hormone replacement therapy (to alleviate the symptoms of menopausal women) and meat and fish production. In commercial fisheries, steroid oestrogens have been exploited to induce complete sex reversal when they offer economic advantages in terms of growth promotion (Pandian and Sheela, 1995). In the USA and many other countries, they are also used in farm animals with low endogenous oestrogen production (calves, lambs, heifers, steers) to enhance growth performance by 5–15 per cent. However, registration of oestrogens and other anabolic hormones is not allowed in the European Union (Meyer, 2001). Oestrogen is also used to induce ovulation in timed artificial insemination of pre-synchronised dairy cattle, to increase the success rates of such programs (Bartolome *et al.*, 2005).

9.3.2.2 Phyto-oestrogens and myco-oestrogens

Phyto-oestrogens occur naturally in plants, and people may select a diet high in these chemicals by selecting particular foodstuffs, enriching foods with supplements containing phyto-oestrogens or using concentrated phyto-oestrogen extracts for their perceived health effects. Human activities may also result in the concentration of the phyto-oestrogens naturally present in the environment; for example, the output of paper mills contains genistein at concentrations of 10–13 μg/L (Kiparissis *et al.*, 2001). Waste-water treatment can mitigate some of these effects, but concentrations in water are liable to be higher in regions where the lack of infrastructure means that waste water receives no, or incomplete, treatment before release into the environment. For example, river water in Brazil was found to contain daidzein, coumestrol and genistein at levels up to 366 ng/L (Kuster *et al.*, 2009).

Livestock may be deliberately exposed to myco-oestrogens. For example, α-zearalanol is used as a growth promoter in the US and Canada, although it was banned in the EU in 1985 (Hartmann *et al.*, 2008). Humans may be exposed to myco-oestrogens through the consumption of food contaminated with moulds, for example grain that has been poorly stored.

9.4 Environmental pathways

9.4.1 Steroid oestrogens

The range of concentrations of oestrone, oestradiol and oestriol in various environmental compartments reported in the scientific literature are summarised in Table 9.3. In many matrices, concentrations of oestrone exceed those of oestradiol, while, when reported, levels of oestriol tend to be much lower.

Natural steroid oestrogens occur in all vertebrates and this is illustrated by the levels detected in meat, eggs and dairy produce (Table 9.3). Although much concern has been expressed in the general media regarding the potential consequences of the presence of female hormones in tap water on male urogenital health, it should be clear that the levels detected in water are several orders of magnitude lower than those found in foods of animal origin (Table 9.3), consumption of which is likely to be the main route of exposure for humans. Genetically improved dairy cows, such as the Holstein, lactate almost throughout pregnancy, when oestrogen levels are highest. Correspondingly, concentration of oestrone sulphate was found to increase from 30 pg/ml in non-pregnant cows to a maximum level of 1000 pg/ml at 220 days of gestation (Maruyama *et al.*, 2010).

Steroid oestrogens are generally recognised as the most significant oestrogenically active substances in domestic sewage effluent. These compounds are mostly excreted in urine as their hydrophilic sulphate and glucuronide conjugates. Some deconjugation takes place in the sewers and then during sewage treatment. *Escheria coli* is excreted in large quantities in the faeces and is able to synthesise the β-glucuronidase enzyme, whilst it has only weak arylsulphatase activity (D'Ascenzo *et al.*, 2003).

Equally, the wastes generated by husbandry practices of the same animals contain high levels of female hormones and should not be overlooked as a route of such compounds into the aquatic environment. This is well illustrated by the concentrations detected in headwater streams, impacted by agriculture but not by sewage effluent (Table 9.3). Some very high levels of steroid oestrogens have been reported for surface water and river sediments (Table 9.3). Another interesting and often neglected source of steroid oestrogens in the aquatic environment is fisheries. Levels of oestrone reported downstream of hatcheries were found to be comparable to those in a river with spawning salmon (Table 9.3) and illustrate the need to better understand the occurrence of hormones and other oestrogenic compounds of natural origin in aquatic environments before the impact of anthropogenic sources can be fully assessed.

Although the aquatic environment has traditionally been thought of as the main environmental compartment, recent reports indicate that synthetic and naturally occurring steroid oestrogens have been detected in the dust generated by large industrialised cattle feedlots in the United States (Smith *et al.*, 2010).

9.4.2 Phyto-oestrogens and myco-oestrogens

Both phyto-oestrogens and myco-oestrogens are naturally present in the environment, due to their natural occurrence in plants and fungi. Environmental levels are also affected by human activities such as farming. The main environmental pathways of myco-oestrogens were investigated experimentally by Hartmann *et al.* (2008). In order to study the distribution of zearalenone in the environment, wheat and corn fields were artificially infected with *Fusarium graminearum* and samples from plant, soil, drainage water, manure, sewage sludge and surface waters were examined. Zearalenone was found to have three main input pathways: (1) washoff from infected plants, (2) plant debris left on soil after harvest and (3) manure. The resulting levels of zearalenone in topsoil reached several grams

Table 9.3 Concentrations of steroid oestrogens measured in various environmental matrices (Adapted from Martin and Voulvoulis, 2009)

	E1	E2	E3	References
Waste waters (μg/l)				
Industry (unspecified)	nd–0.120	nd–0.054		Vethaak *et al.*, 2005
Septic tank	0.049–0.019	0.016–0.074		Swartz *et al.*, 2006
Land sources (μg/kg)				
Soil	0.012–0.025	0.002–0.055	nd–0.001	Beck *et al.*, 2008
Sewage sludge	5.6–25.2	1.7–5.1		Andersen *et al.*, 2003
Swine manure	217–4,728	nd–1,215		Hanselman *et al.*, 2003
Dairy wastes	203–543	113–236		Hanselman *et al.*, 2003
Poultry manure	nd	14–904		Hanselman *et al.*, 2003
Aquatic environment				
Headwater stream (μg/l)	0.0001–0.009	nd–0.0009	nd	Matthiessen *et al.*, 2006
Surface water (μg/l)	nd–0.112	nd–0.200	nd–0.051	Belfroid *et al.*, 1999; Cargouet *et al.*, 2004; Hollert *et al.*, 2005; Huang and Sedlak, 2001; Kolpin *et al.*, 2002; Mouatassim-Souali *et al.*, 2003; Petrovic *et al.*, 2002; Wen *et al.*, 2006; Wicks *et al.*, 2004; Williams *et al.*, 2003; Xiao *et al.*, 2001; Zuehlke *et al.*, 2005
Fish hatchery (μg/l)	nd–0.0007			Kolodziej *et al.*, 2004
River with spawning salmon	0.0004			Kolodziej *et al.*, 2004
Groundwater (μg/l)	nd–0.120	nd–0.045		Swartz *et al.*, 2006; Kolodziej *et al.*, 2004
Tile drainage (μg/l)	up to 0.07	up to 0.02		Kolodziej *et al.*, 2004; Kjaer *et al.*, 2007
Sea (μg/l)	nd–0.002			Atkinson *et al.*, 2003
Rain (μg/l)	nd	nd		Vethaak *et al.*, 2005
Riverine sediments (μg/kg)	nd–11.8	nd		Vethaak *et al.*, 2005; Petrovic *et al.*, 2002; Williams *et al.*, 2003
Marine sediment (μg/kg)	0.05–3.60	nd–0.59	nd	Isobe *et al.*, 2006
Food/biota (μg/kg)				
Earthworms (sewage works)		3 (ww)		Markman *et al.*, 2007
Earthworms (garden)		nd		Markman *et al.*, 2007
Fish muscle	<0.02	<0.03		Fritsche and Steinhart, 1999
Beef	nd–3.96	nd–1.03		Fritsche and Steinhart, 1999
Milk	0.01–0.26	0.01–0.09		Fritsche and Steinhart, 1999
Butter	1.47	nd		Fritsche and Steinhart, 1999
Yoghurt	0.16	nd		Fritsche and Steinhart, 1999
Cheese	0.17	nd–0.03		Fritsche and Steinhart, 1999
Pork	<0.02–26.06	nd–10.56		Fritsche and Steinhart, 1999
Poultry	<0.02–0.51	nd–0.73		Fritsche and Steinhart, 1999
Eggs	0.18–0.89	nd–0.22		Fritsche and Steinhart, 1999
Vegetables	<0.01–0.02	nd		Fritsche and Steinhart, 1999
Drinking water (μg/l)	0.0002–0.0006	nd–0.002	nd	Wen *et al.*, 2006; Kuch and Ballschmiter, 2001; Webb *et al.*, 2003

per hectare. The same study found that surface water levels of other substances, such as deoxynivalenol, a mycotoxin, and formononetin, a phyto-oestrogen, were considerably higher than zearalenone, perhaps suggesting that myco-oestrogen contamination is not a priority for concern. However, the higher potency of zearalenone makes this simple comparison less meaningful, because the likelihood of effect is a function of both potency and exposure.

9.4.2.1 Human health exposure pathways

Food is almost certainly the major route of human exposure to phyto-oestrogens. Dietary exposure results in sustained and substantial exposure to phyto-oestrogens. The amount of phyto-oestrogens in the diet depends on regional or geographical diets, selection of vegetarian or vegan diets and on the type of vegetables typically consumed. Humans are also exposed to foods specifically supplemented with phyto-oestrogens or through the direct use of phyto-oestrogens, perhaps for their perceived health benefits. Human exposure to isoflavones is rising as they are increasingly added to beverages, nutrient bars, yogurt, baked goods, meal replacements and confections (Shu *et al.*, 2009).

Human exposure to myco-oestrogens most probably occurs from the consumption of food prepared from poorly stored grain, which has consequently become fungally infected.

Food levels of phyto-oestrogens and myco-oestrogens are documented in Table 9.1. Genistein and daidzein are considered to be the most prevalent phyto-oestrogens in foodstuffs (COT, 2003). For example, they are the main isoflavonoids in soya, which is an import component of many Asian diets (Adlercreutz *et al.*, 1999). Lignan precursors occur in legumes, whole grains and green and yellow vegetables, which feature prominently in vegetarian diets (Whitten *et al.*, 2001). Phyto-oestrogen levels in plants, and the foods derived from them, can vary significantly, depending on a number of factors, including the species of plant, strain, crop year and geographical location (COT, 2003).

9.5 Effects on humans

9.5.1 Steroid oestrogens

9.5.1.1 Endogenous production of oestradiol

The contribution of consumption of animal produce to circulating levels of oestrogens has to be considered in relation to endogenous production. This question has been addressed by various authorities in relation to the safety of steroid hormones used in beef cattle. The US Food and Drug Administration (FDA) issued guidelines for compounds used in food-producing animals, and their conclusion regarding the use of natural sex steroids was 'that no physiological effect will occur in individuals chronically ingesting animal tissues that contain an increase of endogenous steroid equal to 1 per cent or less of the amount in micrograms produced by daily synthesis in the segment of the population with the lowest daily production' (FDA, 1999). Pre-pubertal boys have the lowest daily production rate for oestradiol, estimated in the guidelines as 6.5 μg/day. The maximum acceptable daily intake (ADI) was therefore estimated to be 65 ng/day. Identical oestradiol production rates were also given in the report of the joint FAO/WHO expert committee on food additives (JECFA) on veterinary drug residues in food, considering the use of hormone implants in cattle (JECFA, 1988). However, this value has been contested on the basis that it would have been calculated using metabolic clearance rates obtained in adults, because of the ethical issues related to the injection of radioactive hormones in healthy children necessary to derive such values experimentally (Andersson and Skakkebaek, 1999). It has been argued that calculating oestradiol production rates in children using adult metabolic clearance rates would overestimate such rates because of the difference in body size. Other factors known to influence metabolic clearance are plasma levels of SHBG and the enzymatic capacity of the liver, which are influenced by hormonal status. Furthermore, there is a high level of uncertainty regarding the plasma levels of sex hormones in pre-pubertal children, because the concentrations are often close to or below the limit of detection of the assays that have been used in the past (Andersson and Skakkebaek, 1999). In fact, reliable estimates of steroid hormone production in pre-pubertal children are still lacking.

9.5.1.2 Dietary contribution to circulating hormone levels

Recent studies have reviewed and reported a positive association between meat or dairy consumption and circulating levels of oestrogens. Maruyama *et al.* (2010) found that serum concentrations of oestrone, and urine excretion of oestrone, oestradiol and oestriol increased in men after intake of cow's milk. This was accompanied by a significant decrease in concentrations of the follicular stimulating hormone, luteinising hormone and testosterone in serum two hours after intake suggesting that oestrogens in milk were absorbed, suppressed gonadotropin secretion in turn decreasing testosterone production. The main oestrogen in milk is oestrone. Oestradiol levels in serum did not change for two hours after the intake, whereas oestradiol urinary excretion was increased, also suggesting that oestrone is converted slowly to oestradiol. Urinary excretion of oestrogens was also significantly increased after milk intake (Maruyama *et al.*, 2010). Although the oral bioavailability of free oestradiol and oestrone is relatively low, that is not true of oestrone sulphate (Ganmaa and Sato, 2005). It is also interesting to note that endometrial, vaginal and mammary tumour tissues have been found to be able to convert precursor oestrogens such as oestrone and its sulphated conjugate very efficiently. Therefore, circulating levels of these precursors may be important contributors of oestradiol in some target tissues (Andersson and Skakkebaek, 1999).

Brinkman *et al.* (2009) found a negative association between total red and fresh red meat consumption and circulating concentrations of SHBG, and a positive association between consumption of dairy products and total and free oestradiol concentrations in naturally postmenopausal women. The Nurses' Health Study showed significantly higher concentrations of total oestrogen and free oestradiol among women in the highest category for consumption of the Western dietary pattern – a diet that included high intakes of red and processed meats. However, these associations appeared to be largely accounted for by the BMI (body mass index) (Fung *et al.*, 2007). This illustrates a major confounding factor in such dietary studies, i.e. that these foods might influence endogenous steroid hormone production indirectly through their nutrient components such as cholesterol, a major substrate for steroid hormone synthesis. Therefore, the reported positive association between high dietary intake of animal products and circulating steroid hormones and negative association with SHBG remains to be fully elucidated.

9.5.1.3 Adverse health effects

As discussed in the previous sections on health hazards, steroid oestrogens have numerous critical physiological functions. In this section, some of the potential risks from exposure to an excess of these hormones in the diet are summarised. Whilst steroid oestrogens have been implicated in the aetiology of various adverse health effects, these should be considered in the context of known beneficial effects such as the prevention of osteoporosis, cardio-vascular diseases or protection against neurodegenerative diseases such as Alzheimer's. The following section focuses on the rationale behind the hypothesis that dietary intake of steroid oestrogens is implicated in the development of some common hormonally responsive tumours.

9.5.1.3.1 Breast cancer Known risk factors for breast cancer include early onset of puberty, late menopause and low parity or later age at first pregnancy, all of which are consistent with an aetiological role for exposure to endogenous oestrogens. Further, some epidemiological evidence has suggested an association between maternal endogenous oestrogen levels during fetal life and the development of breast pathology, including breast cancer, in adulthood (Aksglaede *et al.*, 2006). Other factors associated with breast-cancer risk include a high birth weight, early age at peak growth, a high stature and low BMI at 14 years of age, all of which are consistent with stimulation of pre-pubertal growth by very low levels of oestradiol (Aksglaede *et al.*, 2006).

In addition to the known role played by endogenous oestrogens in mammary carcinogenesis, there is epidemiological evidence of a positive correlation between breast-cancer risk and consumption of meat, milk and dairy products (Ganmaa and Sato, 2005). This has however been attributed to dietary fats, which have long been considered a risk factor for breast cancer, particularly in post-menopausal women. However, it is argued that consumption of whole milk has declined steadily since the 1950s. Indeed, commercially available low-fat milk was found to promote the development of 7,12-dimethylbenz[a]anthracene (DMBA)-induced mammary tumours in rats and tumour promotion by cows' milk was almost comparable to that of 100 ng oestrone sulphate per millilitre (Qin *et al.*, 2004). Milk also contains insulin-like growth factor I (IGF-I), and IGF-I is known to stimulate the proliferation of human oestrogen-sensitive breast cancer cell line MCF-7 (Macaulay, 1992).

9.5.1.3.2 Testicular cancer There is a birth-cohort effect on the age of onset of testicular cancer between men born during the second world war and those born after it, the peak mortality of the latter occurring in their twenties rather than their thirties or forties. This suggests that causative factors operate in early life, possibly *in utero* or in the perinatal or pre-pubertal period. This is also consistent with an influence of maternal nutrition.

In Japan, the intake of milk and dairy products, meat and eggs increased 20-, 9- and 7-fold, respectively, between 1950 and 1998 (Ganmaa *et al.*, 2002). Ganmaa *et al.* (2002) examined the relation between dietary practices throughout the world and the incidence and mortality from testicular cancer. The food that was most closely correlated with the incidence of testicular cancer was cheese, followed by animal fats and milk. Several studies have examined dietary influences on the onset of testicular cancer and found associations with high intake of fat as well as dairy products (Garner *et al.*, 2005). A case-control study in East Anglia, in the United Kingdom, also found that men with testicular cancer had consumed significantly more milk during adolescence than had controls (Davies *et al.*, 1996).

9.5.1.3.3 Prostate cancer Fat intake is a known risk factor for prostate cancer. However, dairy products have been implicated because of their high calcium rather than their fat content. This hypothesis is based on the fact that calcium suppresses the formation of 1,25-dihydroxyvitamin D. This decreases the level of the biologically active form of vitamin D in circulation, which, in turn, is thought to reduce cellular proliferation and enhance cellular differentiation (Ganmaa *et al.*, 2002). Indeed, the World Cancer Research Fund/American Institute for Cancer Research (WCRF/AICR) report on Food, Nutrition, Physical Activity and the Prevention of Cancer (WCRF/AICR, 2007) judged that foods containing calcium are probably a cause of prostate cancer and that there is some evidence suggesting that processed meat and dairy products are causes of this cancer. As mentioned above, consumption of milk increases blood levels of IGF-1, which has been associated with increased prostate cancer risk in some studies. Several meta-analyses of epidemiological studies considering prostate-cancer risk have reported marginally significant risk estimates (Parodi, 2009). A recent, extensive study confirmed that intake of dairy products would more than double the risk of prostate cancer, whilst calcium showed only a borderline risk (Raimondi *et al.*, 2010). There is also some supporting evidence for an oestrogen-imprinting or developmental oestrogenisation role in prostate cancer. During the third trimester of *in utero* development, rising maternal oestradiol and declining fetal androgen production result in an elevated oestrogen/testosterone ratio. This has been shown to stimulate extensive transient squamous metaplasia within the developing prostatic epithelium, which regresses immediately after birth. Although the role of oestrogens for prostatic development is still unclear, it has been proposed that excessive oestrogenisation during prostatic development may predispose to abnormal function and disease in later life (Prins and Korach, 2008). There is some mechanistic *in-vitro* evidence that oestradiol may play a role in prostate cancer; both ERα and ERβ are expressed in normal and malignant prostatic epithelial cells, and physiological concentrations of oestradiol stimulated the growth of human prostatic cells in the LNCaP line (Ganmaa *et al.*, 2002).

In conclusion, the World Cancer Research Fund (WCRF), in collaboration with the American Institute for Cancer Research (AICR), conducted an authoritative review of the state of evidence with regard to cancer risk or prevention by food, nutrition, weight and physical activity (WRCF/AICR, 2007). While testicular cancer is not mentioned in this report, it concluded that red meat was a possible cause of cancers of breast and prostate, and that there is limited evidence that milk and dairy are a cause of prostate cancer. An association with steroid oestrogens occurring in animal produce is generally biologically plausible, particularly in view of the known carcinogenic activity of natural steroid oestrogens. Nonetheless, other compounds contained in these products may be equally biologically active and thus contribute to the aetiology of such cancers. However, mechanistic pathways for the role of dietary steroid oestrogens in these cancers remain to be fully elucidated.

9.5.2 Phyto-oestrogens

9.5.2.1 *Metabolism of phyto-oestrogens*

Phyto-oestrogens occur in various forms in plants and foodstuffs. Isoflavones are often present as glucose conjugates or glucosides, and lignans occur as aglucones or as mono or diglycosides. Glucosides may undergo metabolism in the human gastrointestinal tract prior to absorption, where they can be deconjugated to aglycone or demethylated; for example, biochanin A and formononetin are demethylated to yield genistein and daidzein, respectively. Deconjugation or demethylation by the gut microflora may be essential to permit absorption from the gastrointestinal tract. Gut microflora play an important role in lignan metabolism, converting the lignans present in plants into the forms to which humans are exposed, namely enterodiol and enterolactone. Following absorption, phyto-oestrogens may undergo sulphate conjugation, or glucuronidation followed by excretion. The oestrogenicity of both the parent compound found in plants and the metabolites are of potential relevance to human exposure, and in some cases, the aglycone has been identified as the bioactive form, rather than the glucoside (COT, 2003).

Pharmacokinetic studies of daidzein and genistein have shown that phyto-oestrogens have a large volume of distribution and probably undergo enterohepatic recirculation. Peak plasma concentrations occur around 4–8 hours after dietary intake, and elimination kinetics can depend on the matrix, a solid matrix resulting in slower elimination than a fluid matrix (Setchell, 2000).

An important example of individual variation within the human population is given by the phyto-oestrogen equol. Equol is not found at significant levels in plants, but is produced by the gut bacteria of humans from degradation of other phyto-oestrogens (Setchell *et al.*, 1984). Equol has a higher affinity for oestrogen receptors than many of the phyto-oestrogens from which it can be derived; consequently, equol production can increase the oestrogenic impact of dietary phyto-oestrogens (Setchell *et al.*, 2002). The ability to produce equol varies within the human population, and the proportion of a population that are deemed to be 'equol producers' is strongly affected by ethnicity. In one study of healthy, menopausal women from Taiwan, around half of the population were equol producers. Interestingly, that study found that soya administration reduced menopausal symptoms (hot flushes and excessive sweating) in those women who produced equol, but not in the non-equol producers (Jou *et al.*, 2008). Equol producers may be subject to adverse effects of phyto-oestrogens whilst non-producers are not, and equol status needs to be controlled for in epidemiological studies.

9.5.2.2 *Levels of phyto-oestrogens in human tissues*

Levels of phyto-oestrogens and myco-oestrogens in human tissues are documented in Table 9.1. Plasma levels vary significantly between countries; for example, the levels of total isoflavonoids in plasma were up to 110 times higher in Japanese men than in Finnish men (Adlercreutz *et al.*, 1993). Asian diets that are rich in soya result in isoflavone consumption of around 1 mg/kg/day. Plasma isoflavone concentrations up to 1 μM can be measured in Japanese men. European and American concentrations are much lower, typically below 0.07 μM, though vegetarians can have concentrations reaching 0.4 μM (Whitten *et al.*, 2001). Phyto-oestrogen levels can be measured in urine, which provides a means of monitoring human exposure without invasive sampling (Wolff *et al.*, 2007). However, these levels can be hard to relate to the tissue of interest in which a biological effect may be occurring.

To put adult dietary exposure to phyto-oestrogens into context, some authors have noted that typical Western exposure to oestrogens through diet is lower than exposure from oral contraceptives (North and Golding, 2000). Human tissue levels of phyto-oestrogens would be expected to be markedly affected by diet supplementation. Dietary supplementation with linseed resulted in peak plasma levels of 500 ng/ml (combined enterolactone and enterodiol), and supplementation with soya flour or clover sprouts resulted in levels of 43 (for equol), 312 (for daidzein) and 148 (for genistein) ng/ml (Morton *et al.*, 1994). One study has compared plasma levels resulting from a soya-rich diet with those resulting from consumption of soya-isoflavone tablets; the results depended on the phyto-oestrogen concerned; whilst daidzein levels were similar irrespective of route, genistein levels were highest for dietary exposure (Gardner *et al.*, 2009). This result has implications for attempts to mimic a soya-rich diet with tablets, and shows that pharmacokinetic considerations must be kept in mind in evaluating such data.

Vulnerable groups may have a different exposure profile to phyto-oestrogens, compared to population averages. Population subgroups that are expected to have a higher than average intake of phyto-oestrogens include: vegetarians and vegans (isoflavones and lignans); particular ethnic groups, e.g. Japanese and Chinese (isoflavones); consumers of soya-based foods (isoflavones); consumers of phyto-oestrogen-containing dietary supplements (mostly isoflavones) (COT, 2003). Vegetarian or vegan subjects had phyto-oestrogen levels that were five to fifty times higher than subjects consuming a non-vegetarian diet (Peeters *et al.*, 2007). Infants fed with soya-based formula (SBF) may have an exposure that far exceeds the average adult exposure (Whitten *et al.*, 2001). The concentration of isoflavones in SBF (100–175 μM) is far greater than that of breast milk (10–70 nM) or bovine-based milk formula. The resulting isoflavone consumption of an infant receiving SBF is 6–11 mg/kg/day, compared with 0.3–1.2 mg/kg/day for an adult consuming a soya-rich Asian diet. Infants consuming SBF have isoflavone plasma concentrations of 2.4–6.5 μM; the concentration for infants consuming cows' milk is 0.03 μM and for those consuming breast milk 0.02 μM (Whitten *et al.*, 2001). Fetal exposure to phyto-oestrogens is also possible. Phyto-oestrogens, including genistein, daidzein, coumestrol and equol, can be transferred from the mother to the fetus, and differences in metabolism and elimination may result in phyto-oestrogens being present in the fetal circulation for longer than they would be in the mother (Todaka *et al.*, 2005).

Finally, it is important to note that phyto-oestrogens are not consumed individually; humans are exposed simultaneously to a range of phyto-oestrogens and to a plethora of other chemicals which may have similar effects, such as xeno-oestrogens. Exposure to such mixtures raises the possibility of additive effects when different chemicals have a common biological effect, such as the ability to activate ER(s). The implication of this is that exposure to many weakly acting chemicals could result in an additive effect that would not be adequately addressed by simply considering each chemical in isolation (Kortenkamp, 2007).

9.5.2.3 Health effects

The possibility of endocrine disruption from phyto-oestrogens and myco-oestrogens merits concern because substances such as zearalenone exhibit oestrogenicity with a comparable potency to natural oestrogens, and have a greater potency than other endocrine disrupters, such as bisphenol A, DDT (dichlorodiphenyltrichloroethane) and atrazine (Hartmann *et al.*, 2008). Not all phyto/myco-oestrogens are as potent as zearalenone, and each chemical needs individual consideration; there is also the need to consider mixture effects, as mentioned above.

Endocrine disruption includes a wide range of human disorders, including hormonal cancers, reproductive disorders, impaired neurodevelopment and metabolic dysfunction. The remainder of this section will focus on the association between phyto-oestrogens and three endocrine-disruption conditions, namely: impaired fertility, male reproductive disorders and breast cancer. Finally, the evidence for beneficial health effects will be reviewed briefly.

9.5.2.3.1 Factors affecting epidemiological studies of phyto-oestrogens Epidemiological studies of phyto-oestrogens can be subject to a number of experimental issues. Studies often compare the effect of diet, and then attribute the observed health differences to components of the diet, such as phyto-oestrogens. However the health benefits may come from consumption of a generally healthy diet, high in fibre, fruit and vegetables, and from reduced consumption of risk-associated foods (fat, processed meat etc). Shu *et al.* (2009) identified some of these problems when discussing their study of phyto-oestrogens and breast cancer. For example, their study participants consumed soya foods, and any effects on risk of breast cancer were attributed to the isoflavones known to be present in soya; however soya also contains folate, protein, protease inhibitors, calcium and fibre, which are all also bioactive. Additionally these results from a Chinese population may not be relevant to a Western population or diet, the beneficial effect of soya may not apply to processed soya, such as soya milk or soya supplements, and the linkage between benefit and soya is open to question, since soya consumption may co-vary with a generally healthy lifestyle, including low consumption of processed meat and dairy produce, high consumption of vegetables and fish and higher rates of exercise.

Differences in epidemiological studies between e.g. Eastern and Western populations could be due to different diets, different genetic backgrounds or a combination of both. The failure to specify the equol production status of study participants may have made the results of dietary intervention studies with soya hard to interpret (Setchell *et al.*, 2002). Short-term studies may be meaningless for effects related to a whole human lifespan; for example, rats with long-term exposure to isoflavones or genistein show an elevated response to oestradiol (Moller *et al.*, 2010).

9.5.2.3.2 Impaired fertility Concern regarding human fertility probably arises from the early observations of impaired fertility in animals exposed to phyto-oestrogens or myco-oestrogens. Although Asian diets are high in soya, and therefore also rich in phyto-oestrogens, there has been no obvious reduction in fertility at a population level. A possible role for consumption of soya, and phyto-oestrogen in impaired fertility was examined in a three-month study of 99 males from sub-fertile couples. An inverse association was found between soya intake and sperm concentration, but no association with sperm motility, sperm morphology or ejaculate volume (Chavarro *et al.*, 2008). The study assessed soya intake by means of a questionnaire to measure the consumption of 15 soya-based foods, and the cautions needed when interpreting this study include the use of a questionnaire, rather than a laboratory test, to assess intake, the possibility that exposure to soya, and thus phyto-oestrogens, may be important during a particular developmental window, and the possibility that current exposure may not be relevant for, or related to, risk. Finally it may not be possible to assess the cause of sub-fertility of a couple without considering both partners in the couple.

A recent review of the possible effects of soya and phyto-oestrogens on male reproductive function found that there were experimental indications that phyto-oestrogens could affect reproductive hormone function, spermatogenesis, sperm capacitation and fertility; however, a risk has not yet been demonstrated in humans (Cederroth *et al.*, 2010). Caution suggests that human exposure to phyto-oestrogens should be avoided in the perinatal stages, when the reproductive tract is developing and may be vulnerable (Cederroth *et al.*, 2010). This caution would include not feeding SBF to infants.

9.5.2.3.3 Testicular dysgenesis syndrome (TDS) Male reproductive health is generally considered to be declining, and male reproductive disorders, such as testicular cancer and impaired semen quality, are considered to be increasing in prevalence in many countries (Main *et al.*, 2010). These disorders have been associated with conditions such as congenital cryptorchidism and hypospadias to form a syndrome known as the testicular dysgenesis syndrome (TDS). The increasing prevalence of TDS suggests a role for lifestyle factors and exposure to environmental chemicals. For example, prospective studies have linked perinatal exposure to persistent halogenated compound to cryptorchidism, and phthalates to anti-androgenic effects in newborns (Main *et al.*, 2010).

Hypospadias is a genital malformation in males whereby the urethral orifice is abnormally located. Hypospadias may have genetic, endocrine or environmental causes (Kalfa *et al.*, 2009). Cryptorchidism describes a situation in which the testis does not descend normally and the condition is a risk factor for subsequent sub-fertility and for testis cancer (Pierik *et al.*, 2004). The incidence of hypospadias in the general population is less than 1 per cent and that of cryptorchidism is 1–5 per cent (Pierik *et al.*, 2004).

A maternal vegetarian diet has been associated with an increased risk of hypospadias, which has been attributed to the higher exposure of vegetarians to phyto-oestrogens (North and Golding, 2000). Mothers who had a vegetarian diet during pregnancy had an adjusted odds ratio of 4.99 for giving birth to a boy with hypospadias compared to mothers with an omnivorous diet. This odds ratio indicates an almost 5-fold higher risk of hypospadias if the maternal diet was vegetarian. However, the number of cases was very low, probably due to the overall low incidence rate for hypospadias, and thus may not justify drawing strong conclusions. Alternative explanations for the study's findings include the possibility that a vegetarian diet may indicate exposure to higher levels of pesticides, which are present in fruits and vegetables, or that the vegetarian diet may have lacked one or more essential nutrients. These alternative explanations will need to be evaluated before strong advice can be given to vegetarian mothers. Other studies have found that dietary intake of phyto-oestrogens was not a significant risk factor for hypospadias or cryptorchidism, and instead identified paternal pesticide exposure and paternal smoking as risk factors (Pierik *et al.*, 2004). Other risk factors such as low parity, an atypically young or old maternal age, premature birth, low birth weight and pesticide exposure are also associated with hypospadias (Norgil Damgaard *et al.*, 2002). As mentioned above, epidemiological studies can be difficult to interpret: Akre *et al.* (2008) reported an association between hypospadias and consumption during pregnancy of a diet that was low in fish and meat and, as a consequence, high in vegetables. These authors considered that their results implicate a lack of animal proteins in the causation of hypospadias rather than implicating a vegetable-rich, and thus phyto-oestrogen-rich, diet.

Clearly it is necessary to put any potential risk of phyto-oestrogen consumption in context, and to assess the potential benefits of phyto-oestrogen-rich diets as well. With this aim in mind, approaches such as those used in a study that found an increased oestrogenicity of the serum of newborns with genital abnormalities (Paris *et al.*, 2006) could be useful in identifying and prioritising which of the many potential endocrine disruptors to which humans are exposed are actually responsible for the eventual adverse health outcomes.

9.5.2.3.4 Breast cancer Breast cancer is less common in Asian women than Western women, and it is sometimes suggested that this may be due to their higher consumption of phyto-oestrogens; for example, women native to Japan, Singapore or China have an estimated soya intake that is 10–15 times higher than that of Westerners (Wu *et al.*, 1998; Shu *et al.*, 2009). A role for lifestyle factors, including phyto-oestrogen consumption, over genetic factors is suggested by the observation that migrating Asian women acquire the risk of the host country. However epidemiology has not proved the link between lower risk and higher phyto-oestrogen consumption (Mense *et al.*, 2008). Concerns that phyto-oestrogens could increase the risk of breast cancer come from experimental observations that phyto-oestrogens such as genistein can enhance the proliferation of breast cancer cells *in vitro* and can promotes oestrogen-dependent mammary tumour growth in mice (Allred *et al.*, 2001).

The Shanghai Breast Cancer Survival Study was a large, population-based cohort study of 5042 female breast cancer survivors in China, designed to address three issues: first, to examine the hypothesis that consumption of soya would reduce the risk of breast cancer; secondly, to address the risk that oestrogenicity from the high isoflavone content of soya would result in adverse effects, namely an increased breast-cancer incidence; and thirdly, to examine the risk that soya consumption might interact with a commonly used cancer therapy, tamoxifen (Shu *et al.*, 2009). The study did not find an association of soya with breast-cancer risk, and rather found that moderate soya food intake was safe and potentially beneficial for women with breast cancer. Soya also did not appear to impair the response to tamoxifen. The authors noted that although soy isoflavones may be directly oestrogenic, soya may lower the availability of endogenous oestrogens by a number of effects, including reducing oestrogen synthesis, increasing steroid clearance from the circulation and increasing protein binding (Shu *et al.*, 2009).

The Shanghai study was relatively short (5 years) and could have missed any risk associated with cancer recurrence, for example the study authors note that it may be desirable to keep levels of oestrogens low after treatment for breast cancer to reduce the chance of oestrogen-dependent recurrence (Shu *et al.*, 2009). It is also possible that adverse effects of phyto-oestrogens may be manifest only if exposure occurs during a crucial developmental window, a theory referred to as the developmental/fetal basis of adult disease (Diamanti-Kandarakis *et al.*, 2009).

9.5.2.3.5 Beneficial effects There is a widespread perception that phyto-oestrogens have health benefits, and mechanisms have been proposed whereby they could beneficially affect the skeletal and cardiovascular systems and reduce the occurrence of hot flushes during menopause (Baber, 2010; Beck *et al.*, 2005). An authoritative meta-review of the use of phyto-oestrogens as an alternative to hormone replacement therapy (HRT) for the treatment of menopausal symptoms such as hot flushes found no consistent evidence of effectiveness, though some individual trials had found a slight reduction in the occurrence of symptoms (Lethaby *et al.*, 2007).

Beneficial effects of phyto-oestrogens have been proposed for a number of human diseases (see Table 9.4 for examples). The Phyto-oestrogens working group on the UK's Committee on Toxicity of Chemicals in Food, Consumer Products and the Environment (COT) reviewed the evidence for beneficial effects of dietary phyto-oestrogens and reported a number of difficulties with the available data, in particular the following aspects:

- There are not many direct studies of phyto-oestrogens; usually a soya-rich diet was studied and effects could be due to the presence of other biologically active components.
- Many cases were short-term intervention studies that could not examine the effect of early phyto-oestrogen exposure on disease manifestation later in life.
- Population studies were often carried out on Japanese and Chinese populations, and the extrapolation to UK or Western populations and diets may not be valid, due to possible confounding by differences in lifestyle, diet, gut microflora, genetic make-up and absorption and metabolism.

Given these limitations, the available literature suggests that the evidence for the beneficial effects of phyto-oestrogens remains inconclusive or that, when such effects were detected, they were found to be relatively small. The conclusions of the Committee on Toxicity are given in Table 9.4 (COT, 2003).

9.6 Risk reduction

Risk reduction at source does not at first appear a viable option for compounds produced naturally by such a wide variety of organisms.

This situation reflects fundamental differences in the perception of risk from natural substances occurring in the diet: namely, compounds to which humans have been exposed for millennia should not present a substantial risk to human health. Furthermore, there are inherent distinctions between risk from exposures one feels control over (e.g. how the addictive properties of tobacco smoking come to be perceived as personal freedom) and risk from seemingly occult exposures. Thus, consumers are not necessarily aware of potential exposure to synthetic chemicals from everyday objects and may, as a result, experience such exposures as an intrusion when they realise that these take place without their prior knowledge. This may explain why there has been much more public concern in the general media over very low levels of steroid hormones detected in drinking water while the main route of exposure is undoubtedly via the diet. Another interesting example of this disconnect between scientific evidence

Table 9.4 Summary of COT opinion (COT, 2003)

Condition	Conclusion: state of the evidence
Menopausal symptoms	Inconclusive
Osteoporosis	Small protective effect in lumbar spine
	Other sites: data is equivocal
Cardiovascular disease	Beneficial effect of soya, but same effect not seen for purified phyto-oestrogen
Cancer	Some evidence of benefit in animals studies, human studies less convincing
Breast cancer	Inconclusive
Endometrial and ovarian cancer	No evidence of protection
Prostate cancer	Data too limited
Colorectal cancer	Suggested reduction in risk with non-fermented soya
	Increased risk with fermented soya
Stomach cancer	Inconclusive
Lung cancer	Inconclusive

of a risk or hazard and its perception would be the use of natural or synthetic substances in cosmetics, where any potential risk will be weighed against perceived benefit. It should be evident, therefore, that risk management of naturally occurring compounds is more often a socio-political decision than a rational scientific one.

Nonetheless, some of the concerns raised in this chapter can be related to relatively recent developments in food production and husbandry practices. There is clear evidence that effective risk-management measures can be effectively implemented. Thus, reducing animal exposure to phyto-oestrogens or myco-oestrogens, through changes in grazing practices to avoid phyto-oestrogen-rich pastures, or by halting the use of myco-oestrogens as growth promoters has led to risk reduction, at least for the animals. Other sustainable husbandry practices could be applied in dairy farming to reduce the levels of steroid oestrogens in milk. Risk reduction for humans is more complex, even when only considering the scientific basis for risk. Due at least partially to the inherent limitations of epidemiological studies and the possible complex biological activity of natural products, there is a lack of clear-cut evidence of harm in the general population from exposure to steroid oestrogens, phyto-oestrogens or myco-oestrogens. If human exposure to myco-oestrogens can be managed by food-quality standards and best practices to prevent fungal infections of stored foods and thus contamination of derived foods, steroid oestrogens and phyto-oestrogens cannot be argued to be food contaminants *per se* and risks have to be balanced against the nutritional value and other benefits of these compounds and other co-occurring substances. Indeed, general advice to avoid phyto-oestrogens would be likely to have adverse effects on public health by discouraging the consumption of fruit and vegetables. Ideally, such personally and culturally motivated choices should be guided by adequate information on risk and benefits. In practice, this may not be easy to achieve. Modern developments in intensive agriculture, coupled with increased urbanisation have led to a disconnection between food consumption and production. Information on methods of food production is therefore indirect and not necessarily independent, although the perceptions of concerned consumers are taken seriously by the food industry and can arguably be addressed by certification and labelling. Nonetheless, there are examples of misinformation and miscommunication; for example, in order to address public concern over the use of steroidal hormones, attempts to induce sex reversal in commercial fisheries by administering phyto-oestrogens have been recently reported (Yılmaz *et al.*, 2009).

Pragmatically, attention should be paid to the risk assessment of products directed at or used by vulnerable groups (e.g. SBF for infants) and towards high-exposure situations (e.g. resulting from the use of phyto-oestrogen supplements or the milking of cows during pregnancy). In such cases, consistent criteria should be used for products containing synthetic chemicals and naturally occurring substances. Furthermore, the fact that steroid oestrogens and phyto-oestrogens occur naturally should not be taken to mean there is no cause for concern regarding human exposure.

References

Adams NR. 1995. Detection of the effects of phytoestrogens on sheep and cattle. *J Anim Sci* **73**(5): 1509–1515.

Adlercreutz H, Markkanen H, Watanabe S. 1993. Plasma concentrations of phyto-oestrogens in Japanese men. *Lancet* **342**: 1209–1210.

Adlercreutz H, Yamada T, Wahala K, Watanabe S. 1999. Maternal and neonatal phytoestrogens in Japanese women during birth. *Am J Obstet Gynecol* **180**: 737–743.

Akre O, Boyd HA, Ahlgren M, Wilbrand K, Westergaard T, Hjalgrim H, Nordenskjold A, Ekbom A, Melbye M. 2008. Maternal and gestational risk factors for hypospadias. *Environ Health Perspect* **116**: 1071–1076.

Aksglaede L, Juul A, Leffers H, Skakkebaek NE, Andersson AM. 2006. The sensitivity of the child to sex steroids: possible impact of exogenous estrogens. *Human Reproduction Update* **12**(4): 341–349.

Allred CD, Allred KF, Ju YH, Virant SM, Helferich WG. 2001. Soy diets containing varying amounts of genistein stimulate growth of estrogen-dependent (MCF-7) tumors in a dose-dependent manner. *Cancer Res* **61**: 5045–5050.

Allred KF, Yackley KM, Vanamala J, Allred CD. 2009. Trigonelline is a novel phytoestrogen in coffee beans. *J Nutr* **139**(10): 1833–1838.

Amakura Y, Tsutsumi T, Nakamura M, Kitagawa H, Fujino J, Sasaki K, Toyoda M, Yoshida T, Maitani T. 2003. Activation of the aryl hydrocarbon receptor by some vegetable constituents determined using *in vitro* reporter gene assay. *Biol Pharm Bull* **26**: 532–539.

Andersen H, Siegrist H, Halling-Sorensen B, Ternes TA. 2003. Fate of estrogens in a municipal sewage treatment plant. *Environ Sci Technol* **37**(18): 4021–4026.

Andersson A, Skakkebaek N. 1999. Exposure to exogenous estrogens in food: possible impact on human development and health. *Eur J Endocrinol* **140**(6): 477–485.

Atkinson S, Atkinson MJ, Tarrant AM. 2003. Estrogens from sewage in coastal marine environments. *Environ Health Perspect* **111**(4): 531–535.

Baber R. 2010. Phytoestrogens and post reproductive health. *Maturitas* **66**: 344–349.

Barker DJP. 2004. The Developmental Origins of Adult Disease. *J Am Coll Nutr* **23**(suppl 6): 588S–595.

Bartolome JA, Sozzi A, McHale J, Melendez P, Arteche ACM, Silvestre FT, *et al.* 2005. Resynchronization of ovulation and timed insemination in lactating dairy cows, II: assigning protocols according to stages of the estrous cycle, or presence of ovarian cysts or anestrus. *Theriogenology* **63**(6): 1628–1642.

Basak S, Pookot D, Noonan EJ, Dahiya R. 2008. Genistein down-regulates androgen receptor by modulating HDAC6-Hsp90 chaperone function. *Mol Cancer Ther* **7**: 3195–3202.

Beck J, Totsche KU, Kogel-Knabner I. 2008. A rapid and efficient determination of natural estrogens in soils by pressurised liquid extraction and gas chromatography-mass spectrometry. *Chemosphere* **71**(5): 954–960.

Beck V, Rohr U, Jungbauer A. 2005. Phytoestrogens derived from red clover: an alternative to estrogen replacement therapy? *J Steroid Biochem Mol Biol* **94**: 499–518.

Beck V, Unterrieder E, Krenn L, Kubelka W, Jungbauer A. 2003. Comparison of hormonal activity (estrogen, androgen and progestin) of standardized plant extracts for large scale use in hormone replacement therapy. *J Steroid Biochem Mol Biol* **84**: 259–268.

Belfroid AC, Van der Horst A, Vethaak AD, Schafer AJ, Rijs GBJ, Wegener J, *et al.* 1999. Analysis and occurrence of estrogenic hormones and their glucuronides in surface water and waste water in The Netherlands. *Sci Total Environ* **225**(1–2): 101–108.

Bennetts HW, Uuderwood E, Shier FL. 1946. A specific breeding problem of sheep on subterranean clover pastures in Western Australia. *Aust Vet J* **22**(1): 2–12.

Bergman Jungestrom M, Thompson LU, Dabrosin C. 2007. Flaxseed and its lignans inhibit estradiol-induced growth, angiogenesis, and secretion of vascular endothelial growth factor in human breast cancer xenografts *in vivo*. *Clin Cancer Res* **13**: 1061–1067.

Bilezikian JP, Morishima A, Bell J, Grumbach MM. 1998. Increased bone mass as a result of estrogen therapy in a man with aromatase deficiency. *New Engl J Med* **339**(9): 599–603.

Brinkman MT, Baglietto L, Krishnan K, English DR, Severi G, Morris HA, *et al.* 2009. Consumption of animal products, their nutrient components and postmenopausal circulating steroid hormone concentrations. *Eur J Clin Nutr* **64**(2): 176–183.

Cargouet M, Perdiz D, Mouatassim-Souali A, Tamisier-Karolak S, Levi Y. 2004. Assessment of river contamination by estrogenic compounds in Paris area (France). *Sci Total Environ* **324**(1–3): 55–66.

Cassidy A, Brown JE, Hawdon A, Faughnan MS, King LJ, Millward J, Zimmer-Nechemias L, Wolfe B, Setchell KD. 2006. Factors affecting the bioavailability of soy isoflavones in humans after ingestion of physiologically relevant levels from different soy foods. *J Nutr* **136**: 45–51.

Cederroth CR, Auger J, Zimmermann C, Eustache F, Nef S. 2010. Soy, phyto-oestrogens and male reproductive function: a review. *Int J Androl* **33**: 304–316.

Chavarro JE, Toth TL, Sadio SM, Hauser R. 2008. Soy food and isoflavone intake in relation to semen quality parameters among men from an infertility clinic. *Hum Reprod* **23**: 2584–2590.

Chen JJ, Chang HC. 2007. By modulating androgen receptor coactivators, daidzein may act as a phytoandrogen. *Prostate* **67**: 457–462.

COT working group on phytoestogens. 2000. Chemistry of phytoestrogens and overview of analytical methodology. PEG/2000/06. 2000.

COT. 2003. Phytoestrogens and health. FSA/0826/0503. 2003.

Couse JF, Korach KS. 1999. Estrogen receptor null mice: what have we learned and where will they lead us? *Endocr Rev* **20**(3): 358–417.

Dahlman-Wright K, Cavailles V, Fuqua SA, Jordan VC, Katzenellenbogen JA, Korach KS, Maggi A, Muramatsu M, Parker MG, Gustafsson JA. 2006. International Union of Pharmacology. LXIV. Estrogen receptors. *Pharmacol Rev* **58**: 773–781.

D'Ascenzo G, Di Corcia A, Gentili A, Mancini R, Mastropasqua R, Nazzari M, *et al.* 2003. Fate of natural estrogen conjugates in municipal sewage transport and treatment facilities. *Sci Total Environ* **302**(1–3): 199–209.

Davies T, Palmer C, Ruja E, Lipscombe J. 1996. Adolescent milk, dairy product and fruit consumption and testicular cancer. *Br J Cancer* **74**(4): 657–660.

Diamanti-Kandarakis E, Bourguignon JP, Giudice LC, Hauser R, Prins GS, Soto AM, Zoeller RT, Gore AC. 2009. Endocrine-disrupting chemicals: an Endocrine Society Scientific Statement. *Endocr Rev* **30**: 293–342.

Dioscorides P. 2000. Book three: roots. In: *De Materia Medica* (Osbaldeston TA, ed) Johannesburg, South Africa: Ibidis Press.

Erlund I, Meririnne E, Alfthan G, Aro A. 2001. Plasma Kinetics and urinary excretion of the flavanones naringenin and hesperetin in humans after ingestion of orange juice and grapefruit juice. *J Nutr* **131**: 235–241.

Fang H, Tong W, Branham WS, Moland CL, Dial SL, Hong H, Xie Q, Perkins R, Owens W, Sheehan DM. 2003. Study of 202 natural, synthetic, and environmental chemicals for binding to the androgen receptor. *Chem Res Toxicol* **16**: 1338–1358.

FDA. 1999. Guideline 3, part 2: Guideline for toxicological testing.

Fox EM, Davis RJ, Shupnik MA. 2008. ERbeta in breast cancer – onlooker, passive player, or active protector? *Steroids* **73**: 1039–1051.

Fraser SP, Ozerlat-Gunduz I, Onkal R, Diss JK, Latchman DS, Djamgoz MB. 2010. Estrogen and non-genomic upregulation of voltage-gated Na(+) channel activity in MDA-MB-231 human breast cancer cells: role in adhesion. *J Cell Physiol* **224**: 527–539.

Fritsche S, Steinhart H. 1999. Occurrence of hormonally active compounds in food: a review. *Eur Food Res Technol* **209**(3–4): 153–179.

Fung TT, Hu FB, Barbieri RL, Willett WC, Hankinson SE. 2007. Dietary patterns, the Alternate Healthy Eating Index and plasma sex hormone concentrations in postmenopausal women. *Int J Cancer* **121**(4): 803–809.

Ganmaa D, Li XM, Wang J, Qin LQ, Wang PY, Sato A. 2002. Incidence and mortality of testicular and prostatic cancers in relation to world dietary practices. *Int J Cancer* **98**(2): 262–267.

Ganmaa D, Sato A. 2005. The possible role of female sex hormones in milk from pregnant cows in the development of breast, ovarian and corpus uteri cancers. *Med Hypotheses* **65**(6): 1028–1037.

Gardner CD, Chatterjee LM, Franke AA. 2009. Effects of isoflavone supplements vs. soy foods on blood concentrations of genistein and daidzein in adults. *J Nutr Biochem* **20**: 227–234.

Garner MJ, Turner MC, Ghadirian P, Krewski D. 2005. Epidemiology of testicular cancer: An overview. *Int J Cancer* **116**(3): 331–339.

Gehm BD, McAndrews JM, Chien PY, Jameson JL. 1997. Resveratrol, a polyphenolic compound found in grapes and wine, is an agonist for the estrogen receptor. *Proc Natl Acad USA* **94**: 14138–14143.

Gopalakrishnan A, Xu CJ, Nair SS, Chen C, Hebbar V, Kong AN. 2006. Modulation of activator protein-1 (AP-1) and MAPK pathway by flavonoids in human prostate cancer PC3 cells. *Arch Pharm Res* **29**: 633–644.

Han DH, Denison MS, Tachibana H, Yamada K. 2002. Relationship between estrogen receptor-binding and estrogenic activities of environmental estrogens and suppression by flavonoids. *Biosci Biotechnol Biochem* **66**: 1479–1487.

Hanselman TA, Graetz DA, Wilkie AC. 2003. Manure-borne estrogens as potential environmental contaminants: a review. *Environ Sci Technol* **37**(24): 5471–5478.

Harris DM, Besselink E, Henning SM, Go VLW, Heber D. 2005. Phytoestrogens induce differential estrogen receptor alpha- or beta-mediated responses in transfected breast cancer cells. *Exp Biol Med* **230**: 558–568.

Harris RM, Wood DM, Bottomley L, Blagg S, Owen K, Hughes PJ, Waring RH, Kirk CJ. 2004. Phytoestrogens are potent inhibitors of estrogen sulfation: implications for breast cancer risk and treatment. *J Clin Endocrinol Metab* **89**: 1779–1787.

Hartmann N, Erbs M, Wettstein F, Horger C, Vogelsgang S, Forrer H, Schwarzenbach R, Buchell T. 2008. Environmental exposure to estrogenic and other myco- and phytoestrogens. *Chimia* **62**: 364–367.

Hollert H, Durr M, Holtey-Weber R, Islinger M, Brack W, Farber H, *et al.* 2005. Endocrine disruption of water and sediment extracts in. a non-radioactive dot blot/RNAse protection-assay using isolated hepatocytes of rainbow trout – Deficiencies between bioanalytical effectiveness and chemically determined concentrations and how to explain them. *Environ Sci Pollut Res* **12**(6): 347–360.

Howell A. 2008. The endocrine prevention of breast cancer. *Best Pract Res Clin Endocrinol Metab* **22**: 615–623.

Huang C-H, Sedlak D. 2001. Analysis of estrogenic hormones in wastewater and surface water using ELISA and GC/MS/MS. *Environ Toxicol Chem* **20**(1): 133–139.

IARC. 1979. Sex Hormones (II) Monographs on the Evaluation of Carcinogenic Risk of Chemicals to Humans. Lyon, International Association of Research on Cancer.

Ikeda K, Arao Y, Otsuka H, Nomoto S, Horiguchi H, Kato S, Kayama F. 2002. Terpenoids found in the umbelliferae family act as agonists/antagonists for ER(alpha) and ERbeta: differential transcription activity between ferutinine-liganded ER(alpha) and ERbeta. *Biochem Biophys Res Commun* **291**: 354–360.

Isobe T, Serizawa S, Horiguchi T, Shibata Y, Managaki S, Takada H, *et al.* 2006. Horizontal distribution of steroid estrogens in surface sediments in Tokyo Bay. *Environ Pollut* **144**(2): 632–638.

JECFA. 1988. Evaluation of certain veterinary drug residues in food. Technical Report Series 763. Geneva, Joint Expert Committee on Food Additives.

Jobling S, Nolan M, Tyler CR, Brighty G, Sumpter JP. 1998. Widespread Sexual Disruption in Wild Fish. *Environ Sci Technol* **32**(17): 2498–2506.

Jou HJ, Wu SC, Chang FW, Ling PY, Chu KS, Wu WH. 2008. Effect of intestinal production of equol on menopausal symptoms in women treated with soy isoflavones. *Int J Gynaecol Obstet* **102**: 44–49.

Kalfa N, Philibert P, Sultan C. 2009. Is hypospadias a genetic, endocrine or environmental disease, or still an unexplained malformation? *Int J Androl* **32**: 187–197.

Karkola S, Wahala K. 2009. The binding of lignans, flavonoids and coumestrol to CYP450 aromatase: A molecular modelling study. *Mol Cell Endocrinol* **301**(1–2): 235–244.

Kelly MJ, Levin ER. 2001. Rapid actions of plasma membrane estrogen receptors. *Trends Endocrinol Metab* **12**(4): 152–156.

Kiang DT, Kennedy BJ, Pathre SV, Mirocha CJ. 1978. Binding characteristics of zearalenone analogs to estrogen receptors. *Cancer Res* **38**: 3611–3615.

Kim JY, Han EH, Shin DW, Jeong TC, Lee ES, Woo ER, Jeong HG. 2004. Suppression of CYP1A1 expression by naringenin in murine Hepa-1c1c7 cells [abstract]. *Arch Pharm Res* **27**: 857–862.

Kiparissis Y, Hughes R, Metcalfe C, Ternes T. 2001. Identification of the isoflavonoid genistein in bleached kraft mill effluent. *Environ Sci Technol* **35**: 2423–2427.

Kirk CJ, Harris RM, Wood DM, Waring RH, Hughes PJ. 2001. Do dietary phytoestrogens influence susceptibility to hormone-dependent cancer by disrupting the metabolism of endogenous oestrogens? *Biochem Soc Trans* **29**: 209–216.

Kjaer J, Olsen P, Bach K, Barlebo HC, Ingerslev F, Hansen M, *et al.* 2007. Leaching of estrogenic hormones from manure-treated structured soils. *Environ Sci Technol* **41**(11): 3911–3917.

Koehler KF, Helguero LA, Haldosen LA, Warner M, Gustafsson JA. 2005. Reflections on the discovery and significance of estrogen receptor beta. *Endocr Rev* **26**: 465–478.

Kolodziej EP, Harter T, Sedlak DL. 2004. Dairy wastewater, aquaculture, and spawning fish as sources of steroid hormones in the aquatic environment. *Environ Sci Technol* **38**(23): 6377–6384.

Kolpin DW, Furlong ET, Meyer MT, Thurman EM, Zaugg SD, Barber LB, *et al.* 2002. Pharmaceuticals, hormones, and other organic wastewater contaminants in U.S. streams, 1999–2000: A national reconnaissance. *Environ Sci Technol* **36**(6): 1202–1211.

Kortenkamp A. 2007. Ten years of mixing cocktails: a review of combination effects of endocrine-disrupting chemicals. *Environ Health Perspect* **115** Suppl 1: 98–105.

Kuch HM, Ballschmiter K. 2001. Determination of endocrine-disrupting phenolic compounds and estrogens in surface and drinking water by HRGC-(NCI)-MS in the picogram per liter range. *Environ Sci Technol* **35**(15): 3201–3206.

Kuiper GG, Carlsson B, Grandien K, Enmark E, Haggblad J, Nilsson S, Gustafsson JA. 1997. Comparison of the ligand binding specificity and transcript tissue distribution of estrogen receptors alpha and beta. *Endocrinology* **138**: 863–870.

Kuiper GG, Enmark E, Pelto-Huikko M, Nilsson S, Gustafsson JA. 1996. Cloning of a novel receptor expressed in rat prostate and ovary. *Proc Natl Acad Sci USA* **93**: 5925–5930.

Kuster M, Azevedo DA, Lopez de Alda MJ, quino Neto FR, Barcelo D. 2009. Analysis of phytoestrogens, progestogens and estrogens in environmental waters from Rio de Janeiro (Brazil). *Environ Int* **35**: 997–1003.

Larsson DGJ, Hállman H, Fórlin L. 2000. More male fish embryos near a pulp mill. *Environ Toxicol Chem* **19**(12): 2911–2917.

Le Guevel R, Pakdel F. 2001. Assessment of oestrogenic potency of chemicals used as growth promoter by in-vitro methods. *Hum Reprod* **16**: 1030–1036.

Lethaby AE, Brown J, Marjoribanks J, Kronenberg F, Roberts H, Eden J. 2007. Phytoestrogens for vasomotor menopausal symptoms. *Cochrane Database Syst Rev* CD001395.

Levin ER. 2008. Rapid signaling by steroid receptors. *Am J Physiol Regul Integr Comp Physiol* **295**: R1425–R1430.

Levin ER. 2009. Membrane oestrogen receptor alpha signalling to cell functions. *J Physiol* **587**: 5019–5023.

Levin ER. 2002. Cellular functions of plasma membrane estrogen receptors. *Steroids* **67**(6): 471–475.

Liao QC, Xiao ZS, Qin YF, Zhou HH. 2007. Genistein stimulates osteoblastic differentiation via p38 MAPK-Cbfa1 pathway in bone marrow culture. *Acta Pharmacol Sin* **28**: 1597–1602.

Liehr JG. 2000. Is estradiol a genotoxic mutagenic carcinogen? *Endocr Rev* **21**(1): 40–54.

Lund TD, Munson DJ, Haldy ME, Setchell KD, Lephart ED, Handa RJ. 2004. Equol is a novel anti-androgen that inhibits prostate growth and hormone feedback. *Biol Reprod* **70**: 1188–1195.

Macaulay V. 1992. Iinsulin-like growth-factors and cancer. *Br J Cancer* **65**(3): 311–320.

Main KM, Skakkebaek NE, Virtanen HE, Toppari J. 2010. Genital anomalies in boys and the environment. *Best Pract Res Clin Endocrinol Metab* **24**: 279–289.

Markman S, Guschina IA, Barnsley S, Buchanan KL, Pascoe D, Muller CT. 2007. Endocrine disrupting chemicals accumulate in earthworms exposed to sewage effluent. *Chemosphere* **70**(1): 119–125.

Martin OV, Voulvoulis N. 2009. Sustainable risk management of emerging contaminants in municipal wastewaters. *Philos Trans R Soc Ser A Math Phys Eng Sci* **367**(1904): 3895–3922.

Maruyama K, Oshima T, Ohyama K. 2010. Exposure to exogenous estrogen through intake of commercial milk produced from pregnant cows. *Pediatr Int* **52**(1): 33–38.

Matsumura A, Ghosh A, Pope GS, Darbre PD. 2005. Comparative study of oestrogenic properties of eight phytoestrogens in MCF7 human breast cancer cells. *J Steroid Biochem Mol Biol* **94**: 431–443.

Matthiessen P, Arnold D, Johnson AC, Pepper TJ, Pottinger TG, Pulman KGT. 2006. Contamination of headwater streams in the

United Kingdom by oestrogenic hormones from livestock farms. *Sci Total Environ* **367**(2–3): 616–630.

McCarty MF. 2006. Isoflavones made simple – Genistein's agonist activity for the beta-type estrogen receptor mediates their health benefits. *Med Hypotheses* **66**: 1093–1114.

Mense SM, Hei TK, Ganju RK, Bhat HK. 2008. Phytoestrogens and breast cancer prevention: possible mechanisms of action. *Environ Health Perspect* **116**: 426–433.

Mersereau JE, Levy N, Staub RE, Baggett S, Zogric T, Chow S, Ricke WA, Tagliaferri M, Cohen I, Bjeldanes LF, Leitman DC. 2008. Liquiritigenin is a plant-derived highly selective estrogen receptor [beta] agonist. *Mol Cell Endocrinol* **283**: 49–57.

Meyer HHD. 2001. Biochemistry and physiology of anabolic hormones used for improvement of meat production. *APMIS* **109**(1): 1–8.

Milligan S, Kalita J, Pocock V, Heyerick A, De Cooman L, Rong H, De Keukeleire D. 2002. Oestrogenic activity of the hop phyto-oestrogen, 8-prenylnaringenin. *Reproduction* **123**: 235–242.

Milligan SR, Kalita JC, Heyerick A, Rong H, De Cooman L, De Keukeleire D. 1999. Identification of a Potent Phytoestrogen in Hops (Humulus lupulus L.) and Beer. *J Clin Endocrinol Metab* **84**: 2249.

Milligan SR, Kalita JC, Pocock V, Van De Kauter V, Stevens JF, Deinzer ML, Rong H, De Keukeleire D. 2000. The Endocrine Activities of 8-Prenylnaringenin and Related Hop (Humulus lupulus L.) Flavonoids. *J Clin Endocrinol Metab* **85**: 4912–4915.

Minervini F, Dell'Aquila ME, Maritato F, Minoia P, Visconti A. 2001. Toxic effects of the mycotoxin zearalenone and its derivatives on *in vitro* maturation of bovine oocytes and 17[beta]-estradiol levels in mural granulosa cell cultures. *Toxicol in vitro* **15**: 489–495.

Mizukami Y. 2010. *In vivo* functions of GPR30/GPER-1, a membrane receptor for estrogen: from discovery to functions *in vivo*. *Endocr J* **57**: 101–107.

Moller FJ, Diel P, Zierau O, Hertrampf T, Maa J, Vollmer G. 2010. Long-term dietary isoflavone exposure enhances estrogen sensitivity of rat uterine responsiveness mediated through estrogen receptor alpha. *Toxicology Letters* **196**(3): 142–153.

Morton MS, Chan PS, Cheng C, Blacklock N, Matos-Ferreira A, branches-Monteiro L, Correia R, Lloyd S, Griffiths K. 1997. Lignans and isoflavonoids in plasma and prostatic fluid in men: samples from Portugal, Hong Kong, and the United Kingdom. *Prostate* **32**: 122–128.

Morton MS, Wilcox G, Wahlqvist ML, Griffiths K. 1994. Determination of lignans and isoflavonoids in human female plasma following dietary supplementation. *J Endocrinol* **142**: 251–259.

Mouatassim-Souali A, Tamisier-Karolak SL, Perdiz D, Cargouet M, Levi Y. 2003. Validation of a quantitative assay using GC/MS for trace determination of free and conjugated estrogens in environmental water samples. *J Separation Sci* **26**(1–2): 105–111.

Nelson LR, Bulun SE. 2001. Estrogen production and action. *J Am Acad Dermatol* **45**(3, Suppl 1): S116–S124.

Norgil Damgaard I, Maria Main K, Toppari J, Skakkebaek NE. 2002. Impact of exposure to endocrine disrupters inutero and in childhood on adult reproduction. *Best Pract Res Clin Endocrinol Metab* **16**: 289–309.s

North K, Golding J. 2000. A maternal vegetarian diet in pregnancy is associated with hypospadias. The ALSPAC Study Team. Avon Longitudinal Study of Pregnancy and Childhood. *BJU Int* **85**: 107–113.

Olsen CM, Meussen-Elholm ETM, Hongslo JK, Stenersen J, Tollefsen KE. 2005. Estrogenic effects of environmental chemicals: An interspecies comparison. *Comp Physiol Part C Toxicol Pharmacol* **141**: 267–274.

Overk CR, Yao P, Chadwick LR, Nikolic D, Sun Y, Cuendet MA, Deng Y, Hedayat AS, Pauli GF, Farnsworth NR, vanBreemen RB, Bolton JL. 2005. Comparison of the *in vitro* estrogenic activities of compounds from hops (*Humulus lupulus*) and red clover (*Trifolium pratense*). *J Agric Food Chem* **53**: 6246–6253.

Pandian TJ, Sheela SG. 1995. Hormonal induction of sex reversal in fish. *Aquaculture* **138**(1–4): 1–22.

Pardridge WM. 1981. Transport of protein-bound hormones into tissues *in vivo*. *Endocr Rev* **2**(1): 103–123.

Paris F, Jeandel C, Servant N, Sultan C. 2006. Increased serum estrogenic bioactivity in three male newborns with ambiguous genitalia: a potential consequence of prenatal exposure to environmental endocrine disruptors. *Environ Res* **100**: 39–43.

Parodi PW. 2009. Dairy product consumption and the risk of prostate cancer. *Int Dairy J* **19**(10): 551–565.

Patisaul HB, Jefferson W. 2010. The Pros and Cons of Phytoestrogens. *Front Neuroendocrinol* **31**(4): 400–419.

Pearce ST, Jordan VC. 2004. The biological role of estrogen receptors alpha and beta in cancer. *Crit Rev Oncol Hematol* **50**: 3–22.

Peeters PH, Slimani N, van der Schouw YT, Grace PB, Navarro C, Tjonneland A, *et al.* 2007. Variations in plasma phytoestrogen concentrations in European adults. *J Nutr* **137**: 1294–1300.

Petrovic M, Sole M, Lopez de Alda MJ, Barcelo D. 2002. Endocrine disrupters in sewage treatment plants, receiving river waters, and sediments: Integration of chemical analysis and biological effects on feral carp. *Environ Toxicol Chem* **21**(10): 2146–2156.

Pierik FH, Burdorf A, Deddens JA, Juttmann RE, Weber RF. 2004. Maternal and paternal risk factors for cryptorchidism and hypospadias: a case-control study in newborn boys. *Environ Health Perspect* **112**: 1570–1576.

Possemiers S, Heyerick A, Robbens V, DeKeukeleire D, Verstraete W. 2005. Activation of proestrogens from hops (*Humulus lupulus* L.) by intestinal microbiota; conversion of isoxanthohumol into 8-prenylnaringenin. *J Agric Food Chem* **53**: 6281–6288.

Prins GS, Korach KS. 2008. The role of estrogens and estrogen receptors in normal prostate growth and disease. *Steroids* **73**(3): 233–244.

Qin L-Q, Xu J-Y, Wang P-Y, Ganmaa D, Li J, Wang J, *et al.* 2004. Low-fat milk promotes the development of 7, 12-dimethylbenz(A)anthracene (DMBA)-induced mammary tumors in rats. *Int J Cancer* **110**(4): 491–496.

Rad M, Humpel M, Schaefer O, Schoemaker RC, Schleuning WD, Cohen AF, Burggraaf J. 2006. Pharmacokinetics and systemic endocrine effects of the phyto-oestrogen 8-prenylnaringenin after single oral doses to postmenopausal women. *Br J Clin Pharmacol* **62**: 288–296.

Raimondi S, Ben Mabrouk J, Shatenstein B, Maisonneuve P, Ghadirian P. 2010. Diet and prostate cancer risk with specific focus on dairy products and dietary calcium: a-case control study. *Prostate* **70**(10): 1054–1065.

Ranhotra HS, Teng CT. 2005. Assessing the estrogenicity of environmental chemicals with a stably transfected lactoferrin gene promoter reporter in HeLa cells. *Environ Toxicol Pharmacol* **20**: 42–47.

Rice S, Whitehead SA. 2006. Phytoestrogens and breast cancer – promoters or protectors? *Endocr Relat Cancer* **13**: 995–1015.

Riddle JM. 1991. Oral contraceptives and early-term abortifacients during classical antiquity and the middle ages. *Past Present* **132**(1): 3–32.

Ruh MF, Zacharewski T, Connor K, Howell J, Chen I, Safe S. 1995. Naringenin: a weakly estrogenic bioflavonoid that exhibits anti-estrogenic activity. *Biochem Pharmacol* **50**: 1485–1493.

Safford B, Dickens A, Halleron N, Briggs D, Carthew P, Baker V. 2003. A model to estimate the oestrogen receptor mediated effects from exposure to soy isoflavones in food. *Regul Toxicol Pharmacol* **38**: 196–209.

Schaefer O, Humpel M, Fritzemeier KH, Bohlmann R, Schleuning WD. 2003. 8-Prenyl naringenin is a potent ER[alpha] selective phytoestrogen present in hops and beer. *J Steroid Biochem Mol Biol* **84**: 359–360.

Scudamore KA, Patel S. 2000. Survey for aflatoxins, ochratoxin A, zearalenone and fumonisins in maize imported into the United Kingdom. *Food Addit Contam* **17**: 407–416.

Setchell KD, Borriello SP, Hulme P, Kirk DN, Axelson M. 1984. Nonsteroidal estrogens of dietary origin: possible roles in hormone-dependent disease. *Am J Clin Nutr* **40**: 569–578.

Setchell KD, Brown NM, Lydeking-Olsen E. 2002. The clinical importance of the metabolite equol-a clue to the effectiveness of soy and its isoflavones. *J Nutr* **132**: 3577–3584.

Setchell KD. 2001. Soy isoflavones – benefits and risks from nature's selective estrogen receptor modulators (SERMs). *J Am Coll Nutr* **20**: 354S–362S.

Setchell KD. 2000. Absorption and metabolism of soy isoflavones – from food to dietary supplements and adults to infants. *J Nutr* **130**: 654S–655S.

Shaw I, McCully S. 2002. A review of the potential impact of dietary endocrine disruptors on the consumer. *Int J Food Sci Technol* **37**: 471–476.

Shu XO, Zheng Y, Cai H, Gu K, Chen Z, Zheng W, Lu W. 2009. Soy food intake and breast cancer survival. *JAMA* **302**: 2437–2443.

Shupnik MA. 2007. Estrogen receptor-beta: why may it influence clinical outcome in estrogen receptor-alpha positive breast cancer? *Breast Cancer Res* **9**: 107.

Simpson E, Michael M, Agarwal V, Hinshelwood M, Bulun S, Zhao Y. 1997. Cytochromes P450 11: expression of the CYP19 (aromatase) gene: an unusual case of alternative promoter usage. *FASEB J* **11**(1): 29–36.

Smith PN, Blackwell B, Cai Q, Cobb GP. 2010. Hormone quantification in particulate matter near confined animal feeding operations. In: SETAC Europe: 20th Annual Meeting. Seville, Spain: Society of Environmental Toxicology and Chemistry.

Stob M, Baldwin R, Tuite J, Andrews F, Gillette K, Baldwin RS, *et al.* 1962. Isolation of an anabolic, uterotrophic compound from corn infected with Gibberella zeae. *Nature* **196**: 1318.

Stocco DM, Wang X, Jo Y, Manna PR. 2005. Multiple signaling pathways regulating steroidogenesis and steroidogenic acute regulatory protein expression: more complicated than we thought. *Mol Endocrinol* **19**(11): 2647–2659.

Swartz CH, Reddy S, Benotti MJ, Yin HF, Barber LB, Brownawell BJ, *et al.* 2006. Steroid estrogens, nonylphenol ethoxylate metabolites, and other wastewater contaminants in groundwater affected by a residential septic system on Cape Cod, MA. *Environ Sci Technol* **40**(16): 4894–4902.

Takamura-Enya T, Ishihara J, Tahara S, Goto S, Totsuka Y, Sugimura T, Wakabayashi K. 2003. Analysis of estrogenic activity of foodstuffs and cigarette smoke condensates using a yeast estrogen screening method. *Food Chem Toxicol* **41**: 543–550.

Todaka E, Sakurai K, Fukata H, Miyagawa H, Uzuki M, Omori M, Osada H, Ikezuki Y, Tsutsumi O, Iguchi T, Mori C. 2005. Fetal exposure to phytoestrogens – the difference in phytoestrogen status between mother and fetus. *Environ Res* **99**: 195–203.

Toniolo P, Koenig KL, Pasternack BS, Banerjee S, Rosenberg C, Shore RE, *et al.* 1994. Reliability of measurements of total, protein-bound, and unbound estradiol in serum. *Cancer Epidemiol Biomarkers Prevent* **3**(1): 47–50.

Verma SP, Salamone E, Goldin B. 1997. Curcumin and genistein, plant natural products, show synergistic inhibitory effects on the growth of human breast cancer MCF-7 cells induced by estrogenic pesticides. *Biochem Biophys Res Commun* **233**: 692–696.

Vethaak AD, Lahr J, Schrap SM, Belfroid AC, Rijs GBJ, Gerritsen A, *et al.* 2005. An integrated assessment of estrogenic contamination and biological effects in the aquatic environment of The Netherlands. *Chemosphere* **59**(4): 511–524.

WCRF/AICR. 2007. Food, Nutrition, Physical Activity, and the Prevention of Cancer: a Global Perspective. Washington DC: AICR.

Webb S, Ternes T, Gibert M, Olejniczak K. 2003. Indirect human exposure to pharmaceuticals via drinking water. *Toxicol Lett* **142** (3): 157–167.

Wen Y, Zhou BS, Xu Y, Jin SW, Feng YQ. 2006. Analysis of estrogens in environmental waters using polymer monolith in-polyether ether ketone tube solid-phase microextraction combined with high-performance liquid chromatography. *J Chromatogr A* **1133**(1–2): 21–28.

Wendler A, Baldi E, Harvey BJ, Nadal A, Norman A, Wehling M. 2010. Position paper: Rapid responses to steroids: current status and future prospects. *Eur J Endocrinol* **162**: 825–830.

Whitten PL, Patisaul HB. 2001. Cross-species and interassay comparisons of phytoestrogen action. *Environ Health Perspect* **109** Suppl 1: 5–20.

Wicks C, Kelley C, Peterson E. 2004. Estrogen in a karstic aquifer. *Ground Water* **42**(3): 384–389.

Williams RJ, Johnson AC, Smith JJL, Kanda R. 2003. Steroid estrogens profiles along river stretches arising from sewage treatment works discharges. *Environ Sci Technol* **37**(9): 1744–1750.

Wiseman H, Casey K, Bowey EA, Duffy R, Davies M, Rowland IR, Lloyd AS, Murray A, Thompson R, Clarke DB. 2004. Influence of 10 wk of soy consumption on plasma concentrations and excretion of isoflavonoids and on gut microflora metabolism in healthy adults. *Am J Clin Nutr* **80**: 692–699.

Wolff MS, Teitelbaum SL, Windham G, Pinney SM, Britton JA, Chelimo C, *et al.* 2007. Pilot study of urinary biomarkers of phytoestrogens, phthalates, and phenols in girls. *Environ Health Perspect* **115**: 116–121.

Wu AH, Ziegler RG, Nomura AM, West DW, Kolonel LN, Horn-Ross PL, Hoover RN, Pike MC. 1998. Soy intake and risk of breast cancer in Asians and Asian Americans. *Am J Clin Nutr* **68**: 1437S–1443S.

Xiao X-Y, McCalley DV, McEvoy J. 2001. Analysis of estrogens in river water and effluents using solid-phase extraction and gas chromatography-negative chemical ionisation mass spectrometry of the pentafluorobenzoyl derivatives. *J Chromatogr A* **923**: 195–204.

Yan GR, Xiao CL, He GW, Yin XF, Chen NP, Cao Y, He QY. 2010. Global phosphoproteomic effects of natural tyrosine kinase inhibitor, genistein, on signaling pathways. *Proteomics* **10**: 976–986.

Yılmaz E, Çek Ş, Mazlum Y. 2009. The effects of combined phytoestrogen administration on growth performance, sex differentiation and body composition of sharptooth catfish *Clarias gariepinus* (Burchell, 1822). *Turk J Fish Aquat Sci* **9**: 33–37.

Zeleniuch-Jacquotte A, Adlercreutz H, Akhmedkhanov A, Toniolo P. 1998. Reliability of serum measurements of lignans and isoflavonoid phytoestrogens over a two-year period. *Cancer Epidemiol Biomarkers Prev* **7**: 885–889.

Zhu BT, Conney AH. 1998. Functional role of estrogen metabolism in target cells: review and perspectives. *Carcinogenesis* **19**(1): 1–27.

Zierau O, Hauswald S, Schwab P, Metz P, Vollmer G. 2004. Two major metabolites of 8-prenylnaringenin are estrogenic *in vitro*. *J Steroid Biochem Mol Biol* **92**: 107–110.

Zierau O, Morrissey C, Watson RW, Schwab P, Kolba S, Metz P, Vollmer G. 2003. Antiandrogenic activity of the phytoestrogens naringenin, 6-(1, 1-dimethylallyl)naringenin and 8-prenylnaringenin. *Planta Med* **69**: 856–858.

Zivadinovic D, Gametchu B, Watson CS. 2005. Membrane estrogen receptor-alpha levels in MCF-7 breast cancer cells predict cAMP and proliferation responses. *Breast Cancer Res* **7**: R101–R112.

Zuehlke S, Duennbier U, Heberer T. 2005. Determination of estrogenic steroids in surface water and wastewater by liquid chromatography-electrospray tandem mass spectrometry. *J Separation Sci* **28**(1): 52–58.

10

Airborne particles

Edward Derbyshire[1*], Claire J. Horwell[2], Timothy P. Jones[3] and Teresa D. Tetley[4]

[1]*Centre for Quaternary Research, Royal Holloway, University of London, Egham, Surrey, TW20 0EX*
[2]*Institute of Hazard, Risk and Resilience, Department of Earth Sciences, Durham University, Science Laboratories, South Road, Durham, DH1 3LE*
[3]*School of Earth and Ocean Sciences, Main Building, Cardiff University, Cardiff, Wales, CF10 3YE*
[4]*Section of Pharmacology and Toxicology, National Heart & Lung Institute, Imperial College London, London SW3 6LY*
**Corresponding author, email ed4gsl@sky.com*

10.1 Introduction

In many parts of the world, atmospheric dust palls can reach concentrations that are deleterious to human and animal health (Figure 10.1). They are variously made up of fine, crystalline and fibrous minerals, mineral aggregates, toxic trace and other chemical elements, organic matter, gases and pathogens.

Natural airborne particles consist of dusts released by weathering and erosion of rocks, unconsolidated sediments, dryland soils, biogenic fibres and residues from forest fires and ash ejected during volcanic eruptions. Sea spray produces aerosols containing particles that are blown inland; the particles are commonly of salt, but can also consist of radionuclides such as those released in discharges during the accident at the Sellafield nuclear power station in north-west England in 1957.

Substantial amounts of particulate aerosols from human activities began to be released, initially as a result of burning and clearing of forests for agriculture in Eurasia from about 8000 years ago, but the quantities increased especially after the Industrial Revolution in Britain in the eighteenth century; such aerosols include dusts from mining and quarrying and the combustion of fossil fuel for energy generation. Exposed agricultural soils are another important source, particularly at local and regional scales. Particulate matter released by biomass burning (for example, from forest clearance and agricultural practices), particularly in developing countries, continues to be important.

There are references to the presence of dust particles in the atmosphere in 3000-year-old Chinese records and in classical Greco-Roman documents in relation to the Mediterranean basin. Dust of Saharan origin over the eastern North Atlantic is mentioned in ships' logs from the eighteenth century and written records of Saharan dust falls in Western Europe have increased steadily since the mid-nineteenth century. The 'dust-bowl years' of the 1930s, which affected the prairies of Canada and the panhandle regions of Texas and Oklahoma, as well as adjacent areas of New Mexico and Colorado, greatly increased public awareness of the threat to health and led to considerable research into particulate aerosols. The dust-bowl conditions arose largely from a combination of several years of drought and intensive agricultural practices.

Prolonged inhalation of mineral particles can damage health and has no known health benefits. Potentially toxic natural particulate dust includes several species of crystalline silica and fibrous silicates, notably asbestos minerals (principally chrysotile, crocidolite (riebeckite), anthophyllite, tremolite, actinolite and amosite), erionite, alkaline salts and dusts containing toxic trace elements (e.g. volcanic ash particles, which hold transition metals and other toxic trace metal on their surfaces: Witham *et al.*, 2005). Potentially toxic particulate dust arising from anthropogenic activities includes quartz and other silicates from quarrying and mining, agricultural biomass burning and wild fires and higher-rank coal dust (giving rise to coal-worker's pneumoconiosis) from coal extraction and processing. Other major anthropogenic particle sources include the burning of fossil fuels, combustion of diesel oil and cement manufacture.

Many rocks are made up mainly of silicates and pure silica (predominantly quartz), but understanding the impact of

Pollutants, Human Health and the Environment: A Risk Based Approach, First Edition. Edited by Jane A. Plant, Nikolaos Voulvoulis and K. Vala Ragnarsdottir.

Figure 10.1 Before and during a Mongolian dryland dust pall over Beijing, March 2003. Visibility is ca. 50 km (left) and 1 km (right) (*Photos: Edward Derbyshire*)

prolonged inhalation of such dusts comes almost entirely from occupational medicine. It has been known for centuries that workers employed in industries such as mining, quarrying, sandblasting, silica milling and stone masonry are particularly exposed to fine, crystalline quartz dust and can develop inflammation and fibrosis of the lung (silicosis), which is one of the most studied occupational lung diseases. Crystalline silica is also classed as a human carcinogen (International Agency for Research on Cancer, 1997).

In contrast, the impact of high concentrations of naturally occurring silica-rich dust on human and animal health received little attention until recently. The earliest account, based on studies of a small population in the Sahara Desert, was published only half a century ago (Policard and Collet, 1952), although 'desert lung syndrome' (non-occupational silicosis with asthmatic symptoms including dyspnoea (breathlessness) and fatigue) has been known for over a century; evidence of it has been found in ancient Egyptian mummies (Tapp *et al.*, 1975). In the past half-century this condition has been reported from several dryland regions including Pakistan, California, Ladakh (north India), the Thar Desert of Rajasthan (north-west India) and northern China.

Large quantities of silica and silicates, together with a range of chemicals including potentially toxic trace elements, are released during some volcanic eruptions. Inhaled ash can exacerbate symptoms in people who are susceptible to asthma and respiratory disease. Such an increase in health problems followed the 1980 eruption of Mount St Helens in the United States, when total suspended particulates (TSPs) exceeded 1000 $\mu g/m^3$ for 7 days, compared with a mean ambient 80 $\mu g/m^3$ (Baxter *et al.*, 1983). Increased wheeze symptoms in children have also been reported from Montserrat, in the West Indies, following the Soufrière Hills volcanic eruptions (1995 to the present) (Forbes *et al.*, 2003). It is still not clear, however, if exposure to volcanic ash is associated with chronic diseases such as silicosis (Horwell and Baxter, 2006); given that approximately 9 per cent of the world's population live near a historically active volcano, this is an important question.

Asbestosis is a progressive, incurable chronic lung disease, which has been known for at least a century and is attributable to prolonged exposure to fibrous silicates known collectively as asbestos (see Section 10.5.1). Unfortunately, the important insulation and fireproof properties of asbestos saw its widespread use in many industries, including building construction, ship building and industrial refrigeration plants, despite the known link to serious lung disease. World production and use of asbestos did not decline until the mid-1970s (Wagner, 1997), but asbestos continues to be used quite extensively in the automotive and other transport industries in brake shoes and pads in cars, trains and other vehicles. Asbestiform particulate matter from rocks, sediments and soils may also be transported naturally by wind. Cases of non-occupational asbestosis in people exposed to agricultural soil dusts have been reported in several European countries, including Italy, Turkey and France (Corsica) (Baris *et al.*, 1987; Selçuk *et al.*, 1992; Viallat *et al.*, 1999; Magnani *et al.*, 1995) and in villagers in the Troodos region of Cyprus who had no involvement in the open-cast mining of asbestos on that island (McConnochie *et al.*, 1987). Pleural mesothelioma can occur after brief exposure to relatively low levels of amphibole asbestos such as crocidolite or tremolite. Prolonged exposure is not necessary (see Section 10.5).

Toxic trace elements are commonly found in, or adsorbed on to, dust particles. Their impact on lung tissue is poorly understood, but they are associated with inflammation of lung airways, giving rise to acute conditions such as asthma, bronchitis and rhinitis, as well as chronic conditions including fibrosis. Pathogens, including bacteria, fungi, pollen and spores, have been found in cavities in airborne mineral particles. They have been associated with a wide range of environmental impacts including Saharan dust palls acting as a carrier of the meningitis pathogen (*Neiseria meningitides*) (Molesworth *et al.*, 2003). Other bacterial pathogens found in Middle East dust palls (Lyles *et al.*, 2005) raise the question of the role of dust-transported pathogens in epidemics and pandemics. A particular case associated with Gulf War Syndrome and the health of military veterans involves the use of munitions containing depleted uranium (DU) during conflicts in the Middle East and the Balkans. The concern involves inhalation of DU aerosol particles and the toxicity of the uranium. Inhaled particles can enter the bloodstream by way of the lungs, ultimately affecting kidney function. Alternatives to DU, such as tungsten, may be just as toxic.

The annual release of fine silt and clay particles, commonly with alkali salts, by wind action on seasonally dry and former lake depressions (palaeo-lakes and artificially drained water bodies such as Owens Lake (USA) and the Aral Sea (Uzbek Republic)) also constitutes a health risk, causing acute irritation of the lungs (Gomez *et al.*, 1992).

There are adverse health impacts from the extraction and burning of coal. It has been known for decades that chronic exposure to coal dust can give rise to incurable fibrotic lung diseases. Coal-worker's pneumoconiosis is not a response to a specific mineral; the presence of quartz in the coal was thought to influence the toxicity of coal dust, but this is only true when very high concentrations of quartz are present, when silicosis develops (see Section 10.5.2; McCunney *et al.*, 2009). However, the presence of clay minerals may inhibit the toxicity of quartz in coal dust. Mineralised coal, including much of the coal in China (United States Geological Survey, 2000) may be enriched in toxic trace elements such as arsenic, fluorine, mercury, antimony and thallium as well as radioactive elements such as uranium and its decay products such as radium and radon. Burning such coal in unventilated stoves exposes families to toxic smoke (Finkelman *et al.*, 2003) and combustion in coal-fired power stations exposes both local and regional populations.

Emissions from industrial coal combustion have been greatly reduced in most rich countries, but emission levels in countries such as China and India remain high. Incomplete coal combustion releases inhalable, unburned carbon and other particles potentially containing toxic trace elements (TTEs). The health impacts can be severe, as in the London smog of November 1952 when a temperature inversion trapped fog and particles from burning coal for several days, leading to 4000 excess deaths. The incident led to the UK Clean Air Acts of 1956 and 1968, which legislated for smokeless fuels, relocation of power stations away from urban areas and construction of tall chimneys to disperse pollution.

Emissions from diesel road and rail vehicles release polluting gases and particles that consist mainly of carbonaceous soot. The clustered spherules are ultra-fine (mean diameter 10–80 nm) and, as about 95 per cent of dust exhaust particles are less than 1 μm in diameter, most remain in suspension and are readily inhaled. Exhaust particles make up about 40 per cent of the particulate matter finer than 10 μm (PM_{10}) aerodynamic diameter in the airborne dust fraction in cities such as Los Angeles (Diaz-Sanchez, 1997).

Biomass burning, as a common agricultural practice or as a result of lightning strikes, has a strong impact on temperate, savannah and some equatorial landscapes, to the extent that some ecosystems depend on periodic fires. In addition to gases, particles emitted include organic carbon and silica from the tissues of plant species such as grasses and sugar cane (Le Blond *et al.*, 2008).

The impact of high levels of pollutants in smoke from wild or bush fires, although sometimes intense, is generally short-lived (Morawska and Zhang, 2002), so that healthy people generally recover quickly from acute exposure. The effects can be more serious in people with pre-existing conditions such as pulmonary and cardiovascular diseases, as well as the old and the young.

Whole communities in central and eastern Asia are exposed to the adverse effects of natural airborne dusts at exposure levels encountered elsewhere only in some high-risk industries. The level of exposure to the dust hazard is considerable in China's semi-arid north and west, which includes a range of silt-rich dynamic terrain types, from large river flood plains, where silt is deposited during flash flooding events, to annually replenished sedimentary fans and a plateau covered by extensive loess (wind-lain mineral dust deposited in the past 2.6 million years) with a total area of about 0.75 million km^2. Farmers and herders are at particular risk and, in this huge region in which most domestic dwellings are built of *adobe* (a mixture of fine silty clay and water), indoor mineral dust can reach high levels, usually exacerbated by open-fire smoke from cooking hearths. Burning of low-grade coal, wood and animal dung inside such dwellings increases the impact of airborne particulate matter on human health over large tracts of China, northern India, Tibet and Africa.

The prime target of mineral-related disease is the lungs and lung-lining by inhalation of airborne dust, leading to the respiratory and related diseases that form the focus of this chapter. However, there are other exposure routes, such as absorption of ultra-fine volcanic clay minerals through the feet, leading to podoconiosis, an endemic disease in Africa, characterised by chronic lymphatic irritation, inflammation and collagenesis, leading to obstruction and lymphoedema of the lower limbs (Davies, 2003).

10.2 Hazardous properties

The effect on the human body from inhaling atmospheric particulate matter varies widely in response to many factors, including certain characteristics of the particulate matter and the nature of any pathogens attached to them, the dust concentration, the length of the exposure period and an individual's personal vulnerability.

The hazardous properties of airborne dusts vary with their geochemistry, mineralogy, degree of crystallinity, size, shape, density, solubility and reactivity with human fluids and tissue (Guthrie, 1997). Fine particles may remain suspended in the atmosphere for hours to weeks depending on their size, form and density. They may also be easily re-suspended by natural and human action. All particulate matter finer than 10 μm aerodynamic diameter is referred to as the respirable fraction or PM_{10} within which the fine fraction (<2.5 μm, or $PM_{2.5}$) and the ultrafine fraction (<100 nm) are of particular importance to pollution and public health.

10.2.1 *Crystalline silica*

Crystalline silica, as fine particles of quartz, cristobalite and tridymite, constitutes a major health hazard. Quartz is the commonest and by far the most studied polymorph, but both cristobalite and the rarer tridymite are regarded as being equally or more pathogenic. Crystalline silica is a highly fibrogenic agent in lung tissue, by a process that appears to arise from the reactivity of particle surfaces in contact with lung-lining fluid and cells. Reactivity of quartz surfaces is greatest when particle surfaces are freshly fractured (either by artificial or natural processes, Table 10.1). Non-crystalline (amorphous) silica, present in diatoms, phytoliths and testate amoebas, generally lacks surface radicals and thus is regarded as benign when inhaled. However, there is a minority view that the presence in wind-blown dust of impurities and particles such as soil-derived sponge spicules with aerodynamic diameters of between 5 and 3 μm (thus meeting the definition of hazardous mineral fibre: Skinner *et al.*, 1988) 'raises the possibility that they may present a respirable silica hazard in dust-affected areas' (Clarke, 2003). Potential pathogenicity of crystalline quartz is moderated or nullified by aging of particle surfaces, by the presence of adhering minerals or coatings and by the presence of small amounts of contaminant metals, notably aluminium and iron. The presence of aluminium, commonly in the form of clay minerals such as kaolinite, substantially moderates quartz toxicity.

Crystalline silica is odourless and a non-irritant when inhaled, with no noticeable symptoms, but protracted exposure over long periods (the dose) may lead to progressive thickening and scarring of lung tissue, although radiographically visible symptoms of silicosis may take many years to appear (see Section 10.5.2). Prolonged progression may result in death owing to cardiopulmonary failure. Cell response to silica varies widely, both among and within its polymorphs, such that silica cannot be regarded as a single, discrete hazardous material (Donaldson and Borm, 1998).

The processes by which particles induce pulmonary pathogenicity are complex and incompletely understood. Particles in the lungs must be cleared rapidly if the gas-exchange system is to work effectively. Problems arise when particles such as crystalline silica or asbestos are not cleared, primarily because of complex interaction with cellular components, or because the shape–size characteristics inhibit particle transport within the pulmonary airways. Free-radical release and the formation of reactive oxygen species (ROS) are thought to be primary processes inducing inflammation. In the case of crystalline silica, breaking of the bond between Si and O by homolytic and heterolytic cleavage generates dangling bonds (reactive surface radicals $Si^{\bullet}$ and $SiO^{\bullet}$) and surface charges (Si^{+} and SiO^{-}), respectively (Fubini *et al.*, 1995) (Figure 10.2). Particle-derived ROS from crystalline silica particles in contact with epithelial cells enhance both oxidative stress and inflammation (Fubini, 1998a, 1998b; Fubini and Hubbard, 2003; Figure 10.3).

Traces of iron on crystalline silica surfaces sustain production of free radicals in a series of catalytic stages (the Haber Weiss cycle) involving the Fenton reaction in the lungs (formulas 10.2–10.4).

Table 10.1 Characteristics of particulate matter determined by its origin (Modified after Fubini, 1998b)

Comminution of crystals e.g. grinding through industrial or natural (e.g. volcanic and glacial) processes	Combustion e.g. coal fly ash, diesel particulates	Biogenic e.g. diatoms, sponge spicules
Sharp edges and corners Irregular surface charges Surface radicals Hydrophilic Contamination by grinding processes and co-existence with other minerals	Spherical particles Smooth surface No surface radicals Hydrophobic Contamination by carbon and other component processes	Retention of organism shapes Indented, irregular surfaces High surface area, particularly internal surface area No surface radicals Very hydrophilic Sometimes contains alkaline, alkaline earth and iron ions from original material

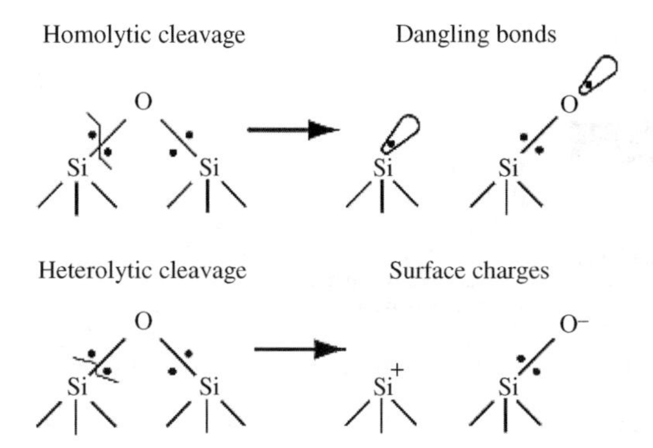

Figure 10.2 Diagram to show homolytic and heterolytic cleavage of silica tetrahedra. Silicon/oxygen radicals and surface charges are generated during cleavage of atomic bonds due to, for example, fragmentation of crystalline silica during quarrying or volcanic eruption (After Fubini *et al.*, 1995)

The hydroxyl radical ($HO^{\bullet}$) is produced through the Fenton reaction (10.4, below) (Fubini and Otero Arean, 1999):

$$Fe^{3+} + \text{reductant} \rightarrow Fe^{2+} + \text{oxidised reductant} \quad (10.1)$$

$$Fe^{2+} + O_2 \rightarrow Fe^{3+} + O_2^{-\bullet} \quad (10.2)$$

$$O_2^{-\bullet} + 2H^+ + e^- \rightarrow H_2O_2 \quad (10.3)$$

$$Fe^{2+} + H_2O_2 \rightarrow Fe^{3+} + OH^- + HO^{\bullet} \quad (10.4)$$

Trivalent Fe on crystalline silica surfaces in contact with H_2O_2 in the body is converted to divalent Fe to generate highly reactive hydroxyl radicals ($HO^{\bullet}$), leading to a cycle of increasing cell damage (Fubini and Otero Areán, 1999). Free-radical generation can cause mutations in DNA, leading to inflammation and cancer (Hardy and Aust, 1995). Particle-induced inflammation is central to a wide range of occupational diseases (Figure 10.4).

The progression from chronic inflammation to fibrosis and silicosis following long-term exposure to fine crystalline silica particles also has deleterious effects on the immune system. For example, the link between silicosis and tuberculosis was established decades ago (Snider, 1978), the risk to silicotic patients of developing tuberculosis being up to 20 times the level found in the general population (Westerholm, 1980). There is a reduction in the ability of the macrophages to inhibit the growth of tubercle bacilli responsible for tuberculosis. Some rheumatic, as well as chronic renal diseases also show higher than average incidence in individuals exposed to silica and it is likely that increased susceptibility to some mycobacterial diseases is due to impaired function of macrophages in silicotic lungs (Snider, 1978).

10.2.2 Asbestos

Asbestos and asbestiform fibres constitute a severe health hazard. Inhalation over a period of years can give rise to fibrosis (asbestosis) in industrial situations and also in some natural environments. Crocidolite is the most pathogenic form of asbestos. Together with chrysotile and amosite, it can induce pulmonary fibrosis, bronchogenic carcinoma, mesothelioma and a number of pleural diseases (Wagner *et al.*, 1960). In addition, amphibole asbestos and particularly crocidolite, causes pleural mesothelioma. The increased toxicity of amphibole asbestos over that of serpentine asbestos (chrysotile) is due in part to the high aspect ratios (i.e. length to width ratio greater than 3) and durability of the amphibole asbestos, where the fibres are long, thin and needle-like and are biopersistent (see Section 10.5.1). Asbestiform compounds have mineral aggregates with the distinctive features of amphibole asbestos, namely discrete, long,

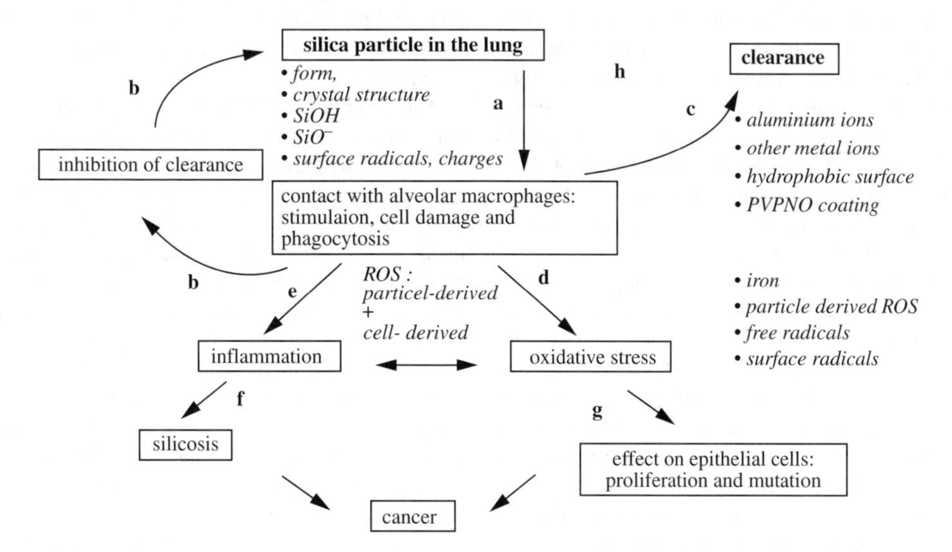

Figure 10.3 Sequence of events leading to quartz pathology, based on animal studies and cells *in vitro* (After Fubini, 1998a, modified from Donaldson and Borm, 1998)

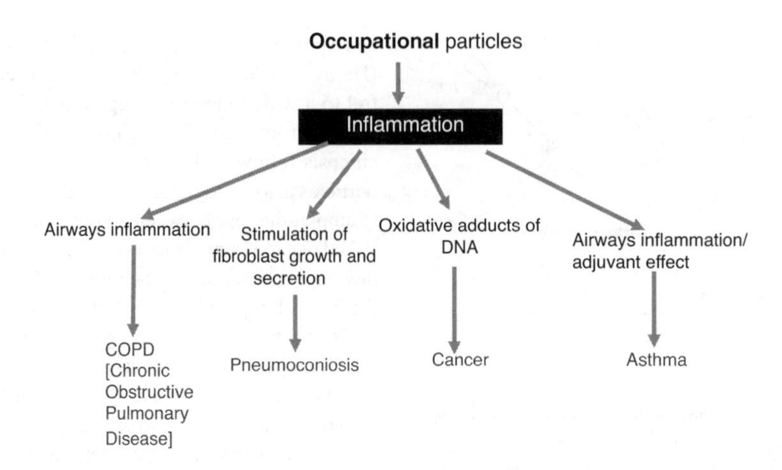

Figure 10.4 The central role of inflammation in a number of occupational diseases (Courtesy of Professor K. Donaldson)

thin, strong and flexible fibres in bunches or mats. Fibrogenesis arising from inhalation of asbestos fibres spreads out along elements of the lung structure, rather than as discrete nodules as occurs with silicosis. Some fibres penetrate tissue and remain in the lungs, lung lining and abdominal cavity. On deposition within the conducting and alveolar regions, their surfaces are modified by adsorption of macromolecules which may enhance cell damage due to generation of free radicals. To count as a 'fibre', the World Health Organisation states that an asbestos particle must have an aspect ratio greater than 3:1, but the most important single property for respirability is fibre diameter, which should be less than 3 μm (World Health Organisation, 1986). Once in the lung, fibre length, surface chemistry, solubility and other physical and chemical properties control biological activity.

Response to grinding varies with species. For example, prolonged grinding deactivates chrysotile but in other asbestos fibres grinding may raise reactivity, releasing the reactive oxygen species that enhance oxidative stress, perhaps because of exposure to small amounts of ferrous iron (Neijari *et al.*, 1993; Fubini, 1997). Oxidative stress induced by asbestos fibres increases with the rate of free-radical production (Ghio *et al.*, 1998). Pathogenicity may be affected by adsorption of metal ions on mineral dusts. For example, the presence of iron on asbestos fibres is thought to influence the catalysing release of free radicals (Hardy and Aust, 1995).

Although not a member of the asbestos group, the fibrous zeolite mineral erionite is regarded as a serious environmental health threat, being between 200 and 1000 times more carcinogenic than asbestos minerals (Hill *et al.*, 1990; Carbone *et al.*, 2002). In the Cappadocia region of Turkey, malignant mesothelioma accounts for more than 50 per cent of deaths in some villages (Dogan *et al.*, 2008), caused by exposure to natural wind-borne erionite-rich particles. Incidence rates can be 10 000 times greater than in the general population, though it has been claimed that epidemiological evidence points to genetic susceptibility as an additional factor (Roushdy-Hammady *et al.*, 2001).

10.2.3 Toxic trace elements

Another group of particles that occur as airborne dust comprises toxic trace elements (TTEs) (see also Chapter 4). These include the well-studied elements mercury, lead, arsenic, cadmium and iron and also fluoride, radioactive elements and other trace metals such as copper and zinc. They may occur adsorbed on airborne mineral dust and as integral constituents of such particles. Most knowledge of the adverse health effects of inhaled and ingested TTEs comes from occupational medicine. Toxic trace elements can cause both upper and lower airway injury and sensitivity (Cook *et al.*, 2005). Arsenic and mercury can cause rhinitis and sinusitis, while mercury and zinc and many other TTEs can cause tracheitis, bronchitis and asthma. Inhalation of elements such as cadmium causes inflammation, oedema and fibrosis of the parenchyma. Iron can act catalytically in free-radical generation, leading to greater oxidative stress and epithelial cell damage. Trace amounts of iron help to generate ROS, leading to DNA damage, cell transformation and pulmonary reaction.

The potential of TTEs to cause damage is related to their solubility: high solubility reflects dissociation and dispersal deep into lung tissues and a degree of insolubility is associated with deposition in the airways (Newman, 1996). A recent study of selected TTEs in a dryland dust pall over Beijing, using an *in vitro* plasmid assay, showed that dust-particle bioreactivity derives mainly from the water-soluble fraction (Shao *et al.*, 2006). Total water-soluble concentrations of aluminium, vanadium, chromium, manganese, iron, cobalt, nickel, copper, zinc, arsenic and lead were found to be higher in the more strongly bio-reactive samples and water-soluble zinc was the element responsible for most plasmid DNA damage. However, the chemical processes by which metals cause pathophysiological responses, especially as compounds, remain poorly understood (Cook *et al.*, 2005).

As mentioned above, iron can act catalytically in free-radical generation, leading to greater oxidative stress and epithelial cell damage. Trace amounts of iron help to generate ROS, leading to

DNA damage, cell transformation and pulmonary reaction. However, bioreactivity varies with the type of iron oxide and other factors. Toxic reaction of lung tissue to dust particles is also proportional to the total water-soluble concentration of trace elements in the dust, a factor that affects oxidative capacity.

As with other natural particles, the potential toxicity of volcanic particles depends not only on mineral composition, grain size and surface area but also on the presence of TTEs and complex factors relating to the interaction of the TTEs, particle surface, lung tissue and cells (i.e. surface reactivity). It is known that basaltic ash produces particularly high numbers of hydroxyl free radicals through interaction of reduced iron on the surface of ash particles with hydrogen peroxide (Horwell *et al.*, 2007). Fluoride is injected into the atmosphere by volcanic activity and other natural processes, as well as from anthropogenic inputs. Airborne fluoride is toxic if high concentrations are inhaled or ingested. It is rapidly absorbed following intake and has a high affinity for calcified tissue, including bones and teeth. Indoor combustion of fluoride-rich coal can lead to high concentrations of particles and fluoride, as is seen in some Chinese rural communities. High concentrations of urinary fluoride have been noted in fluoride-rich coal-burning regions and, in some Chinese villages, all elementary and junior high-school students between 10 and 12 years old were found to have dental fluorosis (Ando *et al.*, 1998).

Manganese (Mn) is an essential trace metal, but excessive exposure is associated with neurotoxicity, notably in mining, the ferroalloy and battery industries and welding (Lucchini *et al.*, 1999). Released manganese-rich dust has been implicated in manganese-induced Parkinsonism (Koller *et al.*, 2004), inhaled Mn dissolving in body fluids to be deposited in the striatum of the brain (involving damage to the basal ganglia structures). Common symptoms are weakness, tremor, gait abnormality and slowness of speech. Manganese-induced Parkinsonism can be differentiated from other forms of Parkinsonism using clinical and imaging techniques (Cersosimo and Koller, 2006).

10.2.4 Diesel fuels

The emissions from the burning of liquid fossil fuels have been causally linked with a large number of adverse health effects. The principal path into the body is respiratory and, once inside the lung, particles can translocate to numerous and diverse locations within that organ. The particles possess the intrinsic hazardous property of being harmful as well as acting as carriers for bioreactive compounds that have condensed on their surfaces, or metals that have been trapped within the carbonaceous structure of the spheres. Exposure to these particles ranges from short-term, high-intensity exposure (usually occupational), which can result in irritation to the eyes and the respiratory tract, to longer-term exposure resulting in compromised lungs, coughing and breathlessness. There is good evidence to support the view that long-term exposure increases the risk of lung cancer, COPD, heart attacks, strokes and numerous other conditions (Neuberger *et al.*, 2007). Human exposure investigations have shown that roadside exposure to traffic fumes results in decreased lung function in asthmatic children (Delfino *et al.*, 2009).

10.2.5 Biomass and wild fires

Air pollution arising from natural fires (or bushfires) and the seasonal burning of vegetation can have a severe impact on large populations locally and sometimes at long distances. Combustion of vegetation releases solids, hydrocarbons, organic compounds and gases, but particulate matter and polycyclic aromatic hydrocarbons (also called polynuclear aromatic hydrocarbons) are of particular concern with respect to human health (World Health Organisation, 1999), especially as amorphous species may be converted into crystalline minerals during combustion. A majority of discrete particles are ultra-fine, with most of the particle mass in the less than 2.5 μm range. The smaller, fine particulate matter consists of up to 70 per cent organic carbon (Morawska and Zhang, 2002).

10.2.6 Micro-organisms

The impact on lung tissue of airborne mineral dust may be exacerbated by the presence within fine dust particles of bacteria, fungi and other micro-organisms (Kellogg *et al.*, 2004). The global extent of the fine-dust transport system has been implicated in damage to plants and animals, including impairment of human health. Although susceptible to destruction by ultraviolet radiation, a proportion of any included micro-organisms may survive in cavities and cracks within suspended dust particles. Dryland dust can act as a carrier for the meningitis pathogen. A study in Mali (Kellogg *et al.*, 2004) found that, of 95 dust-borne bacteria identified, 25 per cent were opportunistic pathogens. No outbreaks of disease attributable to fungi in dust palls have been reported, but the prevalence of fungi in dryland soils and dust storms indicates diverse communities including pathogenic genera and species. More than 40 fungal colony-forming units or spores have been found within dust palls (Griffin, 2007); most of the genera known to contain pathogenic species are derived from the Sahara, Sahel and Middle East. A wide range of human diseases is caused by pathogenic or opportunistic pathogens and some fungi listed include mild to potent allergens. Short-range transmission of infectious human viruses is well documented, but little is known about long-range transmission within dust storms (Griffin, 2007).

10.3 Sources

10.3.1 Natural sources

10.3.1.1 Crustal dust

World dust emissions from arid and semi-arid terrains can reach as much as 5 billion tonnes per year (Schultz, 1980). Sources of natural dust occur on all continents, but the most extensive

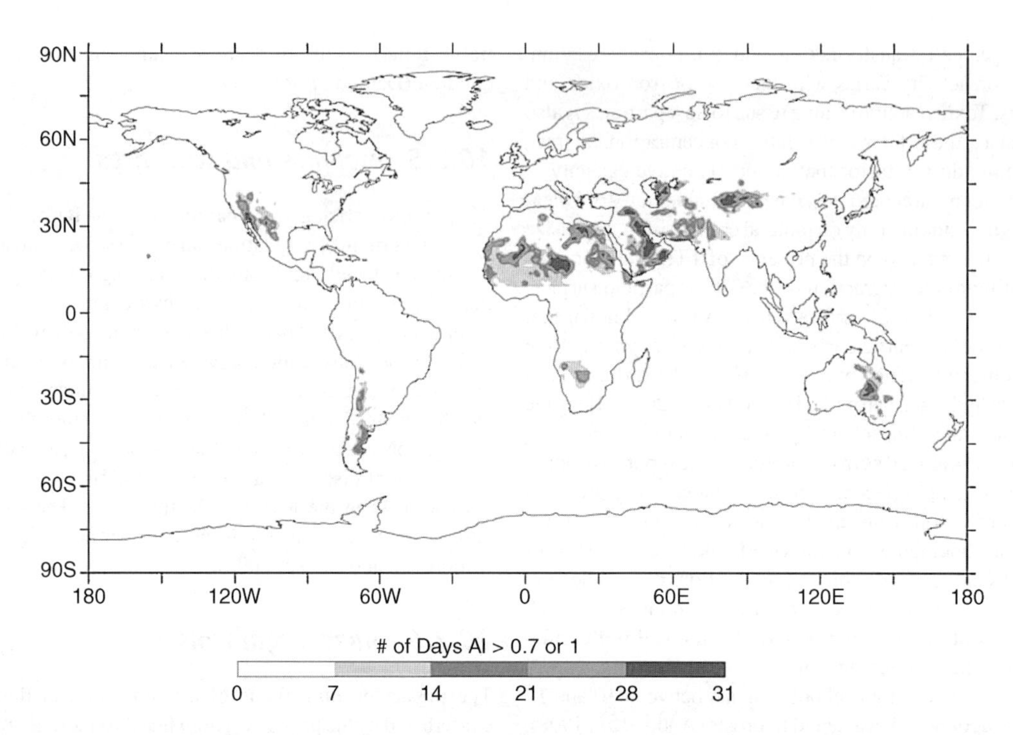

Figure 10.5 Total ozone mapping spectrometer (TOMS) indicating main dust source regions. UV spectral contrast is used as an absorbing aerosol index (AAI). An AAI threshold of 1.0 is used in the 'global dust belt' (West Africa to China) and 0.7 elsewhere (After Prospero *et al.*, 2002, with kind permission)

source region stretches from the western Sahara, through the Middle East and central Asia to the Yellow Sea (Figure 10.5). The distribution of palaeo-lake beds closely matches the major source areas of atmospheric dust (Ginoux *et al.*, 2001; Tegen, 2003: Figure 10.6). The lake deposits in the Bodelé depression (formerly mega-lake Chad, North Africa) include abundant broken particles of diatomite (siliceous shells of a eukaryotic algae: Figure 10.7); and fluxes from this source can

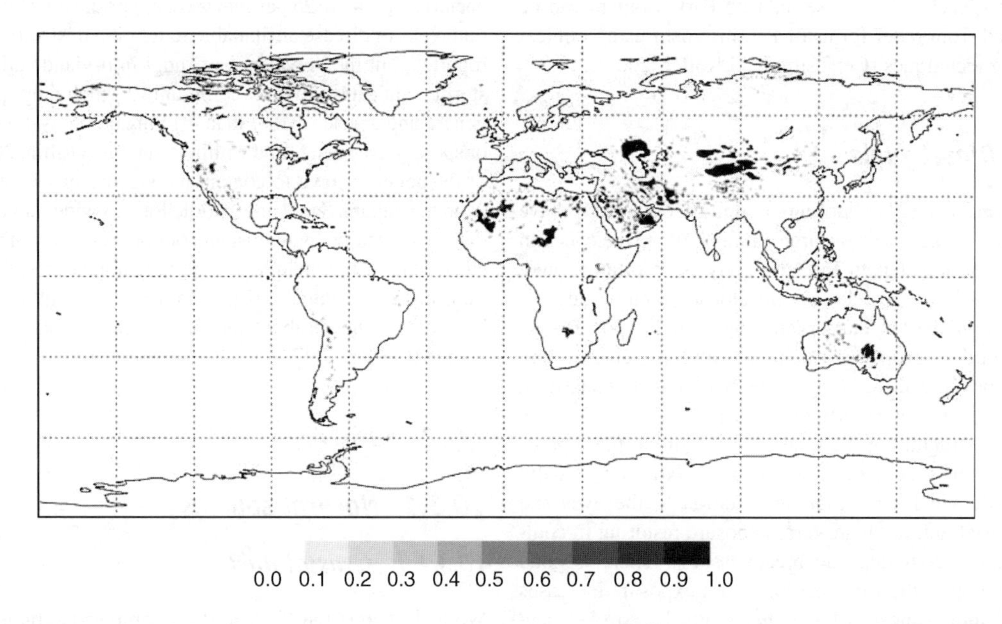

Figure 10.6 Areal coverage of preferential dust sources, calculated from the extent of potential lake areas, excluding actual lakes (After Tegen, 2003)

Figure 10.7 Fragment of a diatom abstracted from a Saharan dust fall on Gran Canaria (Canary Islands) on 4 March 2004 (*Photo Edward Derbyshire*)

reach 1.18 ± 0.45 Tg per day, with an estimated contribution to global dust emissions of between 6 and 18 per cent (Todd *et al.*, 2007).

The massive deposits (>300 m thick) of wind-lain mineral dust known as loess in northern China are impressive evidence of sustained, semi-continuous deposition of the coarser airborne dust fractions (silt and sandy silt) for at least the past 2.6 million years. The loess formation of north-west China, made up of alternating silt (loess proper) and clayey silt palaeosols (fossil soils), marking, respectively, substantial dust depositional rates and reduced dust depositional rates with soil profile development, is supplied by sources lying to the north and west (see Figures 10.12 and 10.13), a natural process that has continued at varying intensity for at least the past 7 million years (Ding *et al.*, 2001). Huge loess units such as the Loess Plateau of north-west China are subject to natural erosion by streams and rivers and mass movements (landslides) as well as to anthropogenic actions such as deforestation, excessive irrigation, intensive agricultural practices and overgrazing. When the loess surface is breached by such actions, it is highly susceptible to wind erosion, making such degraded loess a notable but secondary source of present-day wind-blown dust.

The toxicity of natural dust is greatly influenced by its composition, which varies according to its provenance (Krueger *et al.*, 2004). The percentage of quartz in airborne dust samples from North Africa and China varies from less than 20 per cent to more than 60 per cent, clay minerals from less than 10 per cent to more than 40 per cent and carbonates up to 40 per cent.

There are relatively few studies of natural particulate matter in major dust source regions using systematic sampling and analysis. One study obtained a year's airborne samples from 15 sites across the Middle East: Djibouti, Afghanistan, Qatar, United Arab Emirates, Iraq and Kuwait (Engelbrecht *et al.*, 2009). All of the sites exceeded the WHO guidelines for maximum ambient particulate matter exposure (20 $\mu g/m^3$ for PM_{10} and 10 $\mu g/m^3$ for $PM_{2.5}$) and all contained mixtures of silicate minerals, carbonates, oxides, sulphates and salts. Dust events were found to yield short-term maxima of the major soil-forming elements silicon, aluminium, calcium and manganese, as well as magnesium, potassium, titanium, vanadium, iron, ruthenium, strontium, zirconium and barium. All quartz and other silicate mineral particles were thinly coated with a silicon-aluminium-magnesium layer made up of clay minerals. Despite the high dust concentrations, none of the quartz particles had fractured surfaces; they were partly rounded and coated with clay minerals and iron oxides suggesting that the dust is of low toxicity. No asbestos fibres were found.

10.3.1.2 Asbestos

Sources of asbestos-rich dust occur in both the natural and built environments. Asbestos occurs in metamorphic rocks in orogenic belts around the world. Chrysotile (white asbestos) occurs in ultramafic rocks that have been altered to serpentinite. The other main types of asbestos are the amphibole minerals crocidolite (blue asbestos) found in feldspathoid rocks and amosite (brown asbestos; fibrous anthophyllite) found in metamorphosed basic and ultrabasic igneous rocks. The Jurassic to Cretaceous metamorphic rocks of north-east Corsica, including outcrops of *schistes lustrés,* are rich in amphiboles, serpentinite and chrysotile. There is no history of occupational contact with asbestos, but it is reported that 94.6 per cent of villagers born in north-east Corsica have bilateral plaques in the lung, compared with only 5.4 per cent of subjects in villages in the north-west. The disease has been linked to the presence of abundant airborne chrysotile fibres (Boutin *et al.*, 1986). Erionite, a fibrous zeolite, occurs in volcanic tuffs and agricultural soils in the Cappadocia region, central Turkey. Weathered rock and fine soil particles make up dust palls that affect the health of village communities in dry periods, a situation exacerbated by the use of erionite-bearing rocks and soils as building stone and as a whitewash on buildings.

10.3.1.3 Volcanic ash

Substantial volumes of silica and silicate minerals may be emitted during volcanic eruptions, depending on the type of magma erupted. Basic (e.g. basaltic) eruptions produce silica-poor (<52 wt per cent) ash, which contains mafic minerals such as calcium-rich feldspar, pyroxenes and olivine, whereas acidic (e.g. rhyolitic) eruptions produce silica-rich (>69 wt per cent) ash with high concentrations of felsic minerals such as quartz, potassium-rich feldspar and silica glass. Basaltic volcanoes are usually found above oceanic hot spots such as the Hawaiian island chain and where Earth's crust is thin. Volcanoes erupting more evolved magma, with more silica (such as andesitic, dacitic and rhyolitic volcanoes) are generally found around Earth's subducting plate margins, e.g. the Pacific Ring of Fire which has given rise to the volcanic chains stretching from Chile, Mexico, around the western US coast, eastern Russia, Japan, Philippines, Indonesia and south to New Zealand.

The amount of crystalline silica generated varies substantially depending on the type of volcano, with basaltic volcanoes producing no free silica. The andesitic (~57–63 wt per cent silica) Soufrière Hills volcano, Montserrat, West Indies (part of the Lesser Antilles island arc) generates abundant crystalline silica in the form of cristobalite, formed by vapour-phase crystallisation and devitrification of volcanic glass within its volcanic dome. Superheated, silicon-rich vapours escaping from the volcano pass through pore networks and cracks in the lava, depositing cristobalite as a metastable mineral in voids. At the Soufrière Hills volcano, the ash produced during collapse of the dome contains in the region of 15 wt per cent cristobalite (Horwell *et al.*, 2010).

The quantity of respirable material (<4 μm aerodynamic diameter) varies greatly among different types of volcanic eruptions, related to magma composition, with increasing explosivity correlating with increasing silica content and viscosity. Basaltic eruptions, which are generally effusive, produce coarse particles with <5 vol. per cent respirable material. Andesitic, dacitic or rhyolitic eruptions, which are usually more explosive, produce finer particles. The eruption of Vesuvius, Italy, in AD 79 produced ash with around 17 vol. per cent respirable material. Dome-forming eruptions such as the Soufrière Hills volcano, Montserrat, can generate ash with about 10–15 vol. per cent respirable material during dome collapse (Horwell, 2007).

The crystalline silica content of volcanic ash is the main cause of concern in terms of health hazard but, in recent years, other potential hazards have been identified. For example, volcanic ash can generate substantial quantities of hydroxyl radicals through the Fenton reaction. Iron-rich basaltic ash has greater reactivity than iron-poor ash (Horwell *et al.*, 2007).

10.3.2 Anthropogenic sources

Human activities that emit fine dust include agriculture, mining and quarrying and combustion of fossil fuels for energy generation and transport. The expansion of agriculture over the past 200 years, especially in the world's sub-humid and semi-arid marginal lands, has involved ploughing fragile soils with heavy machinery, leading to the re-suspension of loess and fine soil particles. The problem is exacerbated by the use of fertilisers and pesticides and the large-scale diversion of water for irrigation and domestic use from rivers and lakes, which impacts on the toxicity of the dusts and enhances erosion.

Fertilisers and pesticides contaminate groundwater and some are sorbed on soil particles that are entrained by strong winds and transported for long distances. The addition of trace elements, such as arsenic, cadmium, chromium, copper, molybdenum, nickel, lead, uranium, vanadium and zinc in fertilisers such as phosphates further increases the toxicity of soil-derived dust palls. The Aral Sea in Kazakhstan has been progressively diminished since 1960 by diverting feeder streams and canals for irrigation, with reduction of the water surface area by 80 per cent. The input of fertilisers and pesticides across the Aral basin has made the periodic dust palls more toxic. Morbidity and mortality rates have increased greatly across the region. Over 50 per cent of illnesses reported in children were found to be of respiratory type (O'Hara *et al.*, 2000; Wiggs *et al.*, 2003; Figure 10.8).

The extractive industries, which provide minerals for the metallurgical, energy, aggregate, cement and brick-making industries, generate dusts during crushing and grinding. Dust is

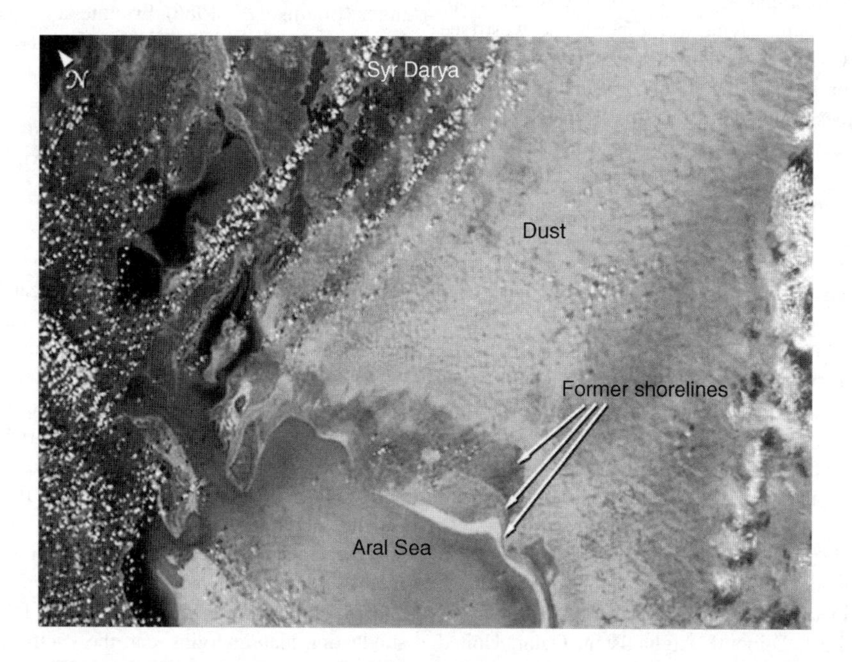

Figure 10.8 The diminishing Aral Sea on 30 June 2001, showing a pall of toxic dust, driven by a westerly wind (Reproduced by kind permission of Earth Sciences and Image Analysis Laboratory, Johnson Space Centre, USA)

also deflated from waste piles (tailings) if they are not kept moist. Such dust palls may represent a high risk to communities nearby and sometimes some distance away if TTEs are present in the tailings dust.

Public awareness of such hazards has increased recently. Following closure of a vermiculite mine in Libby, Montana, USA in 1990, it was found that the vermiculite was contaminated with tremolite-actinolite asbestos and radiographic pleural and interstitial abnormalities were found to be present in 51 per cent of former mine workers, increasing with age and exposure. Abnormalities in non-miner groups (3.8%) were also higher than elsewhere (United States Environmental Protection Agency, 2003).

Coal and oil combustion is also a major source of particulate matter, coal fly ash being widespread in industrial, many urban and some natural environments (Jones *et al.*, 2009). Fly ash can contain a component of unburnt organic matter but, if this exceeds certain levels, it is common practice to re-inject this material into the furnace for secondary combustion. The smaller-sized fly-ash particles, typically around a few microns in diameter, are usually spherical and often display gas-expulsion holes. Larger particles tend to have a more blocky appearance and often have numerous smaller glass spheres adhering to the surface. In modern coal-burning factories or power stations, the vast majority of fly-ash particles are collected post-furnace by electrostatic collectors or in bag rooms. The fly ash is then usually dumped on nearby ash piles; thus, the potential exists for this fly ash to be blown into the atmosphere (Brown *et al.*, 2009). The chemistry of the fly ash is controlled by the original chemistry of the coal (mineral impurities) and the operating conditions at the furnace. Although the coal minerals are completely melted into glass, after solidification there is often a small amount of recrystallisation producing quartz, mullite and haematite. The fly ash itself is not particularly reactive (Wlodarczyk *et al.*, 2008), most having limited pozzalinic properties; usually it has to be mixed with Portland cement to make any sort of building material. However, it does contain a range of TTEs that can be leached out in certain chemical environments. These include arsenic, mercury, lead, barium, cadmium, chromium, copper, fluorine, nickel, selenium, zinc and uranium and its decay products.

The human-health implications of fly ash produced by domestic coal burning are currently a subject of concern. This is especially so in poorer countries where the coal is burnt in poorly ventilated rooms in order to retain as much heat as possible and, consequently, the fumes are also retained. In northern China, for example, the domestic burning of local Permian coals appears to have resulted in clusters of lung cancer (Shao *et al.*, 2009). The relatively few studies of coal fly ash toxicity have yet to provide convincing evidence of human lung inflammation and there is a continuing debate about the relative importance of this material versus its TTE components. However, the absence of fibrosis in subjects exposed to coal fly ash indicates that the glass spheres are less bio-reactive than equivalent doses of crystalline silica (Borm, 1997).

There is compelling evidence that residual oil fly ash (ROFA) may be more harmful to human health than coal fly ash. ROFA is mostly less than 2.5 μm in diameter, chemically complex, largely inorganic and rich in metals, especially vanadium. Under the electron microscope, ROFA is seen to consist mostly of clusters of tiny carbon spheres; it is, therefore, soot produced by the burning of the oil. Lung inflammation, eye and throat irritation, cough, dyspnoea, rhinitis and bronchitis, have all been found in workers (notably in power-generating plants) exposed to high ash concentrations. Given the similarity of the symptoms associated with ROFA and vanadium exposure, transition metals, especially vanadium in ROFA, may be involved in Fenton-like chemical reactions in the lungs. Exposure to vanadium in the mining, steel and chemical industries is mainly by inhalation, however, suggesting the presence of ROFA particles in ambient air (Ghio *et al.*, 2002).

Petrol and diesel-powered vehicles are an important source of gaseous and particulate atmospheric pollution. This is a field that has constantly changing trends, resulting in constantly changing particulate emissions. For petrol engines, the most important change in recent years has been the decrease in atmospheric lead levels following the ban on the use of leaded petrol in most developed countries (Cook and Gale, 2005). For diesel engines, there are a number of important trends that include significant increases in domestic diesel cars, as opposed to petrol cars, the development of the so-called new-generation diesel engines with sophisticated fuel-injection systems and tailpipe after-treatment devices, the promotion of low-sulphur diesel and the introduction of biofuels. The nature of the particles generated by diesel combustion is therefore a function of the type of fuel, the technology employed in the engine and exhaust systems, the running condition of the engine and ambient atmospheric conditions (Maricq, 2007). An overloaded truck straining to get up a steep hill will produce different particles from those released by the same truck driving along a flat road under normal operating conditions.

All types of liquid fossil fuels produce combustion particles (soot). These particles consist of nanometre-scale carbon spheres that can exist individually but, more commonly, rapidly aggregate into larger chains or clusters. Confusion exists as to whether consideration of the behaviour of these particles should take account of the size of the individual spheres (nanoparticles) or the whole clusters that can be PM_{10} or more in diameter. A pragmatic approach is to classify the particles as 'nanostructured', which means that the size of the cluster is considered. However, it is understood that these particles can, both chemically and biologically, behave like the constituent nanoparticles. The overall size of the particles is significant, as it has implications for collection methods as well as health issues such as the ability of the particles to penetrate deeply into the lung. The individual spheres have an onion-like structure of perturbed (turbostratic) small graphitic structures, which can be imaged by high-resolution techniques such as transmission electron microscopy. The spaces between the graphitic molecules in the primary carbon spheres can trap metal sourced from the fuel, engine and exhaust systems. This includes platinum-group metals from the catalytic converters. The surfaces of the carbon spheres act as substrates for the condensation of a large variety of organic species, heavier-end hydrocarbons and inorganic species such as sulphates (Clague *et al.*, 1999). In addition to the carbon-sphere-based particles, other particles are generated by the direct condensation of both organic and inorganic species as the hot emission fumes cool to ambient air temperatures.

10.4 Global pathways

The detachment of mineral dust from the ground surface ('deflation') and its transport by the wind are functions of several variables, including the wind speed (notably the critical wind speed, or threshold velocity, required to dislodge particles), the degree of instability of the atmosphere, the size of the particles, the roughness and moisture content of the land surface and the degree of particle exposure. A recent classification of dust concentrations in relation to wind velocities in north-east Asia claims that sustained wind velocities of 10 m/s or more can generate total suspended solid loads in excess of 6000 μg/m^3 with PM_{10} fractions as high as 5000 μg/m^3 (Song *et al.*, 2007).

Silt-sized particles in the 10–50 μm range are readily picked up by the wind from dry, unvegetated surfaces, but, because of the high inter-particle cohesive forces typical of very fine (colloidal) materials, entrainment of clay particles (<2 μm) usually occurs as cohering silt-size aggregates or as attachments to silt-size grains. Critical wind threshold velocities vary markedly, those for the semi-arid and sub-humid silt-covered terrains of northern China being approximately twice those required to initiate dust storms in the Sahara (Wang *et al.*, 2000). It is important to discriminate between source-proximal and source-distal dust plumes (Pye, 1987; Figure 10.9). The mean size of entrained particles diminishes with transport distance because of fallout of larger and denser particles. Thus, the proportion of a dust plume consisting of the respirable fractions increases progressively with distance from the source, although the absolute mass of the respirable fraction is greatest close to source. Median diameters less than 10 μm characterise dust particles transported over long distances. The finest fractions (<1 μm) settle more slowly under gravitational force when in moving air streams, often staying at high altitudes for weeks. This is true of very fine-grained particulate pollutants, such as asbestos, which has been traced thousands of km from source.

Clearance of lower atmospheric dust by deposition occurs mainly as dry deposition, in which high atmospheric pressure is dominant, but it can also occur as wet deposition when dust particles are drawn down by raindrops in the unstable conditions associated with vigorous cells of low atmospheric pressure (depressions). Wet dust deposition is known regionally as 'loess rain' in eastern China and 'blood rain' in Mediterranean Europe.

The natural processes by which mineral dust is injected into the atmosphere are usually episodic (e.g. as in volcanic eruptions), often strongly seasonal and, in the case of dust deflated from the land surface, located mainly in sub-tropical arid and semi-arid regions. Such major dust sources cover about 30 per cent of the total land area of the Earth. The finest particulates are carried long distances along the following four main pathways (Tanaka *et al.*, 2005):

- the monsoon and westerly systems across Asia and trans-Pacific;
- the sub-tropical easterly system (Trade Winds: Figure 10.10), modified by seasonal westerlies in North Africa and trans-Atlantic;
- the westerlies from the Middle East to East Asia (Figure 10.11);
- the westerly system across North America, South America and Australasia.

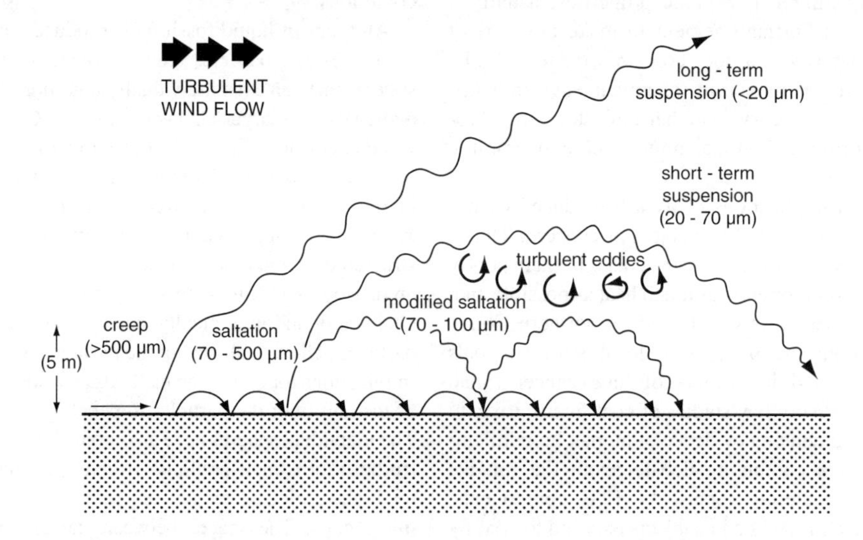

Figure 10.9 Modes of particle transport by the wind. The indicated particle-size ranges in different transport modes are those typically found during moderate wind storms [$\varepsilon = 10^4$–10^5 cm^2/s^1] (After Pye, 1987)

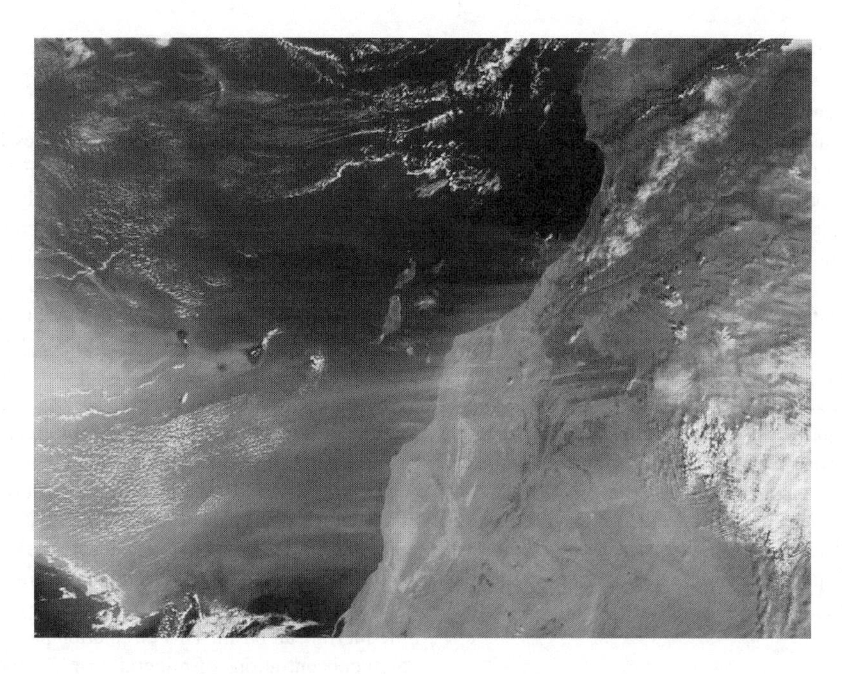

Figure 10.10 Easterly pall of Saharan dust over north-west Africa (southern Morocco, western Algeria, Western Sahara and Mauritania), covering the Canary island archipelago and extending into the eastern Atlantic and beyond in the northern winter (January, 2002) (Reproduced by kind permission of NASA)

In the planet's two most abundant dust-source regions (North Africa and Central and East Asia), dust pathways and deposition are influenced by distinctive and seasonally variable meteorological factors, including the subtropical trade winds, the monsoons and the westerlies. It should also be noted that aeolian dust is a vital source of ocean ecosystem nutrition. The dominant input of iron to oceanic life, for example, is supplied by wind-borne dust from the world's drylands

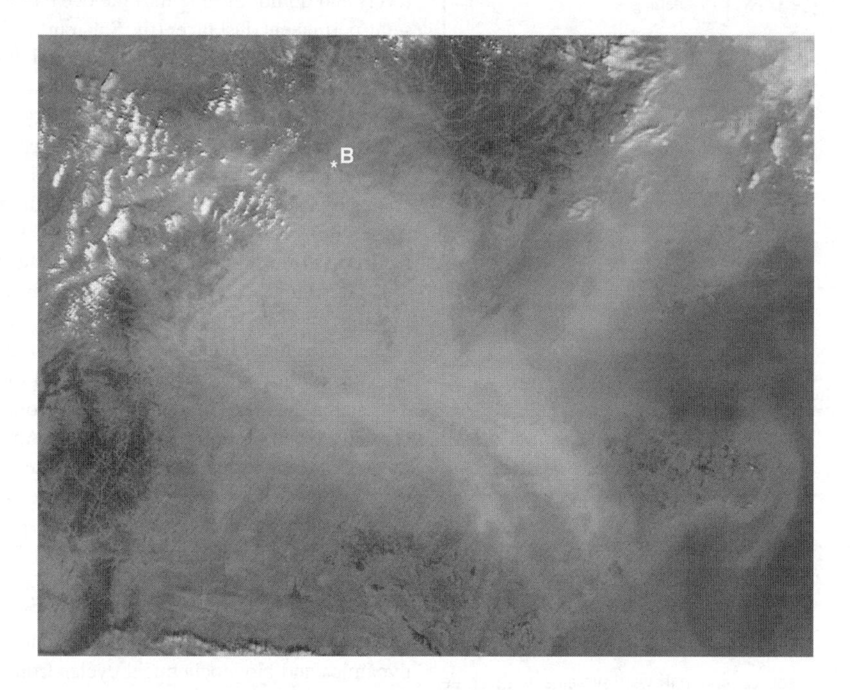

Figure 10.11 Dense outbreak of dust from the drylands of Mongolia and northern China in the northern winter–spring transition period (March 2009). Westerly airflow carries dust eastwards over the north China plain (*B = Beijing) including the Shandong Peninsula, the Bohai Gulf and Yellow Sea and the Pacific Ocean (Reproduced by kind permission of NASA)

(Jickells *et al.*, 2005). Both African and Asian dust pathways are global in distribution.

Dust transport in central and eastern Asia is influenced by both westerly and Asian monsoon climatic systems. An extensive high-pressure system (the Siberian High) dominates the cold, dense but stable atmospheric conditions in northern Eurasian winters; but between late winter and early summer, the Siberian High progressively diminishes in size and stability as westerly-driven depressions and associated low-pressure troughs track along its receding southern margins. Mineral dust in the lower troposphere, driven by westerly frontal systems, is sourced from mid-latitude drylands in Mongolia and western China. Two major source-pathway systems (Mongolia to North China and Taklamakan Desert to the northern margins of Tibet) drawing dust palls in the lower troposphere to Beijing, Nanjing and beyond have recently been determined using Nd-Sr isotopic composition (Chen *et al.*, 2007; Li *et al.*, 2009; Figure 10.12).

The dust deposition sequence in East Asia begins on the north and north-west China plateau regions, with the coarsest fractions falling on the loess lands of Gansu, Shaanxi and Shanxi provinces. Progressively finer dust fractions are transported eastward to the densely populated north China plain, Korea and Japan. The city of Xi'An on the southern margin of the Chinese Loess Plateau in Shaanxi province receives dust palls from seven different pathways (Wang *et al.*, 2006; Figure 10.13).

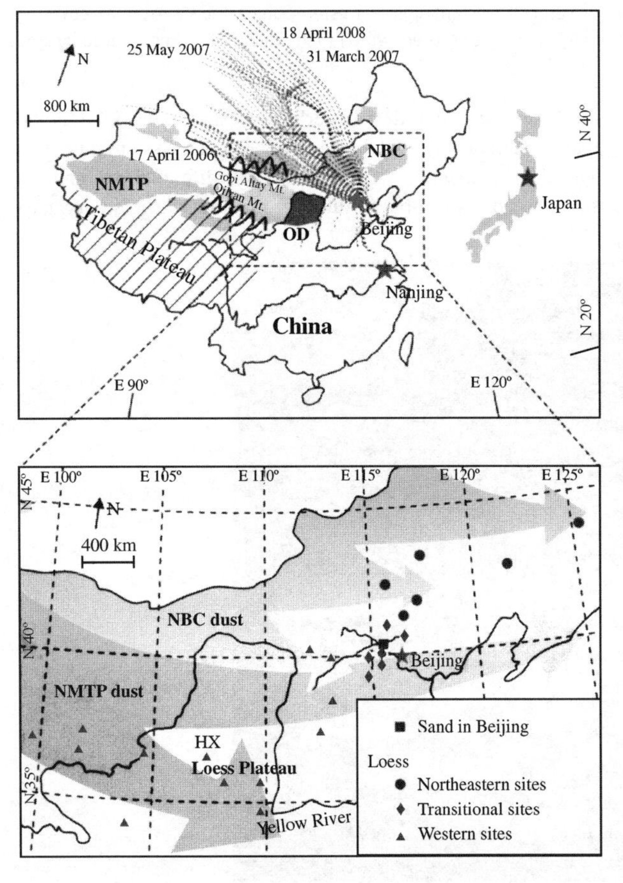

Figure 10.12 Two major source and pathway systems with three distinctive Nd-Sr isotopic signatures across northern and western China. Thirty six hour back trajectories from Beijing were calculated using the HYSPLIT method (Draxler and Rolph, 2003). After Li *et al.* (2009)

Lower-troposphere dust palls usually travel considerable distances beyond China, crossing the Pacific to deposit in North America (McKendry *et al.*, 2007). Both lower and upper troposphere dusts from East Asia have been found in the Greenland ice sheet (Bory *et al.*, 2003) and the French Alps (Grousset *et al.*, 2003). Recently, an upper troposphere dust stream from the Tarim Basin was traced along a $\sim 450°$ meridian circuit crossing the eastern Pacific at the 120°E meridian twice, with a dust transport mass flux decay of an order of magnitude from 75 to 8 Gg (gigagrams: Uno *et al.*, 2009).

In North Africa, the nature and extent of pressure cells over the region vary with the season. Winter outbreaks of Saharan dust (February–March) are associated with the seasonal high pressure system over North Africa and its characteristic mobility, in which major cells can be found from the eastern Atlantic to the Mediterranean basin. This circulation pattern draws prolonged, high concentrations of mineral dust into the lower troposphere from the Sahelian zone south of the Sahara. Outbreaks during summer (June–August) include dust palls of low frequency and density ($\sim$0.75 mg/m^3 of PM_{10}) but high persistence (15–30 days). The thermal low pressure that characterises the Sahara in summer shifts the anticyclone above the 850 hPa isobar level (approximately 1500 m a.s.l.), with the result that summer outbreaks of Saharan dust origin occur at higher tropospheric levels and do not intrude into the oceanic boundary layer. This semi-permanent dust layer (the Saharan Air Layer; Prospero and Carlson, 1972), clearly seen on orbital imagery, sustains an almost continuous westward dust flow across the Atlantic (Figure 10.14). A third, autumn–winter type of outbreak occurs at low frequency in October–November. In this season, the anticyclone is at ground level, the Sahelian surface having being cooled by lower temperatures and more frequent rainfall events, so inhibiting large-scale dust injections. However, locally violent dust storms sometimes raise PM_{10} dust densities of up to 400–500 mg/m^3. In general, these systems undergo dry deposition, but unstable atmospheric conditions involving draw-down of Saharan dust to the surface by intense low-pressure cells offshore of the Western Sahara can lead to wet deposition.

Saharan dust takes about a week to cross the Atlantic Ocean, typically reaching north-eastern South America, including the lower Amazon basin, in the (northern) late winter and spring and the Caribbean, Central America and the south-eastern United States in summer and early autumn (Prospero and Nees, 1986). Depending on season and meteorological conditions, particulate dust advected westwards from the Sahara influences air quality in North America (Prospero, 1999), the Caribbean, South America, Europe, the Middle East and Asia and affects the nutrient dynamics and biogeochemical cycles from northern Europe to South America. It has been estimated that 13 million tonnes of African dust falls on the north Amazon Basin every year (Griffin *et al.*, 2001, 2002). On the Caribbean island of Trinidad during

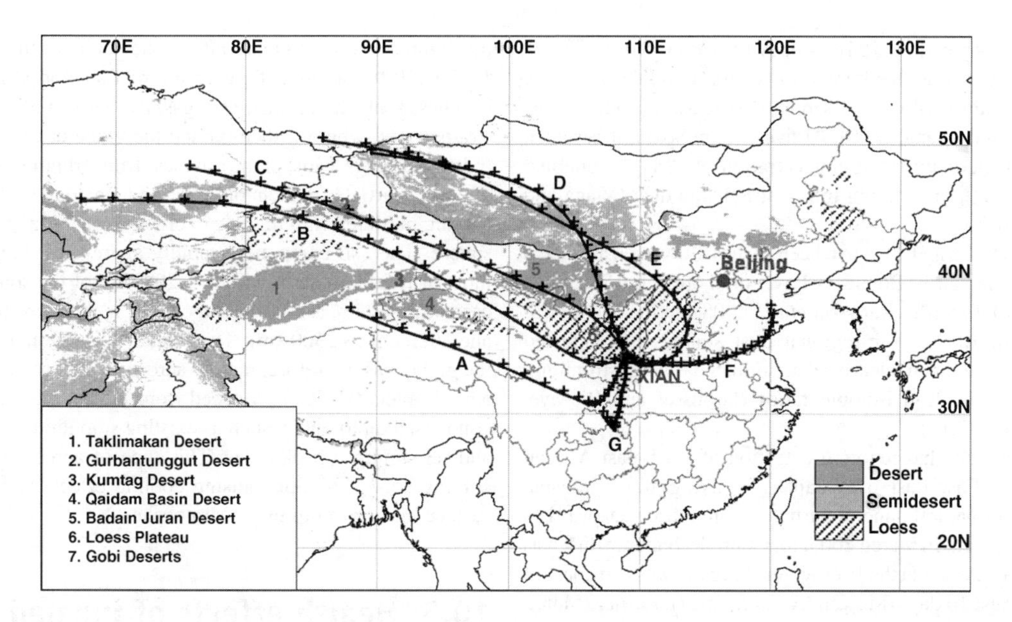

Figure 10.13 Back trajectories (HYSPLIT: Draxler and Rolph, 2003) of pathway clusters of springtime dusts deposited on the city of Xi'An in 2001–2003 (Wang *et al.*, 2006). Pathway clusters A, B and C, traversing the major sources of Asian dust, show mean PM_{10} loadings higher than the three spring months' average (159 μg/m^3); they account for 8 per cent, 16 per cent and 29 per cent of all trajectories depositing at Xi'An. Clusters D and E (the more northerly sources, centred on the Badain Juran desert) and F are less important routes for transport to Xi'An. No dust sources lie along Cluster G, the deposits being anthropogenic dust only. The major dust sources for Xi'An and Beijing are different, with north-westerly sources more important for Xi'An and arid and semi-arid regions in Mongolia more important for Beijing. PM_{10} loadings in springtime at Xi'An are usually less than those observed in Beijing (194 μg/m^3)

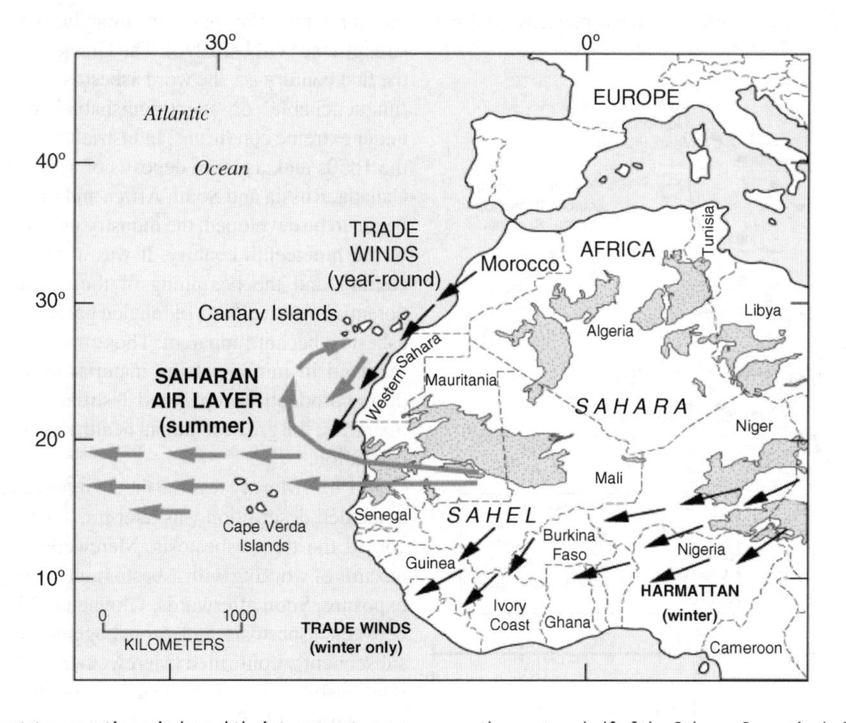

Figure 10.14 Main dust-transporting winds and their transport seasons, over the western half of the Sahara. Grey stippled areas indicate major active sand seas (From US Department of Agriculture, Natural Resources Conservation Service: http://www.soils.usda.gov/use/worldsoils/mapindex/order.html), based on FAO-UNESCO Soil Map of the World (Food and Agriculture Organization of the United Nations, UNESCO, 1974)). After Muhs *et al.* (2010), with kind permission of the first author

influx of Saharan dust, PM_{10} values of 135–149 μg/m^3 were registered (four times the known values for non-Saharan days), leading to dramatically increased paediatric asthma admissions (Rajkumar and Chang, 2000). At five sites in Spain, it has been shown that the daily mean concentration of $PM_{2.5}$ is doubled (from 4 to 11 μg/m^3) during influx of Saharan dust (Viana and Averol, 2007).

Isotopes ^{137}Cs and ^{40}K have been successfully used as tracers of PM_{10} in the Marine Boundary Layer offshore of the Sahara on the island of Tenerife (Karlsson *et al.*, 2008). In the Mediterranean, North African sub-regional dust sources and inferred pathways have been clearly discriminated using strontium, neodymium and lead isotopic ratios (Grousset and Biscaye, 2005; Figure 10.15)

A recent study has reported detection of north-east African and Middle East dust over Japan, implying trans-Eurasian movement of natural mineral particles from Africa eastward as well as westward. In March 2005, for example, Saharan dust was traced across Asia and the Pacific, reaching the western coast of Canada in just 10 days (McKendry *et al.*, 2007). Earlier (March 2003), rainwater samples in Japan contained dust and nanometre-scale particles that differed in both composition and shape from typical Chinese dust. In addition, carbon-bearing nanoparticles in the rainwater were consistent with the crude-oil fires raging in the Middle East at that time (Tanaka *et al.*, 2005).

Volcanic plumes can also travel great distances from the source volcano. The global extent depends on the volcano's location (plumes circulate in the hemisphere within which the volcano is located, following high-level wind patterns) and on the volume of erupted material and the height of the plume. On 15 June 1991, a climactic eruption of Pinatubo volcano, Philippines, sent fine-grained ash, gas and aerosol 35 km into the stratosphere, where it then circled the globe at equatorial latitudes in 22 days. Most of the ash mass from eruptions is made up of large particles which rapidly sediment out of the plume, leaving particles with diameters of less than 2 μm in the stratosphere. The ash component rapidly aggregates, however, and most of it falls to earth within 1–2000 km of the source; in the case of Pinatubo, ash fall was recorded as far away as Malaysia and Vietnam (~2500 km). The Pinatubo plume that circulated the globe was, therefore, mainly aerosol composed of sulphuric acid droplets (H_2SO_4) produced from SO_2 in the plume which condensed on to nuclei such as existing sulphuric acid particles and remaining ash particles. This type of aerosol scatters incoming solar radiation causing cooling effects at the Earth's surface and warming in the stratosphere.

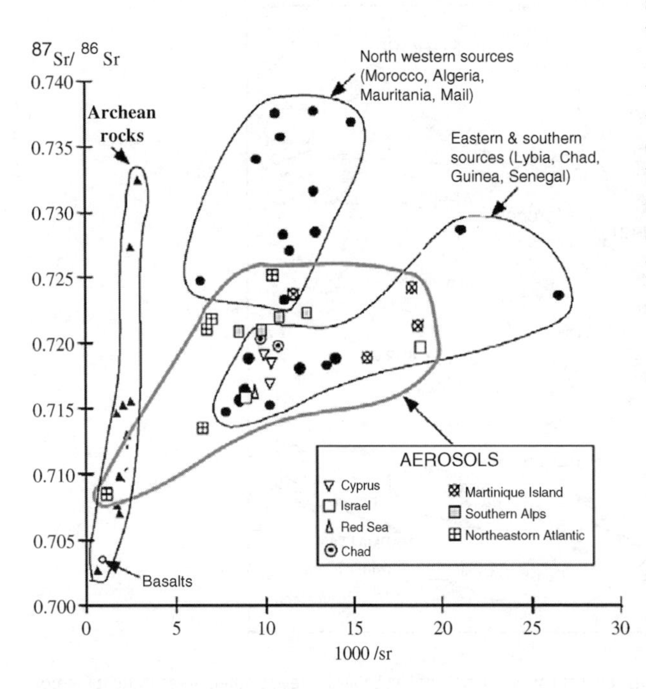

Figure 10.15 Strontium, neodymium and lead isotope ratios of Saharan aerosol dusts and their source regions (After Grousset and Biscaye, 2005)

10.5 Health effects of inhaled particulate material

10.5.1 Asbestos

Asbestos has been of practical use to mankind for over 4,000 years. The special properties of asbestos were recognised in Finland 4500 years ago, where they incorporated anthophyllite fibres into clay pots, presumably to give the vessels added strength (see Gibbs, 1996). The Greeks have used asbestos since the first century BC; the word asbestos is derived from the Greek, 'unquenchable' or 'inextinguishable', reflecting its durability under extreme conditions. Industrial mining of asbestos began in the 1850s and, as large deposits of asbestos were discovered in Canada, Russia and South Africa and as novel asbestos products began to be developed, the industry grew rapidly in the latter half of the nineteenth century. It was at the turn of the nineteenth century and the beginning of the twentieth century that the potential health effects of inhaled particulate material, including asbestos, became apparent. Those most at risk have been, or are, involved in mining of the material as well as being exposed during production, repair and destruction of asbestos-containing materials; the greatest human health risk is to the lungs by way of inhalation.

In 1907, Murray was the first to report a case of asbestosis and a detailed description was later provided by Cooke (1924) who coined the term asbestosis. Merewether (1930) addressed the hazards of working with asbestos and made recommendations on exposure. Soon afterwards, Gloyne (1935) noted a relationship between asbestosis and bronchogenic carcinoma, which was subsequently confirmed (Merewether, 1949; Doll, 1955) and the relationship between exposure to low levels of asbestos and a rare, untreatable form of lung cancer, malignant mesothelioma, was first described by Wagner *et al.* (1960). Asbestos exposure has devastating consequences on the respiratory system, causing pulmonary fibrosis, bronchogenic lung cancer, mesothelioma and pleural plaques. Although legislation to limit exposure was first introduced in Britain in the 1930s, following the report of Merewether and Price, legislation was not introduced into the USA until the 1970s. The exceptional qualities and versatility of asbestos meant that it was widely and increasingly, used for many more years before regulatory control was introduced in the UK in the 1980s (Asbestos Licensing Regulations, 1983; Control of Asbestos at Work Regulations, 1987) and, ultimately, importation to the UK was prohibited in 1992 (Asbestos Prohibition Regulation, 1992; The Control of Asbestos at Work (Amended) Regulations Act, 1992). Owing to the long latent period of asbestos-related diseases (some 30–40 years), the incidence of diffuse malignant mesothelioma, which has the longest latency, is predicted to reach its peak in Europe in 2020 (Peto *et al.*, 1999). Diagnosis of asbestosis continues to rise and, consequently, the deaths from these untreatable diseases are significant.

The nature of the disease depends, at least in part, on the type of asbestos to which individuals are exposed (Berman and Crump, 2008). There are two main types of asbestos, which have distinct physical differences: serpentine (chrysotile) and the amphiboles (crocidolite, amosite, anthophyllite, tremolite, actinolyte) (Table 10.2). The amphiboles are straight and needle-like, or form flakes. These fibres tend to retain their structure within the lung tissue; their needle-like shape and biopersistance is believed to be an important factor in their pathogenicity, penetrating parenchymal lung tissue and accessing the pleural surface (Donaldson, 2009). In contrast, chrysotile asbestos is made up of hundreds of small fibrils that form fibres, which are flexible and fray at the ends. These fibres can wrap around the lung structures, depending on the site of deposition; they are less durable than the amphiboles. The deposition and retention of asbestos in the lung critically depends on fibre size and physicochemical format. Long fibres are trapped in the large airways, where they may be phagocytosed by macrophages or trapped in the mucus and transported to the throat where they will be swallowed or expectorated. Very long particles are sometimes engulfed by many macrophages (frustrated phagocytosis) and develop into asbestos bodies (Figure 10.16), readily seen by light microscopy. These usually contain amphibole asbestos and high levels of asbestos bodies in lung secretions obtained by bronchoalveolar lavage indicate high tissue levels. Smaller, respirable particles less than 10 μm aerodynamic diameter, will reach the peripheral, respiratory zone. Macrophages will phagocytose single particles and particle aggregates in the micron range and will clear the particles from the lung via the mucociliary escalator or the lymphatic system. Particles that escape phagocytosis and the normal clearance mechanisms of the lung will access, and cause injury to, the lung tissue.

The Great Britain Asbestos Survey of 98 117 asbestos workers (Harding *et al.*, 2009) confirmed the known association between exposure to asbestos, asbestos-related lung diseases and the high mortality rates due to asbestosis, lung cancer (e.g. bronchogenic carcinoma), as well as pleural and peritoneal cancers and malignant mesothelioma. Pleural plaques are also common in asbestos workers but are not in themselves a cause

Table 10.2 Chemical formulae of asbestos and some asbestiform minerals. There are many other fibrous zeolites, such as mesolite, mordenite, natrolite, paranatrolite, tetranatrolite, scolecite and thomsonite

Name	Other names	Mineral type	Formula
Asbestos minerals			
Chrysotile	White asbestos;	Serpentine	$Mg_3(Si_2O_5)(OH)_4$
Amosite	Brown asbestos; cummingtonite-grunerite (solid solution series)	Amphibole	$Fe_7Si_8O_{22}(OH)_2$
Crocidolite	Blue asbestos; fibrous riebeckite	Amphibole	$Na_2Fe^{2+}{}_3Fe^{3+}{}_2Si_8O_{22}(OH)_2$
Tremolite		Amphibole	$Ca_2Mg_5Si_8O_{22}(OH)_2$
Anthophyllite		Amphibole	$(Mg,Fe)_7Si_8O_{22}(OH)_2$
Actinolite		Amphibole	$Ca_2(Mg,Fe)_5(Si_8O_{22})(OH)_2$
Some other asbestiform minerals			
Erionite		Zeolite	$(Na_2,K_2,Ca)_2Al_4Si_{14}O_{36}{\cdot}15H_2O$
Fluoro-edenite		Edenite	$NaCa_2Mg_5(Si_7Al)O_{22}F_2$
Richterite		Amphibole	$Na(Ca,Na)(Mg,Fe^{2+})_5(Si_8O_{22})(OH)_2$
Winchite		Amphibole	$(Ca,Na)Mg_4(Al,Fe^{3+})(Si_8O_{22})(OH)_2$
Balangeroite		Serpentine	$(Mg,Fe^{2+},Fe^{3+},Mn^{2+})_{42}Si_{16}O_{54}(OH)_{36}$
Carlosturanite		Serptentine-like	$(Mg,Fe,Ti)_{21}(Si,Al)_{12}O_{28}(OH)_{34}$
Nemalite	Fibrous brucite	Magnesium hydroxide	$Mg(OH)_2$
Palygorskite	Attapulgite	Clay	$(Mg,Al)_2Si_4O_{10}(OH){\cdot}4(H_2O)$
Wollastonite		Pyroxenoid	$CaSiO_3$
Sepiolite		Clay	$Mg_4Si_6O_{15}(OH)_2{\cdot}6H_2O$

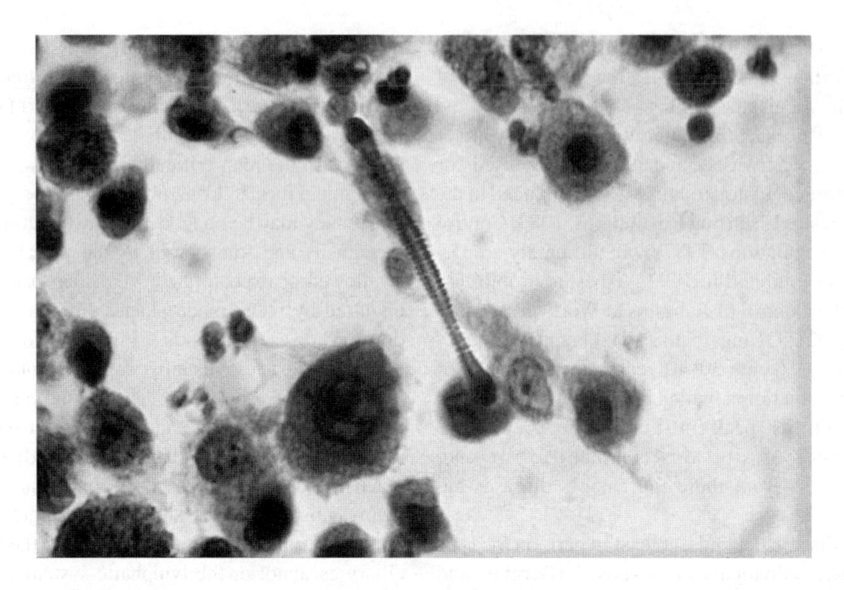

Figure 10.16 Photomicrograph of an asbestos body recovered during diagnostic bronchoscopy. Asbestos bodies consist of a central core fibre of asbestos, coated with a layer of iron-protein-polysaccharide, giving this beaded, dumb bell appearance. The fibre is surrounded by alveolar macrophages and other inflammatory cells, which were recovered during bronchoalveolar lavage (Courtesy of Professor T. D. Tetley)

of mortality (Weiss, 1993). Those who work in the insulation industry were found to be almost 7 times more likely to die from asbestos exposure than those who were in manufacturing; those involved with asbestos removal, a more recent occupation, were also found to be more likely to die than those in the manufacture of asbestos (Harding *et al.*, 2009; Antonescu-Turcu and Shapira, 2010). There is evidence to suggest that smoking is an important additional risk factor in cases of asbestos-related lung cancer and asbestosis, but not for mesothelioma (Harding *et al.*, 2009; Henderson *et al.*, 2004). Smoking has been described as being both additive and synergistic, but its influence probably reflects the complexity of the combination of differing pathogenicity of the asbestos, the type of cigarette smoked, genetic background and other environmental factors. The most recent report on British asbestos workers, the largest survey of its kind, shows a synergistic effect (Harding *et al.*, 2009). Churg and Stevens (1995) suggest that smoking delays clearance of the asbestos fibres from the lung, increasing retention within the airway mucosa and contributing to airways obstruction; increased accumulation of asbestos fibres is also likely to enhance the possibility of carcinogenic changes in the lung.

All types of lung cancer have been related to asbestos exposure, including small-cell and non-small-cell carcinomas. As mentioned above, the effect of concurrent smoking has been shown to be synergistic, with a significant, manifold increase in development of lung cancer. Since it is not possible to distinguish tumours related to asbestos exposure from those induced by tobacco smoke, well-controlled epidemiological studies have been essential in establishing synergy (Henderson *et al.*, 2004). It is suggested that cigarette smoke acts as a tumour promoter (Mossman and Gee, 2010) or co-carcinogen. There is also some debate regarding an association between asbestosis and the incidence of lung cancer. This is further complicated by the difference in relative potency of chrysotile and amphibole asbestos: both can induce lung cancer, but amphibole asbestos has at least six times greater carcinogenic potential (Berman and Crump, 2008). Another important factor is fibre length; the presence in lung tissue of asbestos fibres of 10 μm and above (but not shorter fibres) have consistently been associated with development of lung cancer.

Asbestosis causes an interstitial pneumonitis and fibrosis as a result of fibre inhalation and deposition within the lung (Antonescu-Turcu and Shapira, 2010; Becklake *et al.*, 2007; Gibbs, 1996). The condition develops over many years and initially there are few symptoms; it will be diagnosed only some 20–30 years following the peak exposure, the risk and incidence relating to the cumulative dose of asbestos exposure and the time since the first exposure. Patients present with breathlessness and dry cough, reduced gas diffusion capacity and oxygen desaturation on exertion. Lung function in advanced disease shows restrictive disease and reduced total lung capacity and vital capacity. Lung pathology shows usual interstitial pneumonia; high-resolution tomography shows significant remodelling of the parenchyma and honeycombing in advanced disease. There is no treatment for asbestosis; the condition gets progressively worse, even after exposure has ceased.

Malignant mesothelioma is an aggressive tumour of the pleura with a prognosis of only 6–18 months following diagnosis, depending on the cellular origin of the tumour (Churg, 1998). Again, there is a long latency between exposure and symptoms, up to 50–60 years following exposure and, unlike asbestosis, exposure to very low levels of amphibole asbestos can induce mesothelioma. Thus, exposure to low levels of amosite or crocidolite (but not chrysotile) induces mesothelioma. Fibres above 10 μm are associated with mesothelioma, but short fibres are not (Berman and Crump, 2008). A complication has been that samples of chrysotile asbestos may be contaminated with amphiboles or even erionite, a potent inducer of mesothelioma. Erionite-induced mesothelioma was thought to be confined to a small region of Turkey, but cases have recently been described in the United States (Dogan *et al.*, 2008; Kliment *et al.*, 2009). Unlike bronchogenic cancer and asbestosis, smoking is not a risk factor (Harding *et al.*, 2009). It is difficult to diagnose the condition, as the symptoms are non-specific. The patients report breathlessness, cough and chest pain, possibly associated with weight loss. The tumour can be seen as dense white tissue encasing the lung. Over time, it penetrates the respiratory tissue, destroying the pulmonary architecture and may impact on other structures such as the oesophagus and superior vena cava. Because of the ubiquitous, aggressive nature of the tumour and late diagnosis, there is no effective treatment.

10.5.2 Silica

Since silica is one of the most common constituents of the Earth's crust, people have always been at risk of exposure to pathologically significant levels, particularly in an occupational setting. Silica is derived from the Latin *silex,* which describes a flint, a stone tool containing silica. Pneumoconiosis, including silicosis, has been shown in Egyptian mummies, in sixteenth-century Bohemian miners and in eighteenth-century stone cutters (see Gibbs, 1996). The industrial revolution in Britain and increased, unanticipated occupational diseases, led to serious study of conditions related to occupation, including those generating a dusty atmosphere. Thus, in 1832 Thackrah noted that sandstone workers died prematurely, usually before 40 years of age (Thackrah, 1832). Peacock (1860) and Greenhow (1865) reported silica in the lungs of stone workers. The first to use the term silicosis was Visconti, in 1870 (Rovida, 1871); silicosis was previously known as miner's phthisis, grinder's asthma, potter's rot and other terms relating to occupation. It was often confused with tuberculosis (TB; also termed phthisis), until the TB bacillus was discovered by Koch in 1882.

Numerous reports in the early twentieth century correlated high rates of silicosis amongst South African gold miners, as well as in miners of tin, metal, slate, sandstone and quartz in the United Kingdom and Australia, with exposure to silica dust. These studies were the first to propose that crystalline silica (rather than amorphous silica) was responsible (Lanza, 1938). The early twentieth century also saw the introduction of sand-blasting and the use of abrasive mechanical equipment to produce fine particulate silica dust, often without adequate protection. Of particular relevance was the Hawk's Nest Tunnel disaster at Gauley Bridge, West Virginia, in the 1930s (Madl *et al.*, 2008). The project entailed blasting large natural rock formations to construct a source of hydroelectric power. Many of the workers died of respiratory disease during the project (called 'tunnelitis') and an epidemic of silicosis occurred subsequently. It was discovered that the rock consisted largely of pure quartz which, when processed, resulted in prolonged exposure to high concentrations of pure silica dust. Consequently, the Air Hygiene Foundation of America Inc. was founded and the first occupational exposure limits (OEL) were established. Regulations have continued to be tightly controlled; nevertheless, respiratory disease related to silica inhalation still occurs, particularly in underdeveloped countries.

There is a wide range of industries in which there may be exposure to silica, including mining and milling of metals, iron and steel foundries, quarrying of granite, stone and slate and the glass, pottery and ceramics industries. The construction industry, shipbuilding and maintenance and some agricultural activities are significant activities that involve contact with silica dust at some stage. Consequently, respiratory morbidity and mortality due to inhalation of silica continues, albeit at low levels (Madl *et al.*, 2008). There are now fewer than 30 deaths per year in the UK (Health and Safety Executive UK) and 200–300 deaths per year in the USA (American College of Occupational and Environmental Medicine, 2005).

Silica exists in a number of forms, which have different pathogenicity. Whilst the non-crystalline forms (e.g. diatomaceous earth (diatomite), fume silica, opal and mineral wool) are not considered to be pathogenic, the crystalline forms are. Quartz, cristobalite, moganite, some forms of tridymite, melanphlogite, coesite and stishovite are all crystalline forms of silica; quartz is by far the most common. Piezoelectric properties, where the silica exhibits opposite electrical charges at opposite ends of the crystal under pressure, are believed to contribute to their pathogenicity, particularly that of quartz (Greenberg *et al.*, 2007; Rimal *et al.*, 2005). Furthermore, toxicity is greatest when the particles have been freshly generated, as the resulting increase in redox potential at the surface favours strong reactions with hydrogen, oxygen, carbon and nitrogen. Oxygen free radicals at the surface of freshly mined silica and stimulation of cellular oxidative stress have been implicated in the cellular damage and lung injury that occurs in affected individuals. These processes are discussed in more detail in Section 10.2.

Chronic (nodular, classic) silicosis is the most common of the respiratory conditions caused by exposure to silica. It usually occurs after 20 or more years of exposure to relatively low concentrations of silica; some cases may occur at 5 or 10 years, described as accelerated silicosis, due to exposure to higher concentrations of often freshly generated silica. The lungs contain numerous, firm nodules, which vary in colour depending on the type of silica exposure. The nodules may vary from a few millimetres in diameter to larger regions of fused nodules in the upper lobes of the lung. In progressive massive fibrosis (PMF), a subset of chronic silicosis, there is marked conglomeration of nodules to create a large, hard fibrotic mass, most commonly observed bilaterally in the upper lobes (Gibbs, 1996; Greenberg *et al.*, 2007; Rimal *et al.*, 2005). The nodules contain silica crystals and many have an area of necrosis centrally, due to inadequate blood supply, surrounded by whorls of fibroblasts and peripheral, silica-laden macrophages and other inflammatory cells, including lymphocytes. In PMF, the fibrotic masses may show cavitation due to necrosis.

Clinically, the diagnosis is based on the patient's history of exposure and the presence of nodules on chest radiographs or, preferably, by computerised tomography. Classically, the nodules are approximately one centimetre in diameter; the patients have relatively few symptoms, although there is some evidence of airflow obstruction and cough, sputum production and breathlessness. These may be symptoms of associated conditions, such as bronchitis. In contrast, complicated silicosis, with PMF, shows the larger fibrotic regions and sometimes basilar emphysema, on radiography and computerised tomography. These patients are often hypoxic at rest and may have mycobacterial infections such as tuberculosis. They are breathless on exertion and exhibit both restrictive and obstructive lung disease, as well as decreased diffusing capacity. They also develop spontaneous pneumothoraces and die of respiratory failure. Those with accelerated silicosis present with the same symptoms as those with chronic silicosis and may rapidly progress to complicated silicosis, but are also likely to exhibit a variety of autoimmune disorders.

Acute silicosis most often occurs when the silica dust is extremely fine and affects occupations involving sandblasting and rock drilling. It develops rapidly over 4–5 years and may be fatal. It is very different from chronic silicosis and is also referred to as silicoproteinosis. It is rare, but was prevalent in the Hawk's Nest disaster. Individuals may present with breathlessness, fever, pleuritic pain and weight loss. It is important to diagnose quickly as the condition can progress very rapidly to respiratory failure and death. The airspaces become filled with fluid containing eosinophils and enriched with lipids and proteins. This partly reflects increase in numbers of alveolar epithelial type 2 cells, which line the respiratory units and secrete large amounts of proteins and lipids, being the source of pulmonary surfactant which lubricates and protects the respiratory units under normal conditions, but which is significantly elevated in acute silicosis.

The relationship between silica exposure and cancer has proved difficult to establish conclusively. In 1997, the International Agency for Research on Cancer (IARC) rated silica as a Group 1 carcinogen (with 'sufficient evidence in humans for the carcinogenicity of inhaled crystalline silica in the form of quartz or cristobalite') (International Agency for Research on Cancer, 1997). However, this has been contested by others (Madl *et al.*, 2008; Hessel *et al.*, 2000) who used many of the same studies together with studies that took place subsequently, but applied alternative statistical analysis (Hessel *et al.*, 2000) which did not show an association between exposure to silica and cancer. One problem has been separating silica exposure and silicosis from exposure to other known carcinogens, including cigarette smoke (Madl *et al.*, 2008; American College of Occupational and Environmental Medicine, 2005). Other factors include differences in the socioeconomic status of the subjects and whether the control groups were appropriate. Thus, the risk of developing cancer increases in those with silicosis who smoke (American College of Occupational and Environmental Medicine, 2005), but evidence of a risk of cancer in those without silicosis is inconsistent (Ulm *et al.*, 1999; Checkoway and Franzblau, 2000). It has been suggested that the relationship between exposure to silica, silicosis and lung cancer may never be resolved (Madl *et al.*, 2008).

As alluded to previously, a number of recent studies suggest that occupational exposure to silica may be associated with autoimmune diseases, such as rheumatoid arthritis, scleroderma, progressive systemic sclerosis and lupus erythematosis, as well as kidney disease. However, it is difficult to know the exact risks since these diseases are also dependent on other environmental and genetic factors, as well as being more common in women, so that epidemiological investigations are not easy to control. The evidence has been interrogated (American Thoracic Society Committee of the Scientific Assembly on Environmental and Occupational Health, 1997; NIOSH, 2002) and the conclusion is that there is not enough evidence and more research is needed before reliable conclusions can be drawn. A similar situation exists in relation to silica exposure and kidney disease (NIOSH, 2002).

10.5.3 Coal

Coal has been mined in the UK for over a thousand years. The first mention of coal was in 852, in the Saxon Chronicle of Peterborough (Gibbs, 1996). During the nineteenth century and the industrial revolution, coal became an important source of energy. Coal-workers' pneumoconiosis (CWP) was first described by Laennec (1819). The term anthracosis was introduced (Stratton, 1838) to describe the blackened lungs of coal miners from Fifeshire, Scotland. A particularly high incidence of CWP was found in miners in South Wales, which led to a survey, set up by the Medical Research Council, leading to a series of reports in the early 1940s (Gibbs, 1996). These were followed up by further investigations by the MRC and the National Coal Board, which culminated in setting standards for dust exposure (Hurley *et al.*, 1982). NIOSH have more recently recommended levels of exposure aimed at preventing or, at the least, limiting CWP (NIOSH, 1995). During the latter half of the twentieth century, these controls, together with a decline in coal mining, have resulted in a significant reduction in the incidence of CWP in Europe and North America. However, coal production is increasing in developing countries, where it is an important source of energy and contributes to the economy of the country. In China, coal is used to generate approximately 70 per cent of the electricity (Liu *et al.*, 2009) and the incidence of CWP is high. The predicted increase in mining and use of coal in Asia and Africa is causing concern that there will be a parallel increase in respiratory health problems (McCunney *et al.*, 2009).

CWP was originally believed to be a form of silicosis, but most coal has relatively little free crystalline quartz and there is a poor correlation between quartz content and CWP (McCunney *et al.*, 2009). Thus, the incidence of CWP was found to be low in miners who mined coal with a comparatively high level of silica, whilst there was a high incidence of CWP in collieries with comparatively low levels of silica. A series of epidemiological studies in Britain during the 1970s and 1980s demonstrated that development of CWP was related to exposure to coal dust but did not correlate with the silica content (McCunney *et al.*, 2009; Hurley *et al.*, 1982; Maclaren *et al.*, 1989). Similar findings were found in studies from Europe. Furthermore, although the diseases may have similar radiographic patterns, they are pathologically very different. It is not entirely clear exactly what properties in coal cause CWP. It is well known that the rank of the coal – low, medium or high, reflecting the carbon content – is important. High rank coal is the hardest and oldest, containing the most carbon, the least volatile matter and least silica; it is retained more readily in the lung and is associated with severe CWP. When the incidence of CWP was tracked against the rank of the coal, CWP was found to be 6- to 10-fold higher in those exposed to high-rank coal (>90 per cent carbon) than in those exposed to the low-ranked coal (<80 per cent carbon) (Attfield and Morring, 1992). Furthermore, the amount of bioavailable iron has also been related to development of CWP (Zhang and Huang, 2002). There is no evidence that coal dust induces cancer.

CWP is largely due to inhalation of coal dust during the process of mining. The disease is described as simple CWP (SCWP) or complicated CWP (CCWP), depending on the pathology and severity (Gibbs, 1996; McCunney *et al.*, 2009; Ross and Murray, 2004). As mentioned earlier, anthracosis is the accumulation of the coal pigment/particles in the lung; this is asymptomatic. When the lung becomes overloaded with coal-dust particles, the defence mechanisms are overwhelmed. The phagocytic macrophages are the first line of defence, internalising coal-dust particles and migrating away from the airways to the throat where they are swallowed or expectorated, or to the lymphatic system to remove excessive material. During heavy or chronic exposure, macrophage migration and particle clearance is reduced so that there is an accumulation of macrophages in hot spots of particle deposition in the peripheral respiratory zone; there is relatively little collagen deposition, unlike that seen in silicosis and asbestosis. These focal areas of macrophages and particles are known as coal macules, which are defined as being less than a centimetre in diameter and are a characteristic of the disease. Macules are mostly situated in the upper regions of the lung and many may not be observed radiologically. Accumulation of coal dust and activation of macrophages leads to tissue injury and, consequently, fibrosis in an attempt to repair the damage. Thus, fibrotic nodules may also be present in SCWP, which are larger and far fewer in number than the macules. If the process of nodule enlargement continues and there is fusion of nodules to create fibrotic lesions many centimetres in size, CCWP or PMF occurs; this may involve a whole lobe, or many lobes, of the lung. Emphysema, which is a loss of lung tissue showing enlarged airspaces associated with reduced surface area and elasticity, is often present in the lung tissue surrounding the macules, nodules and fibrotic regions. Emphysema is a component of chronic obstructive pulmonary disease (COPD), a disease that affects approximately 15 per cent of tobacco smokers, but which the evidence suggests can also be one of the effects of exposure to coal dust (Coggon and Newman, 1998; Ross and Murray, 2004).

The diagnosis of CWP is based on history of coal-dust exposure and radiology in the first instance, but the macules and nodules cannot always be visualised so that computed tomography scanning may be performed. SCWP is largely asymptomatic, showing only small changes, if any, in lung function. In those with CCWP who present with cough and breathlessness, the ventilatory capacity is reduced in proportion to the presence of diseased tissue, as is the gas exchange capacity and increasing hypoxemia. There is also evidence that coal miners have symptoms of chronic bronchitis, chronic airflow limitation and emphysema, all features of chronic obstructive pulmonary disease (Coggon and Newman, 1998). The severity of CWP is related to the duration and intensity of exposure.

Once the disease has progressed to PMF it continues to progress regardless of removal from the coal-dust stimulus. Thus, if SCWP has been diagnosed, it is suggested that the individual changes jobs to an environment where there is low, or no exposure to avoid progression to CCWP. As smoking has an additive effect, this is to be discouraged. Depending on the degree of breathlessness and hypoxemia, the patient could have oxygen therapy.

10.5.4 Bioaccessibility and bioavailability of inhaled particulate material

The bioaccessibility and bioavailability of inhaled particles depends on numerous factors. Bioavailability refers to the difference between the amount of airborne particulate matter to which one is exposed and the actual amount that the lung retains. Bioaccessibility refers to subsequent events whereby the particles are processed by body fluids and cells to facilitate absorption by, and assimilation into, the lung. Thus, on exposure to airborne particulate matter, particle size and shape, chemistry, crystallinity, surface properties, biopersistence and durability are all important, impacting on the particle burden and reactivity within the lung and development of lung disease.

10.5.4.1 Particle deposition and retention

Deposition of inhaled particles is determined by their density, as well as their size and aspect ratio, which dictate their aerodynamic diameter. The smaller the aerodynamic diameter and/or density, the greater is the likelihood of deposition in the distal, respiratory region of the lung. In addition, the anatomy of the lung favours deposition under gravity or impaction of particles greater than 5 µm in the large, conducting airways; in contrast, a high proportion of particles smaller than 5 µm can reach the respiratory units and may be deposited on the alveolar epithelium. However, very long fibres with high aspect ratios may be deposited in the deep lung.

The ability to clear deposited particles depends on both physicochemical and physiological mechanisms. Large particles which deposit in the conducting airways are cleared by way of the mucociliary escalator to the throat, where they may be swallowed or expectorated. In addition, airway macrophages will phagocytose particulate material (optimally <5 µm), which will then be cleared by the same route. However, long fibres greater than the diameter of the macrophage (>20 μm), cause 'frustrated phagocytosis', involving the efforts of numerous macrophages to internalise the fibre. These chronically-activated macrophages contribute to generation of reactive oxygen species and other mediators of inflammation (Oberdorster, 2010).

Clearance of particles from the distal, epithelial air-liquid interface of the respiratory region of the lung will also involve macrophage phagocytosis and clearance by the mucociliary escalator. In addition, particles may translocate into the interstitium, due to physical breaching of the alveolar barrier or as a result of active epithelial uptake processes (e.g. endocytosis). Alternatively, particulate material may pass across the epithelial barrier paracellularly, following the flow of water under negative interstitial water pressure. Interstitial particles can enter the lymphatic system under negative pressure and translocate to the pleural surface or to the tracheobronchial lymph nodes, where they may accumulate (Dodson *et al.*, 1991; Miserocchi *et al.*, 2008). The difference between the amount of particulate material that is deposited and the amount that is cleared is the particle retention, which accounts for the lung burden.

10.5.4.2 Interaction with lung-lining fluid

Another significant factor in the bioaccessibility and biopersistence of inhaled particles is their reaction with components of lung-lining fluid. Mineral particles, such as asbestos and metal oxides, were shown to adsorb proteins from serum many years ago (Desai and Richards, 1978); the amount and nature of protein adsorption depends on the surface chemistry and surface area of the particles, as well as the flexibility of the protein. Protein adsorption by nano-sized particles has recently been described as forming a 'corona', where strong protein binding forms a 'hard corona' (Lundqvist *et al.*, 2008). Again, the surface properties (i.e. charge), size and surface area of the particles are important. It is interesting that, to date, there appears to be no predictable pattern or profile of serum protein adsorption based on nanoparticle physicochemistry. Previous studies of ambient environmental particulate matter have shown agglomeration of particles which deposit in human lung-lining liquid due, in part, to alveolar epithelial cell-derived surfactant protein D (SPD) (Kendall *et al.*, 2004, 2002) and very precise interactions with macrophages to induce phagocytosis. In addition, surfactant protein A (SPA) adsorbed to inhaled particles may enhance epithelial uptake and translocation of particles less than 1 μm in diameter that deposit at the epithelial surface (Kemp *et al.*, 2008) and which escape macrophage phagocytosis. Dipalmitylphosphatidylcholine, a major component of lung surfactant, which has high surface reactivity, also binds to inhaled ambient particulate matter (Kendall, 2007); it is believed to be more likely to enhance particle dispersion rather than agglomeration. Thus, adsorption of components of lung-lining liquid may have protective or detrimental effects on particle reactivity, cellular uptake, translation into the pulmonary interstitium and retention in the lung.

10.5.4.3 Asbestos

There are marked differences in lung retention between serpentine (chrysotile) and amphibole asbestos (Churg and Wright, 1994; McDonald, 1998; Bernstein and Hoskins, 2006). Amphibole asbestos is found in the lung in greater quantities than chrysotile asbestos. Particle retention will reflect a mixture of the dose inhaled and rate of clearance from the lung. Clearance is compromised when there is an exposure overload, which, in experimental animals, occurs when a single bolus of a high concentration is introduced into the airways. It is not clear how relevant particle overload is to human exposure. Nevertheless, inhalation studies in experimental animals (i.e. not involving particle overload) show that, for the same exposure dose, chrysotile is cleared from the lung more effectively than amphibole asbestos, the former being cleared in a matter of days ($T_{1/2} + 0.3$–11 days) whereas the latter takes years ($T_{1/2} > 500$ days). Studies in people are less exact, but show that retention of amphibole asbestos, particularly fibres with a high aspect ratio and a longest diameter greater than 20 μm long, is significantly greater than chrysotile asbestos (Bernstein and Hoskins, 2006; Churg and Wright, 1994; McDonald 1998), even when the amphibole asbestos (e.g. tremolite: McDonald and McDonald, 1997) is a small impurity in a predominantly chrysotile exposure dose. Regardless of the significant difference in the exact amounts of chrysotile retained in the lung compared to the amphibole asbestos, there is still a correlation between the concentration of chrysotile and amphibole asbestos with the estimated cumulative dust exposure. It has been suggested that the increased biopersistence (i.e. fibre retention within the lung, despite the lung's physiological clearance mechanisms) and durability (i.e. ability to resist dissolution under a wide range of experimental *in vitro* conditions; Maxim *et al.*, 2006; Muhle and Bellmann, 1995) of the long, thin amphibole asbestos fibres may be one reason why these fibres cause mesothelioma.

An important factor in the rapid clearance of chrysotile asbestos from the lung is its solubility. The crystal structure of serpentine consists of layers of octahedrally coordinated magnesium (brucite sheets) and layers of tetrahedrally coordinated silica; these sheets are mismatched so they curl and form scrolls, with the magnesium on the outside (Bernstein and Hoskins, 2006; Fubini, 1997). The structure is susceptible to environmental pH, in that the outer, magnesium, brucite-like sheets dissolve in mildly acid conditions. The silica matrix is susceptible to acid pH, a situation which exists in the phagolysosomal apparatus and in the immediate vicinity of the cell surface membrane of macrophages (Fubini, 1997; Wypych *et al.*, 2005). Consequently, regardless of the initial length of fibre, chrysotile deposited at the air–liquid interface of the lung disintegrates into fragments that can readily be internalised and cleared by macrophages and other clearance systems in the lung. If this occurs, it is unlikely to be a significant health hazard. In contrast, amphibole asbestos has chemical resistance similar to that of quartz. The fibres, each of which is a double chain of tetrahedral silicate structures, are weakly associated lengthwise by a variety of ions, depending on the host rock. These ions bond the fibres together, but it is at these weakly-bonded surfaces that the fibres will break apart to generate fibres with greater aspect ratios than the original and these are highly insoluble at any pH (see Bernstein and Hoskins, 2006) and may retain their structure in the lung even after many decades. This is why amphibole-asbestos fibres are significantly more biopersistent than chrysotile.

10.5.4.4 Silica and coal

Unlike fibrous asbestos minerals, silica and coal are, for the most part, small, respirable particles finer than 5 μm, often as small as 1 μm, which readily access the deep lung during inhalation. Macrophages play an important role in clearance of both silica and coal and are a key factor in controlling the lung burden. However, phagocytosis of particles and induction of macrophage cell death leads to release of the intracellular particles, which are then re-ingested by fresh macrophages. This continuous cycle of particle phagocytosis and macrophage death contributes to inadequate particle clearance and retention in the lung, which, in turn, perpetuates the inflammatory process. However, the reactivity of coal and various types of silica differs according to their crystallinity (silica) and associated contaminating material, such as clay and pyrite (Donaldson and Borm, 1998).

Inhaled silica deposits in the alveolar ducts and bronchiolar region of the deep lung consist of particles that are, on average, 1.5 μm in diameter. Particles which escape removal by macrophage phagocytosis, as described above, interact with and injure the epithelium, when the particles are then translocated into the lymph nodes (Hemenway *et al.*, 1990). In the Hemenway study, the clearance of cristobalite was far less than that found for quartz, while the inflammatory response was 30 per cent greater, lasting over several months. This coincided with increased lung hydroxyproline, as an indicator of fibrosis. The variability in the bioreactivity of silica is well documented (Donaldson and Borm, 1998; Fubini, 1997) and relates to its surface properties. In addition, particle surfaces may be modified by components of lung-lining liquid, as described above. Orr and colleagues (Orr *et al.*, 2010) recently showed that uptake of nano-sized amorphous silica by macrophages depends on macrophage scavenger receptor A, by way of clathrin-dependent mechanisms and suggested that the physical agglomeration state influenced the mode of cellular trafficking. In this study, the uptake of silica nanoparticles by pulmonary A549 adenocarcinoma cells (often likened to alveolar epithelial type 2 cells) critically depended on protein adsorption (Stayton *et al.*, 2009). Proteins adsorbed to the particles very rapidly and cellular uptake was consequently far slower, approximately one third that of the naked particles. Interestingly, the particles were also slowly expelled from the cells, though at a slower rate than uptake, resulting in cellular retention. Such complex interactions, between silica, macrophages and epithelial cells will impact on the overall bioaccessibility of silica within the lung, at the gas-liquid interface and within the interstitium.

Coal dust is less life-threatening than silica. Exposure to moderately high doses of coal dust, with its significant biopersistence in the lung, does not have the same impact on health as exposure to silica. It is interesting that, although there is a massive accumulation of coal dust in the lungs of miners, related to exposure dose, the health effect, in the absence of exposure to tobacco smoke, is lower than that observed with equivalent exposure to silica. Furthermore, although silica is found in coal dust at a level that might be considered to be detrimental to health, the expected pathogenicity is not observed. Within the lung, factors related to particle deposition, uptake and translocation will apply to all inhaled particulate matter, as discussed above. Clearly, there are some important differences between coal dust exposure and that of mineral fibres and silica (and other dusts described in this chapter). Some clues to these differences lie in the finding that water-soluble extracts of coal dust and both low- and high-aluminium clays (kaolin and attapulgite) inhibit the cellular reactivity of quartz (Stone *et al.*, 2004; Clouter *et al.*, 2001); such events might explain why occupational exposure to coal dust containing less than 20 per cent quartz does not induce classic silicosis. It is not yet clear whether the influence of components of coal dust simply depresses the bioreactivity of silica or whether they influence the bioaccessibilty, relocation and subsequent reactivity of silica.

10.6 Risk reduction and future trends

This chapter has shown that airborne particulate matter includes material from both natural and anthropogenic sources and that the impact on the environment and human health varies in intensity, scale and timing. The hazards of airborne dust have been described and the potential for reduction of the risks of inhaling particles is considered here. The risk takes into consideration the vulnerability and resilience of local populations, the potential economic impact and the likelihood of the hazard occurring.

10.6.1 Natural dust palls

Dust palls arising from natural sources may have local, regional and even global impacts, their periodicity varying with source type and location.

10.6.1.1 Dust storms

A key strategy in risk reduction is control, minimisation or prevention of the hazard. Dust storms of anthropogenic origin are increasing in many parts of the world. Continuing global deforestation and desertification correlate quite closely with increases in the severity and frequency of dust storms. For example, desertification in China increased total desert area by 2 to 7 per cent in the second half of the twentieth century. This resulted in disproportionately large areas of potential sources of soil dust, which have increased troposphere dust loading in the mid-latitudes and generated 10 to 40 per cent more dust storms. Any reversal of the soil-dust load by re-establishing grassland in the sensitive 200–400 mm annual rainfall belt of north-west China would require regional cooperation (Gong *et al.*, 2004) upon which the degree of risk prevention would depend.

Asian dust storms are known to occur at any time of the year but the prime period is spring to early summer. In contrast, dust palls sourced in North Africa occur throughout the year, dust reaching North America and the northern Caribbean in the northern summer and South America and the southern Caribbean in the northern winter. Substantial regional dust palls derived from some of the world's smaller dust source regions may have a lower mean frequency. For example, the major dust storm that originated in Australia's Northern Territory in September 2009 and which extended beyond New Zealand into the Southern Ocean, was described as the largest dust storm event for over 40 years (Figure 10.17).

Such wide-ranging differences in the magnitude and frequency of naturally occurring dust palls have a bearing on human-health impacts. There is a clear link between frequent exposure to high-density airborne particles over several decades and pneumoconiosis and related conditions; this precludes the application of effective risk reduction strategies, a situation made more complex by the distribution of human population density in relation to regional and global dust pathways. For example, up to 50 per cent of the annual dust flux from eastern Asia derived from its two global sources (the Taklamakan Desert in western China and the Mongolia/North China region) is deposited in China, Korea and Japan, which have urban population densities of between 100 and more than 500 people per km^2 (Figure 10.18); some parts of western China have a siliceous pneumoconiosis incidence of 7 per cent, rising to three times that proportion in people over 40 years old (Xu *et al.*, 1993). While available research results are overwhelmingly concerned with occupational pneumoconiosis, there is increasing recognition that non-occupational pneumoconiosis is a serious problem in China (Yin, 2005). Regions in the Caribbean and North and South America affected by Saharan dust have moderately dense populations (>40 per km^2 with some regional centres having higher densities of >100 km), but dust palls are more frequent than the mean for eastern Asia. Saharan regional-scale dust reaches moderately dense populations in Europe, but the limited extent, low density and/or low frequency of many of these incursions reduces their potential health impact.

Given that dust storms have the potential to affect large areas, reduction of risk is most likely to be achieved by adopting local mitigation measures such as small-scale forestation projects, long-term use of dust masks and even relocation of populations. Direct action has been undertaken in China, designed to stem the frequency and magnitude of dust storms around dryland margins; trees have been planted, including, most notably, the Green Great Wall initiative. However, judging from evidence of a

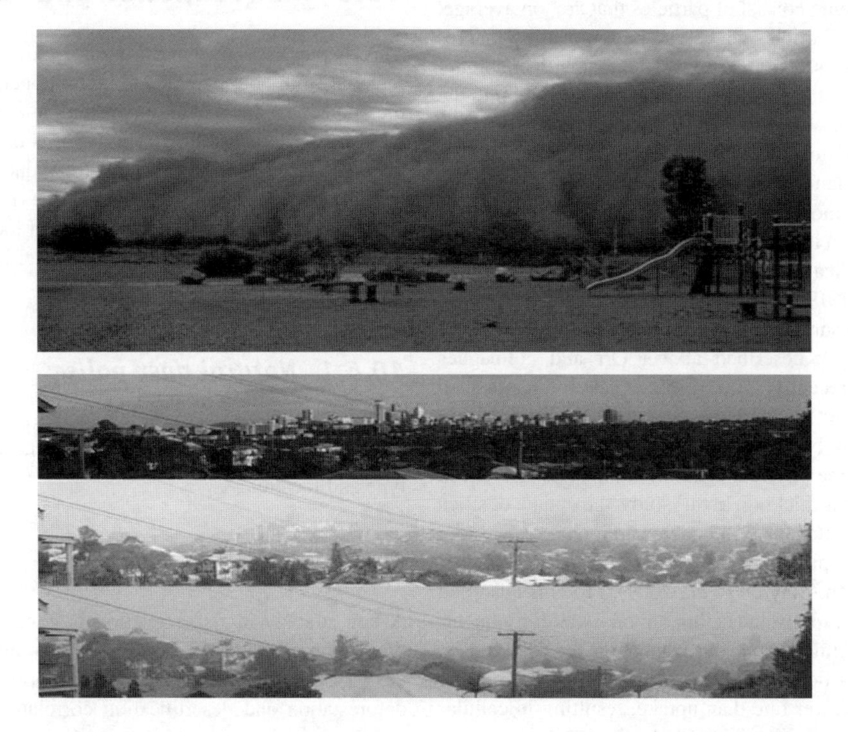

Figure 10.17 Dust storm progression eastwards from a central Australian source in 2009. Top: Advancing dust storm front approaching a children's playground in Alice Springs, 22 September 2009. Photograph source: unknown. Bottom: Changing air quality over the city of Brisbane, Queensland, some 2000 km ESE of source. Upper: Before the storm, early September 2009. Middle: 11.07 a.m. on 23 September. Lower: 11.44 a.m. on 23 September (Set of three photographs courtesy of Ben Garratt, University of South Australia)

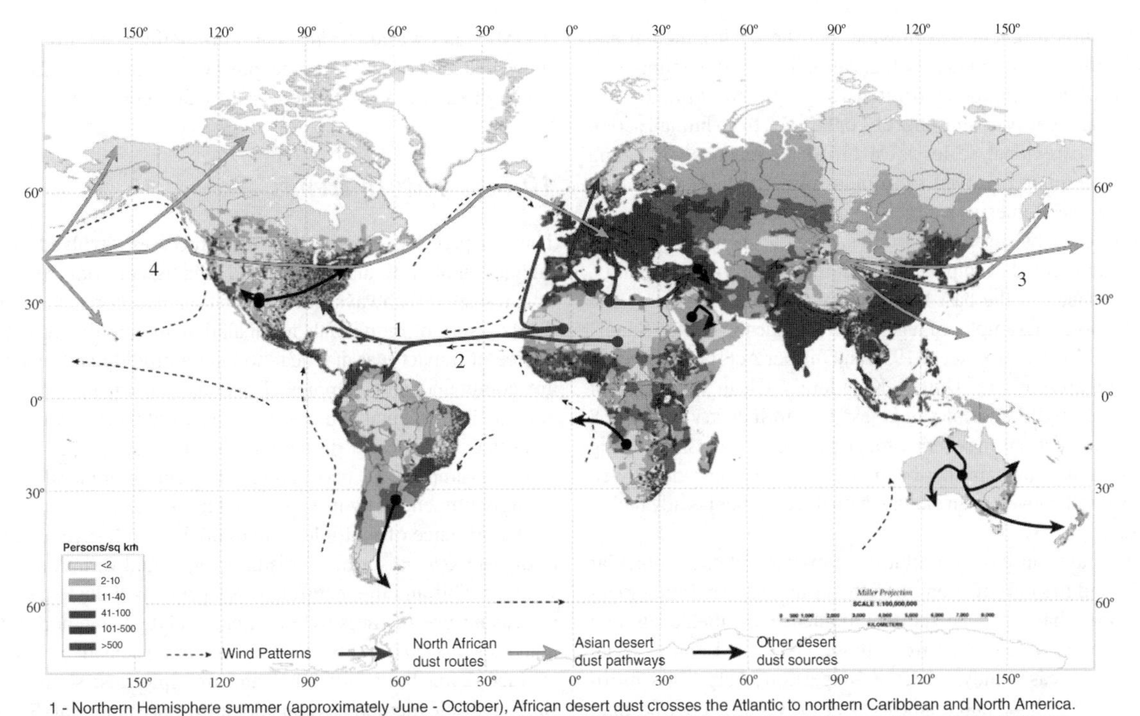
Persons/sq km
<2
2-10
11-40
41-100
101-500
>500
Wind Patterns
North African dust routes
Asian desert dust pathways
Other desert dust sources
1 - Northern Hemisphere summer (approximately June - October), African desert dust crosses the Atlantic to northern Caribbean and North America.
2 - Northern Hemisphere winter (approximately November - May), African desert dust crosses the Atlantic to southern Caribbean and North America.
3 - Asian dust season typically lasts from late February to late April.
4 - Large Asian dust events can travel significant distance in the Northern Hemisphere, including a full circuit

Figure 10.18 Primary sources of mineral dust and their atmospheric pathways (modified after Griffin, 2007) in relation to world population density 1994 (United States Department of Agriculture) (After Derbyshire (in press))

continuing increase in dust storms in the past quarter of a century, the degree of success has been limited. Nevertheless, success in some local communities has been achieved by considerably reducing the number of grazing animals and limiting their grazing range, together with improvement of water conservation practices (Normile, 2007). At the urban scale, rapid growth and increasing dependence on motor traffic, as well as growth of manufacturing, power and other industries, has accelerated a rise in particle-rich air pollution in many cities in eastern China and in settlements in dryland areas of intense farming activity. Most direct action to reduce risk has generally been a matter of personal initiative in response to governmental and company-management advice, notably in the use of cotton face-masks during periods of high dust density.

10.6.1.2 Volcanic plumes

A large volcanic eruption can spew ash into the stratosphere, into the path of jet streams, thereby enabling the ash to circle the globe. Occasionally, the impact of such an eruption can be significant, causing global temperature change which, in turn, can lead to droughts, crop failures and famines. The hazard itself is uncontrollable, so risk reduction can be achieved only through international economic and logistical scenario planning.

More frequently, volcanoes emit substantial quantities of ash which falls on local populations and may remain in the environment for months, years or, occasionally, decades, if the eruption is sustained. For example, the Soufrière Hills volcano, Montserrat, West Indies, started erupting in July 1995 and is still active. The population has been subjected to frequent ash falls over the past 15 years. Those most likely to develop respiratory problems are workers who are occupationally exposed to the ash (e.g. outdoor workers) and detailed risk analyses have been carried out to assess the likelihood of specific occupations developing chronic diseases (Hincks *et al.*, 2006). Assuming continuing volcanic activity, outdoor workers, as the group most at risk, are thought to have a 2–4 per cent risk of silicosis after 20 years of exposure to the cristobalite-rich ash. Risk reduction is also achieved through education of local populations, taking into account cultural issues and vulnerabilities. This includes teaching safe ways to clear ash off roofs, to clean and seal houses and to maintain respiratory health through the use of dust masks.

10.6.2 Anthropogenic dust

Biomass burning and some bush fires, the former being part of the annual agricultural cycle in many parts of the world, can generate major dust palls and smoke plumes with local to regional impact. Fires in several locations in recent years, including those in Indonesian tropical forests and Siberian boreal woodlands, led to dust palls that could be traced over several thousand kilometres. The burning of sugar cane at harvest has been common practice worldwide, but the environmental and human impact has led to new legislation in Brazil, banning pre- and post-harvesting burns from 2010 (Le Blond *et al.*, 2010).

Poor air quality arising entirely, or largely, from human action at local to regional scales is of broad provenance, including particulate matter derived from industrial, transportation, agricultural and domestic sources. Progress in reducing risks to health posed by airborne particles, largely by means of state legislation aimed at imposing measures to improve air quality, has advanced in the past half-century, although there is some way yet to go. Examples of legislation include the United States' Air Pollution Control Act of 1955 and further Acts between 1963 and the present, the United Kingdom's Clean Air Acts of 1956 (demanding use of smokeless fuel in industry) and 1968 and similar actions in several other countries. Progress in setting up air-quality standards has been much slower in less-developed countries, however, in an era in which development is advancing rapidly.

The main cause of air pollution in modern cities is vehicular traffic and the introduction of traffic-reducing or calming measures may have had an indirect influence on the levels and distributions of particles from those vehicles. London is an appropriate case study, where a congestion charge was introduced in February 2003. Atkinson *et al.* (2009) reported that, while there did not appear to be an overall reduction in pollution in London as a direct result of introducing the scheme, there was 'evidence of relative reduction' of PM_{10}. However, within London's charging zone, Beevers and Carslaw (2005) reported that PM_{10} emissions were reduced by 11.9 per cent and by 1.4 per cent on London's inner ring road. They attributed these reductions to increases in vehicle speed, as this is as important in reducing emissions as changes in vehicle numbers. Conversely, traffic calming measures such as road humps and width barriers can have a negative effect on emission levels in the immediate proximity because drivers 'aggressively' brake and accelerate through such systems (Kyoungho and Hesham, 2009).

Legislation designed to encourage decreased use of cars and to create financial advantages by using smaller and less polluting vehicles have had mixed success. High petrol taxes have been justified on the grounds that they would decrease car use and, therefore, emission levels. However, a study in Southern California and Connecticut in the USA revealed that environmental taxes on fuel resulted in only minimal reductions in driving levels (Sipes and Mendelsohn, 2001). There are also significant developments in producing viable electrical and hybrid vehicles and promoting their use. For example, the Irish government has plans for 10 per cent of all vehicles to be powered by electricity by 2020, resulting in significant emission reductions (Brady and O'Mahony, in press). In addition to new technology in the vehicle fleet, the development of new fuels offers the chance of reduced emission levels. The use of biofuels is rapidly increasing and, although there are concerns about decreased engine efficiency using these new fuels, there is generally a marked decrease in gaseous and particulate emissions. A study by Ozsezen and Canakci (2011), using a direct-injection diesel engine fuelled with canola-oil methyl ester (COME) and waste (frying) palm oil methyl ester (WPOME), showed a decrease in engine power and increased fuel use, although smoke opacity decreased by 56–63 per cent.

10.6.3 Future trends

This chapter has shown that the processes involved in the interaction of particulate matter and lung tissue, from the derivation, transport pathways and composition of airborne particles to the exposure of individuals to inhalation and the likelihood and degree of impact that may lead to lung morbidity and mortality, are compound and complex. Uncertainties arising from such complexity, especially those concerned with matters of cause and effect, are present throughout the process.

For example, the geochemical content of mineral dust is strongly influenced by its source and its complexity is increased by the presence of multiple sources arising from several factors as diverse as land surface moisture regime and variations in the weather. This and the manifold re-suspension of dust, has led to increasing use of the single-particle analytical method, rather than those based on bulk samples; as a more sensitive means of characterising both individual and multiple dust sources, the single particle method has considerable future potential (Stevens *et al.*, 2010).

Global diversity in the mode and time-span of dust collection, dust measurement and statistical analysis of pollutant data has also been a limiting factor in comparative analysis between regions and in respect of community sub-groups. Other non-standard practices may include variability of criteria used in diagnosis of certain conditions (Monteil, 2008) and the assessment of the relative importance for human health of local, regional and global components in dust pollution when data on the health effects of global-scale mineral dust are sparse and opinion remains contentious (Monteil and Antoine, 2009). Future long-term, standardised studies are needed at all scales, with collaborative, comparative global programmes on the impact on health of both semi-continuous and seasonal dust palls, providing higher quality background data than is currently available.

At the other end of the spectrum, information on the precise mechanisms involved in mineral-particle pathogenesis is relatively sparse. For example, the condition of quartz-particle surfaces and the presence both upon and within them of toxins affects pathogenicity. However, while metallic iron is known to reduce quartz toxicity and thus lung inflammation, traces of ferric or ferrous iron stimulates oxidative stress on quartz surfaces that leads to damage of epithelium cells (Donaldson and Borm, 1998).

Evidence is accumulating of the presence of cause-and-effect links between far-travelled dust and morbidity arising from micro-organisms, including bacteria, fungi, pollen and viruses attached to soil particles (e.g. Griffin *et al.*, 2001; Kellogg *et al.*, 2004; Griffin, 2007), but the role of airborne pathogens in epidemics remains poorly understood.

Understanding of the role of airborne particles in environmentally linked health conditions has rested to varying degrees upon epidemiological results, while definitive studies of the part played by physico-chemical processes are sparser. These and other gaps in knowledge currently legitimise doubt as to whether or not chronic exposure to naturally derived dust particles, such as those of regional to global scale, is necessarily a *direct* cause of disease.

References

American College of Occupational and Environmental Medicine (2005) Medical surveillance of workers exposed to crystalline silica. *Journal of Occupational and Environmental Medicine*, **48**(1), 95–101.

American Thoracic Society Committee of the Scientific Assembly on Environmental and Occupational Health (1997) Adverse effects of crystalline silica exposure. *American Journal of Respiratory and Critical Care Medicine*, **155**(2), 761–768.

Ando, M., Tadano, M., Asanuma, S., Tamura, K., Matsushima, S., *et al.* (1998) Health effects of indoor fluoride pollution from coal burning in China. *Environmental Health Perspectives*, **106**, 239–244.

Antonescu-Turcu, A. L. and Shapira, R. M. (2010). Parenchymal and airway diseases caused by asbestos. *Current Opinion in Pulmonary Medicine*, **16**(2), 155–161.

Atkinson, R.W., Barratt, B., Armstrong, H.R. Anderson, S.D., Beevers, I.S., *et al.* (2009) The impact of the congestion charging scheme on ambient air pollution concentrations in London. *Atmospheric Environment*, **43**(34), 5493–5500.

Attfield, M.D. and Morring, K. (1992) An investigation into the relationship between coal workers' pneumoconiosis and dust exposure in U.S. coal miners. *American Industrial Hygiene Association Journal*, **53**(8), 486–492.

Baris, I., Simonato, L., Artvinli, M., Pooley, F., Saracci, R., *et al.* (1987) Epidemiological and environmental evidence of the health effects of exposure to erionite fibres: a four-year study in the Cappadocian region of Turkey. *International Journal of Cancer*, **39**, 10–17.

Baxter, P.J., Ing, R., Falk, H. and Plikaytis, B. (1983) Mount St. Helens eruptions: the acute respiratory effects of volcanic ash in a North American community. *Archives of Environmental Health* **38**(3), 138–143.

Becklake, M.R., Bagatin, E., and Neder, J.A. (2007) Asbestos-related diseases of the lungs and pleura: uses, trends and management over the last century. *International Journal of Tuberculosis and Lung Disease*, **11**(4), 356–369.

Beevers, S.D. and Carslaw, D.C. (2005) The impact of congestion charging on vehicle emissions in London. *Atmospheric Environment*, **39**(1), 1–5.

Berman, D.W. and Crump, K.S. (2008) A meta-analysis of asbestos-related cancer risk that addresses fiber size and mineral type. *Critical Reviews in Toxicology*, **38** (Suppl 1), 49–73.

Bernstein, D.M. and Hoskins, J.A. (2006) The health effects of chrysotile: current perspective based upon recent data. *Regulatory Toxicology and Pharmacology*, **45**(3) 252–264.

Borm, P.J.A. (1997) Toxicity and occupational health hazards of coal fly ash (CFA): a review of data and comparison to coal mine dust. *Annals of Occupational Hygiene*, **41**, 659–676.

Bory, A.J.M., Biscaye, P.E. and Grousset, F.E. (2003) Two distinct seasonal Asian source regions for mineral dust deposited in Greenland (NorthGRIP), *Geophysical Research Letters*, **30**, 1167.

Boutin, C., Viallat, J.R., Steinbauer, D.G., Massey, D.G. and Mouries, J.C. (1986) Bilateral pleural plaques in Corsica: a non-occupational asbestos exposure marker. *European Journal of Respiratory Diseases*, **69**, 4–9.

Brady, J. and O'Mahony, M. (in press) Travel to work in Dublin. The potential impacts of electric vehicles on climate change and urban air quality. Transportation Research Part D: Transport and Environment.

Brown, P.D., Jones, T.P., Shao, L. and BéruBé, K. (2009) The geochemistry and geotoxicity of ash from coal-burning power stations. Proceedings of the 11th International Conference on Atmospheric Science and Applications to Air Quality. Ji'nan, China. http://www.asaaq.sdu.edu.cn/asaaq.

Carbone, M., Kratzke, R. A. and Testa, J. R. (2002). The pathogenesis of mesothelioma. *Seminars in Oncology*, **29**(1), 2–17.

Cersosimo, M.G. and Koller, W.C. (2006) The diagnosis of manganese-induced parkinsonism. *NeuroToxicology*, **27**, 340–346.

Checkoway, H. and Franzblau, A. (2000) Is silicosis required for silica-associated lung cancer? *American Journal of Industrial Medicine*, **37**(3), 252–259.

Chen, J., Li, G.J., Yang, J.D., Rao, W.B., Lu, H.Y., Balsam, W., Sun, Y.B. and Ji, J.F. (2007) Nd and Sr isotopic characteristics of Chinese deserts: implications for the provenances of Asian dust. *Geochimica et Cosmochimica Acta*, **71**: 3904–3914.

Churg, A. (1998) Neoplastic induced asbestos-related disease, In; Churg, A. and Green, F.H.Y. (eds.) *Pathology of Occupational Lung Disease*, 2nd edn. Baltimore, MD: Williams and Wilkins, pp. 339–391.

Churg, A. and Stevens, B. (1995) Enhanced retention of asbestos fibers in the airways of human smokers. *American Journal of Respiratory and Critical Care Medicine*, **151**(5), 1409–1413.

Churg, A. and Wright, J.L. (1994) Persistence of natural mineral fibers in human lungs: an overview. *Environmental Health Perspectives*, **102** (Suppl 5), 229–233.

Clague, A.D.H., Donnet, J.B., Wang, T.K. and Peng, J.C.M. (1999) A comparison of diesel engine soot with carbon black. *Carbon*, **37**, 1553–1565.

Clarke, J. (2003) The occurrence and significance of biogenic opal in the regolith. *Earth-Science Reviews*, **60**, 175–194.

Clouter, A., Brown, D., Hohr, D., Borm, P. and Donaldson, K. (2001) Inflammatory effects of respirable quartz collected in workplaces versus standard DQ12 quartz: particle surface correlates. *Toxicological Sciences*, **63**(1), 90–98.

Coggon, D. and Newman, T.A. (1998) Coal mining and chronic obstructive pulmonary disease: a review of the evidence. *Thorax*, **53**(5), 398–407.

Cook, A.G., Weinstein, P. and Centeno, J.A. (2005) Health effects of natural dust: role of trace elements and compounds. In: Schrauzer, G.N. (ed.) *Biological Trace Element Research*, **103**, 1–15.

Cook, D.E. and Gale, S.J. (2005) The curious case of the date of introduction of leaded fuel to Australia: implications for the history of Southern Hemisphere atmospheric lead pollution. *Atmospheric Environment*, **39**(14), 2553–2557.

Cooke, W.E. (1924) Fibrosis of the lungs due to the inhalation of asbestos dust. *British Medical Journal*, **ii**, 147.

Davies, T.C. (2003) Some environmental problems of geomedical relevance in East and Southern Africa. In: Skinner, H.C.W. and Berger, A.R. (eds.) *Geology and Health: Closing the Gap*. Oxford University Press, Oxford and New York, pp 139–144.

Derbyshire, E. (in press) *Global Dust. Encyclopaedia of Natural Hazards*. Springer, Heidelberg.

Desai, R. and Richards, R.J. (1978) The adsorption of biological macromolecules by mineral dusts. *Environmental Research*, **16**(1–3), 449–464.

Delfino, R.J., Chang, J., Wu, J., Ren, C., Tjoa, T., Nickerson, B.D. and Gillen L. (2009) Repeated hospital encounters for asthma in children and exposure to traffic-related air pollution near the home. *Annals of Allergy, Asthma and Immunology*, **102**(2), 138–144.

Diaz-Sanchez, D. (1997) The role of diesel exhaust particles and their associated polyaromatic hydrocarbons in the induction of allergic airway disease. *Allergy*, **52** (Suppl 38), 52–56.

Ding, Z. L., Yang, S. L., Sun, J. M., Liu, T. S. (2001) Iron geochemistry of loess and red clay deposits in the Chinese Loess Plateau and implications for long-term Asian monsoon evolution in the last 7.0 Ma. *Earth and Planetary Science Letters*, **185**, 99–109.

Dodson, R.F., Williams, M.G., Jr., Corn, C.J., Brollo, A. and Bianchi, C. (1991). A comparison of asbestos burden in lung parenchyma, lymph nodes and plaques. *Annals of the New York Academy of Sciences*, **643**, 53–60.

Dogan, A., Dogan, M., and Hoskins, J. (2008) Erionite series minerals: mineralogical and carcinogenic properties. *Environmental Geochemistry and Health*, **30**(4), 367–381.

Doll, R. (1955) Mortality from lung cancer in asbestos workers. *British Journal of Industrial Medicine*, **12**, 81–86.

Donaldson, K. (2009) The inhalation toxicology of p-aramid fibrils. *Critical Reviews in Toxicology*, **39**(6), 487–500.

Donaldson, K. and Borm, P.J.A. (1998) The quartz hazard: a variable entity. *Annals of Occupational Hygiene*, **42**, 287–294.

Draxler, R.R. and Rolph, G.D. (2003) HYSPLIT (Hybrid Single-Particle Lagrangian Integrated Trajectory) Model access via NOAA ARL READY Website (http://www.arl.noaa.gov/ready/hysplit4.html): Silver Spring, Maryland, National Oceanic and Atmospheric Administration Air Resources Laboratory.

Engelbrecht, J.P., McDonald, E.V., Gillies, J.A., Jayanty, R.K.M., Casuccio, G. and Gertler, A.W. (2009) Characterizing mineral dusts and other aerosols from the Middle East—Part 1: Ambient sampling. *Inhalation Toxicology*, **21**(4), 297–326.

Finkelman, R.B., Belkin, H.E., Centeno, J.A. and Zheng, B. (2003) Geological epidemiology: coal combustion in China. In: Skinner, H. C. and Berger, A.R. (eds) *Geology and Health: closing the gap*. Oxford University Press, pp 45–50.

Forbes, L., Jarvis, D., Potts, J. and Baxter, P.J. (2003) Volcanic ash and respiratory symptoms in children on the island of Montserrat, British West Indies. *Occupational and Environmental Medicine*, **60**, 207–211.

Fubini, B. (1997) Surface reactivity in the pathogenic response to particulates. *Environmental Health Perspectives*, **105** (Suppl 5), 1013–1020.

Fubini, B. (1998a) Health effects of silica. In: Legrand, A.P. (ed.) *The Surface Properties of Silicas*. John Wiley & Sons Ltd, Chichester, pp 415–464.

Fubini. B. (1998b) Surface chemistry and quartz hazard. *Annals of Occupational Hygiene*, **42**, 521–530.

Fubini, B., Bolis, V., Cavenago, A. and Volante., M. (1995) Physicochemical properties of crystalline silica dusts and their possible implication in various biological responses. *Scandinavian Journal of Work Environment and Health*, **21** (Suppl 2), 9–14.

Fubini, B. and Hubbard, A. (2003) Reactive oxygen species (ROS) and reactive nitrogen species (RNS) generation by silica in inflammation and fibrosis. *Free Radical Biology and Medicine*, **34**(12), 1507–1516.

Fubini, B. and Otero Aréan, C. (1999) *Chemical aspects of the toxicity of inhaled mineral dusts*. Chemical Society Reviews, **28**, 373–381.

Ghio, A.J., Carter, J.D., Richards, J.H., Brighton, L.E., Lay, J.C. and Devlin, R.B. (1998) Disruption of normal iron homeostasis after bronchial instillation of an iron-containing particle. *American Journal of Physiology*, **274**, L396–403.

Ghio, A.J., Silbajoris, R., Carson, J.K.L. and Samet, J.M. (2002) Biologic effects of oil fly ash. *Environmental Health Perspectives Supplements*, **110** (Suppl 1), 89–102.

Gibbs, A.R. (1996) Occupational lung disease. In: Hasleton, P. S. (ed.), *Spencer's Pathology of the Lung*, 5th edition, McGraw-Hill, pp. 461–506.

Ginoux, P., Chin, M., Tegen, J., Prospero, J.M., Holben, B., Dubovik, O., and Lin, S.J. (2001) Sources of distributions of dust aerosols simulated with the GOCART model. *Journal of Geophysical Research*, **106**, 20255–20274.

Gloyne, S.R. (1935) Two cases of squamous carcinoma of the lung occurring in asbestosis. *Tubercle*, **17**, 5–10.

Gomez, S.R., Parker, R.A., Dosman, J.A. and McDuffie, H.H. (1992) Respiratory health effects of alkali dust in residents near desiccated Old Wives Lake. *Archives of Environmental Health*, **47**, 364–369.

Gong, S.L., Zhang, X.Y., Zhao, T.L. and Barrie, A.L. (2004) Sensitivity of Asian dust storm to natural and anthropogenic factors. *Geophysics Research Letters*, **31**, L07210.

Greenberg, M.I., Waksman, J. and Curtis, J. (2007) Silicosis: a review. *Disease-a-Month*, **53**(8), 394–416.

Greenhow, E.H. (1865) Specimen of diseased lung from a case of grinder's asthma. *Transactions of the Pathological Society London*, **16**, 59.

Griffin, D.W. (2007). Atmospheric movement of microorganisms in clouds of desert dust and implications for human health. *Clinical Microbiology Reviews*, **20**, 459–477.

Griffin, D.W., Kellogg, C.A., Garrison, V.H. and Shinn, E.A. (2002) The global transport of dust. *American Scientist*, **90**, 228–235.

Griffin, D.W., Kellogg, C.A. and Shinn, E.A. (2001) Dust in the wind: long range transport of dust in the atmosphere and its implications for global transport and ecosystems. *Global Change and Health*, **2**, 20–30.

Grousset, F.E. and Biscaye, P.E. (2005) Tracing dust sources and transport patterns using Sr, Nd and Pb isotopes. *Chemical Geology*, **222**, 149–167.

Grousset, F.E., Ginoux, P., Bory, A. and Biscaye, P.E. (2003) Case study of a Chinese dust plume reaching the French Alps. *Geophysical Research Letters*, **30**(6), 23–26.

Guthrie, G.D. (1997) Mineral properties and their contributions to particle toxicity. *Environmental Health Perspectives*, **105** (Suppl 5), 1003–1011.

Harding, A.H., Darnton, A., Wegerdt, J. and McElvenny, D. (2009) Mortality among British asbestos workers undergoing regular medical examinations (1971–2005). *Occupational and Environmental Medicine*, **66**(7), 487–495.

Hardy, J.A. and Aust, A.E. (1995) Iron in asbestos chemistry and carcinogenicity. *Chemical Reviews*, **95**, 97–118.

Health and Safety Executive, UK. Pneumoconiosis and silicosis. http://www.hse.gov.uk/statistics/causdis/pneumoconiosis.

Hemenway, D.R., Absher, M.P., Trombley, L. and Vacek, P.M. (1990) Comparative clearance of quartz and cristobalite from the lung. *American Industrial Hygiene Association Journal*, **51**(7) 363–369.

Henderson, D.W., Rodelsperger, K., Woitowitz, H.J. and Leigh, J. 2004. After Helsinki: a multidisciplinary review of the relationship between asbestos exposure and lung cancer, with emphasis on studies published during 1997–2004. *Pathology*, **36**(6), 517–550.

Hessel, P.A., Gamble, J.F., Gee, J.B., Gibbs, G., Green, F.H., Morgan, W.K. and Mossman, B.T. (2000) Silica, silicosis and lung cancer: a response to a recent working group report. *Journal of Occupational and Environmental Medicine*, **42**(7), 704–720.

Hill, R.J., Edwards, R.E. and Carthew, P. (1990). Early changes in the pleural mesothelium following intrapleural inoculation of the mineral fibre erionite and the subsequent development of mesotheliomas. *Journal of Experimental Pathology*, **71**, 105–118.

Hincks, T.K., Aspinall, W.P., Baxter, P.J., Searl, A., Sparks, R.S.J. and Woo, G. (2006) Long-term exposure to respirable volcanic ash on Montserrat: a time series simulation. *Bulletin of Volcanology*, **68**(3), 266–284.

Horwell, C.J. (2007) Grain size analysis of volcanic ash for the rapid assessment of respiratory health hazard. *Journal of Environmental Monitoring*, **9**(10), 1107–1115.

Horwell, C.J. and Baxter, P.J. (2006) The respiratory health hazards of volcanic ash: a review for volcanic risk mitigation. *Bulletin of Volcanology* **69**(1): 1–24.

Horwell, C.J., Fenoglio, I. and Fubini, B. (2007) Iron-induced hydroxyl radical generation from basaltic volcanic ash. *Earth and Planetary Science Letters*, **261**(3–4), 662–669.

Horwell, C.J., Le Blond, J.S., Michnowicz, S.A.K. and Cressey, G. (2010) Cristobalite in a rhyolitic lava dome: Evolution of ash hazard. *Bulletin of Volcanology*, **72**, 249–253.

Hurley, J.F., Burns, J., Copland, L., Dodgson, J., and Jacobsen, M. (1982) Coalworkers' simple pneumoconiosis and exposure to dust at 10 British coalmines. *British Journal of Industrial Medicine*, **39**(2), 120–127.

International Agency for Research on Cancer (1997) Silica, some silicates, coal dust and para-aramid fibrils. IARC Monograph on the evaluation of carcinogenic risks to humans. Lyon, France, 68, p. 506.

Jickells, T. D., An, Z. S., Andersen, K.K., Baker, A. R., Bergametti, G., Brooks, N., *et al.* (2005) Global iron connections between desert dust, ocean biogeochemistry and climate. *Science*, **308**, 67–71.

Jones, T.P., Wlodarczyk, A., Koshy, L., Brown, P., Longyi, S. and BéruBé, K.A. (2009) The geochemistry and bioreactivity of fly-ash from coal-burning power stations. *Biomarkers*, **14**(1), 45–48.

Karlsson, L., Hernandez, F., Rodriguez, S., López-Pérez, M., Hernandez,-Armas, J., Alonso Pérez, S. and Cuevas, E. (2008) Using ^{137}Cs and ^{40}K to identify natural Saharan dust contributions to PM10 concentrations and air quality impairment in the Canary Islands. *Atmospheric Environment*, **42**, 7034–7042.

Kellogg, C.A., Griffin, D.W., Garrison, V.H., Peak, K.K., Royall, N., Smith, R.R. and Shinn, E.A. (2004) Characterization of aerosolized bacteria and fungi from desert dust events in Mali, West Africa. *Aerobiologia*, **20**, 99–110.

Kemp, S.J., Thorley, A.J., Gorelik, J., Seckl, M.J., O'Hare, M.J., Arcaro, A., Korchev, Y., Goldstraw, P., and Tetley, T.D. (2008) Immortalization of human alveolar epithelial cells to investigate nanoparticle uptake. *American Journal of Respiratory Cell and Molecular Biology*, **39**(5), 591–597.

Kendall, M. (2007) Fine airborne urban particles (PM2.5) sequester lung surfactant and amino acids from human lung lavage. *American Journal of Physiology–Lung Cellular and Molecular Physiology*, **293**(4), L1053–L1058.

Kendall, M., Brown, L. and Trought, K. (2004) Molecular adsorption at particle surfaces: a PM toxicity mediation mechanism. *Inhalation Toxicology*, **16** (Suppl 1), 99–105.

Kendall, M., Tetley, T.D., Wigzell, E., Hutton, B., Nieuwenhuijsen, M. and Luckham, P. (2002) Lung lining liquid modifies PM(2.5) in favor of particle aggregation: a protective mechanism. *American Journal of Physiology–Lung Cellular and Molecular Physiology*, **282**(1), L109–L114.

Kliment, C.R., Clemens, K. and Oury, T.D. (2009) North American erionite-associated mesothelioma with pleural plaques and pulmonary fibrosis: a case report. *International Journal of Clinical and Experimental Pathology*, **2**(4) 407–410.

Koller, W.C., Lyons, K.E. and Truly, W. (2004) Effect of levodopa treatment for parkinsonism in welders. *Neurology*, **62**, 730–733.

Krueger, B.J., Grassian, V.H., Cowin, J.P., and Laskin A. (2004) Heterogeneous chemistry of individual mineral dust particles from different dust source regions: the importance of particle mineralogy. *Atmospheric Environment*, **38**, 6253–6261.

Kyoungho, A. and Hesham, R. (2009) A field evaluation case study of the environmental and energy impacts of traffic calming. *Transportation Research Part D: Transport and Environment*, **14**(6), 411–424.

Laennec, R. T. H. (1819) *Traite de l'Auscultation Mediate ou Traite du Diagnostic des Maladies des Poumons at du Coeur*, 1st edition, Paris: Brosson & Chaudé.

Lanza, A.J. (1938) *Silicosis and Asbestosis*, Oxford University Press.

Le Blond, J.S., Horwell, C.J., Williamson, B.J. and Oppenheimer, C. (2010) Generation of crystalline silica from sugarcane burning. *Journal of Environmental Monitoring*, **12**(7), 1459–1470.

Le Blond, J.S., Williamson, B.J., Horwell, C.J., Monro, A.K., Kirk, C. A. and Oppenheimer, C. (2008) Production of potentially hazardous respirable silica airborne particulate from the burning of sugarcane. *Atmospheric Environment*, **42**(22), 5558–5568.

Li, G., Chen, J., Ji, J., Yang, J. and Conway, T.M. (2009) Natural and anthropogenic sources of East Asian dust. *Geology*, **37**, 727–730.

Liu, H.B., Tang, Z.F., Yang, Y.L., Weng, D., Sun, G., Duan, Z.W. and Chen, J. (2009) Identification and classification of high risk groups for Coal Workers' Pneumoconiosis using an artificial neural network based on occupational histories: a retrospective cohort study. *BMC Public Health*, **9**, 366.

Lucchini, R., Apostoli, P., Perrone, C., Placidi, D., Albini, E. and Migliorati, P., *et al.* (1999) Long-term exposure to "low levels" of manganese oxides and neurofunctional changes in ferroalloy workers, *NeuroToxicology*, **20**, 287–297.

Lundqvist, M., Stigler, J., Elia, G., Lynch, I., Cedervall, T., and Dawson, K.A. (2008) Nanoparticle size and surface properties determine the protein corona with possible implications for biological impacts. *Proceedings of the National Academy of Science USA*, **105**(38), 14265–14270.

Lyles, M.B., Fredrickson, H.L., Bednar, A.J., Fannin, H.B. and Sobecki, T.M. (2005) The chemical, biological and mechanical characterization of airborne micro-particulates from Kuwait. Abstract 8th Annual Force Health Protection Conference, session 2586, Louisville, Kentucky, USA.

Maclaren, W.M., Hurley, J.F., Collins, H.P. and Cowie, A.J. (1989) Factors associated with the development of progressive massive fibrosis in British coalminers: a case-control study. *British Journal of Industrial Medicine*, **46**(9), 597–607.

Madl, A.K., Donovan, E.P., Gaffney, S.H., McKinley, M.A., Moody, E.C., Henshaw, J.L. and Paustenbach, D.J. (2008) State-of-the-science review of the occupational health hazards of crystalline silica in abrasive blasting operations and related requirements for respiratory protection. *Journal of Toxicology and Environmental Health Part B Critical Reviews*, **11**(7), 548–608.

Magnani, C., Terracini, B., Ivaldi, C., Botta, M., Mancini, A. and Andrion, A. (1995) Pleural malignant mesothelioma and non-occupational exposure to asbestos in Casale Monferrato, Italy. *Occupational and Environmental Medicine*, **52**, 362–367.

Maricq, M.M. 2007. Chemical characterisation of particulate emissions from diesel engines: a review. *Aerosol Science*, **38**, 1079–1118.

Maxim, L.D., Hadley, J.G., Potter, R.M. and Niebo, R. (2006) The role of fiber durability/biopersistence of silica-based synthetic vitreous fibers and their influence on toxicology. *Regulatory Toxicology and Pharmacology*, **46**(1), 42–62.

McConnochie, K., Simonoto, L., Mavrides, P., Christofides, P., Pooles, F.D. and Wagner, J.C. (1987) Mesothelioma in Cyprus: the role of tremolite. *Thorax*, **42**, 342–347.

McCunney, R.J., Morfeld, P. and Payne, S. (2009) What component of coal causes coal workers' pneumoconiosis? *Journal of Occupational and Environmental Medicine*, **51**(4), 462–471.

McDonald, J.C. (1998) Mineral fibre persistence and carcinogenicity. *Industrial Health*, **36**(4), 372–375.

McDonald, J.C. and McDonald, A.D. (1997) Chrysotile, tremolite and carcinogenicity. *Annals of Occupational Hygiene*, **41**(6), 699–705.

McKendry, I.G., Strawbridge, K.B., O'Neill, N.T., Macdonald, A.M., Liu, P.S.K., *et al.* (2007) Trans-Pacific transport of Saharan dust to western North America: A case study. *Journal of Geophysical Research*, **112**, D01103.

Merewether, E.R.A. (1949) Annual Report of the Chief Inspector of Factories for the year 1947. HMSO, London, UK.

Merewether, E.R.A. and Price, C.W. (1930) Report on the effects of asbestos dust on the lungs and dust suppression in the asbestos industry. HMSO, London, UK.

Miserocchi, G., Sancini, G., Mantegazza, F. and Chiappino, G. (2008) Translocation pathways for inhaled asbestos fibers. *Environmental Health*, **7**, 4.

Molesworth, A.M., Cuevas, L.E., Connor, S.J., Morse, A.P. and Thomson, M.C. (2003) Environmental risk and meningitis epidemics in Africa. *Emerging Infectious Diseases*, **9**, 1287–1293.

Monteil, M.A. (2008) Saharan dust clouds and human health in the English-speaking Caribbean: what we know and don't know. *Environmental Geochemistry and Health*, **30**, 339–343.

Monteil, M.A. and Antoine, R. (2009) African dust and asthma in the Caribbean: medical and statistical perspectives. *International Journal of Biometeorology*, **53**(5), 379–381.

Morawska, L. and Zhang, J. (2002) Combustion sources of particles. 1. Health relevance and source signatures. *Chemosphere*, **49**, 1045–1058.

Mossman, B.T. and Gee, J.B. (2010). Asbestos-related diseases. *New England Journal of Medicine*, **320**(26), 1721–1730.

Muhle, H. and Bellmann, B. (1995) Biopersistence of man-made vitreous fibres. *Annals of Occupational Hygiene*, **39**(5), 655–660.

Muhs, D.R., Budahn, J., Skipp, G., Prospero, J.M., Patterson, D. and Bettis, E.A. III (2010) Geochemical and mineralogical evidence for Sahara and Sahel dust additions to Quaternary soils on Lanzarote, eastern Canary Islands, Spain. *Terra Nova*, **22**(6), 399–410.

Murray, H. M. (1930) Testimony before the Departmental Committee on Compensation for Industrial Diseases "Minutes of Evidence, Appendices and Index", 1907, p. 127. Cited and summarized in Merewether and Price (1930).

Neijari, A., Fournier, J., Pezerat, H. and Leanderson, P. (1993) Mineral fibres: correlation between oxidising surface activity and DNA base hydroxylation. *British Journal of Industrial Medicine*, **50**, 501–504.

Neuberger, M., Rabczenko, D. and Moshammer H. (2007) Extended effects of air pollution on cardiopulmonary mortality in Vienna. *Atmospheric Environment*, **41**(38), 8549–8556.

Newman, L.S. (1996) Metals. In: Harber, P., Schenker, M.B. and Blames, J.R. (eds.) *Occupational and Environmental Respiratory Disease*. Mosby, St Louis, MO, USA.

NIOSH (1995) *Criteria for recommended standard: Occupational exposure to coal mine dust.*, DHHS (NIOSH), Washington, DC, pp 95–106.

NIOSH (2002) *Health effects of occupational exposure to respirable crystalline silica*. Washington DC: National Institute for Occupational Safety and Health, p 129.

Normile, D. (2007) Getting at the roots of killer dust storms. *Science*, **317**, 314–316.

Oberdorster, G. (2010) Safety assessment for nanotechnology and nanomedicine: concepts of nanotoxicology. *Journal of International Medicine*, **267**(1), 89–105.

O'Hara, S.L., Wiggs, G.F.S., Mamedov, B., Davidson, G. and Hubbard, R.B. (2000) Exposure to airborne dust contaminated with pesticide in the Aral Sea region. *The Lancet*, **355**: 627–628.

Orr, G.A., Chrisler, W.B., Cassens, K.J., Tan, R., Tarasevich, B.J., Markillie, L.M., Zangar, R.C. and Thrall, B.D. (2010) Cellular recognition and trafficking of amorphous silica nanoparticles by macrophage scavenger receptor A. *Nanotoxicology*, epub 2010 September 17.

Ozsezen, A.N. and Canakci, M. (2011) Determination of performance and combustion characteristics of a diesel engine fueled with canola and waste palm oil methyl esters. *Energy Conversion and Management*, **52**(1), 108–116.

Peacock, T.B. (1860) On French millstone makers pthisis. *British Foreign Medical Chair Reviews*, **25**, 214.

Peto, J., Decarli, A., La, V.C., Levi, F. and Negri, E. (1999) The European mesothelioma epidemic. *British Journal of Cancer*, **79**(3–4), 666–672.

Policard, A. and Collet, A. (1952) Deposition of silicosis dust in the lungs of the inhabitants of the Saharan regions. *AMA Archives of Industrial Hygiene and Occupational Medicine*, **5**, 527–534.

Prospero, J. M. (1999). Long-range transport of mineral dust in the atmosphere: Impact of African dust on the environment of the southeastern United States. *Proceedings of the National Academy of Science USA*, **96**, 3396–3403.

Prospero, J.M., Ginoux, P., Torres, O., Nicholson, S.E. and Gill, T.E. (2002) Environmental characterization of global sources of atmospheric soil dust identified with the NIMBUS-7 TOMS Absorbing Aerosol Product. *Reviews of Geophysics*. **40**(1), 1002.

Prospero, J.M. and Nees, R.T. (1986) Impact of the North African drought and El Niño on mineral dust in the Barbados trade winds. *Nature*, **320**: 735–738.

Prospero, J.M. and Carlson, T.N. (1972) Vertical and areal distribution of Saharan dust over the western equatorial North Atlantic Ocean. *Journal of Geophysical Research*, **77**, 5255–5265.

Pye, K. (1987) *Aeolian Dust and Dust Deposits*. Academic Press, London.

Rajkumar, W.S. and Chang, A.S. (2000) Suspended particulate concentrations along the East-West-Corridor. Trinidad, West Indies. *Atmospheric Environment*, **34**, 1181–1187.

Rimal, B., Greenberg, A.K. and Rom, W.N. (2005) Basic pathogenetic mechanisms in silicosis: current understanding. *Current Opinion in Pulmonary Medicine*, **11**(2), 169–173.

Ross, M.H. and Murray, J. (2004) Occupational respiratory disease in mining. *Occupational Medicine (London)*, **54**(5), 304–310.

Roushdy-Hammady, I., Siegel, J., Emri, S., Testa, J.R. and Carbone, M. (2001) Genetic-susceptibility factor and malignant mesothelioma in the Cappadocian region of Turkey. *The Lancet*, **357**, 445–445.

Rovida, C.L. (1871) Un casoi di silicosis del pulmone con analisi chimica. *Ann Chim Appl Med*, **2**, 102–106.

Schultz, L. (1980) Long range transport of desert dust with special emphasis on the Sahara. *Annals of the New York Academy of Sciences*, **338**, 515–532.

Selçuk, Z.T., Cöplü, L., Emri, S., Kalyoneu, A.F., Sahin A.A. and Baris, Y.I. (1992) Malignant pleural mesothelioma due to environmental mineral fiber exposure in Turkey; analysis of 135 cases. *Chest*, **102**, 790–806.

Shao, L., Shi, Z., Jones, T.P., Li, J., Whittaker, A.G. and BéruBé, K.A. (2006) Bioreactivity of particulate matter in Beijing air: results from plasmid DNA assay. *Science of the Total Environment*, **367**, 261–272.

Shao, L., Yang, Y., Wu, M., Xiao, Z. and Zhou, L. (2009) Particle-induced oxidative damage of indoor PM10 from homes in the lung cancer area in Xuanwei, China. Proceedings of the 11th International Conference on Atmospheric Science and Applications to Air Quality. Ji'nan, China. http://www.asaaq.sdu.edu.cn/asaaq

Sipes, K.L. and Mendelsohn, R. (2001) The effectiveness of gasoline taxation to manage air pollution. *Ecological Economics*, **36**, 299–309.

Skinner, H.C.W., Ross, M. and Frondel, C. (1988) *Asbestos and Other Fibrous Materials*. Oxford University Press, New York.

Snider, D.E. (1978) The relationship between tuberculosis and silicosis. *American Review of Respiratory Diseases*, **118**, 455–460.

Song, Z., Wang, J. and Wang, S. (2007) Quantitative classification of northeast Asian dust events. *Journal of Geophysical Research*, **112**, D04211.

Stayton, I., Winiarz, J., Shannon, K., and Ma, Y. (2009) Study of uptake and loss of silica nanoparticles in living human lung epithelial cells at single cell level. *Analytical and Bioanalytical-Chemistry*, **394**(6), 1595–1608.

Stevens, T., Palk, C., Carter, A., Lu, H. and Clift, P.D. (2010) Assessing the provenance of loess and desert sediments in northern China using U-Pb dating and morphology of detrital zircons. *Geological Society of America*, **122**, 1331–1344.

Stone, V., Jones, R., Rollo, K., Duffin, R., Donaldson, K. and Brown, D.M. (2004) Effect of coal mine dust and clay extracts on the biological activity of the quartz surface. *Toxicology Letters*, **149**(1–3), 255–259.

Stratton, T.M.L. (1838) Case of anthrocosis or black infiltration of the whole lungs. *Edinburgh Medicne and Surgery*, **49**, 490–491.

Tanaka, T.Y., Kurosaki, Y., Chiba, M., Matsumura, T., Nagai, T., Yamazaki, A., Uchiyama, A., Tsunematsu, N. and Kai, K. (2005) Possible transcontinental dust transport from North Africa and the Middle East to East Asia. *Atmospheric Environment*, **39**, 3901–3909.

Tapp, E., Curry, A. and Anfield, C. (1975) Sand pneumoconiosis in an Egyptian mummy. *British Medical Journal*, **2**(5965), 276.

Tegen, I. (2003) Modelling the mineral dust aerosol cycle in the climate system. *Quaterary Science Review*, **22**, 1821–1834.

Thackrah, C.T. (1832) *The Effects of Arts, Trades and Professions: And of Civic States And Habits Of Living, on Health And Longevity: With Suggestions'*, 2nd edn. London: Longman, Rees, Orme, Brown, Green and Longman; Simpkin and Marshall.

Todd, M.C., Washington, R., Martins, J.V., Dubovik, O., Lizcano, G., M'Bainayel, S. and Engelstaedter, S. (2007) Mineral dust emission from the Bodélé Depression, northern Chad, during BoDEx 2005. *Journal of Geophysical Research*, **112**, D06207.

Ulm, K., Waschulzik, B., Ehnes, H., Guldner, K., Thomasson, B., Schwebig, A. and Nuss, H. (1999) Silica dust and lung cancer in the German stone, quarrying and ceramics industries: results of a case-control study. *Thorax*, **54**(4), 347–351.

United States Environmental Protection Agency (2003) http://www.epa.gov/region8/superfund/libby/lbybkgd.html.

United States Geological Survey (2000) Health impacts of coal combustion. USGS Fact Sheet FS-094-00, 2pp. Available at: http://pubs.usgs.gov/fs/fs94-00/fs094-00.pdf.

Uno, I., Eguchi, K., Yumimoto, K., Takemura, T., Shimizu, A., Uematsu, M., Liu, Z., Wang, Z., Hara, Y. and Sugimoto, N. (2009) Asian dust transported one full circuit around the globe. *Nature Geoscience*, **2**, 557–560.

Viallat, J.R., Boutin, C., Steinbauer, J., Gaudichet, A. and Dufour, G. (1999) Pleural effects of environmental asbestos pollution in Corsica. *Annals of the New York Academy of Science*, **643**, 438–443.

Viana, M. and Averol, X. (2007) Source apportionment of ambient PM 2.5 at 5 Spanish centres of the European community respiratory health survey (ECRHS II). *Atmospheric Environment*, **41**(7), 1395–1406.

Wagner, J.C., Sleggs, C.A. and Marchand, P. (1960) Diffuse pleural mesothelioma and asbestos exposure in the North Western Cape Province. *British Journal of Industrial Medicine*, **17**, 260–271.

Wagner, G.R. (1997) Asbestosis and silicosis. *The Lancet*, **349**, 1311–1315.

Wang, Z., Ueda, H. and Huang, M. (2000) A deflation module for use in modelling long-range transport of yellow sand over East Asia. *Journal of Geophysical Research*, **105**(D22), 26947–26959 .

Wang, Y.Q., Zhang, X.Y. and Arimoto, R. (2006) The contribution from distant dust sources to the atmospheric particulate matter loadings at Xi'An, China during spring. *Science of the Total Environment*, **368**, 875–883.

Weiss, W. (1993) Asbestos related pleural plaques and lung cancer. *Chest*, **103**, 1854–1859.

Westerholm, P. (1980) Observations on a case register. *Scandinavian Journal of Work and Environmental Medicine*, **9**, 523–531.

Wiggs, G.F.S., O'Hara, S.L., Wegerdt, J., Van Der Meers, J., Small, I. and Hubbard, R.B. (2003) The dynamics and characteristics of aeolian dust in dryland central asia: possible impacts on human exposure and respiratory health in the Aral Sea Basin. *Geographical Journal*, **169**, 142–157.

Witham, C.S., Oppenheimer, C. and Horwell, C.J. (2005) Volcanic ash leachates: a review and recommendations for sampling methods. *Journal of Volcanology and Geothermal Research*, **141**, 299–326.

Wlodarczyk, A., Koshy, L., Jones, T.P. and BéruBé, K. (2008) Reactive oxygen species drives fly ash bioreactivity. In: Proceedings of 11th Annual UK Review Meeting on Outdoor and Indoor Air Pollution Research. Institute of Environment and Health, Cranfield University, Web Report W25, http//www.silsoe.cranfield.ac.uk/ieh/.

World Health Organisation (1986) Asbestos and other natural mineral fibres: Environmental Health Criteria 53. World Health Organisation, Geneva, Switzerland.

World Health Organisation (1999) World Health Organization Health Guidelines for Vegetation Fire Events. World Health Organization, Geneva, Switzerland.

Wypych, F., Adad, L.B., Mattoso, N., Marangon, A.A. and Schreiner, W.H. (2005) Synthesis and characterization of disordered layered silica obtained by selective leaching of octahedral sheets from chrysotile and phlogopite structures. *Journal of Colloid Interface Science*, **283**(1), 107–112.

Xu, X.Z., Cai, X.G. and Men, X.S. (1993) A study of siliceous pneumoconiosis in a desert area of Sunan County, Gansu province, China. *Biomedical and Environmental Science*, **6**: 217–222.

Yin, Y. (2005) A trend analysis on the national pneumoconiosis epidemics in 2003. *Chinese Occupational Medicine*, 2005-05. (in Chinese)

Zhang, Q. and Huang, X. (2002) Induction of ferritin and lipid peroxidation by coal samples with different prevalence of coal workers' pneumoconiosis: role of iron in the coals. *American Journal of Industrial Medicine*, **42**(3), 171–179.

Selected further reading

Dodson, R.F. and Hammar, S.P. (eds) (2006) *Asbestos: Risk Assessment, Epidemiology and Health Effects*, Boca Raton: CRC Press, Taylor and Francis Group, ISBN 13: 9780849328299, ISBN 10: 0849328292.

Durant, A.J., Bonadonna, C. and Horwell, C.J. (2010) Atmospheric and environmental impacts of volcanic particulates. *Elements*, **6**(4), 235–240.

Engelbrecht, J.P. and Derbyshire, E. (2010) Airborne mineral dust. *Elements*, **6**(4), 241–246.

Gehr, P. and Heder, J. (eds.) (2000) *Particle–Lung interactions*, Volume 143 in book series *Lung Biology in Health and Disease,* Marcel Dekker, Inc,. New York-Basel, ISBN 0-8247-9891-0.

Griffin, D.W. (2007) Atmospheric movement of microorganisms in clouds of desert dust and implications for human health. *Clinical Microbiology Reviews*, **20**, 459–477.

Maynard, R.L. and Howard, C.V. (eds.) (1999) Particulate matter: properties and effects upon health. *Bios Scientific Publishers in association with the Royal Microscopical Society*. ISBN 1–85996-172-X.

Plumlee, G.S., Morman, S.A. and Ziegler, T.L. (2006) The toxicological geochemistry of Earth materials: an overview of processes and the interdisciplinary methods used to understand them. *Reviews in Mineralogy and Geochemistry*, **64**, 5–57.

11

Engineered nanomaterials

Superb K. Misra[1], Teresa D. Tetley[2], Andrew Thorley[2], Aldo R. Boccaccini[3] and Eugenia Valsami-Jones[1*]

[1]Natural History Museum, Mineralogy, London SW7 5BD
[2]National Heart and Lung Institute, Imperial College London, London SW3 6LY
[3]University of Erlangen-Nuremberg, Department of Materials Science & Engineering, 91058 Erlangen, Germany
**Corresponding author, email e.valsami-jones@nhm.ac.uk*

11.1 Introduction

A nanometre (nm) is a metric unit of length equal to one billionth of a metre (10^{-9} m), or one thousandth of a micron. The discovery of the unique properties acquired by matter when its dimensions are at the nanoscale has been followed by a technological and scientific explosion in the application of nanomaterials; and the prefix nano (Greek for 'dwarf') is now used for many different products including nano-objects, nanostructures, nanoparticles (NPs), and nanowires (refer to Table 11.1 for definitions of terms). In this chapter emphasis is given to free engineered NPs and nanomaterials, since they may be dispersed and pose a more direct risk to human health and the environment than nanostructured objects such as microprocessor chips or antimicrobial nanosilver particles from the linings of fridges, which are contained or well attached.

Although nanotechnology is a relatively new field of science and technology, naturally occurring NPs have always been present in our environment. The fabrication of nanomaterials is a broad and rapidly evolving field. Recent advances in synthesis and characterisation methods have resulted in a rapid increase in the production of nanostructured materials, and specifically of NPs. The growth of the nanotechnology sector over the past few years indicates the potential it has to develop the commodities market (http://www.nanotechproject.org). The global nanotechnology market was worth $11.6 billion in 2007 and is projected to grow to more than $25 billion by 2013 (http://www.bccresearch.com/pressroom/NAN031C.html).

Artificial nanomaterials can be synthesised by methods which can be broadly grouped into two categories: top-down approach (e.g. grinding or lithography) and bottom-up approach, involving processes such as wet chemical synthesis (Royal Society, 2004). One of the great benefits of manufactured NPs is that they can be synthesised in a wide range of shapes, sizes and with various surface modifications (Figures 11.1 and 11.2), giving a matrix with a large number of different materials with different physical properties. Generally, particles with all three dimensions <100 nm are classified as NPs. In some cases the distinct physicochemical properties associated with the nano domain apply only well below the 100 nm value (Borm *et al.*, 2006).

The significant difference between the behaviour of nanomaterials and that of the bulk material arises from (a) the smaller size of the particles, (b) the surface effects[1] of the particles and (c) the quantum effects (Roduner, 2006). These factors influence the chemical reactivity, mechanical, optical, electrical, and magnetic properties of NPs. In some cases, the changes from the corresponding bulk material are the opposite of what would be predicted. For example, gold is a noble, non-magnetic yellow metal which is a good conductor of heat and electricity, but in the form of small NPs (2–3 nm) it is red, an excellent catalyst, and it exhibits magnetic properties and can behave as an insulator

[1] Surface effects encompasses several parameters such as surface chemistry, specific surface area, surface defects, surface porosity, exposed crystallographic planes, which can all have a significant influence on the properties of nanomaterials.

Pollutants, Human Health and the Environment: A Risk Based Approach, First Edition. Edited by Jane A. Plant, Nikolaos Voulvoulis and K. Vala Ragnarsdottir.

Table 11.1 Common terms in nanotechnology and their definitions from International Organisation for Standardisation (ISO), European Standardisation Committee (CEN), Organisation for Economic Cooperation and Development (OECD), and European Scientific Committee on Emerging and Newly Identified Health Risks (SCENIHR)

Terms	Definitions	Standardisation committee
Nanoscale	Size range from approximately 1 nm to 100 nm	ISO/TC 229
Nano-object	Material with one, two or three external dimensions in the nanoscale	ISO/TC 229
Nanostructure	Having an internal or surface structure at the nanoscale	OECD
Nanoparticle	Nano-object with all three external dimensions in the nanoscale	CEN ISO/TS 27867
Nanowire	Electrically conducting or semi-conducting nanofibre	CEN ISO/TS 27867
Nanomaterial	Material with any external dimension in the nanoscale or having internal structure or surface structure in the nanoscale	ISO/TS 80004–1
Agglomerate	Collection of weakly bound particles or aggregates or mixtures of the two where the resulting external surface area is similar to the sum of the surface areas of the individual components.	ISO/TC 24/SC 4
Aggregate	Particle comprising strongly bonded or fused particles where the resulting external surface area may be significantly smaller than the sum of calculated surface areas of the individual components	ISO/TC 24/SC 4
Nanoplate	Nano-object with one external dimension in the nanoscale and the two other external dimensions significantly larger	CEN ISO/TS 27867
Nanofibre	Nano-object with two similar external dimensions in the nanoscale and the third dimension significantly larger	CEN ISO/TS 27867
Nanotube	Hollow nanofibre	CEN ISO/TS 27867
Nanorod	Solid nanofibre	CEN ISO/TS 27867
Nanoparticulate matter	A substance comprising particles, the substantial majority of which have three dimensions of the order of 100 nm or less	EU SCENIHR
Quantum dots	Quantum dots are crystalline nanoparticles that exhibit size-dependent properties due to quantum confinement effects on the electronic states	CEN ISO/TS 27867

(Roduner, 2006). Compared with microparticles, particles in the nano size range have a very large specific surface area,[2] high particle numbers per unit mass,[3] a higher proportion of their atoms at the surface[4] and a lower binding energy per atom (Roduner, 2006), which significantly enhances the reactivity of NPs. The lower binding energy can affect the physical properties of the material. For example, the melting temperature of 3 nm gold NPs is more than 400 K lower than that of bulk gold (Roduner, 2006).

Quantum dots are increasingly used in biomedical applications (bio-imaging and target-specific gene and drug delivery)

[2] If a cube with 1 cm side is divided into cubes of 1 μm side then the surface area becomes 10,000 times more, and if the cube is divided into cubes of 10 nm a side then the surface area is increased by million times.

[3] By dividing a cube of 1 cm side length to 1 μm side length, the number of particles increases to 10^{12} and if reduced to 10 nm the number of particles amounts to 10^{18}.

[4] The fraction of surface atoms of a 20 μm cubic particle is 0.006 per cent, which increases to 0.6 per cent for a 200 nm particle and almost half of the atoms in the case of a 2 nm particle.

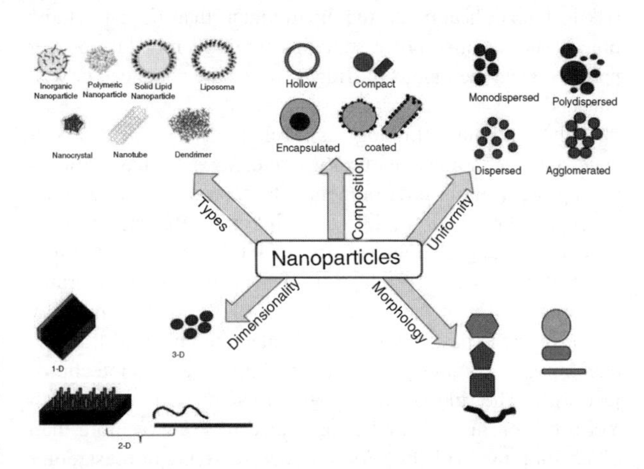

Figure 11.1 Classification of NPs according to their composition, morphology, uniformity, types and dimensionality. The figure is not comprehensive, but provides an indication of the variation in the nanoparticle domain (Buzea *et al.*, 2007; Faraji and Wipf, 2009)

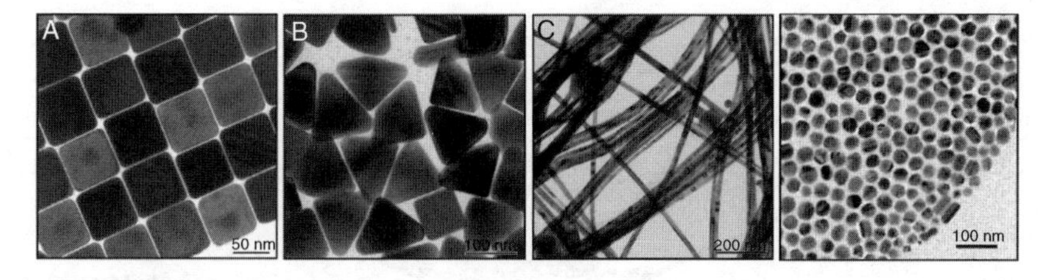

Figure 11.2 Hydrothermally synthesised silver NPs showing the different shapes (nanocubes, nanoprisms, nanowires and nanospheres) that can be produced (Reproduced, with permission, from Journal of Physical Chemistry B (2005) 109, 5497 © American Chemical Society)

and in the electronics industries (LED displays, ultra-high-density data storage, quantum information processing) for their special optical and electrical properties (Hardman, 2006). Due to quantum confinement effects, quantum dots of the same composition (material) but different size can emit light of different colours. Large quantum dots have lower energy and a red fluorescence spectrum, whereas smaller quantum dots have higher energy and emit a blue fluorescence spectrum. Such novel properties in materials, resulting simply from a change in size, have become the driving force for further nanotechnological discoveries, with the potential for breakthroughs in fields as disparate as medicine, energy generation and environmental protection. There is concern, however, about the potential risks of nanotechnology to human health and the environment, which remain unknown. Because of their small size and a large surface area per given mass, nanoparticles are likely to have a direct connection to increased toxicity by reactive atoms and molecules on the surface (Nel *et al.*, 2006). Nanotoxicology is the branch of toxicology that deals with such risks imposed by engineered NPs (Hardman, 2006; Nel *et al.*, 2006; Oberdorster *et al.*, 2007) and combustion-derived nanomaterials (e.g. vehicle emissions, industrial activity). Studies by the Royal Society and the Royal Academy of Engineering in the UK (Royal Society, 2004), the US Environmental Protection Agency (Environmental Protection Agency, 2007) and European Commission (http://cordis.europa.eu/nanotechnology) have encouraged industry and governments to clarify the potential risks posed to human health and the environment.

11.2 Useful and hazardous properties

Knowledge of bulk-material toxicity cannot be extrapolated to that of equivalent nanomaterials. The variety of types of NPs, their properties and the different formulations used for any one nanomaterial suggests that toxicity will depend critically on the final unique product rather than the starting materials and their composition. For any given mass of material there are significantly greater numbers of nano-sized particles than micron-sized particles, and therefore, even at low nanoparticulate concentrations, greater numbers of cells may be exposed. In addition, the chemical(s) remaining in (or on) the NPs following synthesis and formulation may also have cytotoxic effects. Moreover, there are different mechanisms of toxicity exerted by nanomaterials depending on their physicochemical properties (Hoet *et al.*, 2004; Lanone and Boczkowski, 2006; Oberdorster *et al.*, 2005).

The smaller size and greater surface area of nanomaterials would, depending on surface chemistry, confer greater chemical reactivity, induce permeability in biological membranes, or adsorb onto their surfaces macromolecules which, in turn can affect the toxicity of NPs. The increased surface area may also enhance cellular uptake, leading to cytotoxicity from the NPs or from species adsorbed on them which are transported into the cell ('Trojan horse' effect) (Limbach *et al.*, 2007). Various experiments have shown NPs to be toxic to cells, resulting in increased oxidative stress, inflammatory cytokine production, DNA mutation, with the potential to cause major structural damage to mitochondria and cell death (Buzea *et al.*, 2007; Gurr *et al.*, 2005; Risom *et al.*, 2005). Several characteristics, such as chemical composition, size, crystal structure, dopant concentration, shape, surface structure, surface charge, presence/absence of functional groups, aggregation and solubility of NPs combined with multiple exposure routes (Section 11.5), and environmental fates, create a plethora of scenarios that may influence the toxicity of NPs (Buzea *et al.*, 2007; Oberdorster *et al.*, 2005). Hence, it is not possible to generalise the health risks associated with exposure to NPs and each new type of NP must be assessed individually taking account of all its properties. At present, nanomaterials are developed to exploit their unique properties such as their antimicrobial, photocatalytic or electrical activity and in terms of current commercialisation and large-scale production; there are only a handful of such NPs available. In this chapter, emphasis is given to four such key nanomaterials; zinc oxide NPs, silver NPs, fullerenes and carbon nanotubes.

The differences between the properties and behaviour of NPs and those of their bulk counterpart make it imperative to have a systematic and mechanistic understanding of the relationship between the toxicity and physicochemistry of NPs. Although dose, dimension and durability of NPs are considered to be important parameters in determining their adverse health effects, recent studies show a more complex relationship between other physicochemical properties and their associated health effects. Some of the parameters of NPs that may have a bearing on their toxicity are shown in Figure 11.3 and ways of measuring them are shown in Table 11.2 (Buzea *et al.*, 2007; Handy *et al.*, 2008; Hassellov *et al.*, 2008).

Several toxicological studies have related the small size of NPs to toxicity by demonstrating that NPs cause adverse respiratory health effects, typically causing more inflammation than

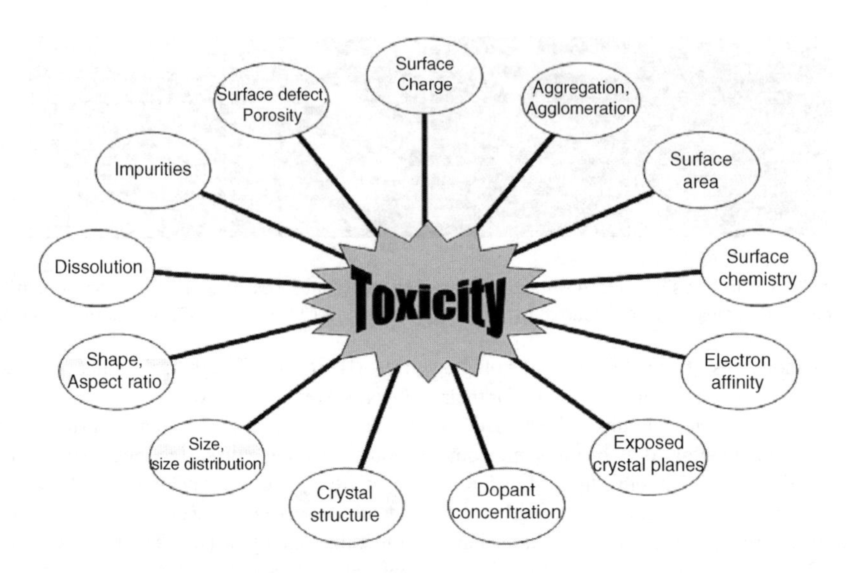

Figure 11.3 Physicochemical properties influencing the toxicity of NPs

larger particles made from the same material (Gurr *et al.*, 2005; Oberdorster *et al.*, 1994; Oberdorster *et al.*, 2005; Yu *et al.*, 2009). *In vivo* studies on rats demonstrated that 20 nm titanium dioxide (TiO_2) NPs were more pro-inflammatory in the lungs than 250 nm TiO_2 particles (Oberdorster *et al.*, 1994, 2005). In the post-exposure period (up to one year), 20 nm TiO_2 particles had significantly prolonged retention and increased translocation to the pulmonary interstitium, and caused greater impairment of alveolar macrophage function (Oberdorster *et al.*, 1994). Along with size, the size distribution and state of aggregation/agglomeration can also play a vital part in its biological role.

Table 11.2 Physicochemical properties of NPs with the potential to affect their toxicity, and analytical methods suitable for their determination

Properties	Techniques available
Size, size distribution	SEM, TEM, AFM, FFF, DLS, DCS, BET, PTA
Shape	SEM, TEM, AFM
Composition and purity	SAXS, EXAFS, ICP-AES, ICP-MS, XRD, FTIR, Raman
Surface chemistry	XPS, FTIR, Raman
Surface charge	Zeta potential
Surface area/porosity	BET
Crystal structure	XRD, TEM
Concentration	ICP-AES, ICP-MS, PTA
Agglomeration	DLS
Redox potential	Cyclic voltammetry, XPS

SEM, scanning electron microscopy; TEM, transmission electron microscopy; AFM, atomic force microscopy; FFF, fluid flow fractionation; DLS, dynamic light scattering; DCS, disk centrifuge system; SAXS, small-angle X-ray scattering; EXAFS, extended X-ray absorption fine structures; ICP-AES, inductively coupled plasma-atomic emission spectroscopy; ICP-MS, inductively coupled plasma-mass spectroscopy; XRD, X-ray diffraction; FTIR, Fourier transform infrared; XPS, X-ray photoelectron spectroscopy; PTA, particle tracking analysis; BET, Brunauer, Emmet, and Teller

The shape-specific toxicity of NPs is also important. In the case of asbestos it has been shown that the aspect ratio and biopersistence of particles is closely related to long-term carcinogenic effects (Lippmann, 1990; Oberdorster, 2002). For example, induction of lung cancer in experimental animals was associated with the presence of asbestos fibres longer than 10 μm, mesothelioma with fibres longer than 5 μm, and asbestosis with fibres longer than 2 μm (Lippmann, 1990). It is difficult for long fibres (longer than 20 μm in the case of humans) to be cleared from the respiratory tract effectively, partly because of the inability of macrophages to phagocytise the fibres (Hoet *et al.*, 2004). Carbon nanotubes which have a high aspect-ratio are a class of nanomaterials that have recently attracted a lot of attention because of their similar morphology to asbestos (Lam *et al.*, 2004; Muller *et al.*, 2005; Shvedova *et al.*, 2005). A recent study by Poland *et al.* (2008) showed that intraperitoneal administration of long, thin carbon nanotubes (i.e. with high aspect ratios: tubes with diameter 85 and 165 nm and length 13 and 56 μm, respectively), resulted in a similar toxic response (pleural granuloma) to long thin asbestos fibres, although this specific toxic response was not observed for short thin nanotubes (diameter 15 and 10 nm and length 5 and 15 μm). Similarly, pulmonary deposition of high-aspect-ratio carbon nanotubes in rodents induced significantly more pulmonary toxicity than spherical particles (amorphous carbon black) (Bottini *et al.*, 2006; Lam *et al.*, 2004; Shvedova *et al.*, 2005).

Particles can have the same composition but different crystalline structures. For example, titanium dioxide has three polymorphs (rutile, anatase and brookite) with different structures; they have different chemical and physical properties and hence potential toxicity. This difference must not be overlooked when using TiO_2 NPs for toxicity studies. Studies comparing the polymorphs of TiO_2 demonstrate that rutile NPs (200 nm) induce oxidative DNA damage in the absence of light, but anatase NPs of the same size do not (Gurr *et al.*, 2005). An important factor to consider in such studies is the ability of NPs to change their crystal structure after interaction with water or liquids. For example, zinc sulphide NPs (3 nm) rearrange their crystal structure in the presence of water and become closer to the structure of bulk zinc sulphide (i.e. sphalerite) (Zhang *et al.*, 2003). Therefore, it is essential to evaluate the surface chemistry and structure of NPs at the same time as determining their toxicity, as these can in turn affect the cellular uptake and sub-cellular localisation, and trigger the production of reactive oxygen species (ROS) (Long *et al.*, 2007; Risom *et al.*, 2005; Donaldson *et al.*, 2003).

The surfaces of the NPs are being modified using capping agents, dispersants, fluorescent labels, proteins and natural organic matter to incorporate numerous functionalities for different applications. The presence of such chemical moieties on the surface of particles changes their surface chemistry and influences their physicochemical properties such as their electrical and optical properties, their chemical reactivity, their aggregation and their surface charge in any given environment, which can in turn affect the cytotoxicity of NPs (Oberdorster *et al.*, 2005; Gupta and Gupta, 2005). Surface coatings can also give pseudo-toxic effects to NPs in certain cases. For example, the specific cytotoxicity of silica is strongly affected by the presence of surface radicals and ROS (Hoet *et al.*, 2004); the formation of blood clots in hamsters has been shown to increase when the surface of polystyrene NPs is modified with amine groups (Nemmar *et al.*, 2002a); the dispersion and biocompatibility of carbon nanotubes can be improved by functionalising the nanotubes with a range of chemicals (Lin *et al.*, 2004); and nickel ferrite particles, with and without oleic acid on the surface, show different cytotoxicity (Yin *et al.*, 2005).

The traditional toxicological approach of correlating observed toxic effects to units of mass might be unsuitable in the case of NPs. For the same mass of material, the number of particles and the corresponding specific surface area is greater than that of micron-sized or bulk material. Therefore, the body's reaction to billions of NPs differs from its reaction to the same mass dose consisting of several microparticles. As mentioned previously, Oberdorster *et al.* (2000) showed that for the same mass dose, 20 nm TiO_2 particles induced a greater inflammatory response than 250 nm TiO_2 particles during an intra-tracheal instillation study on mice. However, for the same surface area dose of TiO_2 in rodent lungs, there were no significant differences between the toxicity caused by the 20 nm and 250 nm TiO_2 particles (Oberdorster *et al.*, 2000). This increase in surface area for NPs also has a direct effect on the surface reactivity of the particles and can be a source of ROS causing intracellular damage (Long *et al.*, 2007; Donaldson *et al.*, 2003; Risom *et al.*, 2005). As shown experimentally, ROS and free-radical production may result in oxidative stress, inflammation, and consequent damage to proteins, membranes and nucleic acids (Nel *et al.*, 2006).

Oxidative stress is a result of an imbalance between production of ROS or reactive nitrogen species (RNS) and the antioxidant defence system of the body. Oxidative stress has been shown to be an important mechanism of particle-induced health effects, and concerted efforts are made to develop screening tests based on oxidative-stress biomarkers to regulate nanomaterials. The NPs can induce oxidative stress through several different processes (Buzea *et al.*, 2007). (1) ROS can be generated directly from the surface of NPs due to the presence of oxidants, free radicals. (2) Dissolution of NPs can release metals, which can then generate ROS by acting as catalysts in Fenton-type reactions (e.g. reduction of hydrogen peroxide with ferrous ion leads to the formation of hydroxyl radical). (3) NPs can enter mitochondria causing physical damage and contributing to oxidative stress. (4) Activation of inflammatory cells induced by phagocytosis can lead to generation of ROS and RNS. However, just as ROS can disturb the normal redox state of cells and cause toxic effects, they can also be used by the immune system to attack or kill pathogens (refer to the case study of C60 in Section 11.2.3.2).

The other side of almost any toxicology is a potential therapy. Just as the unique properties of an NP can make it toxic, the same properties also have huge potential benefits for use of NPs in diagnostics and targeted therapeutics. One such example discussed in this chapter is C60-fullerene whose radical scavenging property is used to develop nanomedicine. Considering that nanotechnology is still at its developing stage, there is a concern that its development should avoid damaging public perception of the field, as has happened with other novel technologies such as genetically modified organisms (GMOs).

11.2.1 Zinc oxide (ZnO) NPs

ZnO NPs are becoming increasingly popular in applications such as sunscreens (Nohynek *et al.*, 2007), water remediation, antimicrobial, bioimaging and gas sensing. High UV absorption efficiency and transparency to visible light makes ZnO NPs ideal for applications as sunscreens, coatings and paints. There are also reports of ZnO NPs exhibiting antimicrobial properties (Liu *et al.*, 2009). Two aspects of ZnO NPs are discussed in this section, dissolution of ZnO NPs and traceability of ZnO NPs.

11.2.1.1 Dissolution of ZnO

To evaluate the toxicology of nanomaterials, it is important to determine whether the observed toxicity is due to the nanomaterial itself or its dissolved species. Dissolution of NPs is therefore important to determine the toxic effects of the NPs as a function of the exposure conditions. For example, if ZnO NPs are soluble, they will deliver a mixed NP/ionic toxic effect until they dissolve completely. This, in turn, will have a major impact in toxicity experiments, where the ratio of nanoparticulate to dissolved zinc will change throughout the experiment and, in some cases (e.g. at low exposure concentrations), complete dissolution of NPs may occur so that only an ionic effect is tested. On the other hand, if ZnO NPs are sparingly soluble, in short-term exposures only the NP effect should be significant. Recent studies with ZnO NPs have shown that zinc ions (Zn^{2+}) can be the source of toxicity. Several researchers[5] have highlighted the differences in dissolved zinc concentrations released from ZnO NPs. For example, dissolved zinc concentrations of 12.35 μg/ml in bronchial epithelial growth medium (Xia *et al.*, 2008), 14.63 μg/ml in DMEM cell culture medium (Xia *et al.*, 2008), 10 μg/ml in cell culture medium (Song *et al.*, 2010), 60 μg/ml in DMEM-FBS media (Horie *et al.*, 2009), 7.6 μg/ml in simulated uterine fluid (Yang and Xie *et al.*, 2006) and 3.7 μg/ml in sea water (Wong *et al.*, 2010).

Franklin *et al.* (2007) reported dissolution of ZnO NPs (ca 30 nm, aggregated) and bulk ZnO in water to be 8 μg/ml by 6 h and 16 μg/ml (i.e. 19 wt per cent) after 72 h. There was no significant difference in the rate of dissolution and saturation solubility of bulk ZnO and ZnO NPs. Further toxicity experiments on fresh-water algae with the particles showed that ZnO NPs, bulk-ZnO and ionic Zn had comparable toxicity (72 h IC50 value 60 μg Zn/l), attributed solely to dissolved zinc. Brunner *et al.* (2006) also inferred that toxic effects of ZnO NPs on MSTO or 3T3 cells may be attributed to dissolved zinc. Deng *et al.* (2009) showed that ZnO NPs and zinc chloride ($ZnCl_2$) had the similar toxic effect on mouse neural stem cells, without measuring the dissolution of ZnO NPs, which could have possibly contributed to the toxicity. Song *et al.* (2010) showed that the dissolved Zn^{2+} concentrations reached equilibrium in cell culture medium at concentrations of higher than 40 μg/ml of ZnO NPs, with the equilibrium concentrations of dissolved Zn^{2+} at 10 μg/ml. At this equilibrium concentration, dissolved zinc ions severely reduced cell viability by 50 per cent, and the toxic effect was similar to that of $ZnCl_2$ (IC50 of 13.33 μg/ml). For dosage less than 40 μg/ml the toxicity was mainly due to dissolved zinc ions, and for concentrations greater than 40 μg/ml the increase in the toxicity was attributed to increased concentration of ZnO NPs.

11.2.1.2 Traceability of ZnO

An important aspect of the toxicological evaluation of NPs is to determine the bioaccumulation of NPs for an organism, which will then determine the toxicological impact a nanoparticle may or may not have for the organism. A US Environmental white paper on nanotechnologies published in 2007 recommended the need to include careful tracking of uptake and disposition of nanoparticles to understand toxicity as a function of dose at the site of action. Differentiating between background concentrations of elements and those from NPs is difficult at the low levels of concentration that are environmentally relevant. There are a few ways of labelling nanoparticles (fluorescent markers, radioactive labelling and stable isotope labelling) (Gulson and Wong, 2006). However, each of these techniques has certain limitations, ranging from radioactivity to disassociation of markers from the NPs. Dybowska *et al.* (2011) showed for the first time that stable isotopically modified ^{67}ZnO[6] NPs can be synthesised and used for bioaccumulation studies. By feeding diatoms spiked with ^{67}ZnO NPs to *Lymnaea stagnalis* (a fresh-water snail), the sensitivity to measure the Zn concentration at environmentally realistic concentration is vastly improved. For example, if no tracer is used, Zn can be detected in an organism only at concentrations greater than 5000 μg/g but it can be detected at 15 μg/g if isotopically modified ^{67}ZnO NPs is used. Although the importance of isotopically modified NPs is that they allow ecotoxicological tests of NPs to be conducted at environmentally realistic exposure conditions, there has been very limited research into the use of stable-isotope NPs because of the lack of suitable isotopic precursors for NP synthesis and isotopically modified NPs.

11.2.2 Silver (Ag) NPs

Silver has been amongst the most common materials used by humans. Applications include ornaments, currencies, dental materials, photography, electronics and optics (Edward-Jones, 2009). Silver has also been exploited for its medicinal properties, due to its broad-spectrum toxicity to bacteria. Since the time of Hippocrates (known as the 'father of medicine'), who wrote about its beneficial healing properties (Magner, 1992). Since then, the germicidal effect of Ag has been used in storage (e.g. Ag coins were put in milk bottles and water barrels to prevent spoiling) and for surgical use (e.g. the use of Ag in amalgam for dental applications, and of Ag compounds to prevent infection in the first world war). The use of Ag and its compounds has changed from the simple use of silver sulphadizane to treat serious burns and of Ag nitrate in eye drops to more sophisticated applications such as Ag-coated wound dressings,[7] Ag-alloy catheters and Ag-coated endotracheal breathing tubes. While consumption of colloidal Ag suspensions (in homeopathic solutions) as health supplements has been common, excessive consumption can lead to argyria (characterised by pigmentation or discoloration of the skin, nails, eyes, mucous membranes or internal organs) (Drake and Hazelwood, 2005). The widespread use of Ag for medicinal and biomedical

[5] Note that the ZnO NPs used in the studies referred to here were not all of same size; and that different experimental set-ups were used for solubility studies.

[6] ^{67}Zn is one of the five stable isotopes of zinc, with a natural abundance of 4.1 per cent.

[7] A review published by Cochrane Collaboration found insufficient evidence to recommend the use of Ag-treated dressings to treat infected wounds (Lo *et al.*, 2008).

applications is primarily due to the oligodynamic effect of Ag ions and Ag compounds, which show toxic effects on bacteria, viruses and algae, but lower toxicity to humans. Although the mechanism of Ag toxicity is not clearly understood, recent results suggest that Ag ions are the primary cause of Ag toxicity to bacteria. These ions interact and form complexes with thiol groups present in enzymes and proteins, and react with amino, carboxyl, phosphate and imidazole groups, resulting in inactivation of the respiratory enzymes leading to the production of ROS (Hwang *et al.*, 2007; Kim *et al.*, 2007). Silver ions also inhibit DNA replication by interfering with DNA unwinding. Silver has been reported to induce oxidative stresses at the cell wall, which may affect bacterial growth (Kim *et al.*, 2007). Silver toxicity in higher organisms is thought to be similar to that in bacteria, by depleting energy reserves as a result of the effects on cell metabolism and DNA synthesis.

Despite the fact that Ag has been used for centuries, the need for nanosized Ag stems from the effectiveness of the Ag NPs to deliver ionic silver. The increased specific surface area and the dispersibility of Ag NPs give an advantage over bulk Ag for antimicrobial applications. Additionally, Ag NPs can be readily incorporated in present-day commercial products, where they optimise the duration and rate of release of Ag ions. Due to the long-established toxicity of Ag to bacteria and its apparent limited toxicity to humans, commercial products generating Ag ions or containing Ag NPs are becoming increasingly common in consumer products (see Table 11.5). The assumption behind the widespread use of Ag NPs in consumer goods derives from the observation (arising from earlier use of Ag and its compounds) that although Ag is taken into the body it is not known to be toxic to humans except at very high concentrations. However, extrapolation of this behaviour of bulk Ag to Ag NPs needs to be verified experimentally, and more research is required to understand the mechanism of biointeraction of Ag NPs. Recent studies have demonstrated that Ag ions and Ag NPs can be toxic (Lubick, 2008) and that Ag NPs can act as Trojan horses, whereby Ag NPs pass through the cell membrane and, once inside the cell, release Ag ions, which can be toxic (Limbach *et al.*, 2007; Luoma, 2008). Additionally, in the light of the widespread use of Ag NP embedded substances, it remains to be seen whether the concentration of Ag to which humans are exposed will have a similar effect to that of bulk Ag. Choi *et al.* (2008) investigated the inhibitory effects of Ag NPs and other important Ag species (silver nitrate ($AgNO_3$), silver chloride (AgCl) colloids) on microbial growth. Their results demonstrated that nitrifying bacteria are inhibited more by Ag NPs than by Ag ions or AgCl, suggesting that Ag NPs could have detrimental effects on the micro-organisms in waste-water treatment.

Recent technological advances in NP synthesis allow the use various capping agents to modify the surface of the Ag NPs and their surface-related properties (Figure 11.4), and it should be possible to initiate a time-dependent release of Ag ions with an important role in their fate and the toxicity they may cause. At this stage it is imperative to analyse each NP format individually and not to oversimplify or generalise the results deduced from experiments using a certain type of NP. Some of the other important research findings involving capped Ag NPs are summarised in Table 11.3.

11.2.3 Fullerene

A fullerene is any molecule composed entirely of carbon. Spherical fullerenes are called buckyballs, and cylindrical ones are called carbon nanotubes (discussed in the following section) or buckytubes. Buckminsterfullerene (also known as C60) was the first molecule discovered from the fullerene family by Kroto *et al.* (1985) and is a spherical molecule (diameter of 0.7 nm) with 60 linked carbon atoms in a highly stable icosahedron structure, consisting of 60 vertices and 32 faces (12 pentagonal and 20 hexagonal) (Figure 11.5). The unique chemistry of fullerenes leads to high electroconductivity, high tensile strength and unique thermal and optical properties, which allow it to be used for applications in material science, electronics (optics and semiconductors) and medicine (drug delivery). Recently, some cosmetic products have also been reported to use C60 (Lens, 2009). With its small size, C60 resembles many pharmaceutical molecules and is better able to quench various free radicals (acting as a free-radical sponge) to its cage structure than conventional antioxidants. Conversely, C60 can also generate singlet oxygen and other reactive oxygen species (C60 absorbs strongly in the UV and moderately in the visible regions of the spectrum) that can cause cell damage (Bosi *et al.*, 2003; Markovic and Trajkovic, 2008). The physical chemistry of C60-mediated ROS generation and radical quenching is comprehensively discussed in several reviews (Markovic and Trajkovic, 2008; Nielsen *et al.*, 2008) and is not discussed here. However, its ROS-generating and free-radical-quenching ability has allowed C60 to be used for various biomedical applications, as discussed in the next section.

11.2.3.1 Anticancer and antimicrobial activity

Fullerenes in general can photosensitise the transition of molecular oxygen to highly reactive ROS, which could make them useful for photodynamic killing of cancer cells (Mroz *et al.*, 2007a). In addition to singlet oxygen producing capacity, degree of cell membrane incorporation and cellular uptake, they might also influence the phototoxicity of C60-based agents. There are several ways of using C60 for anticancer applications. (1) Selective tumour-specific activation of photosensitising agents by highly focused light beam (e.g. optical fibres) delivered to tumour region (Brown *et al.*, 2004). (2) Linking of fullerenes with other photosensitisers (porphyrin), which enables the C60-porphyrin dyads to generate ROS-mediated cytotoxicity, even if oxygen levels are low (Alvarez *et al.*, 2006). Several studies have shown the efficient photodynamic action of C60 derivatives against different types of cancer cell lines and *in vivo* tumours (Mroz *et al.*, 2007a). The observed anticancer activity of derivatised fullerene appears to depend on the generation of both singlet oxygen and superoxide anions, and is

Table 11.3 Summary of some of the toxicity results for Ag NPs using different capping agents

Type	Size (nm)	Capping agent	Cells/Organism	Comments	References
S	9–21	PVA	Nitrifying bacteria	5 nm Ag NPs were found to be more toxic to bacteria	Choi and Hu, 2008
C	1–100	Carbon matrix	Gram negative bacteria	NPs (<10 nm) mainly attach to the surface of the cell membrane and are able to penetrate inside the bacteria and cause further damage	Morones *et al.*, 2005
S	1–10	PVP	HIV-I virus	Ag NPs in the range 1–10 nm specifically bind to the virus and inhibit the virus from binding to host cells	Elechiguerra *et al.*, 2005
S	27–40	Tween-80 PEG PVP	Eukaryotic organism	At concentrations below 25 mg/l no toxicity was observed due to exposure of Ag NPs, whereas ionic Ag showed toxicity even at 0.4 mg/l	Kvitek *et al.*, 2009
C	7–10	Polyethylenimine	Human hepatoma cells	At low doses (<0.5 mg/l) Ag NPs accelerated cell proliferation. However, at high doses (>1.0 mg/l) there was significant cytotoxicity	Kawata *et al.*, 2009
S	6–20	Starch	Lung fibroblast cells	Ag NPs reduced ATP content in cells, caused damage to mitochondria and increased ROS production in a dose-dependent manner	Asharani *et al.*, 2009
S	25	Uncoated and polysaccharide coated	Mouse embryonic stem cells and fibroblasts	The presence of Ag NP induced cell death, up regulation of p53 protein and DNA damage protein. Polysaccharide-coated NPs showed more damage than uncoated NPs	Ahamed *et al.*, 2008
S	5–20	Starch and Bovine serum albumin	Zebrafish embryos	Embryos exhibited phenotypic defects, altered physiological functions, namely bradycardia, axial curvatures and degeneration of body parts Exposure of Ag showed presence of NPs in the brain of the embryos and also resulted in accumulation of blood in different parts of the body, thereby causing edema and necrosis	Asharani *et al.*, 2008

S, synthesised; C, commercial

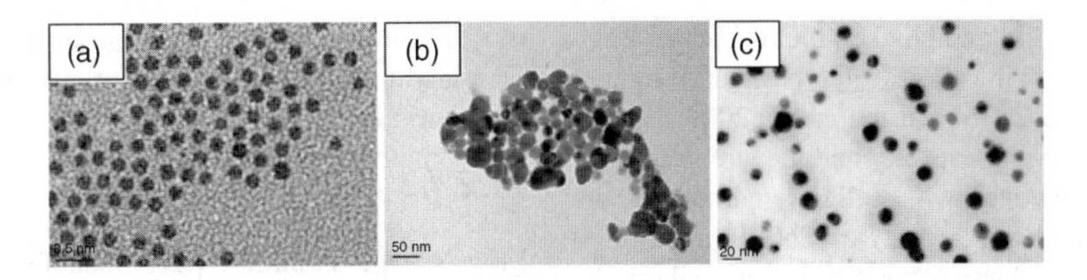

Figure 11.4 Silver NPs prepared using different capping agents (a) dodecylamine, (b) citrate and (c) starch (© Superb Misra)

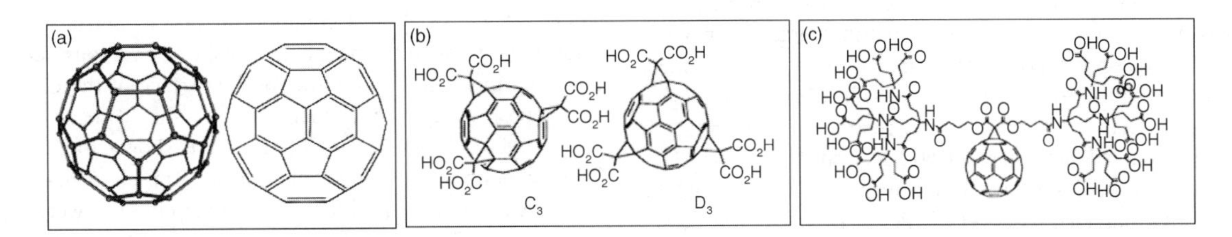

Figure 11.5 The structure of (a) C60, (b) carboxyfullerenes and (c) water-soluble dendrimeric fullerene (Reproduced, with permission, from Biomaterials (2010) 29, 3561 © Elsevier; Reproduced, with permission, from European Journal of Medicinal Chemistry (2003) 38, 913 © Elsevier)

inversely correlated with the extent of derivatisation of the fullerene cage (Alvarez *et al.*, 2006; Mroz *et al.*, 2007b). The photodynamic anti-tumour action of water-soluble C60 derivatives apparently involves induction of Type I apoptotic cell death (Alvarez *et al.*, 2006; Mroz *et al.*, 2007b). Recent results have also indicated the possibility of C60 exerting its antiproliferative and pro-apoptotic action, not only by producing cell-damaging ROS, but also through antioxidant effects (Gloire *et al.*, 2006). For example, solvent exchange-prepared THF/C60 NPs exerted oxidative stress-mediated necrotic cell death (Harhaji *et al.*, 2008) and at low doses might still affect tumour cells by inducing Type II cell cycle through arrest and autophagy (Harhaji *et al.*, 2007).

Along with killing tumour cells, C60 has also demonstrated its antibacterial properties on various bacterial strains through photodynamic inactivation of bacteria (Mroz *et al.*, 2007a). The bactericidal effects of C60 were more pronounced in Gram positive bacteria than in Gram negative bacteria, suggesting that the antibacterial effect could be dependent on C60 interaction with the microbial cell wall and altering the membrane lipid composition and membrane fluidity in response to oxidative stress (Fang *et al.*, 2007; Tsao *et al.*, 2002). Cationic C60 derivatives have been shown to be more effective than neutrally or negatively charged ones, presumably due to stronger interaction with bacterial cell walls (Tang *et al.*, 2007). Pure C60 and C60 derivatives have also been reported to inactivate viruses (belonging to Togaviridae, Rhabdoviridae and Flaviviridae families) and some bacteriophages, through either photodynamic ROS-dependent or ROS-independent actions (Kasermann and Kempf, 1997; Badireddy *et al.*, 2007; Lin *et al.*, 2000). For example, incorporation of specifically tailored C60 derivatives into the active sites of HIV protease has been shown to inhibit HIV replication (Friedman *et al.*, 1998).

11.2.3.2 Antioxidant and cytoprotective activity

Oxidative stress related cellular injuries are prevalent in many pathological conditions, such as autoimmunity, atherosclerosis, diabetes, and neurodegenerative disorders (such as Alzheimer's disease, Parkinson's disease, amyotrophic lateral sclerosis and HIV-associated dementia) (Droge, 2002). The diminished capacity of the central nervous system to prevent oxidative damage makes therapies targeting against ROS generation a promising approach to minimise oxidative stress-mediated neurodegenerative injuries. C60 and derivatised C60 (such as fullerols or carboxyfullerenes) have been suggested for cytoprotective antioxidant treatment because of their ability to associate with mitochondria and to quench ROS more efficiently than conventional antioxidants (Markovic and Trajkovic, 2008). The mechanism for ROS deactivation-dependent cytoprotective action of fullerol and carboxyfullerenes involves interference with the oxidative stress-mediated induction of apoptotic cascades (Chirico *et al.*, 2007; Harhaji *et al.*, 2008). Carboxyfullerenes are reported not only to reduce age-associated oxidative stress in the brain, but also to extend the lifespan and prevent age-related cognitive impairment in mice (Quick *et al.*, 2008). Fullerol (polyhydroxylated C60) has been shown to protect rat lungs from ischemia-reperfusion injury and reduce cell death triggered by proinflammatory cytokine TNF, hydrogen peroxide or nitric oxide (NO) (Chen *et al.*, 2004; Isakovic *et al.*, 2006; Harhaji *et al.*, 2008), while carboxy-C60 protected zebrafish embryos exposed to ionising radiation and blocked the oxidative stress-mediated apoptosis induced in various cell types by UV light or cytokines (Daroczi *et al.*, 2006; Chirico *et al.*, 2007; Harhaji *et al.*, 2008). ROS scavenging capacity and the strength of the interaction of C60 derivatives with cellular membranes are essential for the cytoprotective activity of C60. In addition, it has been shown that fullerol can prevent cell death induced by high concentrations of NO (Mirkov *et al.*, 2004).

The cytoprotective activity of C60 NPs has been well reported in the literature (Xiao *et al.*, 2005; Takada *et al.*, 2006). For example polyvinylpyrrolidone (PVP)/C60 NPs reduced articular cartilage degeneration in a rabbit model of osteoarthritis, suppressed production of extracellular matrix-degrading enzymes (matrix metalloproteinases) and reduced apoptosis in human chondrocytes *in vitro* (Yudoh *et al.*, 2007). Although the cytoprotective activity of C60 NPs has been reported in the literature, it has been observed that different C60 NPs display different ROS-scavenging properties in terms of the quenching capacity and the type of ROS targeted (Xiao *et al.*, 2005; Huang *et al.*, 2003). Whether or not these differences have an effect on the cytoprotective potency of the C60 NPs remains to be investigated. Despite the excellent prospects of using fullerenes in biomedicine, their use depends on the availability of sufficient material and an adequate understanding of structure-function relationships.

11.2.4 Carbon nanotubes (CNTs)

CNTs possess unique structural, mechanical, thermal, electrical and chemical properties, which have led to a multitude of proposed applications in the field of manufacturing, electronics, and biomedical sciences (Sinha and Yeow, 2005). They are well-ordered, high-aspect-ratio allotropes of carbon and can essentially be thought of as sheets of carbon one atom thick (known as graphene) rolled up into a cylinder. There are two main forms: single-walled carbon nanotubes (SWCNTs) and multi-walled carbon nanotubes (MWCNTs). They differ in the arrangement of their graphene cylinders, whereby SWCNTs have only one single graphene sheet (0.4–2 nm in diameter) and MWCNT consist of multiple concentric graphene cylinders of increasing diameter (2–100 nm). The promising properties of CNTs have to be weighed against the potential toxicity they may cause and also their fate *in vivo*. Although, several toxicological, and biocompatibility studies have been conducted on modified and raw CNTs, there are several discrepancies in the published results, which create uncertainty. While several *in vitro* studies have shown CNTs to be cytotoxic, some have reported them to be an excellent substrate for cell growth. Table 11.4 gives selected experimental results for the biological response of CNTs. One of the main reasons for concern is the potential of CNTs to be manufactured in large quantities and also to be synthesised with a very high aspect ratio. As discussed earlier, recent studies have even indicated that high-aspect-ratio CNTs may behave like asbestos and induce pulmonary carcinomas (Poland *et al.*, 2008). Whilst considering the apparently contradictory results on the toxicity of CNTs it is important to consider some of the reasons that may lead to differences in the observed results.

11.2.4.1 Impurities

Most of the common CNT manufacturing techniques use metal catalysts for CNT growth. As a result, residual metal catalysts (iron and nickel) and amorphous carbon can be present in the CNT samples as contaminants, and these can be a major factor in inducing oxidative stress and cytotoxic response (Pulskamp *et al.*, 2007). Acid treatment of CNTs (Dillon *et al.*, 1999) is a common method of purification, to oxidise and remove the metal catalysts, carbonaceous deposits and other impurities and allow for the surface of the CNTs to be appropriately functionalised in the process (Lin *et al.*, 2004). Kagan *et al.* (2006) have shown that impure CNTs containing 26 weight per cent of iron can cause more oxidative stress than purified CNTs containing 0.2 weight per cent of iron. This highlights the need to identify and measure the concentration of any impurities in CNTs that are used for toxicological studies.

11.2.4.2 Functionalisation

CNTs are functionalised for a wide variety of reasons, ranging from improving their dispersive properties to improving their biocompatibility (Lin *et al.*, 2004). For example, CNTs functionalised with glycopolymers can show a high degree of *in vitro* cytocompatibility. CNTs, being hydrophobic, tend to aggregate and are therefore functionalised to improve their dispersive properties. Aggregation is an important property, as experiments have shown that mortality caused by intratracheal instillation of SWCNTs into rats is due to agglomeration of the CNTs in the major airways, instead of just the inherent toxicity of SWCNTs (Warheit *et al.*, 2004). Similarly, higher bacterial cytotoxicity is observed when CNTs are uncapped, de-bundled, short, and dispersed in solution (Kang *et al.*, 2008). Therefore, careful consideration must be given while performing toxicological studies on functionalised CNTs. Finally, while designing the exposure experiment it is important to use a physiologically relevant concentration of NPs, as it is known that high concentration of NPs can lead to particle agglomerations or aggregations and therefore reduce or alleviate the toxic effects compared to lower concentrations (Gurr *et al.*, 2005; Takenaka *et al.*, 2001; Warheit *et al.*, 2004).

11.2.4.3 Exposed vs unexposed

In the field of biomaterials, CNTs are increasingly used as fillers and coatings in polymeric and ceramic matrices. As reviewed by Smart *et al.* (2005), experiments conducted on polymer and CNT composites have shown favourable cellular responses without significant cytotoxicity. However, such experiments do not measure the toxicity of CNTs (they

Table 11.4 Selected experimental results obtained for biological response of CNTs

Samples	Cells	Observation	References
SWCNT and MWCNT	Alveolar macrophages	Increase in cytotoxicity after 6 h of exposure SWCNT significantly impaired phagocytosis at low doses, showed dose-dependent cytotoxicity MWCNT at dosage of 3 mg/cm^2 resulted in necrosis and degeneration Phagocytic ability reduces with increase in dosage	Jia *et al.*, 2005
0.23 wt% Fe containing purified SWCNT vs 26 wt% Fe containing unpurified SWCNT	Human macrophages	26 wt% Fe containing SWCNT showed significantly more toxicity than 0.23 wt% Fe containing SWCNT Neither type of SWCNT produced superoxides within the cells. Cell death through oxidative stress	Kagan *et al.*, 2006
MWCNT	Human fibroblasts	Dose-dependent effect of gene expression At 0.6 mg/ml serious impact on the cellular functions in maintenance, growth and differentiation occurred Cell apoptosis occurring with increasing dose	Ding *et al.*, 2005
Carbon black (14 nm)	Human type II alveolar epithelial cells	Increased in vitro oxidative stress Increased murine alveolar macrophage migration in fetal calf serum Long-term, sub-chronic inhalation can cause the development of pulmonary tumours in rats	Barlow *et al.*, 2005; Stone *et al.*, 1998; Gallagher *et al.*, 2003
SWCNT	Epidermal keratinocytes	Produces oxidative stress and cellular toxicity. Cell death through oxidative stress Time and dose dependent reduction in cell viability Concentration of dead cells increases with time and dose upon exposure of SWCNTs. Also showed DNA damage Time and concentration dependent release of cytokine IL-8	Shvedova *et al.*, 2003 Manna *et al.*, 2005
MWCNT	Epidermal keratinocytes	Presence of CNTs in cell cytoplasm Concentration-dependent platelet aggregation	Monteiro-Riviere *et al.*, 2005
SWCNT, MWCNT,	Human blood platelets and rats	Infusion of both types of CNTs accelerates the rate of carotid artery thrombosis development	Radomski *et al.*, 2005
Carbon black, MWCNTs	Human T-lymphocytes	Toxicity was found in the order of oxidised CNT>MWCNT>Carbon black	Bottini *et al.*, 2006

(Continued)

Table 11.4 (*Continued*)

Samples	Cells	Observation	References
Polyurethane/carbon nanofibre composites	Astrocytes osteoblasts	Decreased astrocytes formation and greater adhesion of osteoblasts with increasing carbon nanofibre concentration in the composite. The composites were recommended for neural and orthopaedic applications	Webster *et al.*, 2004
Lithography patterned functionalised and non-functionalised CNT surfaces	Neurons	Neurons localised in CNT-rich regions and formed an interconnected network that replicated the CNT template. No cytotoxicity was observed	Hu *et al.*, 2004
		Functionalised MWCNT provided a substrate for neurite extension, and the neurite extension was loosely based on the ionic charge. No cytotoxicity towards neuronal cells was observed	Gabay *et al.*, 2005
PLA/MWCNT composites	Osteoblasts	Under current stimulation the composites with MWCNTs showed increased cell proliferation and increased extracellular calcium deposition	Supronowicz *et al.*, 2002
		Extensive growth, spreading and attachment of the cells was observed, with no cytotoxic effects	Correa-Duarte *et al.*, 2004
3-D substrate with functionalised MWCNT	L929 Mouse fibroblast cells	Cells attach and survive on MWCNT substrates without penetrating the network. The proliferation was not comparable to planar control (HOPG)	George *et al.*, 2006
MWCNTs grown on quartz substrate	Lung epithelial cells and osteoblast cells	Osteoblasts sense and react to MWCNT constructs, having nanotube diameters ranging from 13 to 53 nm. Cells adhered and survived on the MWCNT constructs	Mwenifumbo *et al.*, 2007
Polyurethane/MWCNT composites	Osteoblasts	Increasing CNT fraction did not cause osteoblast cytotoxicity and did not retard the differentiation or mineralisation Vascular endothelial growth factor increased in proportion to CNT	Jell *et al.*, 2008
SWCNT, asbestos, carbon nanofibres (CNF)	Lung fibroblast, macrophages	Cytotoxicity tests revealed a concentration- and time-dependent loss of V79 cell viability after exposure to all tested materials in the following sequence: asbestos>CNF>SWCNT DNA damage and micronucleus induction were found after exposure to all tested materials with the strongest effect seen for CNF	Kisin *et al.*, 2011

measure the cell response to the CNT-based composites) because CNTs are embedded within the matrix and may not be exposed to the cells directly during the duration of the cytotoxicity studies.

11.2.4.4 Interference with assays

Recent reports have revealed that CNTs and other carbon-based nanomaterials may interact with components of the most commonly used cytotoxicity assays and thus interfere with absorption and fluorescence data used to evaluate cytotoxicity (Worle-Knirsch *et al.*, 2006). Casey *et al.* (2007) showed that the indicator dyes (MTT assay, Alamar Blue, Commassie Blue, and Neutral Red) were not appropriate for quantitative toxicity assessment of SWCNTs and recommended the need for developing new screening techniques to evaluate their toxicity.

11.3 Sources of NPs

11.3.1 Natural sources

Concerns about the health effects of exposure to NPs are recent, although natural nanoparticulates have existed on our planet for millennia through geological processes (Figure 11.6). Many natural events such as volcanic activity, dust storms, forest fires, weathering and erosion can produce vast quantities of nanoparticulates (Buzea *et al.*, 2007). Forest and grass fires, which occur globally, increase the amount of airborne particulate matter, including NPs (Sapkota *et al.*, 2005). A single volcanic eruption can release a vast amount of volcanic ash containing micron and nano-sized particles such as crystalline silica (cristobalite) (Taylor, 2002). In addition to naturally occurring inorganic NPs, some of the smallest microbes including bacteriophages and viruses are nano-sized, whilst many organisms are known to produce inorganic nanoparticulates through intracellular and extracellular processes. For example, magnetite NPs and other metallic NPs are synthesised by micro-organisms (Narayanan and Sakthivel, 2010; Thakkar *et al.*, 2010), siliceous nanomaterials are produced by diatoms (Buzea *et al.*, 2007), and nano-sized apatite crystals are the main component of vertebrate bones. NPs are known to occur widely in the Earth's atmosphere, from natural sources, as well as anthropogenic sources, such as car exhaust fumes. NPs from fires, spillage and erosion can be contained locally, but natural phenomena such as dust storms can transport mineral dust ranging in size from 100 nm to several microns and anthropogenic pollutants over distances spanning countries and continents and have the potential to carry NPs (Shi *et al.*, 2005; Taylor, 2002).

11.3.2 Anthropogenic sources

Nanomaterials created by humans (Figure 11.6) have existed for centuries, in the form of the by-products of combustion and, more recently, through various manufacturing processes such as ore refining, welding, and power generation (Rogers *et al.*, 2005; Seames *et al.*, 2002). Nano and micro-particles from diesel and automobile exhausts provide a common and ever-increasing supply of anthropogenic NPs (Buzea *et al.*, 2007; Westerdahl *et al.*, 2005). Even inside our homes we are exposed to nanomaterials, released through cooking, smoking, cleaning and combustion, including of candles (Afshari *et al.*, 2005; Ning *et al.*, 2006). Tobacco smoke and smoke from incense sticks contain nanomaterials (10–700 nm) (Afshari *et al.*, 2005). All of these sources have been contributing to the atmospheric burden of NPs for generations, but current concern is focused mainly on manufactured nanomaterials (nanomaterials that are intentionally produced to have specific compositions or specific properties). Manufactured nanomaterials are now used in a range of consumer products (Table 11.5) that can directly or indirectly expose humans and the environment to them. There are several ways of preparing manufactured NPs (Royal Society, 2004; Faraji and Wipf, 2009; Thakkar *et al.*, 2010) with specific properties ranging across any given spectrum. This makes it imperative to study the risk such manufactured NPs pose to human health and the environment.

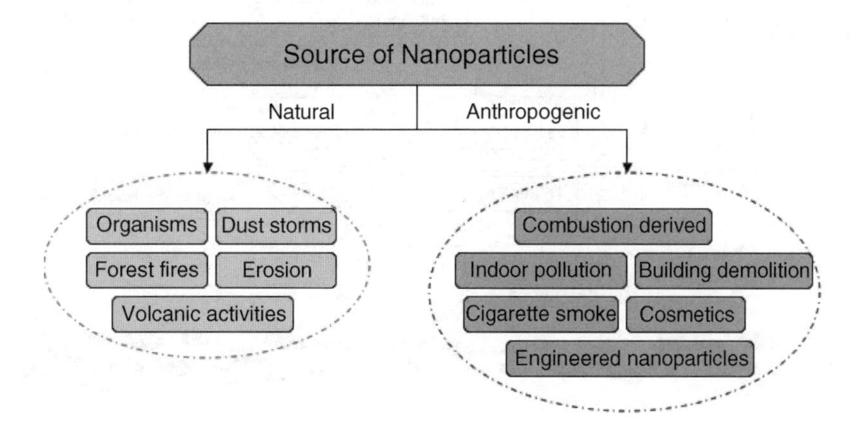

Figure 11.6 Classification and common examples of nanoparticle sources

Table 11.5 Anthropogenic nanomaterials used in consumer products

Type	Use
Silver	Garments, toothpastes, wound dressings, air purifiers, kitchenware, pet products, baby toys, vacuum cleaners, fabric softeners, paints, personal hygiene products, water purifiers, cosmetics, catheters, silver-impregnated paints, silver coatings on hospital fittings and fixtures
Gold	Pregnancy test kits, cosmetics, personal hygiene products, health supplements
Zinc oxide	Sun screens, filter cartridges
Titanium dioxide	Sun screens, self-cleaning windows, computer accessories, air purifiers
Carbon-based (CNTs, C60s)	Sporting goods, air filters, cosmetics, catalysts, organic solar cells, water filtration systems
Silicon and silica	Paints, cosmetics, electronic goods, sports goods, fabric softeners
Ceria	Fuel additives

11.4 Environmental pathways

The increased use of nanomaterials in consumer products through direct exposure (e.g. sunscreens containing TiO_2 as rutile, or socks containing silver) or indirect exposure (e.g. in buildings with self-cleaning glass windows coated with titanium dioxide as anatase NPs) and imperfect waste treatment of nanomaterials and their derived products will inevitably lead to release of NPs within the key environment compartments (i.e. sediment, soils, water, air and biota). Furthermore, this could lead to bioaccumulation of NPs in natural systems, for example via waste-water discharges, since current waste-water treatment does not deal with nanomaterials, and this poses a potential environmental risk (Powell *et al.*, 2008). The release of nanomaterials into the environment can lead to organism exposure. While the release of nanomaterials in the air and any subsequent potential uptake by receptors has been addressed in some recent studies, and is a key part of occupational-exposure assessment, exposure via sediments, soils and water systems has not been adequately addressed to date. The role of natural nanomaterials (such as clay minerals, oxides and hydroxides of aluminium and iron in soils) in the environment is reasonably well understood, but this is not the case for anthropogenic nanomaterials, which can enter the environment in varying quantities through several pathways and either stay suspended in the relevant water system or aggregate and sorb onto particles of sediment and soil. Once the NPs are available in the ecosystem they can be transported through organisms and eventually move through the food chain (Figure 11.7). The degradability and persistence of nanomaterials will play an important role in their ecological fate. For example, the bioaccumulation of Ag NPs will be different from that of CNTs, which have low biodegradability.

NPs in the soil can often have a coating of organic compounds through their interaction with enzymes and humic substances, and this can make such NPs accessible to various micro-organisms and facilitate their travel through the soil pores. On the other hand, due to the high reactivity of NPs they can be sorbed to soil particles and can also form larger agglomerates, restricting their movement through soil. The nature of the soil (parameters such as pH, ionic strength, composition, presence of natural organic matter) can play a huge role in the stability of NPs. Fang *et al.* (2009) used saturated homogenous soil column experiments to show that TiO_2 NPs are stable in soil suspensions (even after 10 days), resulting in a higher mobility of the NPs through soil layers. Jaisi and Elimelich (2009) showed that SWCNTs have very limited movement through soils, and a low deposition-rate

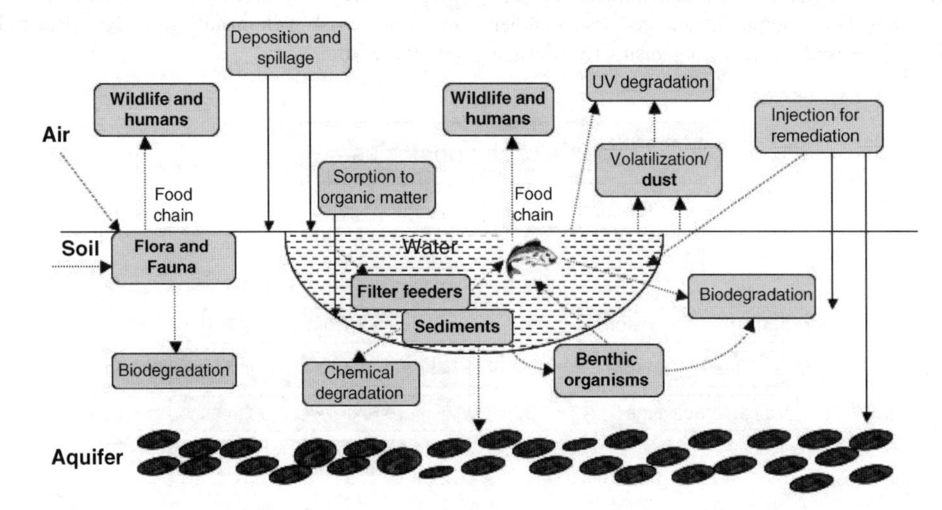

Figure 11.7 Possible routes of exposure, uptake, distribution, and degradation of NPs in the environment. Bold lettering indicates possible sinks and sources (Reproduced, with permission, from Environmental Health Perspectives (2005), 113, 823 © U.S. National Institute of Environmental Health Sciences)

coefficient in soil. Due to the bundled and agglomerated nature of SWCNTs they can also be retained in the soil. Despite concern about the ecotoxicology of nanomaterials, NPs are being investigated for deliberate use in remediation (e.g. zerovalent iron), which could have a significant effect on benthic and soil flora and fauna. Not only the nanomaterials themselves, but also the production process, can lead to considerable damage to the environment and be energy intensive. As an example, it has been reported that for manufacturing fullerenes, only 10 per cent of material is usable and the rest is sent as waste to landfill (Royal Commission on Environmental Pollution, 2008).

In recent years, a rapid increase in the use of Ag NPs for their antibacterial properties has given rise to a variety of Ag-impregnated consumer products, and the mass of Ag released and dispersed into the environment from such new products is likely to become substantial. This has led to recent risk evaluation being performed on such items (Benn and Westerhoff, 2008; Geranio *et al.*, 2009). Although Ag has beneficial properties, USEPA (US Environmental Protection Agency) has classified it as a primary pollutant in natural waters due to it being toxic, persistent and bioaccumulative under certain circumstances to certain forms of life (Luoma, 2008). However, the paradox of this scenario is that nearly one-third of Ag NP embedded products currently sold have the potential to release and disperse Ag into the environment (Ag concentrations in natural waters can range from 0.03 to 500 ng/l) (Luoma, 2008). In response to the increasing use of Ag NPs in fabrics, a recent study considered the amount of Ag released into water from commercially available socks and determined its fate in waste-water treatment plants (Benn and Westerhoff, 2008). Their analysis showed that the socks contained up to a maximum of 1360 μg of Ag per g of sock and leached as much as 650 μg of Ag in 500 ml of distilled water (Benn and Westerhoff, 2008). On microscopic examination of the wash water, Ag NPs in the size range 10–100 nm were identified. Gottschalk *et al.* (2009) calculated the predicted environmental concentrations (PEC) for all environmental compartments based on a probabilistic material flow analysis from a life-cycle perspective of products containing TiO_2 NPs, ZnO NPs, Ag NPs, CNTs, and fullerenes for US, Europe and Switzerland (Table 11.6). They showed that the risk quotient from Ag, TiO_2, and ZnO NPs to aquatic organisms may increase in sewage treatment effluents and from Ag NPs in surface waters.

11.5 Regulation and effects on human receptors

11.5.1 Regulation

Due to limited toxicological data and the huge variety of engineered NPs, there are practically no operational exposure limits (OELs) specific to nanomaterials that have been adopted or circulated by authoritative standards and guidance organisations. The vast range of nanomaterials that can be generated and their very diverse properties limit the number of specific OELs that can be developed. The variety of physicochemical properties of nanomaterials may result in a similar variety of toxic potential. For example, the tendency for NPs to agglomerate will influence the amount of time they remain airborne and their inhalability. Nevertheless, recent toxicological data on dose-response of NPs from *in vitro* and/or *in vivo* studies for certain specific nanomaterials might help in the formulation of such OELs for nanomaterials. With more development in the field of nanotechnology, it is becoming clearer that the existing OELs based on bulk materials may not be appropriate for the same material at the nanoscale, which makes the task of developing OELs for distinct nanomaterials impractical. Therefore, the way forward may be to develop OELs for specific categories of nanomaterials with common hazard potential based on their physicochemical properties and *in vivo* studies (Schulte *et al.*, 2010). For example, developing OELs for nanomaterials with certain aspect ratios or the ability to generate oxidative stress. As the information on the route of exposure and mechanism of toxicity for NPs becomes available, a quantitative risk assessment method incorporating pharmacokinetic models can be used to extrapolate the dose administered in animal studies to humans (Kuempel *et al.*, 2006). Marchant *et al.* (2008) have proposed an incremental regulatory pyramid for nanomaterials, starting from information gathering followed by self-regulation and eventually leading to hard regulatory actions and legislation. Figure 11.8 lists the proposed OELs by various health organisations for certain anthropogenic nanomaterials.

11.5.2 Effect on human receptors

11.5.2.1 Bioaccessibility

As discussed earlier, the increasing use of engineered NPs has raised concerns about its effect on human health. There is more than one route of exposure of NPs to humans and to the environment (Borm *et al.*, 2006; Buzea *et al.*, 2007; Maynard and Kuempel, 2005; Oberdorster *et al.*, 2005). The main characteristics that determine the pathogenicity of NPs are:

- particle dimensions – the dimensions of the NP determine its respirability and potential for translocation inside the body;
- biopersistence – high biopersistence is likely to increase the toxicity of particles; particles that dissolve or disaggregate can be phagocytosed and metabolised and therefore cleared more easily;
- reactivity or inherent toxicity of NPs – the relative toxicity of NPs will finally depend on their chemical components. It is clear that the behaviour of NPs inside the organism is likely to be a complex process, and several factors must be considered in designing *in vitro* studies.

Most micron-sized particulate matter is cleared by biological processes in the body, but some NPs may evade the

Table 11.6 Predicted environmental concentration from simulation results for (a) various nanomaterials showing the mode, lower (Q0.15) and upper (Q0.85) quantiles for Europe, US and Switzerland (Gottschalk *et al.*, 2009) and (b) predicted environmental concentration for Ag and TiO2 NPs and CNTs showing the realistic scenario (RE) and high-emission scenario (HE) in Switzerland (Mueller and Nowack, 2008)

(a)	Europe			US			Switzerland			
	Mode	$Q_{0.15}$	$Q_{0.85}$	Mode	$Q_{0.15}$	$Q_{0.85}$	Mode	$Q_{0.15}$	$Q_{0.85}$	
					nano-TiO_2					
soil	1.28	1.01	4.45	0.53	0.43	2.13	0.28	0.21	1.04	Δμg/kg · y
sludge treated soil	89.2	70.6	310	42.0	34.5	170				Δμg/kg · y
surface water	0.015	0.012	0.057	0.002	0.002	0.010	0.021	0.016	0.085	μg/l
STP effluent	3.47	2.50	10.8	1.75	1.37	6.70	428	3.50	16.3	μg/l
STP sludge	136	100	433	137	107	523	211	172	802	mg/kg
sediment	358	273	1409	53	44	251	499	426	2382	Δμg/kg · y
air		<0.0005			<0.0005		0.001	0.0007	0.003	μg/m^3
					nano-ZnO					
soil	0.093	0.085	0.661	0.050	0.041	0.274	0.032	0.026	0.127	Δμg/kg · y
sludge treated soil	3.25	2.98	23.1	1.99	1.62	10.9				Δμg/kg · y
surface water	0.010	0.008	0.055	0.001	0.001	0.003	0.013	0.011	0.058	μg/l
STP effluent	0.432	0.340	1.42	0.3	0.22	0.74	0.441	0.343	1.32	μg/l
STP sludge	17.1	13.6	57.0	23.2	17.4	57.7	21.4	16.8	64.7	mg/kg
sediment	2.90	2.65	51.7	0.51	0.49	8.36	3.33	3.30	56.0	Δμg/kg · y
air		<0.0005			<0.0005			<0.0005		μg/m^3
					nano-Ag					
soil	22.7	17.4	58.7	8.3	6.6	29.8	11.2	8.7	41.2	Δng/kg · y
sludge treated soil	1581	1209	4091	662	526	2380				Δng/kg · y
surface water	0.764	0.588	21.6	0.116	0.088	0.428	0.717	0.555	2.63	ng/l
STP effluent	42.5	32.9	111	21.0	16.4	74.7	38.7	29.8	127	ng/l
STP sludge	1.68	1.31	4.44	1.55	1.29	5.86	1.88	1.46	6.24	mg/kg
sediment	952	978	8593	195	153	1638	1203	965	10184	Δng/kg · y
air	0.008	0.006	0.02	0.002	0.0020	0.0097	0.021	0.017	0.074	ng/m^3
					CNT					
soil	1.51	1.07	3.22	0.56	0.43	1.34	1.92	1.44	3.83	Δng/kg · y
sludge treated soil	73.6	52.1	157	31.4	23.9	74.6				Δng/kg · y
surface water	0.004	0.0035	0.021	0.001	0.0006	0.004	0.003	0.0028	0.025	ng/l
STP effluent	14.8	11.4	31.5	8.6	6.6	18.4	11.8	7.6	19.1	ng/l
STP sludge	0.062	0.047	0.129	0.068	0.053	0.147	0.069	0.051	0.129	mg/kg
sediment	241	215	1321	46	40	229	229	176	1557	Δng/kg · y
air	0.003	0.0025	0.007	0.001	0.00096	0.003	0.008	0.006	0.017	ng/m^3
					Fullerenes					
soil	0.058	0.057	0.605	0.024	0.024	0.292	0.026	0.019	0.058	Δng/kg · y
sludge treated soil	2.2	2.1	22.2	1.01	1.0	12.2				Δng kg · y
surface water	0.017	0.015	0.12	0.003	0.0024	0.021	0.04	0.018	0.19	ng/l
STP effluent	5.2	4.23	26.4	4.6	4.49	32.66	3.82	3.69	25.1	ng/l
STP sludge	0.012	0.0088	0.055	0.0 1	0.0093	0.068	0.0107	0.0101	0.068	mg/kg
sediment	17.1	6.22	530	2.5	1.05	91.3	20.2	8.2	787	Δng/kg · y
air		<0.0005			<0.0005			<00005		ng/m^3

Table 11.6 *(Continued)*

(b)		nano-Ag		nano-TiO_2		CNT	
	unit	RE	HE	RE	HE	RE	HE
air	$\mu g/m^3$	1.7×10^{-3}	4.4×10^{-3}	1.5×10^{-3}	4.2×10^{-2}	1.5×10^{-3}	2.3×10^{-3}
water	$\mu g/l$	0.03	0.08	0.7	16	0.0005	0.0008
soil	$\mu g/kg$	0.02	0.I	0.4	4.8	0.01	0.02

clearance mechanism and if they remain in contact with cell membranes there is a greater chance of damage. Skin, lungs, and the gastrointestinal tract are the most likely points of entry for natural or anthropogenic NPs (apart from those that are injected or implanted) and their small size enables them to translocate readily into the circulatory and lymphatic systems. Once the NPs have entered the circulatory system, depending on the duration of exposure and their stability, long-term translocation to organs (liver, heart, spleen, kidney etc.) is possible (Geiser *et al.*, 2005; Oberdorster *et al.*, 2005).

Studies have shown that NPs have the potential to access various intracellular locations (such as cytoplasm, mitochondria, lipid vesicles) (Garcia-Garcia *et al.*, 2005a; Stefani *et al.*, 2005; Xia *et al.*, 2006) in a number of cell types (such as alveolar macrophages, endothelial, pulmonary epithelial and gastrointestinal epithelial cells, red blood cells, and nerve cells) (Hoet *et al.*, 2004; Takenaka *et al.*, 2001). Inhaled NPs have been shown to translocate into the nervous system (through the olfactory nerves and/or blood–brain barrier) (Borm *et al.*, 2006; Oberdorster *et al.*, 2004; Peters *et al.*, 2006) and gastrointestinal tract (via the mucociliary escalator, lymphatic and circulatory systems) (Liu J *et al.*, 2006; Semmler *et al.*, 2004). Broken skin and hair follicles can also act as a route of entry for NPs. Particles between diameters of 750 nm and 6 μm can selectively penetrate the skin at hair follicles, and larger particles (up to 7 μm) can enter through broken skin (Toll *et al.*, 2004). The risk of NPs crossing the epidermis layer and interacting with other cells is high, due to the presence of blood, lymph vessels and nerve endings in the dermis layer. For example, translocation of soil particles through the skin into the lymphatic system has been shown to cause podoconiosis (Blundell *et al.*, 1989; Buzea *et al.*, 2007) and Kaposi's sarcoma (Montella *et al.*, 1997; Mott *et al.*, 2002).

Recent studies have demonstrated that NPs are able to enter the nervous system (Garcia-Garcia *et al.*, 2005b) and the olfactory nerves and the blood–brain barrier could possibly be the route (Borm *et al.*, 2006; Oberdorster *et al.*, 2004; Peters *et al.*, 2006). *In vivo* studies on rats have shown that about 7 per cent of the ingested 50 nm latex NPs and 4 per cent of the ingested 100 nm latex NPs translocated to the liver, spleen, blood and bone marrow; particles larger than 100 nm did not reach the bone marrow and those larger than 300 nm were absent from the blood (Jani *et al.*, 1990). Although the exact order of translocation of NPs from the gastrointestinal tract to organs and the blood is not known, a preliminary study on dental prosthesis porcelain

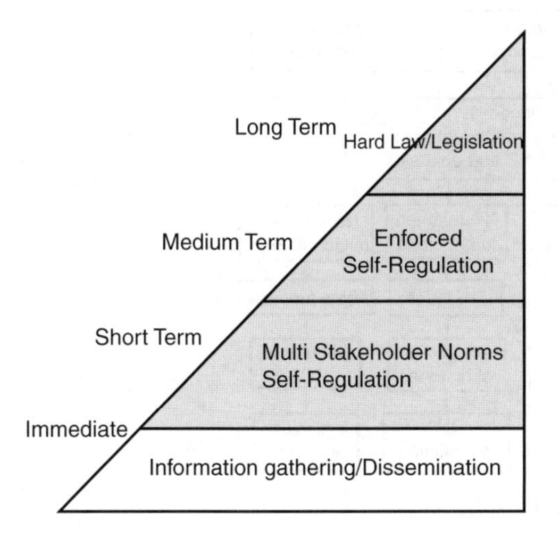

Nonomaterial	Parameter	OEL
General	0.004% risk level	Mass-based OEL: 15
Titanium dioxide	0.1 risk level particles < 100 nm	$0.1 mg/m^3$
General dust		$3 mg/m^3$
Photocopier Toner	Tolerable risk	$0.6 mg/m^3$
	2009 acceptable risk	$0.06 mg/m^3$
	2018 acceptable risk	$0.006 mg/m^3$
Biopersistent granular materials (metal oxides, others)	Density > 6,000 kg/m^3	20,000 $particles/cm^3$
Biopersistent granular materials	Density < 6,000 kg/m^3	40,000 $particles/cm^3$
CNTs	Exposure ristk ratio for asbestos	$0.01 f/cm^3$
Nanoscale liquid		Mass-based OEL
Fibrous	3:1; length 75,000 nm	$0.01 f/cm^3$
CMAR[a]		Mass-based OEL: 10
Insoluble	Not fibrous	Mass-based OEL: 15
Soluble	Not fibrous	Mass-based OEL: 10
	Not CMAR	
MWCNT	Bayer product only	$0.05 mg/m^3$
MWCNT	Nanocyl product only	$0.0025 mg/m^3$

[a] Carcinogenic, mutagenic, asthmagenic, and reproductive toxicants

Figure 11.8 Regulatory pyramid for nanotechnology and nanomaterials, and proposed OELs for engineered NPs (Reproduced from Journal of Nanoparticle Research (2010), 12, 1971 © Springer; Reproduced, with permission, from Safety Science (2010) 48, 957 © Elsevier)

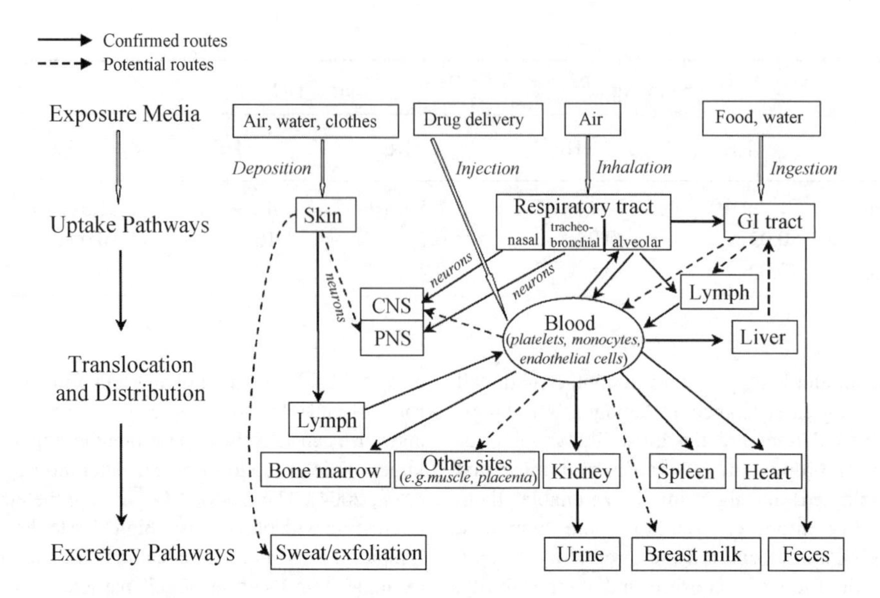

Figure 11.9 Uptake and translocation routes for NPs (Reproduced, with permission, from Environmental Health Perspectives (2005), 113, 823, © U.S. National Institute of Environmental Health Sciences)

debris suggests that intestinal absorption of particles is followed by liver clearance before reaching the general circulation and the kidneys (Ballestri *et al.*, 2001). Figure 11.9 shows some uptake and translocation routes for NPs and Figure 11.10 is a representation of their toxic effects at the cellular level.

11.5.2.2 Human health effects

The toxicity of NPs will depend on their route of entry to the body. Some NPs may be well tolerated in some organ systems but highly toxic in others. The focus of this section is the potential

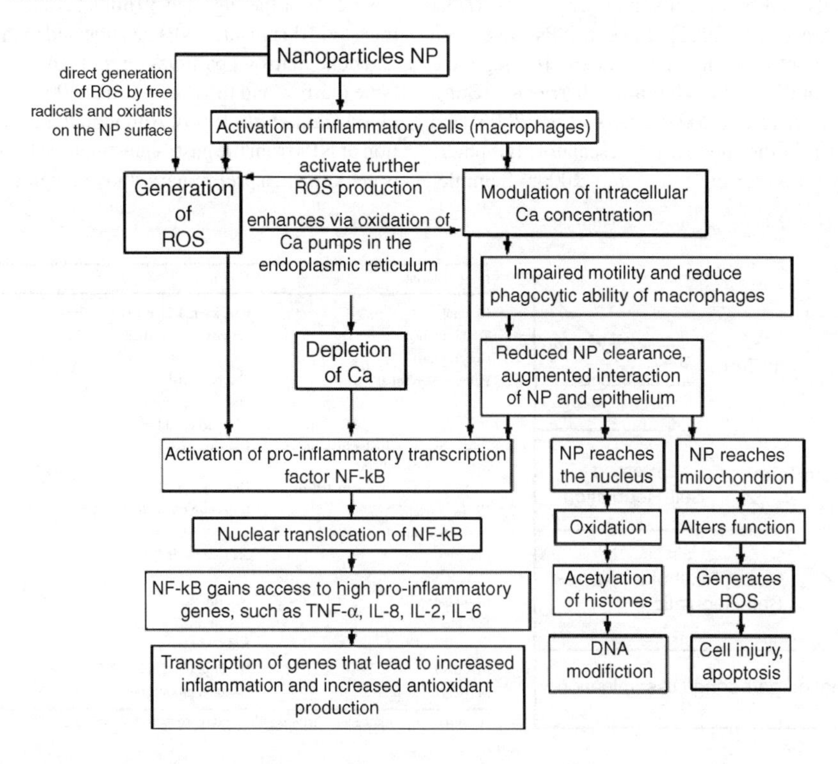

Figure 11.10 Representation of the events by which NPs exert their toxic effects at the cellular level (Reproduced, with permission, from Biointerphases (2007), 2, MR17 © Elsevier)

toxic effects of NPs on the respiratory tract, skin, lymphatic and circulatory system, gastrointestinal system and nervous system, since these represent the major routes of entry to, and possible translocation around, the body.

Table 11.7 gives a selection of recent publications on the ecotoxicity and toxicity of NPs.

11.5.2.2.1 Cellular uptake NPs can be classified into two categories: (1) NPs without any specific receptors on the outer surface and (2) NPs with specific receptors on the outer surface. For NPs without receptors, cellular uptake is passive (adhesive interaction), initiated by van der Waals forces, electrostatic charges, steric interactions, or interfacial tension without the formation of vesicles. A NP can interact with the exterior of the plasma membrane of a cell and enter into the cell through a process termed 'endocytosis'. Endocytosis is a multistage process (Figure 11.11), whereby NPs are engulfed to form membrane-bound vesicles, known as endosomes (Sahay *et al.*, 2010a). Once the NPs are taken into the cell, the endosomes deliver them to various specialised vesicular structures. Eventually the NPs are delivered to intracellular compartments, recycled to the extracellular environment or delivered across cells (transcytosis). Endocytosis can be divided into phagocytosis (uptake of large particles) and pinocytosis (the uptake of fluids and solutes), with pinocytosis further classified on the basis of the different proteins involved in the endocytic pathway, as shown in Figure 11.11. However, very small NPs, such as C60 molecules with a diameter of 0.7 nm, are able to penetrate cells via a different mechanism than phagocytosis, probably through ion channels or via pores in the cell membrane (Porter *et al.*, 2006). A recent study (Bhabra *et al.*, 2009) also showed that NPs may cause damage to DNA and chromosomes across an intact cellular barrier through a mechanism involving pannexin and connexion hemichannels and gap junctions and purinergic signalling.

Phagocytes such as macrophages, neutrophils, monocytes and dendritic cells are primarily responsible for performing phagocytosis in four distinct steps (Figure 11.11b). Step 1 consists of recognition of particles by opsonisation. Opsonisation of particles occurs through adsorption of proteins (e.g. immunoglubulins, blood serum proteins) onto the surface of particles, which may affect the net charge and composition, and the cell-trafficking mechanism (Sahay *et al.*, 2010a). In Step 2 opsonised particles adhere to the macrophage cell membrane through specific receptors (e.g. Fc receptor, complement receptors can bind to the immunoglobulins or complement molecules adsorbed onto the NPs). The receptor–ligand interaction results in actin rearrangement and formation of a phagosome, which constitutes Step 3. The particles are then ingested into the phagosome. Eventually in Step 4 the phagosomes mature, fuse with lysosomes and become enzyme-rich phagolysosomes where the particles are prone to degradation.

Another form of endocytosis, clathrin-mediated endocytosis, is a common route for uptake of essential nutrients into cells (e.g. cholesterol carried by low-density lipoprotein (LDL) via the LDL receptor, or iron carried by transferrin (Tf) via the Tf receptor) (Sahay *et al.*, 2010a). Clathrin-mediated endocytosis is also a multistage process involving engulfment of receptors associated with their ligands to a coated pit, followed by pulling the assembled vesicle from the plasma membrane. Once inside the cell, actin defines spatial regulation and movement of the endocytic vesicle towards the interior of the cells, where the vesicles fuse with the early endosomes. Here they are sorted to late endosomes and lysosomes, to trans-Golgi network or to be transported back to plasma membrane (Pucadyil and Scmid *et al.*, 2009; Sahay *et al.*, 2010a). Clathrin-mediated endocytosis has been shown in some cases to be a preferred trafficking pathway for cellular entry for nanomaterials (Sahay *et al.*, 2010a). Other forms of endocytosis (viz caveolae-mediated endocytosis, clathrin and caveolae-independent endocytosis, and macropinocytosis) are thoroughly discussed by Sahay *et al.* (2010a). A distinguishing feature of caveolae-mediated endocytosis pathways is that in some cases they can bypass lysosomes and prevent lysosomal degradation (Carver and Schnitzer *et al.*, 2003; Sahay *et al.*, 2010a). Therefore current research is focusing on using specific molecules (e.g. CTB, Shiga toxin) on NPs that will help the particle to undergo caveolae-mediated endocytosis for cellular delivery of proteins and DNA (Rejman *et al.*, 2006; Doherty and McMahon, 2009). However, these markers are not restricted to caveolae and can also enter through clathrin- and caveolae-independent pathways.

The available research suggests that the transport of nanomaterials in cells depends on structure, physicochemical characteristics of nanomaterials (size, shape, charge, hydrophobicity, surface chemistry, etc.), biointeractions between biological moieties on nanomaterials with cells, and the endocytic pathway of the cell (for example, the endocytic pathway appears to be different between normal and tumour cells; Sahay *et al.*, 2010a, b). Gratton *et al.* (2008) used three different sizes (2–5 μm, 1 μm and 100–400 nm) and shapes (cubic microparticles, cylindrical microparticles and cylindrical nanomaterials) of particles made from cross-linked polyethylene glycol (PEG)-based hydrogels and showed that even the largest micron-sized particle was internalised in cervical cancer cell lines (HeLa) through macropinocytosis, while the NPs entered more rapidly than the micron-particles, using clathrin-mediated endocytosis and caveolae-mediated endocytosis. The same study showed that by inversing the surface charge of the particles, their entry into the cells became negligible. In another study, Lai *et al.* (2008) showed that 43 nm polystyrene NPs can enter HeLa cells through clathrin-mediated endocytosis and reside in the endolysosomal compartment, whereas for 24 nm polystyrene NPs the entry was through a clathrin- and caveolae-independent pathway. A recent study by Greulich *et al.* (2011) showed that for 80 nm PVP-coated Ag NPs the mesenchymal stem cells use clathrin-dependent endocytosis and macropinocytosis to ingest the Ag NP and that the ingested nanoparticles subsequently occur as agglomerates in the perinuclear region. Although shape effect of micron-sized particles on phagocytosis has been described (Champion and Mitragotri, 2006) no such shape effect has been thoroughly investigated for pinocytosis. This could

Table 11.7 Selection of recent publications on ecotoxicity and toxicity of NPs

Type	Size (nm)	Cells/Organism	Observation	References
Cobalt-chromium	29±5	Human fibroblasts	NPs may cause damage to DNA and chromosomes across an intact cellular barrier Damage caused by a novel mechanism involving pannexin and connexion hemi-channels, gap junctions and purinergic signalling, instead of NPs passing through the barrier	Bhabra *et al.*, 2009
MWCNTs		Mice (inhalation)	CNTs reach the subpleura after a single inhalation exposure of 30 mg/m for 6 h. Nanotubes were embedded in the subpleural wall and within subpleural macrophages Subpleural fibrosis unique to this form of nanotubes increased after 2 and 6 weeks following inhalation None of these effects was seen in mice that inhaled carbon black NPs or a lower dose of nanotubes (1 mg/m)	Ryman-Rasmussen *et al.*, 2009
PEG-functionalised SWCNTs		Mice (i.v.)	Survival, clinical and laboratory parameters revealed no evidence of toxicity over 4 months Functionalised SWCNTs persisted within liver and spleen macrophages for 4 months without apparent toxicity	Schipper *et al.*, 2008
MWCNTs		Tomato seeds	CNTs can penetrate the tomato seed coat, support water uptake inside seeds and increase the germination and growth rates	Khodakhovskaya *et al.*, 2009
ZnO	15–25	Ryegrass	In the presence of ZnO NPs, ryegrass biomass significantly reduced, root tips shrank, and root epidermal and cortical cells highly vacuolated or collapsed ZnO NPs greatly adhered onto the root surface. Individual ZnO NPs were observed present in apoplast and protoplast of the root endodermis and stele	Lin and Xing, 2008
Ag	10–15	Blood platelets (in vitro) and mice (i.v.)	Ag NPs have an innate antiplatelet property and effectively prevent integrin-mediated platelet responses (in vivo and in vitro) in a concentration-dependent manner Ultrastructural studies show that nanosilver accumulates within platelet granules and reduces interplatelet proximity Ag NPs suppressed the aggregation of platelets obtained from patients with type 2 diabetes mellitus	Shrivastava *et al.*, 2009

ZnO CeO_2 TiO_2	13 8 11	Macrophage and epithelial	ZnO induced toxicity and led to the generation of ROS, oxidant injury, excitation of inflammation, and cell death CeO_2 nanoparticles were taken up intact into positive endosomal compartments, without inflammation or cytotoxicity. CeO_2 nanoparticles suppressed ROS production and induced cellular resistance to an exogenous source of oxidative stress TiO_2 NPs were processed by the same uptake pathways as CeO_2 but did not elicit any adverse or protective effects	Xia *et al.*, 2008
Ni	30, 60, 100	Zebrafish embryos	Ni NPs are equal to or less toxic than soluble Ni With each Ni NP exposure, thinning of the intestinal epithelium first occurs around the LD10 continuing into the LD50	Ispas *et al.*, 2009
TiO_2 (anatase) (rutile)	7–10 15–20	Modular immune in vitro construct	TiO_2 NPs in the system led to increased maturation and expression of costimulatory molecules on dendritic cells TiO_2 NPs generate an immune response through the confluence of increased proinflammatory cytokine production and ROS generation	Schanen *et al.*, 2009
CdTe/ZnS core/shell quantum dots	5 nm	Human osteosarcoma cells	The cellular targeting of NPs is inherently imprecise due to the randomness of nature at the molecular scale. Statistical analysis of NP loaded endosomes indicates that particle capture is described by an over-dispersed Poisson probability distribution whereas partitioning of NPs in cell division is random and asymmetric, following a binomial distribution	Summers *et al.*, 2011
TiO_2 (75% anatase and 25% rutile)	21 nm	Wild type mice	NPs induce clastogenicity, genotoxicity, oxidative DNA damage, and inflammation in vivo in mice. Utero exposure to NPs results in an increased frequency in DNA deletions in the fetus	Trouiller *et al.*, 2009
Carbon-coated iron carbide	30 nm	Blood	Therapeutic nanomagnets were able to extract metal ions (Pb2þ), steroid drugs (digoxin, cardiac drug), and proteins (interleukin-6 (IL-6), inflammatory mediator) from the blood	Hermann *et al.*, 2010

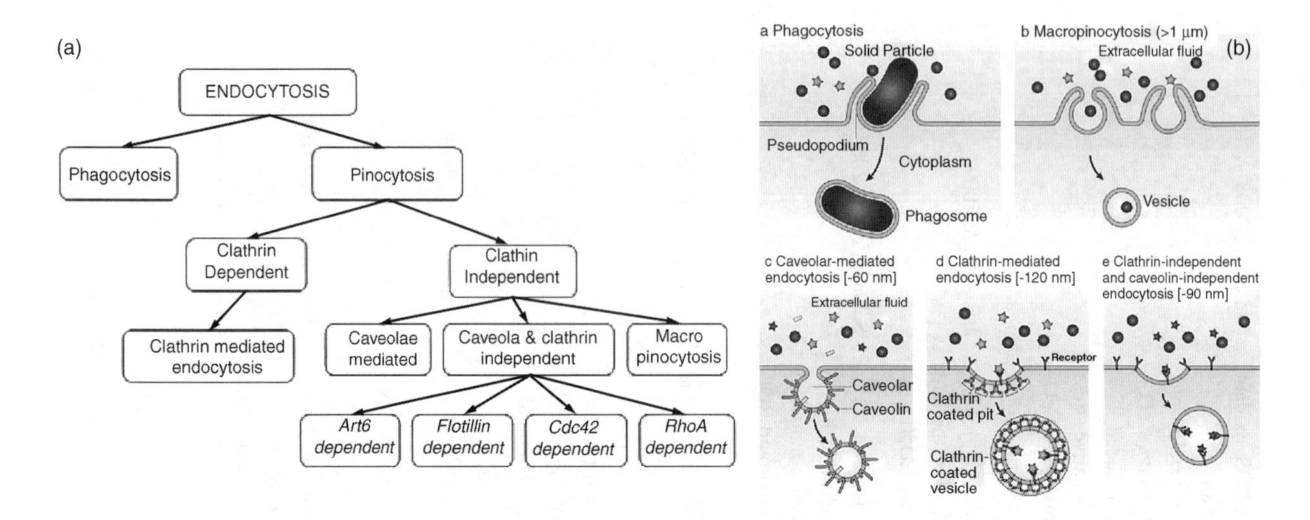

Figure 11.11 (a) Classification of endocytosis based on endocytosis proteins interactions with entry of particles and solutes (Sahay *et al.*, 2010a). (b) Modes of cellular internalisation for NPs (Reproduced, with permission, from Nature Reviews Drug Discovery (2010) 9, 615 © Nature Publishing)

be partially due to the lack of availability of shape-specific monodispersed NPs.

Although the small size of NPs may facilitate rapid entry into cells, there is evidence that surface chemistry plays a more significant role than size. The effect of surface modifications on the uptake of polylactic acid (PLA) and PEG-co-PLA NPs (100 nm) in fibroblast and epithelial cells showed that the NPs, when altered by cationic surfactant stearylamine, were more readily accumulated in cells than unmodified anionic NPs (Harush-Frenkel *et al.*, 2007, 2008). For polarised MDCK epithelial cells, the NPs used clathrin-mediated endocytosis irrespective of their charge. For non-polarised HeLa cervical-cancer cells, anionic particles used multiple pathways (clathrin-mediated endocytosis and caveolae-mediated), while cationic particles appeared to be restricted to clathrin-mediated endocytosis as well as to macropinocytosis (Harush-Frenkel *et al.*, 2007, 2008). In addition, for epithelial cells, the cationic particles were routed for transcytosis and the anionic particles reached the lysosomes (Harush-Frenkel *et al.*, 2007, 2008). Recent studies on PEG, PEG-amine or poly(acrylic acid) modified CdSe/ZnS core-shell ellipsoid quantum dots (6 nm and 12 nm), differing in their surface charge, showed that negatively charged particles displayed the greatest uptake in skin cells through caveolae-mediated endocytosis, while the neutral and positive particles did not appear to use this pathway (Zhang *et al.*, 2009). Experiments conducted on template-synthesised silica nanotubes (50 nm in diameter and 200 nm long) functionalised with positively charged aminosilane groups were shown to internalise via clathrin-mediated endocytosis and reach lysosomes in cancer cells (Nan *et al.*, 2008). Similarly, mesoporous silica-based nanoparticles (ca 110 nm) were internalised into mesenchymal stem cells and fibroblasts via a clathrin-mediated endocytosis but not through caveolae (Chung *et al.*, 2007). Vasir and Labhasetwar (2008) showed that cellular uptake of negatively charged poly (lactic-co-glycolic acid) (PLGA) particles (300 nm) are both cell-type and surface-charge dependent. In the vascular smooth muscle cells these particles entered predominantly through clathrin-mediated endocytosis, and in rat corneal epithelial cells the particles entered via clathrin- and caveolae-independent pathways (Panyam and Labhasetwar, 2003; Qaddoumi *et al.*, 2004). Some 85 per cent of these particles were recycled to the cell surface from the early endosomes and 15 per cent of the particles escaped the endosomes to reach the cytosol (Panyam and Labhasetwar *et al.*, 2003; Qaddoumi *et al.*, 2004). The PLGA particles when coated with cationic polymer, poly(l-lysine), led to a five-fold increase in the interaction force between the functionalised particles and the cell-plasma membrane (Vasir and Labhasetwar, 2008). Currently, researchers are deliberately using specific surface ligands to promote clathrin-mediated endocytosis, but the cellular trafficking of these modified NPs differs from their ligands alone, as shown in the case of Tf-functionalised NPs and Shiga-toxin-modified NPs (Tekle *et al.*, 2008). Constant development of receptor-modified NPs and an understanding of their mechanism of intracellular trafficking are vital to the use of NPs as carriers for drugs, as therapeutic agents are targeted not only to a specific cell population but also to a specific intracellular compartment. Nanomedicine is aiming to exploit the physicochemical properties of NPs to direct them towards selected intracellular compartments through use of specific cellular-trafficking machinery(ies).

11.5.2.2.2 Respiratory tract Inhaled NPs are deposited throughout the respiratory tract, but the alveolar region is known to be a site of significant deposition. Particles smaller than 10 μm can reach the alveoli, whereas larger particles tend to be deposited higher up the respiratory tract (Figure 11.12) (Hoet *et al.*, 2004; Lippmann, 1990; Oberdorster, 2001). Particle residence time in the respiratory tract could also induce a range of toxic responses, even at low concentrations. *In vivo* rodent studies investigating the respiratory toxicity of CNTs demonstrate

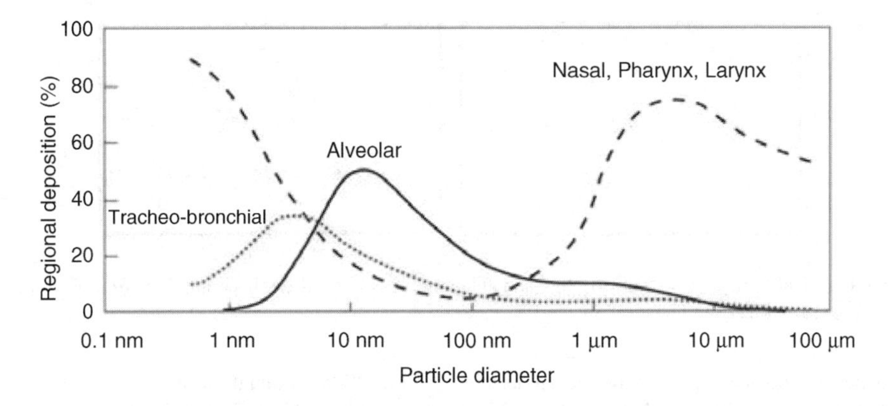

Figure 11.12 Deposition of inhaled particles in the human respiratory tract for different particle sizes (Reproduced from International Archives of Occupational and Environmental Health (2001), 74, 1 © Springer)

that 81 per cent of the carbon nanotubes instilled into the lung were still present 60 days from exposure (Muller *et al.*, 2005), although the dose under investigation was not physiological.

The particles insoluble in intracellular or extracellular fluids are eliminated much more slowly than soluble particles (Borm *et al.*, 2004; Takenaka *et al.*, 2001). The main route of clearance in the upper airways is via the mucociliary escalator, which clears particulate matter towards the oesophagus for swallowing (Ferin, 1994). Inhaled NPs can interact with lung-lining fluid, which can displace them towards the epithelial surface and facilitate their uptake by epithelial cells (Gehr and Schurch, 1992). The lung is also protected by antioxidants that are secreted into the lung-lining liquid. If large numbers of oxidative particles are inhaled the protective antioxidants can be overwhelmed, and this can lead to pulmonary inflammation and toxicity (Wright *et al.*, 1994). Particles that reach the lower airways are primarily cleared through macrophage phagocytosis (Buzea *et al.*, 2007). Human alveolar macrophages can efficiently engulf particles in the 1–5 μm size range, but become less effective for particles that are much larger or smaller. Experimental data show that NPs smaller than 100–200 nm are capable of evading alveolar macrophage phagocytosis (Peters *et al.*, 2006). Particles that can evade macrophage phagocytosis are likely to enter extrapulmonary sites, via interactions with, and penetration of, epithelial cells (Oberdorster *et al.*, 2005; Takenaka *et al.*, 2001). Although individual NPs can evade phagocytosis, at higher concentrations they are more likely to agglomerate to a size that can be phagocytosed by alveolar macrophages. In inhalation studies, low concentrations of Ag NPs (15 nm) were shown to translocate to other parts of the body, whereas inhalation of high concentrations caused NPs to agglomerate and be phagocytosed by alveolar macrophages (Takenaka *et al.*, 2001). Studies have shown that labelling of NPs with antibodies or complement molecules can increase the speed of phagocytosis, whereas coating particles with hydrophilic polymers (e.g. polyethylene glycol) decreases the probability of being phagocytosed (Garnett and Kallinteri, 2006). Furthermore, it has been suggested that the electric charge plays a role in activating the scavenger-type receptors for certain types of NPs (TiO_2, iron oxide, quartz), whereas toll-like receptors are responsible for the recognition of uncharged NPs (carbon-based NPs) (Kobzik, 1995).

Through these complex interactions with cells, NPs may also modify the host response to microbial infection. Macrophages exposed to zirconium dioxide (ZrO_2) NPs have enhanced expression of some viral receptors, making them hyper-reactive to viral infections and leading to excessive inflammation, whereas exposure to silicon dioxide (SiO_2) and TiO_2 NPs led to a decrease in the expression of some viral and bacterial receptors, resulting in a lower resistance to microbial infections (Inoue *et al.*, 2006). Therefore, it seems that the adverse effect of inhaled NPs on the lungs critically depends on the rate of particle deposition and clearance, the physiochemical properties of the NPs and their residence time in the lungs.

11.5.2.2.3 Dermal system NPs are currently used in cosmetic products because of (1) their antimicrobial properties (e.g. Ag) (2) their UV-absorption properties (e.g. TiO_2, ZnO) and (3) the presence of scavenging free radicals (e.g. fullerene). ZnO and TiO_2 (rutile) NPs are used in sunscreens for their strong UV-absorbing capabilities and their resistance to discolouration under UV radiation. The TiO_2 NPs that reflect and scatter UV rays most efficiently are in the size range 60–120 nm (Popov *et al.*, 2005); for ZnO NPs the optimal size is 20–30 nm (Cross *et al.*, 2007). However, concern about using these NPs in cosmetics arises because of (1) the possibility that they might penetrate through the skin and (2) the ability of some NPs to create radicals while undergoing photocatalytic reaction. TiO_2 (particularly anatase) has been shown to generate ROS when added to aqueous media, which can have serious implications for cellular toxicity. Recent studies suggest that photoactive nanomaterials are generally classified into two groups: photosensitisers (e.g. fullerenes), and semiconductors (e.g. TiO_2 NPs). The distinct mechanism of ROS generation by TiO_2 NPs and fullerenes are shown in Figure 11.13. A recent study by Murray *et al.* (2009) showed that exposure to unpurified SWCNTs induced free-radical generation and inflammation, thus causing dermal toxicity. Some studies have shown that

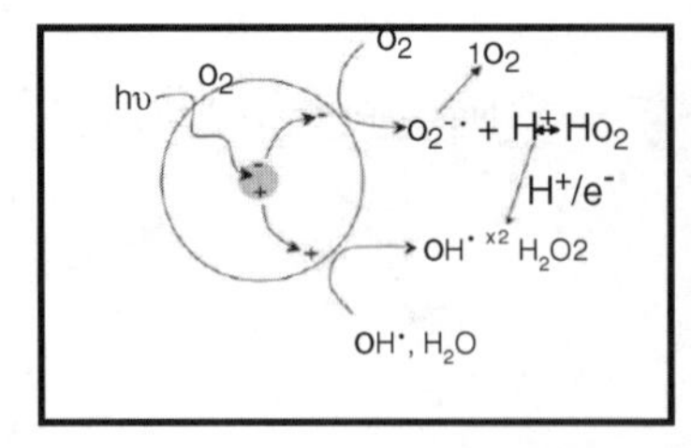

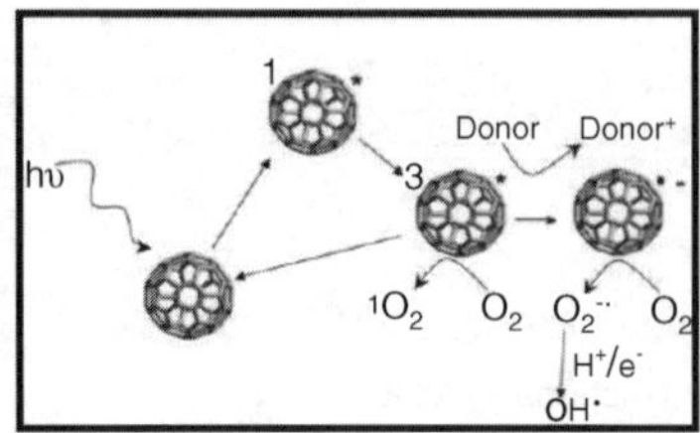

Figure 11.13 Mechanism of ROS generation for TiO_2 NPs and fullerene (Reproduced, with permission, from Environmental Science and Technology (2009), 43, 4355 © American Chemical Society)

NPs are able to penetrate the outer portion of the epidermis (Blundell *et al.*, 1989; Borm *et al.*, 2006; Toll *et al.*, 2004), which is a 10 μm-thick keratinised layer of dead cells with a scaly and porous microstructure. It has been shown *in vitro* that multi-walled CNTs can be internalised by human epidermal keratinocytes into cytoplasmic vacuoles and can induce the release of pro-inflammatory mediators (Monteiro-Riviere *et al.*, 2005). Table 11.8 summarises the results of some skin adsorption studies conducted on TiO_2 NPs.

11.5.2.2.4 Lymphatic and circulatory system Considerable attention is currently given to understanding the translocation of NPs to lymph nodes for drug delivery, tumour imaging, and cancer treatment (Liu J *et al.*, 2006). Studies have shown that colloidal particles (0.7–2 μm) after intrapleural administration were taken up in the regional lymphatic system and are an appropriate size for intrapleural lymphatic targeting to regional lymph nodes (Liu J *et al.*, 2006). As well as direct translocation into the lymphatic system, NPs may also be transported to lymph nodes by phagocytes. Studies have shown that carbon black NPs that have been intratracheally instilled into the lungs of mice can be detected in phagocytes residing in the lymph nodes within 24 hours (Win-Shwe *et al.*, 2005). Although the adverse health effects of NP uptake by the lymphatic system are not clear, the oxidative stress created by certain types of NPs could lead to damage of lymphocytes and lymph nodes (Buzea *et al.*, 2007).

Studies of NP inhalation and intratracheal instillation in healthy animals showed that metallic NPs less than 30 nm in dimension pass rapidly into the circulatory system, while non-metallic NPs with dimensions in the range 4–200 nm have a very low permeability (Takenaka *et al.*, 2001; Geiser *et al.*, 2005; Wiebert *et al.*, 2006; Chen *et al.*, 2006; Liu J *et al.*, 2006; Mills *et al.*, 2006). Epidemiological studies on the effects of ambient particulate matter suggest that subjects suffering from cardio-respiratory diseases may have an increased possibility of NPs translocating from lungs to the circulatory system due to pulmonary inflammation and increased microvascular permeability (Chen *et al.*, 2006, Wiebert *et al.*, 2006), but this hypothesis has yet to be proved. It has been shown that 20 nm Ag particles are cleared from the lungs and enter the circulatory/lymphatic system faster than 100 nm Ag particles because of their ability to evade phagocytosis (Takenaka *et al.*, 2001). Clinical and laboratory experimental evidence shows that inhalation of nano and microparticles can cause cardiovascular effects (Schulz *et al.*, 2005). Inhalation studies on rats showed that a small proportion of 22 nm TiO_2 NPs could pass across the pulmonary epithelium, enter the underlying pulmonary architecture and translocate to the heart (Geiser *et al.*, 2005). Intratracheally instilled gold NPs (30 nm) in rats have been also found in platelets inside pulmonary capillaries (Berry *et al.*, 1977). Evidence for translocation of inhaled nano-sized particles (technetium-labelled particles) across the human epithelial barrier remains equivocal (Mills *et al.*, 2006; Nemmar *et al.*, 2001).

It is clear from studies mentioned here that the uptake of NPs by specific blood cells (red blood cells, white cells and platelets) depends on the physicochemistry of NPs. NP uptake by red blood cells has been shown to be dictated by size, whereas NP uptake by platelets is influenced by the charge of the NP (Peters *et al.*, 2006; Nemmar *et al.*, 2002a; Rothen-Rutishauser *et al.*, 2006). As well as entering blood cells, NPs may also affect the clotting capability of blood. Studies using polystyrene NPs demonstrated that uncharged NPs do not have an effect on blood-clot formation, whereas negatively charged NPs significantly inhibit thrombin formation and positively charged NPs enhance platelet aggregation and thrombosis (Nemmar *et al.*, 2002a). Therefore, it seems possible that NPs may interact with the charged platelets and aggregate due to the reduction of the surface charge (Gatti *et al.*, 2005, 2004). Intratracheally or intravenously instilled polystyrene (60 nm, charged) and diesel exhaust particles (20–50 nm) in hamsters also reveal a clear dose-dependent response between the pollutant (quantity) and the thrombus size (Vermylen *et al.*, 2005; Nemmar *et al.*, 2002a, 2002b).

11.5.2.2.5 Gastrointestinal tract The two primary sources of NPs in the gastrointestinal tract are directly ingested material (e.g. pharmaceuticals, water and toothpaste) and the respiratory tract, where they have been cleared by the mucocilliary escalator and swallowed (Lomer *et al.*, 2004). The gastrointestinal tract is highly developed to facilitate the absorption of ingested materials. The physicochemical properties can influence particle absorption in the gastrointestinal tract and influence or facilitate site-specific targeting (Hoet *et al.*, 2004). It has been shown that the uptake of particles decreases with increasing particle size. Studies on the uptake of polystyrene NPs following gavage, demonstrated that 6.6 per cent of 50 nm NPs were absorbed, compared to 5.8 per cent of 100 nm particles, 0.8 per cent for 1 μm particles, and 0 per cent for 3 μm particles

Table 11.8 Summary of some of the skin adsorption studies conducted on TiO_2 NPs

TiO_2 details	Study	Observation	References
Silica and methicone-coated needle-shaped NPs ($30–60 \times 10$ nm).	In vitro: Franz diffusion cells with porcine skin	Aggregate size (200–1000 nm) TiO_2 particles were not able to penetrate through porcine stratum corneum No titanium was found in the receptor fluid at any sampling time Most of the applied material was recovered in the first five tape strips A simple washing procedure was sufficient to remove TiO_2 particles from the skin	Gamer *et al.*, 2006
Microfine and nanosized particles	In vivo: human foreskin grafts transplanted to immunodeficient mice	NPs penetrated into the corneocyte layers of stratum corneum (confirmed using TEM and via their chemical fingerprint in PIXE) No particle was observed in the cytoplasm of the granular cells	Kertesz *et al.*, 2005
9 nm (anatase)	In vivo: human foreskin grafts transplanted to immunodeficient mice In vitro: keratinocytes, dermal fibroblasts	In the case of intact epidermal barrier, TiO_2 nanoparticles in vivo do not penetrate the intact epidermal barrier stratum corneum In vitro exposure of NPs, exerts significant and cell-type dependent effects (viability, proliferation, apoptosis and differentiation) Particles were internalised by fibroblasts and melanocytes but not by epidermal keratinocytes and sebocytes	Kiss *et al.*, 2008
20 nm hydrophobic coated; 10 nm amphiphilic Al_2O_3 and SiO_2 coated; 100 nm hydrophilic coated	In vivo: human volunteers' skin biopsies	Microscopy examination revealed that TiO_2 pigments were located exclusively on the outermost layer of the human stratum corneum The particles were deposited on the outermost surface of the stratum corneum and were not detected in deeper stratum corneum layers, the human epidermis and the dermis Neither the surface characteristics, the particle size nor the shape of the particles resulted in any dermal absorption of TiO_2 particles	Schulz *et al.*, 2002
Four formulations of commercial products containing TiO_2 particles	In vivo: pig-skin biopsies	TiO_2 particles penetrated through the stratum corneum into the underlying stratum granulosum via intercellular space No TiO_2 found in the hair follicles and particle concentration in the stratum spinosum was below the minimum detection limit of about 1 particle/lm^2	Menzel *et al.*, 2004

(Jani *et al.*, 1990). In addition, the smaller particles cross the gastrointestinal mucus layer quicker than the larger particles. In studies investigating uptake of latex NPs it was shown that 14 nm latex NPs crossed the epithelial barrier within 2 min and 415 nm NPs translocated within 30 min, whereas, 1000 nm particles did not cross this barrier (Hoet *et al.*, 2004). Particles that penetrate the mucus reach the enterocytes and are able to translocate further (viz enter the lymphatic system and capillaries). Since the mucus lining of the gastrointestinal tract is negatively charged it can be perceived that positively charged particles can get trapped whilst negatively charged particles can diffuse across the mucus layer and become available for interaction with epithelial cells, as shown for latex NPs *in vitro* (Szentkuti, 1997). However, it is important to note that due to the presence of a complex mixture of compounds inside the gastrointestinal tract (such as enzymes, food, bacteria, proteins), NPs can interact with, and can be modified by, these organic compounds and thus may have reactivity and toxic responses that are very different from those of the unmodified NPs.

Studies in rats have shown that up to 98 per cent of ingested fullerenes are excreted in the faeces within 48 h and most of the remainder was excreted via the urine (Yamago *et al.*, 1995). Once NPs have crossed the gastrointestinal epithelial barrier, they can be transported to various systemic sites. They have been found in colon tissue of subjects affected by cancer, Crohn's disease, and ulcerative colitis (Gatti, 2004). It has recently been suggested that there is an association between high levels of dietary nanomaterials (100 nm to 1 μm) and Crohn's disease (Lomer *et al.*, 2002). Macrophages located in lymphoid tissue can internalise 100–200 nm TiO_2 (anatase) nanomaterials from food additives (Powell *et al.*, 1996).

11.5.2.2.6 Nervous system Whilst the route of entry of NPs to the nervous system may not be immediately obvious, it has been demonstrated that inhaled NPs can travel along the olfactory nerves and cross the blood–brain barrier (Borm *et al.*, 2006; Oberdorster *et al.*, 2004; Peters *et al.*, 2006). Studies using silver-coated gold NPs (50 nm) showed that, following intranasal instillation, particles could translocate to the olfactory nerves and bulb of monkeys (De Lorenzo, 2008). This has also been demonstrated in rat inhalation studies, which have shown that 30 nm magnesium oxide and 20–30 nm carbon NPs translocate to the olfactory bulb (Elder *et al.*, 2006; Oberdorster *et al.*, 2002). Since the diameter of the olfactory axons is typically 100–200 nm, micron-sized particles are not able to enter the olfactory nerve (Oberdorster *et al.*, 2005), but NPs are able to translocate into deeper brain structures, as has been shown for viruses in the 20–200 nm size range (Oberdorster *et al.*, 2005; Takenaka *et al.*, 2001). Another route of uptake of NPs is through the blood–brain barrier, a physical anionic barrier that selectively restricts the access of certain substances, only allowing the passage of particular types of charged molecules (Borm *et al.*, 2006; Lockman *et al.*, 2004). The blood–brain barrier allows more cationic NPs to pass than neutral or anionic particles (Lockman *et al.*, 2004). The integrity of the blood–brain barrier may, however, be compromised by diseases such as hypertension, brain inflammation and respiratory-tract inflammation. These may increase the blood–brain barrier permeability, which may facilitate the entry of NPs (Peters *et al.*, 2006; Borm *et al.*, 2006). Accumulation of high concentrations of copper, aluminium, zinc or iron in the brain and oxidative stress are common factors associated with initiation and promotion of neurodegenerative diseases, such as Alzheimer's, Parkinson's, and Pick's (Liu G *et al.*, 2006; Campbell *et al.*, 2005; Miu and Benga, 2006), and thus it will be important to evaluate whether the presence of metals in the brain of patients suffering from neurodegenerative diseases may be due to NPs translocating to the brain, or their soluble compounds. Apart from influencing neurological diseases, NPs are also being used to cure these disorders by using the antioxidative properties of certain types of NPs (Schubert *et al.*, 2006; Bosi *et al.*, 2003) Functionalised fullerenes and other NPs (cerium oxide (CeO_2) and yttrium oxide (Y_2O_3)) have strong antioxidant properties, and fullerols, or polyhydroxylated fullerenes, have shown encouraging results as neuroprotective agents due to their high solubility and ability to cross the blood–brain barrier (Schubert *et al.*, 2006; Bosi *et al.*, 2003).

11.6 Future trends and risk reduction

Nanomaterials have exceptional electronic, chemical and mechanical properties with important current applications and the potential to dominate future markets (especially in the fields of drug delivery and human therapeutics). Despite the proven benefits of nanotechnology, its future smooth implementation depends critically on successful demonstration that health concerns have been addressed. This chapter demonstrates the need to evaluate the physicochemical properties of NPs and adapt them into an experimental protocol for toxicological studies. Considerable effort needs to be directed to the development of appropriate reference nanomaterials and controls (both positive and negative) to enable toxicity studies to be standardised. The need for reliable tracers or labelled particles is also a research priority, as is the need to improve our ability to detect NPs, particularly in contaminated or complex matrices (soil, food). It will be important that future studies assess whether NP uptake and potential health effects depend on genetic susceptibility and health status. To advance research into the hazards and risks associated with NPs, both statistical and mechanical models are required that offer generic predictions of the relative safety of different classes of nanomaterials. For the success of research in the field of nanotoxicology, it is essential that the researchers involved are prepared to work in an interdisciplinary environment, rather than focus exclusively on their narrow area of expertise. Important comprehensive, well-documented and thoughtfully implemented work in the field is now beginning to emerge, but the field is still lacking major risk-assessment and risk-reduction studies, partly because not all new lab-based studies have, as yet, fed into appropriate models. Until such risk-assessment studies become available and whilst current legislation does not require declaration when nanomaterials are present in any consumer products, application of the precautionary principle is the only appropriate risk-reduction strategy.

References

Afshari, A., Matson, U., Ekberg, L. E. (2005) Characterization of indoor sources of fine and ultrafine particles: a study conducted in a full-scale chamber. *Indoor Air*, **15**, 141.

Ahamed, M., *et al.* (2008) DNA damage response to different surface chemistry of silver nanoparticles in mammalian cells. *Toxicology and Applied Pharmacology*, **233**, 404.

Alvarez, M. G., Prucca, C., Milanesio, M. E., Durantini, E. N., Rivarola, V. (2006) Photodynamic activity of a new sensitizer derived from porphyrin-C60 dyad and its biological consequences in a human carcinoma cell line. *International Journal of Biochemistry and Cell Biology*, **38**, 2092.

Asharani, P. V., Wu, Y. L., Gong, Z., Valiyaveettil, S. (2008) Toxicity of silver nanoparticles in zebrafish models. *Nanotechnology*, **19**, 255102.

Asharani, P. V., Kah Mun, G. L., Hande, M. P., Valiyavettil, S. (2009) Cytotoxicity and Genotoxicity of silver nanoparticles in human cells. *ACS Nano*, **3**, 279.

Badireddy, A. R., Hotze, E. M., Chellam, S., Alvarez, P. J., Wiesner, M. R. (2007) Inactivation of bacteriophages via photosensitization of fullerol nanoparticles. *Environmental Science and Technology*, **41**, 6627.

Ballestri, M., *et al.* (2001) Liver and kidney foreign bodies granulomatosis in a patient with malocclusion, bruxism, and worn dental prostheses. *Gastroenterology*, **121**, 1234.

Barlow, P. G., Donaldson, K., MacCallum, J., Clouter, A., Stone, V. (2005) Serum exposed to nanoparticle carbon black displays increased potential to induce macrophage migration. *Toxicology Letters*, **155**, 397.

Benn, T. M., Westerhoff, P. (2008) Nanoparticle silver released into water from commercially available sock fabrics. *Environmental Science and Technology*, **42**, 4133.

Berry, J. P., Arnoux, B., Stanislas, G., Galle, P., Chretien, J. (1977) A microanalytic study of particles transport across the alveoli: role of blood platelets. *Biomedicine*, **27**, 354.

Bhabra, G., *et al.* (2009) Nanoparticles can cause DNA damage across a cellular barrier. *Nature Nanotechnology*, **4**, 876.

Blundell, G., Henderson, W. J., Price, E. W. (1989) Soil particles in the tissues of the foot in endemic elephantiasis of the lower legs. *Annals of Tropical Medicine and Parasitology*, **83**, 381.

Borm, P. J. A., Schins, R. P., Albrecht, C. (2004) Inhaled particles and lung cancer, part B: paradigms and risk assessment. *International Journal of Cancer*, **110**, 3.

Borm, P. J. A., *et al.* (2006) The potential risks of nanomaterials: a review carried out for ECETOC. *Particle and Fibre Toxicology*, **3** (11).

Bosi, S., *et al.* (2003) Fullerene derivatives: an attractive tool for biological applications. *European Journal of Medicinal Chemistry*, **38**, 913.

Bottini, M., *et al.* (2006) Multi-walled carbon nanotubes induce T lymphocyte apoptosis. *Toxicology Letters*, **160**, 121.

Brown, S. B., Brown, E. A., Walker, I. (2004) The present and future role of photodynamic therapy in cancer treatment. *Lancet Oncology*, **5**, 497.

Brunet, L., *et al.*, (2009) Comparative photoactivity and antibacterial properties of C60 fullerene and TiO2 NPs. *Environmental Science and Technology*, **43**, 4355.

Brunner, T. J., *et al.* (2006) *In vitro* cytotoxicity of oxide nanoparticles: comparison to asbestos, silica, and the effect of particle solubility. *Environmental Science and Technology*, **40**, 4374.

Buzea, C., Pacheco, I. I., Robbie, K. (2007) Nanomaterials and nanoparticles: Sources and toxicity. *Biointerphases*, **2**, MR17.

Campbell, A., *et al.* (2005) Particulate matter in polluted air may increase biomarkers of inflammation in mouse brain. *Neurotoxicology*, **26**, 133.

Carver, L. A., Schnitzer, J. E. (2003) Caveolae: mining little caves for new cancer targets. *Nature Review Cancer*, **3**, 571.

Casey, A., *et al.* (2007) Spectroscopic analysis confirms the interactions between single walled carbon nanotubes and various dyes commonly used to assess cytotoxicity. *Carbon*, **45**, 1425.

Champion, J. A., Mitragotri, S. (2006) Role of target geometry in phagocytosis. *Proceedings of the National Academy of Sciences*, **103**, 4930.

Chen, J., *et al.* (2006) Quantification of extrapulmonary translocation of intratracheal-instilled particles *in vivo* in rats: effect of lipopolysaccharide. *Toxicology*, **222**, 195.

Chen, Y. W., Hwang, K. C., Yen, C. C., Lai, Y. L. (2004) Fullerene derivatives protect against oxidative stress in RAW 264.7 cells and ischemia-reperfused lungs. *American Journal of Physiology Regulatory Integrative and Comparative Physiology*, **287**, R21.

Chirico, F., *et al.* (2007) Carboxyfullerenes localize within mitochondria and prevent the UVB-induced intrinsic apoptotic pathway. *Experimental Dermatology*, **16**, 429.

Choi, O., Hu, Z. (2008) Size dependent and reactive oxygen species related nanosilver toxicity to nitrifying bacteria. *Environmental Science and Technology*, **42**, 4583.

Choi, O., *et al.* (2008) The inhibitory effects of silver nanoparticles, silver ions, and silver chloride colloids on microbial growth. *Water Research*, **42**, 3066.

Chung, E. H., *et al.* (2007) The effect of surface charge on the uptake and biological function of mesoporous silica nanoparticles in 3 T3-L1 cells and human mesenchymal stem cells. *Biomaterials*, **28**, 2959.

Correa-Duarte, M. A., *et al.* (2004) Fabrication and biocompatibility of carbon nanotube-based 3D networks as scaffolds for cell seeding and growth. *Nano Letters*, **4**, 2233.

Cross, S. E., Innes, B., Roberts, M. S., Tsuzuki, T., Robertson, T. A., McCormick, P. (2007) Human skin penetration of sunscreen nanoparticles: *In vitro* assessment of a novel micronised zinc oxide formulation. *Skin Pharmacology and Physiology*, **20**, 148.

Daroczi, B., Kari, G., McAleer, M. F., Wolf, J. C., Rodeck, U., Dicker, A. P. (2006) *In vivo* radioprotection by the fullerene nanoparticle DF-1 as assessed in a zebrafish model. *Clinical Cancer Research*, **12**, 7086.

De Lorenzo, A. J. D. (2008) The olfactory neuron and the blood–brain barrier. In: Wolstenholme G.E.W. and Knight, J. (eds), *Ciba Foundation Symposium on Taste and Smell in Vertebrates*, John Wiley & Sons, Chichester. Chapter 9.

Deng, X., *et al.* (2009) Nanosized zinc oxide particles induce neural stem cell apoptosis. *Nanotechnology*, **20**, 115101.

Dillon, A. C., *et al.* (1999) A simple and complete purification of single walled carbon nanotube materials. *Advanced Materials*, **11**, 1354.

Ding, L., *et al.* (2005) Molecular characterization of the cytotoxic mechanism of multiwall carbon nanotubes and nano-Onions on human skin fibroblast. *Nano Letters*, **5**, 2448.

Donaldson, K., *et al.* (2003) Oxidative stress and calcium signaling in the adverse effects of environmental particles (PM_{10}). *Free radical Biology and Medicine*, **34**, 1369.

Doherty, G. J., McMahon, H. T. (2009) Mechanisms of endocytosis, *Annual Review of Biochemistry*, **78**, 857.

Drake, P. L., Hazelwood, K. J. (2005) Exposure-related health effects of silver and silver compounds: A review. *Annals of Occupational Hygiene*, **49**, 575.

Droge, W. (2002) Free radicals in the physiological control of cell function. *Physiological Reviews*, **82**, 47.

Dybowska A., *et al.* (2011) Synthesis of isotopically modified nanoparticles and their potential as nanotoxicity tracers. *Environmental Pollution*, **159**, 266.

Edward-Jones, V. (2009) The benefits of silver in hygiene, personal care and healthcare. *Letters in Applied Microbiology*, **49**, 147.

Elder, A., *et al.* (2006) Translocation of inhaled ultrafine manganese oxide particles to the central nervous system. *Environmental Health Perspectives*, **114**, 1172.

Elechiguerra, J. L., *et al.* (2005) Interaction of silver nanoparticles with HIV-I J. *Nanobiotechnology*, **3**, 6.

Environmental Protection Agency (2007) EPA Nanotechnology White Paper. Washington DC, United States.

Fang, J., Lyon, D. Y., Wiesner, M. R., Dong, J., Alvarez, P. J. (2007) Effect of a fullerene water suspension on bacterial phospholipids and membrane phase behavior. *Environmental Science and Technology*, **41**, 2636.

Fang, J., Shan, X., Wen, B., Lin, J., Owens, G. (2009) Stability of titania nanoparticles in soil suspensions and transport in saturated homogeneous soil columns. *Environmental Pollution*, **157**, 1101.

Faraji, A. H., Wipf, P. (2009) Nanoparticles in cellular drug delivery. *Bioorganic and Medicinal Chemistry*, **17**, 2950.

Ferin, J. (1994) Pulmonary retention and clearance of particles. *Toxicology Letters*, **72**, 121.

Franklin, N. M., Rogers, N. J., Apte, S. C., Batley, G. E., Gadd, G. E., Casey, P. S. (2007) Comparative toxicity of nanoparticulate ZnO, Bulk ZnO and $ZnCl_2$ to a freshwater microalga (Pseudokirchneriella subcapita): The importance of particle solubility. *Environmental Science and Technology*, **41**, 8484.

Friedman, S. H., Ganapathi, P. S., Rubin, Y., Kenyon, G. L. (1998) Optimizing the binding of fullerene inhibitors of the HIV-1 protease through predicted increases in hydrophobic desolvation. *Journal of Medicinal Chemistry*, **41**, 2424.

Gabay, T., Jakobs, E., Ben-Jacob, E., Hanein, Y. (2005) *Engineered self-organization of neural networks using carbon nanotube clusters. Physica A*, **350**, 611.

Gallagher, J., *et al.* (2003) Formation of 8-oxo-7, 8-dihydro-2-deoxyguanosine in rat lung DNA following subchronic inhalation of carbon black. *Toxicology and Applied Pharmacology*, **190**, 224.

Gamer, A. O., Leibold, E., Van Ravenzwaay, B. (2006) The *in vitro* absorption of microfine zinc oxide and titanium dioxide through porcine skin. *Toxicology In vitro*, **20**, 301.

Garcia-Garcia, E., Andrieux, K., Gil, S., Couvreur, P. (2005a) Colloidal carriers and blood–brain barrier (BBB) translocation: A way to deliver drugs to the brain? *International Journal of Pharmaceutics*, **298**, 274.

Garcia-Garcia, E., *et al.* (2005b) A relevant *in vitro* rat model for the evaluation of blood–brain barrier translocation of nanoparticles. *Cellular and Molecular Life Sciences*, **62**, 1400.

Garnett, M. C., Kallinteri, P. (2006) Nanomedicines and nanotoxicology: some physiological principles. *Occupational Medicine*, **56**, 307.

Gatti, A. M. (2004) Biocompatibility of micro- and nano-particles in the colon. *Part II. Biomaterials*, **25**, 385.

Gatti, A. M., Montanari, S., Gambarelli, A., Capitani, F., Salvatori, R. (2005) In-vivo short- and long-term evaluation of the interaction material-blood. *Journal of Materials Science: Materials in Medicine*, **16**, 1213.

Gatti, A. M., *et al.* (2004) Detection of micro- and nano-sized biocompatible particles in the blood. *Journal of Materials Science: Materials in Medicine*, **15**, 469.

Gehr, P., Schurch, S. (1992) Surface forces displace particles deposited in airways toward the epithelium. *News in Physiological Sciences*, **7**, 1.

Geiser, M., *et al.* (2005) Ultrafine particles cross cellular membranes by nonphagocytic mechanisms in lungs and in cultured cells. *Environmental Health Perspectives*, **113**, 1555.

George, J. H., Shaffer, M. S., Stevens, M. M. (2006) Investigating the cellular response to nanofibrous materials by use of a multi-walled carbon nanotube model. *Journal of. Experimental Nanoscience* **1**, 1.

Geranio, L., Heuberger, M., Nowack, B. (2009) The behaviour of silver nanotextiles during washing. *Environmental Science and Technology*, **43**, 8113.

Gloire, G., Legrand-Poels, S., Piette, J. (2006) NF-kB activation by reactive oxygen species: fifteen years later. *Biochemical Pharmacology*, **72**, 1493.

Gottschalk, F., Sonderer, T., Scholz, R. W., Nowack, B. (2009) Modeled environmental concentrations of engineered nanomaterials (TiO_2, ZnO, Ag, CNT, Fullerene) for different regions. *Environmental Science and Technology*, **43**, 9216.

Gratton, S. E. A., *et al.* (2008) The effect of particle design on cellular internalization pathways. *Proceedings of the National Academy of Sciences*, **105**, 11613.

Greulich, C., Diendorf, J., Simon, T., Eggeler, G., Epple, M., Köller, M. (2011) Uptake and intracellular distribution of silver nanoparticles in human mesenchymal stem cells. *Acta Biomaterialia*, **7**, 347.

Gulson, B., Wong, H. (2006) Stable isotopic tracing – a way forward for nanotechnology. *Environmental Health Perspective*, **114**, 1486.

Gupta, A. K., Gupta, M. (2005) Cytotoxicity suppression and cellular uptake enhancement of surface modified magnetic nanoparticles. *Biomaterials*, **26**, 1565.

Gurr, J. R., Wang, A. S. S., Chen, C. H., Jan, K. Y. (2005) Ultrafine titanium dioxide particles in the absence of photoactivation can induce oxidative damage to human bronchial epithelial cells. *Toxicology*, **213**, 66.

Handy, R. D., *et al.* (2008) The ecotoxicology and chemistry of manufactured nanoparticles. *Ecotoxicology*, **17**, 287.

Hardman, R. (2006) A toxicologic review of Quantum dots: Toxicity depends on physicochemical and environmental factors. *Environmental Health Perspectives*, **114**, 165.

Harush-Frenkel, O., Debotton, N., Benita, S., Altschuler, Y. (2007) Targeting of nanoparticles to the clathrin-mediated endocytic pathway. *Biochemical and Biophysical Research Communications*, **353**, 26.

Harush-Frenkel, O., Rozentur, E., Benita, S., Altschuler, Y. (2008) Surface charge of nanoparticles determines their endocytic and transcytotic pathway in polarized MDCK cells. *Biomacromolecules*, **9**, 435.

Harhaji, L., *et al.* (2008) Modulation of tumor necrosis factor-mediated cell death by fullerenes. *Pharmaceutical Research*, **25**, 1365.

Harhaji, L., *et al.* (2007) Multiple mechanisms underlying the anticancer action of nanocrystalline fullerene. *European Journal of Pharmacology*, **568**, 89.

Hassellov, M., Readman, J. W., Ranville, J. F., Tiede, K. (2008) Nanoparticle analysis and characterisation methodologies in environmental risk assessment of engineered nanoparticles. *Ecotoxicology*, **17**, 344.

Hermann, I. K., *et al.* (2010) Blood purification using functionalised core/shell nanomagnets. *Small*, **6**, 1388.

Hoet, P. H. M., Bruske-Hohlfield, I., Salata, O. V. (2004) Nanoparticles – known and unknown health risks. *J. Nanobiotechnology*, **2**, 12.

Horie, M., *et al.* (2009) Protein adsorption of ultrafine metal oxide and its influence on cytotoxicity toward cultured cells. *Chemical Research in Toxicology*, **22**, 543.

Hu, H., Ni, Y., Montana, V., Haddon, R. C., Parpura, V. (2004) Chemically functionalized carbon nanotubes as substrates for neuronal growth. *Nano Letters*, **4**, 507.

Huang, H. L., Fang, L. W., Lu, S. P., Chou, C. K., Luh, T. Y., Lai, M. Z. (2003) DNA-damaging reagents induce apoptosis through reactive oxygen species-dependent Fas aggregation. *Oncogene*, **22**, 8168.

Hwang, M. G., Katayama, H., Ohgaki, S. (2007) Inactivation of *Legionella pneumophila* and *Pseudomonas aeruginosa*: Evaluation of the bactericidal ability of silver cations. *Water Research*, **41**, 4097.

Inoue, K., *et al.* (2006) The role of toll-like receptor 4 in airway inflammation induced by diesel exhaust particles. *Archives of Toxicology*, **80**, 275.

Isakovic, A., *et al.* (2006) Distinct cytotoxic mechanisms of pristine versus hydroxylated fullerene. *Toxicological Sciences*, **91**, 173.

Ispas, C., *et al.* (2009) Toxicity and developmental defects of different sizes and shape nickel nanoparticles in zebrafish. *Environmental Science and Technology*, **43**, 6349.

Jaisi, D. P., Elimelech, M. (2009) Single walled carbon nanotubes exhibit limited transport in soil columns. *Environmental Science and Technology*, **43**, 9161.

Jani, P., Halbert, G. W., Langridge, J., Florence, A. T. (1990) Nanoparticle uptake by the rat gastrointestinal mucosa: quantitation and particle size dependency. *The Journal of Pharmacy and Pharmacology*, **42**, 821.

Jell, G., *et al.* (2008) Carbon nanotube-enhanced polyurethane scaffolds fabricated by thermally induced phase separation. *Journal of Material Chemistry*, **18**, 1865.

Jia, G., *et al.* (2005) Cytotoxicity of carbon nanomaterials: single-wall nanotube, multi-wall nanotube, and fullerene. *Environmental Science and Technology*, **39**, 1378.

Kagan, V. E., *et al.* (2006) Direct and indirect effects of single walled carbon nanotubes on RAW 264.7 macrophages: Role of iron. *Toxicology Letters*, **165**, 88.

Kang, S., Mauter, M. S., Elimelech, M. (2008) Physicochemical determinants of multiwalled carbon nanotube bacterial cytotoxicity. *Environmental Science and Technology*, **42**, 7528.

Kasermann, F., Kempf, C. (1997) Photodynamic inactivation of enveloped viruses by buckminsterfullerene. *Antiviral Research*, **34**, 65.

Kawata, K., Osawa, M., Okabe, S. (2009) *In vitro* toxicity of silver nanoparticles at noncytotoxic does to HepG2 Human Hepatoma cells. *Environmental Science and Technology*, **43**, 6046.

Kertesz, Z., *et al.* (2005) Nuclear microprobe study of TiO_2 penetration in the epidermis of human skin xenografts. *Nuclear Instruments and Methods in Physics Research B*, **231**, 280.

Khodakovskaya, M., *et al.* (2009) Carbon nanotubes are able to penetrate plant seed coat and dramatically affect seed germination and plant growth. *ACS Nano*, **3**, 3221.

Kim, J. S., *et al.* (2007) Antimicrobial effects of silver nanoparticles. *Nanomedicine: Nanotechnology, Biology and Medicine*, **3**, 95.

Kisin, E. R., Murray, A.R., Sargent, L. *et al.* (2011) Genotoxicity of carbon nanofibers: are they potentially more or less dangerous than carbon nanotubes or asbestos? *Toxicology and Applied Pharmacology*, **252**: 1–10.

Kiss, B., *et al.* (2008) Investigation of micronized titanium dioxide penetration in human skin xenografts and its effects on cellular functions of human skin-derived cells. *Experimental Dermatology*, **17**, 659.

Kobzik, L. (1995) Lung macrophage uptake of unopsonized environmental particulates. Role of scavenger-type receptors. *Journal of Immunology*, **155**, 367.

Kroto, H. W., Heath, J. R., O'Brien, S. C., Curl, R. F., Smalley, R. F. (1985) C60: Buckminsterfullerene. *Nature*, **318**, 162.

Kuempel, E. D., Tran, C. L., Castranova, V., Bailer, A. J. (2006) Lung dosimetry and risk assessment of nanoparticles: evaluating and extending current models in rats and humans. *Inhalation Toxicology*, **18**, 717.

Kvitek, L., *et al.* (2009) Initial study on the toxicity of silver nanoparticles against Paramecium caudatum. *Journal of Physical Chemistry C*, **113**, 4296.

Lai, S. K., Hida, K., Chen, C., Hanes, J. (2008) Characterization of the intracellular dynamics of a non-degradative pathway accessed by polymer nanoparticles, *Journal of Controlled Release*, **125**, 107.

Lam, C. W., James, J. T., McCluskey, R., Hunter, R. L. (2004) Pulmonary toxicity of single-wall carbon nanotubes in mice 7 and 90 days after intratracheal instillation. *Toxicological Sciences*, **77**, 126.

Lanone, S., Boczkowski, J. (2006) Biomedical applications and potential health risks of nanomaterials: Molecular mechanisms *Current Molecular Medicine*, **6**, 651.

Lens, M. (2009) Use of fullerenes in cosmetics. *Recent Patents on Biotechnology*, **3**, 118.

Limbach, L. K., *et al.* (2007) Exposure of engineered nanoparticles to human lung epithelial cells: Influence of chemical composition and catalytic activity on oxidative stress. *Environmental Science and Technology*, **41**, 4158.

Lin, D., Xing, B. (2008) Root uptake and phytotoxicity of ZnO nanoparticles. *Environmental Science and Technology*, **42**, 5580.

Lin, Y., *et al.* (2004) Advances towards bioapplications of carbon nanotubes. *Journal of Materials Chemistry*, **14**, 527.

Lin, Y. L., Lei, H. Y., Wen, Y. Y., Luh, T. Y., Chou, C. K., Liu, H. S. (2000) Light-independent inactivation of dengue-2 virus by carboxyfullerene C3 isomer. *Virology*, **275**, 258.

Lippmann, M. (1990) Effects of fiber characteristics on Lung deposition, retention and disease. *Environmental Health Perspectives*, **88**, 311.

Liu, G., *et al.* (2006) Nanoparticle iron chelators: A new therapeutic approach in Alzheimer disease and other neurologic disorders associated with trace metal imbalance. *Neuroscience Letters*, **406**, 189.

Liu, J., Wong, H. L., Moselhy, J., Bowen, B., Yu Wu, X., Johnston, M. R. (2006) Targeting colloidal particulates to thoracic lymph nodes. *Lung Cancer*, **51**, 377.

Liu, Y., He, L., Mustapha, A., Li, H., Hu, Z. Q., Lin, M. (2009) Antibacterial activities of zinc oxide nanoparticles against Escherichia coli O157:H7. *Journal of Applied Microbiology*, **107**, 1193.

Lo, S. F., Hayter, M., Chang, C. J., Hu, W. Y., Lee, L. L. (2008) A systematic review of silver-releasing dressings in the management of infected chronic wounds. *Journal of Clinical Nursing*, **17**, 1973.

Lockman, P. R., Koziara, J. M., Mumper, R. J., Allen, D. D. (2004) Nanoparticle surface charges alter blood–brain barrier integrity and permeability. *Journal of Drug Targeting*, **12**, 635.

Lomer, M. C., *et al.* (2004) Dietary sources of inorganic microparticles and their intake in healthy subjects and patients with Crohn's disease. *British Journal of Nutrition*, **92**, 947.

Lomer, M. C., Thompson, R. P., Powell, J. J. (2002) Fine and ultrafine particles of the diet: influence on the mucosal immune response and association with Crohn's disease. *Proceedings of the Nutrition Society*, **61**, 123.

Long, T. C., *et al.* (2007) Nanosize titanium dioxide stimulates reactive oxygen species in brain microglia and damages neurons *in vitro*. *Environmental Health Perspectives*, **115**, 1631.

Lubick, N. (2008) Nanosilver toxicity: ions, particles or both? *Environmental Science and Technology*, **42**, 8617.

Luoma, S. N. (2008) Silver nanotechnologies and the environment: Old problems or new challenges. The Project on Emerging Technologies. Washington DC, Woodrow Wilson International Center for Scholars.

Magner, L. N. (1992) Hippocrates and the Hippocratic tradition. In: *A History of Medicine*. Marcel Dekker, pp 66–68.

Manna, S. K., *et al.* (2005) Single-walled carbon nanotube induces oxidative stress and activates nuclear transcription factor-κB in human keratinocytes. *Nano Letters*, **5**, 1676.

Marchant, G. E., Sylvester, D. J., Abbott, K. W. (2008). Risk management principles for nanotechnology. *Nanoethics* **2**, 43.

Markovic, Z., Trajkovic, V. (2008) Biomedical potential of the reactive oxygen species generation and quenching by fullerenes. *Biomaterials*, **29**, 3561.

Maynard, A. D., Kuempel, E. D. (2005) Airborne nanostructured particles and occupational health. *Journal of Nanoparticle Research*, **7**, 587.

Menzel, F., Reinart, T., Vogt, J., Butz, T. (2004) Investigations of percutaneous uptake of ultrafine TiO_2 particles at the high energy ion nanoprobe LIPSION. *Nuclear Instruments and Methods in Physics Research B*, 219–220, 82.

Mills. N. L., *et al.* (2006) Do inhaled carbon nanoparticles translocate directly into the circulation in humans? *American Journal of Respiratory and Critical Care Medicine*, **173**, 426.

Mirkov, S. M., *et al.* (2004) Nitric oxide-scavenging activity of polyhydroxylated fullerenol, $C60(OH)_{24}$. *Nitric Oxide*, **11**, 201.

Miu, A. C., Benga, O. (2006) Aluminum and Alzheimer's disease: a new look. *J. Alzheimer's Diesease*, **10**, 179.

Monteiro–Riviere, N. A., Nemanich, R. J., Inman, A. O., Wang, Y. Y., Riviere. J. E. (2005) Multi-walled carbon nanotube interactions with human epidermal keratinocytes. *Toxicology Letters*, **155**, 377.

Montella, M., *et al.* (1997) Classical Kaposi sarcoma and volcanic soil in southern Italy: a case-control study. *Epidemiologia e prevenzione*, **21**, 114.

Morones, J. R., *et al.* (2005) The bactericidal effect of silver nanoparticles. *Nanotechnology*, **16**, 2346.

Mott, J. A., *et al.* (2002) Wildland forest fire smoke: health effects and intervention evaluation, Hoopa, California, 1999. *The Western Journal of Medicine*, **176**, 157.

Mroz, P., Tegos, G. P., Gali, H., Wharton, T., Sarna, T., Hamblin, M. R. (2007a) Photodynamic therapy with fullerenes. *Photochemical and Photobiological Sciences*, **6**, 1139.

Mroz, P., *et al.* (2007b) Functionalized fullerenes mediate photodynamic killing of cancer cells: type I versus type II photochemical mechanism. *Free Radical Biology and Medicine*, **43**, 711.

Mueller, N. C., Nowack, B. (2008) Exposure modeling of engineered nanoparticles in the environment. *Environmental Science and Technology*, **42**, 4447.

Muller, J. *et al.* (2005) Respiratory toxicity of multi-wall carbon nanotubes. *Toxicology and Applied Pharmacology*, **207**, 221.

Murray, A. R., *et al.* (2009) Oxidative stress and inflammatory response in dermal toxicity of single-walled carbon nanotubes. *Toxicology*, **257**, 161.

Mwenifumbo, S., Shaffer, M. S., Stevens, M. M. (2007) Exploring cellular behaviour with multi-walled carbon nanotube constructs. *Journal of Material Chemistry*, **17**, 1894.

Nan, A., Bai, X., Son, S. J., Lee, S. B., Ghandehari, H. (2008) Cellular uptake and cytotoxicity of silica nanotubes, *Nano Letters*, **8**, 2150.

Narayanan, K. B., Sakthivel, N. (2010) Biological synthesis of metal nanoparticles by microbes. *Advances in Colloid and Interface Science*, **156**, 1.

Nel, A., Xia, T., Madler, L., Li, N. (2006) Toxic potential of Materials at the nanolevel. *Science*, **311**, 622.

Nemmar, A., *et al.* (2002a) Ultrafine particles affect experimental thrombosis in an *in vivo* hamster model. *American Journal of Respiratory and Critical Care Medicine*, **166**, 998.

Nemmar, A., Nemery, B., Hoylaerts, M. F., Vermylen, J. (2002b) Air pollution and thrombosis: an experimental approach. *Pathophysiology of Haemostatis and Thrombosis*, **32**, 349.

Nemmar, A., *et al.* (2001) Passage of intratracheally instilled ultrafine particles from the lung into the systemic circulation in hamster. *American Journal of Respiratory and Critical Care Medicine*, **164**, 1665.

Nielsen, G. D., Roursgaard, M., Alstrup, K., Steen, J., Poulsen, S., Larsen, S. T. (2008) *In vivo* biology and toxicology of fullerenes and their derivatives. *Basic and Clinical Pharmacology and Toxicology*, **103**, 197.

Ning, Z., Cheung, C. S., Fu, J., Liu, M. A., Schnell, M. A. (2006) Experimental study of environmental tobacco smoke particles under actual indoor environment. *Science of the Total Environment*, **367**, 822.

Nohynek, G. J., Lademann, J., Ribaud, C., Roberts, M. S., (2007) Grey goo on the skin? Nanotechnology, cosmetic and sunscreen safety. *Critical Reviews in Toxicology*, **37**, 251.

Oberdorster, G. (2001) Pulmonary effects of inhaled ultrafine particles. *International Archives of Occupational and Environmental Health*, **74**, 1.

Oberdorster, G. (2002) Toxicokinetics abd effects of fibrous and nonfibrous particles. *Inhalation Toxicology*, **14**, 29.

Oberdorster, G., Ferin, J., Lehnert, B. E. (1994) Correlation between particle size, *in vivo* particle persistence, and lung injury. *Environmental Health Perspectives*, **102**, 173.

Oberdorster, G., *et al.* (2000) Acute pulmonary effects of ultrafine particles in rats and mice. *Health Effects Institute*, **96**, 5.

Oberdorster, G., *et al.* (2002) Extrapulmonary translocation of ultrafine carbon particles following whole body inhalation exposure of rats. *Journal of Toxicology and Environmental Health, Part A*, **65**, 1531.

Oberdorster, G., *et al.* (2004) Translocation of inhaled ultrafine particles to the brain. *Inhalation Toxicology*, **16**, 437.

Oberdorster, G., Oberdorster, E., Oberdorster, J. (2005) Nanotoxicology: an emerging discipline evolving from studies of ultrafine particles. *Environmental Health Perspectives*, **113**, 823.

Oberdorster, G., Stone, V., Donaldson, K. (2007) Toxicology of nanoparticles: A historical perspective. *Nanotoxicology*, **1**, 2.

Panyam, J., Labhasetwar, V. (2003) Dynamics of endocytosis and exocytosis of Poly(D, L lactide-co-Glycolide) nanoparticles in vascular smooth muscle cells. *Pharmaceutical Research*, **20**, 212.

Peters, A., *et al.* (2006) Translocation and potential neurological effects of fine and ultrafine particles a critical update. *Particle and Fibre Toxicology*, **3**, 13.

Petros, R. A., DeSimone, J. M. (2010) Strategies in the design of nanoparticles for therapeutic applications. *Nature Reviews Drug Discovery*, **9**, 615.

Poland, C. A., *et al.* (2008) Carbon nanotubes introduced into the abdominal cavity of mice show asbestos-like pathogenicity in a pilot study. *Nature Nanotechnology*, **3**, 423.

Popov, A. P., Lademann, J., Priezzhev, A. V., Myllyla, R. (2005). Effect of size of TiO_2 nanoparticles embedded into stratum corneum on ultraviolet-A and ultraviolet-B sun-blocking properties of the skin. *Journal of Biomedical Optics*, **10**, 1.

Porter, A. E., Muller, K., Skepper, J., Midgley, P., Welland, M. (2006) Uptake of C60 by human monocyte macrophages, its localization and implications for toxicity: Studied by high resolution electron microscopy and electron tomography. *Acta Biomaterialia* **2**, 409.

Powell, J. J., *et al.* (1996) Characterisation of inorganic microparticles in pigment cells of human gut associated lymphoid tissue. *Gut*, **38**, 390.

Powell, M. C., Griffin, M. P. A., Tai, S. (2008) Bottom-up risk regulation? How nanotechnology risk knowledge gaps challenge federal and state environmental agencies. *Environmental Management*, **42**, 426.

Pucadyil, T. J., Schmid, S. L. (2009) Conserved functions of membrane active GTPases in coated vesicle formation. *Science*, **325**, 1217.

Pulskamp, K., Diabate, S., Krug, H. F. (2007) Carbon nanotubes show no sign of acute toxicity but induce intracellular reactive oxygen species in dependence on contaminants. *Toxicology Letters*, **168**, 58.

Qaddoumi, M. G., Ueda, H., Yang, J., Davda, J., Labhasetwar, V., Lee, V. H. (2004) The characteristics and mechanisms of uptake of PLGA nanoparticles in rabbit conjunctival epithelial cell layers. *Pharmaceutical Research*, **21**, 641.

Quick, K. L., Ali, S. S., Arch, R., Xiong, C., Wozniak, D., Dugan, L. L. (2008) A carboxyfullerene SOD mimetic improves cognition and extends the lifespan of mice. *Neurobiology of Aging*, **29**, 117.

Radomski, A., *et al.* (2005) Nanoparticle-induced platelet aggregation and vascular thrombosis. *British Journal of Pharmacology*, **146**, 882.

Rejman, J., Conese, M., Hoekstra, D. (2006) Gene transfer by means of lipo- and polyplexes: role of clathrin and caveolae-mediated endocytosis. *Journal of Liposome Research*, **16**, 237.

Risom, L., Moller, P., Loft, S. (2005) Oxidative stress-induced DNA damage by particulate air pollution. *Mutation Research*, **592**, 119.

Roduner, E. (2006) *Size matters: why nanomaterials are different*. Chemical Society *Reviews*, **35**, 583.

Rogers, F., *et al.* (2005) Real-time measurements of jet aircraft engine exhaust. *Journal of Air and Waste Management Association*, **55**, 583.

Rothen-Ruthishauser, B. M., Schurch, S., Haenni, B., Kapp, N., Gehr, P. (2006) Interaction of fine particles and nanoparticles with red blood cells visualized with advanced microscopic techniques. *Environmental Science and Technology*, **40**, 4353.

Royal Commission on Environmental Pollution (2008) Novel materials in the environment: The case of nanotechnology. 27th Report.

Royal Society (2004) Nanoscience and nanotechnologies: opportunities and uncertainities. London, The Royal Society and The Royal Academy of Engineering.

Ryman-Rasmussen, J. P., *et al.* (2009) Inhaled carbon nanotubes reach the subpleural tissue in mice. *Nature Nanotechnology*, **4**, 747.

Sahay, G., Alakhova, D. Y., Kabanov A. V. (2010a) Endocytosis of nanomedicne. *Journal of Controlled Release*, **145**, 182.

Sahay, G., Kim, J. O., Kabanov, A. V., Bronich, T. K. (2010b) The exploitation of differential endocytic pathways in normal and tumor cells in the selective targeting of nanoparticulate chemotherapeutic agents. *Biomaterials*, **31**, 923.

Sapkota, A., *et al.* (2005) Impact of the 2002 Canadian forest fires on particulate matter air quality in Baltimore city. *Environmental Science and Technology*, **39**, 24.

Savolainen, K., *et al.* (2010) Nanotechnologies, engineered nanomaterials and occupational health and safety – A review. *Safety Science*, **48**, 957.

Schanen, B. C., *et al.* (2009) Exposure to titanium dioxide nanomaterials provokes inflammation of an *in vitro* human immune construct. *ACS Nano*, **3**, 2523.

Schipper, M. L., *et al.* (2008) A pilot toxicology study of single-walled carbon nanotubes in a small sample of mice. *Nature Nanotechnology*, **3**, 216.

Schubert, D., Dargusch, R., Raitano, J., Chan, S. W. (2006) Cerium and yttrium oxide nanoparticles are neuroprotective. *Biochemical and Biophysical Research Communications*, **342**, 86.

Schulz, H., *et al.* (2005) Cardiovascular effects of fine and ultrafine particles. *Journal of Aerosol Medicine*, **18**, 1.

Schulz, J., *et al.* (2002) Distribution of sunscreens on skin. *Advanced Drug Delivery Reviews, 54 Supplement* **1**, S157.

Schulte, P. A., Murashov. V., Zumwalde, R., Kuempel, E. D., Geraci, C. L. (2010) Occupational exposure limits for nanomaterials: state of the art. *Journal of Nanoparticle Research*, **12**, 1971.

Seames, W. S., Fernandez, A., Wendt, J. O. L. (2002) A study of fine particulate emissions from combustion of treated pulverized municipal sewage sludge. *Environmental Science and Technology*, **36**, 2772.

Semmler, M., *et al.* (2004) Long-term clearance kinetics of inhaled ultrafine insoluble iridium particles from the rat lung, including transient translocation into secondary organs. *Inhalation Toxicology*, **16**, 453.

Shi, Z., Shao, L., Jones, T. P., Lu, S. (2005) Microscopy and mineralogy of airborne particles collected during severe dust storm episodes in Beijing, China. *Journal of Geophysical Research*, **110**, D01303.

Shrivastava, S., *et al.* (2009) Characterisation of antiplatelet properties of silver nanoparticles. *ACS Nano*, **3**, 1357.

Shvedova, A. A., *et al.* (2003) Exposure to carbon nanotube material: Assessment of nanotube cytotoxicity using human keratinocyte cells. *Journal of Toxicology and Environmental Health Part A*, **66**, 1909.

Shvedova, A. A., *et al.* (2005) Unusual inflammatory and fibrogenic pulmonary responses to single-walled carbon nanotubes in mice. *American Journal of Physiology - Lung Cellular and Molecular Physiology*, **289**, L698.

Sinha, N., Yeow, J. T. W. (2005) Carbon nanotubes for biomedical applications. *IEEE Transactions on Nanoscience*, **4**, 180.

Smart, S. K., Cassady, A. I., Lu, G. Q., Martin, D. J. (2005) The biocompatibility of carbon nanotubes. *Carbon*, **44**, 1034.

Song, W., *et al.* (2010) Role of the dissolved zinc ion and reactive oxygen species in cytotoxicity of ZnO nanoparticles. *Toxicology Letters*, **199**, 389.

Stefani, D., Wardman, D., Lambert, T. (2005) The implosion of the Calgary General Hospital: ambient air quality issues. *Journal of Air and Waste Management Association*, **55**, 52.

Stone, V., *et al.* (1998) The role of oxidative stress in the prolonged inhibitory effect of ultrafine carbon black on epithelial cell function. *Toxicology in vitro*, **12**, 649.

Summers, H. D., *et al.* (2011) Statistical analysis of nanoparticle dosing in a dynamic cellular system. *Nature Nanotechnology*, **6**(3), 170–174.

Supranowicz, P. R., *et al.* (2002) Novel-current conducting composite substrates for exposing osetoblasts to alternating current stimulation. *Journal of Biomedical Material Research-A*, **59**, 499.

Szentkuti, L. (1997) Light microscopical observations on luminally administered dyes, dextrans, nanospheres and microspheres in the pre-epithelial mucus gel layer of the rat distal colon. *Journal of Control Release*, **46**, 233.

Takada, H., Kokubo, K., Matsubayashi, K., Oshima, T. (2006) Antioxidant activity of supramolecular water-soluble fullerenes evaluated by beta-carotene bleaching assay. *Bioscience Biotechnology and Biochemistry*, **70**, 3088.

Takenaka, S., *et al.* (2001) Pulmonary and systemic distribution of inhaled ultrafine silver particles in rats. *Environmental Health Perspectives*, **109**, 547.

Tang, Y. J., *et al.* (2007) Charge associated effects of fullerene derivatives on microbial structural integrity and central metabolism. *Nano Letters*, **7**, 754.

Taylor, D. A. (2002) Dust in the wind. *Environmental Health Perspectives*, **110**, A80.

Tekle, C., Deurs, B., Sandvig, K., Iversen, T. G. (2008) Cellular trafficking of quantum dot ligand bioconjugates and their induction of changes in normal routing of unconjugated ligands. *Nano Letters*, **8**, 1858.

Thakkar, K. N., Mhatre, S. S., Parikh, R. Y. (2010) Biological synthesis of metallic nanoparticles. *Nanomedicine*, **6**, 257.

Toll, R., *et al.* (2004) Penetration profile of microspheres in follicular targeting of terminal hair follicles. *The Journal of Investigative Dermatology*, **123**, 168.

Trouiller, B., Reliene, R., Westbrook, A., Solaimani, P., Schiestl, R. H. (2009) Titanium dioxide nanoparticles induce DNA damage and genetic instability *in vivo* in mice. *Cancer Research*, **69**, 8784.

Tsao, N., *et al.* (2002) *In vitro* action of carboxyfullerene. *Journal of Antimicrobial Chemotherapy*, **49**, 641.

Vasir, J. K., Labhasetwar, V. (2008) Quantification of the force of nanoparticle-cellmembrane interactions and its influence on intracellular trafficking of nanoparticles. *Biomaterials*, **29**, 4244.

Vermylen, J., Nemmar, A., Nemery, B., Hoylaerts, M. F. (2005) Ambient air pollution and acute myocardial infarction. *Journal of Thrombosis and Haemostatis*, **3**, 1955.

Warheit, D. B., *et al.* (2004) Comparative pulmonary toxicity assessment of single-wall carbon nanotubes in rats. *Toxicological Sciences*, **77**, 117.

Webster, T. J., Waid, M. C., McKenzie, J. L., Price, R. L., Ejiofor, J. U. (2004) Nano-biotechnology: carbon nanofibres as improved neural and orthopaedic implants. *Nanotechnology*, **15**, 48.

Westerdahl, D., Fruin, S., Sax, T., Fine, P. M., Sioutas, C. (2005) Mobile platform measurements of ultrafine particles and associated pollutant concentrations on freeways and residential streets in Los Angeles. *Atmospheric Environment*, **39**, 3597.

Wiebert, P., *et al.* (2006) No significant translocation of inhaled 35-nm carbon particles to the circulation in humans. *Inhalation Toxicology*, **18**, 741.

Win-Shwe, T. T., Yamamoto, S., Kakeyama, M., Kobayashi, T., Fujimaki, H. (2005) Effect of intratracheal instillation of ultrafine carbon black on proinflammatory cytokine and chemokine release and mRNA expression in lung and lymph nodes of mice. *Toxicology and Applied Pharmacology*, **209**, 51.

Worle-Knirsch, J. M., Pulskamp, K., Krug, H. F. (2006) Oops they did it again! Carbon nanotubes hoax scientists in viability assays. *Nano Letters*, **6**, 1261.

Wright, D. T., Cohn, L.A., Li, H., Fischer, B., Li, C. M., Adler, K. B. (1994) Interactions of oxygen radicals with airway epithelium. *Environmental Health Perspectives*, **102**, 85.

Wong, S. W. Y., Leung, P. T. Y., Djurisic, A. B., Leung, K. M. Y. (2010). Toxicities of nano zinc oxide to five marine organisms: influences of aggregate size and ion solubility. *Analytical and Bioanalytical Chemistry*, **396**, 609.

Xia, T., *et al.* (2006) Comparison of the abilities of ambient and manufactured nanoparticles to induce cellular toxicity according to an oxidative stress paradigm. *Nano Letters*, **6**, 1794.

Xia, T., *et al.* (2008) Comparison of the mechanism of toxicity of zinc oxide and cerium oxide nanoparticles based on dissolution and oxidative stress properties. *ACS Nano*, **2**, 2121.

Xiao, L., Takada, H., Maeda, K., Haramoto, M., Miwa, N. (2005) Antioxidant effects of water-soluble fullerene derivatives against ultraviolet ray or peroxylipid through their action of scavenging the reactive oxygen species in human skin keratinocytes. *Biomedicine and Pharmacotherapy*, **59**, 351.

Yamago, S., *et al.* (1995) *In vivo* biological behavior of a water-miscible fullerene: ^{14}C labeling, absorption, distribution, excretion and acute toxicity. *Chemistry and Biology*, **2**, 385.

Yang, Z., Xie, C. (2006) Zn^{2+} release from zinc and zinc oxide particles in simulated uterine solution. *Colloids and Surfaces. B* **47**, 140.

Yin, H., Too, H. P., Chow, G. M. (2005) The effects of particle size and surface coating on the cytotoxicity of nickel ferrite. *Biomaterials*, **26**, 5818.

Yu, D., Wing-Wah Yam, V. (2005) Hydrothermal induced assembly of colloidal silver spheres into various nanoparticles on the basis of HTAB-modified silver mirror reaction. *Journal of Physical Chemistry B*, **109**, 5497.

Yu, K. O., *et al.* (2009) Toxicity of amorphous silica nanoparticles in mouse keratinocytes. *Journal of Nanoparticle Research*, **11**, 15.

Yudoh, K., *et al.* (2007) Water-soluble C60 fullerene prevents degeneration of articular cartilage in osteoarthritis via down-regulation of chondrocyte catabolic activity and inhibition of cartilage degeneration during disease development. *Arthritis and Rheumatism*, **56**, 3307.

Zhang, H., Gilbert, B., Huang, F., Banfield J. F. (2003) Water-driven structure transformation in nanoparticles at room temperature. *Nature*, **424**, 1025.

Zhang, L. W., Monteiro-Riviere, N. A. (2009) Mechanisms of quantum dot nanoparticle cellular uptake. *Toxicological Sciences*, **110**, 138.

Conclusions: pollutants, risk and society
Sustainable systems

Richard Owen[1*], Jane A. Plant[2], K. Vala Ragnarsdottir[3] and Nikolaos Voulvoulis[4]

[1]*University of Exeter Business School, Streatham Court, Rennes Drive, Exeter, EX4 4PU; European Centre for Environment and Human Health, Peninsula College of Medicine and Dentistry, Royal Cornwall Hospital, Truro, TR1 3HD*
[2]*Centre for Environmental Policy and Department of Earth Science and Engineering, Imperial College London, Prince Consort Road, London SW7 2AZ*
[3]*Faculty of Earth Sciences, School of Engineering and Natural Sciences, Askja, University of Iceland, Reykjavik 101, Iceland*
[4]*Centre for Environmental Policy, Imperial College London, Prince Consort Road, London SW7 2AZ*
Corresponding author, email r.j.owen@exeter.ac.uk

At the beginning of the twenty-first century, society is becoming increasingly concerned with the global issues of climate change and sustainability in which the role of humanity, although contested, is ever more certain (IPCC, 2001, 2007). There is growing awareness of the finite nature of the resources on which modern society is critically dependent (Graedel and Klee, 2002; Ragnarsdottir, 2007) and of the rising pressure from populations that are growing exponentially and at the same time increasingly aspire to a Western-style material standard of living, although some scientists suggest that the limits to growth will be determined more by the planet's ability to absorb waste than by material shortages (Bloodworth *et al.*, 2010).

Important though it is to be aware of the harmful effects of chemicals on the environment and human health, we must consider chemicals in a wider context. They clearly have a potentially important role in helping us adapt to and mitigate the effects of climate change (for example, as catalysts in carbon capture). The risks and benefits of chemical substances also need to be considered in relation to the sustainability and security of food, water, energy and other material supplies.

The issue of sustainability in a world of finite resources is likely to become ever more important in considering chemicals in the twenty-first century. For example, 10 cal of oil are used to produce each 1 cal of conventionally produced food (Soil Association, 2008). The oil provides fuel for agricultural machinery and energy for transport, cooling and processing, as well as being used in the manufacture of agricultural machinery, fertilisers and pesticides. Hence most food is highly energy intensive, as well as containing synthetic chemicals with uncertain effects on health. The risks and benefits of modern food production must be considered in a situation where many experts consider that we are approaching peak oil (Hubbert, 1966, 1972, 1982; Arleklett, 2003, 2007) – where 50 per cent of the geologically produced oil will have been extracted, resulting in demand exceeding supply. Beyond this, energy-intensive farming could become prohibitively expensive and thus less competitive than low-energy-input agriculture. Hence, chemicals are linked to the environment, health, economics, society and, ultimately, sustainability.

Moves towards more sustainable living, coupled with improved understanding of the risks of chemicals to the environment and human health, suggest that in the future a more precautionary[1] approach should be adopted, emphasising the

[1] The **precautionary principle** is discussed in Chapters 1 and 2. It states that if an action or policy has a suspected risk of causing harm to the public or to the environment, in the absence of scientific consensus that the action or policy is harmful, the burden of proof that it is *not* harmful falls on those taking the action. In some legal systems, for example that of the European Union, the application of the precautionary principle is a statutory requirement (UN, 1992; Santillo *et al.*, 2001; Wiener and Rogers, 2002; Gee and O'Riordan, 2010).

Pollutants, Human Health and the Environment: A Risk Based Approach, First Edition. Edited by Jane A. Plant, Nikolaos Voulvoulis and K. Vala Ragnarsdottir.

responsible development and use of chemicals. One example is the approach known as 'green chemistry' (USEPA, 2010). Indeed, some environmentalists describe the twentieth century as the technological age and suggest that the twenty-first century will be the ecological age (Head, 2008), where the emphasis is placed on learning from nature. Emergent fields such as 'biomimicry' (Benyus, 2002), for example, seek to demonstrate that nature uses thousands of sustainable, non-toxic processes, which can be copied in the design of industrial processes. McDonough and Braungart 2002 suggest that we should aim to make the outflow water from factories cleaner than the inflow in their 'cradle to cradle' concept. These authors propose that industry should develop clean cycles that are built on natural processes whereby waste from one process is feed for another (Figure A.1).

For example, carbon dioxide captured from combustion processes might be converted by artificial photosynthesis to fuels. Key to the cradle-to-cradle concept is the idea that all manufactured products, whether aircraft, mobile phones, cars or computers, be designed from the outset for recycling 'technical nutrients'.

McDonough and Braungart (2002) advocate a separate cycle for biological nutrients, but this concept needs some modification because human and animal wastes are frequently contaminated with toxic trace elements and other hazardous substances. We need better systems for dealing with sewage. Many sewage systems were designed in the nineteenth and twentieth centuries to take out pathogenic organisms and large molecules but cannot remove small molecules such as those in many modern drugs (Jobling and Owen, 2010). Moreover, it has been suggested that if nutrient cycles are not closed, rock phosphorus will run out (Sverdrup *et al.*, 2011), and therefore sewage needs to become a source of fertilisers for the future.

The disposal of sewage sludge represents a growing political and environmental problem (Lottermoser and Morteani, 1993) and its re-use for agriculture is diminishing across Europe. Around big cities, not enough farmland is available and farmers increasingly refuse to apply sewage sludge, often under pressure from supermarkets, reflecting consumer concerns about chemical residues entering the food chain. Hence dewatered sludge is increasingly incinerated at temperatures of 850 °C, releasing gases which are widely perceived as harmful.

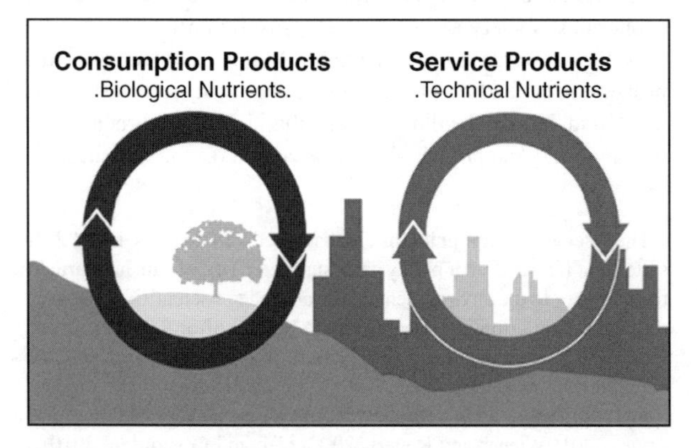

Figure A.1 The cradle-to-cradle concept (After McDonough and Braungart, 2002) (Reproduced with permission from EPEA GmbH, www.epea-hamburg.org)

Under the Braungart cradle-to-cradle model, sewage sludge would be considered as a source of energy and chemicals, using new integrated sewage-treatment processes. Such processes could incorporate many of those presently used to recover different substances in different countries.

For example, anaerobic digestion (AD), which uses naturally occurring bacteria in the absence of air, can be used to treat sewage sludge produced by the aerobic treatment of waste water, generating methane-rich biogas, and stabilised pathogen-free biosolids. AD is increasingly widely used in the water industry in the UK (DEFRA, 2009) and in many other EU countries, including France, Germany, Italy, Sweden and Denmark (AD-Net, 2010).

The biogas produced can be used directly for heating or converted into combined heat and (electrical) power (CHP). Biogas can also be used as vehicle fuel. AD will play an important role in enabling the UK to meet its targets for reducing carbon emissions under the Climate Change Act and the EU renewable energy directive (New Energy Focus, 2010).

The stabilised biosolids produced can be excellent soil conditioners and fertilisers, but there is concern that high concentrations of organic chemicals and toxic trace elements could be taken up by plants and reach humans through the food chain, despite scientific evidence that the use of biosolids poses no threat to the environment or humans (USEPA, 1992, 2003; Smith, 2000; Bright and Healey, 2003). Switzerland, for example has banned the addition of biosolids to agricultural land.

Further treatment of sewage sludge could help solve the problem. For example, phosphate from detergents and from human, animal and food waste has long been considered a contaminant of concern in sewage, but is now being recovered as calcium phosphate ($Ca_3[PO_4]_2$) (Van Der Houwen and Valsami-Jones, 2001) and struvite ($MgNH_4PO_4 \cdot 6H_2O$) (Jaffer, *et al.*, 2002) from sewage sludge in Australia, USA, Netherlands, Canada, Italy and Sweden (Tillotson, 2006; Liberti *et al.* 1986; Petruzelli *et al.*, 2001), helping to conserve scarce phosphate resources and reduce greenhouse-gas emissions. Methods for the recovery of urea as a source of nitrogen are also being developed and applied. The challenge for modern sewage-treatment systems is to shift the perspective from the removal of nutrients to their recovery (Kirchmann *et al.*,2006).

Precious metals are readily recycled from sewage; gold recovery is outstandingly high, with > 98 per cent recovered, on average (Sverdrup *et al.*,2011b). Gold and platinum are being extracted from sewage sludge in Japan and Germany (Jackson *et al.*,2010; Lottermoser, 2001; Mineweb, 2009). Techniques for extracting precious metals and, variably, removing toxic trace elements such as cadmium, antimony, mercury and chromium from a historical sewage-sludge stockpile near Melbourne, Australia have also been demonstrated, although the process consumed considerable quantities of reagents (Reeves and

Plimer, 2001). Such processes could not only increase the supply of metals and reduce greenhouse-gas emissions from mining but also enable the sludge to be used in agriculture, improving soil quality and displacing the carbon footprint of synthetic fertiliser. Any residue that could not be returned to land because of its content of hazardous substances could be used as filler in building or road surfacing.

Regulation and governance

In the past, the identification of hazards and risk to human health and the environment of substances ranging from organochlorine pesticides to tetraethyl lead and asbestos has been met with denial, obfuscation and mendacity on the part of industry and trade bodies, who have frequently lobbied governments to maintain their products on the basis of their safety, despite overwhelming scientific evidence to the contrary. Moreover, following the banning of such substances in the USA and Europe, companies have frequently switched their markets to developing countries causing great harm to human health and the environment.

Since the 1970s the number of policies, and regulatory agencies, for the protection of the environment has increased dramatically, reflecting growing public environmental consciousness. Earth day 1970, which was held almost a decade after the publication of *Silent Spring*, is sometimes cited as the beginning of the environmental revolution (Dunlap and Mertig, 1992; Anderson, 1994).

Early environmental legislation developed in response to high-profile events such as the widely reported occurrence of Minamata and itai-itai disease in Japan related to mercury and cadmium poisoning, respectively. Initially it was aimed mainly at protecting human health but it now covers environmental health generally. Policy has also evolved from the prevention of local pollution to preventing pollution on a wider scale, and recently to holistic management of environmental quality. For example, water-protection policy in the USA, has shifted since the early 1970s (Figure A.2), from a programme-by-programme, source-by-source, and pollutant-by-pollutant approach to a watershed-based strategy, emphasising the protection of healthy waters and restoring polluted ones (Cao, 2006). In the EU there has been a similar change in water-protection policy, from a fragmented approach involving different sectors to integrated water-resource management, covering pollution control, agriculture, hydropower, flood control and navigation (Bone *et al.,* 2011).

Air-quality policy and regulation has developed along similar lines to that for water in the USA and Europe. The Clean Air Act of 1956 in the UK, in response to the Great Smog of 1952, demonstrates how initially the reaction by legislators was catalysed by serious high-profile problems that affected human health. Similar legislation was introduced in 1962 in the USA (Figure A.3). As with water, air-quality legislation is now directed towards achieving integrated management of environmental quality (Bone *et al.,* 2011).

Recently, impact-based policies have been developed by the UN and the European Economic Commission (UN/ECE). Their protocols are based on critical loads, which are estimates of how much of a pollutant nature can withstand without damage. Critical loads of acidity for nitrogen and sulphur species have been developed and adopted in Europe under the UN/ECE Long-range Transboundary Air Pollution Convention (LRTRP) (UN-ECE, 1999). As an example, the 1999 Gothenburg protocol to the LRTAP Convention takes into account acidification (of surface waters and soils), eutrophication of soils and ground-level ozone and the emissions of sulphur dioxide, ammonia, nitrogen oxide and non-methane volatile organic compounds (NMVOC). The critical loads concept was used for acidification and eutrophication. Lead, mercury and cadmium emissions have also been limited using LTRAP (e.g. Hettelingh *et al.,* 2001; Sverdrup, 2001).

There is increasing implementation of chemicals regulation, underpinned by risk assessment, at the national, regional and international scale. One example is the Stockholm Convention on Persistent Organic Pollutants (2001).

Risk assessment has been used to support the development of a wide range of legal standards for compliance, covering the workplace and wider environment. Environmental Quality

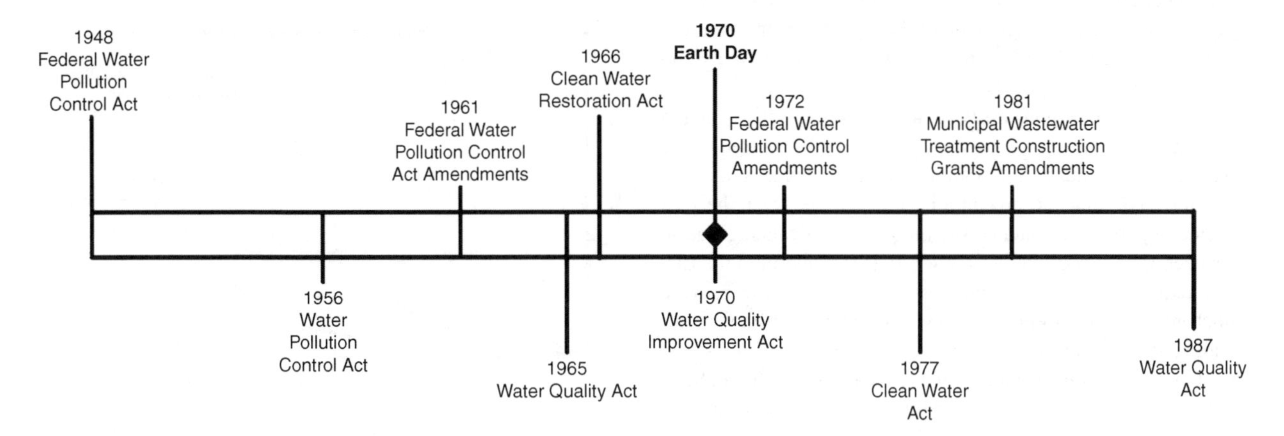

Figure A.2 Major US water-quality policy and amendments timeline (From Bone *et al.,*2011)

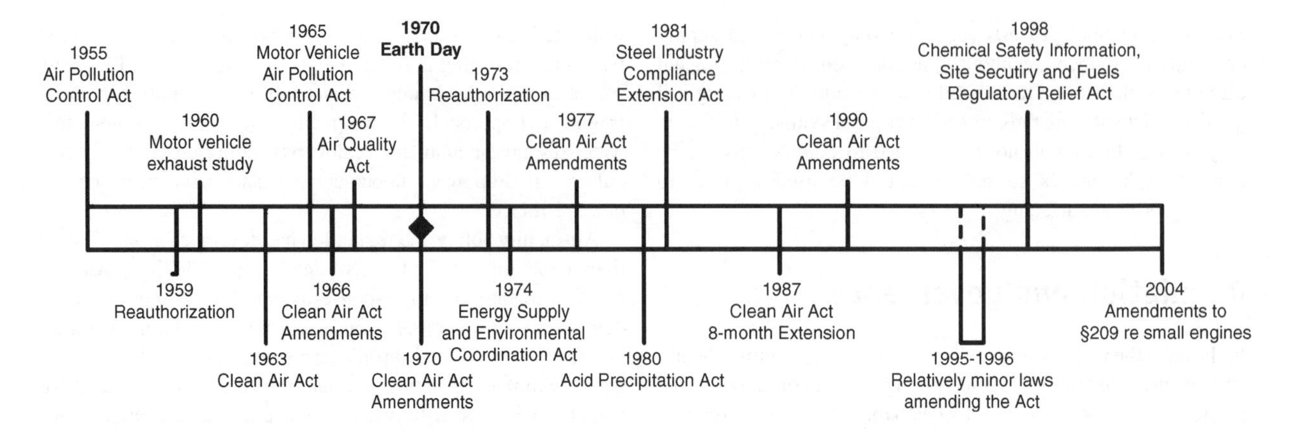

Figure A.3 Major US air-quality policy and amendments timeline

Standards and Occupational Exposure Limits have been set for specific chemicals that are known to be toxic and/or persistent and bioaccumulative. New legislation introduced over the past 20 years also shows a more holistic approach to environmental protection. This still has a foundation in risk assessment but incorporates life-cycle analysis of products from cradle to grave, with an appreciation of the effects of mixtures and consideration of both biological and chemical components. For example, under the EU Water Framework Directive (Directive 2000/60) 'good ecological status' at an ecosystem or catchment scale is considered along with 'good chemical status'. The move away from a substance-specific to an 'integrated risk assessment' approach recognises the complex multifactorial nature of environmental risks and apportions them to specific pressures such as diffuse agricultural pollution and point-source discharges from industrial complexes or sewage treatment works, for example. In recent decades there has been a further drive to apportion observed effects in humans and wildlife to specific chemicals in the environment (e.g. EU Environment and Health Action Plan), with the advent of a range of complementary approaches, from biomonitoring and epidemiology (e.g. Galloway *et al.*, 2010) to health-impact assessment, which reduce associated uncertainties and support interventions and risk management. Risk-based management of European water resources was developed in the EC funded RiskBase project (Brils and Harris, 2009).

The burden of responsibility for understanding and managing the risks of chemicals has been progressively transferred to the manufacturers and users, with increasing emphasis on 'the polluter pays' and the precautionary principle. Over the past decade in particular, 'data before market' chemicals legislation has been introduced, such as the REACH legislation in the EU. This requires chemical manufacturers or importers to provide much more information than previously about substances, including their hazardous properties, exposure and risk assessment, depending on the volume of chemical manufactured. Such legislation is also applied to existing substances and must be completed before authorisation is given to commence marketing them. In Europe, the burden of responsibility for risk assessment has shifted away from governmental authorities to industry, which is required to be more accountable and efficient.

Chemicals regulation based on risk assessment has promoted a more responsible culture and contributed significantly to improvements in the environment and human health; this will continue to be important to chemicals governance in the future. There is, however, growing unease about reliance on this form of governance alone. One reason for this is awareness of the long time delay, often of the order of many decades, between the preparation of new chemicals and their applications and products, and an understanding of their wider health, environmental and social impacts and associated risks (Figure A.4). There is often an even longer delay before the enactment, amendment or development of an appropriate regulatory response (RCEP, 2008). Engineered nanomaterials, one of the most important group of emerging chemical substances, illustrate this well (Karinen and Guston, 2010): it has been estimated that toxicity testing of manufactured nanoparticles currently available in the US will cost between \$249 million and \$1.18 billion and take 34–53 years to complete (Choi *et al.*, 2009). Meanwhile, the field of engineered nanomaterials is changing

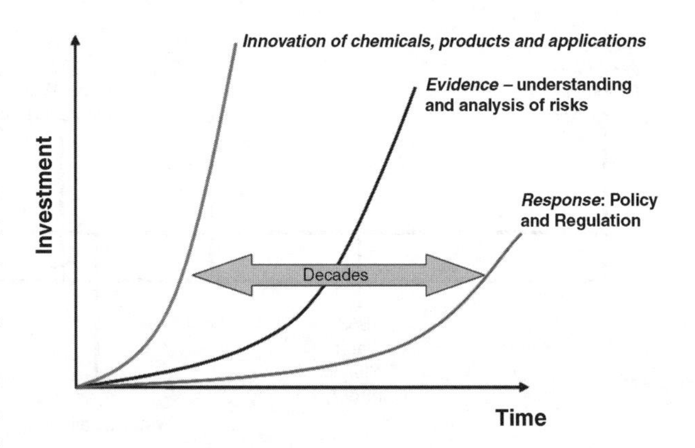

Figure A.4 Time lags between innovation and governance (Adapted from Owen et al., 2009)

rapidly, from passive nanoparticles to active structures and evolutionary systems (IRGC, 2006), and this is increasingly at the convergence of disciplines, as seen with synthetic biology.

Although it is interpreted and enacted in different ways (Wiener and Rogers, 2002), the adoption in some jurisdictions of the precautionary principle reflects the need for decision making to proceed even in the absence of full scientific certainty, and in particular where irreversible, harmful effects of substances are reported.

The problem of time lags between innovation and governance is compounded by globalisation, whereby the manufacturing and distribution of modern chemicals cuts across geographical and regulatory borders and is characterised by complex networks, so that chemicals produced in one part of the world are changed into products in another and shipped across the globe to consumers elsewhere.

In response, the twenty-first century has seen the emergence of concepts of *anticipatory and adaptive governance* and *co-responsibility* (RCEP, 2008; Owen and Goldberg, 2010; Guston, 1999; Mitcham, 2003), which go beyond traditional downstream regulation and reflect the distributed nature of risk and responsibility in modern society. These require that all those involved in chemicals and other technology-based innovation, including research scientists and their sponsors, understand their roles and responsibilities in managing the wider impacts of innovation at the same pace as this takes place. Recent years have seen the proliferation of various non-regulatory codes of conduct for chemicals research, development and use. These include the 2008 European Commission code for responsible nanosciences and nanotechnologies research (Maynard *et al.,* 2006; EC, 2008). Among other things, this recommends that research should be safe, ethical and anticipatory with regard to potential impacts on health and the environment and should be guided by principles of openness and transparency. It also asks for accountability, such that 'researchers and research organisations remain accountable for the social, environmental and human health impacts that their research may impose on present and future generations'.

Recent decades have also seen the advent of approaches such as real-time and constructive technology assessment, which facilitate identification and discussion of the wider impacts of innovation from the outset of the research and development process, feeding this back into decisions about the trajectories of innovation in real time (Guston and Sarewitz, 2002; Fisher *et al.,* 2006; Owen *et al.,* 2009). Such 'governance from within' has also been complemented by a drive to open up dialogue and participation in risk assessment and decision making (e.g. RCEP, 1998), with a move away from a purely technocratic approach to a more interdisciplinary and participatory approach (RCEP, 1998) which stresses engagement and dialogue (Wilsdon *et al.,* 2005; Guston and Sarewitz, 2006).

The field of environmental policy has had a long history of public interest and participation (Renn *et al.,* 1995; Abelson *et al.,* 2003). Future environmental policy is likely to involve the public in the decision-making processes even more, with interactions explicitly prescribed in policy documents. New and emerging policies aim to improve public understanding of environmental problems and involvement in the legislative process (European Parliament Council, 2000; Council of the European Union, 2009). The past 40 years have seen an increase in the collection of environmental data, a need for better communication of findings to the public, and in some cases collection of such data by the public (Silvertown, 2009). With an increase in environmental protection and consciousness, from both public and policy spheres, there is potential to improve public engagement through participation in environmental analysis and better management of environmental quality.

The future: sustainability and individual responsibility

Much has changed in the decades since Rachel Carson's *Silent Spring* was published, but the beginning of the twenty-first century has brought a mounting sense of urgency about the impacts of massive human overpopulation, lack of sustainability and global environmental change (Brown and Kane, 2004; Brown 2009a, 2009b; Ehrlich, 1968; Ehrlich *et al.,* 1992; Ehrlich and Goulder, 2007; Meadows *et al.,* 1972, 1993, 2005). The risks of chemicals to health via the environment must be considered in this context – and also their potential benefits, for example for maintaining supplies of food and materials and helping to mitigate climate change. In the middle of UNESCO's Education for Sustainable Development decade (UNESCO, 2005), the world education framework is shifting towards sustainability being at the core of the curriculum at all levels. Research supported by funding agencies around the world is also focusing on the development of sustainable communities and a campaign has been launched by the Optimum Population Trust (2010) to help limit population growth. Many around the world are already taking more personal responsibility for transforming their way of life to more sustainable practices. Hundreds of communities, for example, are now following the twelve steps of the Transition Towns to develop sustainable communities (Hopkins, 2008). Other organisations that are on a similar track include the Global Ecovillage Network (GEN), the founders of which (e.g. Findhorn, Scotland (www.findhorn.org) and Damanhur, Italy (www.damanhur.org)) have more than 30 years' experience of moving towards sustainable living. More recently, villages in the developing world have been changing traditional villages into ecovillages. One example is the Ministry of Ecovillages in Senegal (www.gouv.sn/spip.php?article989), which aims to change 14 000 villages in Senegal to ecovillages. Such communities will not stop using chemicals, but will consider their benefits and risks in the overall context of sustainability. The drive to a society which is more literate and aware of sustainability rather than one focused on unsustainable and unequal wealth creation with deleterious effects on the environment may in turn drive industrial practices, government policy and individual behaviour that support sustainability and the use of chemicals in this context (Stibbe, 2009).

Rather than relying on the burdensome centralised regulation of the past, the drive to sustainable living could also see increased focus on our actions as individuals, and require us to consider our own responsibilities, behaviour and values. The challenges for reducing the impacts of pollutants on human health and the environment are twofold. First, careful consideration needs to be given to the risks and benefits of chemicals in a debate in which all members of society can participate. In the UK, the Royal Commission on Environmental Pollution pioneered the principle of public access to information, and this has become the template for European practice (Fisk, 1996). This commission no longer exists, but it is important that government and the public continue to receive expert advice from sources that are independent of government: the universities, learned societies and the Royal Society are well placed to carry out this function. Secondly, this must be translated into the behaviour, responsibilities and actions of us all as individuals, as a driving force for change. In order to achieve this, more information on chemicals in the environment is needed, based on sound science, and this should be made available to a wider public in accessible language. We hope that this book will make such a contribution by introducing and informing students in tertiary education about the risks of pollutants in the environment and their potential impacts on human health.

References

Abelson, J., Forest, P. G., Eyles, J., Smith, P., Martin, E., Gauvin, F. P. (2003) Deliberations about deliberative methods: issues in the design and evaluation of public participation processes. *Social Science and Medicine* **57**, 239–251.

AD-Net (2010) The European Anaerobic Digestion Network. Available online [http://www.adnett.org/] Accessed 1 November 2010

Anderson, J. L. The Environmental Revolution at Twenty-Five. *Rutgers L. J.* 1994, **26**, 395–430.

Arleklett, K. (2003) The Peak and Decline of World Oil and Gas Production, *Journal of Minerals and Energy* **18**, 5–20.

Arleklett, K. (2007) Peak oil and the evolving strategies of oil importing and exporting countries. Discussion paper 17. Joint Transport Research Centre, OECD, International Transport Forum.

Benyus, J. M. (2002) *Biomimicry: Innovation Inspired by Nature*. Harper Perennial.

Bloodworth, A., Gunn, G., Lusty, P. (2010). Malthus revisited? www.bgs.ac.uk/about/bgs175/docs/019ABloodworth_Malthus/index.htm, www.bgs.ac.uk/downloads/start.cfm?id=1785

Bone, J., Head, M., Jones, D., Barraclough, D., Archer, M., Schieb, C., Flight, D., Eggleton, P., Voulvoulis, N. (2011) From chemical risk assessment to environmental quality management: the challenge for soil protection. *Environmental Science and Technology* **45**(1), 104–110.

Bright, D.A., Healey, N. (2003) Contaminant risks from biosolids land application: contemporary organic contaminant levels in digested sewage sludge from five treatment plants in Greater Vancouver, British Columbia. *Environmental Pollution* **126**, 39–49.

Brils J., Harris B. (Eds) (2009) *Towards Risk-Based Management of European River Basins*. Key-findings and recommendations of the RISKBASE project. EC FP6 reference GOCE 036938, December 2009, Utrecht, The Netherlands.

Brown, L. R. (2009a) Plan B 4.0. Mobilizing to save civilization Earth Policy Institute, Norton and Company, London.

Brown, L. B. (2009b) Could food shortages bring down civilization? *Scientific American* **301**, 38–45.

Brown, L. R., Kane, H. (1994) *Full House: Reassessing the Earth's Population Carrying Capacity*. Norton, NewYork.

Cao, Y. S. (2006) Evolution of Integrated Approaches to Water Resource Management in Europe and the United States. Some Lessons from Experience. Background Paper No. 2, pp 1-39.

Carson, R. (1962) *Silent Spring*. Penguin Classics, London 323pp.

Choi, J., Ramachandran, G., Kandlikar, M. (2009) The impact of toxicity testing costs on nanomaterial regulation. *Environmental Science and Technology* **43**: 3030–3034.

Colborn, T., Dumanoski, D., Myers, J. P. (1996) *Our Stolen Future: Are We Threatening Our Fertility, Intelligence, and Survival? A scientific detective story*. New York, Dutton.

Council of the European Union (2009). Proposal for a Directive of the European Parliament and of the Council establishing a framework for the protection of soil—Political agreement) Presidency proposal, 10387/09, pp 1-47.

DEFRA (2009) Water industry shows the way in turning waste into energy and fertiliser. Department of Environment, Food and Rural Affairs. Avaliable online [http://www.water.org.uk/home/news/press-releases/defra-anaerobic-digestion] Accessed 19 November 2010

Dunlap, R. E., Mertig, A. G., (1992) *American Environmentalism: The US Environmental Movement, 1970-1990*. Crane Russak & Co. USA, Vol. **1**, p 134.

EC (2008) European Commissions Code of Conduct for Responsible Nanotechnologies Research. Available at http://ec.europa.eu/nanotechnology/pdf/nanocode-rec_pe0894c-en.pdf

EEA (1998) Chemicals in the European Environment: Low Doses, High Stakes? The EEA and UNEP Annual Message 2 on the State of Europe's Environment 1998.

Ehrlich, P. R., (1968) *The Population Bomb*. Ballantine books, New York.

Ehrlich, P. R., Goulder, L. H., (2007) Is current consumption excessive? A general framework and some indications for the United States. *Conservation Biology* **21**, 1145–1154.

Ehrlich, P. R., Ehrlich, A. H., Daily, G. C. (1992) Population, ecosystem services, and the human food supply. *Morrison Institute for Population and Resource Studies Working Paper No. 44*. Stanford University, Stanford, CA.

European Parliament Council (2000) Directive 2000/60/EC of the European Parliament and of the Council of 23 October 2000 establishing a framework for Community action in the field of water policy. 2000/60/EC, pp 1-72.

EU Water Framework Directive (2000) Integrated River Basin Management for Europe. Directive 2000/60/EC (2000).

Fisher, E., Mahajan, R. L., Mitcham, C., (2006) Midstream modulation of technology: Governance from within. *Bulletin of Science, Technology and Society* **26**, 485–496.

Fisk, D. J., (1996) Opening address on behalf of the secretary of state for the environment. *Radiation Protection Dosimetry* **68**, 1–2.

Galloway, T., Cipelli, R., Guralnik, J., Ferrucci, L., Bandinelli, S., Corsi, A. M., Money, C., Paul, M., Melzer, D., (2010) Daily Bisphenol A excretion and associations with sex hormone concentrations: results from the InCHIANTI adult population study. *Environmental Health Perspectives* **118**(11), 1603–1608.

Gee, D., O'Riordan, T., (2010) *The Precautionary Principle in the 20th Century. Late Lessons from Early Warnings*. Earth Scan Publications Ltd.

Graedel, T. E., Klee, R. J., (2002) Getting serious about sustainability. *Environmental Science and Technology* **36**(15), 3455–3456.

Guston, D. H., (1999) Stabilizing the boundary between politics and science: the role of the Office of Technology Transfer as a boundary organization. *Social Studies of Science* **29**: 87–112.

Guston, D., Sarewitz, D., (2002) Real-time technology assessment. *Technology in Society* **24**, 93–109.

Guston, D. H., Sarewitz, D. R., (2006) *Shaping Science and Technology Policy: the next generation of research*. Madison, Wisconsin, The University of Wisconsin Press.

Head, P., (2008) Entering the Ecological Age: The Engineers Role. Royal College of Engineering Brunel Lecture. London.

Hettelingh, J.-P., Posch, M., de Smet, P.A.M., (2001). Multi-effect critical loads used in multi-pollutant reduction agreements in Europe. *Water and Air Soil Pollution* **130**(1-4): 1133–1138.

Hopkins, R., (2008) *The Transition Handbook: From Oil Dependency to Local Resilience*. Totnes, Devon, Green Books.

Hubbert, M. K. (1966) History of Petroleum Geology and Its Bearing Upon Present and Future Exploration. Presented to the 15th annual convention of the Gulf Coast Association of Geological Societies, Houston, Texas, October 28, 1965. Washington: U.S. Geological Survey. Published under same title in *AAPG Bulletin* **50**, 2504–2518.

Hubbert, M. K. (1972) Estimation of Oil and Gas Resources. In U.S. Geological Survey, Workshop on Techniques of Mineral Resource Appraisal, 16-50. Denver, U.S. Geological Survey.

Hubbert, M. K. (1982) Techniques of Prediction as Applied to Production of Oil and Gas, United States Department of Commerce, NBS Special Publication 631, May 1982.

IPCC (2001) *Climate Change 2001. The Scientific Basis. Intergovernmental Panel on Climate Change*. Cambridge University Press.

IPCC (2007) *Climate Change 2007. The Physical Science Basis. Intergovernmental Panel on Climate Change*. Cambridge University Press.

IRGC (2006), (International Risk Governance Council) White Paper on Nanotechnology Risk Governance. Accessed at www.irgc.org

Jackson, M. J., Prichard, H. M., Sampson, J. (2010) Platinum-group elements in sewage sludge and incinerator ash in the United Kingdom: Assessment of PGE sources and mobility in cities. *Science of the Total Environment* **408**(6), 1276–1285.

Jaffer, Y., Clark, T.A., Pearce, P., Parsons, S.A. (2002) Potential phosphorus recovery by struvite formation. *Water Research* **36**, 1834–1842.

Jobling, S., Owen, R. (2010) Ethinyl oestradiol: bitter pill for the precautionary principle, In Gee. D (ed). *Late Lessons from Early Warnings II*. European Environment Agency.

Karinen, R., Guston, D. H. (2010). Towards anticipatory governance: the experience with nanotechnology. In: Kaiser, M., Kurath, M., Maasen, S., Rehmann-Sutter, C. (Eds). *Nanotechnology and the Rise of an Assessment Regime*. Dordrecht, Netherlands, Springer, p. 217–232.

Kirchmann, H., Nyamangara, J., Cohen, Y. (2006) Recycling municipal wastes in the future: from organic to inorganic forms? *Soil Use and Management* **21**(s1) 152–159.

Liberti, L., Limoni, N., Lopez, A., Passino, R., Boari, G. (1986) The $10\,m^3\,h^{-1}$ RIM NUT demonstration plant at west Bari for removing and recovering N and P from waste-water. *Water Research* **20**, 735–739.

Lottermoser, B.G. (2001) Gold in municipal sewage sludges: a review on concentratons, sources and potential extraction. *The Journal of Solid Waste Technology and Management* **27**(2).

Lottermoser, B.G., Morteani, G. (1993) Sewage sludges: toxic substances, fertilisers or secondary metal resources? *Episodes* **16**(1-2) 329–333.

Maynard, A. D., Aitken, R. J., Butz, T., Colvin, V., Donaldson, K., Oberdorster G,*et al.* (2006) Safe handling of nanotechnology. *Nature* **444**: 267–269.

McDonough, W., Braungart, M. (2002) *Cradle to Cradle: Remaking the Way We Make Things*. North Point Press.

Meadows, D. H., Meadows, D. L., Randers, J., Behrens, W. (1972) *Limits to Growth*. New York, Universe Books.

Meadows, D. H., Meadows, D. L., Randers J. (1993) *Beyond the Limits: Confronting Global Collapse, Envisioning a Sustainable Future*. Chelsea Green Publishing Company:

Meadows, D. H., Randers, J., Meadows, D. L. (2005) *Limits to Growth, the 30 Year Update*. Sterling VA, Earthscan.

Mills, C. F. (1969) Trace element metabolism in animals and man. *British Medical Journal* **3**, 352–353.

Mineweb (2009) Japan discovers new source of mineral wealth. Available online [http://www.mineweb.com/mineweb/view/mineweb/en/page34?oid=77714&sn=Detail] Accessed 19 November 2010

Mitcham, C. (2003) Co-responsibility for research integrity. *Science and Engineering Ethics* **9**(2), 273–290.

New Energy Focus (2010) UK's first injection of biogas into gas network set for 'early 2011'. Avaliable online [http://www.newenergyfocus.com/do/ecco/view_item?listid=1&listcatid=125&listitemid=2741] Accessed 1 November 2010

Optimum Population Trust (2010) Optimum Population Trust. Available online [REFhttp://www.optimumpopulation.org/stopattwo.html] Accessed 1 November 2010

Owen, R., Crane, M., Deanne, K., Handy, R. D., Deanne, K., Linkov, I., Depledge, M. H. (2009) Strategic approaches for the management of environmental risk uncertainties posed by nanomaterials. In: Linkov, I. (ed.), *Nanotechnologies: Risks and Benefits*. Springer.

Owen, R., Goldberg, N. (2010) Responsible innovation: a pilot study with the UK Engineering and Physical Sciences Research Council. *Risk Analysis* **30**(11), 1699–1707.

Ragnarsdottir, K. V. (2007) Rare metals getting rarer. *Nature Geoscience* **1**, 720–721.

Reeves, S. J., Plimer, I. R. (2001) Exploitation of gold in an historic sewage sludge stockpile, Werribee, Australia: resource evaluation, chemical extraction and subsequent utilisation sludge. *Journal of Geochemical Exploration* **72**, 77–79.

Petruzzelli, D., Volpe, A., Limoni, N., Passino R. (2000) Coagulants removal and recovery from water clarifier sludge. *Water Research* **34**(7), 2177–2182.

Renn, O., Webler, T., Wiedemann, P. M. (1995) *Fairness and Competence in Citizen Participation: Evaluating Models for Environmental Discourse*. Springer, USA, Vol. **1**, p 400.

RCEP (1998) Royal Commission on Environmental Pollution, Report no 21: Setting Environmental Standards (1998). Available at www.rcep.org.uk

RCEP (2008) Royal Commission on Environmental Pollution, Report no 27: Novel Materials in the Environment: The case of Nanotechnology (2008). Available at www.rcep.org.uk

Santillo, S., Johnston, P., Langston, W. J. (2001) Tributyltin (TBT) antifoulants: a tale of ships, snails and imposex. In: *Late Lessons from Early Warnings: the Precautionary Principle 1896-2000*. EEA, pp 135–143.

Silvertown, J. (2009) A new dawn for citizen science. *Trends in Ecology and Evolution*, **24**, 467–471.

Smith, A.H., Lingas, E. O., Rahman, M. (2000) Contamination of drinking-water by arsenic in Bangladesh: a public health emergency. *Bulletin of the World Health Organization*, Vol **78**, No. 9. Geneva, WHO.

Smith, S. (2000) Are controls on organic contaminants necessary to protect the environment when sewage sludge is used in agriculture? *Progress in Environmental Science* **2**, 129–146.

Soil Association (2008) *Soil not Oil*. Bristol, Soil Association.

Stibbe, A. (2009) *The Handbook of Sustainability Literacy: Skills for a Changing World*. Dartington, Green Books.

Stockholm Convention on Persistent Organic Pollutants (2001) Available at http://chm.pops.int/Convention/tabid/54/language/en-GB/Default.aspx

Sverdrup, H. (2001) Setting critical limits for mercury, cadmium and lead to be used in calculation of critical loads for different receptors. In: Report from the Ad Hoc Expert meeting in Bratislava, Slovakia, 93-100, J. Curlik, P. Sefcik, Z. Viechova, (Eds.). Peer reviewed through a series of official UN/ECE workshops held in Schwerin, Berlin and Bratislava. Published by the Slovak Ministry of Environment, Bratislava, 2001.

Sverdrup, H. U., Koca, D., Ragnarsdottir, K. V., Haraldsson, H. V., Robert, K.-H. (2011a) Population, society and phosphate consumption. Assessing sustainability and planetary boundaries using a systems dynamics assessment model. *AMBIO*, submitted.

Sverdrup, H., Koca, D., Granath, C. (2011b). Modelling the gold market, explaining the past and assessing the physical and economical sustainability of future scenarios. Submitted for publication.

Tillotson, S. (2006) Phosphate removal: an alternative to chemical dosing. *Filtration & Separation*, **43**, 10–12.

UN (1992) Precautionary Principle. Principle 15 of the UN Rio Declaration on Environment and Development, accessed at: www.unep.org/Documents.Multilingual/Default.asp?DocumentID=78&ArticleID=1163

UNECE (1999). Protocol to the 1979 convention on long-range transboundary air pollution to abate acidification, eutrophication and ground-level ozone. United Nations Economic Commission for Europe http://www.unece.org/env/lrtap

UNESCO (2005) Education for Sustainable Development. http://www.unesco.org/en/esd/

USEPA (1992) Technical support document for land application of sewage sludge, Volume II. Eastern research Group, Lexington.

USEPA (2003) Final action not to regulate dioxins in land-applied sewage sludge. Retrieved 4 February, 2010, from http://www.epa.gov/waterscience/biosolids/dioxinfs.html

USEPA (2010) Twelve principles of green chemistry. Avaliable online [http://www.epa.gov/greenchemistry/pubs/principles.html] Accessed 19 November 2010

Van Der Houwen, J.A.M., Valsami-Jones, E. (2001) The application of calcium phosphate precipitation chemistry to phosphorus recovery: the influence of organic ligands. *Environmental Technology* **22**, 1325–1335.

Wiener, J. B., Rogers, M. D. (2002) Comparing Precaution in the United States and Europe. Available online [http://scholarship.law.duke.edu/faculty_scholarship/1191] Accessed 1 November 2010

Wilsdon J., Wynne B., Stilgoe J. (2005) The Public Value of Science. Demos, pp 67. Available at www.demos.co.uk

Index

Pollutants, Human Health and the Environment: A Risk Based Approach, First Edition. Edited by Jane A. Plant, Nikolaos Voulvoulis and K. Vala Ragnarsdottir.